Microwave Radiometry and Remote Sensing of the Environment

MICROWAVE RADIOMETRY AND REMOTE SENSING OF THE ENVIRONMENT

EDITOR:

D. SOLIMINI

Università Tor Vergata
Rome, Italy

///VSP///

Utrecht, The Netherlands, 1995

VSP BV
P.O. Box 346
3700 AH Zeist
The Netherlands

First published in 1995

ISBN 90-6764-189-8

CIP-DATA KONINKLIJKE BIBLIOTHEEK, DEN HAAG

Microwave

Microwave radiometry and remote sensing of the environment
/ ed.: D. Solimini. - Utrecht : VSP
Papers presented at the Specialist meeting held in Rome,
Italy, on 14-17 February, 1994. - With index.
ISBN 90-6764-189-9 bound
NUGI 814
Subject headings: microwave radiometry / remote sensing

Printed in The Netherlands by Koninklijke Wöhrmann bv, Zutphen

PREFACE

This volume contains a collection of refereed papers presented at the Specialist Meeting on Microwave Radiometry and Remote Sensing of the Environment held in Rome, Italy, on 14-17 February, 1994. This conference was the fourth of its kind, following those held in Rome, Italy, in 1983, in Florence, Italy, in 1988, and in Boulder, Colorado, in 1992. The 11-year timespan since the first conference has marked a period of steady advances and new developments in the observation of the terrestrial environment by passive microwave sensors. Both ground-based and satellite-borne systems have improved their accuracy, stability and spatial resolution and are providing a wealth of quantitative data that are being increasingly employed in application-oriented projects. The technological progress has gone together with the development of realistic electromagnetic models which now provide sophisticated means to stimulate the emission from the Earth's surface and atmosphere and to analyze the sensitivity of radiometric measurements to the geophysical parameters. The planned integration of microwave radiometers into the Earth Observing System is further boosting the advancement of passive remote sensing to monitor the Earth system dynamics.

This volume covers the different fields of application of microwave radiometry, the various observation and retrieval techniques, and the recent technological developments. It is subdivided into four sections, regarding measurement of atmospheric water vapor and cloud liquid, measurement of rain, observation of the surface, and new radiometric systems.

All papers published in this volume were reviewed by an editorial board. A special acknowledgment goes to the reviewers' recommendations.

The colleagues in the scientific steering committee were of great help in devising and organizing the conference.
I am particulary indebted to both Ed Westwater, who circulated the announcements and managed the collection of manuscripts from the U.S.A., and Paolo Pampaloni, for their valuable advice and substantial support.

I wish to express my gratitude to the rector of the Tor Vergata University, Professor Aldo Brancati, for the financial support.

Finally, special thanks go to Antonio Perrone for his willing cooperation and hard work in the organization of this volume.

Domenico Solimi
Roma, September 1994

SCIENTIFIC STEERING COMMITTEE

Giovanni d'Auria, Università la Sapienza, Rome, Italy
Jin A. Kong, M.I.T., Cambridge, MA, U.S.A.
Paolo Pampaloni, IROE/CNR, Florence, Italy
Domenico Solimini, Università Tor Vergata, Rome, Italy
Calvin T. Swift, University of Massachusetts, Amherst, MA, U.S.A.
Ed R. Westwater, NOAA/ETL, Boulder, CO, U.S.A.

LIST OF REVIEWERS

G. d'Auria
K.H. Ding
G. Elgered
P. Ferrazzoli
N.C. Grody
D. Hiltbrunner
N. Kämpfer
J.A. Kong
C. Mätzler
P. Pampaloni
R. Peter
G. Schiavon
E.R. Westwater

CONTENTS

Preface v

1. RADIOMETRIC SENSING OF ATMOSPHERIC WATER VAPOR AND CLOUD LIQUID

1.1 Water vapor

Clear-air observations of water vapor by ground-based microwave radiometers and Raman lidar
Y. Han, J.B. Snider, E.R. Westwater, S.H. Melfi and R.A. Ferrare 3

Observed and theoretical atmospheric emission at 20, 30, and 90 GHz: Recent results from land- and ocean-based locations
J.B. Snider 13

Forward modelling for AMSU
P.J. Rayer 23

Profiling of atmospheric water vapor with the millimeter-wave imaging radiometer
J.R. Wang, P. Racette and L.A. Chang 33

Assimilation of satellite derived moisture fields for numerical weather prediction
G. Deblonde, L. Garand, P. Gauthier and C. Grassotti 43

Estimation of errors in the two-beamwidth-antenna method for microwave measurement of atmospheric water vapor
S. Hashimoto 53

Improved GPS vertical surveying & GPS sensing of atmospheric water vapor
R. Ware, F. Solheim, C. Rocken, T. van Hove and C. Alber 65

Using microwave radiometry and space geodetic systems for studies of atmospheric water-vapor variations
G. Elgered, J.M. Johansson, and J.L. Davis 69

Influence of tropospheric electrical path delay on interferometric observations in millimeter radioastronomy
L. Olmi 79

1.2 Cloud liquid

The retrieval of tropospheric water vapor and cloud liquid with an iterative non-linear algorithm
R. Peter 95

Observations of total column precipitable water vapor and cloud liquid water using a dual-frequency microwave radiometer
J.C. Liljegren 107

Retrieval of atmospheric water vapour and liquid water using a single frequency microwave radiometer
P.O.J. Jarlemark 119

Effectiveness of brightness temperature ratios as indicators of the atmospheric path conditions
A.V. Bosisio and C. Capsoni 129

Spatial features of the sky brightness temperature at Roma
F. Barbaliscia, E. Fionda and P. Masullo 139

Horizontal variability of water vapor and cloud liquid water as derived from space borne observations
S.A. Tjemkes and M. Visser 147

Estimation of cloud liquid water contents from SSM/I and METEOSAT observations: Application to the SOFIA-ASTEX experiment
C. Prigent and G. Seze 155

An improved total precipitable water algorithm for cloudy situations
R. Furhop and E. Ruprecht 165

Spectral and temperature dependencies of the millimeter and centimeter wave absorption in clouds
B.G. Kutuza 175

2. RADIOMETRIC SENSING OF RAIN

2.1 Models of precipitation

Effects of hydrometeor shape and orientation upon passive microwave brightness temperature measurements
J. Turk and J. Vivekanandan 187

Influence on cloud and rainfall characteristics on brightness temperatures of the Earth measured by satellite
B.G. Kutuza, A. Hornbostel and A. Schroth 197

A quantitative comparison between 3-D and plane parallel microwave radiative transfer codes applied to horizontally and vertically structured precipitating clouds
L. Roberti and C. Kummerow 209

3-D radiation transfer effects of a raining cloud
Q. Liu and C. Simmer 219

A microwave radiometry characterization of precipitating clouds
P. Basili, P. Ciotti, G. d'Auria, F.S. Marzano and N. Pierdicca 229

2.2 Retrieval of rain

Polarimetric measurements and model calculations of downwelling rain brightness temperatures
A. Hornbostel, A. Schroth and B.G. Kutuza 239

Precipitation profile retrieval from airborne microwave radiometers: A case study over ocean during CaPE
F.S. Marzano, A. Mugnai, N. Pierdicca, E.A. Smith, J. Turk and J. Vivekanandan 253

Sensitivity of SSM/I rain rate algorithms to variations in water vapor and cloud liquid water: A modeling study
B.A. Burns and F. Flender 265

Retrieval of liquid and ice water content in atmosphere using Special Sensor Microwave Imager (SSM/I)
F. Weng and N.C. Grody 281

Review of the SSM/I-based algorithms submitted for the GPCP-AIP/2
G.L. Liberti 297

3. RADIOMETRIC SENSING OF THE EARTH'S SURFACE

3.1 Snow and ice

Identification of snowcover and precipitation using the Special Sensor Microwave Imager (SSM/I)
N.C. Grody, R.R. Ferraro and A.N. Basist 309

Multi-temporal aspects in snow retrieval using spaceborne radiometers
J. Noll, J.P.V. Poiares Baptista, M. Borgeaud and A. Rogne 321

Retrieval of snow and sea ice parameters from SSM/I data
J. Grandell, L. Kurvonen and M. Hallikainen 333

Airborne microwave radiometer measurements of Baltic sea ice
L. Kurvonen and M. Hallikainen 347

3.2 Ocean surface

Validation of liquid water path and surface windspeed retrievals from an airborne microwave radiometer
S.J. English, D.C. Jones and R.W. Saunders 357

How to describe the ocean roughened surface in microwave emissivity models?
C. Guillou, C. Prigent and S.J. English 369

On the use of different ocean surface models in radiative transfer calculations
M. Schrader and Q. Liu 379

3.3 Crops and forest

Microwave dielectric models of leaves
C. Mätzler 389

Potential of multifrequency techniques in microwave radiometry of crops
P. Ferrazzoli, L. Guerriero, S. Paloscia and P. Pampaloni 391

A model and experiments for microwave radiometry of forests
L. Guerriero and C. Mätzler 401

Microwave radiometry as a tool for forest fire detection: Model analysis and preliminary experiments
G. Luzi, P. Coppo, P. Ferrazzoli, S. Gagliani and T. Mazzoni 411

Polarimetric scattering and emission from a layer of random clusters of small spheroids and dense spheres
Y. Jin 419

Microwave radiometry of vegetation: Recent advances
G. Macelloni, S. Paloscia, P. Pampaloni, R. Ruisi and C. Susini 429

4. NEW RADIOMETRIC SYSTEMS

4.1 Synthetic aperture radiometers

Progress in remote sensing with thinned array radiometers
D.M. Le Vine, C.T. Swift, T.J. Jackson, A. Griffis, M. Kao and P. Gaiser 441

The TUD synthetic aperture radiometer demonstration model
B. Laursen, H.M. Pedersen and N. Skou 451

MIRAS: Preliminary concept of a two-dimensional L-band aperture synthesis radiometer
J.M. Goutoule, U. Kraft and M. Martin-Neira 463

4.2 MIMR

Multifrequency Imaging Microwave Radiometer: Design, calibration and expected performance
R. Bordi, M. L'Abbate and P. Spera 469

Radiometer absolute calibration: The solution implemented in MIMR
E. Battistelli and R. Bordi 485

Antenna electrical design for the Multifrequency Imaging Microwave Radiometer (MIMR)
S. Contu, F.M. Marinelli and P. Rinous 499

4.3 AMSU

AMSU-B antenna test results
T.J. Hewison 509

The radiometric characterization of AMSU-B
R.W. Saunders, T.J. Hewison, S.J. Stringer and N.C. Atkinson 519

4.4 Ground-based

A Multi-frequency mm-wave Radiometer (MFR): instrument description and retrieval algorithm
E. Battistelli, C. Capitani, A. Culebras, F. Del Frate, G. Schiavon, B. Arbesser-Rastburg and J.P.V. Poiares Baptista 529

New designs for portable microwave temperature profilers and for water vapor profilers
F. Solheim and S.J. Keihm 537

Author index 547

Subject index 549

1. RADIOMETRIC SENSING OF ATMOSPHERIC WATER VAPOR AND CLOUD LIQUID

OVERVIEW

- **Water vapor**

The simultaneous operation of microwave radiometers and other sensors, such as lidars, has significant synergy in studies of atmospheric absorption and emission. In fact, this combined technique, as well as extensive radiometric measurements at several frequencies and in a variety of climatic conditions, can contribute to assess the accuracy of the available absorption models, which are crucial in understanding the basic relations between the meteorological variables and the corresponding emission. From the operational point of view, the studies strongly suggest the soundness of microwave radiometric techniques both for large scale vapor profiling and for deriving moisture fields usable in numerical weather prediction.

The integrated water vapor content of the atmosphere is a key quantity in climatology. Moreover, since it causes a non negligible propagation delay, it may introduce relevant errors in measurements used in a variety of applications, including space geodesy, VLBI radioastronomy, and high-accuracy GPS surveying. The susceptibility of these systems to the amount of vapor offers an additional means for the measurement of this latter at the various time scales of interest.

- **Cloud liquid**

  The liquid water content of clouds is another key quantity in climatology, meteorology and telecommunications.

  The combined retrieval of vapor and liquid is typically carried out by dual-channel radiometers, but the use of a single frequency instrument, which exploits the diverse spatial and temporal behaviors of vapor and liquid may be an alternative cheaper method. Multi-channel systems for the accurate retrieval of temperature and moisture profiles are also under development, and nonlinear inversion algorithms are being devised and worked out, both to enhance the accuracies of the estimates and to release the retrieval procedure from particular observing geometries and site climatology.

  Radiometric observations have been used by the radiopropagation community since a long time to infer path attenuation. However, the approach is now gradually shifting from the "frequency scaling" towards a more complete observation of the physical status of the troposphere, from which the relevant propagation parameters can be obtained. The angular inhomogeneities of water vapor and, especially, cloud liquid are detected by ground-based scanning radiometers and their characterization can lead to the evaluation of the parameters of interest along different slant paths from measurements taken at a single elevation.

  The spatial variability of atmospheric water content over different regions is of importance in characterizing the hydrological cycle. The statistical properties of the moisture fields are determined from spaceborne platforms, like the Defense Meteorological Satellite Programme Special Sensor Microwave/Imager (*DMSP SSM/I*), which observe large spatial domains almost instantaneously. These radiometric systems not only are effective in producing maps of integrated water vapor and cloud liquid, but also are promising for cloud classification. The retrieval algorithms rely upon radiative transfer models which must take into account both temperature and cloud effects. Indeed, the enhancement of the retrieval accuracy, which is crucial to the performance of the weather prediction models, demands sound theoretical foundations of the dependence of absorption by liquid droplets on cloud temperature and frequency.

Clear-Air Observations of Water Vapor by Ground-Based Microwave Radiometers and Raman Lidar

YONG HAN[1], J.B. SNIDER[1], E.R. WESTWATER[1], S.H. MELFI[2], and R.A. FERRARE[3]
[1]*NOAA/ERL/Environmental Technology Laboratory, 325 Broadway, Boulder, CO 80303 USA*
Laboratory for Atmospheres, NASA/Goddard Space Flight Center
Greenbelt, MD 20771 USA
[3]*Hughes-STX Corp., Lanham, MD 20706 (under contract at NASA/GSFC)*

Abstract In November-December 1991, a substantial number of remote sensor and in situ instruments were operated together in Coffeyville, Kansas, USA, during the climate experiment FIRE II. Included in the suite of instruments were (a) the NOAA Environmental Technology Laboratory's (ETL) three-channel microwave radiometer (20.6, 31.65, and 90.0 GHz); (b) the NASA/GSFC Raman lidar; (c) ETL's Radio Acoustic Sounding System (RASS); and (d) frequent research-quality radiosondes. In this paper we compare radiometric observations with calculations of brightness temperature based on Raman lidar data and contemporary absorption models. We also compare measurements of water vapor profiles by the Raman lidar with measurements of water vapor (total amount and profiles) by the microwave radiometer. The experiments clearly demonstrate the potential of simultaneous operation of radiometers and Raman lidars for fundamental physical studies of water vapor.

Key Words: absorption models, microwave radiometer, precipitable water vapor, Raman lidar

1. INTRODUCTION

In radiometric remote sensing of the atmosphere, the ability to calculate radiances from underlying state variables is fundamental. For example, to infer temperature and water vapor profiles from satellites, one must be able to calculate radiance emerging from the top of the Earth's atmosphere from underlying profiles of temperature and water vapor. Similar needs occur when measuring precipitable water vapor from ground-based radiometers. Traditionally, such "forward model" calculations have coupled radiosonde observations of the state variables with detailed absorption and radiative transfer models to compare with radiance observations. However, for a variety of reasons, radiosondes are not always satisfactory, especially during low humidity conditions, or when there are large horizontal or temporal gradients in the humidity structure. An alternative method of measuring humidity profiles during clear conditions at night is Raman lidar [1].

In November-December 1991, a substantial number of remote sensor and in situ instruments were operated together in Coffeyville, Kansas, USA, during the climate experiment FIRE II. Included in the suite of instruments were (a) the NOAA Environmental Technology Laboratory's (ETL) three-channel microwave radiometer; (b) the NASA/Goddard Space Flight Center's (GSFC) Raman lidar; (c) ETL's Radio Acoustic Sounding System (RASS) [2]; and (d) frequent research-quality radiosondes. The Raman lidar operated only at night and our focus was on clear conditions. Here, we present results of simultaneous microwave radiometer measurements with collocated Raman lidar measurements of water vapor. Information on temperature profiles was obtained from composite data from surface meteorological instruments, radiosondes and RASS.

2. DESCRIPTION OF EQUIPMENT

A: NOAA Transportable Three-Channel Microwave Radiometer

The NOAA transportable microwave radiometer is a three-frequency system which is used to measure the column abundances of liquid water in clouds and water vapor in the atmosphere. The system contains three independent microwave radiometers that operate at 20.6, 31.65, and 90.0 GHz. Due to the substantially different frequency dependence of emission by liquid water and water vapor, a simple inversion algorithm [3] permits the two water phases to be separated. Inversion is further simplified since all three channels have the same beamwidth of 2.5°. An important feature of the electronics is that the gain is continuously monitored by switching between two reference blackbody loads.

The three radiometers are coupled into an antenna that is steerable in both azimuth and elevation. Here, we use only observations at zenith. The brightness temperatures T_b of the three channels were calibrated by the so-called "tipping curve" calibration procedure [3], a technique that is completely independent of either radiosondes or Raman lidar. Based on a substantial set of special research observations, the analysis of our tipping curve data [4] leads to 99% confidence limits for the absolute accuracy of our measurements at 20.6, 31.65, and 90.0 GHz of (±0.5, ±0.9, ±1.4 K). The corresponding rms values for the sensitivities, at our 2-minute averaging time, are (0.03, 0.02, 0.11 K).

B: RAMAN Lidar

During the FIRE-II experiment, a new NASA/GSFC Raman water vapor lidar system was deployed [5]. The characteristics of this lidar will be discussed here briefly. The system uses a XeF laser to transmit light at 351 nanometers (nm). A telescope collects the combined aerosol and molecular backscattered light at the laser wavelength as well as Raman scattered light from water vapor (403 nm), nitrogen (383 nm), and oxygen (372 nm) molecules. In normal operation, data are recorded as one-minute profiles with a range resolution of 75 meters; the lowest range is 185 meters above ground level (AGL) and the upper is nominally about 8.5 km.

Profiles of water vapor mixing ratio are computed from the ratio of the Raman water vapor to Raman nitrogen return signals. The lidar water vapor mixing ratio profiles are calibrated using a weighted least squares regression of the lidar ratios to the water vapor mixing ratios measured by coincident radiosonde launches at the lidar site. Radiosonde data below 30% relative humidity are not used because of potentially unreliable radiosonde moisture measurements in dry conditions [6]. During the three-week FIRE-II experiment period, there were 41 radiosonde humidity profiles used for calibration. A small correction (<7% for altitudes below 12 km) is applied to account for the differential atmospheric transmission between these two Raman wavelengths. A penetration distance of about 10 meters in a liquid cloud generally limits water vapor measurements to altitudes below cloud base. Since the lidar also measures scattering by aerosols and clouds, these cloud base altitudes are easily identified from the lidar data.

3. COMPARISON OF BRIGHTNESS TEMPERATURE MEASUREMENTS AND CALCULATIONS

Previously, comparisons of measured and calculated T_b's have been reported [7, 8]. In these works, clear-sky measurements of zenith T_b were compared with calculations of the Radiative Transfer Equation (RTE) [3] that are based on atmospheric absorption models. Radiosonde soundings of temperature T, relative humidity RH, and pressure P provided the input profiles to the RTE. Recently, both water vapor and oxygen absorption models have been updated. In this section we compare measurements of T_b's with calculations based on several models. For convenience we will refer to the models as: RTE76 [9] for H_2O, [10] for O_2; RTE87 [11] for H_2O, [10] for O_2; and RTE93 [12] for H_2O, [13] for O_2. Here, in contrast to previous calculations that were based on radiosondes, we use two-minute measurements of water vapor profiles provided by the Raman lidar, together with temperature profiles provided by RASS and radiosondes. There were a total number of 2506 coincident radiometric and Raman lidar measurements available when there were clear-sky conditions during 10 nights of the experiment. Details of our data-editing procedures are given by [4]. The Raman lidar also allowed us to eliminate cloudy data.

In Figure 1, we show typical results of comparing measured and calculated brightness temperatures at 20.6 GHz for a 12 hour time period. Although, there is an overall offset between measurements and calculations using RTE76, RTE87, and RTE93, it is encouraging that excellent correlation between measurements is obtained on the finest temporal scale allowed - two minutes. It is also evident that at 20.6 GHz, the RTE76 model gives the best agreement with measurements.

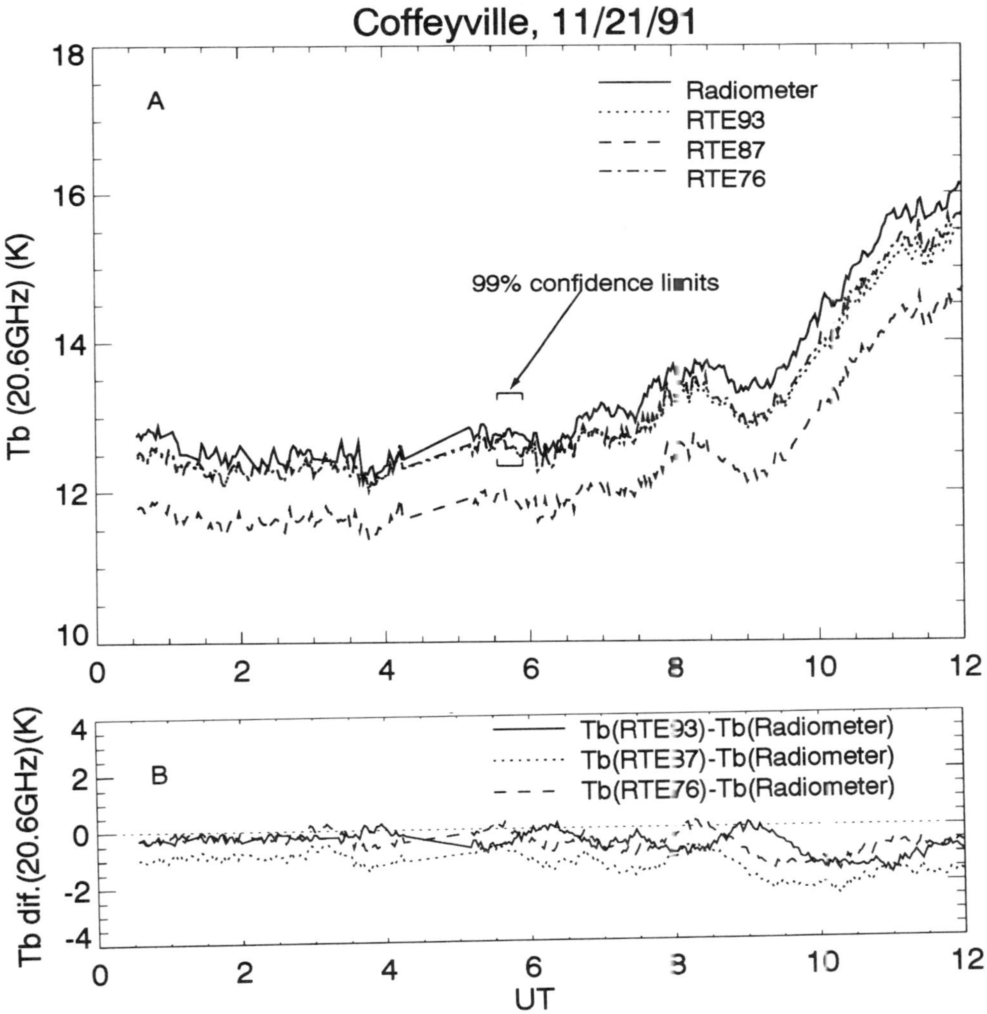

Figure 1. (A) Time series of measured and modeled T_b at 20.6 GHz on 11/21/91. (B) Time series of difference between modeled and calculated T_b.

We also show on the figure our estimate of the absolute accuracy in the T_b measurement. We supplement the results shown in Figure 1 by a statistical analysis carried out over the entire data set of clear and quality-controlled measurements. We present in Table 1 a summary of our linear regression results. These statistical results show: the 20.6 GHz results are better with RTE76, the 31.65 GHz results are better with RTE93, and the 90 GHz results show no clear winner between the RTE87 and the RTE93 models, although both are clearly better than RTE76. The latter results are consistent with those of [7] which also showed poor agreement of the RTE76 at 90.0 GHz. The recent changes to the Liebe model RTE93 produce better agreement at 20.6 and 31.65 GHz than do previous models used in RTE87. This better agreement results from improvements in the line parameters associated with the 22.235 GHz water vapor line. The rather mixed results at 90 GHz indicate that the recent water vapor continuum model (RTE93) may not be as accurate as that of the earlier RTE87. Certainly, measurements in more humid environments would be useful in addressing these modeling questions. A companion paper [14] compares measurements and calculations using the same absorption models used here, but with radiosondes providing ground truth.

Table 1.
Comparison of measured and calculated brightness temperatures at 20.6, 31.65, and 90.0 GHz. N = 2506. After Han et al., 1994.

20.6 GHz	rms diff. (K)	bias (K) (meas.-cal.)	slope	intercept (K)
RTE76	0.40	0.27	0.99	-0.10
RTE87	1.48	1.41	0.90	0.28
RTE93	0.63	0.53	0.94	0.46
31.65 GHz				
RTE76	1.30	1.27	0.88	0.53
RTE87	1.96	1.93	0.83	0.70
RTE93	0.76	0.73	0.94	0.23
90.0 GHz				
RTE76	7.66	-7.38	1.27	-2.36
RTE87	1.95	1.77	1.00	-1.75
RTE93	1.48	-0.04	1.18	-6.42

4. COMPARISON OF PRECIPITABLE WATER VAPOR RETRIEVALS

The total amount of water vapor in a vertical column through the atmosphere per unit cross-section, the precipitable water vapor V, is given by

$$V = \int_0^\infty \rho_V(h)\, dh \ , \tag{1}$$

where

ρ_V = density of water vapor (g m^{-3})

and

h = height (m).

Since the density of liquid water is approximately 1 g/cm^3, the units of V in g/cm^2 or in cm are practically interchangeable. Other commonly used units of V are mm (1 mm = 0.1 cm) and kg/m^2 (1 kg/m^2 = 1 mm). From polar regions to the tropics, V can vary from 0.01 cm to 7.5 cm.

Precipitable water vapor can be derived directly from the Raman lidar data, given a suitable dry air density profile, and from the radiometer data, by means of a retrieval algorithm. The radiometric retrieval algorithm that we used was the linear statistical method [3], conditioned on clear sky conditions; all three channels were used in the retrievals. In addition, we removed radiometric - RTE model offsets by adjusting the radiometer T_b data to be consistent with the RTE93 calculations, using radiosonde data as ground-truth profile data. A priori statistics were obtained from Oklahoma City, Oklahoma, a station about 200 km from Coffeyville. We also used coincident radiosonde data to estimate the non-diagonal covariance matrix of experimental errors S_ϵ.

Typical V comparisons are shown in Figure 2, where Raman lidar, microwave and radiosonde measurements are shown. As is evident, there is excellent correlation on short time scales between Raman lidar and microwave data, and, due to the radiometric offset adjustments, there is only a small bias in V. Note the sometimes poor agreement with radiosonde measurements, the best agreement being with the CLASS radiosondes. Over a data set of 2506 comparisons, we achieved a rms difference of 0.03 cm. These results can be compared with the values of 0.17 cm (radiometer vs. radiosonde) reported by [3]. The much better agreement between Raman and radiometer arises from at least four reasons: (a) the remote sensors are both simultaneously viewing the same zenith direction, although with different beamwidths (Raman beamwidth = 0.5 x 0.7 mrad, microwave beamwidths ≈ 2.5^0.); (b) these data are confined to clear sky conditions, and the radiometer retrieval coefficients are determined for these conditions; (c) three, rather than two, channels were used in the radiometric retrievals; and (d) a non-diagonal experimentally-determined covariance matrix S_ϵ was used in the construction of retrieval coefficients.

5. COMPARISON OF WATER VAPOR PROFILE RETRIEVALS

The principal products from ground-based microwave radiometers are precipitable water vapor and integrated cloud liquid. However, if surface measurements of water vapor density are available, low vertical resolution water vapor profiles can be derived [15]; [3]. Although these derived profiles do not approach the 75 meters vertical resolution of the Raman lidar, they may have use in some radiative transfer and meteorological applications. Here, we will compare derived profiles, obtained during clear air conditions, with the Raman lidar measurements.

The methodology of deriving profiles from microwave radiometer measurements, surface measurements of absolute humidity, and an a priori ensemble of radiosondes profiles is described in [3]. Two representative retrievals are compared with the Raman lidar and with radiosonde soundings in Figure 3. Excellent agreement between the lidar and the radiosonde is seen; the microwave retrievals give a smoothed, but reasonable representative picture of the vapor profiles. Statistical differences between the Raman and microwave retrievals are given in Figure 4. The large differences between the microwave and Raman retrievals near the surface are probably due to inaccuracies in the in situ measurements of surface absolute humidity, which radiometers use as a statistical predictor.

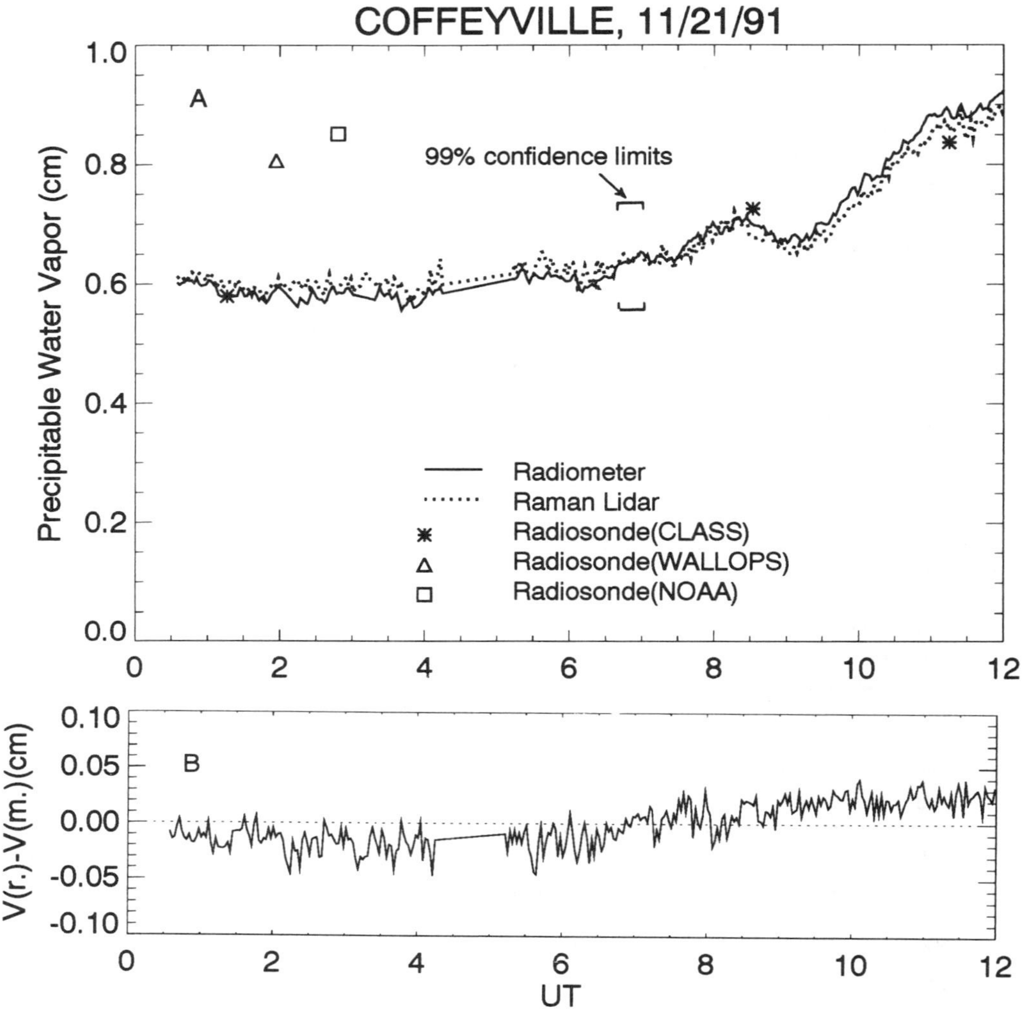

Figure 2. (A) Time series of precipitable water vapor V from the radiometer, Raman lidar and radiosonde on 11/21/91. (B) Time series of the difference between V determined from Raman lidar (V(r)) and radiometer (V(m)).

6. CONCLUSIONS

The use of Raman lidar data for detailed studies of tropospheric water vapor absorption and emission has significant potential. The vertical resolution of 75 meters and the temporal resolution of 1 minute have significant advantages over radiosondes which frequently are displaced from a radiometer's observing volume and which take about 30 minutes to traverse the major portion of the water vapor overburden. Another advantage is that the lidar can indicate the presence or absence of clouds, so that clear-sky radiative transfer calculations can be confidently made. On the negative side, the Raman lidar soundings reach to ≈ 8.5 km altitudes only during the night, and the lidar usually does not penetrate clouds. Another negative feature is that the Raman lidar requires in situ profile measurements for calibration and thus may not be independent of radiosonde sensors.

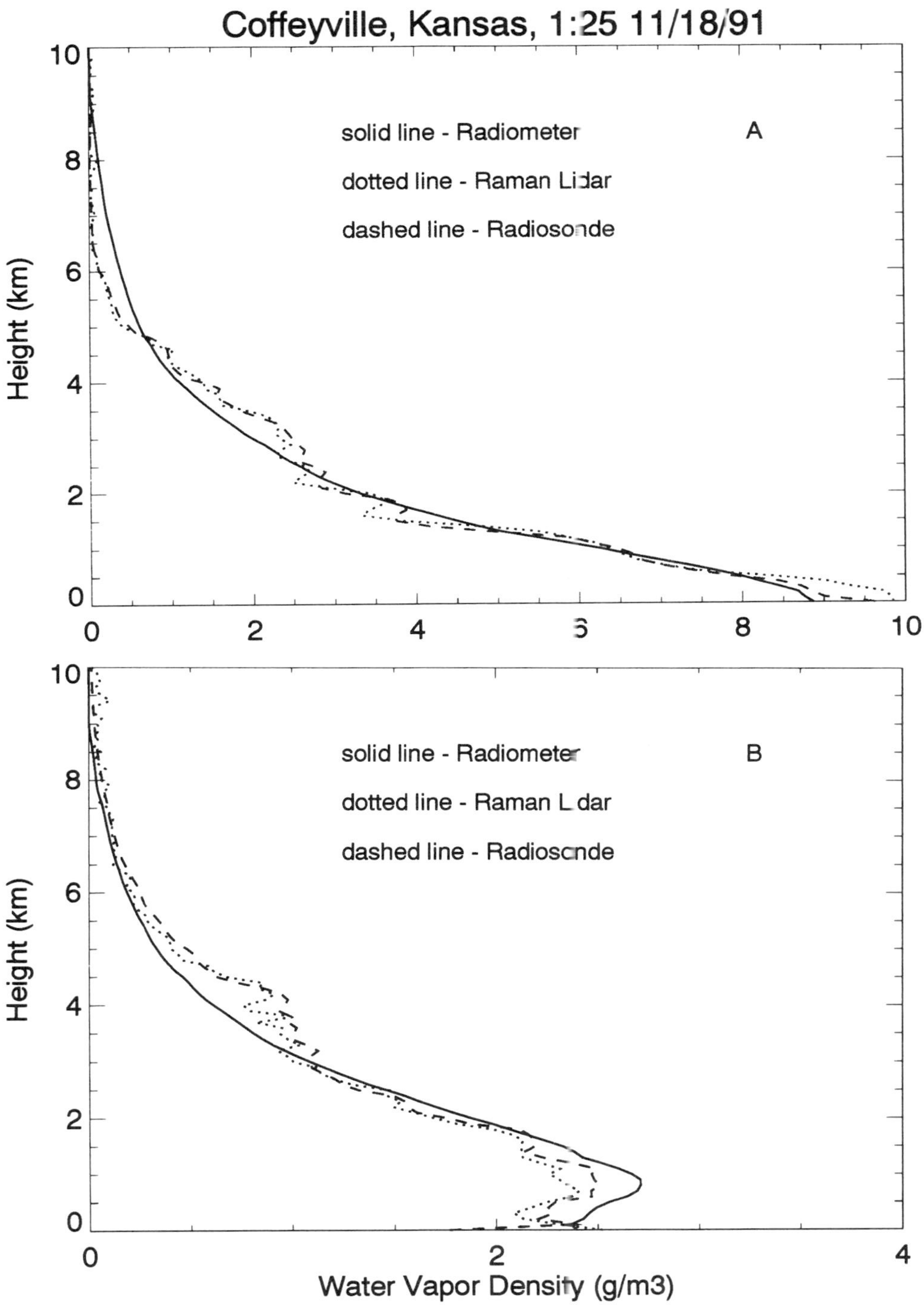

Figure 3. Examples of comparisons of radiometric retrievals of water vapor profiles with measurements by the Raman lidar and by radiosondes; data taken on (A) 11/18/91 at 1:25 UT, and (B) 12/■4/91 at 2:58 UT.

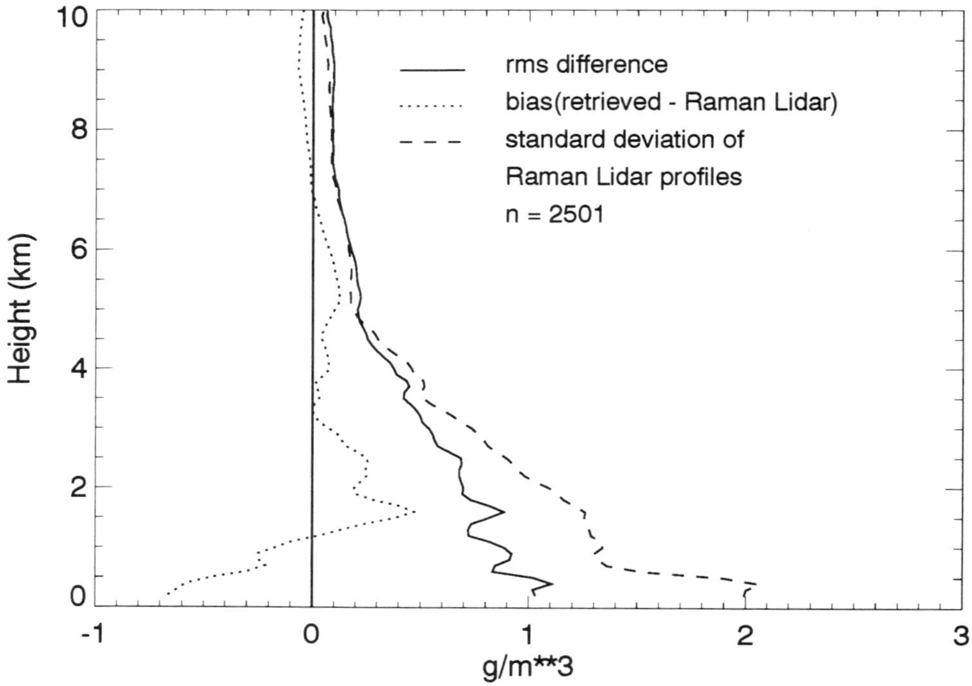

Figure 4. Rms comparisons of radiometric retrievals of water vapor profiles with Raman lidar, N = 2501.

We have examined three absorption models here. At the lower frequencies of 20.6 and 31.65 GHz and for the relatively dry Coffeyville, Kansas, environment, the calculations based on RTE76 and RTE93 produced similar results, with rms differences between measurements and calculations of ≈0.5K. However, the results at 90 GHz using RTE93 were not substantially better than the earlier RTE87, and may be worse. It was previously established that RTE76 poorly modeled 90 GHz emission. Since 90 GHz emission is not affected by changes in the 22.235 water vapor line parameters, the differences could be due to changes in the parameterization of the water vapor continuum or also in the description of the O_2 far wing behavior.

Definitive experiments using collocated Raman lidar and microwave radiometers could be conducted in humid locations to address these issues. The excellent temporal correlations between the Raman lidar's and the microwave radiometer's determination of precipitable water vapor confirms the ability of both of the instrument's to follow rapid changes in moisture. Temporal resolutions of 30 seconds could easily be obtained for both instruments. The already demonstrated agreement in profile determination between the Raman lidar and radiosondes lends further confirmation to this instruments ability as a meteorological research tool and as ground truth for passive remote sensors.

Acknowledgements
The authors thank Dr. H. J. Liebe for providing us with copies of his latest MPM software, as well as discussing several aspects of his work. The authors also thank Dr. G. Elgered for his useful comments on the manuscript. This work was partially supported by the Department of Energy's Atmospheric Radiation Measurements Program.

REFERENCES

1. England, M.N., R.A. Ferrare, S.H. Melfi, D.N. Whiteman, and T.A. Clark, 1992.Atmospheric Water Vapor Measurements: Comparison of Microwave Radiometry and Lidar, J. Geophys. Res.--Atmospheres, 97, 899-916.

2. May, P.T., R.G. Strauch, and K.P. Moran, 1988.The altitude coverage of temperature measurements using RASS with wind profiler radars, Geophys. Res. Lett., 15, 1381-1384.

3. Westwater, E.R., 1993."Ground-based remote sensing of meteorological variables," in Atmospheric Remote Sensing by Microwave Radiometry, M.A. Janssen, ed., 145-213, John Wiley and Sons, Inc.

4. Han, Yong, J.B. Snider, E.R. Westwater, S.H. Melfi, and R.A. Ferrare, 1994.Observations of water vapor by ground-based microwave radiometers and Raman lidar, J. Geophys. Res.--Atmospheres, in press.

5. Ferrare, R.A., S.H. Melfi, D.N. Whiteman, and K.D. Evans, 1992.Raman lidar measurements of Pinatubo aerosols over southeastern Kansas during November-December 1991, Geophys. Res. Letters, 19, 1599-1602.

6. Garand, L., C. Grassotti, J. Halle, and G. Klein, 1992.On differences in radiosonde humidity-reporting practices and their implications for numerical weather prediction and remote sensing, Bull. Amer Meteor. Soc., 73, 1417-1423.

7. Westwater, E.R., J.B. Snider, and M.J. Falls, 1990.Ground-based radiometric observations of atmospheric emission and attenuation at 20.6, 31.65, and 90 GHz: A comparison of measurements and theory, IEEE Trans. Antennas Prop., AP-38, 1569-1580.

8. England, M.N., F.J. Schmidlin, and J.M. Johansson, 1993.Atmospheric moisture measurements: A microwave radiometer-radiosonde comparison. IEEE Trans. Geosci. Remote Sensing, 31, 389-398.

9. Waters, J.W., 1976."Absorption and emission of microwave radiation by atmospheric gases," Methods of Experimental Physics, vol 12B, M.L. Meeks, ed., Radio Astronomy, Sect. 2.3.

10. Rosenkranz, P.W., 1988.Interference coefficients for overlapping oxygen lines in air, J Quant. Spectrosc. Radiat. Transfer, 39, 287-297.

11. Liebe, H.J., 1989.MPM - An atmospheric millimeter wave propagation model, Int. J. Infrared and Millimeter Waves, 10, 631-650.

12. -----, G.A. Hufford, and M.G. Cotton, 1993.Propagation modeling of moist air and suspended water/ice particles below 1000 GHz, AGARD 52nd Spec. Meet. of the Electromagnetic Wave Propagation Panel, Palma de Mallorca, Spain, 17-21 May, 1993, 3-1 - 3-10.

13. -----, P.W. Rosenkranz, and G.A. Hufford, 1992.Atmospheric 60-GHz oxygen spectrum: New laboratory measurements and line parameters, J. Quant. Spectros. Radiat. Transfer, 48, 629-643.

14. Snider, J.B., 1994.Observed and theoretical atmospheric emission at 20, 30, and 90 GHz: Recent results from land- and ocean-based locations. This Proceedings Volume.

15. Askne, J.I.H. and E.R. Westwater, 1986.A review of ground-based remote sensing of temperature and moisture by passive microwave radiometers, IEEE Trans. Geoscience and Remote Sensing, GE-24, 340-352.

Observed and Theoretical Atmospheric Emission at 20, 30, and 90 GHz: Recent Results from Land- and Ocean-based Locations

JACK B. SNIDER

National Oceanic and Atmospheric Administration
R/E/ET5
325 Broadway
Boulder, CO 80303-3328
U. S. A.

Abstract The Environmental Technology Laboratory (ETL) operated a number of microwave radiometers at various domestic and international locations during the period 1987 to 1993. The ability of these systems to operate continuously and unattended for extended periods of time has provided significant information on cloud liquid and water vapor for climate-related studies. The radiometers have also provided new information for the evaluation of absorption models, which relate measurements and theory. Because absorption models are often developed partly from laboratory observations, it is important for remote sensing that their applicability in the atmosphere be determined. In this paper, measurements of atmospheric emission from 20 to 90 GHz are compared with values predicted by theoretical models. Agreement between observations and theory is examined in terms of geographical location and differences in radiosonde humidity sensors.

Keywords: Atmospheric absorption, emission, microwave radiometry, remote sensing

INTRODUCTION

In remote sensing of atmospheric variables using passive techniques, it is necessary to understand the relationship between the amount of the atmospheric variable and the absorption (hence, emission) produced. Customarily, these relationships are predicted through absorption models. However, as discussed by Liebe [1], the purpose of a model is to describe a phenomenon with a limited number of parameters; it thus may not identify all factors present in the operating environment. For the past several years, the Environmental Technology Laboratory (ETL), has participated in a number of observational programs related to climate and global change. We have employed microwave radiometers operating at frequencies from 20.6 to 90.0 GHz to observe atmospheric water vapor and liquid content of clouds. These experiments provided a variety of supporting data including radiosondes which have been launched on a systematic basis for time periods ranging from one week to two months. The combination of microwave radiometer and simultaneous radiosonde data provided the opportunity to evaluate a number of microwave absorption models at a variety of locations. This paper presents a comparison of observed brightness temperatures with those predicted by three absorption models.

DESCRIPTION OF INSTRUMENTATION

ETL Microwave Radiometers
Microwave instrumentation included the following radiometers: a three-channel system operating at 20.6, 31.65, and 90.0 GHz, and two dual-channel radiometers operating (1) at 20.6 and 31.65 GHz, and (2) at 23.87 and 31.65 GHz. Supporting data included narrowband infrared radiometers (wavelength 10.7 μm), which provided an indication of cloudy conditions and whether the clouds contained liquid or ice. All microwave radiometers employ triple Dicke switching and offset parabolic antennas with beamwidths ranging from 2.5 to 4.0 deg. The absolute accuracy of the brightness temperature measurements varies from about ± 0.5 K at 20.6 GHz, ± 0.9 K at 31.65 GHz, to ± 1.5 K at 90 GHz. During the observations, the radiometers operated continuously with antenna beams directed to the zenith.

Microwave radiometers were calibrated using "tipping curves" where atmospheric absorption is determined from antenna elevation scans in clear weather [2]. The brightness temperature of the atmosphere is calculated from the absorption and the mean radiating temperature of the atmosphere which is determined from climatological data. Finally, constants are derived which relate the atmospheric brightness temperature to the radiometer output voltages and internal temperatures.

Radiosondes
The microwave data were supplemented by radiosondes launched at three- to six-hour intervals, depending on the experiment. As shown in Table 1, most soundings employed the Vaisala type RS80 radiosonde with HUMICAP capacitive humidity sensors. At two locations, the VIZ radiosonde, which is the standard unit employed by the U. S. National Weather Service, was used. The humidity sensor used in the latter unit is the Accu-lok hygristor -- a resistive design.

Table 1. Type and number of radiosondes used at each measurement location.

Location	Total Soundings	Type of Radiosonde	
		Vaisala	VIZ
Wallops Island, VA.	24	20 (Note 1)	24
Coffeyville, KS	112	74	38
Porto Santo Island, Portugal	99	99	
R/V Malcolm Baldrige	67	67	
Kavieng, Papua New Guinea	110	110	

Note 1: Twenty soundings were made with both Vaisala and VIZ humidity sensors attached to a common balloon.

Measurement Locations
Measurement locations in this study ranged from the central United States in the temperate zone, to the North Atlantic, to the tropical Pacific region. Moisture amounts varied from extremely dry to excessively wet. Table 2 shows the measurement locations along with the mean precipitable water vapor, V (in cm), measured during each experiment. Note that V expressed in cm is equivalent to V in g m^{-2} [11]. The NOAA Research Vessel (R/V) Malcolm Baldrige cruised the region of the North Atlantic Ocean shown in Table 2; other locations were fixed.

Table 2. Measurement locations employed for the absorption study.

Location	Latitude (deg)	Longitude (deg)	Elevation (m)	Mean Precipitable Water Vapor (cm)	Period of Measurements
Wallops Island, VA	37.51 N	-75.29 W	4	1.56	11-18 April 1989
Coffeyville, KS	37.10 N	-95.57 W	227	1.66	13 November - 7 December 1991
Porto Santo Island, Portugal	33.08 N	-16.35 W	97	2.36	1-28 June 1992
R/V Malcolm Baldrige	28 to 38 N	-18 to -27 W	6	2.47	7-28 June 1992
Kavieng, Papua New Guinea	-2.58 S	150.8 W	3	5.26	15 January - 28 February 1993

ABSORPTION MODELS

Three water vapor absorption models were employed in this comparison of theory and measurement. Two are models developed by Liebe [3], [4] while the third was developed by Waters [5]. Two absorption models for O_2 are combined with the vapor absorption models as discussed in the following brief description of each model. The models are designated as Radiative Transfer Equation (RTE) followed by the year of publication expressed in two digits.

RTE87 Absorption Model
This model, by Liebe [3], uses the known properties of all H_2O lines below 1000 GHz and the Van Vleck-Weisskopf line shape to calculate the absorption. A "continuum" term, which accounts for the effect of lines above 1000 GHz and possible dimer contributions, is added to the line absorption. The continuum term was determined empirically from laboratory measurements. The RTE87 model employs the O_2 model described in [6].

RTE93 Absorption Model
Recently, Liebe [4] revised both his 1987 model and the previously used O_2 model [7]. The revisions include changes to both H_2O and O_2 parameters as well as the continuum term.

RTE76 Absorption Model
The Waters model [5] uses the kinetic line shape of Zhevakin-Naumov-Gross [8],[9] to fit the 22.235 GHz water vapor line and an empirical term for the wing absorption below 100 GHz. In the calculations described in the next section, the O_2 absorption model [6] is combined with the H_2O model of Waters [5].

COMPARISON OF OBSERVATIONS AND MODEL PREDICTIONS

Radiosonde profiles of temperature, relative humidity, and atmospheric pressure were input to the three absorption models to calculate downwelling brightness temperatures. Radiometric measurements of brightness temperature at the two minute period coincident with the radiosonde launch time were compared to the model calculations. To avoid contamination of the comparisons caused by emission from liquid water, comparisons were limited to clear or cirrus cloud conditions. Data from the narrowband infrared (IR) radiometer were employed to filter the microwave data. For example, data with IR brightness temperatures greater than -20° C were not considered.

A typical scatter plot of calculated versus measured brightness temperatures is shown in Fig. 1. This example is for Porto Santo Island, Portugal and shows results at 20.6 GHz for each absorption model. A more detailed comparison of results at Porto Santo for all three frequencies and models is given in Tables 3 through 5. The mixed nature of the comparisons is readily apparent. At 20.6 GHz, all three models produce similar results with RTE93 yielding a slight improvement in agreement over RTE87. At 31.65 GHz, RTE76 and RTE87 give comparable agreement although RTE93 gives the best overall agreement. In general, at 20 and 30 GHz, all models tend to overpredict the brightness temperature at low values of V and underpredict the brightness at higher V values. At 90 GHz, RTE87 produces the best agreement but generally underpredicts the brightness temperature. Although RTE93 shows a fairly large bias of about 5 K in brightness, the offset is nearly constant. The same is true for RTE76 but the offset is about 13 K.

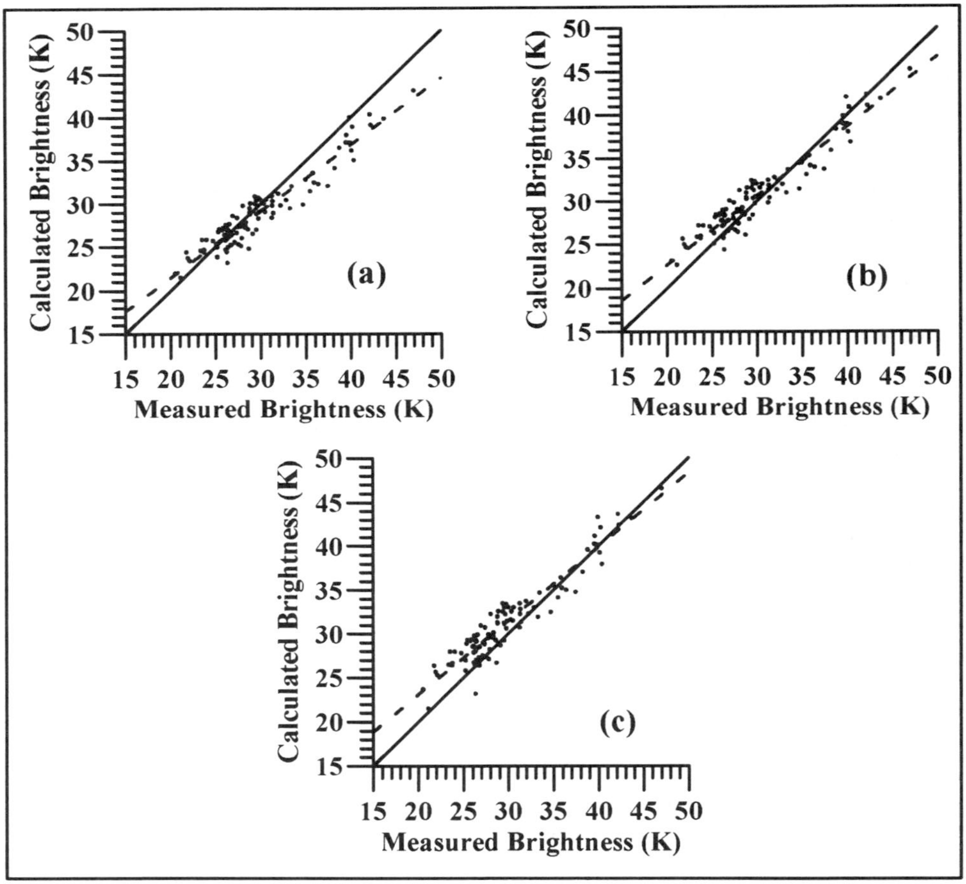

Figure 1. Comparison of measured and calculated brightness temperatures at 20.6 GHz for three absorption models: (a) RTE87, (b) RTE93. and (c) RTE76. Location: Porto Santo Island, Portugal, June 1992. Solid lines represent perfect agreement; dashed lines are the linear least squares fit to the data.

Table 3. Differences in measured and calculated brightness temperatures for three absorption models as a function of geographic location. Frequencies: 20.6, 23.87 GHz.

Location	Frequency (GHz)	Model	Sample Size	MeanΔT (K) Note 1	RMSΔT (K)	Fit Intercept	Fit Slope	Fit Correlation
Wallops I.	20.6	RTE87	24	-0.228	1.018	1.959	0.904	0.992
		RTE93		-1.258	1.533	2.247	0.945	0.992
		RTE76		-1.646	1.848	1.739	0.995	0.992
Coffeyville	20.6	RTE87	112	1.402	1.600	0.260	0.911	0.994
		RTE93		0.459	0.794	0.434	0.952	0.994
		RTE76		1.182	1.272	-0.078	0.998	0.994
Porto Santo	20.6	RTE87	99	0.853	2.035	6.075	0.769	0.954
		RTE93		-0.649	1.888	6.595	0.802	0.954
		RTE76		-1.478	2.309	6.255	0.841	0.949
R/V Baldrige	20.6	RTE87	67	0.248	1.171	3.535	0.878	0.982
		RTE93		-1.338	1.716	3.937	0.916	0.982
		RTE76		-2.305	2.521	3.820	0.951	0.982
Kavieng	23.87	RTE87	110	-2.017	3.563	6.215	0.942	0.856
		RTE93		-5.378	6.173	7.277	0.974	0.856
		RTE76		4.640	5.379	6.469	0.847	0.858

Note 1: Mean ΔT = Measured - Calculated Brightness Temperature

The trends observed at Porto Santo are not necessarily observed at the other locations. A brief summary of the details of Tables 3 - 5 follows:

20.6 and 23.87 GHz
The RTE87 model results in the smallest mean and RMS differences at Wallops Island, R/V M. Baldrige, and Kavieng. However, the RTE93 model gives better agreement at Coffeyville and Porto Santo.

31.65 GHz
The smaller differences at Wallops I., Coffeyville, and Porto Santo are obtained with RTE93. RTE76 gives the best agreement at R/V Baldrige and shows reasonable agreement at Kavieng. RTE87 results in the best agreement at Kavieng although the comparison is not greatly different than for RTE76.

90.0 GHz
RTE87 gives the best agreement at Wallops and Porto Santo, while RTE93 is superior at Coffeyville. The RTE76 model shows the poorest agreement at all three locations where 90.0 GHz measurements were made. This behavior is consistent with results obtained in California and Colorado in 1987 [10].

The lack of a consistent trend in agreement between the three models is somewhat surprising. Differences in the comparison can be due to either errors in the brightness measurements or errors in the radiosonde data that are input to the absorption models. However, data from Wallops I. suggest that inaccuracies in the radiosonde humdity sensor are likely to be the source of the differences. During the Wallops I. observations, both VIZ Accu-lok and Vaisala HUMICAP humidity sensors were flown on a common balloon on 20 flights. Fig. 2 presents a comparison of precipitable water vapor measured by the two types of sensor. The mean difference is 0.14 cm which translates to a difference in brightness temperature at 20.6 GHz, for example, of approximately 1.5 K. Therefore, differences in individual humidity sensors could possibly mask a systematic variation in model performance.

Table 4. Differences in measured and calculated brightness temperatures for three absorption models as a function of geographic location. Frequency: 31.65 GHz.

Location	Frequency (GHz)	Model	Sample Size	MeanΔT (K) Note 1	RMSΔT (K)	Fit Intercept	Fit Slope	Fit Correlation
Wallops I.	31.65	RTE87	24	1.606	1.732	1.047	0.841	0.987
		RTE93		0.179	0.527	0.618	0.952	0.987
		RTE76		0.741	0.903	0.697	0.914	0.988
Coffeyville	31.65	RTE87	112	1.932	2.005	0.350	0.858	0.989
		RTE93		0.642	0.754	-0.121	0.968	0.990
		RTE76		1.182	1.272	0.429	0.900	0.989
Porto Santo	31.65	RTE87	99	2.203	2.342	2.205	0.805	0.958
		RTE93		0.152	0.751	2.163	0.897	0.958
		RTE76		1.284	1.521	2.988	0.811	0.954
R/V Baldrige	31.65	RTE87	67	1.304	1.492	2.359	0.839	0.982
		RTE93		-0.865	1.078	2.247	0.939	0.973
		RTE76		0.326	0.753	3.311	0.840	0.977
Kavieng	31.65	RTE87	110	-0.538	1.751	7.071	0.812	0.770
		RTE93		-4.315	4.669	7.952	0.896	0.770
		RTE76		-0.863	1.739	9.859	0.742	0.789

Note 1: Mean ΔT = Measured - Calculated Brightness Temperature

Table 5. Differences in measured and calculated brightness temperatures for three absorption models as a function of geographic location. Frequency: 90.0 GHz.

Location	Frequency (GHz)	Model	Sample Size	MeanΔT (K) Note 1	RMSΔT (K)	Fit Intercept	Fit Slope	Fit Correlation
Wallops I.	90.0	RTE87	24	0.053	1.985	0.427	0.988	0.986
		RTE93		-2.777	4.039	-3.461	1.158	0.986
		RTE76		-11.253	11.984	0.048	1.283	0.987
Coffeyville	90.0	RTE87	112	1.936	2.425	-1.974	1.001	0.991
		RTE93		-0.432	2.469	-5.786	1.162	0.992
		RTE76		-8.254	8.900	-0.602	1.231	0.987
Porto Santo	90.0	RTE87	99	1.470	3.462	5.663	0.890	0.946
		RTE93		-4.915	5.945	4.277	1.010	0.946
		RTE76		-13.557	14.269	12.020	1.024	0.912

Note 1: Mean ΔT = Measured - Calculated Brightness Temperature

A companion paper by Han et al. [11] shows that V measured by the three channel microwave radiometer and a RAMAN lidar agree to within 0.03 cm RMS. This result is considered additional evidence that the inaccuracy of the radiosonde humidity sensor is a major contributor to the difference between measurements and theory.

However, when all data are combined a more systematic picture emerges. This may mean that differences in individual radiosondes tend to average out when considered in large quantities. Fig. 3 shows the agreement between measurements and calculations at 20.6 GHz for each

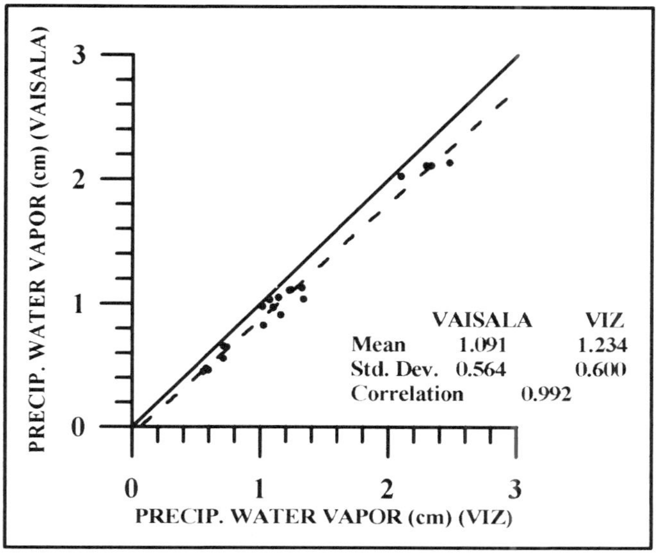

Figure 2. Comparison of precipitable water vapor measured by two humidity sensors flown on same radiosonde balloon. Data shown are for 20 soundings at Wallops Island, VA, April 1989. The solid line represents perfect agreement; the dashed line is the linear least squares fit to the data.

model when data from four locations are combined. Note the improvement relative to the data of Fig. 1. Table 6 contains a detailed comparison of mean and RMS differences, and fit information for the combined data. Considering Table 6, the RTE93 absorption model gives the best agreement at 20.6 and 31.65 GHz with mean differences of less than 0.5 K. The RTE76 and RTE87 models models perform similarly at 20.6 GHz, while RTE76 gives the better agreement at 31.65 GHz. The 31.65 GHz data from Kavieng were not included with the other locations because the relationship between measurements and calculations is different there than at other locations. At 90 GHz, the best fit is provided by the RTE87 model which may indicate that the earlier version of the Liebe model has a better representation of the continuum contribution than does the later model.

Table 6. Differences in measured and calculated brightness temperatures for three absorption models when data are combined. Frequencies: 20.6, 31.65, and 90.0 GHz.

Location	Frequency (GHz)	Model	Sample Size	MeanΔT (K) Note 1	RMSΔT (K)	Fit: Intercept	Fit: Slope	Fit: Correlation
WAL, COF, PSO, BAL	20.6	RTE87	302	0.836	1.641	0.282	0.955	0.985
		RTE93		-0.439	1.497	0.470	0.999	0.985
		RTE76		-1.083	1.891	0.021	1.042	0.985
WAL, COF, PSO, BAL	31.65	RTE87	112	1.856	2.005	-0.603	0.937	0.983
		RTE93		0.110	0.822	-1.105	1.050	0.985
		RTE76		0.990	1.245	-0.259	0.963	0.984
WAL, COF, PSO	90.0	RTE87	99	1.548	2.873	-0.953	0.988	0.990
		RTE93		-2.560	4.411	-4.220	1.136	0.990
		RTE76		-10.794	11.755	2.096	1.175	0.984

Note 1: Mean ΔT = Measured - Calculated Brightness Temperature

Much different results were obtained at Kavieng (Tables 3 and 4) where the best agreement was obtained with the RTE87 model. Further, the mean bias is greater and the correlation is considerably lower than at other locations. The stability of the radiometers was comparable to that experienced at the other locations and is believed not to be a factor in the poorer agreement. Some difficulty with solar heating of the humidity elements prior to balloon launch was experienced at Kavieng. This effect was corrected during editing of the data by the National Center for Atmospheric Research. It is tempting to relate the different behavior at Kavieng to the tropical climate where the effect of the extremely high moisture content upon the radiosonde humidity sensor is unknown. However, the poor agreement relative to the other sites must be considered an anomaly at this time.

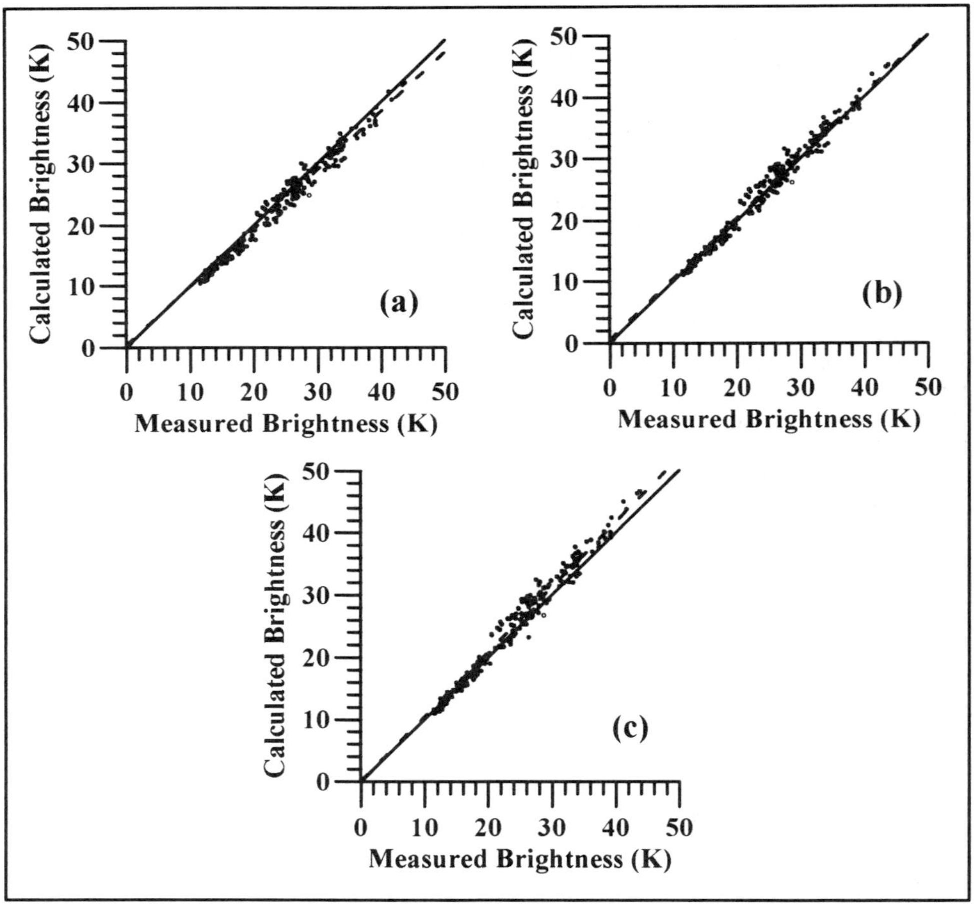

Figure 3. Comparison of measured and calculated brightness temperatures at 20.6 GHz for three absorption models: (a) RTE87, (b) RTE93, and (c) RTE76. Combined data are from Wallops Island, Coffeyville, Porto Santo Island, and R/V Malcolm Baldrige. Solid lines represent perfect agreement; dashed lines are the linear least squares fit to the data.

SUMMARY AND CONCLUSIONS

Brightness temperature measurements near 20, 23, 31, and 90 GHz performed during a series of climate-related campaigns have been compared to brightnesses calculated from simultaneous radiosonde data using three absorption models. Measurement locations ranged from the temperate zone of the central United States and North Atlantic Ocean to the tropical, equatorial Pacific Ocean. As in earlier observations [10], it was found in this work that differences in brightness temperature caused by the radiosonde humidity sensor are sometimes comparable to brightness differences caused by absorption models. For the temperate zone data, it was found that the updated model of Liebe [4] generally results in better agreement with the observed brightness temperatures at 20, 31, and 90 GHz than the previous Liebe model [3]. However, in the tropical region, all three models resulted in poorer agreement than was found in the temperate zone. Whether this is the result of problems with the models or inaccuracies in the humidity sensor when operated in the moist tropical region is unclear.

Acknowledgement
The author is grateful to Dr. H. J. Liebe for providing copies of his latest Millimeter Wave Propagation (MPM) software and to Dr. Yong Han for implementing the changes in ETL software. The author also wishes to acknowledge the outstanding work of the following members of his group who made the radiometric observations: D. A. Hazen, A. J. Francavilla, W. D. Madsen, M. D. Jacobson, and R. H. Beeler. Radiosonde data were provided by Colorado State University, the National Center for Atmospheric Research, Dr. C. Fairall of ETL, and by Frank Schmidlin of NASA/Goddard. Finally, the author thanks Drs. G. Elgered and E. R. Westwater for useful suggestions on the manuscript.

REFERENCES

1. Hans J. Liebe, 1985. An updated model for millimeter wave propagation in moist air, *Raaio Science*, **20**, 1069-1089.

2. D. C. Hogg, F. O. Guiraud, J. B. Snider, M. T. Decker, and E. R. Westwater, 1983. A steerable dual-channel microwave radiometer for measurement of water vapor and liquid in the troposphere, *J. Appl. Meteorol.*, **22**, 789-806.

3. H. J. Liebe and D. H. Layton, 1987. Millimeter-wave properties of the atmosphere: Laboratory studies and propagation modeling, Nat. Telecom. and Inform. Admin., Boulder, CO, NTIA Rep. 87-24.

4. H. J. Liebe, G. A. Hufford, and M. G. Cotton, 1993. Propagation modeling of moist air and suspended water/ice particles below 1000 GHz, *AGARD 52nd Spec. Meet. of the Electromagnetic Wave Propagation Panel,* Palma de Mallorca, Spain, 17-21 May, 1993.

5. J. W. Waters, 1976. Absorption and emission of microwave radiation by atmospheric gases, in *Methods of Experimental Physics,* **12,** sec 2.3, pt. B, Radio Astronomy, M. L. Meeks, ed., Academic Press, NY, 142-176.

6. P. W. Rosenkranz, 1988. Interference coefficients for overlapping oxygen lines in air, *J. Quant. Spectrosc. Radiat. Transfer,* **39,** 287-297.

7. H. J. Liebe, P. W. Rosenkranz, and G. A. Hufford, 1992. Atmospheric 60-GHz oxygen spectrum: New laboratory measurements and line parameters, *J. Quant. Spectros. Radiat. Transfer,* **48,** 629-643.

8. S. A. Zhevakin and A. P. Naumov, 1963. On the absorption coefficient of electromagnetic waves by water vapour in the 10 μ-2 cm band, *Izv. VUZ Radiofiz.,* **6,** 674-695.

9. E. P. Gross, 1955. Shape of collision broadened spectral lines, *Phys. Rev.,* **97,** 395-403.

10. E.. R. Westwater, J. B. Snider, and M. J. Falls, 1990. Ground-based radiometric observations of atmospheric emission and attenuation at 20.6, 31.6, and 90.0 GHz: A comparison of measurements and theory, *IEEE Trans. Antennas Prop.,* **AP-38,** 1569-1580.

11. Han, Yong, J. B. Snider, E. R. Westwater, S. H. Melfi, and R. A. Ferrare, 1994. Clear-air observations of water vapor by ground-based microwave radiometers and RAMAN lidar, this Proceedings Volume.

FORWARD MODELLING FOR AMSU

P.J. RAYER

Remote Sensing Instrumentation
U.K. Meteorological Office
Building Y70, D.R.A.
Farnborough, Hants GU14 6TD
U.K.
tel:+44-252-395727 fax:+44-252-515523
e-mail: pjrayer@email.meto.govt.uk

Abstract. The operational retrieval scheme used by the UK Meteorological Office includes the on-line calculation of radiative transfer along a first-guess profile using the fast forward model RTTOV. The transmittance stage of this model requires a set of instrument-specific coefficients derived by regression. In the past, the UKMO has carried out spectral calculations covering the HIRS and MSU channels using a modified version of the line-by-line model GENLN2 for a diverse set of profiles, and has performed the regressions to derive the RTTOV coefficients for the instruments aboard NOAA-11 and NOAA-12. New spectral calculations have been carried out for the planned AMSU instrument, and are being used to determine regression coefficients for a new version of the fast forward model. The salient parts of the new forward scheme are described, with some discussion of the spectral modelling and the regression predictors.

Keywords: AMSU, GENLN2, RTTOV, modelling, microwave, predictors

INTRODUCTION

The UK Meteorological Office (UKMO) has a continuing need to incorporate satellite data into the assimilation stage of its numerical weather prediction (NWP) models. At present, radiances from the HIRS (High resolution InfraRed Sounder) and MSU (Microwave Sounding Unit) are used to retrieve meteorological variables for input to the NWP data assimilation suite. The first step in the retrieval process is to simulate the radiances observed by the instrument using a first-guess (NWP) atmospheric profile. Therefore, a very fast forward model is required.

MSU is soon to be replaced by AMSU (Advanced Microwave Sounding Unit) which contains two separate modules (see Table 1). Firstly there is the 15-channel temperature sounder called AMSU-A with channels 1 and 2 at 23.8 and 31.4 GHz, channels 3 to 9 in the oxygen band between 50 to 58 GHz, narrow channels 10 to 14 with pass bands clustered around 57 GHz, and a channel at 89 GHz. The second module is the 5-channel humidity sounder AMSU-B which has channel 16 also at 89 GHz, channel 17 at 150 GHz, and channels 18 to 20 with pass bands clustered around the 183 GHz water vapour line.

For AMSU we propose to use essentially the same retrieval scheme discussed above until longer term plans for the assimilation of raw radiances come to fruition. A detailed calculation of transmittance spectra covering the AMSU channels for a diverse global set of radiosonde profiles has been made using the GENLN2[1] line-by-line model, as modified for use in the microwave. Prior to each new satellite instrument being launched, we will convolve these spectra with the channel response functions and regress the resulting set of channel transmittances onto the original set of profiles to obtain regression coefficients for the transmittance stage of the fast radiative transfer model RTTOV[2]. As part of the retrieval algorithm, the fast model is then available to perform the forward radiative transfer calculation for the lifetime of the instrument.

Table 1. A summary of the AMSU channel characteristics. The applications are numbered according to the following key: (1) water vapour; (2) precipitation; (3) cloud liquid water; (4) surface parameters; (5) temperature.

	Channel	Central Frequency	No. of Pass Bands	Retrieval Application
AMSU-A				
	1	23.8	2	1-4
	2	31.4	2	1-4
	3	50.3	2	5
	4	52.8	2	5
	5	53.596	2	5
	6	54.40	2	5
	7	54.94	2	5
	8	55.50	2	5
	9-10	57.290344	2 each	5
	11-14	57.290344	4 each	5
	15	89.0	2	1-4
AMSU-B				
	16	89.0	2	1-3
	17	150.0	2	1-3
	18-20	183.31	2 each	1-3

In preparation for AMSU, both GENLN2 and RTTOV are being modified. For GENLN2, the recommendations from the current microwave exercise of ITRA (Intercomparison of Transmittance and Radiance Algorithms) will be implemented, and additional in-house inter-model comparisons are also in progress, with some validation being available from aircraft measurements. For RTTOV, several proposed changes to improve the prediction scheme for water vapour transmittance are currently being examined for the HIRS $7\,\mu m$ water channels and will be extended to AMSU.

GENLN2 IN THE MICROWAVE

For AMSU the main active gases are water vapour and oxygen, although there is some ozone absorption near the AMSU-B water vapour channels. To obtain suitable RTTOV coefficients, the line-by-line model GENLN2 has been run in transmittance mode to create a spectral archive. For each of the sample set of 32 profiles, the transmittance from each of the 40 standard pressure levels has been calculated at 5 viewing angles and for 4 gas combinations, *viz.* water and oxygen together, oxygen alone, water alone, ozone alone, these combinations allowing more flexibility in the structure of the fast model. A fresh calculation will be undertaken this year to allow for improvements, but this fundamental transmittance archive is then expected to require only occasional updating to take account of significant improvements in line data.

For water vapour, we propose to use Version 3 of GENLN2 which will access the HITRAN-1992[3] line database and incorporate the pre-calculated water 'continuum' due to Clough[4]. Thus the line contribution is truncated at $25\,cm^{-1}$ from the line centre and each line beyond this is assumed to contribute to the pre-calculated continuum which is reconstructed for the given path conditions during run-time from a tabulation of continuum coefficients.

The effect of pressure broadening on a band of spectral lines is determined by the so-called relaxation matrix whose diagonal elements are the line halfwidths. For O_2 the off-diagonal elements which signify line coupling are important. Since there is no allowance for this in Version 3 of GENLN2, we have in the past made modifications for use in MSU simulations based on the Rosenkranz 1975[5] treatment in which the line shape is modified by an additional frequency-dependent term governed by a so-called *y*-coefficient. Pressure-normalised values for these were precalculated for a subset of the oxygen lines at several temperatures and fitted to temperature curves. The fitting coefficients were then tabulated with line identifiers to allow the path-adjusted *y*-coefficients for each oxygen line to be reconstructed during a GENLN2 run for incorporation into the line shape calculation.

We have used this scheme for the current ITRA microwave exercise and part of our preparation for AMSU will be to consider whether it should be updated in the light of ITRA and other comparisons. For instance, in moving from the limited range of MSU to the wider spectral coverage of AMSU, we will need to take some account of the collisionally-induced nitrogen contribution to the spectrum between the strong lines at 118 and 183 GHz.

We also plan to take account of reported work[6] on the oxygen band by Liebe, Rosenkranz, and others in recent years. For instance, the Liebe Millimeter Propagation Model (MPM[7]) now utilises new spectral measurements and a line coupling scheme based on Rosenkranz's 1988[8] treatment. This differs from the present GENLN2 scheme in that the relaxation matrix is not itself calculated, but the *y*-coefficients are obtained instead from the measured spectrum using a retrieval algorithm.

As part of the preparation for AMSU, further in-house modifications are being installed in GENLN2 which will allow it to incorporate this new work as an option. For this purpose, the published MPM spectroscopic parameters have been merged with HITRAN data to provide an alternative for the oxygen calculation. With respect to line coupling, this option replaces the temperature fitting tabulation in the old scheme by one which incorporates the published MPM coefficients.

LINE-BY-LINE MODEL COMPARISONS

Besides our contribution to the ITRA exercise which will not be reported here, work has begun along two main avenues in order to assess how well the modified GENLN2 models the AMSU channel measurements.

Comparison of Simulations with Measurements

Radiances measured at 89 and 157 GHz by an airborne radiometer have been compared with calculations using atmospheric profiles measured in-flight during three experiments from different climate zones: SAAMEX (Arctic), MASTEX (Mediterranean), and FATE (Tropical). The Microwave Airborne Scanning System (MARSS[9]) is a single polarisation, in-flight calibrated instrument with channels at 89 and 157 GHz. It is mounted on the side of a C-130 aircraft and rotates continually to enable measurements to be made along nine upward-looking and nine downward-looking viewing angles every 3 seconds.

For the nominal zenith view (12°) a set of 12 clear-air measurements from these three experiments was selected for study. The measured radiances at 89 and 157 GHz were simulated both by GENLN2 and by the Liebe MPM-89 model[10], this last being a model specifically developed for use in the microwave region which uses an in-stream set of line parameters and incorporates an empirically determined water 'continuum'. Since GENLN2 works in absorber partial pressure instead of absolute absorber amount, it was necessary to add upper levels for that model from US Standard Atmosphere 1976 to avoid incorrect scaling above the flight ceiling.

Results from one profile in each of the three experiments are plotted in Figure 1 (89 GHz) and Figure 2 (157 GHz). These experiments correspond to Flight H990 (Arctic), Flight A014 (Mediterranean), and Flight A144 (Tropical), where the integrated water amounts were 7.56, 18.11, 40.38 kg/m^2 respectively. The Figures show measured down-welling radiances (symbol Δ) from space to each profile level, together with the simulations based on GENLN2 (solid line) and MPM (dashed line). The measurement error bars are also shown.

At 89 GHz both models are in good agreement with each other except in the Mediterranean case of Figure 1(b) where the model values are divergent in the boundary layer, though lying generally within the measurement error.

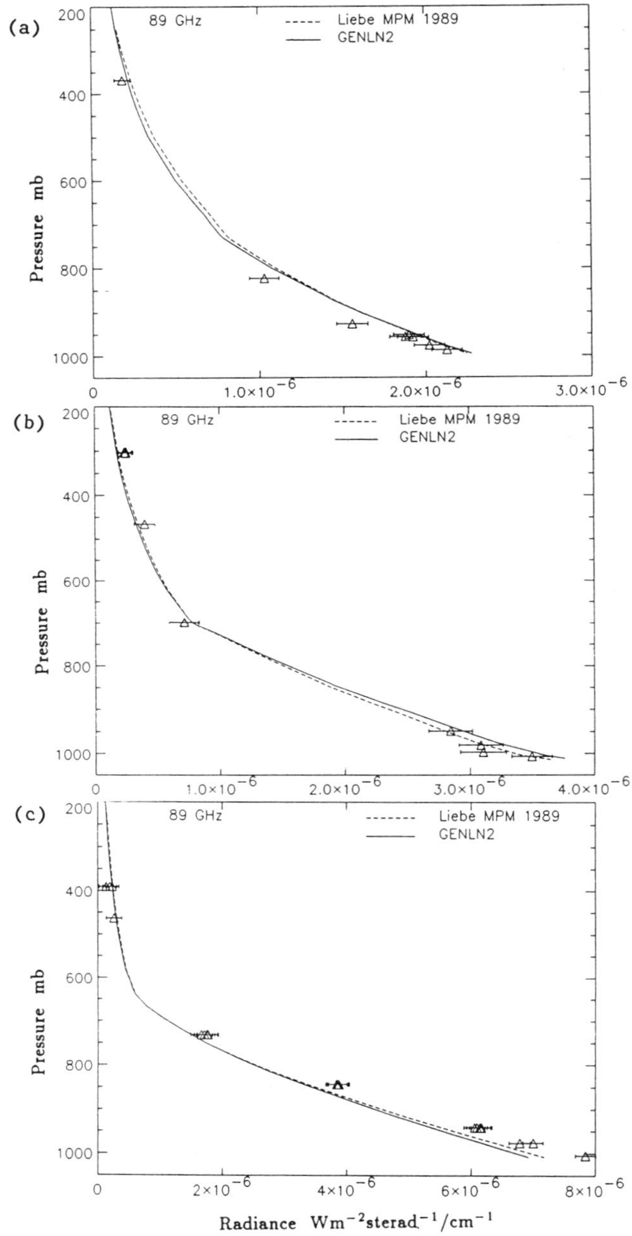

Figure 1. A comparison of radiance simulations using MPM-89 and GENLN2 with MARSS measurements (Δ) at 89 GHz. (a) Arctic profile from flight H990 from SAAMEX; (b) Mediterranean profile from flight A014 from MASTEX; (c) Tropical profile from flight A144 from FATE. Note that the radiance scales differ.

However, in the Arctic case of Figure 1(a), the values for both models are too high between 800 and 900 mb, whereas in the Tropical case of Figure 1(c) the values for both are too low.

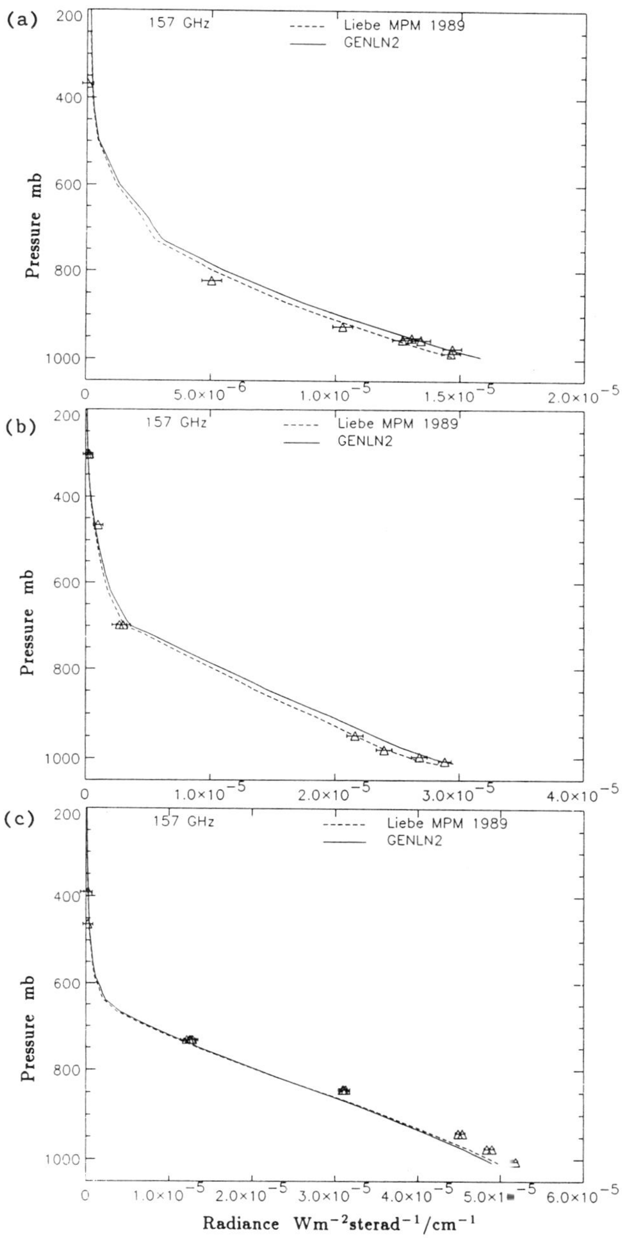

Figure 2. A comparison of radiance simulations using MPM-89 and GENLN2 with MARSS measurements (Δ) at 157 GHz. (a) Arctic profile from flight H990 from SAAMEX; (b) Mediterranean profile from flight A014 from MASTEX; (c) Tropical profile from flight A 44 from FATE. Note that the radiance scales differ.

Since the first profile is very dry and the second is very moist this result may be due to modelling problems for oxygen and the water 'continuum', respectively. In general, the MPM model is closer to the aircraft measurements.

At 157 GHz (see Figure 2), similar characteristics are shown, although there is less agreement between the models below 700 mb in both the Arctic case Figure of 2(a), and the Mediterranean case Figure 2(b).

Comparison between Models

The second study involves no measurements, but compares the results of the same two models (GENLN2 and MPM-89) in calculating the transmittance to space from pressure levels at 1000, 700, 500, 300, 100, 10, and 1 mb. A diverse sample of 36 profiles from a global set of radiosonde measurements, each with 40-levels, was used to provide nadir paths for oxygen and for water vapour. For these initial tests, spectra of transmittance and optical depth at a fairly low resolution (0.03 cm^{-1}) of about 1 GHz were calculated and examined over 6 spectral intervals chosen to cover the AMSU channels.

In each of these spectral intervals, transmittance differences (MPM – GENLN2) were computed for 5 profiles: the average of the original global set, two of the driest (Arctic), and two of the wettest (Tropical). In addition the mean and standard deviation of the differences over all 36 profiles were obtained.

Table 2. Differences in transmittance (MPM – GENLN2) from several pressure levels to space in the region of AMSU channels 15 and 16. Values are given as pecentages of unit transmittance for three cases: (a) all gases, (b) oxygen only, (c) water vapour only; for MPM-89, 'all gases' includes the nitrogen continuum. Two arctic, two tropical and a global mean profile were considered, and the mean and standard deviations of the differences over all 36 profiles in the sample are also shown.

P (mb)	Very Dry	Dry	Global Average	Moist	Very Moist	Mean	Standard Deviation
(a)							
100	-0.01	-0.01	-0.01	-0.01	-0.01	-0.01	0.00
300	-0.08	-0.08	-0.07	-0.08	-0.07	-0.07	0.01
500	-0.19	-0.21	-0.10	-0.24	-0.10	-0.13	0.06
700	-0.30	-0.32	0.00	-0.75	-0.91	-0.23	0.34
1000	-0.41	-0.40	0.30	-2.98	-3.60	-0.75	1.20
(b)							
100	-0.01	-0.01	-0.01	-0.01	-0.01	-0.01	0.00
300	-0.08	-0.08	-0.07	-0.08	-0.07	-0.07	0.01
500	-0.20	-0.21	-0.13	-0.29	-0.22	-0.18	0.05
700	-0.31	-0.34	-0.24	-0.70	-0.66	-0.35	0.16
1000	-0.46	-0.52	-0.54	-1.80	-1.86	-0.83	0.48
(c)							
100	0.00	0.00	0.00	0.00	0.00	0.00	0.00
300	0.00	0.00	0.00	0.00	0.00	0.00	0.00
500	0.00	0.01	0.03	0.05	0.11	0.04	0.04
700	0.01	0.02	0.24	-0.13	-0.33	0.11	0.23
1000	0.06	0.13	0.83	-1.98	-2.64	-0.12	1.05

Table 2 shows the results for the interval near AMSU channels 15 and 16 (89 GHz), for which the central frequency coincides with one of the MARSS channels from the previous comparison just discussed. Note that all transmittance differences have been multiplied by 100 and are therefore expressed as percentages of unit transmittance. For GENLN2 the oxygen path is consistently more transparent and the effect increases somewhat with warmer (moister) profiles. The average difference over all profiles is less than 1% of unit transmittance, and the standard deviation of the differences is about half this. This pattern is fairly representative of all 6 spectral intervals at this resolution.

Table 3 summarises the mean optical depths for the MPM and GENLN2 models over all 36 profiles, optical depths being a better measure for comparison when absorption is high. As before, the GENLN2 oxygen path is more transparent than that of MPM over these spectral intervals.

Table 3. Optical depths averaged over six spectral intervals using the MPM-89 and GENLN2 line-by-line models; the 'all gases' column for MPM-89 includes the nitrogen continuum. The values listed are averages over all 36 profiles for the path from 1000 mb to space.

Interval (GHz)	No. of points averaged	Model	Optical depths All gases	Oxygen	Water
22.4- 24.3	3	MPM-89	0.133	0.016	0.116
		GENLN2	0.134	0.014	0.120
30.6- 32.4	3	MPM-89	0.065	0.027	0.038
		GENLN2	0.059	0.023	0.036
49.5- 57.6	10	MPM-89	7.292	7.224	0.063
		GENLN2	7.244	7.177	0.067
85.4- 92.6	9	MPM-89	0.237	0.056	0.180
		GENLN2	0.224	0.047	0.177
148.4-152.0	5	MPM-89	0.606	0.017	0.589
		GENLN2	0.593	0.010	0.583
174.5-191.6	20	MPM-89	8.760	0.010	8.750
		GENLN2	8.705	0.006	8.699

THE RTTOV PREDICTORS

In its present structure, RTTOV deals with the constant (mixing ratio) gases and the variable (mixing ratio) gases separately so, for the present, the GENLN2 oxygen-and-water gas combination in the transmittance archive is not used. For the constant gases (here oxygen), the RTTOV regression uses a set of ten predictors based primarily on the Taylor expansion of the slant path transmittance about a nadir reference profile. They are, in the main, simple combinations of the deviations of layer temperature, water amount, and viewing angle from the reference.

With respect to water vapour, some allowance has been made for the large variation in water absorption encountered in practice — both in the layers of a single profile and in total burden over a global distribution. Here, therefore, the predictors which enter the regression are pre-scaled by the layer slant-path water amount. RTTOV at present treats ozone, the other variable gas, in a scheme based on total column amounts. This means that the GENLN2 archived ozone-alone transmittances are not used.

The regression on the GENLN2 transmittance archive supplies, for each predictor in each layer, one coefficient for mixed gas and one for water so that, by linear combination, RTTOV can very rapidly generate the respective transmittances whose product is required for the radiance calculation.

We are currently studying ways to improve the existing RTTOV prediction scheme for water vapour. In the mixed gas scheme, the basic predictors are the low order terms from the expansion mentioned above, but there are additional predictors to address the finite width of the instrument channel, and the deviation of the viewing angle from nadir. In the original RTTOV model, the angle predictors have been dropped from the water scheme to allow some of the remaining predictors to appear twice. This allows more freedom in the water scaling factor already mentioned. To take account of the 'weak' and 'strong' limits of integrated line absorption, some predictors are scaled linearly and others by the square root of the layer water amount along the path. However, although the water 'continuum' has a 'self' broadening component which varies quadratically with the water amount, none of the RTTOV predictors are scaled to this power.

Preliminary results point to a significant, almost universal, improvement for the HIRS and MSU channels when, as for mixed gases, the water scheme is provided with angle predictors. Two of the 'weak line' second-order predictors in temperature and water amount can be replaced by first and second order deviations of the zenith angle from nadir, provided that the linear scaling in slant path water amount is retained. However, no improvement was found when including the quadratic continuum path factor among the predictors.

Table 4. The difference in channel averaged transmittance (RTTOV – GENLN2) for each of the nineteen infrared channels of HIRS and the four microwave channels of MSU from each of the 40 standard pressure levels to space. Level 1 represents the top of the atmosphere and the values are given as percentages of unit transmittance. (a) standard deviation of the differences taken over 32 diverse profiles and 5 viewing angles (160 cases); (b) the table in (a) after subtracting a similar table calculated with the new RTTOV predictors. In (b) positive numbers indicate improvement.

(a)

1	2	3	4	5	6	7	8	9	10	11	12	13	14	15	16	17	18	19	1	2	3	4	
0.19	0.03	0.06	0.04	0.03	0.01	0.00	0.00	0.00	0.00	0.00	0.00	0.00	0.00	0.01	0.03	0.00	0.00	0.00	0.00	0.00	0.00	0.00	1
0.14	0.02	0.06	0.04	0.03	0.01	0.01	0.00	0.00	0.00	0.00	0.01	0.00	0.00	0.01	0.04	0.00	0.00	0.00	0.00	0.00	0.00	0.00	2
0.15	0.02	0.05	0.04	0.03	0.01	0.01	0.00	0.00	0.00	0.00	0.00	0.00	0.00	0.01	0.04	0.00	0.00	0.00	0.00	0.00	0.00	0.00	3
0.15	0.03	0.05	0.04	0.03	0.01	0.01	0.00	0.00	0.00	0.00	0.00	0.00	0.00	0.01	0.04	0.00	0.00	0.00	0.00	0.00	0.00	0.00	4
0.14	0.03	0.06	0.04	0.03	0.01	0.01	0.00	0.00	0.00	0.00	0.00	0.00	0.00	0.01	0.04	0.00	0.00	0.00	0.00	0.00	0.00	0.00	5
0.14	0.03	0.06	0.04	0.03	0.01	0.01	0.00	0.00	0.00	0.00	0.00	0.00	0.00	0.01	0.04	0.00	0.00	0.00	0.00	0.00	0.00	0.00	6
0.13	0.03	0.05	0.04	0.03	0.01	0.01	0.00	0.00	0.00	0.00	0.01	0.00	0.00	0.01	0.04	0.00	0.00	0.00	0.00	0.00	0.00	0.00	7
0.13	0.03	0.05	0.04	0.04	0.01	0.01	0.00	0.00	0.00	0.00	0.01	0.00	0.00	0.01	0.04	0.00	0.00	0.00	0.00	0.00	0.00	0.00	8
0.12	0.03	0.05	0.04	0.04	0.01	0.01	0.00	0.00	0.00	0.00	0.01	0.00	0.00	0.01	0.04	0.00	0.00	0.00	0.00	0.00	0.00	0.00	9
0.11	0.03	0.05	0.04	0.04	0.01	0.01	0.00	0.00	0.00	0.00	0.01	0.00	0.00	0.01	0.04	0.00	0.00	0.00	0.00	0.00	0.00	0.00	10
0.10	0.03	0.05	0.04	0.04	0.02	0.01	0.00	0.00	0.00	0.01	0.01	0.00	0.00	0.01	0.04	0.00	0.00	0.00	0.00	0.00	0.00	0.00	11
0.09	0.03	0.05	0.04	0.04	0.02	0.01	0.00	0.00	0.00	0.01	0.01	0.00	0.00	0.01	0.04	0.00	0.00	0.00	0.00	0.00	0.00	0.01	12
0.07	0.03	0.05	0.04	0.04	0.02	0.02	0.00	0.00	0.00	0.01	0.01	0.00	0.00	0.01	0.04	0.00	0.00	0.00	0.00	0.00	0.00	0.01	13
0.06	0.03	0.05	0.04	0.04	0.02	0.02	0.00	0.00	0.00	0.01	0.01	0.00	0.00	0.01	0.04	0.00	0.00	0.00	0.00	0.00	0.00	0.01	14
0.06	0.03	0.04	0.04	0.04	0.02	0.02	0.00	0.00	0.00	0.01	0.01	0.00	0.00	0.01	0.04	0.00	0.00	0.00	0.00	0.00	0.00	0.02	15
0.03	0.03	0.04	0.04	0.04	0.02	0.02	0.00	0.00	0.00	0.02	0.02	0.00	0.01	0.02	0.04	0.00	0.00	0.00	0.00	0.00	0.00	0.05	16
0.02	0.03	0.04	0.04	0.04	0.02	0.02	0.00	0.00	0.00	0.02	0.03	0.00	0.01	0.02	0.04	0.00	0.00	0.00	0.00	0.00	0.01	0.04	17
0.02	0.02	0.04	0.04	0.04	0.03	0.03	0.00	0.00	0.00	0.02	0.04	0.00	0.01	0.02	0.04	0.00	0.00	0.00	0.00	0.01	0.01	0.03	18
0.01	0.02	0.04	0.04	0.05	0.03	0.03	0.00	0.00	0.00	0.03	0.04	0.01	0.01	0.02	0.04	0.00	0.00	0.00	0.00	0.01	0.01	0.02	19
0.01	0.02	0.03	0.05	0.05	0.03	0.03	0.00	0.00	0.00	0.03	0.05	0.01	0.01	0.02	0.04	0.00	0.00	0.00	0.00	0.01	0.01	0.02	20
0.00	0.02	0.03	0.05	0.05	0.03	0.03	0.00	0.00	0.00	0.04	0.17	0.01	0.01	0.03	0.04	0.00	0.00	0.00	0.00	0.01	0.01	0.01	21
0.00	0.01	0.03	0.05	0.05	0.03	0.04	0.00	0.00	0.01	0.10	0.46	0.01	0.01	0.03	0.04	0.00	0.00	0.00	0.00	0.01	0.01	0.01	22
0.00	0.01	0.02	0.05	0.05	0.03	0.04	0.00	0.00	0.01	0.15	0.66	0.01	0.01	0.03	0.03	0.00	0.00	0.00	0.00	0.01	0.02	0.01	23
0.00	0.00	0.02	0.04	0.05	0.03	0.04	0.00	0.00	0.01	0.25	1.07	0.01	0.02	0.03	0.03	0.00	0.00	0.00	0.00	0.01	0.02	0.00	24
0.00	0.00	0.01	0.04	0.05	0.03	0.04	0.00	0.00	0.01	0.31	1.22	0.01	0.02	0.03	0.02	0.00	0.00	0.00	0.00	0.01	0.02	0.00	25
0.00	0.00	0.01	0.04	0.06	0.04	0.04	0.00	0.00	0.02	0.40	1.33	0.02	0.02	0.03	0.02	0.00	0.00	0.01	0.00	0.01	0.02	0.00	26
0.00	0.00	0.01	0.05	0.07	0.05	0.06	0.00	0.00	0.04	0.60	1.42	0.02	0.02	0.03	0.01	0.00	0.00	0.01	0.00	0.02	0.01	0.00	27
0.00	0.00	0.01	0.05	0.08	0.06	0.08	0.01	0.01	0.08	0.89	1.52	0.03	0.02	0.03	0.01	0.00	0.00	0.03	0.01	0.02	0.01	0.00	28
0.00	0.00	0.01	0.05	0.08	0.07	0.10	0.01	0.01	0.10	0.97	1.46	0.03	0.02	0.03	0.01	0.00	0.00	0.05	0.01	0.02	0.01	0.00	29
0.00	0.00	0.00	0.04	0.07	0.08	0.13	0.02	0.01	0.14	1.10	1.56	0.03	0.03	0.03	0.01	0.01	0.01	0.07	0.02	0.02	0.01	0.00	30
0.00	0.00	0.00	0.04	0.07	0.08	0.15	0.03	0.02	0.17	1.15	1.42	0.04	0.03	0.03	0.00	0.01	0.01	0.09	0.02	0.02	0.01	0.00	31
0.00	0.00	0.00	0.03	0.07	0.11	0.22	0.07	0.05	0.28	1.15	1.37	0.06	0.04	0.02	0.00	0.02	0.02	0.17	0.05	0.03	0.01	0.00	32
0.00	0.00	0.00	0.03	0.07	0.12	0.26	0.12	0.08	0.36	1.17	1.49	0.07	0.04	0.02	0.00	0.03	0.04	0.24	0.07	0.03	0.01	0.00	33
0.00	0.00	0.00	0.03	0.07	0.13	0.29	0.17	0.12	0.43	1.17	1.30	0.09	0.05	0.02	0.00	0.05	0.05	0.33	0.10	0.04	0.01	0.00	34
0.00	0.00	0.00	0.02	0.07	0.13	0.31	0.21	0.16	0.48	1.16	1.16	0.11	0.05	0.02	0.00	0.06	0.07	0.38	0.12	0.04	0.01	0.00	35
0.00	0.00	0.00	0.01	0.07	0.16	0.39	0.40	0.30	0.66	1.12	0.71	0.14	0.07	0.01	0.00	0.10	0.11	0.54	0.19	0.04	0.00	0.00	36
0.00	0.00	0.00	0.01	0.07	0.16	0.42	0.55	0.43	0.78	1.11	0.41	0.17	0.07	0.01	0.00	0.14	0.17	0.67	0.25	0.05	0.00	0.00	37
0.00	0.00	0.00	0.01	0.07	0.16	0.47	0.78	0.60	0.93	1.12	0.34	0.20	0.07	0.01	0.00	0.21	0.24	0.82	0.34	0.05	0.00	0.00	38
0.00	0.00	0.00	0.01	0.06	0.16	0.50	0.89	0.69	1.01	1.11	0.66	0.21	0.08	0.01	0.00	0.24	0.28	0.90	0.38	0.05	0.00	0.00	39
0.00	0.00	0.00	0.00	0.06	0.16	0.54	1.07	0.85	1.14	1.09	1.21	0.23	0.08	0.01	0.00	0.30	0.36	1.03	0.46	0.05	0.00	0.00	40

(b)

1	2	3	4	5	6	7	8	9	10	11	12	13	14	15	16	17	18	19	1	2	3	4	
0	0	0	0	0	0	0	0	0	0	0	0	0	0	0	0	0	0	0	0	0	0	0	1
0	0	0	0	0	0	0	0	0	0	0	+0.01	0	0	0	0	0	0	0	0	0	0	0	2
0	0	0	0	0	0	0	0	0	0	0	0	0	0	0	0	0	0	0	0	0	0	0	3
0	0	0	0	0	0	0	0	0	0	0	0	0	0	0	0	0	0	0	0	0	0	0	4
0	0	0	0	0	0	0	0	0	0	0	0	0	0	0	0	0	0	0	0	0	0	0	5
0	0	0	0	0	0	0	0	0	0	0	0	0	0	0	0	0	0	0	0	0	0	0	6
0	0	0	0	-0.01	0	0	0	0	0	0	+0.01	0	0	0	0	0	0	0	0	0	0	0	7
0	0	0	0	0	0	0	0	0	0	0	+0.01	0	0	0	0	0	0	0	0	0	0	0	8
0	0	0	0	0	0	0	0	0	0	0	+0.01	0	0	0	0	0	0	0	0	0	0	0	9
0	0	0	0	0	0	0	0	0	0	0	+0.01	0	0	0	0	0	0	0	0	0	0	0	10
0	0	0	0	0	0	0	0	0	0	+0.01	+0.01	0	0	0	0	0	0	0	0	0	0	0	11
0	0	0	0	0	0	0	0	0	0	+0.01	+0.01	0	0	0	0	0	0	0	0	0	0	0	12
0	0	0	0	0	0	0	0	0	0	+0.01	+0.01	0	0	0	0	0	0	0	0	0	0	0	13
0	0	0	0	0	0	0	0	0	0	+0.01	+0.01	0	0	0	0	0	0	0	0	0	0	0	14
0	0	0	0	0	0	0	0	0	0	+0.01	+0.01	0	0	0	0	0	0	0	0	0	0	0	15
0	0	0	0	0	0	0	0	0	0	+0.02	+0.02	0	0	0	0	0	0	0	0	0	0	0	16
0	0	0	0	0	0	0	0	0	0	+0.02	+0.02	0	0	0	0	0	0	0	0	0	0	0	17
0	0	0	0	0	0	0	0	0	0	+0.02	+0.03	0	0	0	0	0	0	0	0	0	0	0	18
0	0	0	-0.01	0	0	0	0	0	0	+0.02	+0.03	0	0	0	0	0	0	0	0	0	0	0	19
0	0	0	0	0	0	0	0	0	0	+0.02	+0.04	0	0	0	0	0	0	0	0	0	0	0	20
0	0	0	0	0	0	-0.01	0	0	0	+0.02	+0.05	0	0	0	0	0	0	0	0	0	0	0	21
0	0	0	0	0	0	0	0	0	0	+0.04	+0.11	0	0	0	0	0	0	0	0	0	0	0	22
0	0	0	0	0	0	0	0	0	0	+0.05	+0.16	0	0	0	0	0	0	0	0	0	0	0	23
0	0	0	0	0	0	0	0	0	0	+0.04	+0.10	0	0	0	0	0	0	0	0	0	0	0	24
0	0	0	0	0	-0.01	0	0	0	0	+0.03	+0.07	0	0	0	0	0	0	0	0	0	0	0	25
0	0	0	-0.01	0	0	-0.01	0	0	+0.01	+0.04	-0.02	0	0	0	0	0	0	+0.01	0	0	0	0	26
0	0	0	0	0	0	+0.01	0	0	+0.02	+0.04	+0.11	0	0	-0.01	0	0	0	0	0	0	0	0	27
0	0	0	0	+0.01	0	+0.01	0	+0.01	+0.03	+0.21	+0.28	0	0	0	0	0	0	+0.02	+0.01	0	0	0	28
0	0	0	+0.01	+0.01	+0.01	+0.02	0	+0.01	+0.04	+0.23	+0.14	0	0	0	0	0	0	+0.03	0	0	0	0	29
0	0	0	0	0	+0.02	+0.03	0	0	+0.06	+0.21	+0.03	0	+0.01	0	0	+0.01	+0.01	+0.05	+0.01	0	0	0	30
0	0	0	0	0	+0.02	+0.04	+0.01	+0.01	+0.06	+0.20	-0.02	+0.01	+0.01	0	0	+0.01	+0.01	+0.06	+0.01	0	0	0	31
0	0	0	0	+0.01	+0.03	+0.06	+0.03	+0.03	+0.09	+0.24	+0.01	+0.02	+0.01	0	0	+0.01	+0.01	+0.12	+0.03	+0.01	0	0	32
0	0	0	0	+0.01	+0.03	+0.07	+0.06	+0.05	+0.12	+0.21	0	+0.03	+0.01	0	0	+0.02	+0.03	+0.17	+0.04	+0.01	0	0	33
0	0	0	+0.01	+0.01	+0.04	+0.09	+0.09	+0.07	+0.15	+0.16	-0.04	+0.04	+0.02	0	0	+0.03	+0.03	+0.23	+0.06	+0.01	+0.01	0	34
0	0	0	0	+0.01	+0.04	+0.10	+0.12	+0.10	+0.18	+0.14	-0.08	+0.06	+0.02	+0.01	0	+0.04	+0.04	+0.26	+0.07	+0.01	+0.01	0	35
0	0	0	0	+0.02	+0.05	+0.11	+0.15	+0.15	+0.16	+0.11	-0.11	+0.07	+0.04	0	0	+0.06	+0.07	+0.36	+0.12	+0.02	0	0	36
0	0	0	0	+0.03	+0.06	+0.15	+0.24	+0.23	+0.26	+0.11	-0.01	+0.09	+0.04	0	0	+0.09	+0.11	+0.42	+0.16	+0.03	0	0	37
0	0	0	+0.01	+0.03	+0.07	+0.21	+0.41	+0.34	+0.41	+0.15	+0.07	+0.11	+0.04	0	0	+0.14	+0.16	+0.50	+0.23	+0.03	0	0	38
0	0	0	+0.01	+0.02	+0.07	+0.23	+0.50	+0.41	+0.46	+0.16	+0.15	+0.11	+0.05	0	0	+0.16	+0.19	+0.54	+0.25	+0.03	0	0	39
0	0	0	0	+0.03	+0.07	+0.25	+0.61	+0.51	+0.52	+0.17	+0.15	+0.12	+0.04	0	0	+0.20	+0.25	+0.60	+0.31	+0.03	0	0	40

Table 4 shows how the inclusion of angle predictors improves the standard deviation of the water scheme when taken over 5 selected viewing angles for each of a set of 32 diverse radiosonde profiles (160 cases).

The standard deviations in Table 4(a) relate to the statistic (RTTOV – GENLN2) for simulated transmittances over the nineteen HIRS infrared channels and the four MSU microwave channels for the original RTTOV; each value represents the percentage of unit transmittance. A similar tabulation was produced with angle predictors included in the RTTOV water scheme, and this has been subtracted from the original in Table 4(b). There is a dramatic improvement in the 'window' channels HIRS 8, 18, 19 and MSU 1, more than halving the standard deviation at some levels. In the HIRS 'water' channels, there is improvement at every level in channel 11, and an overall gain in channel 12.

At present we hope to use the same fast model for both HIRS and AMSU. It will therefore be important to extend this analysis beyond the oxygen region covered by MSU, and in particular to the AMSU-B channels near the 183 GHz water line.

CONCLUSIONS

The fast forward model adopted by the UK Met.Office retrieval scheme for HIRS and AMSU instruments on future satellites will be RTTOV or a derivative, using coefficients generated either by a microwave version of GENLN2, or perhaps the Liebe MPM model. Improvements to GENLN2 will include options to use more recent line data and an updated line coupling scheme in conjunction with the latest version of the HITRAN database. Improvements in the transmittance stage of RTTOV will primarily involve the scheme for water vapour, and the inclusion of angle predictors. Other improvements to RTTOV, which will lie in the radiance stage of the calculation, have not been addressed here.

Acknowledgements

Thanks is due to Stephen English (UK Met.Office) who supplied the aircraft data and ran the Liebe MPM model, and to Phil Watts (Rutherford Appleton Laboratory) and John Eyre (European Centre for Medium-Range Weather Forecasting) for collaboration on the fast model predictors.

REFERENCES

1. Edwards D.P., 1988. Atmospheric Transmittance and Radiance Calculations using Line-by-Line Computer Models, *SPIE Modeling of the Atmosphere,* **928**, 94-116.
2. Eyre J.R., 1991. A Fast radiative Transfer Model for Satellite Sounding Systems, *ECMWF Technical Memorandum No.176,* (Available from ECMWF, Shinfield park, Reading, UK)
3. Rothman L.S., R.R. Gamache, R. Tipping, C.P. Rinsland, M.A.H. Smith, D. Chris Benner, V. Malathy Devi, J.-M. Flaud, C. Camy-Peyret, A. Perrin, A.Goldman, S.T. Massie, L.R. Brown, and R.A. Toth, 1992. The HITRAN molecular database: editions of 1991 and 1992. *J.Quant.Spectrosc.Radiat.Transfer.* **48** Nos. 5/6,469-507.
4. Clough S.A., F.X. Kneizys, and R.W. Davies, 1989. Line Shape and the Water Vapor Continuum. *Atmospheric Research.* **23**, 229-241.
5. Rosenkranz P.W., 1975. Shape of the 5mm Oxygen Band in the Atmosphere. *IEEE Trans. Antennas Propag.,***23** No. 2, 498-506.
6. Liebe H.J., P.W. Rosenkranz, and G.A. Hufford, 1992. Atmospheric 60 GHz Oxygen Spectrum: New Laboratory Measurements and Line Parameters. *J.Quant.Spectrosc. Radiat.Transfer.*, **48** Nos. 5/6, 629-643.
7. Liebe H.J., G.A. Hufford, and M.G. Cotton, 1993. Propagation Modeling of Moist Air and Suspended Water/Ice Particle at Frequencies Below 1000 GHz. *AGARD 52nd Specialists' Meeting of the Electromagnetic Wave Propagation Panel, Spain.*

8. Rosenkranz P.W., 1988. Interference Coefficients for Overlapping Oxygen Lines in Air. *J.Quant.Spectrosc.Radiat.Transfer.*, **39** No. 4, 287-297.
9. Jones D.C., 1991. Microwave Airborne Radiometer Scanning System: Calibration and Initial Performance Assessment. *UKMO RSI Branch Memorandum No.3*, (Available from National Meteorological Library, Bracknell, UK.)
10. Liebe H.J., 1989. MPM - An Atmospheric Millimeter-Wave Propagation Model. *International Journal of Infrared and Millimeter Waves*, **10** No.6, 631-650.

PROFILING OF ATMOSPHERIC WATER VAPOR WITH THE MILLIMETER-WAVE IMAGING RADIOMETER

J. R. WANG[1], P. RACETTE[1], and L. A. CHANG[2]
[1]*NASA/Goddard Space Flight Center, Greenbelt, Maryland 20771.*
[2]*Futuretech Corporation, Gaithersburg, Maryland 20878.*

Abstract Measurements of atmospheric water vapor were made by the Millimeter-wave Imaging radiometer (MIR), Raman lidar and rawinsondes. Two types of rawinsonde sensor packages (AIR and Vaisala) were carried by the same balloon. The MIR measured brightness temperatures were used to retrieve water vapor profiles. The retrieved profiles were compared with those measured by the Raman lidar and rawinsondes. The results of these comparisons suggest that MIR is capable of retrieving useful water vapor profiles.

Keywords: Water vapor, remote sensing, millimeter-wave radiometry.

1. INTRODUCTION.

Water vapor is one of the key components affecting the atmospheric processes over a wide range of temporal and spatial scales from global climate to micrometeorology [1-3]. Measurements of this parameter vary from single locations (e.g., rawinsondes, ground-based Raman lidar) to aircraft flights at regional scales, and to satellite coverage over the entire globe. A number of different methods have been used for these measurements. The satellite large-scale measurements have been made by either infrared (e.g., TOVS, the Tiros Operational Vertical Sounder) or microwave sensors (e.g., SSM/I, the Special Sensor Microwave/Imager). The TOVS measurements could provide water vapor concentrations at several altitude levels, but the SSM/I measurements could only give the distribution of total precipiatable water over the ocean areas [4]. More recently, a five-channel microwave humidity sounder (SSM/T-2, Special Sensor Microwave/Temperature-2) was placed aboard the DMSP (Defense Meteorological Satellite Project) F-11 satellite to estimate water vapor profiles globally [5]. At aircraft altitudes, a number of measurements have been made previously with a microwave humidity sounder [6-9] to develop algorithms for the retrieval of water vapor profiles. These studies based on the aircraft measurements have shown a reasonable success and strongly suggest the soundness of both infrared and microwave approaches for large scale water vapor profiling.

In this paper we compare the results of near simulataneous measurements from a ground-based Raman lidar system, a 6-channel airborne Millimeter-wave Imaging Radiometer (MIR), and two different rawinsonde sensor packages (AIR, Vaisala). The lidar system has been in operation for the past several years at a few different locations [10-11], while the MIR was a newly developed instrument [12] that was flown for the first time in California aboard the NASA ER-2 aircraft in May 1992. The rawinsonde observations with different sensor packages were quite often not in a good agreement, as previously reported by [11]. The inaccuracy in the rawinsonde data could cause severe prolems for the users. The results of using the different rawinsonde data sets for radiative transfer calculations and comparison with measurements from various MIR frequency channels will be examined and discussed in detail.

2. MEASUREMENTS.

The GSFC Raman lidar is a trailer based system which uses an XeF excimer laser to transmit light at 351.2 nm. When pointed vertically, the lidar can provide one-minute profiles of water vapor from near the surface to an altitude of about 7 km with a vertical resolution of 75 meters [10-11]. A steerable elliptical flat provides 180 degree scan capability within a single scan plane. For this experiment, the lidar was oriented to scan over the rawinsonde launch site which was located on the coast about 3 km from the lidar.

The MIR is a newly built six-frequency (89, 150, 183.3±1, 183.3±3, 183.3±7, and 220 GHz) cross-track scanning total-power radiometer mounted on board the NASA ER-2 aircraft. The design and fabrication of the instrument are joint efforts between Goddard Space Flight Center and Georgia Institute of Technology [12]. The sensor has a 3-dB beamwidth of about 3.5° at all frequency channels and covers an angular swath of ±50° with respect to nadir. In every scanning cycle, it also views two external calibration targets; one of them is heated to a temperature of ~ 330 K and another remains at the ambient temperature which, at the cruising altitude of the ER-2 aircraft, will be ~ 240 K. The temperatures of these calibration targets are closely monitored to within ±0.1 K [12]. The temperature sensitivity of this sensor is on the order of 0.4 K and the calibration accuracy is better than ±1 K in the brightness temperature range of 240-300 K [12]. The measurement accuracy below 240 K is somewhat uncertain; based on the calibration studies in the laboratory, the accuracy near the liquid nitrogen temperature of 77 K is estimated to be ±2 K [12].

Two types of rawinsonde sensor packages (AIR AND Vaisala) were launched during the lidar/radiometer measurements. Balloons carrying both AIR and Vaisala sondes were launched in 3-hour intervals during the measurement period of 0100-0700 UTC on each of these five nights. The humidity sensors from the AIR sondes were made of carbon resistive strips which had a lowest measurement limit near 20% because of the algorithm used. Those on the Vaisala sondes were made of capacitive thin-films; the relative humidities reported by the Vaisala sensors were generally lower than those provided by the other two sensors.

The simultaneous measurements of all these sensors were conducted on five nights between 9 p.m. and 5 a.m. local time (0100-0900 UTC) during the period of July 29-August 6, 1992 at Goddard Space Flight Center/Wallops Flight Facility (WFF). In the first four days of measurements on July 29-30 and August 2-3, the sky was mostly clear; while on the last day of measurements on August 6, there was heavy cloud cover at WFF. During the nearly 8-hour operation, the Raman lidar was operated nearly continuously and rawinsonde launches were generally made at 9 p.m., 12 a.m. and 3 a.m. local time. The ER-2 aircraft spent about 2 hours over the lidar and rawinsonde sites in each of these 5 nights. The flight pattern over the WFF area was the same for all five nights. After reaching the cruising altitude of ~

20 km, the ER-2 aircraft flew four flight lines. Two of the flight lines were over water just off the coast and parallel to the coast line along the direction of south-southwest and north-northeast. The other two flight lines were over the lidar site and perpendicular to the coast line along the direction of west-northwest and east-southeast. MIR performed well in each of these flights.

3. RESULTS..

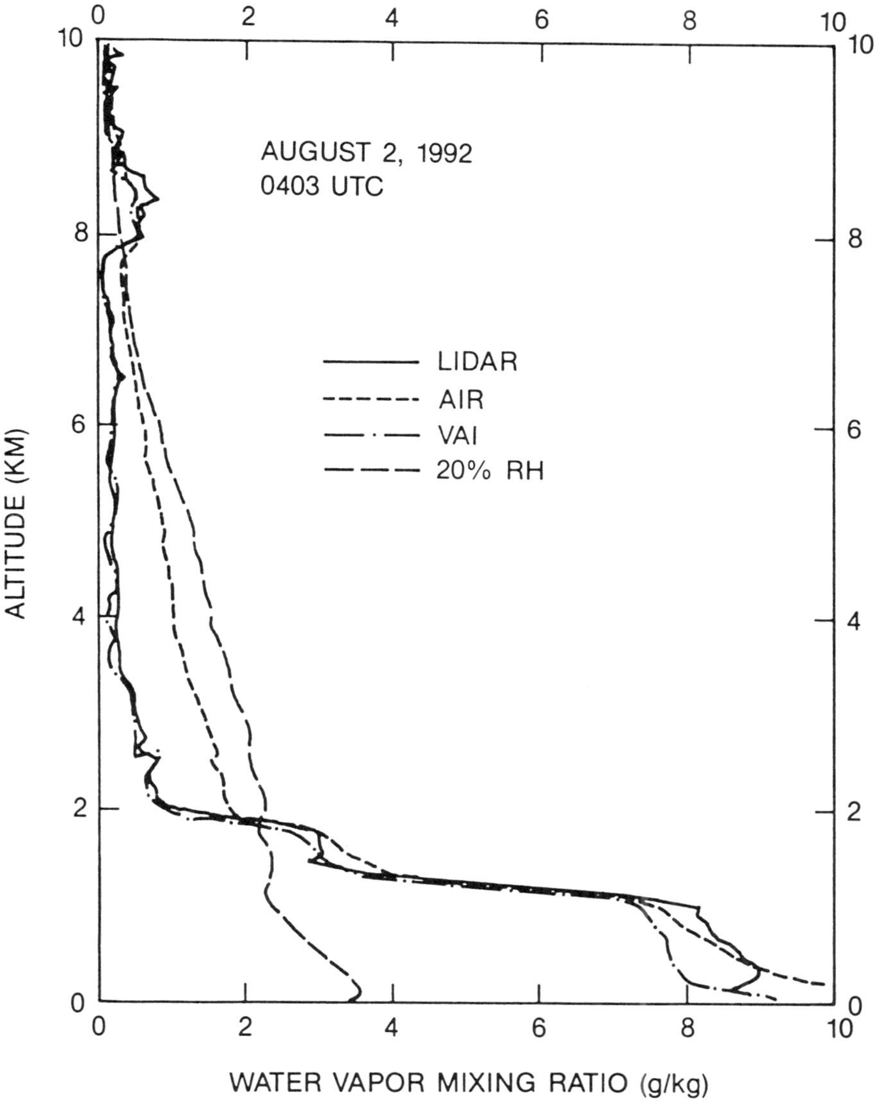

Figure 1. The variations with height of the water vapor mixing ratios measured by the ground-based Raman lidar system and two different rawinsondes on the same balloon.

The results of typical water vapor measurements from the ground-based Raman lidar and the AIR and Vaisala sondes are compared in Figures 1. The figure shows the altitude variation of the water vapor mixing ratios measured from surface to 10-km altitude on the night of August 2, 1992. A curve showing the mixing ratio (ρ) for 20% relative humidity is also included in the figure for comparison. The sky over the region of WFF was clear and there were no observable patches of clouds within the range of the Raman lidar (a sphere of about 10 km in radius). At altitudes < 2 km, the lidar and AIR sonde mixing ratios are in a good agreement, while the Vaisala mixing ratios are generally lower at altitudes < 1 km. The measurements from Raman lidar and Vaisala sondes agree very well at altitudes > 2 km, while that from the AIR sondes gives generally higher ρ values because of the AIR sonde's limitation in sensitivity to low relative humidities.

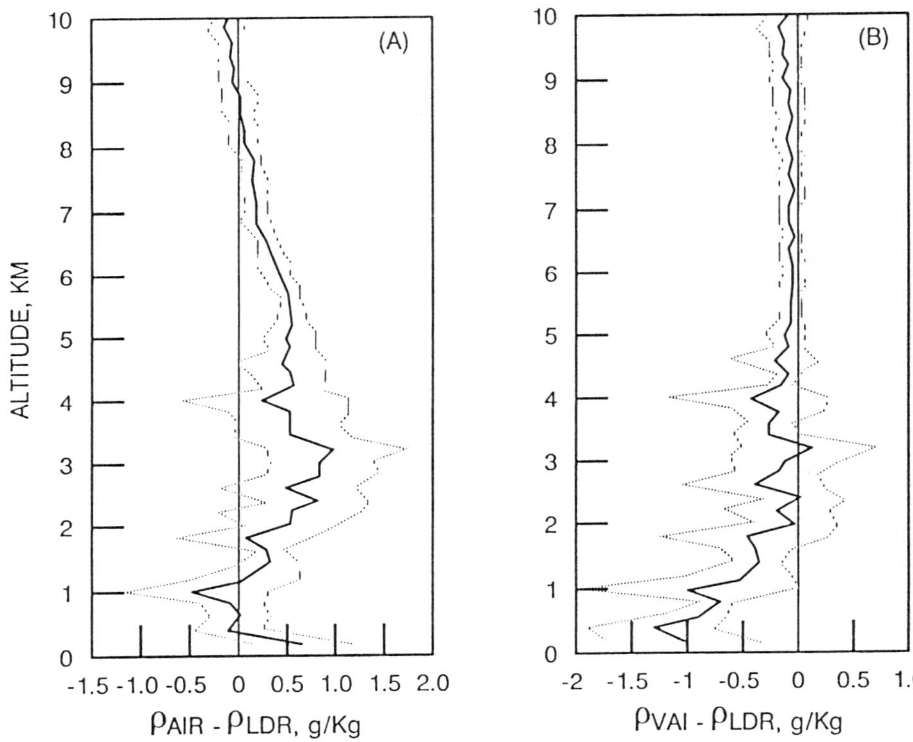

Figure 2. The variations with height of the average mixing ratio differences (solid curves) and their standard deviations (dotted curves) derived from measurements of (A) Raman lidar and AIR sondes and (B) Raman lidar and Vaisala sondes.

The discrepancies in the measurements by the different sensors displayed in Figure 1 are not unique just for the night of August 2, 1992. They are persistent for all five nights of observations. To illustrate this, the differences in the pairs of simultaneous ρ measurements by AIR and Vaisala sondes and by Vaisala sondes and Raman lidar are taken and their averages and standard deviations calculated. The results are shown in Figure 2, a and b respectively, for the height variations of the averages (solid curves) and standard deviations (dotted curves) of the mixing ratio differences ρ_{air}-ρ_{ldr} and ρ_{vai}-ρ_{ldr}, where the subscripts identify the type of measurements (AIR, Vaisala or lidar). At altitudes < 2 km, the mean ρ_{air}-ρ_{ldr} value is close to 0, while the mean ρ_{vai}-ρ_{ldr} is ~ -0.5-1.0 g/kg. At altitudes > 2 km, the mean ρ_{air}-ρ_{ldr} is ~ 0.5 g/kg, while the mean ρ_{vai}-ρ_{ldr} is close to 0. These features reflect the persistent differences in the measurements among the different humidity sensors, which turn out to be important in the radiative transfer calculations and the comparison with the MIR measurements.

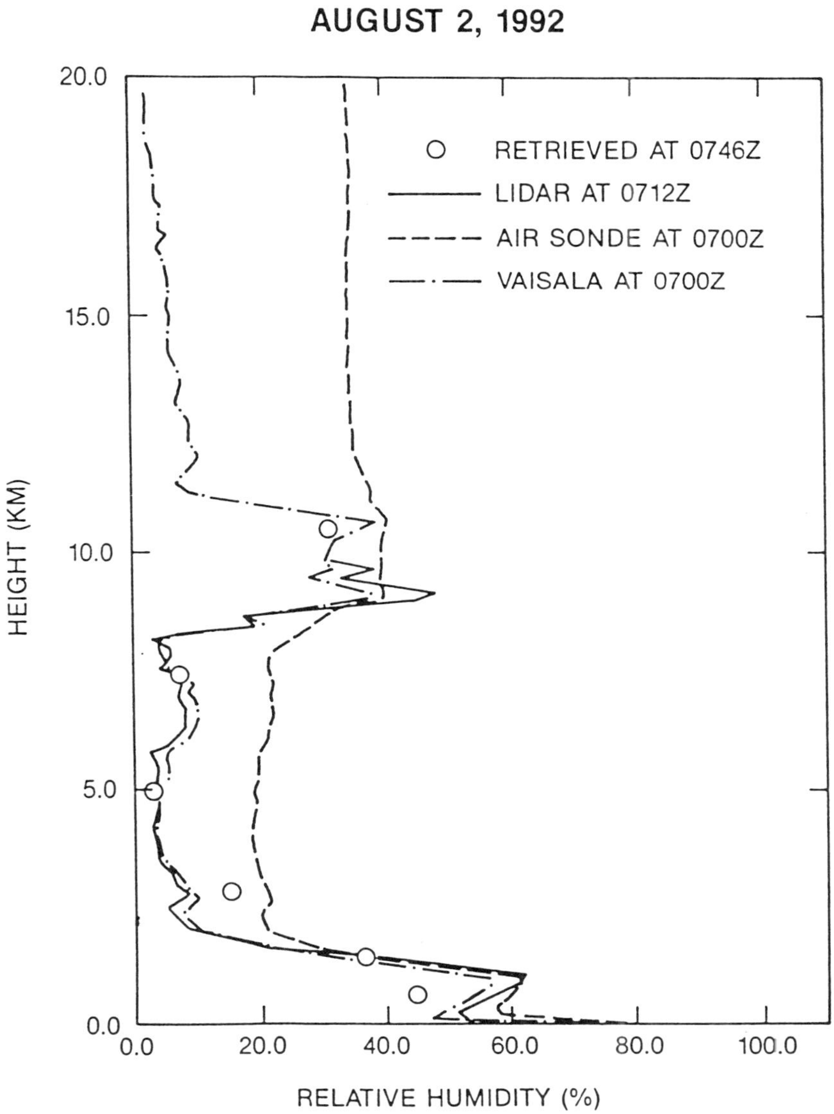

Figure 3. A compariosn of the MIR-retrieved profile with those measured by the Raman lidar, AIR and Vaisala sondes on August 2, 1992.

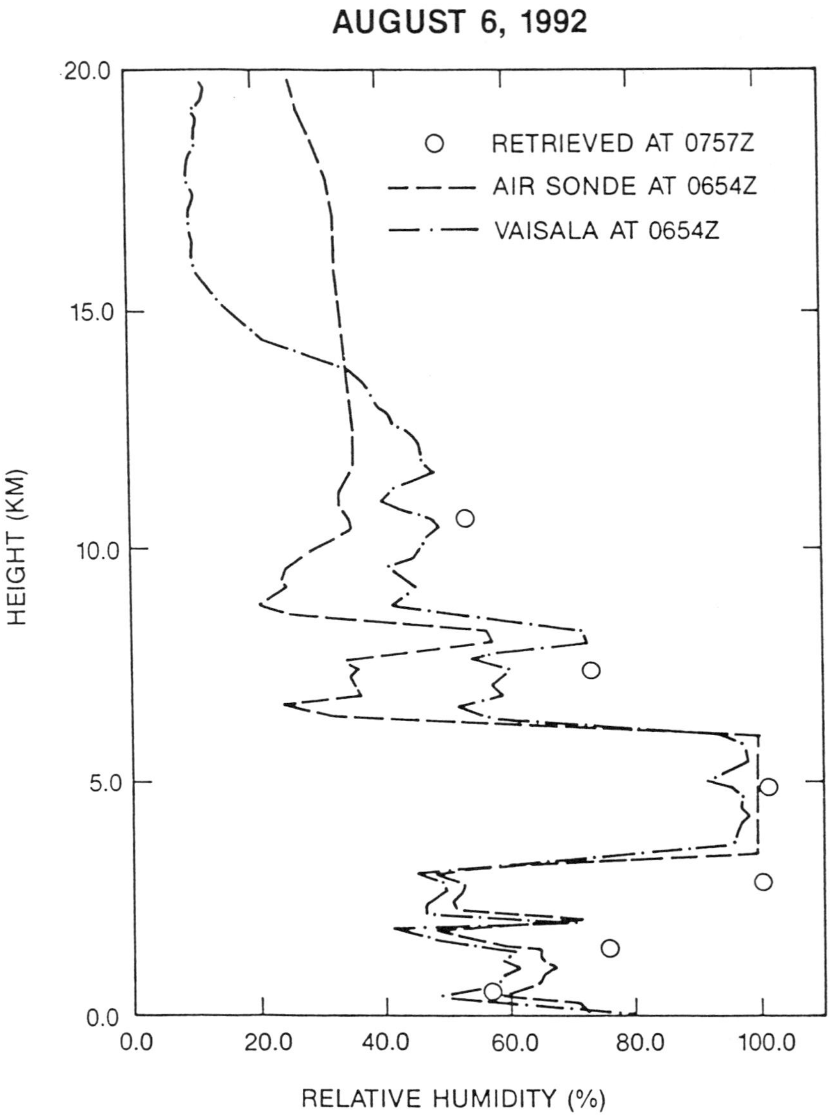

Figure 4. A comparison of the MIR-retrieved profile with those measured by the AIR and Vaisala sondes on August 6, 1992. Raman lidar measurements were not included in the figure because of the cloud cover.

Several different retrieval techniques for water vapor profiling using the microwave radiometric measurements near 90 and 183 GHz have been explored in the past years [6-9, 13]. Both approaches of Lutz et al. [13] and Wang and Chang [7] also allowed for an estimation of cloud liquid water at the level where the estimated relative humidity exceeded 95%. The approach of Lutz et al. employed the algorithm developed by Wilheit [14] in his simulation studies, while that of Wang and Chang [7] was the modified version of a Kalman-Bucy filter discussed by Ledsham and Staelin [15] and used by Wang et al. [6]. The former gave estimated water vapor values at a few altitude levels supposedly most reflective of the radiometric measurements; consequently, the number and the altitudes of the levels might change from one location of measurements to another. The latter [7] provided estimation of water vapor at a few selected altitude levels normally equal to the channel number of the radiometric measurements. This latter approach was used in this paper.

All six channels of MIR measurements over ocean surface closest to the locations of Raman lidar and rawinsonde observations were used in the retrieval. The temperature profiles recorded from the Vaisala sondes at the times closest to the MIR flights were input to the algorithm. The retrieval also adopted the atmospheric millimeter-wave propagation model by Liebe [16]. Six levels of retrieval were selected at the altitudes of 0.6, 1.6, 3.0, 5.0, 7.6, and 10.6 km. Typical results from the retrieval, expressed in relative humidity, were compared with those derived from the Raman lidar and rawinsondes in Figures 3 and 4 for August 2 and 6, 1992 respectively. The relative humidity at 20 km altitude was assumed to be zero and that at the surface was linearly extrapolated from the values retrieved at 0.6 km and 1.6 km altitudes. As reflected in the rawinsonde data, the atmosphere near WFF was quite dry during the night of August 2 and modestly wet on August 6. The retrieved relative humidity profiles from the MIR measuements also suggested a more humid air on August 6 than on August 6. There was a modest cloud cover in the proximity of WFF during the time of the measurements on August 6. The Raman lidar measurements on that day were valid only up to the altitudes of 2-3 km and were not included in Figure 4. The 0654Z AIR sonde data between the altitudes of 3.4 and 6 km were missing and a 100% relative humidity was assumed in the same figure.

The relative humidity profiles from the rawinsondes and Raman lidar on August 2 (Figure 3) show a moderate broad maximum centered at the height of about 10 km and small values between the heights of 2 and 8 km. The corresponding profile retrieved from the MIR data also displays these features. On the other hand, the profiles derived from the rawinsondes and Raman lidar on August 6 (Figure 4) show fluctuations between the heights of 0-3 km and 6-10 km, which are not observed from the MIR-retrieved profile. This could be due to the poor vertical resolution inherent in the microwave moisture sounding - a characteristic displayed by broad peaks in the weighting functions [6,17]. The cloud cover on August 6 apparently occurred in the altitude interval of 3.5-6.0 km, based on the rawinsonde data. The MIR-retrieved profile gives a 100% relative humidity in the altitude interval of 2.8-4.8 km. A total cloud liquid water of 20 mg/cm^2 is required in the 2.0 km layer to arrive at a convergent retrieval.

Table 1. The averages and standard deviations of ρ_{ldr} and of the differences between measured and retrieved ρ values at different altitudes.

Altitude Level (km)	$\rho_{ldr}-\rho_{mir}$ (g/kg)	$\rho_{air}-\rho_{mir}$ (g/kg)	$\rho_{vai}-\rho_{mir}$ (g/kg)	ρ_{ldr} (g/kg)
0.6	0.23±0.59	-0.43±0.99	-0.70±1.06	12.36±2.69
1.6	0.26±1.89	0.86±1.92	0.55±1.92	8.29±1.79
3.0	-0.37±0.84	0.54±1.16	-0.63±1.33	3.87±2.13
5.0	-0.19±0.79	0.12±0.94	-0.57±1.14	1.24±0.83
7.6	-0.05±0.27	0.22±0.27	-0.09±0.16	0.35±0.11
10.6	-0.02±0.02	0.02±0.02	-0.02±0.02	0.29±0.11

To arrive at the statistics of the comparison between the ρ's retrieved from the MIR data and those measured by Raman lidar and rawinsondes, differences of ρ_{ldr}-ρ_{mir}, ρ_{air}-ρ_{mir}, and ρ_{vai}-ρ_{mir} were tabulated and their averages and standard deviations computed. Table 1 summarizes these averages and standard deviations in terms of atmospheric levels. The ρ's were tagged with different subscripts for identifying water vapor densities from four different sources of MIR, Raman lidar, AIR and Vaisala sondes. The averages and standard deviations for ρ_{ldr} at the same altitudes are also entered in the table for comparison. There were 16 pairs of the samples for the calculations of the differences in ρ values and their standard deviations, and a total of 35 measurements for ρ_{ldr}. The averages and standard deviations from the table can be compared with the magnitudes and variations of ρ's measured by any one of the three methods (Raman lidar, AIR and Vaisala sondes) during the experimental period. For example, the standard deviations of ρ_{ldr}'s for the six levels from the surface to 10.6 km altitude are found from the fifth column of the table to be ± 2.69, ±1.79, ±2.13, ±0.83, ±0.11, and ±0.15 g/kg. The values in second column of the table suggests that the MIR could estimate these ρ_{ldr}'s to within ±0.59, ±1.89, ±0.84, ±0.79, ± 0.27, and ±0.02 g/kg, respectively, for the six levels from the surface to 10.6 km altitude. With the exception of the 7.6 km level, the accuracy of the estimation is better than the variation (as represented by the standard deviation) of ρ_{ldr}'s during the experimental period.

4. CONCLUSION.

Simultaneous measurements of atmospheric water vapor profiles were conducted with MIR, Raman lidar, and rawinsondes at NASA Goddard Space Flight Center's Wallops Flight Facility, Wallops Island, Virginia, during July 29-August 6, 1992. A similar measurement was made previously by England et al. [18] with the ground based microwave radiometers at the frequencies of 20.7, 22.7, and 31.4 GHz. The water vapor line at 22.235 GHz is quite weak, and the comparison of the microwave and lidar measurements was made by these authors in terms of total precipitable water only. The analysis presented above showed the measurements by the newly built MIR using the strong water vapor line at 183.3 GHz, which could provide estimated mixing ratios to compare with those observed by Raman lidar and rawinsondes at several different levels in the atmosphere.

Retrieval of water vapor profiles was made from the MIR measured brightness temperatures. Comparisons were made between the retrieved profiles and those measured by Raman lidar, AIR and Vaisala sondes. The results from these comparisons suggest that the MIR radiometric measurements are adequate for water vapor profiling.

REFERENCES.

1 Lindzen, R. S., 1990, Some coolness concerning global warming, Bull. Amer. Meteor. Soc., 71(3), 288-2992.

2. Haydu, K. J., and T. N. Krishnamurti, 1981: Moisture analysis from radiosonde and microwave spectrometer data. J. Appl. Meteor., 20, 1177-1191.

3. Elliott W. P., and D. J. Gaffen, 1991: The utility of radiosonde humidity archives for climate studies," Bull. Amer. Meteor. Soc., 72, 1507-1520.

4. Alishouse, J. C., S. A. Snyder, J. Vongsathorn, and R. R. Ferraro, 1990, Determination of oceanic total precipitable water from the SSM/I, IEEE Trans. Geosci. Remote Sensing, GE-28(5), 811-816.

5. Falcone, V. J., M. K. Griffin, R. G. Isaacs, J. D. Pickle, J. F. Morissey, A. J. Jackson, A. Bussey, R. Kakar, J. Wang, P. Racette, D. J. Boucher, B. H. Thomas, and A. M. Kishi, 1993, DMSP F11 SSM/T-2 calibration and validation. PL-TR-92-2293 Environmental Research Paper No. 1111, Phillips Laboratory, Hanscom Air Force Base, MA 01731-5000.
6. Wang, J. R., J. L. King, T. T. Wilheit, G. Szejwach, L. H. Gesell, R. A. Nieman, D. S. Niver, B. M. Krupp, and J. A. Gagliano, 1983, Profiling atmospheric water vapor by microwave radiometry, J. Clim. Appl. Meteor., 22(5), 779-788.
7. Wang, J. R., and L. A. Chang, 1990, Retrieval of water vapor profiles from microwave radiometric measurements near 90 and 183 GHz, J. Appl. Meteor., 29(10), 1006-1013.
8. Kakar, R. K., 1983: Retrieval of clear sky moisture profiles using the 183 GHz water vapor line. J. Clim. Appl. Meteor., 22, 1282-1289.
9. Kakar, R. K., and B. H. Lambrigtsen, 1984, A statisical correlation method for the retrieval of atmospheric moisture profiles by microwave radiometry, J. Clim. Appl. Meteor., 23, 1110-1114.
10. Whiteman, D. N., S. H. Melfi, and R. A. Ferrare, 1992: Raman lidar system for the measurements of water vapor and aerosols in the earth's atmosphere," Applied Optics, 31(16), 3068-3082.
11. Melfi, S. H., D. Whiteman, and R. Ferrare, 1989, Observation of atmospheric fronts using Raman lidar moisture measurements. J. Appl. Meteor., 28, 789-805.
12. Racette, P., R. F. Adler, D. M. Jackson, A. J. Gasiewski, J. R. Wang and D. S. Zacharias, 1994, An airborne millimeter-wave imaging radiometer for cloud, precipitation, and atmospheric water vapor studies, submitted to J. Atoms. Ocean. Tech.
13. Lutz, R., T. T. Wilheit, J. R. Wang, and R. K. Kakar, 1991, Retrieval of atmospheric water vapor profiles using radiometric measurements at 183 and 90 GHz. IEEE Trans. Geosci. Remote Sens., GE-29(4), 602-609.
14. Wilheit, T. T., 1990: Analgorithm for retrieving water vapor profiles in clear and cloudy atmospheres from 183 GHz radiometric measurements : simulation studies. J. Appl. Meteor, 29(6), 508-515.
15. Ledsham, W. H. and D. H. Staelin, 1978, An extended Kalman-Bucy filter for atmospheric temperature profile retrieval with a passive microwave sounder, J. Appl. Meteor., 17, 1023-1033.
16. Liebe, H. J., 1989, MPM-an atmospheric millimeter-wave propagation model. Intern. J. infrared and Millimeter Waves, 10(6), 631-650.
17. Schaerer, G., and T. T. Wilheit, 1979, A passive microwave technique for profiling atmospheric water vapor, Radio Sci., 14, 371-375.
18. England, M. N., R. A. Ferrare, S. H. Melfi, D. N. Whiteman, and T. A. Clark, 1992: Atmospheric water vapor measurements: comparison of microwave radiometry and lidar. J. Geophys. Res., 97(D1), 899-916.

Assimilation of satellite derived moisture fields for numerical weather prediction

GODELIEVE DEBLONDE[1], LOUIS GARAND, PIERRE GAUTHIER and CRISTOPHER GRASSOTTI[2]

1 Atmospheric Environment Service, Dorval, Québec, Canada.
2 Now with Atmospheric and Environmental Research Inc., Cambridge, MA 02139, USA.

Abstract. Integrated water vapour content (V) retrieved from Special Sensor Microwave / Imager (SSM/I) brightness temperatures and specific humidity retrieved from Geostationary Operational Environmental Satellite (GOES) radiances are assimilated for numerical weather prediction using a one-dimensional variational analysis technique which minimizes the error variance of the analyzed field. The root mean square difference of the analyzed field of V with collocated radiosondes is reduced by 1 kg m^{-2} with respect to the 6-h numerical weather prediction forecast when SSM/I V is assimilated and 0.6 kg m^{-2} when GOES retrievals of specific humidity are assimilated. In the upper levels of the troposphere (above 600 mb), the reduction in root mean square difference of specific humidity is largely due to the assimilation of GOES retrievals whereas in the lower atmosphere (850mb and 700mb) the reduction is mostly due to the assimilation of SSM/I V. This emphasizes the complementarity of using multisensor retrievals.

Key words: SSM/I, GOES, numerical weather prediction, analysis, integrated water vapour content, specific humidity.

INTRODUCTION

Satellite retrievals of moisture fields have been obtained from both microwave and infrared (IR) observations. Microwave sensors have a higher transparency to clouds than IR sensors and thus make them a better tool to retrieve water vapor. Microwave sensors on board of polar-orbiters see the same location about twice a day. Over the ocean surface, due to the low surface emissivity values, microwave brightness temperatures at the top of the atmosphere appear warm with respect to the ocean surface. Over land, the surface emissivity is higher and more variable and thus the retrieval of atmospheric water vapor is made more difficult. Retrievals of water vapor obtained with IR sensors are usually limited to cloud-free areas (techniques that obtain humidity profiles based on cloud-classification schemes are an exception to this) but are also available over land and have a higher sampling frequency (30 minutes) if the sensor is on board of a geostationary satellite. Retrieval techniques based on IR usually have similar or higher rms errors (rmse: root mean square difference between field in question and corresponding radiosondes) on integrated water vapour content than the microwave techniques [1]. Thus, ideally, a combination of both retrieval techniques would give access to the largest observation data set. Both sensor types are used in our study to retrieve the moisture field. Integrated water vapour content (V) retrievals are obtained from SSM/I brightness temperatures using an empirical relation. The IR retrieval technique [2] employed and named Humsat is based on a cloud-classification scheme. The technique provides vertical profiles of specific humidity (q) at six pressure levels: 1000, 850, 700, 500, 400 and 300 mb.

The study is divided into two parts, first, an evaluation of the accuracy of the retrievals of V and specific humidity profiles is done by comparing them with collocated upper air radiosondes. These retrievals are also compared with 6-h forecasts which are also referred to as *trial fields* and are obtained with an operational numerical weather prediction model (horizontal resolution of ≈ 250 km at the equator [3]). Secondly, the satellite retrieved SSM/I V and Humsat q are assimilated via a one-dimensional variational analysis technique [4]. The analyzed field is obtained by combining a trial field with the retrieved satellite data (both with known error covariances) by minimizing the analysis error variance.

METHODOLOGY

Integrated water vapour content was computed from DMSP (Defense Military Satellite Project) SSM/I brightness temperatures for 2 periods of 12 days each (on a 0.3° x 0.3° grid). The first data set extended from June 1 to June 12, 1991 and was available for the two satellites F08 and F10. The second data set extended from March 1 to March 12, 1992 and was available for the F10 and F11 satellites. Three regressions equations (from Alishouse and Ferraro in [5]) for the retrieval of V from brightness temperatures (T_B) over the ocean were compared. The equations (linear, logarithmic and non-linear respectively) are:

$$V = c_0 + c_1 T_{B,22v} \qquad (1)$$

where c_0 =-132.642, c_1= 0.713, V is in kg m^{-2} and T_B in K,

$$V = c_0 + c_1 \log_{10}(290.0 - T_{B,22v}) \qquad (2)$$

where c_0=205.844, c_1= -99.8898, and

$$V = c_0 + c_1 T_{B,19v} + c_2 T_{B,22v} + c_3 T_{B,37v} + c_4 (T_{B,22v})^2 \qquad (3)$$

where c_0=232.89393, c_1=-0.148596, c_2=-1.829125, c_3=-0.36954, and c_4=0.006193.

A precipitation screen [6] was applied prior to computing V. Based on the computation of the rmse and bias, the logarithmic and non-linear equations have a similar performance, whereas the linear equation has a lower performance. Fig. 1 illustrates the regressions for one of the data sets used in this study. The nonlinear regression equation was chosen for this study since it was selected as a calibration/validation algorithm [6] to retrieve V from SSM/I brightness temperatures over the ocean. For the dependent data set that was used to develop the regression, V had a rmse of 3.0 kg m^{-2} with a bias of 0.0 kg m^{-2}. A rmse of 4.2 kg m^{-2} with a bias of -0.7 kg m^{-2} was obtained for an independent data set [6].

Collocations between SSM/I V pixels and radiosondes (launched from small islands) were performed to estimate the accuracy of the retrievals. Collocations were global but restricted to be located between 60°N and 60°S. A SSM/I V pixel was chosen as a collocation pixel provided that both the distance to the radiosonde (< 300 km) and the difference in observation times (< 3 hours) were minimized. Vertical profiles of specific humidity for the trial fields were interpolated at the location of the radiosondes. For the trial fields and the radiosondes, V was computed from vertical profiles of q as follows:

$$V(\text{kg m}^{-2}) = \frac{1}{g}\int_{1000\text{mb}}^{300\text{mb}} q \, dp \qquad (4)$$

where g is the acceleration due to gravity and p is pressure. The integral was discretized using the trapezoidal rule.

The GOES retrieval technique (Humsat) is based on a cloud-classification scheme. Three channels of full disk GOES-7 images are employed: the IR (10.5-12.0 μm) channel, the visible (0.55-0.75 μm) channel when available and the water vapor (6.7 μm) channel also when

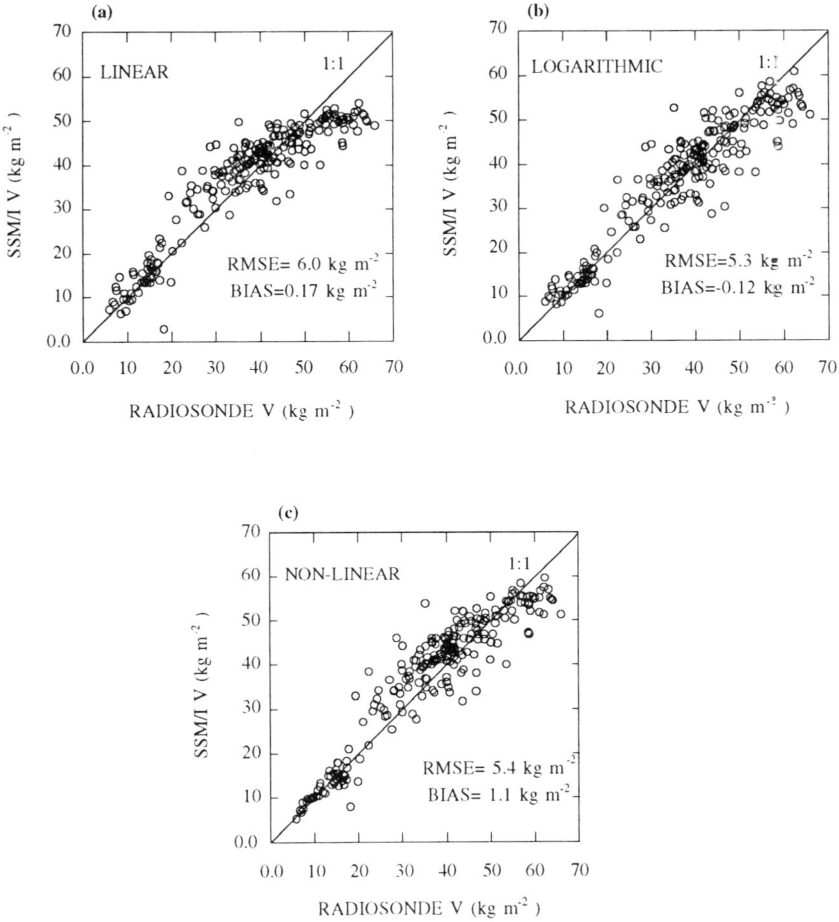

Figure 1: Comparison of the use of three different regression equations to obtain retrievals of integrated water vapour content from SSM/I brightness temperatures. The JUNE 1991 F10 data set was used, see Table 1. **a)** Linear relationship between integrated water vapour content and brightness temperatures (equation 1), **b)** Same as in a) but for a logarithmic relationship (equation 2) and **c)** Same as in a) but for a non-linear relationship (equation 3).

available. The IR channel alone or a combination Visible/IR (when the visible was available) was used to obtain the cloud classes. The features analyzed by the cloud-classification scheme are mean cloud albedo, cloud top pressure and cloud fraction. As a result of the classification, to each cloud class is assigned a mean profile of dew point depression (DPD, DPD is equal to the temperature minus the dew point temperature) which may be considered as a first guess to the retrieval. The first guess is improved on by applying class dependent empirical relationships which significantly reduce the rmse. First, the water vapor brightness temperature when available is used to estimate the humidity above 600 mb (the weighting function of the water vapor channel peaks near 400 mb). Secondly, an empirical relation between DPD and the temperature trial field at 1000 mb is developed. The dynamic range of this relation is too low in the tropics, and hence the relation is only used in the midlatitudes and over desert areas. Vertical profiles of DPD thus obtained are converted to specific humidity using the temperature trial field.

Humsat retrievals of q were available from 29 May 1991 to 12 June 1991 and 26 February 1992 to 12 March 1992 and overlap with the SSM/I data sets. Retrievals were performed in the -60° and 60° latitude and 190° to 320° longitude window. The profiles were obtained on a 1°x1° grid and collocations were performed over both land and oceans. A collocation study with radiosondes showed that a bias corrected rmse of 4.5 kg m^{-2} for V was obtained for an independent data set [2] with a V mean of 22.9 kg m^{-2} (midlatitude data set) and a standard deviation of 11.4 kg m^{-2}.

The aim of the one-dimensional variational analysis is to provide a best estimate for the vertical profile of q by combining different sources of observations and a background field chosen to be the trial field (6-h forecast of the numerical weather prediction model). If one assumes that observations and background error have Gaussian error distributions, then minimizing the cost function J(q) gives an estimate that minimizes the error variance of the analyzed field [4]. J(q) is given by:

$$J(q) = \frac{1}{2}(q - q_b)^T P^{-1}(q - q_b) + \frac{1}{2}(H(q) - z)^T O^{-1}(H(q) - z) \quad \textbf{(5)}$$

where z are the observations to be *assimilated* and H(q) is the so-called "forward model" for the variable z. q_b is the background field of specific humidity. P and O are the error autocovariance matrices of the background field and observations respectively. Since the error matrices vary with latitude and season, they were computed for four latitude bands (midlatitudes: 30° to 60° N or S and tropics: 0.0° to 30° N or S) and for the March 1992 and June 1991 data sets separately. If H(q) is linear in q suth that H(q)=Hq as is the case in our study, an analytic solution for the minimum of equation 5 may be found. If q_a represents the field that minimizes the error variance of the analysis, then one has:

$$q_a = q_b + K(z - H(q_b)) \text{ where } K = PH^T(O + HPH^T)^{-1} \quad \textbf{(6)}$$

In this study, the two sources of observations are: SSMI/I V and Humsat q and it follows that z, O and H may be written as:

$$z = \begin{bmatrix} z_m \\ q_h \end{bmatrix}; \; O = \begin{bmatrix} O_m & 0 \\ 0 & O_h \end{bmatrix}; \; H(q) = \begin{bmatrix} L_m q \\ q \end{bmatrix} \quad \textbf{(7)}$$

where z_m= SSM/I V, q_h is the Humsat retrieved specific humidity, O_m is the error variance for the SSM/I V and O_h is the matrix of the error autocovariance for Humsat retrievals of q. $L_m q$ is the vertical integral of q and L_m is given by applying the trapezoidal rule (see equation 4). q_a will henceforth be referred to as the analyzed field.

SATELLITE RETRIEVALS OF MOISTURE FIELDS

To evaluate the quality of moisture retrievals, three collocations studies with radiosondes were performed. The first study derived the rmse of V for collocations of SSM/I V and radiosondes, the second study derived the rmse for Humsat retrievals and radiosondes and finally, collocations were performed between SSM/I V, Humsat q and radiosondes. In all 3 cases, statistics were also available for the trial fields interpolated from the model grid to the location of the radiosondes.

Table 1 summarizes the collocation analysis for the SSM/I V retrievals (oceanic cases). In Table 1, the total number of collocations (sample size) and the number of different radiosonde

Table 1. SSM/I, trial field and analyzed integrated water vapour content collocation statistics (kg m^{-2}). The analyzed field results from the assimilation of SSM/I integrated water vapour content (see equation 6).

Sample Size / # of Stations	Radiosonde Mean	Radiosonde S.D.	Technique	S.D.	RMSE	BIAS
Data Set: June 91 --F08						
163/27	39.0	12.9	SSM/I	12.1	4.3	1.0
			Trial Field	12.7	4.7	1.1
			Analyzed	12.2	3.7	1.1
Data Set: June 91 --F10						
242/37	37.0	15.0	SSM/I	14.3	5.4	1.1
			Trial Field	15.0	4.6	1.0
			Analyzed	14.4	4.4	1.1
Data Set: March 92 --F10						
182/35	31.2	14.0	SSM/I	13.0	4.8	-0.63
			Trial Field	13.8	5.7	1.7
			Analyzed	13.1	4.4	0.25
Data Set: March 92 --F11						
95/27	36.9	12.3	SSM/I	12.7	4.4	-1.3
			Trial Field	12.6	5.2	1.1
			Analyzed	12.3	3.7	-0.41

launch sites (# of stations) are also listed. The radiosonde mean and standard deviation are listed in the second and third column. The column entitled "technique" lists the statistics for the SSM/I V and the trial field. The mean of these fields may be obtained by adding the bias to the radiosonde mean. The magnitude of the rmse (4.3 to 5.7 kg m^{-2}) is similar to that obtained in other studies (e.g. [7]). The highest rmse values are obtained with the F10 satellite. Data merging for different sensors were not performed because equations for the intercalibration between the different satellites were not available. However, differences between satellites were estimated by comparing near coincident pixels for different satellites [5]. The differences in the mean between F10 and F08 are less than 0.5 K for channels 19.3, 22.2 and 37 GHz. The differences between the F10 and F11 satellites are 1.6 K for the 22.2 GHz channel and less than 0.5 K for the 19.3 and 37.0 GHz channels. Using equation 1, a difference of 1.6 K for the 22.2 GHz channel would correspond to a V difference of ~3% for a V value of 35.0 kg m^{-2}. The rmse for the SSM/I V are smaller than for the trial fields except for one of the data sets (Table 1). This shows a higher performance of the SSM/I. The trial field bias of V is positive for all data sets and thus is over predicting moisture.

The rmse of V obtained with Humsat are listed in Table 2. Since more collocations were available for Humsat (considered both land and ocean cases), the data sets were split by latitudinal bands. The rmse for Humsat V is lowest for the midlatitudes and is comparable to that of the trial fields (5.1 kg m^{-2} for the former compared with 4.5 kg m^{-2} for the latter and for the June data set; these values are 4.0 kg m^{-2} compared with 4.2 kg m^{-2} for the March data set). For the tropical data sets, the Humsat V rmse are 6.7 kg m^{-2} for the June data set and 6.9 kg m^{-2} for the March data set. The trial field performs slightly better than Humsat in the tropics (Table 2). The bias of the trial field is positive in the midlatitudes and negative in the tropics. The retrieval errors obtained for the midlatitudes are similar to the ones reported in [2] and summarized above. In the tropics, the mean V is larger (Table 2) and thus a larger rmse is to be expected. The explained variance (1 minus the ratio of the retrieval error variance to the observed variance was 82% and 77% for the midlatitudes (June and March respectively) and 47% and 63% for the tropics (June and March respectively). Thus, in the midlatitudes, the performance of the retrievals is superior to that in the tropics. This may be explained in part by the fact that the empirical equation that related surface DPD to the temperature trial field is not applicable in the tropics (discussed above).

The number of samples obtained for a collocation study between SSM/I V, Humsat q and trial fields was substantially reduced and as a result quantitative conclusions could not be obtained from this data set. An alternative to this problem would be to use longer data sets.

Table 2. Humsat, trial field and analyzed integrated water vapour content collocation statistics (kg m^{-2}). The analyzed field results from the assimilation of Humsat (GOES) specific humidity (see equation 6).

Sample Size/ # of Stations	Radiosonde Mean	Radiosonde S.D.	Technique	S.D.	RMSE	BIAS
Data Set: 29 May 1991 to 12 June 1991: Midlatitudes						
809/80	24.8	11.9	Humsat	10.6	5.1	-0.37
			Trial Field	11.3	4.5	0.90
			Analyzed	10.8	4.1	0.33
		:Tropics				
238/34	40.6	9.3	Humsat	7.6	6.7	0.49
			Trial Field	8.1	6.4	-0.24
			Analyzed	7.3	5.5	-0.25
Data Set: 26 Feb 1992 to 12 March 1992: Midlatitudes						
251/29	13.5	8.3	Humsat	9.2	4.0	0.56
			Trial Field	9.1	4.2	2.3
			Analyzed	8.8	3.5	1.5
		:Tropics				
231/38	39.8	11.3	Humsat	10.4	6.9	-0.89
			Trial Field	11.3	5.4	-0.84
			Analyzed	10.5	5.1	-0.86

Even if Humsat V statistics are the least performant, statistics of q by level (Fig. 2, q_h) show that Humsat retrievals of q have lower rms errors than the trial fields for the levels of 500, 400 and 300 mb. The reason that Humsat performs better there is in major part due to the water vapor channel. For the midlatitude data sets, the retrieval of q at 1000 mb performs better than the trial field. Again, this is probably due to the fact that the empirical equation developed at the surface that relates DPD and the temperature trial field is not applicable in the tropics. Thus, the least accurate Humsat retrievals are for the levels of 700 and 850 mb. A complementarity between SSM/I and Humsat humidity retrievals exists in the sense that Humsat retrievals of q are better than the trial fields at the higher levels whereas for V, SSM/I retrievals perform as well as the trial fields and better than Humsat. Most of the atmospheric humidity is concentrated in the lowest 2 km of the troposphere and therefore the contribution of the SSM/I is largest for the lowest levels.

ASSIMILATION OF SATELLITE DERIVED MOISTURE FIELDS

Satellite retrievals of SSM/I V and Humsat q were assimilated by solving equation 6. Assimilation of SSM/I V, reduced the V rmse with respect to that of the trial field by between 0.2 and 1.5 kg m^{-2} (Table , analyzed field). Fig. 3 illustrates the rmse reduction of q as a function of pressure level due to the assimilation. The largest reductions occur for the 700 and 850 mb levels. The specific humidity at 1000 mb does not improve much. The rmse of q for the background field is lower at the surface than for the 850 and 700 mb levels. This is typical for profiles over the oceans. Since the error statistics (P matrix, see equation 5) were developed for a hybrid surface type data set (land and ocean), the errors at the surface for the background field are probably overestimated and does not allow for a further correction at the surface. This suggest that the error statistics for q should be available separately for land and ocean surfaces.

Assimilation of Humsat q leads to a reduction of the V rmse by between 0.3 and 0.9 kg m^{-2} (Table 2). This is about one half of that due to assimilation of SSM/I V. However, for the assimilation of Humsat, both ocean and land surfaces are considered. Only a small percentage of the collocation sample is over ocean surfaces (14% of collocations for March and 26% for collocations for June). The q rmse profiles for the background field, the analyzed field and Humsat q are illustrated in Fig. 2. The error of the background q at the surface is the largest as is typical over land surfaces. The errors of the analyzed field are reduced at all levels but by a small amount. The largest relative corrections occur for the levels above 600 mb as was to be expected

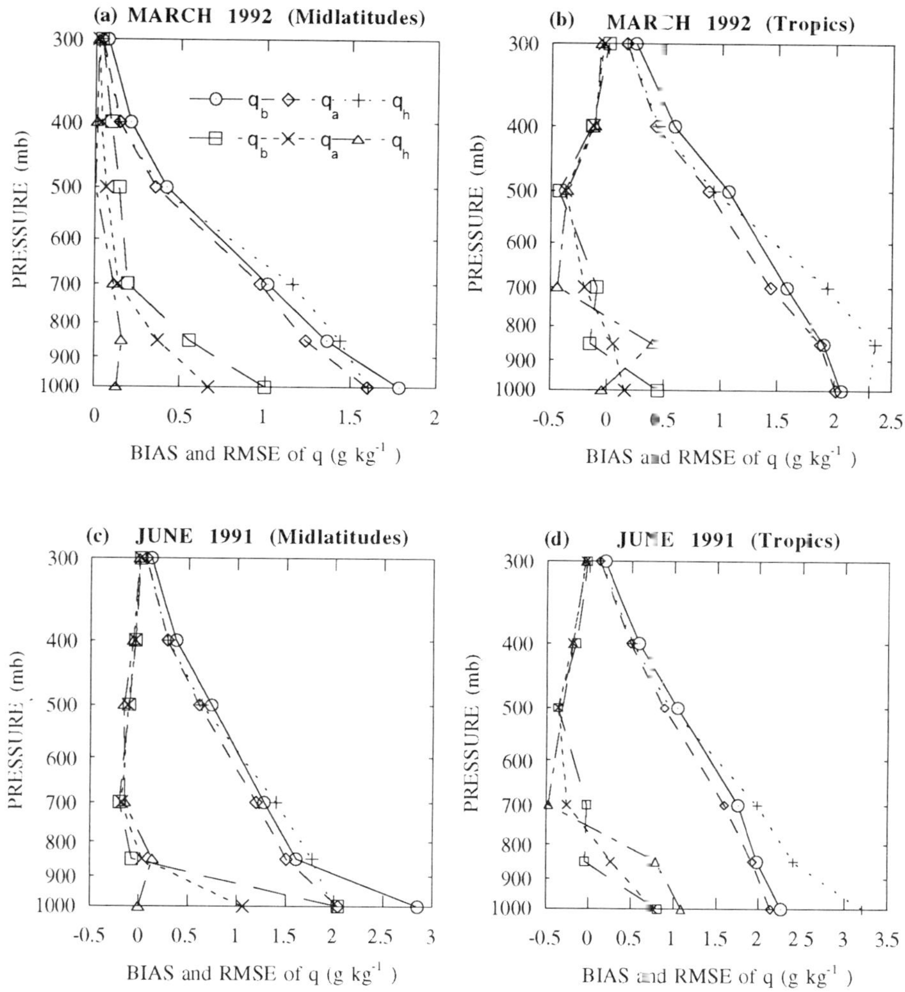

Figure 2: Assimilation of Humsat (or GOES) specific humidity. Both the bias (3 left-most curves in each graph) and the rmse of the specific humidity are illustrated (3 right-most curves in each graph). q_b represents the background (trial field) specific humidity, q_a represents the analyzed field and q_h are the values obtained with the Humsat retrieval scheme. **a)** Midlatitudes, March 1992 data set, **b)** Tropics , March 1992 data set, **c)** Midlatitudes , June 1991 data set, and **d)** Tropics, June 1991 data set.

from the contribution of the water vapor channel. The largest absolute corrections occur at the surface in the midlatitudes.

CONCLUSIONS

A variational analysis study was performed by concentrating on a one-dimensional variational analysis technique for specific humidity (q) and integrated water vapour content (V). First, the accuracy of the satellite retrieved fields was estimated by computing the error using collocated radiosondes. The retrieval of SSM/I V over the oceans had a rmse of ~4.7 kg m^{-2}.

This is comparable (somewhat better) to the rmse of the trial field or 6-h forecast from the numerical weather prediction model (~5.0 kg m^{-2}). The retrieval of q obtained from a

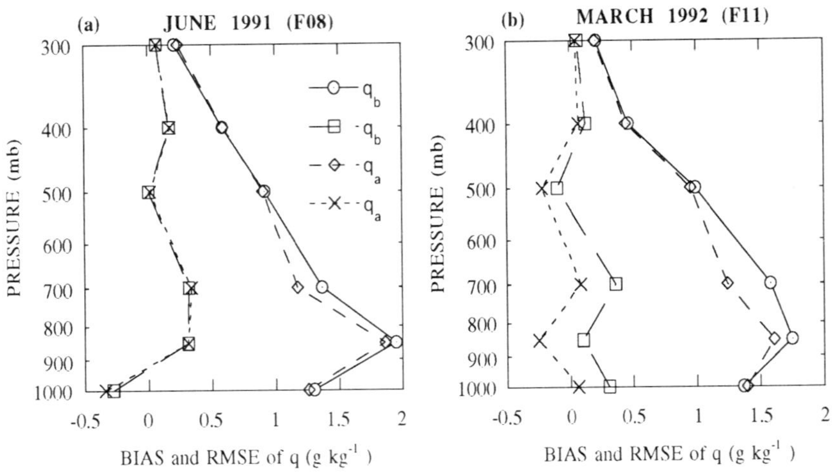

Figure 3: Assimilation of SSM/I integrated water vapour content. Both the bias (2 left-most curves in each graph) and the rmse of the specific humidity are illustrated (2 right-most curves in each graph). q_b represents the background (trial field) specific humidity and q_a represents the analyzed field. **a)** June 1991 data set obtained with the F08 satellite, **b)** March 1992 data set obtained with the F11 satellite. Results for the June 1991 (F10) and March 1992 (F10) are similar to those illustrated in figures a and b respectively.

combination of GOES radiances and ancillary data (Humsat technique) which was subsequently integrated to give V had a similar error of ~4.6 kg m^{-2} while trial field V errors were ~4.4 kg m^{-2} for a midlatitude data set. In the tropics the rmse was higher and respectively equal to ~6.8 and ~5.9 kg m^{-2}. These results are not in disagreement with prior findings (e.g [1]). The rmse of the vertical profiles of q was smaller at the surface than at 700 mb and 850 mb over oceanic surfaces but not over land.

Assimilation of SSM/I V and Humsat q was performed singly at the collocations with radiosondes. The results showed a reduced rmse for the integrated water vapour content of ~1 kg m^{-2} when SSM/I V (limited to ocean surfaces) was assimilated and around 0.6 kg m^{-2} when Humsat q was assimilated. The reduction in V rmse for the SSM/I occurred mainly at the levels of 700 and 850 mb. For the Humsat q assimilation, a small reduction in the rmse of q occurred everywhere but with the largest relative error correction above 600 mb. This was mainly due to the effect of the water vapor channel in the retrievals. Thus the SSM/I mainly determines humidity at the lower levels where most of the atmospheric water vapor is located while Humsat provides important information at the higher levels and non negligible information at the surface in the midlatitudes and at other lower levels when SSM/I data are not available. It has also been shown that Humsat q retrievals in the tropics should be improved and a new retrieval algorithm will be developed by taking into account the specifications of future GOES-NEXT satellites.

REFERENCES

1. Schmetz, J. and L. van de Berg, 1991: Estimation of precipitable water from Meteosat infrared window radiances over sea, *Beitr. Phys. Atmosph.*, **64**, 93-102.
2. Garand, L., 1993: A pattern recognition technique for retrieving humidity profiles from Meteosat or GOES imagery, *J. Appl. Meteor.*, **32**, 1592-1607.
3. Ritchie, H., 1991: Application of the semi-Lagrangian method to a multilevel spectral primitive-equations model, *Quart.J. Roy. Meteor .Soc.*, **117**, 91-106.
4. Lorenc, A.C., 1988: Optimal nonlinear objective analysis,*Quart.J. Roy. Meteor .Soc.*, **114**, 205-240.
5. Isaacs , R.G., E. Kalnay, G. Ohring and R.A. McClatchey, 1992: Proceedings of the first NMC/NESDIS/DOD conference on DMSP retrieval products, 14-15 April 1992, 266pp.
6. Hollinger, J.P., 1991: DMSP Special Sensor Microwave/Imager Calibration/Validation. Final Report, Vol. II, Naval Research Laboratory, Washington.
7. Colton, M.C., 1993: Proceedings of the Shared Processing Network DMSP SSM/I Algorithm Symposium, 8-10 June 1993, Fleet Numerical Oceanography Center, Monterey, California.

Estimation of Errors in the Two-Beamwidth-Antenna Method for Microwave Measurement of Atmospheric Water Vapor

SHINOBU HASHIMOTO†

Electrotechnical Laboratory, Umezono 1-1-4, Tsukuba, Ibaraki 305, Japan.

†*The author is now with the Laboratory of Mechanical Engineering for Fishing, Faculty of Fisheries, Hokkaido University, Minatomachi 3-1, Hakodate, Hokkaido 041, Japan.*

Abstract Estimation of errors in the ground-based microwave measurement of the atmospheric water vapor using the Two-Beamwidth-Antenna method. By selecting frequency pair, two excellent composite weighting functions for the measurement of integrated water vapor were fund, which were used to perform computer simulation for six model atmospheres including cloud to estimate measurement error. Relations between radiometer system parameter and measurement error were defined to provide radiometer design criterion.

key words: Atmospheric water vapor measurement, Tropospheric, Microwave, Solar extinction measurement, Two-Beamwidth-Antenna method, Radiometer system parameter and measurement error.

INTRODUCTION

The solar extinction measurement, though it has limitation of only daytime operation, can provide higher accuracy than the thermal emission measurement in estimation of contribution of the water vapor under cloudy conditions. Problems in this method are the fluctuation of the solar radiation and the thermal emission of the atmosphere.

The author has proposed a new Two-Beamwidth-Antenna method to simplify the measurement procedure and improve accuracy in the solar extinction measurement. The method consists of two ideas. The first is that the temperature difference between two antennas having different beamwidths provides automatic cancellation of the atmospheric thermal emission component in the path of solar radiation, and the second is use of relative intensity between two frequency points to remove solar fluctuation. The relative intensity measured with selected frequency pair can be used as a measure of water vapor concentration, which is based on the specific altitude variation of "the composite weighting function" defined at each frequency pair [1],[2].

This paper describes brief introduction of the Two-Beamwidth-Antenna method, selection of optimum composite weighting functions for the measurement of integrated water vapor, and the results of computer simulation for six model atmospheres using the selected composite weighting functions, and also relations between radiometer system parameter and measurement error defined to provide radiometer design criterion.

TWO-BEAMWIDTH-ANTENNA METHOD

(1) *Relative intensity and opacity difference of the atmospheric water vapor absorption*

The temperature difference between two antennas having different beam solid angles Ω_{A1}, Ω_{A2} and receiving the solar radiation is given by

$$T_m(f, \theta) = \Omega_S \left[\frac{\Omega_{A1} - \Omega_{A2}}{\Omega_{A1} \cdot \Omega_{A2}} \right] T_S \exp\left[-\tau(f) \sec\theta \right], \quad (1)$$

where Ω_S and T_S are the solid angle and the brightness temperature of the sun, respectively, as shown in Fig. 1. τ (f) is the zenith opacity at frequency f, and θ is the zenith angle. In

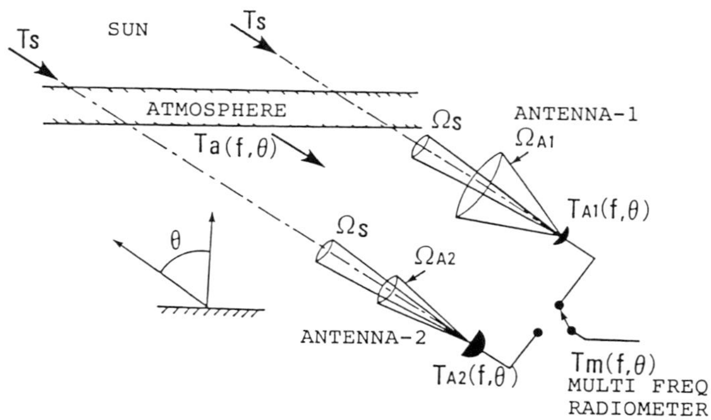

Figure 1. Two different beamwidth antennas observing the sun. T_S : solar brightness temperature, Ta(f, θ) : thermal emission of the atmosphere, Ω_S and Ω_A : solid angles of the solar disk and the antenna beams, respectively.

this expression, horizontally stratified atmosphere is assumed. The contribution of thermal emission of the atmosphere Ta(f, θ) in Fig. 1 is approximately the same for each antenna except when $\theta \sim 90^\circ$, so that these components in both antennas are automatically canceled out in the measurement of antenna temperature difference $T_m(f, \theta)$.

To compensate for the fluctuation of solar radiation T_S, the ratio of the $T_m(f, \theta)$ measured at two frequency points in the absorption spectrum is used as the measure of the water vapor concentration. The relative intensity $R(f_1/f_2)$ is given by

$$R(f_1/f_2) = \frac{T_m(f_1,\theta)}{T_m(f_2,\theta)}$$

$$= C(f_1,f_2)\ \exp[\ [\tau(f_2) - \tau(f_1)] \sec\theta\ ] \tag{2}$$

where $C(f_1,f_2)$ is the frequency correction factor of the antenna beamwidths, and given by

$$C(f_1,f_2) = \frac{\Omega_{A1}(f_2)\Omega_{A2}(f_2)[\Omega_{A1}(f_1) - \Omega_{A2}(f_1)]}{\Omega_{A1}(f_1)\Omega_{A2}(f_1)[\Omega_{A1}(f_2) - \Omega_{A2}(f_2)]}, \tag{3}$$

The opacity difference $[\tau(f_2) - \tau(f_1)]$ in (2) can be written as

$$\tau(f_2) - \tau(f_1) = \int_0^\infty \rho(z)\,[\,W(f_2,z) - W(f_1,z)]dz, \tag{4}$$

using the weighting function defined by $W(f,z) = \alpha(f,z)/\rho(z)$, where $\rho(z)$ is the water vapor density at altitude z, and $\alpha(f,z)$ is the water vapor absorption coefficient at altitude z and frequency f. The opacity difference given by (4) includes small contribution of oxigen absorption, specifically, the contribution of wide extended tail of oxigen absorption at low altitude. This term is important at low humidity condition. For 1 g/m^3 of water vapor density at ground level and standard distribution, for example, the contribution of this term attains 15.9% and 12% for the frequence pair (23.9-20.6)GHz and (24.5-18.0)GHz, respectively. This term, however, can be estimated and compensated satisfactory by theoretical calculation using the stability of oxigen mixing ratio in the atmosphere and measured atmospheric temperature at ground level. Hence it is assumed that the total opacity difference in the rest of this paper represents the one due only to the water vapor and cloud absorption.

As described later, if measurement frequencies f_1 and f_2 are adequately selected, we can find some composite weighting functions $[W(f_i,z)-W(f_k,z)]$ which have enough constant value against altitude, thus, the opacity difference in (4) becomes directly proportional to the integrated water vapor $\int \rho(z)dz$.

(2) *Estimation of water vapor absorption under cloudy conditions*

Clouds add a nonresonant absorption that is proportional to the square of frequency over the whole frequency region of microwave water vapor spectrum [3]. Cloud contribution in opacity $\tau_{cl}(f)$ at any frequency can be related as

$$\tau_{cl}(f_i) = \left[\frac{f_i}{f_k}\right]^2 \tau_{cl}(f_k). \tag{5}$$

If we can find two pairs of frequencies with which the altitude variations of the composite weighting functions are proportional to each other by a constant **A**,

$$[\,W(f_i, z) - W(f_k, z)]_{H_2O} = \mathbf{A}[\,W(f_m, z) - W(f_n, z)]_{H_2O}, \quad \mathbf{(6)}$$

the opacity difference without cloud effect can be estimated from the total opacity difference measured under cloudy conditions. Making measurement of two sets of opacity differences using frequency pair f_1 and f_2, and another pair f_3 and f_4, one of which is enough apart from the other frequencies, and rewriting total opacity difference in (4) by separation of water vapor and cloud contribution as

$$[\,\tau(f_2) - \tau(f_1)] = [\,\tau(f_2) - \tau(f_1)]_{H_2O} + [\,\tau(f_2) - \tau(f_1)]_{cl}, \quad \mathbf{(7)}$$
$$[\,\tau(f_4) - \tau(f_3)] = [\,\tau(f_4) - \tau(f_3)]_{H_2O} + [\,\tau(f_4) - \tau(f_3)]_{cl}, \quad \mathbf{(8)}$$

where subscripts H2O and cl show the opacity difference of water vapor absorption and that of cloud absorption, respectively.

Using (5) and (6), the opacity difference due only to cloud absorption is given by

$$[\,\tau(f_2) - \tau(f_1)]_{cl} = (C_2 - C_1)\,[\,\mathbf{A}[\,\tau(f_2) - \tau(f_1)] - [\,\tau(f_4) - \tau(f_3)\,], \quad \mathbf{(9)}$$

where C_1 and C_2 are constants given by

$$C_1 = \frac{1}{A\left[\left(\frac{f_2}{f_1}\right)^2 - 1\right] - \left[\left(\frac{f_4}{f_1}\right)^2 - \left(\frac{f_3}{f_1}\right)^2\right]}, \quad \mathbf{(10)}$$

and

$$C_2 = \frac{1}{A\left[1 - \left(\frac{f_1}{f_2}\right)^2\right] - \left[\left(\frac{f_4}{f_2}\right)^2 - \left(\frac{f_3}{f_2}\right)^2\right]} \quad \mathbf{(11)}$$

Substituting (9) into (7), the opacity difference without cloud absorption is given by

$$[\,\tau(f_2) - \tau(f_1)]_{H_2O} = [\,\tau(f_2) - \tau(f_1)]$$
$$- (C_2 - C_1)[\,\mathbf{A}[\,\tau(f_2) - \tau(f_1)\,] - [\,\tau(f_4) - \tau(f_3)\,]\,]. \quad \mathbf{(12)}$$

SELECTION OF OPTIMUM FREQUENCY PAIRS FOR THE MEASUREMENT OF INTEGRATED WATER VAPOR

From the measurement procedure described above, it is important to find the composite weighting function which has a constant value against altitude, and to find a pair of composite weighting functions whose values are proportional to each other at any altitudes, and whose one of frequency is enough apart from the other.

Calculated weighting functions in the whole range of 22 GHz water vapor absorption are shown in Fig. 2, where (a) and (b) show those of frequency range below and above 22.2 GHz, respectively. The variations of weighting functions at lower altitudes vary from the center frequency of absorption (22,235 GHz) to lower frequency and to higher frequency.

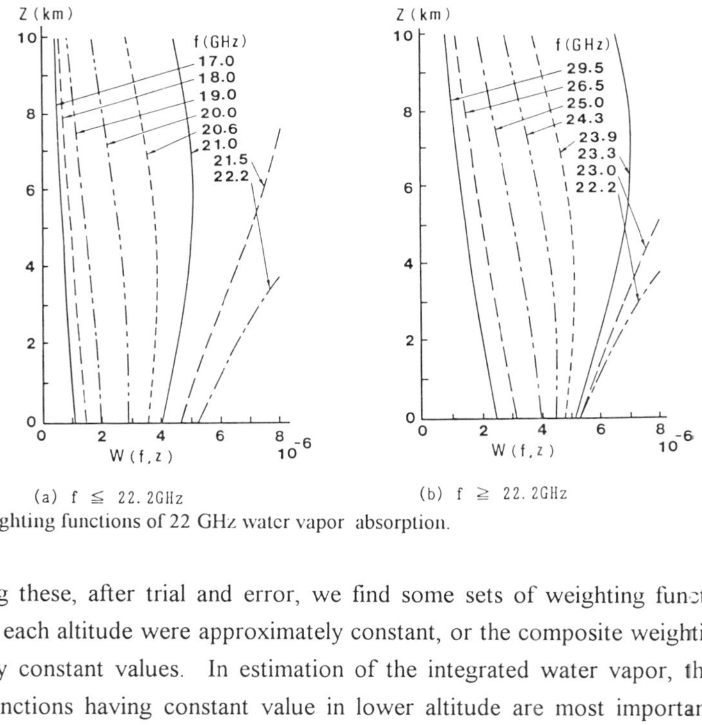

Figure 2. Weighting functions of 22 GHz water vapor absorption.

Among these, after trial and error, we find some sets of weighting functions, whose difference at each altitude were approximately constant, or the composite weighting functions having nearly constant values. In estimation of the integrated water vapor, the composite weighting functions having constant value in lower altitude are most important. Here, we calculate the weighted mean of each composite weighting function using the exponential decay of the standard water vapor distribution as the weight, and use the least deviation from the weighted mean as a selection rule. The weighted mean of weighting function $\overline{W}$ is given by,

$$\overline{W} = \frac{\int_0^\infty \exp(-z/Hs)\left[W(f_2,z)-W(f_1,z)\right]dz}{\int_0^\infty \exp(-z/Hs)dz}, \tag{13}$$

where Hs is the scale height.

Fig. 3 shows three composite weighting functions and their weighted mean. The best one is the frequency pair of (23.9-20.6) GHz, and the next is that of (24.5-18.0)GHz which is generally better than the pair of (20.2-18.0)GHz in the layer below 4 km, and specifically at around ground level.

Fig. 4 shows the stability of the composite weighting functions against water vapor density variation, where the range of density variation at each altitude corresponds to the relative humidity H change from 10% to 100%. The composite weighting functions with frequency pairs of (23.9-20.6)GHz and (24.5-18.0)GHz are excellent also in the stability against density variations. These two composite weighting functions are to be used in the computer simulations for evaluation of accuracies of water vapor measurement.

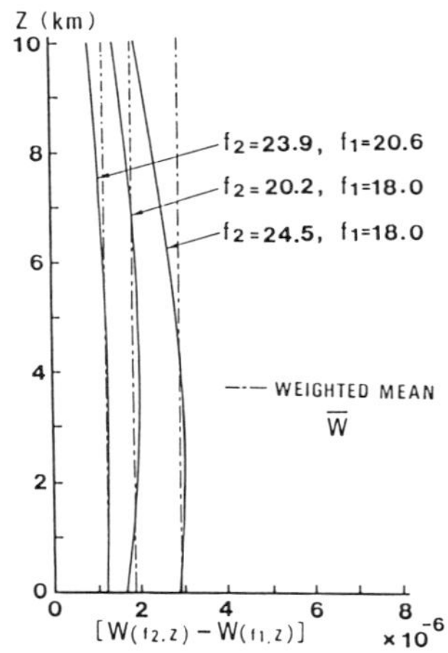

Figure 3. Altitude variations of composite weighting functions.

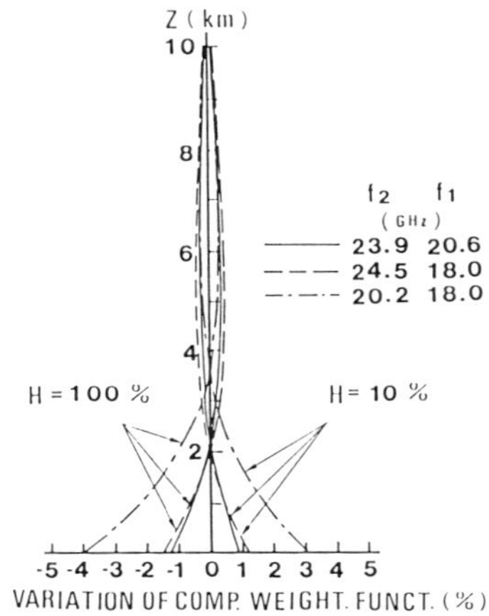

Figure 4. Water vapor density variations of composite weighting function for total variation of relative humidity H from 10% to 100% at each altitude.

With opacity difference given by (12) and weighted mean weight $\overline{W}$, the integrated water vapor estimate V* is given by

$$V^* = [\, \tau(f2) - \tau(f1)\,] / \overline{W}. \tag{14}$$

COMPUTER SIMULATION OF INTEGRATED WATER VAPOR ESTIMATION

In performing computer simulations, six models of atmosphere are assumed. These are given in Table 1. The two clear sky models labeled "E" and "B" are standard atmosphere with the ground level water vapor density ρ_0 = 8 g/m^3, temperature T_0 = 282 K, and ρ_0 = 30 g/m^3, T_0 = 303 K, respectively. The ground level pressure in both models is 1,013 mb. The four cloudy models were made by adding Low-Lying Stratus (1,500-3,000 ft) and Cumulus Congestus (3,000-9,000 ft) taken from the standard cloud model [4] to two clear sky models. Using the six model atmospheres, and using relations in (7), (8), (12) and (13), computer simulations were carried out for the estimation of opacity differences without cloud absorption, integrated water vapors and the errors. The results are shown in Table 2, where the opacity differences without cloud absorption are shown by abbreviated the suffix H_2O. **A** value in the table is the constant used in (6) and in the estimation of cloud opacity $\tau_{cl(f)}$. The value marked *1 is determined by the ratio of two weighted means of the composite weighting

Table 1: Model atmospheres used in computer simulation

Model Atmosphere	Descriptive Atmospheric Condition
"E"	STD distribution with ρ_0 = 8 g/m^3, T_0= 232 K, and P_0 = 1,013 mb at ground level
"B"	STD distribution with ρ_0 = 30 g/m^3, T_0 = 303 K, and P_0 = 1,013 mb at ground level
"LA"	"E" plus Low-Lying Stratus(1,500-3,000ft)
"LB"	"B" plus Low-Lying Stratus(1,500-3,000ft)
"CC"	"E" plus Cumulus Congestus(3,000-9,000ft)
"DC"	"B" plus Cumulus Congestus(3,000-9,000ft)

Cloud Name	Cloud Height Base (m)	Top (m)	Mass Density (g/m^3)	Mode Radius (μm)	Principal Composition
Low-Lying Stratus (1,500-3,000ft)	500	1,000	0.25	10.0	Water
Cumulus Congestus (3,000-9,000ft)	1,000	1,200	0.30	10.0	Water
	1,200	1,600	0.50	15.0	Water
	1,600	2,000	0.80	20.0	Water
	2,000	2,500	1.00	20.0	Water
	2,500	3,000	0.50	20.0	Water

Definition "Water": Cloud with liquid water, 100% humidity, and a continuous drop-size distribution.

functions used. For some model atmospheres, a little shifted **A** value from the value marked *1, for example, **A** = 0.42160 provides better results in estimation of integrated water vapor. It is considered that this value depends considerably upon the accuracy of cancellation of water vapor opacity term in (7) and (8) using (6), and also depends upon the water vapor distribution. Optimum determination of the A value should be made using wide variety of possible atmospheric conditions.

Table 2. Effect of the constant **A** on the estimation of opacity difference without cloud absorption and the integrated water vapor. In the table, $\tau(f_4)-\tau(f_3) = [\tau(f_4)-\tau(f_3)]_{H_2O}$ for f_4 = 23.9GHz, f_3 = 20.6GHz, and *1 mark indicates that the **A** value is determined by the ratio of weighted mean of composite weighting functions.

(1) CLEAR SKY

MODEL ATM.	"E"	"B"
TRUE VALUE :		
τ(f4)−τ(f3)	2.7130×10^{-3}	8.1392×10^{-2}
τ_{cl} (neper)	0	0
V (g/m^2)	2,209.2	66,277.1

ESTIMATED INTEGRATED WATER VAPOR V* USING $\bar{W}$ = 1.22805×10^{-6}

A	τ(f4)−τ(f3)	V*(g/m^2)	ε (%)	τ(f4)−τ(f3)	V*(g/m^2)	ε (%)
0.421856 *1	2.720×10^{-3}	2,215.2	0.27	7.943×10^{-2}	64,679.7	-2.4
0.42160	2.712×10^{-3}	2.208.7	-0.03	7.920×10^{-2}	64,490.0	-2.7
0.42120	2.700×10^{-3}	2,198.6	-0.48	7.884×10^{-2}	64,195.3	-3.1
0.42080	2.688×10^{-3}	2,188.6	-0.94	7.848×10^{-2}	63,902.8	-3.6

(2) CLOUDY : LOW-LYING STRATUS (1,500-3,000 ft)

MODEL ATM.	"LA"	"LB"
TRUE VALUE :		
τ(f4)−τ(f3)	2.2262×10^{-2}	8.2842×10^{-2}
τ_{cl} (neper)	1.4780×10^{-2}	8.193×10^{-3}
V (g/m^2)	18,128.1	67,458.1

ESTIMATED INTEGRATED WATER VAPOR V* USING $\bar{W}$ = 1.22805×10^{-6}

A	τ(f4)−τ(f3)	V*(g/m^2)	ε (%)	τ(f4)−τ(f3)	V*(g/m^2)	ε (%)
0.421856 *1	2.230×10^{-2}	18,158.4	0.17	8.086×10^{-2}	65,845.3	-2.4
0.42160	2.223×10^{-2}	18,105.1	-0.13	8.062×10^{-2}	65,651.4	-2.7
0.42120	2.213×10^{-2}	18,022.4	-0.58	8.025×10^{-2}	65,351.4	-3.1
0.42080	2.203×10^{-2}	17,940.3	-1.0	7.989×10^{-2}	65,053.6	-3.6

(3) CLOUDY : CUMULUS CONGESTUS (3,000-9,000 ft)

MODEL ATM.	"CC"	"DC"
TRUE VALUE :		
τ(f4)−τ(f3)	2.2885×10^{-2}	8.7438×10^{-2}
τ_{cl} (neper)	1.716×10^{-1}	9.510×10^{-2}
V (g/m^2)	18,635.4	71,200.5

ESTIMATED INTEGRATED WATER VAPOR V* USING $\bar{W}$ = 1.22805×10^{-6}

A	τ(f4)−τ(f3)	V*(g/m^2)	ε (%)	τ(f4)−τ(f3)	V*(g/m^2)	ε (%)
0.421856 *1	2.298×10^{-2}	18,708.8	0.39	8.577×10^{-2}	69,841.9	-1.9
0.42160	2.291×10^{-2}	18,654.0	0.10	8.552×10^{-2}	69,637.0	-2.2
0.42120	2.280×10^{-2}	18,568.7	-0.36	8.513×10^{-2}	69,318.9	-2.6
0.42080	2.270×10^{-2}	18,484.1	-0.81	8.474×10^{-2}	69,003.0	-3.1

REQUIREMENTS FOR THE RADIOMETER RESOLUTION AND ANTENNA DIMENSION

With the radiometer resolution ΔT and antenna diameters D_1, D_2, errors in the measurement of the relative intensity ε_R and estimation error of opacity difference ε_τ are given by the following relations. Assuming that the radiometer resolution of all frequency channels are the same, and knowing that $\Delta T/T_m(f, \theta) << 1$,

$$R(f_k/f_i)_{meas} = \frac{T_m(f_k, \theta) \pm \Delta T}{T_m(f_i, \theta) \pm \Delta T}$$

$$= R(f_k/f_i)\left[1 \pm \left\{\frac{\Delta T}{T_m(f_k,\theta)} + \frac{\Delta T}{T_m(f_i,\theta)}\right\}\right]. \tag{15}$$

The measurement error in the relative intensity is defined by

$$\varepsilon_R = \Delta T\left[\frac{1}{T_m(f_k,\theta)} + \frac{1}{T_m(f_i,\theta)}\right]$$

$$= \frac{\pi}{\Omega_S T_S}\frac{\Delta T}{(D_2^2 - D_1^2)}\left[\frac{\lambda_i^2}{\exp\left[-\tau(f_i)\sec\theta\right]} + \frac{\lambda_k^2}{\exp\left[-\tau(f_k)\sec\theta\right]}\right]. \tag{16}$$

where, λ is microwave wavelength, and antenna solid angle with aperture diameter D is approximated by $\Omega_A(f) = \pi(\lambda/D)^2$.

With error ε_R, the estimated opacity difference is given by

$$[\tau(f_i) - \tau(f_k)]_{est} = \ln\left[\frac{R(f_k/f_i)(1+\varepsilon_R)}{C(f_k,f_i)}\right]\frac{1}{\sec\theta}$$

$$= \left[\tau(f_i) - \tau(f_k)\right]\left[1 + \frac{\ln(1+\varepsilon_R)}{\left[\tau(f_i) - \tau(f_k)\right]\sec\theta}\right]. \tag{17}$$

Using approximation $\ln(1+\varepsilon_R) \approx \varepsilon_R$, the opacity estimation error can be written as

$$\varepsilon_\tau = \frac{\varepsilon_R}{\left[\tau(f_i) - \tau(f_k)\right]\sec\theta}$$

$$= \frac{\pi}{\Omega_S T_S}\frac{\Delta T}{(D_2^2 - D_2^2)}\left[\frac{\lambda_i^2}{\exp\left[-\tau(f_i)\sec\theta\right]} + \frac{\lambda_k^2}{\exp\left[-\tau(f_k)\sec\theta\right]}\right]$$

$$\times \frac{1}{[\tau(f_i)-\tau(f_k)]\sec\theta}. \tag{18}$$

The radiometer system parameter $\Delta T/D_2^2[1-(D_1/D_2)^2]$ determines the accuracy of the estimation of opacity difference. If antenna diameter ratio is taken to be larger than 1:3, practically only the larger antenna diameter is important.

For four model atmospheres in Table 1, and taking the solar radiation temperature as 8,000 K, the calculated results of opacity difference error versus radiometer system parameter are shown in Fig. 5. This figure provides a design criterion for radiometers.

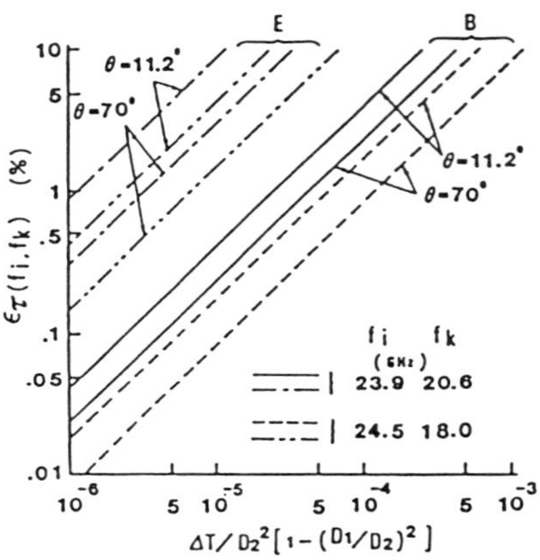

(a) for model atmosphere "E" and "B".

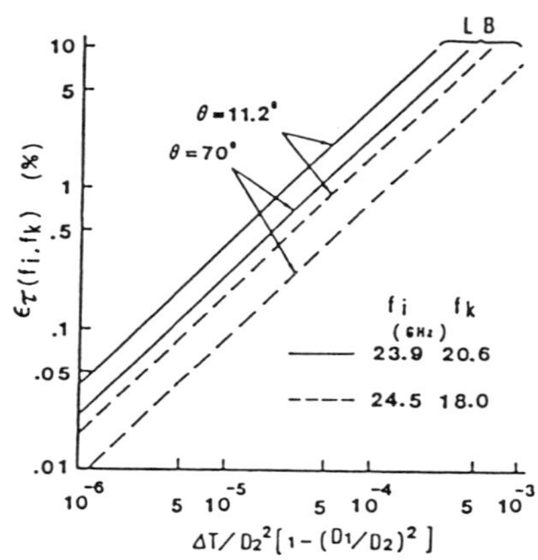

(b) for model atmosphere "LB".

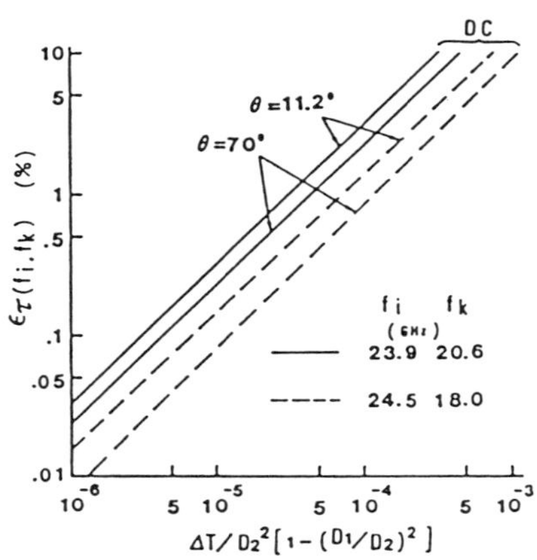

(c) for model atmosphere "DC".

Figure 5. Errors in estimation of opacity difference ε_τ due to radiometer resolution ΔT and antenna diameters D1, D2.

The restriction of the smaller antenna dimension is determined by the errors in cancellation of thermal emission component in the measurement of temperature difference Tm(f, θ). This error arises at large zenith angle where thermal emission increases rapidly with angle, and broad beam antenna measures mean temperature around observing direction. The beamwidth of the smaller antenna should be selected less than 5° to make this error negligible.

CONCLUDING REMARKS

For the purpose of integrated water vapor measurement using Two-Beamwidth-Antenna method, selection of optimum composite weighting functions and computer simulation results for the measurement of integrated water vapor under cloudy condition were described. One of important properties of this method is to be able to obtain excellent constant composite weighting functions against altitude by selection of frequency pairs, so that accurate estimation of integrated water vapor can be expected. Computer simulation of integrated water vapor estimation including cloudy condition shows usefulness of this method. Use of high temperature of the solar radiation as the source would make it possible to measure water vapor through thick cloud. To clarify errors arisen for various possible atmospheric conditions requires future investigation.

The relation between the radiometer system parameter and measurement error were defined to provide radiometer design criterion.

REFERENCES

1. Hashimoto, S., "A Compact Sun Tracking Radiometer with Two Different Beamwidth Antennas for Detection of Atmospheric Water Vapor," IEEE Trans. Instrumentation and Measurement, vol. 38, no. 2, pp. 619-623, April 1989.
2. Hashimoto, S., "A New Simplified Water Vapor Measurement Using a Multifrequency Microwave Radiometer with Two- Different- Beamwidth Antennas," IEEE Trans. Geosci. Remote Sensing, vol. 30, no. 4, pp. 832-837, July 1992.
3. Staelin, D. H., "Measurements and Interpretation of the Microwave Spectrum of the Terrestrial Atmosphere Near 1-centimeter Wavelength," J.Geophys. Res., vol. 71, no.12, pp. 2875-2881, June 1966.
4. Reifenstein, E. C., and N. E. Gaut, Microwave Properties of Clouds in the Spectral Range 30-40 GHz, Tech. Rept. no. 12, Environ. Res. & Tech., Inc., 1971.

Improved GPS Vertical Surveying & GPS Sensing Of Atmospheric Water Vapor

RANDOLPH WARE, FREDRICK SOLHEIM, CHRISTIAN ROCKEN,
TERESA VAN HOVE, CHRIS ALBER

University Navstar Consortium
POB 3000, Boulder, CO 80307 USA

Abstract. Atmospheric water vapor causes Global Positioning System (GPS) signal delay. Wet delay is a major source of error in high-accuracy vertical GPS surveying. Improved vertical GPS surveying has applications in geodesy, earthquake studies, volcanology, sea level measurements, airborne and ground-based gravimetry, and GPS orbit determination. GPS can also be used to sense integrated water vapor, which is important in weather forecasting and atmospheric research. We discuss here recent results and ongoing activities in these two related research areas.

Keywords: Global Positioning System, Atmospheric Water Vapor, Water Vapor Radiometry.

IMPROVED GPS SURVEYING

University Navstar Consortium (UNAVCO) staff have been working to improve vertical GPS surveying accuracy. UNAVCO is a national facility that assists university investigators using GPS technology for geosciences research. UNAVCO recently reported results from a GPS survey using water vapor radiometers (WVRs) pointed toward GPS satellites[1]. GPS receivers and WVRs collected data at both ends of a 50-km baseline from Boulder to Platteville, Colorado. The WVRs were pointed sequentially by a personal computer to each of the GPS satellites being observed. The WVRs measure integrated water vapor which can be converted to wet delay for correction of GPS observations. Typical pointed WVR data, mapped to the zenith, are shown in Figure 1.

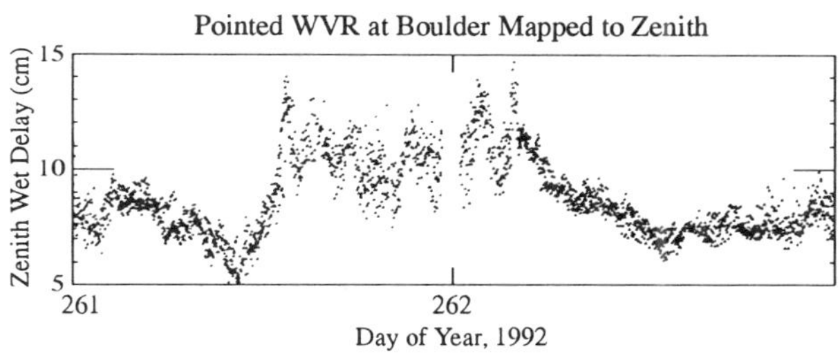

Figure 1. Wet delay measured by a pointed water vapor radiometer at Boulder, mapped to the zenith. Azimuthal variation in wet delay is indicated by scatter in the data.

Wet delays measured by two WVRs separated by 30 m are shown in Figure 2. Agreement is 3 mm rms, indicating the ability to correct for GPS wet delay at this level of accuracy. The tripod-mounted WVRs were designed for use in GPS surveying, and were loaned by Radiometrics, Inc., of Boulder, Colorado.

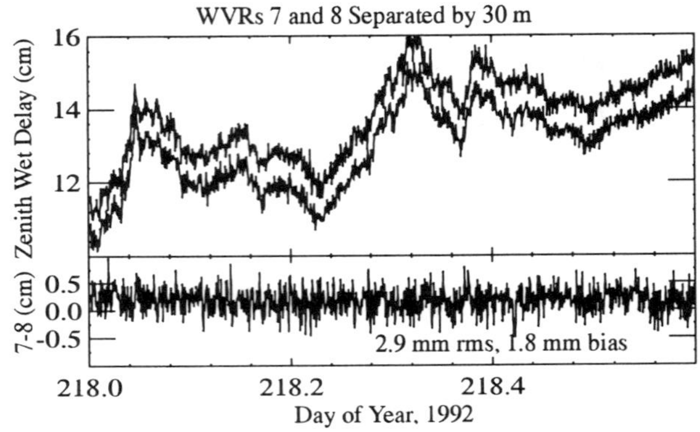

Figure 2. Zenith wet delay observations by the two Radiometrics™ WVRs used at Boulder and Platteville, separated by 30 m. The trace for WVR 8 is shifted upward by 1 cm to show detail. The difference of the two is shown in the lower plot.

The best vertical repeatability was 2.6 mm using pointed WVR corrections, as shown in Figure 3. Pointed WVRs account for azimuthal variations in wet delay, while the other correction methods account only for zenith wet delay.

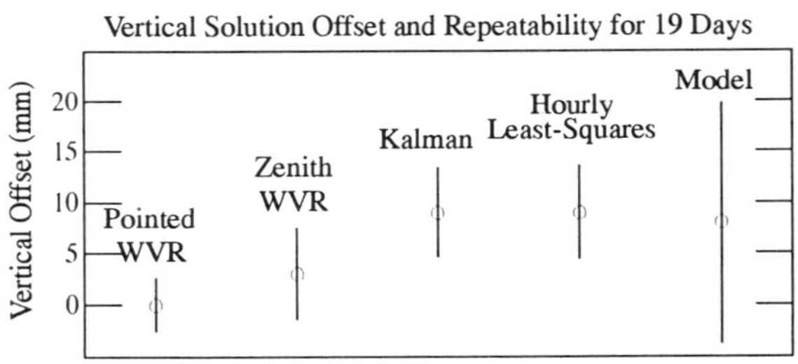

Figure 3. Repeatability for 19 days of vertical solutions for the 50-km Boulder to Platteville baseline. Solutions with pointed water vapor radiometer, zenith WVR, Kalman filter, hourly least squares, and modeled corrections for wet delay are shown.

GPS SENSING OF ATMOSPHERIC WATER VAPOR

Wet delay can be measured directly using GPS receivers and analysis software developed for high-accuracy geodetic surveying[2]. Wet delay can then be converted to integrated water vapor[3] (IWV). More widespread IWV data could be used to improve precipitation forecasts[4]. Currently, weather forecasts are based on radiosonde data collected twice daily from 600 sites worldwide. GPS-sensed IWV measured by permanent GPS arrays could be a valuable source of information on water vapor.

GPS sensing of IWV was recently reported. Agreement between half-hour GPS and WVR-sensed IWV was better than 1 mm for several weeks of observations during September to December, 1992[5]. Typical comparisons are shown in Figure 4.

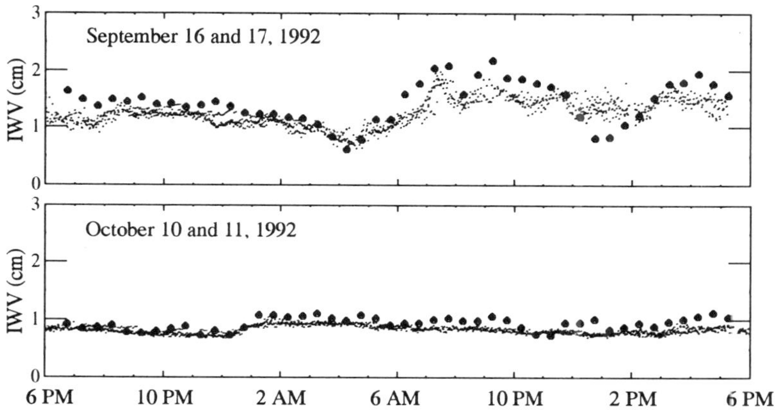

Figure 4. Typical active and quiet days of GPS and water vapor radiometer (WVR)-sensed integrated water vapor (IWV) at Boulder, Colorado. The small dots represent WVR measurements along the line-of-sight to GPS satellites, mapped to the zenith. The circles represent 30-minute averages of IWV calculated from least-squares fits of GPS wet delay. Agreement over 29 days of observations was better than 1 mm rms.

PLANS

Scientists from UNAVCO and North Carolina State University are exploring the use of GPS technology in weather forecasting. They recently gathered GPS and WVR data from National Oceanic and Atmospheric Administration (NOAA) Wind Profiler sites and a Department of Energy Atmospheric Radiation Monitoring (ARM) site in Oklahoma, and from sites in Colorado. Improvement in weather forecasting using GPS-sensed IWV from these sites will be explored.

In a separate project, scientists from NOAA's Forecast Systems Laboratory in Boulder, Colorado, are collecting GPS-sensed IWV data and evaluating its use for weather forecasting. UNAVCO scientists are assisting by analyzing the GPS data and converting it into IWV.

Increasing numbers of high-accuracy GPS receivers are in continuous operation worldwide for geodesy, surveying, orbit determination and other applications. Data from most of these sites could be accessed in near real time via Internet or telephone. This presents a new, cost-effective opportunity to improve precipitation forecasts using GPS sensing of water vapor.

Acknowledgments

Work on improving GPS vertical surveying accuracy and GPS sensing of atmospheric water vapor has been supported by grants from the Earth Sciences Division of the National Science Foundation. Additional work on measuring atmospheric water vapor with GPS for weather forecasting is being conducted under a grant from the Atmospheric Science Division of NSF.

References

1. Ware, R., C. Rocken, F. Solheim, T. Van Hove, C. Alber, and J. Johnson, *Pointed Water Vapor Radiometer Corrections for Accurate Global Positioning System Surveying*, **Geophys. Res. Lett., 20**, 2635-2638, 1993.
2. Tralli, D., S. Lichten, T. Herring, *Comparison of Kalman filter Estimates of Zenith Atmospheric Path Delays using GPS and VLBI*, **Radio Science, 27**, 999-1007, 1992.

3. Bevis, M., S. Businger, T. Herring, C. Rocken, R. Anthes, and R. Ware, *GPS Meteorology: Remote Sensing of Atmospheric Water Vapor Using the Global Positioning System*, **J. Geophys. Res., 97**, 15, 787-15, 801, 1992.
4. Kuo, Y. H., Y. R. Guo, and E. Westwater, *Assimilation of Precipitable Water Measurements into a Mesoscale Numerical Model*, **Monthly Weather Review, 121**, 1215-1238, 1993.
5. Rocken, C., R. Ware, T. Van Hove, F. Solheim, C. Alber, J. Johnson, M. Bevis, and S. Businger, *Sensing Atmospheric Water Vapor with the Global Positioning System*, **Geophys. Res. Lett., 20**, 2631-2634, 1993.

Using microwave radiometry and space geodetic systems for studies of atmospheric water-vapor variations

GUNNAR ELGERED[1], JAN M. JOHANSSON[1], and JAMES L. DAVIS[2]

[1] *Onsala Space Observatory, Chalmers University of Technology S-439 92 Onsala, Sweden*
[2] *Harvard-Smithsonian Center for Astrophysics 60 Garden St., CAMBRIDGE, MA 02138, USA*

Abstract. We use three independent methods to study variations in the propagation delay due to atmospheric water vapor: estimation using Very-Long-Baseline Interferometry (VLBI), Global Positioning System (GPS), and dual-frequency (21.0 and 31.4 GHz) microwave radiometer data. For 20 VLBI experiments, each 24 hour long, we find systematic mean differences between the three techniques with a typical value of about 1 cm. The weighted root mean square (wrms) scatter of the differences about these mean values is also of the order of 1 cm. The best agreement is obtained between estimates from microwave radiometer data and VLBI data where we have a mean offset of -0.3 cm and an wrms scatter about this value of 1.0 cm for the entire data set.

Key words: Microwave Radiometry, Atmospheric Water Vapor, Excess Propagation Path, GPS Meteorology, Very-Long-Baseline Interferometry

INTRODUCTION

It is well known that the varying amount of atmospheric water vapor affects the propagation time of a radio signal. The common method for dealing with this excess propagation time in radio based space geodetic techniques is to estimate an atmospheric correction using the geodetic data themselves. This correction may be modelled, for example, as a constant, a piece-wise linear function, or a stochastic process. In this study we compare estimates of the wet propagation delay obtained from Very-Long-Baseline Interferometry (VLBI), Global Positioning System (GPS), and ground-based microwave radiometry. The model used in both the VLBI and the GPS analysis is a Markov process or more specifically a random walk process.

In addition to studying the influence of the wet delay on the accuracy of VLBI and GPS another goal of this work is to evaluate the use of GPS networks for meteorological measurements (see, e.g., [1]).

INSTRUMENTATION AND ESTIMATION TECHNIQUES

In this study, the maximum distance between the 20 meter radio telescope used for VLBI observations, the GPS antenna and the Water Vapor Radiometer (WVR) is approximately 80 m. The intersection of azimuth and elevation axis of the VLBI antenna is 22 m above the mean sea level, whereas both the GPS and WVR are 9 m above sea level. This means that for a standard atmosphere with a scale height equal to 2 km the VLBI antenna senses a wet delay which is 0.6% less than the other two methods. This effect is of the order of 1 mm and will be ignored in the following.

Twenty geodetic VLBI experiments involving Onsala Space Observatory were carried out during the period June–December 1992 each one 24 hours long. Many of these were dedicated "mobile" experiments and included the mobile VLBI antenna (MV–2) at several sites in northern Europe. Because of the lower sensitivity of the mobile VLBI antenna, typically 180 observations were carried out during one 24 hour experiment. For a comparison, the Research & Development experiments during 1993 (coordinated by NASA Goddard Space Flight Center), using larger antennas, typically include 300 observations at each site.

The GPS receiver, a Rogue SNR-800 with a Dorne-Margolin antenna operated continuously during this time period; data from this receiver have also been used in the International GPS Service for Geodynamics (IGS).

The WVR, measuring the atmospheric emission at 21.0 and 31.4 GHz was operated either in a "sky mapping mode" with a cycle of approximately 8 minutes or in a synchronized mode, with the VLBI observations pointing at each source during data acquisition and performing elevation scans during the time between the VLBI observations.

The VLBI estimates use observations down to elevation angles of approximately 7°. The hydrostatic mapping function is used to calculate the actual hydrostatic delay from the equivalent zenith value based on ground pressure observations [2]. The wet delay is estimated using a special wet delay mapping function [3]. The VLBI data analysis is carried out with the SOLVK software [4] which models the wet delay as a random walk process. The variance between values separated by the time Δt for a random walk process scales linearly with the separation

$$< (x(t+\Delta t) - x(t))^2 >= \sigma^2 \, \Delta t. \tag{1}$$

The value of the random walk variance, σ^2 in (1), is estimated for each site using the delay rate residuals. For these 20 experiments the estimated values of σ^2 varied from 0.03 ps^2/s to 0.90 ps^2/s and had a mean value of 0.26 ps^2/s.

The GPS estimates use observations down to an elevation angle of 15°. One mapping function [5] is used to estimate the total propagation delay in the GIPSY software [6]. The standard value used for the data in this study was 0.32 ps^2/s. The wet delay is, thereafter, calculated by subtracting the zenith value of the hydrostatic delay based on the ground pressure observations.

The microwave radiometer data are presented as equivalent zenith values of actual observations at different elevation angles using the simple cosecant law. In this data set the radiometer was always observing above an elevation angle of 23°. The wet delay estimates from the GPS data were updated with 5 minute intervals which is of the same order as the time between VLBI observations towards different sources. Therefore, we smoothed the WVR data with a gaussian window with a full-width-half-maximum of 5 minutes.

RESULTS

The estimated wet delays using the three different methods have been compared. We present the results obtained for each VLBI experiment and for the entire data set. The values for the means and wrms residuals of the differences are given in Table 1. Figures 1 and 2 show some examples of the data. We see in both Figures that the WVR data show more short-term variations. It should also be noted that the GPS results in Figure 2 were obtained by processing the data in 24 hour long data sets (0–24 UT) independently. This may be a problem occasionally, such as during the large increase in the wet delay around midnight July 25–26. Such problems will, however, not occur if data sets are processed with an overlap in time. This was actually done for the data shown in Figure 2 for which the processing of the GPS data for each day also included the last three hours from the previous day and the first three hours from the next day. The VLBI processing was, however, made for each experiment independently.

DISCUSSION

The result for the whole data set show over all biases (or mean differences) well below 1 cm between all the three methods. Especially interesting is the bias of 3 mm between WVR and VLBI results. An earlier comparison [7] using more than 100 VLBI experiments from the period 1980–1988 resulted in an unexplained bias of 1 cm. We know that a slight modification of the mapping function can cause differences of many mm in the estimated wet delay using VLBI data [8] but these two studies have used identical

Table 1. Mean differences (biases) and weighted rms residuals about the mean differences for equivalent zenith wet delays estimated using WVR, GPS, and VLBI data.

VLBI experiment	Start time 1992 (UT)	WVR–GPS mean (cm)	WVR–GPS wrms[a] (cm)	WVR–VLBI mean (cm)	WVR–VLBI wrms[a] (cm)	GPS–VLBI mean (cm)	GPS–VLBI wrms[a] (cm)
Mobile	Jun. 25, 8	-0.3	1.1	-0.3	0.6	0.0	1.1
Mobile	Jun. 26, 10	-0.7	1.1	-1.9	0.4	-1.2	1.1
Mobile	Jun. 28, 6	-0.6	1.9	-1.3	1.1	-0.8	1.6
POLAR-N1	Jun. 29, 8	-1.5	0.7	-1.8	0.8	-0.4	1.2
IRIS-A 748[b]	Jul. 06, 8	-1.3	1.4	-1.1	0.6	0.3	1.4
Europe-5	Jul. 07, 12	-1.4	0.9	-0.3	0.7	1.1	1.3
IRIS-A 750[b]	Jul. 20, 8	-1.8	1.3	-0.8	0.7	1.1	1.8
Mobile	Jul. 25, 4	-0.1	1.5	-0.2	0.7	0.0	1.6
Mobile	Jul. 26, 6	-1.5	0.9	-1.0	0.9	0.5	1.3
IRIS-A 751[b]	Jul. 27, 8	-1.5	1.2	-0.5	1.1	1.0	1.9
Mobile	Aug. 12, 18	-0.6	1.3	-0.1	0.8	0.5	1.5
Mobile	Aug. 25, 10	0.7	1.2	0.2	1.0	-0.5	1.2
Mobile	Aug. 26, 12	0.2	1.0	-0.1	0.8	-0.2	1.0
Mobile	Aug. 28, 20	-0.1	1.4	-0.2	1.0	-0.2	1.3
Mobile	Aug. 29, 22	-0.4	0.9	0.0	0.8	0.4	1.0
IRIS-A 761	Oct. 05, 8	-0.3	0.7	0.2	0.8	0.3	0.8
IRIS-A 765	Nov. 02, 8	-0.7	1.1	-0.1	1.2	0.5	0.7
Europe-7	Nov. 03, 12	-0.7	0.7	-0.4	0.6	0.3	0.7
IRIS-A 769	Nov. 30, 8	-0.1	0.9	0.4	0.8	0.5	0.5
Europe-8	Dec. 01, 12	-0.3	1.0	-0.3	1.1	-0.4	1.1
All	–	-0.6	1.3	-0.3	1.0	0.1	1.3

[a] Weighted root mean square residual about the estimated mean difference.
[b] These IRIS experiments include the MV–2 antenna and consist, therefore, of fewer observations than the regular IRIS experiments.

mapping functions. It therefore, seems reasonable to assume that the WVR instrumental upgrade made during 1991 has had some impact on the overall WVR accuracy as well as the fact that the VLBI data have become more accurate.

It is also worth noting that the relatively small biases between the WVR data and the wet delays estimated from VLBI or GPS data is an

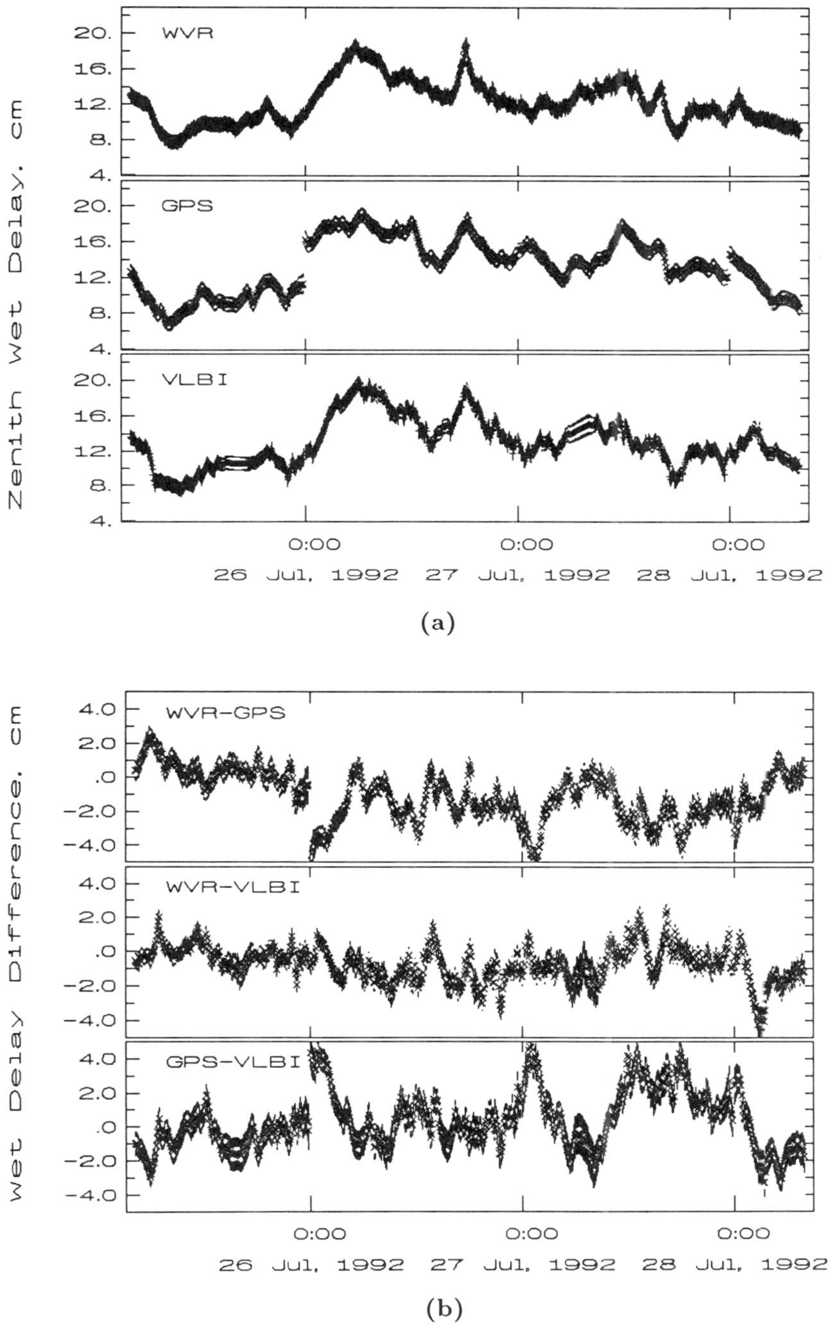

Figure 1. Estimated equivalent zenith wet delays using data from three contiguous VLBI experiments (a). The corresponding differences are shown in (b).

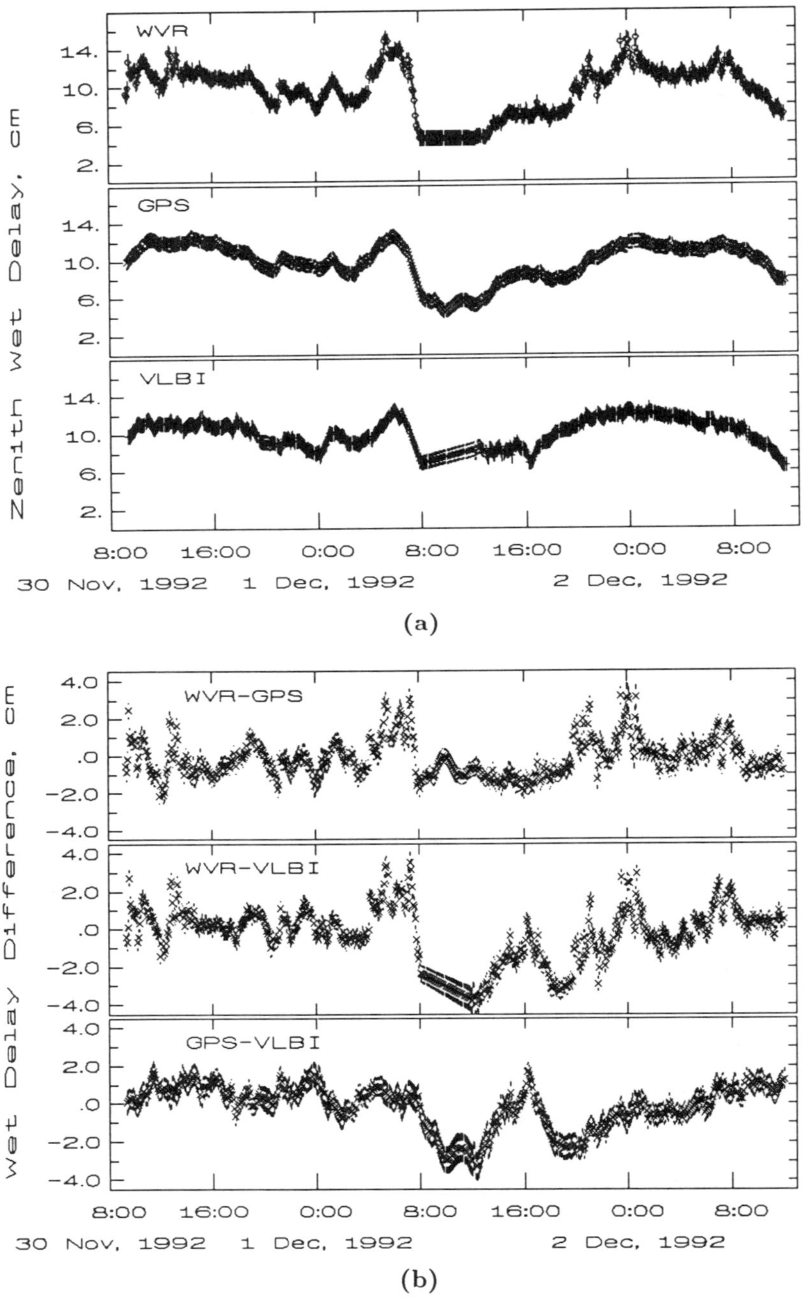

Figure 2. Estimated zenith wet delays from two contiguous VLBI experiments (a) and the corresponding differences (b). Note that there are neither WVR nor VLBI data from the period 8–12 UT on December 1. These interpolated data are, however, not included in the calculations for Table 1.

important indicator of the accuracy of the attenuation coefficient used to relate the observed sky emission to the water vapor content and the wet delay. In this WVR data set we used the equations defining the "Millimeter-wave Propagation Model", version 1992 [9]. The wet delay estimates using VLBI and GPS data are of course not affected by any error in the atmospheric attenuation coefficient. The mean wet delay of the whole data set was 11 cm implying that the overall difference between the WVR and the VLBI results was approximately 3%. Of course this is not a strict test because instrumental biases in the WVR and biases due to incorrect mapping functions may be of similar magnitudes and will also influence the final over all bias. It is also worth noting that the difference between the MPM–92 and the frequently used model presented by Waters [10] is only 2% when a wet delay of $\approx$11 cm is inferred from radiometer observations at 21.0 and 31.4 GHz. Radiometric observations just above the center of the water vapor line at 22 GHz are better suited for a comparison between these two models [11].

The WVR data are less accurate when there are large amounts of liquid water in the atmosphere. An advantage in this technique is, however, that the integrated liquid water content is estimated simultaneously with the wet delay using the WVR data. In this presentation we have discarded all data for which the estimated integrated liquid water content is larger than 0.7 mm (this is equivalent to 0.7 kg/m^3). In the evening of July 26 the rapid increase suggested by the WVR data is supported by the VLBI data. (The liquid water content at the peak is 0.7 mm.) On the other hand events with a liquid water content of the order of 0.7 mm have been seen in other data sets where the WVR data seem to overestimate the wet delay compared to the GPS and the VLBI estimates.

CONCLUSIONS AND FUTURE WORK

Variations in the wet propagation delay are in general detected by all three methods but slowly varying differences at the centimeter level exist. The typical mean biases for a 24 hour period are about 1 cm with a weighted rms scatter about this value of the same order. When we study the whole data set the best agreement is between estimates from WVR data and VLBI data with a mean offset of -0.3 cm and an wrms scatter about this value of 1.0 cm. The corresponding values for the WVR–GPS comparison are -0.6 cm in bias and an additional wrms scatter of 1.3 cm.

This type of comparison has proven valuable for detecting errors (such as ignored clock-breaks) in the VLBI and the GPS data analysis. As more data are being analyzed the goal is to search for systematic differences in the wet delay estimates and for the possibility for improving the time resolution of both the GPS and VLBI estimates. In particular, we plan to evaluate the potential of the Swedish GPS network (SWEPOS) [12], shown in Figure 3, for continuous observations of the atmospheric water vapor content by inferring it from wet delay estimates. Several examples of such time series are shown in Figure 4.

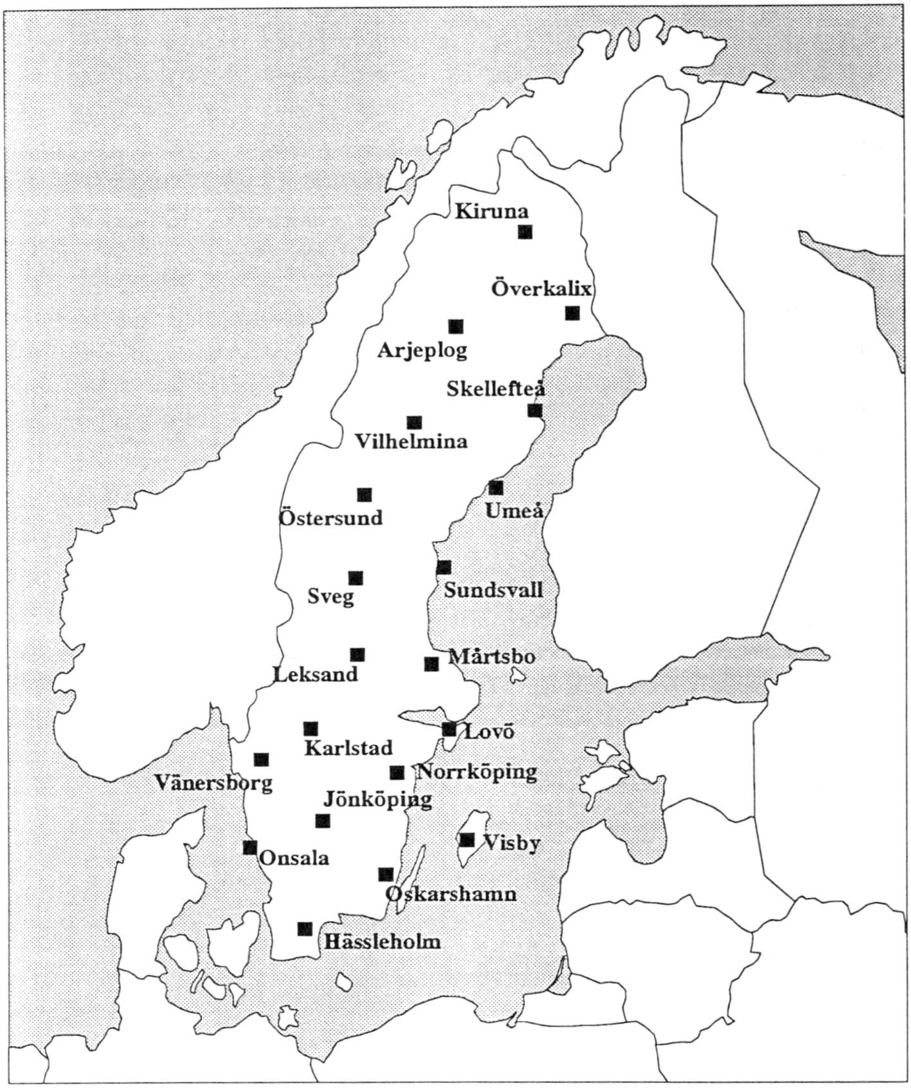

Figure 3. The continuously operating Swedish GPS network. Presently we have Rogue receivers at the following sites: Kiruna, Arjeplog, Skellefteå, Vilhelmina, Östersund, Umeå, Sveg, Sundsvall, Leksand, Mårtsbo, Karlstad, Jönköping, Onsala, Visby, Oskarshamn, and Hässleholm.

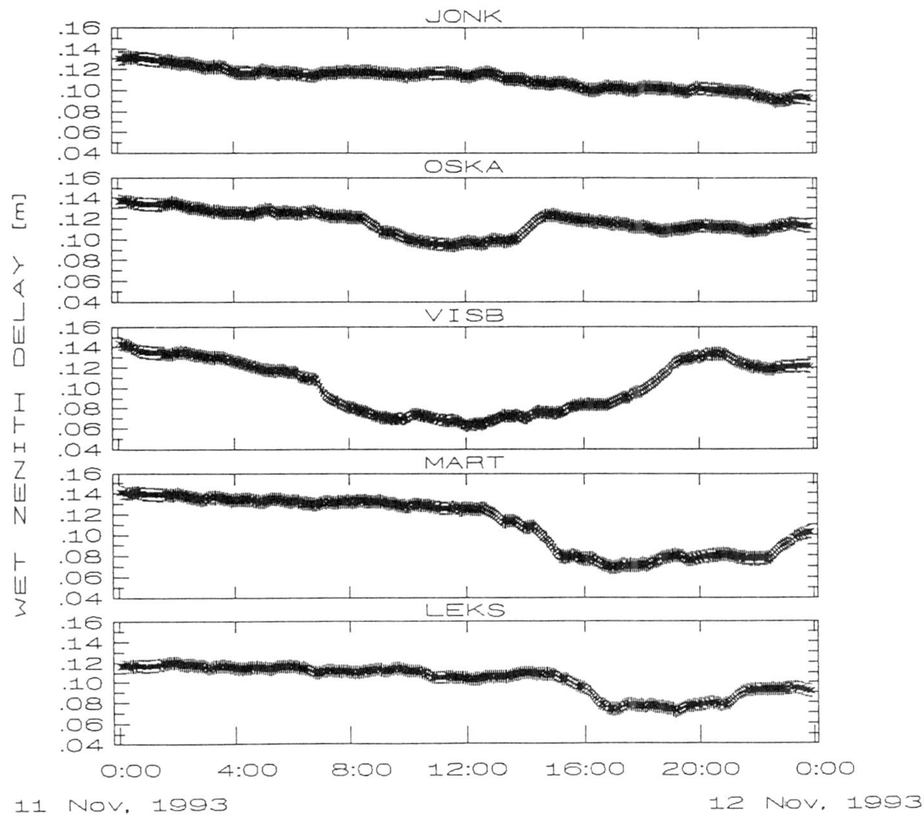

Figure 4. Example of estimated zenith wet delays at a few SWEPOS stations (from the top): Jönköping, Oskarshamn, Visby, Mårtsbo, and Leksand. Stations far to the east (see Figure 3) indicate a larger and longer decrease in the wet delay. The internal consistency indicates that this observed variation is real. Independent evaluation methods such as comparisons with radiosonde and WVR data are planned for the future.

REFERENCES

1. Bevis, M., S. Businger, T.A. Herring, C. Rocken, R.A. Anthes, and R.H. Ware, 1992. GPS Meteorology: Remote Sensing of Atmospheric Water Vapor Using the Global Positioning System, *J. Geophys. Res.*, **97**, 15787–15801.
2. Davis, J.L., T.A. Herring, I.I. Shapiro, A.E.E. Rogers, and G. Elgered, 1985. Geodesy by Radio Interferometry: Effects of Atmospheric Modeling Errors on Estimates of Baseline Length, *Radio Sci.*, **20**, 1593–1607.
3. Chao, C.C., 1972. A model for tropospheric calibration from daily surface and radiosonde balloon measurements, *Tech. Mem. 391-350*, Jet Propulsion Lab., Pasadena, California.

4. Herring, T.A., J.L. Davis, and I.I. Shapiro, 1990. Geodesy by radio interferometry: The application of Kalman filtering to the analysis of very long baseline interferometry data, *J. Geophys. Res.*, **95**, 12,561–12,581.
5. Lanyi, G., 1984. Tropospheric delay effects in radio interferometry, *TDA progress report 42–78*, Jet Propulsion Laboratory, Pasadena, California.
6. Lichten, S.M., and J.S. Border, 1987. Strategies for high-precision Global Positioning System orbit determination, *J. Geophys. Res.*, **92**, 12751–12762.
7. Elgered, G., J.L. Davis, T.A. Herring, and I.I. Shapiro, 1991. Geodesy by Radio Interferometry: Water Vapor Radiometry for Estimation of the Wet Delay, *J. Geophys. Res.*, **96**, 6541–6555.
8. Kuehn, C.E., G. Elgered, J.M. Johansson, T.A. Clark, and B.O. Rönnäng, 1993. A Microwave Radiometer Comparison and Its Implication for the Accuracy of Wet Delays, *Contributions of Space Geodesy to Geodynamics: Technology*, D.E. Smith, and D.L. Turcotte eds., AGU Geodynamics Series, **25**, 99–114.
9. Liebe, H.J., 1992. Atmospheric Spectral Properties Between 10 and 350 GHz: New Laboratory Measurements and Models, *Proc. Specialist Meeting on Microwave Radiometry and Remote Sensing Applications*, Ed. E.R. Westwater, Boulder.
10. Waters, J.W., 1976. Absorption and Emission by Atmospheric Gases, in *Methods of Experimental Physics*, **12B**, ed. M.L. Meeks, Academic Press, New York, 142–176.
11. Keihm, S.J., 1992. Atmospheric Absorption from 20–32 GHz: Radiometric Constraints on the Vapor and Oxygen Components, *Proc. Specialist Meeting on Microwave Radiometry and Remote Sensing Applications*, Ed. E.R. Westwater, Boulder.
12. Johansson, J.M., R.T.K. Jaldehag, T.R. Carlsson, T.M. Carlsson, G. Elgered, P.O.J. Jarlemark, B.I. Nilsson, B.O. Rönnäng, H.-G. Scherneck, J.L. Davis, P. Elosegui, I.I. Shapiro, and J.X. Mitrovica, Fennoscandian Postglacial Land Uplift: Results From Seven Months of Continuous GPS Measurements, *in prep.*, 1994.

Influence of tropospheric electrical path delay on interferometric observations in millimeter radioastronomy

L. OLMI[1 2]

[1] Cornell University, NAIC, Arecibo Observatory, P.O. Box 995, Arecibo, Puerto Rico 00613
[2] Osservatorio Astrofisico di Arcetri, Largo E. Fermi 5, I-50125 Firenze, Italy

Abstract

We describe the effects of the troposphere in limiting the performances of a millimeter radio interferometer, used to observe astronomical sources. We show that at mm wavelengths water vapor is the main cause for fluctuations in the refractive index, n, which in turn are responsible for the degradations in the visibility function and partly for the random fluctuations in the power measured by a ground-based microwave radiometer. We also discuss, both qualitatively and quantitatively, the differences between the radio and the optical fluctuations of n. We then report measurements with the IRAM interferometer of tropospheric phase fluctuations at 86 GHz, on baselines 24 — 288 m. The data are used to derive phase structure function vs. baseline length, Allan variance and one-dimensional temporal phase power spectra. On average, the phase variance has a (baseline)$^{1.42}$ dependence. The one-dimensional phase power spectra vary with temporal frequency f as $f^{-2.1}$. Comparisons are made with a recent statistical model of water vapor fluctuations and with Kolmogorov turbulence. From this analysis, we estimate limits to use of the IRAM interferometer at higher frequencies and longer baselines.

Key words: atmospheric effects — instrumentation: interferometers — techniques: interferometric

1 Introduction

Radio propagation in the troposphere is mainly influenced by water vapor and oxygen. The water vapor is not well mixed, and its fluctuations affect the phase of centimeter and millimeter waves received by interferometers. Because the atmosphere is essentially nondispersive at radio wavelengths, the phase variations increase with operating frequency and become especially pronounced at millimeter wavelengths. In the centimeter range, observations agree with the Kolmogorov turbulence model (e.g., Armstrong and Sramek [1]; Kasuga *et al* [2], [3]; Sramek [4]). It is of great interest to obtain more information in the millimeter range. In this paper, we describe tropospheric phase fluctuations observed with the IRAM interferometer at a wavelength of 3.4 mm. The data, taken in a wide variety of weather conditions and elevation angles, are summarized as phase structure functions vs. baseline length, Allan standard deviation vs. sampling time, and one-dimensional phase power spectra. These results show how tropospheric turbulence affects astronomical observing with the IRAM interferometer and enable us to predict its effect at still higher frequencies.

2 Propagation effects and the visibility function

2.1 Phase fluctuations

It is well known that propagation through a medium whose refractive index varies in an irregular fashion gives rise to diffractive intensity and phase fluctuations. These fluctuations will affect the magnitude and phase of the complex visibility function measured by an interferometer. If we indicate with $V_{obs}(\mathbf{b}_\lambda)$ the observed visibility and with $V(\mathbf{b}_\lambda)$ the visibility that would be measured outside the Earth atmosphere with an ideal interferometer, we have:

$$V_{obs}(\mathbf{b}_\lambda) = V(\mathbf{b}_\lambda)\, e^{j\phi_{rnd}}$$

where $V(\mathbf{b}_\lambda) =\mid V(\mathbf{b}_\lambda) \mid \exp[j\Phi_V]$. The random phase ϕ_{rnd} takes into account the contribution of all the atmospheric and instrumental (noisy) effects. The total phase measured by the correlator for a *point source* located in the phase centre of the interferometer ($\Phi_V = 0$), will thus be:

$$\phi_{ij} = \Phi_{ij} + \phi_{rnd}$$

where the term Φ_{ij} incorporates both the phase delay corresponding to the difference in the optical paths to the two antennas i and j, and any other inserted instrumental phase term. In terms of the excess path delay L, defined by

$$L = c\Delta t = \int [n(l) - 1]\, dl$$

the total phase ϕ_{ij} can also be written as:

$$\phi_{ij} = \Phi_{ij} + \frac{2\pi}{\lambda} \int [n(l_i) - 1]\, dl_i - \frac{2\pi}{\lambda} \int [n(l_j) - 1]\, dl_j \tag{1}$$

where we see that the measured phase at the output of the correlator contains a term due to the *differential* path delay in the optical paths to the two distinct antennas. Since in a radio interferometer the instrumental term Φ_{ij} can be put to zero, from eq. 1 an interferometer can thus directly measure the differential excess path delay.

Therefore, the phase difference in signals received at two antennas has information about the source position, the length and orientation of the baseline, and the phase noise due to refractive index variations between the source and the antennas. Phase noise in radio interferometers at cm and mm wavelengths is mainly caused by water vapor fluctuations. Small scale ($\leq$ 1 km) water vapor variations limit the angular resolution of interferometers with a small number of elements, but there can be at least three other contributions to phase fluctuations: scattering of radiation by irregularities in the "dry" troposphere, scattering by plasma irregularities, and instrumental noise, which are described elsewhere (Olmi and Downes [5]; OD, hereafter), but are anyway negligible compared to water vapor turbulence.

2.2 The refractive index structure parameter

Since only the real part of the refraction index determines the phase delay in an electromagnetic wave propagating through the atmosphere, eq. 1 can be written as:

$$L = \int (n_R - 1)\, dl = 10^{-6} \int N_R\, dl$$

where N_R represents the real part of the refractivity, and the factor 10^{-6} is introduced for practical reasons. Since at radio and optical wavelengths, refractivity is a relatively simple function of temperature, T, humidity, Q, and pressure, P, one can write the fluctuation of n_R (or N_R), δn_R, as a function of the fluctuations in T, Q and P (Hill *et al* [6]), so that one can express the refractive index structure parameter, $C_{nR}^2 \equiv \langle(\delta n_R)^2\rangle/(\delta r)^{2/3}$, as:

$$C_{nR}^2 = A_T^2 \frac{C_T^2}{\langle T\rangle^2} + A_Q^2 \frac{C_Q^2}{\langle Q\rangle^2} + 2A_T A_Q \frac{C_{TQ}}{\langle Q\rangle\langle T\rangle} \simeq A_T^2 \frac{C_T^2}{\langle T\rangle^2}(1 \pm R_A)^2 \tag{2}$$

where the coefficients C_T^2, C_Q^2 are defined in a similar fashion as C_{nR}^2, and $C_{TQ} \equiv \langle(\delta Q\,\delta T)\rangle/(\delta r)^{2/3}$ (Hill *et al* [6]). The plus or minus sign in the second equality is due to the fact that temperature and humidity are often well correlated, although they may be positively or negatively correlated. It is useful to compare the relative importance of the refractive index fluctuations in the optical, infrared and radio regimes (Olmi [7]). We have introduced the parameter $R_A = R_C(A_Q/A_T)$, where

$$R_C \equiv \frac{C_Q}{\langle Q\rangle} / \frac{C_T}{\langle T\rangle} \tag{3}$$

If we now assume a typical order of fractional fluctuations as

$$\frac{\delta P}{\langle P\rangle} \ll \frac{\delta T}{\langle T\rangle} \leq \frac{\delta Q}{\langle Q\rangle} \tag{4}$$

and also that $R_C \simeq 100$ for intense humidity fluctuations (Hill and Clifford [8]), we get the following ratios for the refractive index structure parameters in the various regimes (Olmi [7])

$$\begin{cases} \frac{C_{nR}^2(rad)}{C_{nR}^2(vis)} \simeq 1065 \\ \frac{C_{nR}^2(rad)}{C_{nR}^2(mir)} \simeq 34 \\ \frac{C_{nR}^2(mir)}{C_{nR}^2(vis)} \simeq 31 \end{cases}$$

where *mir* stands for mid–infrared (these values are computed for the region $7 - 25\,\mu$m), *vis* for the visible region, *rad* for the radio regime, and R_A has been calculated by using the curves for A_Q and A_T given by Hill *et al* [6]. From the previous ratios one can see that in case of intense humidity fluctuations, the radio regime undergoes the strongest refractive index fluctuations, and the value of C_{nR}^2 increases rapidly in the sub–mm range.

The same procedure described above, can be applied to the definition of the term C_{nI}^2, the refractive index structure parameter for the imaginary part, n_I, of the refractive index. It describes the random fluctuations in the power extinction of an electromagnetic wave propagating through the atmosphere, and it is thus related to the measurements of a radiometer. In the case of an interferometer, we are mainly interested in C_{nR}^2, that will be indicated simply as C_n^2, hereafter.

2.3 Correlation lengths

A random field can also be characterized through its correlation function which, in the case of the normalized complex amplitude of the perturbed wave $\Psi(x,t) = \exp(j\,\phi(x,t))$, can be written as

$$B_\Psi(x) = \langle\Psi(b)\Psi^*(b+x)\rangle = \exp\left[-\left(\frac{\rho}{\rho_c}\right)^{5/3}\right]$$

where, in the second equality, we have assumed a Kolmogorov turbulence. One can show that $\rho_c \simeq r_\circ/2$, where both the Fried parameter, $r_\circ$ (that we shall introduce later), and thus ρ_c, called the correlation length, depend on the value of C_n^2. From the previous section, one can easily show that

$$C_n^2(rad) \simeq A_Q^2 \frac{C_Q^2}{\langle Q \rangle^2}$$

$$C_n^2(vis) \simeq A_T^2 \frac{C_T^2}{\langle T \rangle^2}$$

Assuming a "frozen" turbulence pattern (see next sect.) is moving past the antennas then, in the radio regime, ρ_c characterizes the typical scale length of the irregularities in the distribution of the *water vapor* in the troposhere; instead, at optical wavelengths, the irregularities in the *temperature* field are characterized by the correlation length. In the transition region of the FIR($\lambda > 50\,\mu$m)/sub–mm, ρ_c cannot be naively associated with only one of the previous fields, since both humidity and temperature fluctuations contribute to C_n^2 (Olmi [7]).

2.4 Radiometric sensing of tropospheric turbulence

As stated in sect. 2.2, tropospheric turbulence can also give rise to random fluctuations in the power measured by a radiometer, due to variations in the *imaginary* part of the refractive index structure parameter. Then, the equation of transport gives the formal description of the atmospheric emission and absorption processes. It can be used either to directly measure the water vapor content of the atmosphere (see, e.g., Elgered *et al* [10]), or to physically characterize the tropospheric turbulence (Andreani *et al* [11], and references therein).

Given the ability of ground–based microwave radiometers to retrieve the precipitable water vapor, they can be used, at least in principle, to correct the atmospheric path length fluctuations in interferometric observations (Resch *et al* [12]). This can be accomplished either by placing the radiometers on the array antennas, or by using different (and smaller) dishes placed next to the main antennas. Difficulties with this technique arise from the small variations in the brightness temperature (e.g., a few tens of mK at 90 GHz) that one has to measure in a reasonable short time (a few seconds).

Amplitude measurements of the incident radiation at each of the antennas of an interferometer, can also provide a data set for a correlation analysis of the turbulence that gave rise to the *amplitude* fluctuations in the incoming electromagnetic wave. This is known in geophysics as the correlation analysis of Moving Random Patterns (see, e.g., Briggs [13], and references therein), and can be used to retrieve some useful parameters (such as the mean wind velocity, the caharacteristic length and lifetime) of a drifting random pattern that moved past the antennas of an astronomical interferometer (see Olmi [7] for an initial discussion of this method applied to the IRAM data).

3 Observations with the IRAM interferometer

In order to test how much the effects described in the previous sections could affect the astronomical observations performed with a millimeter interferometer, we carried on a series of observation with the IRAM interferometer, located at an altitude of 2550 m on Plateau de Bure, France (the data sample is described in OD). To interpret the data we used phase structure functions, Allan variance and temporal phase power spectra. Analytic relations among these functions may be used if the random turbulent field is locally isotropic and if Taylor's hypothesis of frozen turbulence is valid. The latter means temporal variations of any quantity measured at a point are caused by uniform atmospheric motion past the point at the mean wind speed, and internal motions of this moving spatial pattern may be neglected on the measurement time scale.

Table 1. Parameters of the structure function in eq.(8), from linear fits to our data sample, selected by two different criteria. N_r represents the minimum number of scans in any of the nine 10° bins in which the data were grouped, according to their elevation angle. P_o is the integral probability that the data used could have come from an uncorrelated parent population. Units of c^2 are $\mathrm{rad}^2\,\mathrm{km}^{-1.34}$ for $N_r = 3$, and $\mathrm{rad}^2\,\mathrm{km}^{-1.49}$ for $N_r = 6$.

	Fit parameters		correlation
	intensity	index	coefficient
Selection criteria	$\log\ c^2$	$\alpha - 2$	r
$N_r = 3, \quad P_o = 0.1$	0.72 ± 0.03	1.34 ± 0.02	0.74
$N_r = 6, \quad P_o = 0.1$	0.77 ± 0.05	1.49 ± 0.04	0.70

3.1 Phase Structure Functions

The structure function of the phase fluctuations (e.g., Tatarski [9]) is the variance of the tropospheric component of the interferometric phase residual — the difference between the observed and expected interferometric phase — as a function of baseline length between antennas. Structure functions are often used in analyzing wave propagation in turbulent media, as the random processes involved can often be regarded as stationary. This also occurs for radio propagation in the troposphere, where the mean values of several meteorological variables change slowly and smoothly with time.

Adopting the same notation and analysis of OD, we can write the expressions for the structure function for observations at an arbitrary zenith angle, D_ϕ, and that one for observations at the zenith, $D_{\phi o} = \langle(\phi(x) - \phi(x+b))^2\rangle$ (where x is a space coordinate perpendicular to the propagation direction)

$$D_\phi(b) = c^2\, b^{\alpha-2} \sec z \; = D_{\phi o}(b) \sec z \tag{5}$$

(Coulman [14]) where z is the zenith angle to the source. The relevant size of an atmospheric irregularity is now not the physical baseline b but its projection $b_p = b \cos z$ on a plane perpendicular to the line of sight.

In the previous equation, we modeled the structure function as a power law (see, e.g., Armstrong and Sramek [1]). On the other hand, Coulman and Vernin [15] pointed out that in the range $5 \leq b \leq 1200$ m, that the authors call the "spectral gap", no power law is valid for either the spectrum or the structure function of refractive index fluctuations. The "spectral gap" is of special interest as the IRAM interferometer and most existing and proposed mm and sub-mm interferometers fall in just this range of baselines. Treuhaft and Lanyi [17] integrated $D_{\phi o}(b)$ as a function of baseline b, and found the phase structure function $D_{\phi o}(b)$ does not obey a strict power law; instead, the exponent $(\alpha - 2)$ in eq.(7) changes continuously between 2/3 and 5/3. Fig. 1 and Fig. 2 shows the phase structure function derived from IRAM interferometer data on 29 baselines from 24 to 288 m over 49 days in February – August 1990. The phase structure function referred to zenith $D_{\phi o}(b) = D_\phi(b) \cos z$ is plotted vs. projected baseline.

A least squares fit of the form

$$\log\ D_{\phi o}(b_p) = (\alpha - 2) \log\ b_p + \log\ c^2 \tag{6}$$

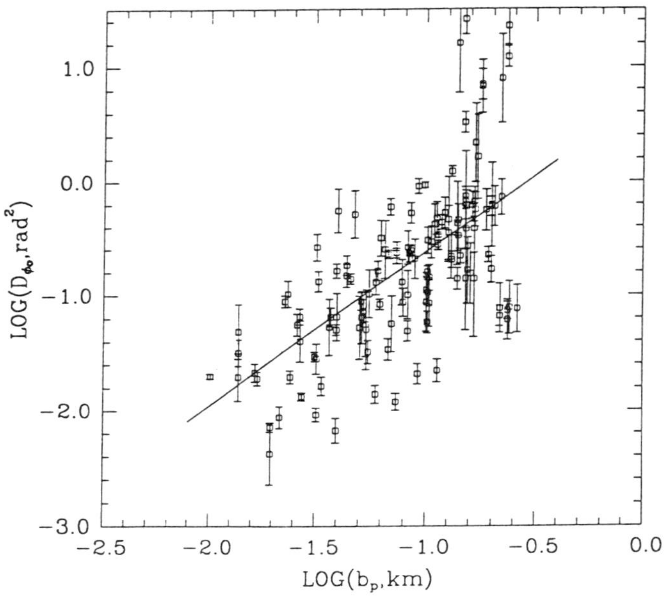

Figure 1. Phase structure function of 86 GHz IRAM data taken on 49 days in February – August 1990 on baselines 24 – 288 m. The data were taken in a variety of weather conditions, elevation angles, and times of day. We show only the data selected with N_r = 3 and P_o = 0.1 (see text and Table 1). The solid line is a fit to the data with exponent 1.34 ± 0.02 .

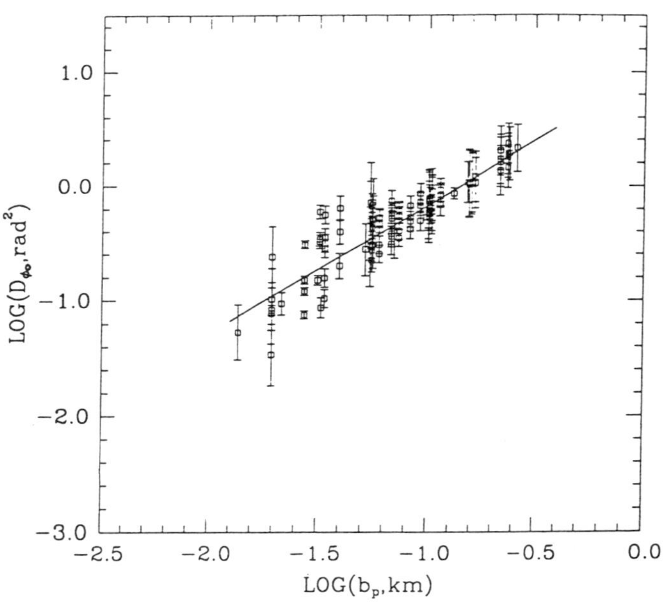

Figure 2. The phase structure function of the same sample as Fig. 1, normalized at a baseline of 144 m. The scatter of Fig. 1 has been considerably reduced, but the slope of the linear fit is different, 1.12 ± 0.03 .

Table 2. Parameters of the structure function, normalized at a baseline of 144 m, as derived from linear fits to our data selected with the same criteria as in Table 1. Units of c^2 are $\mathrm{rad^2\,km^{2-\alpha}}$.

Selection criteria	Fit parameters intensity $\log c^2$	index $\alpha - 2$	correlation coefficient r
$N_r = 3, \quad P_o = 0.1$	0.95 ± 0.04	1.12 ± 0.03	0.87
$N_r = 6, \quad P_o = 0.1$	0.77 ± 0.04	0.93 ± 0.03	0.90

was applied to all samples (for more details, see OD and Table 1) that led to $(\alpha - 2) = 1.42 \pm 0.07$ and $\log c^2 = 0.75 \pm 0.03$ (1σ errors). Although corrected for varying elevations, the plot still has a large scatter due to the changes in seeing quality over this period. The distribution for both the power law index α and $\log c^2$ are shown in Fig. 3. Coulman and Vernin normalized to reduce scatter in the structure function plots. Using a similar procedure, OD obtained the plot shown in Fig. 2, and one can see that the scatter is reduced, but the slope is different (see Table 2). The difference in slopes of the normalized and unnormalized structure functions is probably due to the uncertainty in linear fits with only a few sampling points per day. For a discussion of the behaviour of the structure function with the time of day, see again OD.

3.2 Tropospheric turbulence at the IRAM interferometer at higher frequencies and longer baselines

We now discuss how phase fluctuations can degrade both the resolution and the measured flux density at the frequencies and baselines that may be used in a future extension of the IRAM interferometer, with four to six dishes on baselines to ~ 450 m, covering the $80 - 115$ and $205 - 270$ GHz bands. Our results indicate the performance of such an extended interferometer will be severely limited by tropospheric fluctuations.

There are two effects: visibility is reduced by the factor $w(b) = \exp^{-D_\phi/2}$, (Thompson *et al* [16]), and resolution is limited by the radio seeing disk. The final map is the convolution of true source distribution, synthesized beam, and Fourier transform of the atmospheric weighting function $w(b)$. Hence the synthesis map that would be obtained outside the atmosphere is convolved with a gaussian beam of width θ_s, which for Kolmogorov turbulence is

$$\theta_s = 1.08\ 10^6 (C_n^2 L)^{3/5} \lambda^{-1/5} \quad , \tag{7}$$

where C_n^2 is assumed constant up to a height L and zero thereafter. The wavelength λ and height L are in meters, C_n^2 in $\mathrm{m}^{-2/3}$, θ_s in arcsec. $C_n^2 L$ characterizes the strength of the phase fluctuations and is related to the phase structure coefficient of eq.(7) by

$$c^2 = 2.91 \left(\frac{2\pi}{\lambda}\right)^2 C_n^2 L \quad . \tag{8}$$

This is because the two structure functions, D_ϕ and $D_n \equiv \langle(\delta n_R)^2\rangle$, can be related to each other (see Treuhaft and Lanyi [17]), and by using the definition of the refractive index structure parameter, one finally finds the previous relationship. From the measurements of c^2 we can calculate $C_n^2 L$ and the expected seeing disk at present and future baselines

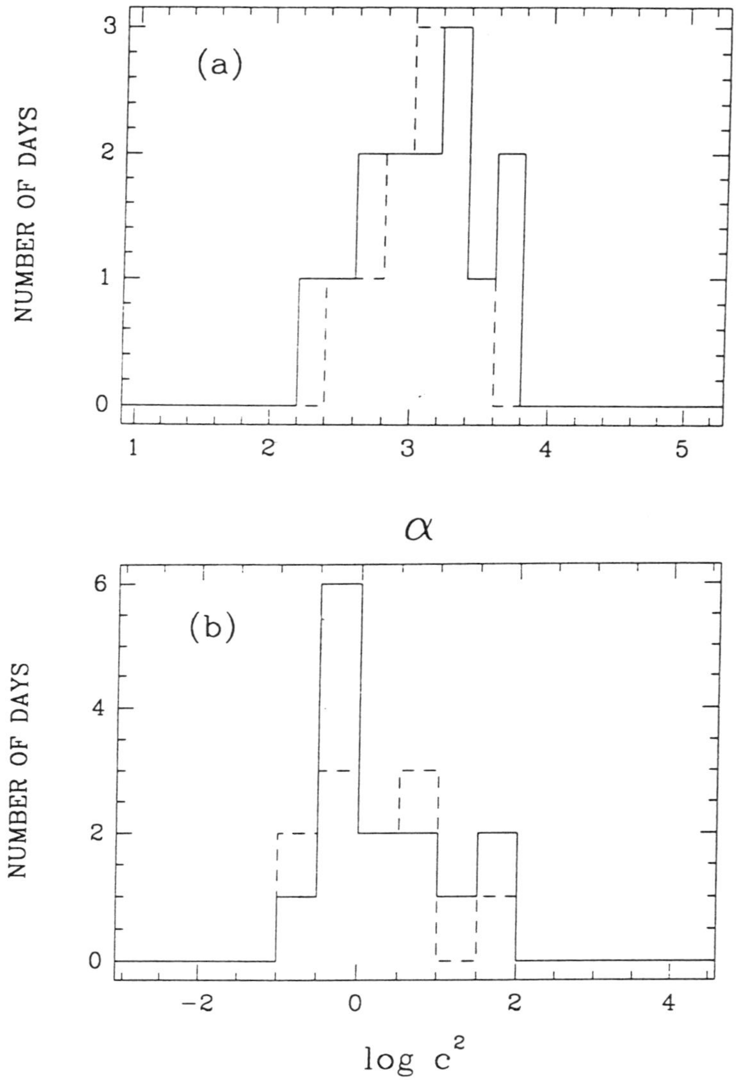

Figure 3. (**a**) Histogram of index α (α = structure function index + 2) derived from eq.(3) for projected baselines on days with $N_r = 3$, $P_o = 0.1$ (solid line) and $N_r = 6$, $P_o = 0.1$ (dashed line). Weighted means from the histograms for $N_r = 3$ and 6 are $\alpha_c = 3.07$ and 3.06, respectively and corresponding means from fits to the structure function (Fig. 1) are $\alpha = 3.34$ and 3.49. (**b**) Histogram of $\log c^2$, as in (a), units of c^2 as in Table 1. For samples with $N_r = 3$ and 6, the respective weighted means are $\log c^2 = 0.3$ and 0.1 .

Table 3. Summary of fits to the phase structure function for the IRAM interferometer

Parameters of IRAM data	range of each day [c]	average selected values	range for $N_r = 3$ $P_\circ = 0.1$	range for $N_r = 6$ $P_\circ = 0.1$
α	2.0 to 3.7	3.42 ± 0.07	2.3 to 3.7	2.5 to 3.5
c^2[86 GHz] [a]	0.14 to 184	31	2.3 to 370	0.3 to 126
c^2[5 GHz] [b]	0.05 to 62	10	0.8 to 130	0.1 to 43

[a] units of $10^{-5}\mathrm{rad}^2\mathrm{m}^{2-\langle\alpha\rangle}$
[b] units of 10^{-7} $\mathrm{rad}^2\mathrm{m}^{2-\langle\alpha\rangle}$
[c] for a selection with $N_r = 3$ and no limits on P_c; best fit $\langle\alpha\rangle = 3.49 \pm 0.01$

and wavelengths. From eq.(10) and the data for the structure coefficient c^2 in Table 3, we get $C_n^2 L = 3\ 10^{-13}$ $\mathrm{m}^{1/3}$ to 10^{-10} $\mathrm{m}^{1/3}$ for the sample selected with $N_r = 6$. Equation (9) then yields the seeing disk size, which we compare with the resolution at present and future (450 m) baselines. Table 4 shows that for the IRAM array's present baselines and frequency coverage, the synthesized beam is larger than the average seeing disk, so beam broadening by the troposphere is not too harmful, except for the worst conditions. At higher frequencies, however, the seeing disk becomes comparable with the beam, and is more than four times the beam size in the worst case. At these frequencies the resolution of the IRAM interferometer will be limited to $1'' - 2''$ or worse.

Table 4 also lists the Fried length, $r_\circ$, defined as the separation of two points along the wavefront for which the differential phase deviation is 2.6 radians r.m.s. . It can be also be associated to the scale length of the irregularities in the distribution of water vapor (sect. 2.3):

$$r_\circ = 0.186\ \lambda^{6/5}(C_n^2 L)^{-3/5} \tag{9}$$

(Thompson *et al* [16]). For the IRAM array at higher frequencies and longer spacings, $\langle r_\circ\rangle$ will become comparable with or smaller than the baselines expected to be used (Table 4).

OD also estimated the signal reduction by the tropospheric turbulence, by computing the atmospheric weighting function $w(b)$ for the present and proposed extension of the IRAM interferometer. The results are shown in Table 5 and discussed in the main paper.

3.3 Power Spectra

The phase variation with time can also be characterized by its power spectrum. To verify our assumptions of a locally isotropic random field and frozen-in turbulence, we derived the phase power spectrum in two ways, from the spatial structure functions and from the Fourier transform of the phase time series. By using the definition of phase structure function, it is possible to show that the one-dimensional temporal phase power spectrum can be written as (Armstrong and Sramek [1]; OD)

$$P_1(f) = c^2 f^{1-\alpha} v^{\alpha-2} B(\alpha) \tag{10}$$

where $B(\alpha)$ is a numerical coefficient depending on α only, f is the temporal frequency in Hz corresponding to the spatial frequency $q_x = 2\pi f/v$, and v is the wind speed in m/s. $P_1(f)$ has dimensions of $\mathrm{rad}^2/\mathrm{Hz}$. Using α and c^2 from the measurements of D_ϕ, we can

Table 4. Seeing disk θ_s (eq. 9) and Fried length $r_\circ$ (eq. 11) for present and future configurations of the IRAM interferometer

	frequencies (GHz) — present —		— proposed —	
	86	115	230	270
Seeing disk:				
$\theta_s{}^a$ [arcsec]	0.1 to 4	0.1 to 4.2	0.1 to 4.7	0.2 to 4.8
$\langle\theta_s\rangle$ [b] [arcsec]	1.6	1.7	2	2
Synthesized beams for $b_{max} = 288$ m:				
θ_{HPW} [arcsec]	2.1	1.6	0.8	0.7
Synthesized beams for $b_{max} = 450$ m:				
θ_{HPW} [arcsec]	1.4	1.1	0.55	0.47
Fried length:				
$r_\circ$ [m]	170 — 5800	130 — 4200	55 — 1800	45 — 1500
$\langle r_\circ\rangle$ [b] [m]	420	300	130	110

[a] from data yielding $C_n^2 L = 3.2\ 10^{-13}$ to $1.3\ 10^{-10}$ m$^{1/3}$, ($N_r = 6, P_\circ = 0.1$)
[b] from data yielding $\langle C_n^2 L\rangle = 3.3\ 10^{-11}$ m$^{1/3}$, from $\langle c^2\rangle$ of Table 3

Table 5. Atmospheric weighting function $w(b)$ (see text) for present and future baselines of the IRAM interferometer, derived from $D_{\phi\circ}(b)$ for observations near the zenith, with the average selected parameters of Table 3.

Baseline	Atmospheric weighting function $w(b)$ present 86 GHz	115 GHz	proposed 230 GHz	270 GHz
$b = 24$ m	1.0	1.0	0.9	0.9
$b = 288$ m	0.6	0.4	0.03	0.008
$b = 450$ m	0.4	0.2	0.002	0.0001

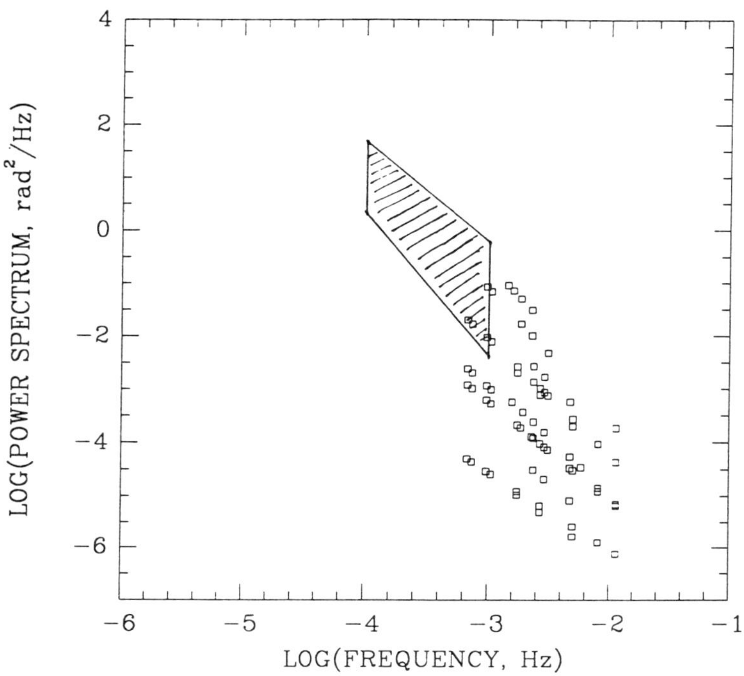

Figure 4. One dimensional temporal phase power spectrum for the IRAM data, vs. temporal frequency $f = v/2\pi b_p$ (v = wind speed, b_p = projected baseline), converted to an equivalent 5 GHz phase spectrum and scaled to a pattern velocity $v = 1$ m/s for comparison with temporal spectra from the VLA (Armstrong and Sramek [1], shaded box). A best fit to the IRAM data has a slope of -2.1 ± 0.4.

derive $P_1(f)$ (Fig. 4). This way of calculating $P_1(f)$ requires phase fluctuation data at several baselines, so interferometers with only a few antennas have to change configurations to build up the sample, and thus the data will not be acquired simultaneously.

We also computed the power spectrum of phase variations by Fourier transforming the phase time series. This approach requires continuous data sampling, which we usually couldn't do since our scans switched between phase calibrators and other sources. So, for this analysis we used two days on which phase calibrators were tracked continuously, 1990 August 23 and July 30 (Fig. 5). See OD for comments about the difference in the two methods.

3.4 Allan variance

We also use the Allan variance (Allan [18]) to describe the atmospheric phase fluctuations. The Allan variance of the frequency stability $y(t) = \delta\nu(t)/\nu_o$ is

$$\sigma_y^2(T) = \frac{1}{2(\pi\nu_o T)^2}\left\langle\left[\frac{\Phi(t+2T)+\Phi(t)}{2} - \Phi(t+T)\right]^2\right\rangle \tag{11}$$

where $\Phi(b,t) = \phi(x,t) - \phi(x+b,t)$ is the interferometric phase residual, ν_o is the observing frequency, and we assume equal sampling and averaging times (no dead time between measurements).

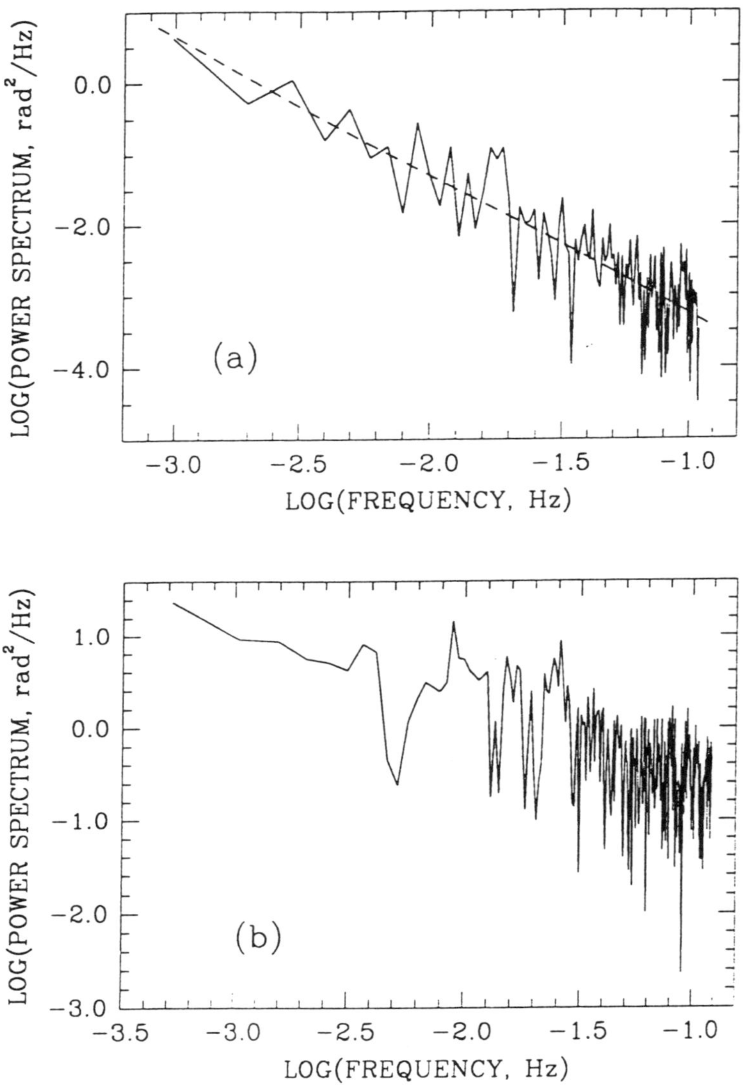

Figure 5. **(a)** Phase power spectrum, as a function of the Fourier temporal frequency, for the phase time series between 03^h30^m and 03^h45^m on 1990 Aug. 23 on a 288 m baseline. The fit has a slope of -2.0 ± 0.1, which agrees with that of Fig. 4, within the uncertainties. **(b)** Phase power spectrum for data taken between 22^h00^m and 22^h35^m. The slope on the lower frequency side is less steep than in (a).

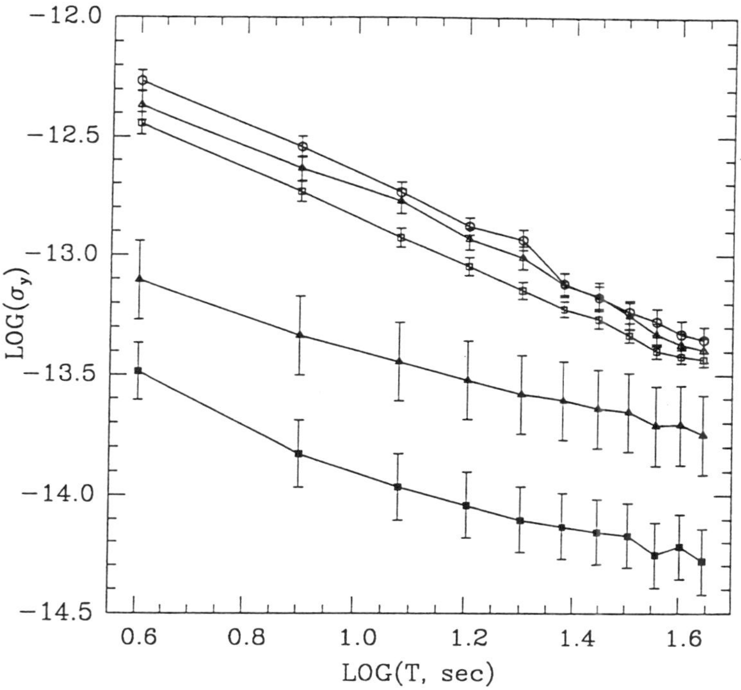

Figure 6. Allan deviations, σ_y, for baselines 24 to 241 m, for 1990 July 22 (two baselines), June 24, and Aug. 1 and 2 (bottom to top). Error bars are standard deviations of σ_y, averaged on each day.

The Allan variance is related to the one-dimensional temporal phase power spectrum $P_1(f)$ (e.g., Armstrong and Sramek [1]; Thompson *et al* [16]), and thus we expect the Allan deviation to vary as (averaging time)$^\beta$, with $\beta = (0.5\alpha - 2)$. The α of the structure function of the IRAM data is 2.0 to 3.7 (Table 3), so the Allan deviation should have a power law index β of -1.0 to -0.15 . Thus our measured range of -1.3 to -0.6 and mean value of -1.0 ± 0.2 for the power law index of the Allan deviation $\sigma_y(T)$ (see Fig. 6) are consistent with the predictions, although the mean is somewhat lower. Possible explanations for this slight discrepancy are given in OD.

4 Conclusions

We have shown that tropospheric turbulence can severely affect the visibility amplitudes and resolution of a mm radio interferometer. The main physical parameter describing these effects is the refractive index structure parameter, C_n^2. At radio wavelengths, C_n^2 is mainly determined by the fluctuations of the water vapor content of the troposphere while, at optical wavelengths, its value depends on the fluctuations in the temperature field. In the intermediate region, and especially at FIR/sub–mm wavelengths, both types of turbulence contribute to the global "seeing".

We also present measurements at 86 GHz of the phase structure function, temporal phase power spectrum and Allan deviation associated with tropospheric scintillations on slant paths. The observations were sensitive to fluctuation scales ~ 24 to 288 m. The phase structure function shows an average (baseline)$^{1.42}$ dependence, which, for the baselines for the IRAM data, is consistent with the predictions of Treuhaft and Lanyi, ~

1.42 to 1.63 . Tropospheric turbulence can have severe effects at the future frequencies and baseline extensions proposed for the IRAM interferometer, possibly preventing the use of resolutions better than 1″. The atmospheric phase fluctuations can strongly attenuate the visibility amplitudes, to as much as $\leq 1\%$ of its normal value.

Using the frozen turbulence model, we found that the one-dimensional phase power spectra vary with temporal frequency f as $f^{-2.1}$, and that the Fourier transform of the phase time series on a single baseline can tell us the approximate phase behaviour on other baselines. The Allan deviations for times of 4 sec to ~1 min show an average slope of -1.0 ± 0.2, close to the slopes of -1.0 to -0.15 expected from the structure function. It may be possible to explain this slight discrepancy by modeling the tropospheric turbulence with nonfrozen flows.

References

[1] Armstrong J.W., Sramek R.A., 1982, Observation of tropospheric phase scintillation at 5 GHz on vertical paths, *Radio Science*, **17**, 1579

[2] Kasuga T., Kanzawa T., Ishiguro M., 1990, Studies of the atmospheric phase fluctuations using the Nobeyama millimeter array, in *Radio Astronomical Seeing*, J.E. Baldwin and Wang Shouguan, eds., Pergamon Press

[3] Kasuga T., Kawabe R., Ishiguro M., 1986, Interferometric measurements of tropospheric phase fluctuations at 22 GHz on antenna spacings of 27 m to 540 m, *IEEE Trans. Antennas Propagat.*, **AP-34**, 797

[4] Sramek, R.A., 1990, Atmospheric phase stability at the VLA, in *Radio Astronomical Seeing*, J.E. Baldwin & Wang Shouguan, eds., Pergamon, Oxford, p. 21

[5] Olmi L. and Downes D., (OD in the text) 1992, Interferometric measurement of tropospheric phase fluctuations at 86 GHz on antenna spacings of 24 m to 288 m, *A&A*, **262**, 634

[6] Hill R.J., Clifford S.F., Lawrence R.S., 1980, Refractive-index and absorption fluctuations in the infrared caused by temperature, humidity, and pressure fluctuations, *JOSA*, **70**, N:10, 1192

[7] Olmi L., 1991, Analisi e misura del seeing astronomico a λ mm e sub–mm, *Ph.D. Thesis* (not published), Università di Firenze, Italy

[8] Hill R.J. and Clifford S.F., 1981, Contribution of water vapor monomer resonances to fluctuations of refraction and absorption for submillimeter through centimeter wavelengths, *Radio Science*, **16**, N:1, 77

[9] Tatarski V.I., 1961, in *Wave Propagation in a Turbulent Medium*, Dover, New York

[10] Elgered G., Ronnang B.O., Askne J.I.H., 1980, Remote sensing of atmospheric water vapor and cloud liquid, *Research Report* N:**141**, Research Laboratory of Electronics and Onsala Space Observatory

[11] Andreani P., Dall'Oglio G., Martinis L., Piccirillo L., Pizzo L., Rossi L., 1990, Millimetric flux densities as a test of atmospheric turbulence, *Infr. Phys.*, **30**, N:6, 479

[12] Resch G.M., Hogg D.E., Napier P.J., 1984, Radiometric correction of atmospheric path length fluctuations in interferometric experiments, *Radio Science*, **19**, N:1, 411

[13] Briggs B.H., 1968, On the analysis of moving patterns in geophysics–I. Correlation analysis, *Journal of Atm. and Terr. Phy.*, **30**, 1777

[14] Coulman C.E., 1991, *A&A*, **251**, 743

[15] Coulman C.E., Vernin J., 1991, Significance of anisotropy and the outer scale of turbulence for optical and radio seeing, *Applied Optics*, **30**, 118

[16] Thompson A.R., Moran J.M., Swenson G.W.JR., 1986, in *Interferometry and Synthesis in Radio Astronomy*, J. Wiley & Sons, New York

[17] Treuhaft R.N., Lanyi G.E., 1987, The effect of dynamic wet troposphere on radio interferometric measurements, *Radio Science*, **22**, 251

[18] Allan D.W., 1966, Proc. IEEE, **54**, 221

The retrieval of tropospheric water vapor and cloud liquid with an iterative non-linear algorithm

R. PETER

Institute of Applied Physics
University of Bern
Sidlerstr. 12
3012 Bern, Switzerland
Tel.: +41 31 6318911

Abstract An iterative retrieval scheme for integrated water vapor and cloud liquid from ground-based multi-channel measurements is presented. The algorithm estimates profiles of water vapor and temperature and adjusts them according to the difference between calculated and measured brightness temperatures.
A retrieval error for water vapor of 0.5 mm (rms) and 0.2 mm (mean deviation) with an instrument noise of 0.5 K can be achieved for zenith as well as lower angles. The iterative profiling algorithm can be used for variable observing conditions (site or elevation) without adaption of the retrieval procedure.

Key Words : Water vapor, cloud liquid, ground-based, radiometer, iterative algorithm

INTRODUCTION

The continuous observation of atmospheric water vapor and liquid water is very important for the propagation of electromagnetic waves and therefore of high relevance for communication and remote sensing. In order to complement already used linear statistical algorithms, a new kind of a non-linear profiling algorithm has been developed. The method and performance of such an algorithm for a ground-based instrument operating at 23.87, 31.65, 22.235, 52.85 and 54.95 GHz, which is currently beeing constructed under an ESA contract,will be reported in this work.

There are two possible approaches for the derivation of atmospheric quantities from microwave sky emission measurements. One is the linear statistical method, which is commonly used [1]-[3]. As the name says, it relates linearly the integrated water vapor (IWV) and the integrated liquid water (ILW) to the linearized brightness temperatures at the different frequencies of the radiometer [4]. The coefficients for the linear retrieval scheme are obtained from correlation of radiosonde measurements with brightness temperature measurements or with brightness temperature calculations. However, a drawback of this algorithm is the fact that the coefficients are only valid for a given observing geometry, site and season due to the strong variability of water vapor.

Although the first method is more elegant and rapid in the calculation, a more complicated approach with larger computation times is proposed in this work. It is a profiling algorithm [5], [6] which uses a complete model of the atmosphere and its related radiative processes. With the Millimeter-wave Propagation Model MPM [7], [8] and a parameterized representation of water vapor and liquid water (clouds), brightness temperatures are calculated and compared to

the measured ones, resulting in an iterative adjusting of the derived atmospheric profiles until agreement between calculation and measurement has been achieved.
Such a non-linear approach has the advantage that it is not site and season dependent and that it can be used for any elevation angle (e.g. time varing elevation angle due to non-stationary satellite). Additionally, each absorption contribution (dry, vapor, liquid) can be calculated separately and can be evaluated at other frequencies of interest due to the validity of MPM up to 1000 GHz.

ABSORPTION MODEL

The microwave absorption or emission, respectively, is based on MPM [7], [8]. The most recent version contains an improved formulation for the permittivity of liquid water [9], a new set of oxygen overlapping coefficients at 60 GHz [10], a stronger H_2O line intensity at 22.235 GHz as suggested by [11] and a modified dry air non-resonant term.
Input parameters for MPM are profiles of barometric pressure, temperature, relative humidity, suspended droplet concentration and rainfall rate with an altitude resolution of 100 m up to 20 km (i.e. 200 layers). The model calculates the complex refractivity, which consists of several additive terms as the contribution of all relevant H_2O and O_2 lines up to 1000 GHz, the continuum absorption (empirical) due to far-wing contribution of strong H_2O lines above 1000 GHz and contributions of non-resonant O_2 and N_2, the absorption due to small water droplets < 50 μm in clouds or haze and the empirically determined extinction due to rain. Integration over all altitudes yields the path opacity and total delay.

ATMOSPHERIC MODEL

A set of atmospheric parameters corresponding to profiles with a realistic behaviour are generated and used as input for the calculation of the complex refractivity (absorption and delay). As a first guess of the temperature profile, standard profiles (as in our case COSPAR International Reference Atmosphere CIRA86 [12]) interpolated for the corresponding latitude and season are used as initial profile. A better estimation of T(z) will certainly be the monthly mean of radiosonde measurements from a meteorological station near the observing site. Due to the high variability of the temperature, especially near the actual surface , an exponential fit of the initial profile to the surface value is applied. The temperature matches therefore at ground level the surface value and approaches with increasing altitude the initial profile.
In order to improve the temperature profile estimation, two channels in the oxygen band (50 - 60 GHz) can be used as this is proposed in [13]. At 54 GHz $T_{B,Meas}$ is sensitive to the temperature near the surface, while at the lower frequency (52 GHz), $T_{B,Meas}$ has a considerable sensitivity to temperature variations at higher altitudes [14].
Therefore T(z) can be corrected with respect to the weighting function at the corresponding frequency, by the use of the deviations of the calculated and measured brightness temperature, i.e. T(z) will be adjusted by the altitude dependent weighting function multiplied by a tuning factor ΔT determined by $T_{B,Meas} - T_{B,Calc}$. Maximum weighting for the 54 GHz channel is given at an altitude of 1.5 km above ground. For the 52 GHz channel a high sensitivity was chosen higher than 3 km above ground. Although both channels will be most sensitive at ground level, a higher altitude has to be used due to the surface temperature matching. The left graph on Figure 1 illustrates a few temperature profiles with different temperature biases.
The detection of an inversion layer is in principle possible, but due to the poor altitude resolution with a single 52/54 GHz measurement it is only a rough estimation of a realistic temperature profile.

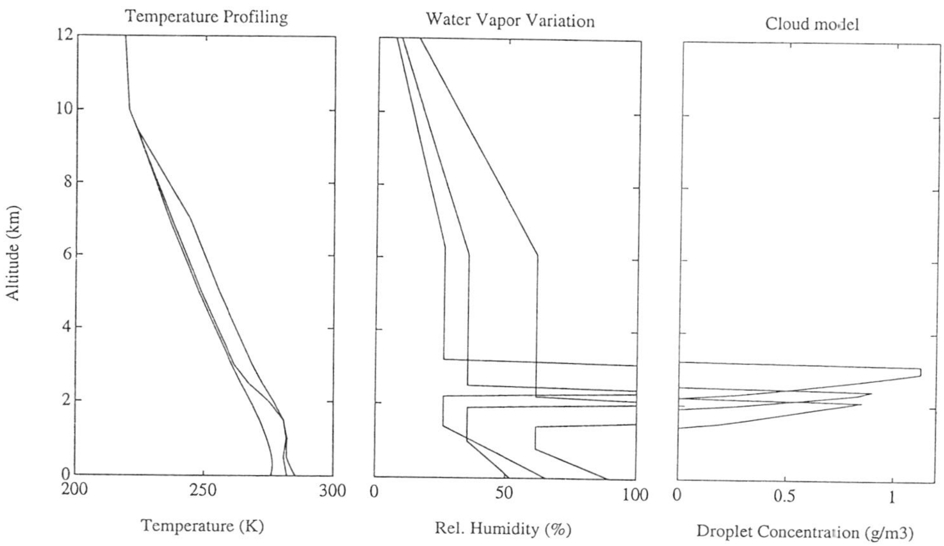

Figure 1 : Examples of generated temperature, humidity and liquid water profiles for cloudy conditions. The first plot shows temperature variations for different surface and bias values, the second one humidity profiles with 100 % relative humidity in the assumed cloud layer and the third one shows the corresponding cloud liquid profile.

The pressure can be well described by a exponential decay as in the case of the temperature profile using a measured surface value and a standard profile in mbar.

The water vapor profile is best expressed in terms of relative humidity since this is a more tractable function and allows a more realistic estimation of the water vapor content. For example in a cloud layer the relative humidity will be 100 % and at ground level $RH(z_0)$ will be the surface value provided by a built-in meteorological instrument. At about 10 km altitude (tropopause) where very little water vapor is present the relative humidity RH_{Top} is set to 0 %. The altitude of the tropopause is determined by the minimum temperature level of the used temperature profile.

A tuning parameter RH_{Ref} for water vapor is introduced in such a manner that the relative humidity RH(z) in the altitude range of approximately 2 - 6 km is equal to RH_{Ref}. The altitude range where RH(z) is equal to RH_{Ref} is slightly dependent on the altitude of the observing site and the cloud top altitude.

The humidity profile is therefore represented by a three-piece linear function defined by interpolation between the endpoints $RH(z_0)$, RH_{Ref} and RH_{Top}. Examples of the generated humidity profiles are given in Figure 1.

Clouds are modeled with a liquid water density profile and a relative humidity of 100 %. The density w(z) increases with altitude from the cloud base to the top of the cloud due to the decrease of the saturation pressure of water vapor with altitude. The difference between $\rho(z)$ in the cloud and ρ_{Base} at the cloud base, is the condensed water vapor of a rising air parcel, called the adiabatic liquid water content. Thus w(z) in a cloud can be represented by a linear increase of w with altitude inside the cloud.

The proportionality constant C of the adiabatic liquid water content is the tuning factor for the algorithm and can vary due to different cloud types [15]. Studies of the cloud structure [16] showed that C is typically about half of the adiabatic liquid water content. The entire cloud consists of several 100 m cloud layers at the adiabatic liquid water content multiplied by C. For simplicity the top of the cloud is determined by the -3 degree level, since above this level droplets have been frozen out by about 50 %. Ice in clouds which occur above this level do not contribute significantly to microwave absorption and are neglected for our model. Thus only the liquid phase in the cloud can be detected. Some liquid water profiles are shown in Figure 1. Another source of liquid water are hygroscopic aerosols (w_A) occuring at relative humidities between 80 % and 99.9 %. They can contribute up to 0.1 g/m3 to the liquid water density, particularly in smog conditions and for maritime climate. MPM includes an aerosol model for calculating the growth of droplets under haze conditions. The input parameters here are the aerosol concentration at 80 % humidity, climate type (rural,urban,sea, sae and wind) and relative humidity.

Rain is described by the rainrate profile R(z) assuming a constant value from ground to the top of the cloud. The instantaneous suspended liquid water concentration $w_R(z)$ can be obtained by integrating over a standard dropsize distribution for a given point rain rate R(z) multiplied with the volume of the droplets. For low rain rates w_R given in g/m3 is approximately one tenth of the rain rate given in mm/hr. The calculation for an exact conversion from the assumed Marshall-Palmer dropsize distribution to the liquid water concentration is performed after [17].

It has to be mentioned that care has to be taken in the interpretation of retrieved rainrates or liquid water under rainy conditions, since increased scattering or contamination of the measurements by water droplets on the antenna, can degrade the retrieval performance significantly.

With this definition of a simplified vertical structure of our model atmosphere the following parameters have to be changed during the iterative retrieval algorithm:

- $\mathbf{H_{Cloud}}$, the cloud base altitude in steps of 100 m indicating the thickness of the layer containing water droplets
- $\mathbf{RH_{Ref}}$,the reference humidity determining the amount of water vapor for a given temperature profile
- **C**, the ratio of the assumed droplet concentration to the adiabatic value allowing a continuous adjustment of the total amount of liquid water in combination with H_{Cloud}
- **R**, the point rainrate asuming a constant value from ground to the top of the cloud
- $\mathbf{\Delta T_{Low}}$, the temperature profile correction indicating the change in temperature for low altitudes
- $\mathbf{\Delta T_{High}}$, the temperature profile correction for high altitudes

ALGORITHM

Figure 2 shows the flow chart of the algorithm. The way that the atmospheric parameters have to be changed is not easy to describe, since it is based on a condition-tree for various situations. Nevertheless, some of the principles are discussed in this section.

Since the sky emission of all channels depends on gaseous and liquid water and the temperature profile, although with different sensitivities, each change of a parameter will influence the calculated T_B at the operating frequencies with different amounts. In order to adjust a parameter it should be known how a small change of RH_{Ref}, H_{Cloud}, C and ΔT would change $T_B(\nu_i)$ for a given observing situation. During the iteration each variation of the parameters and $T_B(\nu_i)$ is recorded for the calculation of the sensitivity of brightness temperatures on parameter variation

At 23 and 22 GHz T_B will be dominated by the value of RH_{Ref}. In a first approach RH_{Ref} is adjusted until T_B is equal to$T_{B,Meas}$ at 23 GHz within 0.5 K. In the next step the cloud parameters C, H_{Cloud} and R are then tuned until the calculated and measured brightness temperatures at 31 GHz do agree. Changing any of the cloud parameters results again in a new value for all other brightness temperatures, which means that we have to proceed as in the previous step. If any of the atmospheric parameters reaches the limits of the allowed range of the parameter, some of the other parameters or the temperature profile has to be changed in an appropriate way, which is one of the reasons for the complexity of this program.

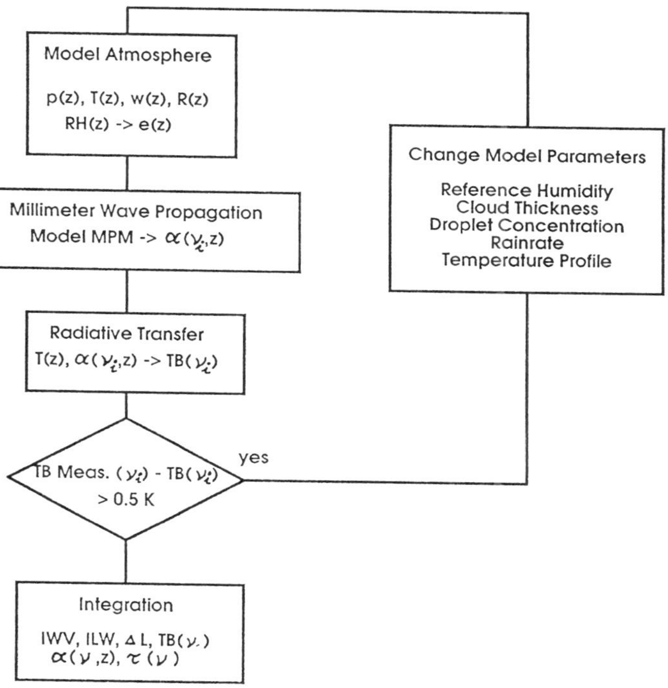

Figure 2 : Principle of the iterative profiling algorithm.

After the humidity channel and liquid water channel do agree, the temperature channel for close surface temperature at 54 GHz and the one for mid-altitudes at 52 GHz are examined, resulting in a change of the temperature profile. Each change might result in a new adjustement of humidity and liquid water, so that we have to proceed at the first step again.
Several (typically 20 -100) iterations are necessary until the exit criterion is fulfilled for the four channels, which means that one retrieval takes a few seconds up to one minute on a 486 PC/33 MHz. The Channel at 22 GHz is not used in the algorithm since it can cause contradictionary parameter changes with respect to the 23 GHz channel, particularly for brightness temperature offsets due to calibration errors. However, the use of the 22 GHz channel instead of the 23 GHz channel might be more advantageous during cloudy conditions due to the stronger signal.
In some cases it is difficult to reach this minimum deviation criterion due to unsuitable profile variation, particularly under rainy conditions. Therefore the allowed deviation can be weighted in a way that after a maximum of 150 iterations the parameters of the minimal weighted deviation will be taken for the retrieved atmospheric state. In this case the exit criterion can be higher than 1 K for the different channels according to their weighting.
Once an appropriate tropospheric condition is found, other quantities as sky brightness temperatures or zenith opacity for an arbitrary elevation and frequency up to 1000 GHz could be estimated using the capabilities of MPM. With the total tropospheric absorption along a

slant path, obtained by integrating $\alpha(\nu,z)$ over all layers, the calculation of the absorption and emission of the troposphere can be performed [6].

A separation of the different contributions due to the liquid and vapor phases in the calculation of $\alpha(\nu,z)$ yields a set of coefficients for the linearized relations between IWV, ILW and $T_B(\nu_i)$ or $\tau(\nu_i)$. Therefore a quick online retrieval using the calculated retrieval coefficients in the linear approach, will be possible without the use of radiosonde statistics.

ACCURACY AND PERFORMANCE OF THE PROFILE ALGORITHM

The overall error of IWV, ILW, $T_B(\nu)$ and $\tau(\nu)$ is primary dependent on the error of the absorption model and the algorithm error including radiometer accuracy, deviation of the assumed profiles from the real profiles and the deviation T_B and $T_{B,Meas}$ due to the exit condition. For the MPM absorption model error an upper limit of about 5 % is claimed by own verification studies [18], [6] and other researchers [11], [19].

The simplest way to quantify the remaining algorithm error is a simulation of $T_{B,Meas}(\nu_i)$ with known profiles of T(z), RH(z) and w(z) for different seasons and climates. Gaussian noise of 0, 0.5 and 1 K rms, respectively, is then added to the calculated $T_{B,Meas}(\nu_i)$ for a mid-latitude site at sea level, which are the input for the retrieval algorithm. Furthermore the calculations have been performed for two elevations (30 degrees and zenith). In order to estimate systematic errors (calibration offsets), a bias of 1 K has been added to each channel separately. Thus the rms deviation of the retrieved IWV, ILW, $T_B(\nu)$ and $\tau(\nu)$ from the initial values will represent the algorithm error.

Table 1: Standard deviation (rms error) of retrieved water vapor at different radiometer noise levels, elevation angles and weather conditions

Clear		Cloudy		Rain		Noise	Elevation
mm	%	mm	%	mm	%	K	
0.12	0.97	0.40	2.24	1.23	5.74	0	90
0.12	0.95	0.49	2.76	1.61	7.53	0	30
0.20	1.62	0.48	2.74	1.20	5.51	0.5	90
0.15	1.24	0.51	2.90	1.76	8.19	0.5	30
0.42	3.33	0.61	3.45	1.21	5.67	1	90
0.24	1.93	0.57	3.23	1.68	7.75	1	30

In order to have a wider variation of the initial profiles and to avoid site specific values as in radiosonde databases, a random generation of realistic profiles was favored. The mean values for the simulated IWV are 12.5, 17.6 and 21.5 mm for clear, cloudy and rainy conditions with a standard deviation of about 7 mm,respectively. For the ILW mean values of 0.4 and 1.9 mm with a standard deviation of 0.4 mm have been used for cloudy and rainy conditions.

The simulation has been done for clear, cloudy and rainy weather separately. The clear and cloudy condition calculations included 400 simulations each, while under precipitating conditions only 200 data points are used. In the statistical analysis also data points where the exit criterion has not been fulfilled (i.e. brightness temperature deviation > 0.5 K) are used.

Table 1 shows for each condition the rms deviation of the retrieved IWV from the initial IWV (square root of the sum of the squared differences). It is evident that increasing the noise of the input brightness temperature will result in a larger error. However, for clear and cloudy conditions, a higher radiometer noise is less significant for the retrieval error at low elevation angles due to the higher brightness temperatures. The analysis shows that the retrieval of IWV under non-precipitating conditions is significantly below 1 mm, even for a radiometer noise of 1 K.

In Table 2 the mean deviation between input and retrieved IWV is indicated. The offsets are smaller than the rms deviations for clear and cloudy conditions. However, for rainy weather there is an overestimation of IWV of about 1mm.

Table 2: Offsets of the retrieved water vapor from its input value. A negative sign means that the retrieved value is lower with respect to the input value. The bias of +1 K has been applied only at the frequency indicated in the parenthesis.

Clear		Cloudy		Rain		Noise	Bias	Elevation
mm	%	mm	%	mm	%	K	K	
-0.09	-0.72	0.14	0.79	0.53	2.41	0	0	90
-0.06	-0.50	0.23	1.30	1.16	5.43	0	0	30
-0.12	-0.96	0.16	0.92	0.45	2.09	0.5	0	90
-0.07	-0.55	0.19	1.11	1.26	5.85	0.5	0	30
-0.17	-1.38	0.12	0.71	0.49	2.28	1	0	90
-0.09	-0.75	0.19	1.09	1.19	5.49	1	0	30
0.65	4.94	1.10	5.92	1.56	6.83	0	+1(23)	90
-0.47	-3.86	-0.34	-1.99	-0.37	-1.72	0	+1(31)	90
-0.09	-0.74	0.06	0.37	0.95	4.33	0	+1(52)	90
-0.13	-1.02	0.14	0.76	0.57	2.55	0	+1(54)	90

The shift due to the increase of each input brightness temperature by 1 K (including the inherent offset of the algorithm) is given in the lower four lines of Table 2. The pure shift, which is the difference between the zero bias and the 1 K bias value of Table 2, is given in Table 3. As expected the increase of IWV is about 5 % caused by the 23 GHz channel and a decrease of about the half for the 31 GHz due to the different sensitivity to water vapor. The temperature channels have a low influence on the retrieved IWV, which means that the accuracy of these channels is not critical for the IWV retrieval.

The results of Table 3 together with typical spectra can also be used to infer requirements for frequency stability and accuracy of the instrument. For example the required instrument accuracy for the IWV retrieval, which is the most critical one, must be 0.5 K at 23 GHz and 1 K at 31 GHz in order to have a lower systematic error than the rms deviation of 0.5 mm for 1 K instrument noise level and zenith direction. Assuming a brightness temperature slope of 10 K/GHz around 23 GHz, the frequency stability can be estimated to be better than 50 MHz for the 23 GHz channel in order not to overpass the required accuracy by a unknown frequency shift. It has also to be mentioned that elevation angle errors larger than 0.2 degrees particularly at low elevation angles contribute in the same way to systematic errors. Therefore not only exact calibration but also an accurate knowledge of these instrument parameters is necessary in order to get a sufficient absolute accuracy for the IWV retrieval.

Table 3: Influence of increasing the brightness temperature by 1 K (calibration error) at the corresponding frequency on the IWV and ILW retrieval. A negative sign means that the retrieved value is lower than the retrieved one with no bias. The shifts in IWV and ILW (mm) reported in this table are the differences between the zero bias and 1 K biased values of Tables 2 and 4 and do therefore not contain inherent algorithm offsets.

Bias				Clear		Cloudy		Rain	
23	31	52	54	IWV	ILW	IWV	ILW	IWV	ILW
+1	0	0	0	+0.74	0.000	+0.96	-0.005	+1.03	0.030
0	+1	0	0	-0.38	+0.016	-0.48	+0.025	-0.80	+0.002
0	0	+1	0	+0.01	+0.001	-0.08	-0.002	+0.42	-0.150
0	0	0	+1	+0.04	0.000	0.00	+0.005	+0.04	+0.036

The wet path delay ΔL is strongly correlated with the IWV (integral over water vapor density) since ΔL is propotional to the integral of the water vapor density divided by the temperature (*Wu* 1987). A comparison between the two quantities lead to the relation

$$\Delta L \approx 6.6 \text{ IWV}$$

A similar value has also been obtained by [20]. The rms between path delay and IWV is 1.7 % which is in the same order than the rms deviation of the retrieved IWV. Therefore the path delay error will be similar to those reported in Table 1 and 2.

Table 4: Standard deviation (rms error) of retrieved liquid water at different radiometer noise levels, elevation angles and weather conditions

Clear		Cloudy		Rain		Noise	Elevation
mm	%	mm	%	mm	%	K	
0.001	-	0.061	17.5	0.421	22.2	0	90
0.002	-	0.085	22.4	0.413	21.6	0	30
0.003	-	0.060	16.8	0.373	19.6	0.5	90
0.002	-	0.078	20.4	0.427	22.3	0.5	30
0.006	-	0.061	17.9	0.380	19.9	1	90
0.003	-	0.073	18.9	0.443	23.0	1	30

Table 4 and 5 show the rms deviation and the mean difference of the ILW retrieval for three noise levels and two elevations. The rms error is about 20 % and is higher at lower elevation

angles, where it is nearly independent of the noise level. It has to be mentioned that even with no additional radiometric noise, some 'noise' is introduced due to the non-perfect agreement of calculated and measured brightness temperature i.e. exit criterion ($T_B(\nu) < 0.5$ K).
Furthermore the algorithm tends to underestimate the ILW. At lower elevation angles this deviation is higher than at zenith. A serious problem occurs under precipitating conditions, where the ILW is underestimated by about 20 % in the average. This is due to a insufficient temperature profile knowledge inside the cloud and rain altitude range and the rainrate profile. However, scattering and saturation will anyway limit the use of such measurements to non-precipitating or at least to weak rain conditions, where the algorithm works well.
Although these errors in the order of 20 %, which are also obtained by other retrieval methods, seem to be quite high, it has to be emphasized that it is impossible to monitor ILW with an other method than microwave radiometry (ground-based or space-borne). Therefore these measurements are still valuable informations about the state of the atmosphere.

Table 5: Offsets of the retrieved liquid water from its input value. A negative sign means that the retrieved value is lower with respect to the input value. The bias of +1 K has been applied only at the frequency indicated in the parenthesis.

Clear		Cloudy		Rain		Noise	Bias	Elevation
mm	%	mm	%	mm	%	K	K	
0.001	-	-0.006	-1.8	-0.317	-20.1	0	0	90
0.001	-	-0.018	-4.8	-0.407	-21.3	0	0	30
0.001	-	-0.006	-1.6	-0.336	-17.6	0.5	0	90
0.001	-	-0.016	-4.3	-0.395	-20.7	0.5	0	30
0.003	-	-0.006	-1.6	-0.343	-18.0	1	0	90
0.002	-	-0.016	-4.2	-0.404	-20.9	1	0	30
0.000	-	-0.011	-3.2	-0.287	-18.0	0	+1(23)	90
0.017	-	0.019	5.1	-0.315	-19.9	0	+1(31)	90
0.002	-	-0.008	-2.3	-0.467	-32.2	0	+1(52)	90
0.001	-	-0.001	-0.2	-0.281	-17.6	0	+1(54)	90

An improvement for the IWV and ILW is clearly obtained by the use of further channels in the 50 - 60 GHz band [13]. On the other hand, for the temperature profile itself, particularly above 3 km, only little improvement can be gained for fixed elevation measurements at two frequencies in the oxygen band, which are used in this algorithm. Due to the shape of the temperature weighting functions showing a high sensitivity near ground a good estimation is only obtained in the lower layers [21]. Since both are similar, only a weak altitude resolution can be obtained.
For the determination of ILW, an uncorrect temperature profile up to 4 - 5 km might be the reason for the high ILW retrieval error as discussed in the previous section.

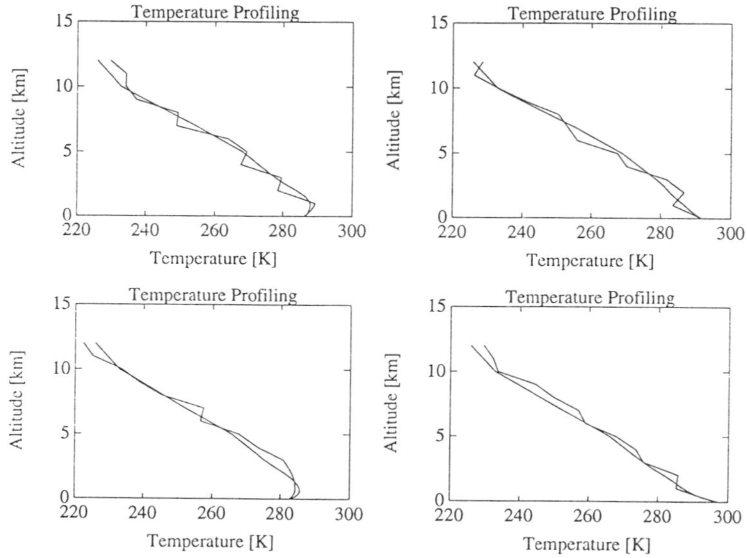

Figure 3 : Examples of the temperature retrieval of the iterative non-linear algorithm. The smoother curves are the retrieved temperature profiles, while the intended lines represent the initial noisy temperature profiles.

SUMMARY AND CONCLUSIONS

The profile algorithm together with the millimeter-wave absorption model MPM allows the estimation of slant path attenuation (or delay) for frequencies up to 1000 GHz as well as the total water vapor and liquid content in the troposphere. While the linear algorithm uses linearized brightness temperatures and statistically determined coefficients, the non-linear algorithm represents a complete modelling of the troposphere and the related radiative processes.

The implicitly contained model atmosphere allows better adaption to the observing site without the use of radiosonde databases and gives additional information about atmospheric profiles. The only local information is surface temperature and humidity as well as standard temperature and pressure profiles for the corresponding latitude. Furthermore no parameters for the retrieval have to be adapted to different seasons and meteorological conditions (e.g. high cloud liquid) as this is the case for the linear approach because the profile algorithm takes into account implicitely seasonal variation and nonlinearities generated by high cloud liquid. In addition the non-linear algorithm can provide an updating of the coefficients in the linear retrieval algorithm.

IWV can be measured under non-precipitating conditions to approximately 0.5 mm (rms) and 0.2 mm (mean deviation) with an instrument having a noise of 1 K and a systematic error (accuracy) of 0.5 K at 23 GHz. The systematic error of 0.5 K at 23 GHz or 1 K at 31 GHz causes an additional offset of the IWV by 0.5 mm in the average. The measurement of cloud liquid (ILW) is more difficult and can be determined to 20 % (rms) and 5 % (mean deviation). Each offset of the instrument by 0.5 K (accuracy) will result in a 5 - 8 % systematic shift.

Opacities for all frequencies used in the retrieval can be determined to 0.01 dB (rms) for 22, 23 and 31 GHz and to 0.05 dB and 0.1 dB (rms) at 52 and 54 GHz, respectively. The mean deviation is generally about half of the rms deviation. The systematic errors due to the limited intrument accuracy (0.5 K at 23 GHz and 1 K at 31 GHz) are of the same magnitude as the

rms error. These errors do not include systematic errors originating from the absorption model, which are in the range of 2 - 6 %.
Verification of a similar algorithm using two frequencies has been done by various methods as radiosonde measurements, IR-spectrometer measurements and inter-frequency comparison at 142 GHz and 204 GHz [6], [18].

Acknowledgements : This work has been carried out within a contract of the European Space Agency (ESA), monitored by B. Arbesser-Rastburg and the Swiss National Science Foundation NF contract no 23330/12/13

REFERENCES

1. Elgered G., B.O. Rönnäng and J.I.H. Askne, 'Measurement of atmospheric water vapor with microwave radiometry', Radio Science, Vol. 17, pp. 1258-1264, Sept./Oct. 1982.
2. Guiraud F.O, J. Howard and D.C. Hogg, 'A dual-channel microwave radiometer for measurement of precipitable water vapor and liquid', IEEE Trans. Geosci. El. , Vol GE-17, pp. 129-136, Oct. 1979.
3. Westwater E.R., 'Ground-based microwave remote sensing of meteorological variables', Chapter 4 in Atmospheric Remote Sensing by Microwave Radiometry, M.A. Janssen (Ed.), John Wiley & Sons Inc., pp. 145-213, 1993.
4. Wu S.-C., 'Optimum frequencies of passive microwave radiometer for tropospheric path length correction', IEEE Trans. Ant. and Prop., Vol. AP-27, pp. 233-239, March 1979.
5. Robinson S.E., 'The profile algorithm for microwave delay estimation from water vapor radiometer data', Radio Science, Vol. 23, pp. 401-408, May-June 1988.
6. Peter R. and N. Kämpfer,' Radiometric determination of water vapor and liquid water and its validation with other techniques', J. Geophys. Res., Vol. 97, pp. 18173-18183, Nov. 1992.
7. Liebe H.J., 'An updated model for millimeter wave propagation in moist air', Radio Science, Vol. 20, pp. 1069-1089, Sept./Oct. 1985.
8. Liebe H.J., 'MPM - An Atmospheric Millimeter-Wave Propagation Model', Int. J. of Infr. and MM Waves, Vol. 10, pp. 631-650, June 1989.
9. Liebe H.J., G.A. Hufford and T. Manabe, 'A model for the complex permittivity of water at frequencies below 1 THz', Int. J. Infr. and MM Waves, Vol. 12, pp. 659-675, July 1991.
10. Liebe H.J., P.W. Rosenkranz and G.A. Hufford, 'Atmospheric 60 GHz oxygen spectrum: New laboratory measurements and line parameters', J. Quant. Spectrosc. Radiat. Transfer, Vol. 48, pp. 629-643, 1992.
11. Keihm S.J., 'Atmospheric absorption from 20 - 32 GHz: Radiometric constraints on the vapor and oxygen components', Proc. Specialist Meeting on Microwave Radiometry and Remote Sensing μλRad 92, Boulder, Colorado, pp. 211-218, Jan. 1992.
12. COSPAR Int. Ref. Atmosphere (CIRA) : 1986, II in Middle Atmosphere Models, edited by D. Rees, J.J. Barnett and K. Labitzke, Adv. Space Res., Vol 10, pp 517, 1990.
13. Schiavon G., D. Solimini and E.R. Westwater, 'Performance analysis of a multifrequency radiometer for predicting atmospheric propagation parameters', Radio Science, Vol. 28, pp. 63-76, Jan./Feb. 1993.
14. Rodgers C.D. 'Retrieval of atmospheric temperature and composition from remote measurements of thermal radiation', Rev. Geophysics Space Phys., Vol. 14, pp. 609-624, 1976.
15. Rogers R.R and M.K. Yau, A short course in cloud physics, Int. Series in Nat. Philosophy, vol. 113, Pergamon Press, 1989.
16. Slobin S.D. 'Microwave noise temperature and attenuation of clouds: Statistics of these effects at various sites in the United States, Alaska and Hawaii', Radio Science, Vol. 17, pp. 1443-1454, Dec. 1982.
17. Olsen R.L., D.V. Rogers and D.B. Hodge, 'The aR^b relation in the calculation of rain attenuation', IEEE Trans. Ant. and Prop., Vol. AP-26, pp. 318-329, March 1978.
18. Peter R. and B. Schmid, 'Comparison of columnar water vapor determined from microwave emission and solar transmission measurements', Proc. CO-MEAS'93, March 22-25 1993, Albuquerque, pp.193-198, New Mexico.
19. Westwater E.R., J.B. Snider and M.J. Falls, 'Ground-based radiometric observations of atmospheric emission and attenuation at 20.6, 31.65, and 90 GHz: A comparison of measurements and theory', IEEE Trans. Ant. and Prop., Vol. 38, pp. 1569-1579, Oct. 1990.
20. Askne J. and H. Nordius, 'Estimation of tropospheric delay for microwaves from surface weather data', Radio Science, Vol. 22, pp. 379-386, May/June 1987.
21. Askne J.I.H. and E.R. Westwater, 'A review of ground-based remote sensing of temperature and moisture by passive microwave radiometers', IEEE Trans. Geosc. and Rem. Sens., Vol. 24, pp. 340-352, May 1986.

Observations of Total Column Precipitable Water Vapor and Cloud Liquid Water Using a Dual-Frequency Microwave Radiometer

J.C. LILJEGREN
Pacific Northwest Laboratory[1]
P. O. Box 999
Richland, Washington, 99352
U.S.A.

Abstract The performance of the WVR-1100 microwave radiometer deployed by ARM at the Oklahoma CART site central facility to provide time series measurements precipitable water vapor (PWV) and liquid water path (LWP) is presented. The instrument is shown to be durable and reliable in continuous field operation since June, 1992. The accuracy of the PWV is demonstrated to achieve the limiting accuracy of the statistical retrieval (0.07 cm) under clear sky conditions, degrading with increasing LWP. Uncertainties in the LWP during clear sky conditions are less than the theoretical accuracy of the retrieval (0.035 mm). The sensitivity of the PWV and LWP measurements for a 1-second integration is approximately an order of magnitude better than the accuracy. Improvements have been made to address moisture accumulation on the Teflon window, as well as to identify the presence of clouds with LWP at or below the retrieval uncertainty.

Keywords: water vapor, cloud liquid water, microwave radiometer, remote sensing, statistical retrieval, ARM

INTRODUCTION

The Atmospheric Radiation Measurement (ARM) Program is focused on improving the treatment of radiation transfer in models of the atmospheric general circulation, as well as on improving parameterizations of cloud properties and formation processes in these models [1]. To help achieve these objectives, ARM has deployed several two-channel microwave radiometers at the Cloud and Radiation Testbed (CART) site in Oklahoma for the purpose of obtaining long time series observations of total precipitable water vapor (PWV) and cloud liquid water path (LWP), defined as

$$\mathrm{PWV} = \frac{1}{\rho_L}\int_0^{\infty} \rho_V(z)\,dz, \quad \mathrm{(cm)}, \tag{1}$$

$$\mathrm{LWP} = \frac{1}{\rho_L}\int_0^{\infty} w(z)\,dz, \quad \mathrm{(mm)}, \tag{2}$$

where ρ_L is the density of liquid water, $\rho_V(z)$ is the density of water vapor and $w(z)$ is the cloud liquid water content (g m^{-3}).

---

[1] Pacific Northwest Laboratory is operated for the U.S. Department of Energy by Battelle Memorial Institute under Contract DE-AC06-76RLO 1830.

Daily averages of PWV from 9 October 1992 through 21 July 1993 are presented in Figure 1. This data reveals the annual trend of low moisture content in the winter (PWV ≈ 0.5 cm) increasing to over 4 cm in the summer. Superimposed on the annual trend are the synoptic variations associated with frontal passages.

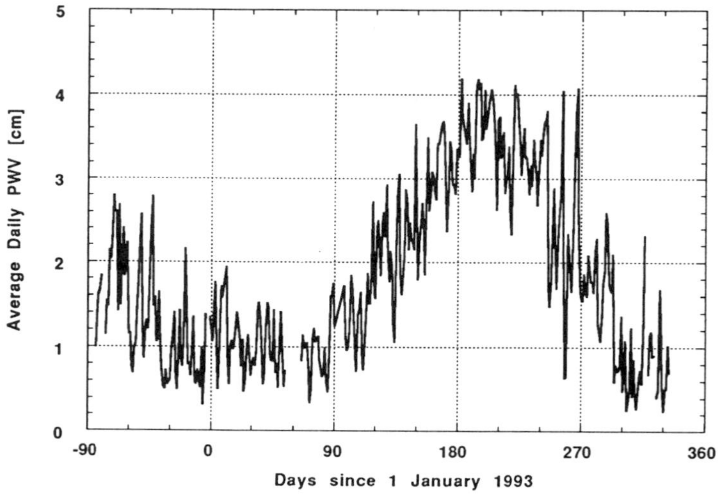

Figure 1. Daily averages of PWV from 9 October 1992 through 30 November 1993 at the ARM Cloud and Radiation Testbed (CART) central facility near Lamont, OK.

Background

Both PWV and LWP have been widely used to understand cloud and radiation interactions. For example, Stephens [2] has shown that LWP is a useful surrogate for the cloud optical depth, $\delta = 3/2\ LWP/r_e$, where r_e is the effective droplet radius. This relationship has been exploited to parameterize cloud radiative properties and to study cloud structure; see for example Stephens [3], Cahalan and Snider [4], Cahalan and Wiscombe [5]. PWV is also used to parameterize atmospheric radiation transfer. For example Li *et al.* [6] used PWV to relate net solar radiation at the surface to satellite-derived fluxes. Kuo *et al.* [7] have shown that short-range precipitation forecasts and predictions of the vertical distribution of specific humidity are improved when PWV is assimilated into numerical weather prediction models.

The use of PWV and LWP for developing and testing radiation and cloud models imposes significant performance requirements for accuracy, sensitivity and long-term stability of the measurements. In this paper the performance of one of the ARM microwave radiometers with respect to these requirements is presented.

THE MICROWAVE RADIOMETER

The microwave radiometers (MWRs) that ARM has acquired are Radiometrics Model WVR-1100. These are dual-frequency instruments operating at 23.8 and 31.4 GHz. (See Table 1.) The 23.8 GHz channel is about three times as sensitive to water vapor as the 31.4 GHz channel; the 31.4 GHz channel is about twice as sensitive to liquid water as the 23.8 GHz channel. The two channels use a common antenna and waveguide system with individual Gunn diode oscillators for each frequency. This design has two implications. First, simultaneous observations at the two frequencies are not possible - only one Gunn diode can be on at a time; and second, the fields of view of each channel are not identical. At 2 km altitude this translates into fields of view of 192 m (at 23.8 GHz) and 161 m (at 31.4 GHz). Considering the differences in the temporal and spatial scales of PWV and LWP, as well as the differences in the sensitivity of the channels to vapor and liquid, this difference in field of view is not particularly significant.

Table 1. WVR-1100 Specifications.

Vapor Sensing Channel	
Frequency	23.8 GHz
Bandwidth	0.4 GHz
Beamwidth (3dB)	5.5 degrees
Liquid Sensing Channel	
Frequency	31.4 GHz
Bandwidth	0.4 GHz
Beamwidth	4.6 degrees
Pointing Characteristics	
Angular coverage	4π sr
Angular slew rate	> 90 degrees/s, elevation
Physical Characteristics	
Dimensions	58 x 28 x 74 cm
Weight	21 kg
Power	120 W max. (winter)
Voltage	90-135 VAC, 47-440 Hz

Radiometric Measurements

The MWR measures the incident microwave energy and reports it as an equivalent blackbody brightness temperature, T_{sky} (K) for each channel:

$$T_{sky} = T_{ref} + \frac{V_{sky} - V_{ref}}{G}. \tag{3}$$

T_{ref} (K) is the actual temperature of the ambient blackbody reference target, V_{sky} (counts) is the signal observed when viewing the sky, V_{ref} (counts) is the signal observed when viewing the reference target, and G is the system gain (counts K^{-1}). The gain is extremely sensitive to the temperature of the microwave hardware. The microwave hardware is mounted on an aluminum plate in an insulated enclosure and thermally stabilized to 323 $\pm$0.25 K to ensure gain stability. Even so, to achieve the necessary accuracy the gain is measured during each observing cycle, defined below. This is accomplished by switching on a "hot load" in the form of a calibrated noise diode that injects microwave energy directly into the system. The gain is determined as

$$G = \frac{V_{ref+nd} - V_{ref}}{T_{nd}}. \tag{4}$$

V_{ref+nd} is the signal when viewing the reference target with the noise diode on, V_{ref} is the signal viewing the target with the noise diode off, and T_{nd} is the noise diode output (K) determined by prior calibration. (See below.) Combining Eqs (3) and (4) gives:

$$T_{sky} = T_{ref} + T_{nd} \frac{V_{sky} - V_{ref}}{V_{ref+nd} - V_{ref}}. \tag{5}$$

During each observing cycle, the three signals in Eq (5) are obtained for each channel. The mirror is rotated to view the sky; the Gunn diode oscillators for each channel are then sequentially energized and the signals recorded. The mirror is then rotated to view the black-body target; for each channel the signals are recorded with the noise source off and on. The target temperature is measured and Eq (5) applied to determine the sky brightness temperature.

The above observing cycle acquires a 1-second "snapshot" of the sky every 20 seconds. This high sampling rate is necessary because the LWP measurements are being used to develop and test new models that describe the inhomogeneous nature of clouds. The sampling rate must be high in order to resolve the fine scale structure of clouds. Averaging over a longer period would tend to smear out the details. Sampling faster (by shortening the observing cycle, if possible) would not necessarily provide additional independent observations: for a 10 m s^{-1} wind speed at

2 km altitude, at least 16 seconds are required for a scene to advect through the field of view of the liquid sensitive channel.

Calibration

Absolute calibration is achieved by use of "tipping curves." (See Decker and Schroeder [8].) Essentially, the atmospheric optical depth at several elevation angles is measured using an old or estimated calibration, and a straight line is fitted against the cosecant of the elevation angle. The slope of the line is the true zenith optical depth, from which can be computed the true zenith brightness temperature. Tipping curves can only be performed on days when the sky is completely free of clouds as the technique assumes that the atmosphere is plane-parallel and horizontally homogeneous.

Combined with measurements of the blackbody reference target, both with the noise diode switched off and on, the noise diode output can be calibrated as well. First, the gain is determined for each tipping curve by solving Eq (3)

$$G = \frac{V_{sky} - V_{ref}}{T_{sky} - T_{ref}}. \tag{6}$$

Equation (4) is then solved for T_{nd} and the gain computed from the tipping curve used to determine the noise diode output T_{nd}:

$$T_{nd} = \frac{V_{ref+nd} - V_{ref}}{G} . \tag{7}$$

The amount of microwave energy injected into the antenna system by the noise diode is not constant; rather ($V_{ref+nd} - V_{ref}$) is a random variable with a Gaussian distribution. To prevent adding noise into the calibration, tipping curves are performed at 90-second intervals over a period of hours and the mean value of T_{nd} for each channel is used. The distributions of T_{nd} resulting from the tipping curves are shown in Figure 2 for the 23.8 GHz (vapor-sensitive) channel and in Figure 3 for the 31.4 GHz (liquid-sensitive) channel.

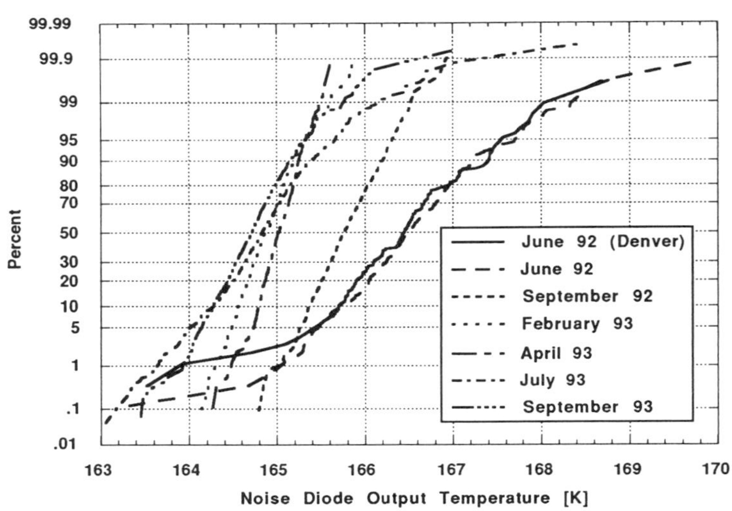

Figure 2. Noise diode output calibration history for the 23.8 GHz channel.

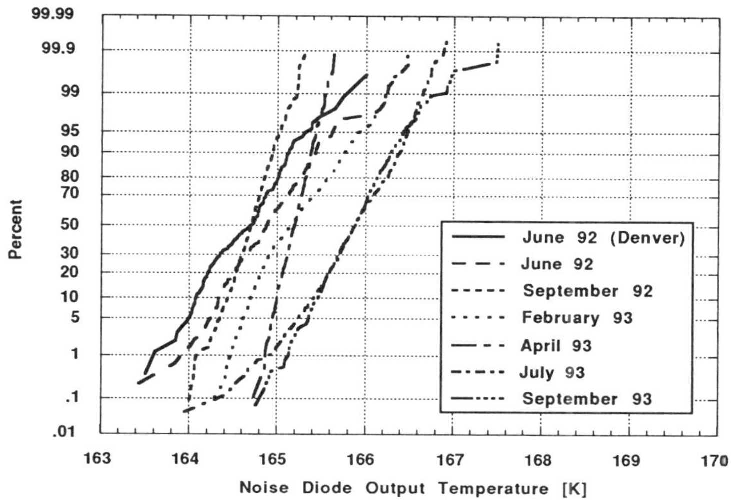

Figure 3. Noise diode output calibration history for the 31.4 GHz channel.

Calibration stability is important for long-term observations. Even though the tipping curves can be performed automatically by the MWR, collecting tipping curve data interrupts normal data collection. Fortunately, the interruptions occur during the least interesting periods, i.e. during clear sky conditions. The MWR calibrations exhibit a 1% overall variation over a period of more than a year. However, it is apparent that the aging of the noise diode is not linear in time. The mean value of T_{nd} for the 23.8 GHz channel appears to have reached a stable value; the mean value for the 31.4 GHz channel changed noticeably between the April and July calibrations, but remained constant from July to September.

CALCULATION OF PWV AND LWP

PWV and LWP are determined from the MWR brightness temperatures using a statistical retrieval technique developed by Westwater [9].

Statistical Retrieval

Essentially, historical radiosonde data representing the climatology where the MWR will be operated are input to a model of microwave radiation transfer [10]. This model computes the zenith optical depths, δ, atmospheric mean radiating temperatures, T_{mr}, and the expected microwave brightness temperatures corresponding to each sounding:

$$T_{sky} = T_c\, e^{-\delta} + T_{mr}(1-e^{-\delta}). \qquad (8)$$

T_c is the cosmic radiating temperature equal to 2.75 K.

For each sounding in the climatological data base the PWV is computed by integrating along the trajectory of the sounding and the LWP is computed by inserting clouds in regions of the sounding where the relative humidity exceeds a preset threshold value. The PWV and LWP are then related to the optical depths using multiple linear regressions of the form

$$PWV = a_0 + a_1\, \delta_{23.8} + a_2\, \delta_{31.4},$$

$$LWP = b_0 + b_1\, \delta_{23.8} + b_2\, \delta_{31.4}. \qquad (9)$$

The retrieval coefficients for Eq (9) have been computed for each calendar month by E.R. Westwater and M.J. Falls of NOAA/Environmental Testing Laboratory (ETL) using ten years of NWS radiosonde data from Oklahoma City, OK. Note that the observed brightness temperatures are related to the optical depths by solving Eq (8)

$$\delta = \ln\left(\frac{T_{mr} - T_c}{T_{mr} - T_{sky}}\right) \tag{10}$$

using the climatological average values of T_{mr} for each month. The residual uncertainty in the regressions, i.e. the variation in the data that is not explained by the regression, is referred to as the theoretical accuracy. That is, even if the observed optical depths were exactly correct some uncertainty in the retrieved values of PWV and LWP would result from the residual uncertainty in the regressions.

Tuning Functions

Because the model used to compute optical depths and brightness temperatures does not capture all of the physics, "tuning functions" must be developed to linearly relate the actual brightness temperatures observed with the MWR, T_{sky}, to those computed with the model, T_{model} [11]. Because MWRs operated by NOAA/ETL used 20.6 GHz and 31.65 GHz [12] rather than 23.8 and 31.4 GHz, these tuning functions had to be developed after the deployment of the first MWR at the CART site in Oklahoma. This was accomplished by implementing the NOAA/WPL microwave radiation transfer model into the CART data system as a Quality Measurement Experiment or QME [13]. After the data from each CART Balloon-Borne Sounding System (BBSS) launch was received by the CART data system, the WPL model was run automatically in order to generate model-calculated brightness temperatures and compute the PWV from the sounding for comparison with the MWR. The results of this proceedure are given in Table 2.

Table 2. Slopes and intercepts for tuning functions, $T_{model} = m\ T_{sky} + b$.

	slope	intercept	R^2
23.8 GHz	0.904	1.307	0.994
31.4 GHz	0.910	1.192	0.980

PERFORMANCE

Accuracy of PWV

The results of the QME comparing the PWV from the MWR with that computed from the BBSS for the time period of Figure 1 are presented as a scatter plot in Figure 4. The general impression from this plot is that the agreement for clear sky conditions is superior to cloudy sky conditions. This impression is borne out by comparing the distributions of differences ($PWV_{MWR} - PWV_{BBSS}$) presented in Figure 5. For clear sky conditions the mean difference (i.e. the bias) is essentially zero whereas there is a significant negative bias for cloudy skies. Also, the standard deviation of the clear sky differences is half of that for cloudy skies; the clear sky value equals the theoretical accuracy of the PWV retrieval, so this is as good as the agreement is likely to be.

The performance of the MWR in terms of PWV accuracy under relatively extreme conditions of high PWV and LWP is presented in Figure 6. The accompanying LWP data are presented in Figure 7. The mean and Root-Mean-Square (RMS) differences between MWR- and BBSS-derived PWV agree with the annual statistics for cloudy skies presented above.

In order to try to understand the relationship between PWV accuracy and cloud cover, the cumulative distribution of RMS differences in PWV as a function of LWP is presented in Figure 8. The cumulative number of BBSS soundings is also plotted; about 25% of the soundings are for clear sky conditions, another 50% occur for LWP less than 0.1 mm; only 25% span the range from 0.1 to 1.5 mm.

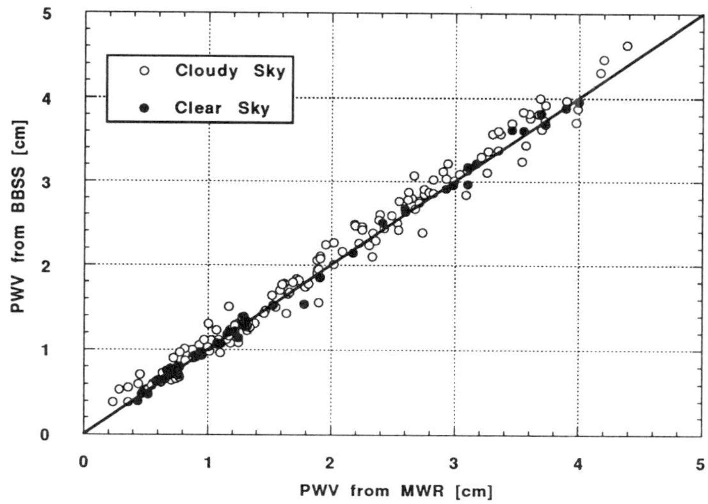

Figure 4. Scatter plot of PWV measured with the MWR against PWV derived from BBSS soundings for clear and cloudy sky conditions though June 1993.

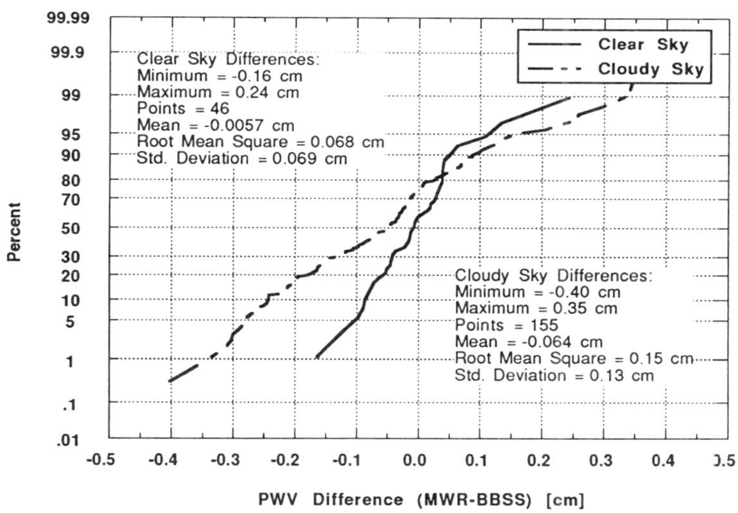

Figure 5. The distribution of differences in PWV between the MWR and BBSS using the data from Figure 4.

Figure 8 reveals a steady increase in the RMS difference with increasing LWP until a value of LWP ≈ 0.3 mm. Several possibilities exist for this increase. First, the horizontal distribution of water vapor may become more inhomogeneous with increasing cloudiness. Because the MWR and BBSS do not sample the same volume of the atmosphere, differences between them should increase as horizontal homogeneity breaks down. Second, Schmidlin [14] has shown that radiosondes can exhibit anomalous behavior upon passing from a cloud into clear air. Third, the number of soundings for which the LWP exceeds 0.2 mm is quite limited. If this trend holds generally, then large LWP events represent extremes in the climatology upon which the retrievals are based; i.e., the farther from the statistical average conditions, the greater the error in the retrieval [12]. The latter may be most apparent in the mean radiating temperature, T_{mr}.

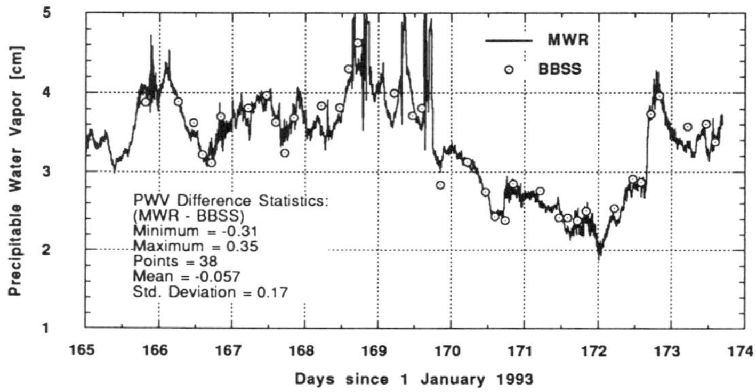

Figure 6. Time series of PWV for 15-23 June 1993.

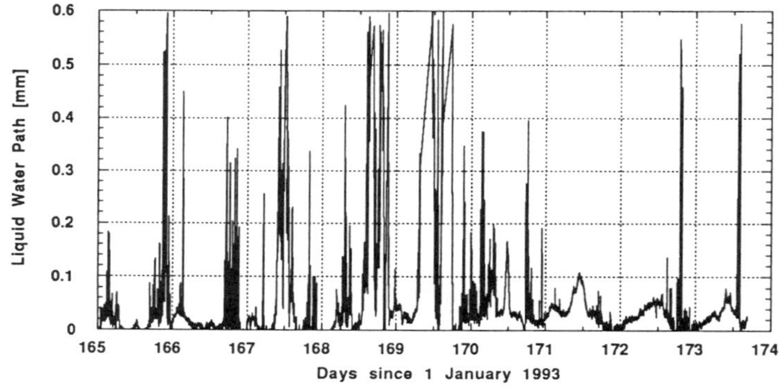

Figure 7. Time series of LWP for 15-23 June 1993.

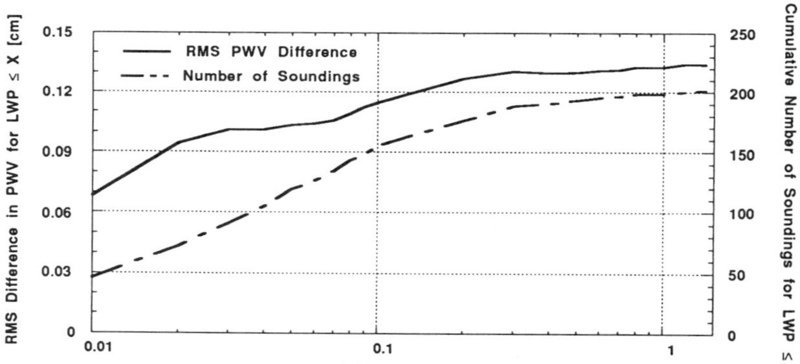

Figure 8. Cumulative distribution of RMS differences in PWV between MWR and BBSS for increasing LWP. The cumulative distribution of soundings according to LWP is also indicated.

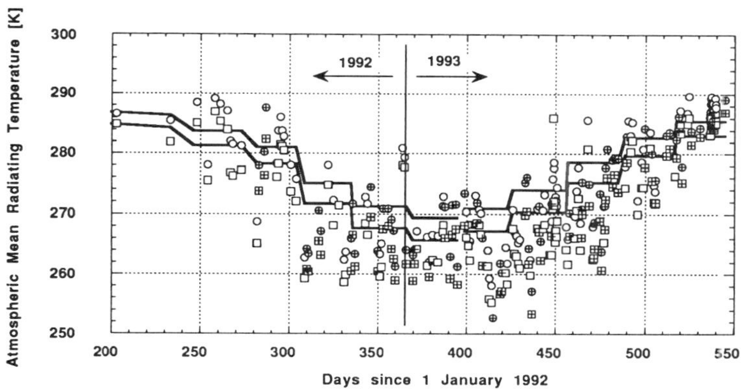

Figure 9. Mean radiating temperatures computed from soundings for 23.8 GHz (circles) and 31.4 GHz (squares) under cloudy sky (open symbols) and clear sky (symbols with '+'s) conditions. The monthly mean values used in the retrievals are indicated as solid lines. ($T_{mr, 23.8} > T_{mr, 31.4}$)

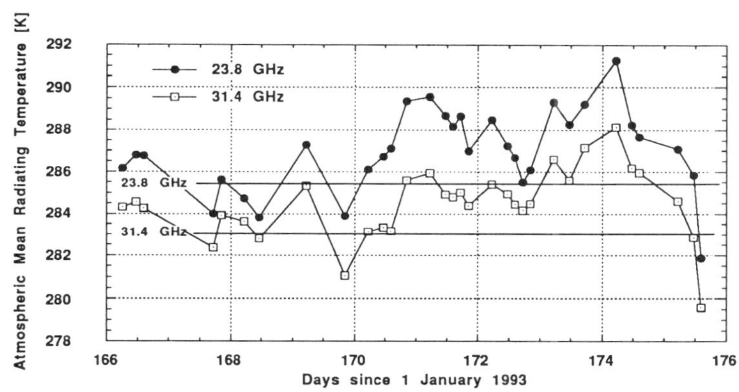

Figure 10. T_{mr} computed from soundings for 16-26 June 1993. Climatological average values for June are indicated as solid lines.

The values of T_{mr} used in the present retrieval method represent climatological averages for each month derived from the same historical data as the retrieval coefficients. In Figure 9 these climatological averages are compared against the values computed for each CART BBSS sounding as part of the QME. In general the BBSS-derived values follow the climatological trend well, with the largest discrepancies occurring during the winter months. This may be due, at least in part, to the known difficulties of the NWS sondes (upon which the climatological values are based) at low humidities [15]. However, T_{mr} is strongly affected by the amount of liquid water present. To illustrate this, the values of T_{mr} computed from BBSS soundings for the same period as Figures 6 and 7 are plotted in Figure 10. The climatological averages for June are also plotted for reference. The variation in T_{mr} during this period is approximately half of the annual variation exhibited by the climatological values. Fortunately, the retrieved value of PWV is relatively insensitive to variations in T_{mr} of the order of ± 5 K; unfortunately, the LWP is much more sensitive to T_{mr}.

Accuracy of LWP

Because the BBSS does not measure liquid water content in clouds, a direct comparison against the MWR-derived values as presented above for PWV is not possible. However, indirect evaluations are possible. For example, Daum [16] used a parameterization developed by Derr *et al.* [17] to relate cloud transmittance to LWP and found that the predicted and observed transmittances could not be reconciled without adjusting the observed LWP for a positive bias ranging from 0.01 to 0.05 mm. This was most apparent on cloud-free days because LWP was greater than zero, indicating the presence of clouds. As discussed above, this bias was corrected by adjusting the tuning functions. It should be noted however that the theoretical accuracy of the LWP retrieval is 0.035 mm. This means that there will be some uncertainty under conditions of low LWP as to whether thin clouds are present. (See "Future Improvements" below.)

Daum also pointed out that even after correcting for the bias, the LWP oscillated about zero (± 0.01 mm). This is believed to be due to short term variations in the mean radiating temperature that cannot not be accounted for with the climotological average monthly values.

Sensitivity

The application of LWP observations to develop and test models of inhomogeneous cloud structure demand a high sampling rate with a short averaging period. Conflicting with this requirement is the need to achieve a sufficient signal-to-noise ratio to distinguish cloud structure. One way to achieve a high signal-to-noise ratio is to average. However this also smears out the cloud structure. The only solution is to use an instrument with sufficiently low noise characteristics (i.e. high sensitivity) to support a high sampling rate.

Additionally, typical analyses of LWP and PWV time series include measures of *changes* in the data such as spectrum analysis; see Hogg, *et al.*, 1983 or Cahalan and Snider, 1989 for examples. These require high-sensitivity observations of the phenomena.

To determine the sensitivity of the MWR, cloud-free periods were selected from two days: 22 October 1992 during which the PWV varied from 2.5 to 3.0 cm, and 13 January 1993 having a variation in PWV from 0.5 to 1.0 cm. The PWV, LWP and brightness temperatures from these periods (1-second integrations at 20-second intervals) were high-pass filtered using a recursive digital filter having a 2-minute time constant. The 2-minute time constant was determined empirically as that which produced uncorrelated samples. That is, the autocorrelation of the output of the high-pass filter equaled zero after one lag.

This process removed the atmospheric contributions to the signal, leaving behind only the instrument noise. (This is why clear-sky conditions were selected for study; clouds contribute to the signal at much higher frequencies.) The RMS of the high-passed signal is thus the 1-second sensitivity of the instrument. These values are summarized in Table 3.

Table 3. Summary of MWR Performance

Precipitable Water Vapor (PWV)	
Theoretical Accuracy	0.07 cm
Measured Accuracy (RMS Difference from BBSS)	
Clear sky	0.07 cm
Cloudy sky	0.15 cm
Sensitivity (1-s RMS noise)	0.01 cm
Liquid Water Path (LWP)	
Theoretical Accuracy	0.03 mm
Sensitivity (1-s RMS noise)	0.003 mm
Brightness Temperature	
23.8 GHz Sensitivity (1-s RMS)	0.11 K
31.4 GHz Sensitivity (1-s RMS)	0.14 K
Calibration Stability	
Annual Average	1 x 10^{-3}month^{-1}

As the values indicate, the sensitivity is an order of magnitude better than the accuracy.

FUTURE IMPROVEMENTS

The most vexing problem with the MWR has been associated with the accumulation of water on the Teflon window covering the mirror. Dew, rain, or melting snow on the Teflon window cause the MWR signal to saturate, effectively "drowning" the atmospheric signal. To try to prevent the accumulation of water and to promote evaporation of water that does accumulate, the MWR has a blower that directs a stream of air over the Teflon window. Unfortunately, in condensing environments (e.g. when the temperature falls to the dew point at night or after a heavy rainfall) the air stream is fully saturated and the blower does little good until the relative humidity falls.

In order to identify and overcome such situations a solid-state moisture sensor has been mounted horizontally, downstream of the Teflon window from the blower. This sensor controls a 500 W heater which has been mounted in the blower housing.

In order to identify sky conditions (i.e. clear or cloudy) more accurately during conditions of low LWP, a zenith-viewing infrared thermometer with a spectral response in the 10-11 μm region and a narrow field of view has been acquired for installation on the MWR.

CONCLUSIONS

The performance of the WVR-1100 microwave radiometer deployed by ARM at the Oklahoma CART site central facility to provide time series measurements precipitable water vapor (PWV) and liquid water path (LWP) has been presented. The instrument has proven to be durable and reliable in continuous field operation since June, 1992.

The accuracy of the PWV has been demonstrated to achieve the limiting accuracy of the statistical retrieval under clear sky conditions, degrading with increasing LWP. A bias in the LWP has been corrected; uncertainties in the LWP during clear sky conditions are less than the theoretical accuracy of the retrieval. The sensitivity of the PWV and LWP measurements for a 1-second integration is approximately an order of magnitude better than the accuracy.

Improvements have been made to address moisture accumulation on the Teflon window, as well as to identify the presence of clouds with LWP at or below the retrieval uncertainty.

Acknowledgements
The author would like to acknowledge many profitable discussions with E.R. Westwater with respect to this work. Discussions with W.J. Wiscombe and F.S. Solheim were also very helpful. Implementation of the QME by N.E. Miller was essential. This work was supported by the U.S. Department of Energy under Contract DE-AC06-76RLO 1830.

REFERENCES

1. U.S. Department of Energy, 1990: Atmospheric Radiation Measurement Program Plan. DOE/ER-0441.
2. Stephens, G.L., 1978: Radiative properties of extended water clouds. Parts I and II. *J. Atmos. Sci.*, **35**, 2111-2132.
3. Stephens, G.L., 1984: The parameterization of radiation for numerical weather prediction and climate models. *Mon. Wea. Rev.*, **112**, 826-867.
4. Cahalan, R.F. and J.B. Snider, 1989: Marine stratocumulus structure. *Remote Sens. Environ.*, **28**, 95-107.
5. Cahalan, R.F. and W.J. Wiscombe, 1991: Plane Parallel Albedo Bias. *Proceedings of the Second Atmospheric Radiation Measurement (ARM) Science Team Meeting*. Denver, CO. USDOE CONF-9110336.

6. Li, Z., H.G. Leighton and R.D. Cess, 1993: Surface net solar radiation estimated from satellite measurements: comparisons with tower observations. *J. Climate*, **6**, 1764-1772.

7. Kuo, Y-H., Y-R. Guo and E.R. Westwater, 1993: Assimilation of precipitable water measurements into a mesoscale numerical model. *Mon. Wea. Rev.*, **121**, 1215-1238.

8. Decker, M.T. and J.A. Schroeder, 1991: Calibration of ground-based microwave radiometers for atmospheric remote sensing. NOAA Tech Memo ERL WPL-197.

9. Westwater, E.R., 1993: Ground-based microwave remote sensing of meteorological variables. *Remote Sensing by Microwave Radiometry*, M.A. Janssen, ed. John Wiley & Sons.

10. Schroeder, J.A. and E.R. Westwater, 1991: User's guide to WPL microwave transfer software. NOAA Tech Memo ERL WPL-213.

11. Westwater, E. R., J. B. Snider and M. J. Falls, 1990: Ground-based radiometric observations of atmospheric emission and attenuation at 20.6, 31.65 and 90.0 GHz: a comparison of measurements and theory. *IEEE Trans. Antenn. and Propag.*, **38**, 1569-1580.

12. Hogg, D. C., F. O. Guiraud, J. B. Snider, M. T. Decker and E. R. Westwater, 1983: A steerable dual-channel microwave radiometer for measurement of water vapor and liquid in the troposphere. *J. Clim. Appl. Meteorol.*, **22**, 789-806.

13. Miller, N.E., J.C. Liljegren, T.R. Shippert, S.A. Clough, and P.D. Brown 1994: Quality Measurement Experiments within the Atmospheric Radiation Measurement Program. Fifth Symposium on Global Change Studies, *74th AMS Annual Meeting*. Nashville, TN.

14. Schmidlin, F.J., 1988: WMO international radiosonde comparison, phase II. *Instruments and Observing Methods Report No. 29*. WMO/TD-312.

15. Nash, J. and F.J. Schmidlin, 1987: WMO international radiosonde comparison, phase I. *Instruments and Observing Methods Report No. 30* WMO/TD-195.

16. Daum, P., 1993: SGP/MWR Bias in LWP measurements. ARM Problem Identification Form PIF930610.2.

17. Derr, V.E., R.S. Stone, L.S. Fedor and H.P. Hanson, 1990: A parameterization for the shortwave transmissivity of stratiform water clouds based on empirical data and radiative transfer theory. *J. Atmos. Sci.*, **47**, 2774-2783.

Retrieval of atmospheric water vapour and liquid water using a single frequency microwave radiometer

PER O. J. JARLEMARK

Onsala Space Observatory, Chalmers University of Technology
S-439 92 Onsala, Sweden

Abstract. Four data sets of water vapour radiometer (WVR) measurements at the Onsala Space Observatory have been analysed in order to investigate the differences in the wet propagation delay derived using either only one or two channels. This was done in order to evaluate the possibility of improving the single channel estimate using statistical methods. By recognizing the more rapid variations in the linearized sky brightness temperature, time intervals with large liquid water content were identified. A Kalman filter whose parameters depend on the presence of liquid water was applied to the data. The filtered one channel estimates of the wet delay had significantly improved performance when compared to the unfiltered. In many cases the deviation from the two channel estimates was reduced to less than one half of the deviation of the unfiltered estimates.

Key words: Microwave Radiometry, Atmospheric Water Vapour, Excess Propagation Path, Space Geodesy, Time Series Analysis

INTRODUCTION

In many applications, inexpensive methods of retrieving both the water vapour and the liquid water content from atmospheric measurements are desired. This is, for instance, the case in space geodetic measurements, where simultaneous monitoring of the excess propagation delay due to the atmospheric water vapour, "wet delay", at several different sites can improve the estimates of geodetic quantities. In connection with this, the performance of a water vapour radiometer (WVR) utilizing only one channel becomes interesting. A single channel WVR will also be less affected by instrumental noise.

The lack of information when the two unknown quantities, water vapour and liquid water, are to be retrieved from the single observable, the sky brightness temperature, can be relieved to some extent by utilizing the differences in the spatial and temporal behaviour of water vapour and liquid water. By assigning linear models for their temporal variations, it is possible to design Kalman filters which update estimates of both quantities based on a sequence of measurements.

DATA ANALYSIS

We have analysed four data sets of WVR measurements from spring and summer 1988 in order to investigate the difference between the wet delay estimated using one and two channels. The goal was to identify periods with high liquid water content using one channel only. This may be done by recognizing the more rapid variations in the sky brightness temperature during these periods. These rapid variations are caused by fluctuations in the liquid water content, e.g. due to the turbulence associated with cloud convection. Variations in the drop–size distribution, which violates the Rayleigh approximation, may also contribute.

A Kalman filter which could reduce the effect of the overestimated wet delay during liquid water conditions was developed.

Wet Delay Retrieval Algorithms

The wet delay of the four data sets was calculated from the sky brightness temperature measurements according to the algorithm

$$\Delta L_w = C_{2f}\left[\left(\frac{31.4}{21.0}\right)^2 T'_{21.0} - T'_{31.4} - T_{bg,ox,2f}\right] \tag{1}$$

where $T'_{21.0}$ and $T'_{31.4}$ are the two "linearized" sky brightness temperatures [1] at 21.0 and 31.4 GHz, respectively. $T_{bg,ox,2f}$ is a linear combination of the contributions from the cosmic background radiation and the atmospheric oxygen at the two frequencies. The parameter C_{2f} is modelled using ground meteorological data [2]. The quantity in brackets in (1) is independent of the liquid water content when drop sizes are significantly smaller than the wavelengths sensed by the radiometer.

The wet delay was also estimated by an algorithm which utilizes one frequency only

$$\Delta L_w = C_{1f}\ [T'_{21.0} - T_{bg,ox,1f}] \tag{2}$$

This algorithm works well under "liquid-free" conditions, but breaks down in the presence of too much liquid water. This can be seen in Figure 1, which shows the zenith wet delay during July 22–27 1988 estimated from the two–channel algorithm (lower curve) and the one–channel algorithm (upper curve). The general bias between the curves, which remains also during periods with no liquid water present, is mainly due to errors in

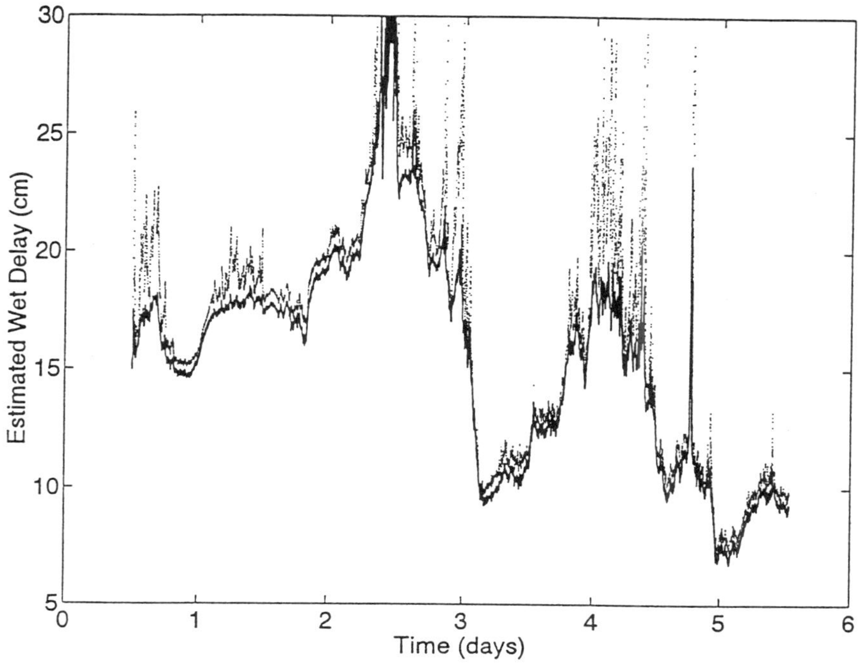

Figure 1. The zenith wet delay from July 22–27 1988, the lower solid curve retrieved using a two channel algorithm and the upper dotted curve using a one channel algorithm.

the attenuation models used in retrieving the parameters C_{1f} and C_{2f}, and their dependence on meteorological data, which are not modelled well enough. This bias will remain until new parameters, C_{1f} and C_{2f}, has been derived.

Liquid Water Detection

In order to use the increased variations in the sky brightness temperature to detect the presense of liquid water, each data set was divided into smaller intervals in which the standard deviation was calculated. It proved to be a trade–off between longer and shorter time intervals. If long time intervals (a couple of hours) were used, good correlation between the standard deviation of $T'_{21.0}$ and the average difference between the one and the two channel estimates of the wet delay was found. On the other hand, the sensitivity to shorter liquid periods was reduced.

In this trade–off, we choose 15 minutes intervals. Since the average sampling rate of the WVR was 23 s, there were, on the average, 39 measurements in each interval. Figure 2 shows the liquid water content in these intervals, here represented by the difference between the one–channel and

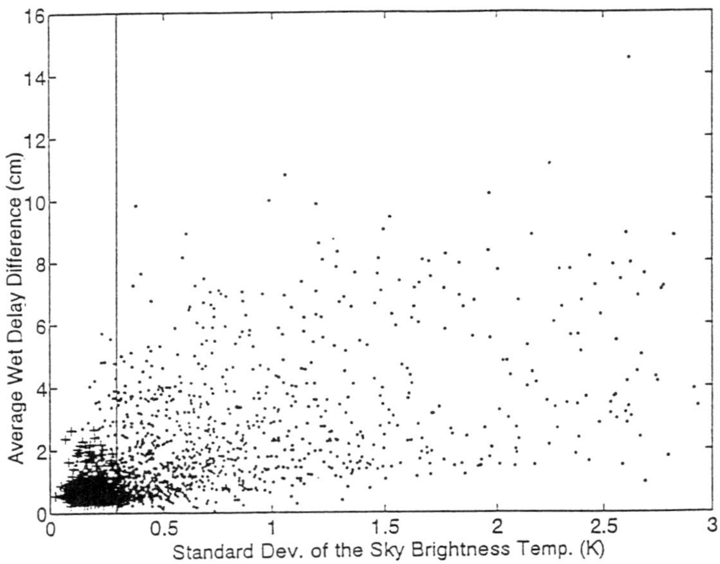

Figure 2. The average difference between the zenith wet delays derived from the one and the two channel algorithms plotted against the standard deviation of the zenith linearized sky brightness temperature within 15 minute intervals. The line indicates the limit between "liquid-free" intervals (to the left), and intervals considered containing liquid water. "+" indicates intervals that will still be considered liquid-free after the removal of neighbour intervals (see text).

the two–channel wet delay estimates, plotted against the standard deviation of the 21 GHz sky brightness temperature. Based on this plot we considered liquid water detected if the standard deviation within an interval was more than 0.3 K. It was found that many of the intervals which had less standard deviation but where the wet delay error was still large were "neighbour" intervals to an interval where liquid water had been detected. Thus, all "neighbour" intervals were also included in the "detect" class.

32% of the intervals were considered "liquid–free" and 0.5% of them still had a wet delay error of 2 cm or more. As can be seen from Figure 2, these criteria are not very strict, but probably good enough as a tool for choosing the model in a Kalman filter.

Kalman Filtering

In Kalman filters, a process is modelled as a number of states at each time epoch and the dynamics of their transitions into the next epoch. Figure 3 shows schematically the steps of the Kalman filter loop.The dynamics are used by first predicting the new states X_p, based on the present states X, and the transition matrix F. The final new states are thereafter calculated

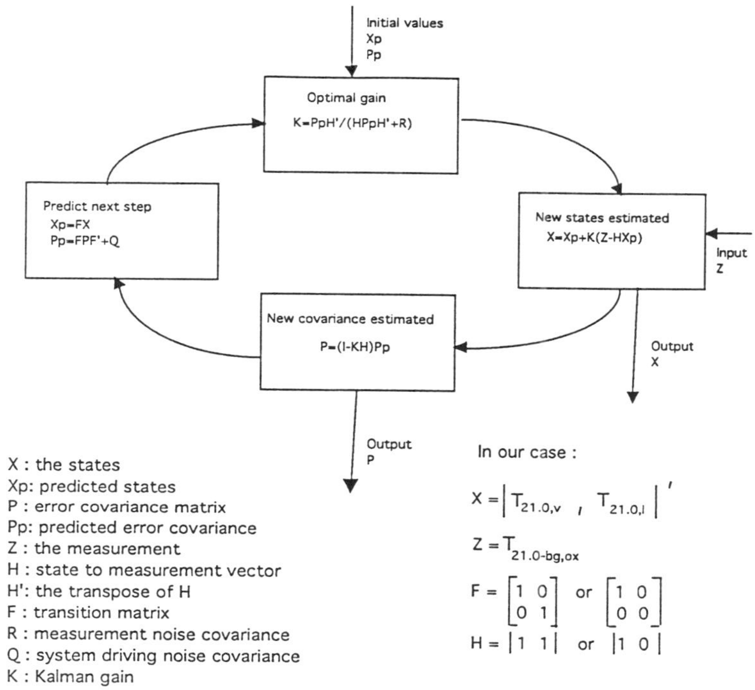

Figure 3. The Kalman filter. In our case the prediction (to the left) also includes a selection of parameters depending on whether liquid water has been detected or not.

as a linear combination of predicted new states and the new measurement Z. The optimal linear combination is associated with a Kalman gain K, which is derived from the predicted covariance matrix of the states P_p, and a model of the measurement noise R. The involved equations are shown in Figure 3.

The linearized sky brightness temperature reduced by the almost constant contribution from the background radiation and oxygen as in (2) can be seen as the sum of a contribution from water vapour and liquid water,

$$T'_{21.0-bg,ox} = T'_{21.0,v} + T'_{21.0,l} \tag{3}$$

where we want to keep the water vapour term in (2) for calculating the wet delay. These two terms were used as our state variables.

The wet delay has successfully been modelled as a random walk process in space geodetic data analysis [3]. This model was then used for $T'_{21.0,v}$. A random walk process is characterized by its variance rate (the diagonal elements of the matrix Q in Figure 3). A variance rate of $2.5 \cdot 10^{-3}$ K^2/s, derived from the data sets, was used under liquid free conditions.

The variance rate used during time intervals where liquid water was detected was varied between different simulations, but the final choice was $2.25 \cdot 10^{-2}$ K^2/s.

A random walk process was also used in modeling $T'_{21.0,l}$. The variance rates were calculated from the data sets using the difference between the two-frequency and the one-frequency wet delay estimates in the intervals where liquid water were detected. Figure 4 shows that the variance rates for each data set calculated for different time lags are fairly constant, thus supporting the choice of random walk processes. The variations from one data set to another is, however, large. The average variance rate was computed to be $2.25 \cdot 10^{-2}$ K^2/s and used in the filters.

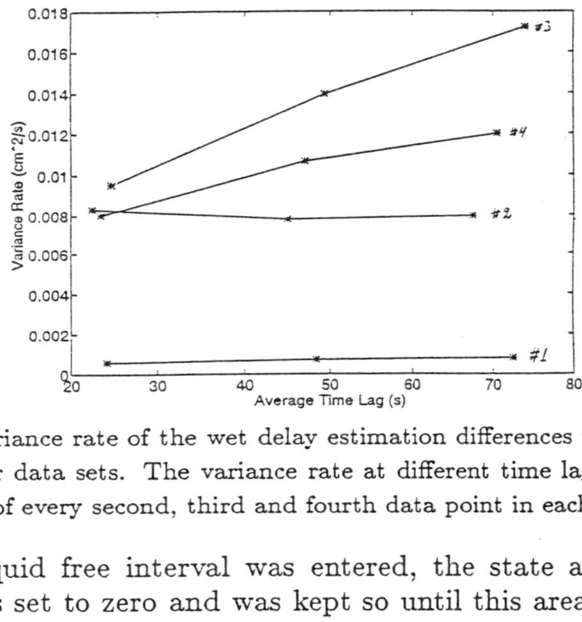

Figure 4. The variance rate of the wet delay estimation differences under liquid conditions for the four data sets. The variance rate at different time lags was computed from comparisons of every second, third and fourth data point in each data set.

When a liquid free interval was entered, the state associated with liquid water was set to zero and was kept so until this area was left. The accuracy in the estimated states was supposed to be low after a long period in liquid containing intervals where both states were "running free". This was compensated by running the filter both forward and backward [4, 5]. The final estimates were computed as a linear combination of the results of these two runs.

RESULTS

Figure 5 gives an example of the performance of the filtered one–channel solution. A large fraction of the less accurate estimates of the wet delay has now been removed. On the other hand, Figure 6, which shows the

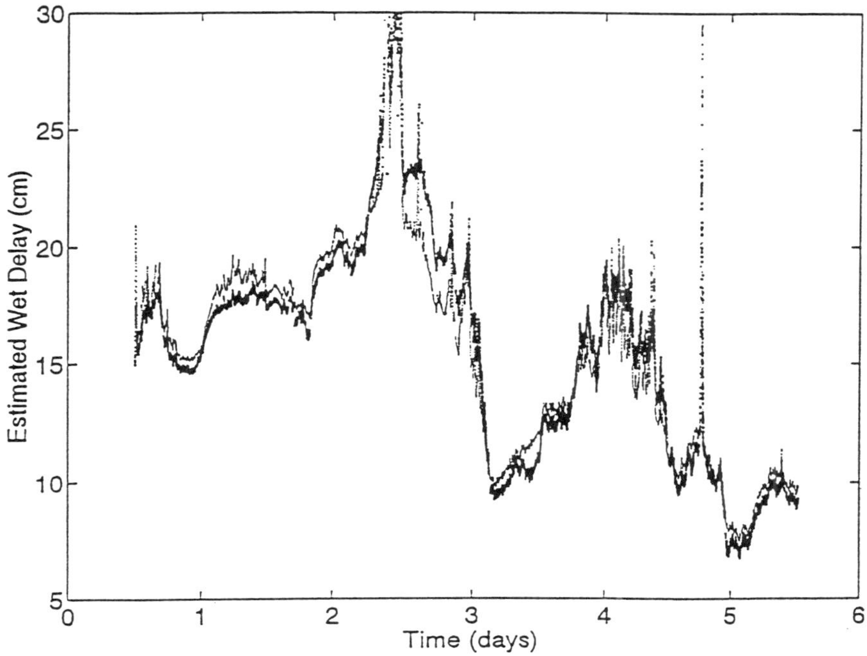

Figure 5. The zenith wet delay from July 22–27 1988, the lower solid curve retrieved using a two channel algorithm and the dotted curve using a one channel algorithm after Kalman filtering (to be compared with Figure 1.).

differences between the two–channel and the one–channel estimates, indicates that during the two long periods with large liquid content our filter generally underestimates the wet delay. However, we are aware that the two–channel solution is also unreliable during conditions with high liquid content. In order to reduce this type of uncertainty, one could (in the future) ignore all data points where $T'_{21.0}$ is larger than a certain value, say 50 K in the zenith direction for the Swedish west coast climate. The two large peaks in the difference plot would have been removed then.

Table 1 presents the performance of the filtered and unfiltered one–channel estimates when compared to the two–channel estimates. For comparison, the performace where points with $T'_{21.0}$ larger than 50 K have been removed are also presented in the table. We can see that, in many cases, the deviation of the filtered estimates from the two channel estimates is less than half of the deviation before filtering. There are exceptions, for instance the data set from July 22–27 when points with $T'_{21.0}$ larger than

50 K had been removed. In this case the "standard" filter yielded a standard deviation of 0.86 cm, which is larger than the deviation of the unfiltered data, 0.79 cm. However, another choice of filter parameters yielded a standard deviation of 0.57 cm for this data set.

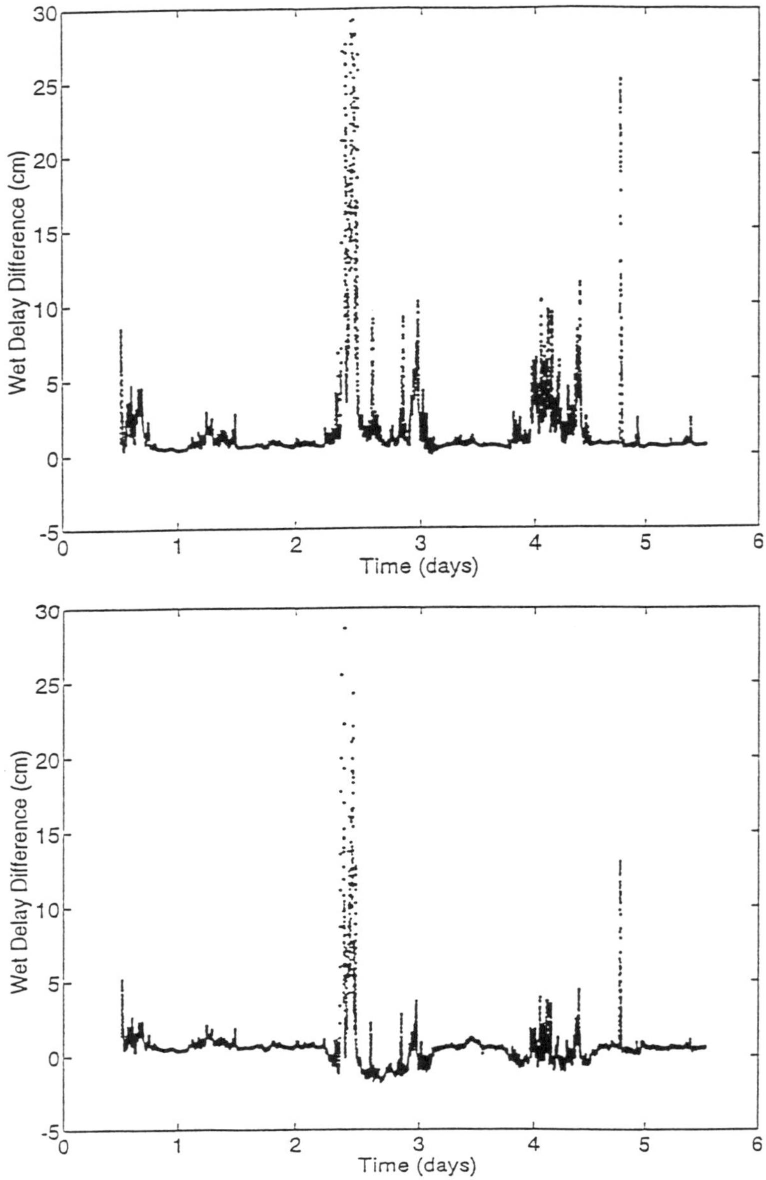

Figure 6. The upper plot shows the difference between the two curves in Figure 1. The lower plot shows the difference between the curves in Figure 5.

Table 1. Standard deviation of the differences between the wet delays derived from the one and the two channel algorithms. The one channel estimates are calculated from filtered and unfiltered sky brightness temperatures. In the right part of the table, the data points where the zenith linearized sky brightness temperatures exceed 50 K are discarded.

Data Set	All Data Points		$T'_{21.0} > 50$ K Removed	
	Unfiltered σ [cm]	Filtered σ [cm]	Unfiltered σ [cm]	Filtered σ [cm]
March 22–26	1.91	0.74	1.72	0.64
April 15–19	3.23	1.38	1.57	1.24
June 30–July 7	3.52	1.13	0.73	0.55
July 22–27	2.79	1.20	0.79	0.86
All Data Sets	2.87	1.09	1.31	0.80

DISCUSSION AND CONCLUSIONS

We have seen that by applying a relatively simple filter to data from a one–channel WVR, its performance is significantly improved under conditions where the atmospheric liquid content is high. We have also evaluated a criterion to identify "good" measurements.

The resulting algorithms may, however, suffer from being optimized for these four data sets. It was, for instance, noticed that the variance–rate of the liquid–containing component of the linearized sky brightness temperature showed large variations between different data sets. Thus, further studies are needed in order to design a robust general filter, or site and weather condition dependent filters.

We used a random walk process as a model for the variations in the component of the sky brightness temperature due to the liquid water. We are not by any means convinced that this is an optimal model, but merely a first test. The choice of state variables (brightness temperatures) that are, by definition, larger than zero may not be optimum for the Kalman gain calculations since Kalman filters assume zero mean variables. This may require the transformation of the variables before entering the Kalman filter.

REFERENCES

[1] Westwater, E.R., 1993. Tropospheric Radio Path Delay from Ground-Based Microwave Radiometry in, *Atmospheric Remote Sensing by Microwave Radiometry*, M. Janssen ed., Wiley and Sons, pp.145–213.

[2] Johansson, J.M., G. Elgered and J.L. Davis, 1993. Wet Path Delay Algorithms Using Microwave Radiometer Data, *Contributions of Space Geodesy to Geodynamics: Technology*, D.E. Smith, and D.L. Turcotte eds., AGU Geodynamics Series, **25**, 81–98.

[3] Herring, T.A., J.L. Davis, and I.I. Shapiro, 1990. Geodesy by radio interferometry: The application of Kalman filtering to the analysis of very long baseline interferometry data, *J. Geophys. Res.*, **95**, 12,561–12,581.

[4] Brown, R.G. and P.Y.C. Hwang, 1992. Introduction to Random Signals and Applied Kalman Filtering, Wiley and Sons.

[5] Fraser, D.C. and J.E. Potter, 1969. The Optimum Linear Smoother as a Combination of Two Optimum Linear Filters, *IEEE Trans. on Auto Control, AC–14*, **387**.

Effectiveness of brightness temperature ratios as indicators of the atmospheric path conditions

A.V. BOSISIO and C. CAPSONI
Dipartimento di Elettronica e Informazione - Politecnico di Milano
P.zza Leonardo da Vinci, 32 - 20133 Milano (I)

Abstract. Atmospheric attenuation is a quantity that may be derived by the scenario temperature under suitable hypotheses. A set of these temperatures, measured at different frequencies, allows to gain information about the amount of integrated water vapour and liquid content and the presence of clouds or precipitation in the atmosphere. In this framework the authors have focused their attention on the ratios between brightness temperatures, since these quantities are more suitable to point out the different atmospheric path condition than the temperatures themselves.
Analysis on a daily basis as well as on a yearly basis of the ratios between the pair temperatures 30-20 GHz and 30-13 GHz have confirmed its use for the estimation of the atmospheric state

Keywords: radiometry, propagation, telecommunications.

INTRODUCTION

Since a long time radiometers are used in the radiopropagation community as a useful tool to perform, although indirectly, attenuation measurements along a satellite path in absence of a satellite beacon. These measurements are based on the property of the matter to absorb and emit radiation (electromagnetic energy) according to the same physical mechanism and therefore such instruments can be used to infer extra attenuation due to the presence along the path of gases, fog, non precipitating clouds and also rain. For an atmosphere in local thermodynamic equilibrium the brightness temperature can be related to the absorptive properties of the medium and then to attenuation.
In practice at Ku band the effect of clear air gases like oxygen and water vapour and that of clouds is negligible which limits the use of radiometer to infer only rain attenuation. At higher frequencies, above 20 GHz, the opposite is true i.e. rain attenuation easily exceeds the maximum predictable value (about 10-12 dB), and absorption due to the other atmospheric constituents is accurately detected by the equipment. The instrument supplies the so called "clear sky reference level" which represents the extra attenuation in non rainy and cloudy condition and is used to calibrate beacon receivers.
The importance of these "minor" atmospheric constituents is becoming increasingly larger with the advent of new telecommunication services like business VSATs where the requirements for availability are less stringent (down to 1% or even less) in order to greatly reduce costs: VSATs systems are designed, in fact, to face only attenuation during non rainy periods or, at most during drizzle.

THE EXPERIMENTAL SETUP

At the experimental station of the Centro di Studi per le Telecomunicazioni Spaziali (CSTS) of the Italian National Research Council sited at Spino d'Adda , 30 km east of Milan, a large set-up of instruments of the Italian Space Agency (ASI) has been installed to "measure" the status of the lower atmosphere.
The actual experimental campaign is part of the Olympus and Italsat propagation experiments whose goal is that of providing exhaustive information on the characteristics of the radio channel from 10 to 50 GHz for present and future telecommunication applications.
The station set-up consists of two satellite receivers for Olympus and Italsat respectively, one S band meteorological radar, three radiometers plus some ancillary instruments.
One radiometer operates at 13 GHz and provides indirect measurements of attenuation to be compared with direct Olympus beacon fade measurements at (almost) the same frequency.
Another one is a dual frequency radiometer working in the 20-30 GHz bands exclusively devoted to the Olympus experiment with the main goal of supplying the clear sky reference level. In particular, the operating frequencies are 23.8 and 31.65 GHz, which represent a proper choice for the retrieval of the total integrated vapour (V) and liquid water content (L). In fact attenuation in non rainy condition may be expressed as a linear function of V and L and then two attenuation values (derived by brightness temperatures) are sufficient to calculate them [1]. From the knowledge of V and L, one may calculate the attenuation at different frequencies by using proper scaling coefficients [2], [3].
A third radiometer devoted to the Italsat propagation experiment and operating in the 20, 30 and 50 GHz bands has been acquired recently.
The three equipments have almost the same characteristics as summarized in Table 1.

Table 1- Radiometers technical characteristics

Frequency (GHz)	13	22.2	23.8	31.65	50.2
integration time (s)	1-32	1-32	1-32	1-32	1-32
resolution (K)	0.5	0.5	0.5	0.5	0.5
accuracy (K)	1	1	1	1	1
beamwidth (deg)	1.8	2	2	2	2.05
receiver bandwidth (MHz)	150	150	150	200	250

This paper refers to a preliminary study carried out on the radiometric data at 13, 23.8, and 31.65 GHz collected during 1992 in the framework of the Olympus experiment with the aim of identifying a tool for the automatic detection of the status of the lower atmosphere seen by the antenna(s) and for the quality check of the data. Due to the behaviour of the receiver outputs which have a different sensitivity to the atmospheric constituents we have envisaged in the ratios between pairs of measured quantities a possible way to get our goal.
The 13 GHz frequency is almost no affected by the presence of oxygen, vapour and liquid water content in the atmosphere, and results appropriate for estimating attenuation during rain events up to a fade level of about 10-12 dB. The 20-30 GHz radiometer, not suitable to predict attenuation during medium or heavy rain events because of the pronounced saturation effect, is appropriate to estimate the amount of the most important atmospheric constituents for atmospheric propagation: the 23.8 Ghz channel is particularly sensitive to water vapour absorption, while the 31.65 GHz channel is sensitive to liquid water.[4]
Among the six possible brightness temperature ratios (from the 13, 23.8 and 31.65 GHz channels) we have focused our attention on the 31.65/23.8 and the 31.65 /13 GHz ones which have shown at a preliminary analysis to be the most adequate to our purpose.[5]

PHYSICAL BACKGROUND

First of all we have investigated the problem from the "theoretical" point of view in order to find out a proper relation between brightness temperature and the amount of V and L present along the propagation path. In this context synthetic brightness temperatures at the three above frequencies have been calculated from one year radiosounding profiles collected worldwide (about 100000 samples)[1]. The quantities were computed by solving the radiative transfer equation with the specific attenuation coefficients obtained through Liebe's model as an input [6]. A study performed on the subset of these data without clouds has shown a very tight linear relation between the brightness temperatures and the water vapour integrated content [7] expressible in the form:

$$T(f_i) = a_i V + b_i \quad \textbf{(1)}$$

where a_i [K mm^{-1}] and b_i [K] are frequency and elevation dependent coefficients, V [mm] is the total water vapour integrated content and f_i the chosen frequency expressed in GHz.
The correlation factor of this regression is about 0.99 at 23.8 GHz, 0.9 at 31.65, 0.7 at 50.2 GHz. In this last case the oxygen dependence on the height of the location may induce a considerable spread on the results. However the linearity is strongly assured. [7]. The physical meaning of the two coefficients is evident: in clear sky the brightness temperature is mainly due to the contribution of oxygen (b_i) and of the amount of water vapour (through a_i).
It follows that:

$$T(30) = k_0 \cdot T(20) + k_1 \quad \textbf{(2.a)}$$

$$T(30) = k_0^* \cdot T(13) + k_1^* \quad \textbf{(2.b)}$$

where:

$$k_0 = \frac{a_{30}}{a_{20}} \qquad k_0^* = \frac{a_{30}}{a_{13}}$$

$$k_1 = \frac{b_{30} - a_{30} \cdot b_{20}}{a_{20}} \qquad k_1^* = \frac{b_{30} - a_{30} \cdot b_{13}}{a_{13}}$$

This behaviour has been confirmed by the results obtained by Fondazione Ugo Bordoni [8].
In the case of cloudy sky the presence of liquid water content along the path affects significantly and of an amount that highly depends on frequency the brightness temperatures. Consequently their ratios change and show a quite large spread, as it can be seen in Figs. 1,2 for the 23.8-31.65 GHz pair.
The third meteorological quantity we have to deal with is rain: in that case, brightness temperatures are governed by attenuation due to rain along the path (A) at least for rainfall rates greater than some mm/h, according to the relation:

$$T = T_m - (T_m - T_c) \cdot 10^{A/10} \quad \textbf{(3)}$$

where T_m is the average medium temperature [K] ant T_c the backgroung cosmic noise temperature (2.7 K) [9]. With increasing rain intensity the measured temperatures increase very rapidly and first T(30) and then T(20) and T(13) saturate.

[1]Radiosonde database kindly provided by CETP - Velizy (France)

Inspection of the results show that the ratios T(30/20) and T(30/13) change with changing the status of the atmosphere, but the same ratio value can be ascribed to clear, cloudy or even rainy weather and therefore they cannot be directly usable; on the contrary, a proper modification of variables makes things effective.

In practice the following approach is suggested:

a) to introduce the modified temperatures T*(f):

$$T^*(30) = T(30) - k_1 \tag{4.a}$$

$$T^*(13) = T(13) - k_1^{'} \tag{4.b}$$

obtained by subtracting the value of the intercepts in (eq. 2) from the brightness temperatures T(30) and T(13)

b) to define the new quantities:

$$R1 = T^*(30) / T(20)$$

$$R2 = T(30) / T^*(13)$$

which have the following properties

	R1	R2
clear sky	0.33-0.42	5
cloudy sky	0.42-1.05	3-5
rainy conditions	0.95-1.31	1.02-4

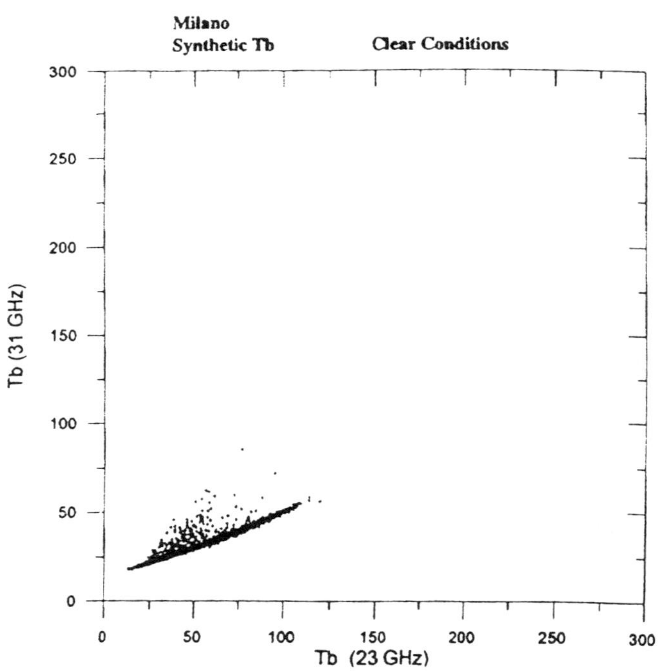

Figure 1 - Scatterplot between brightness temperatures collected in clear sky at two frequencies: data derived from radiosoundings and projected along a 30 deg. inclined path. By courtesy of FUB (Italy) .

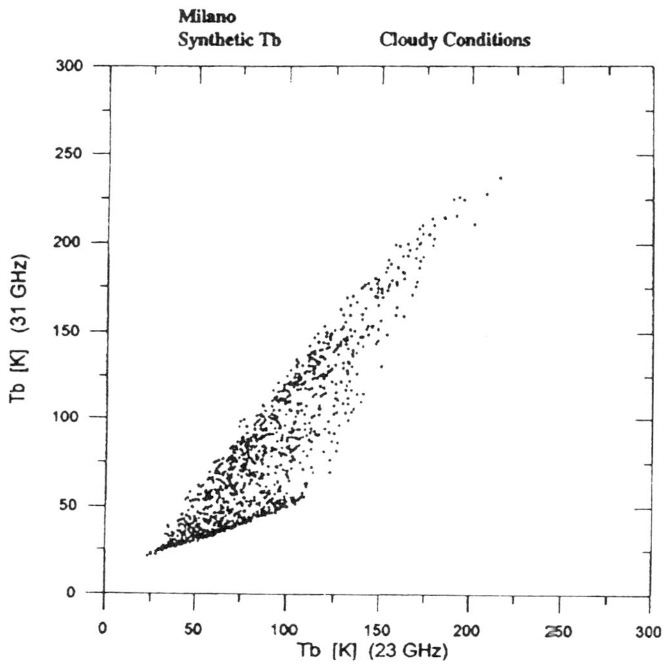

Figure 2 - Scatterplot between brightness temperatures collected in cloudy sky at two frequencies: data derived from radiosoundings and projected along a 30 deg. inclined path. By courtesy of FUB (Italy)

It has to be remembered that these values apply to a 30 deg. inclined path and that obviously they cannot be interpreted as sharp boundaries between the different meteorological situations (in fact intervals overlap a bit each other) as R1 and R2 vary in the continuum.
As it can be easily noted both the values of R1 and R2 change with the atmospheric conditions seen by the antenna of the radiometer, but the behaviour of R2 is less monotone and therefore it is expected to be more difficult to interpret.

NUMERICAL RESULTS

Following the considerations reported in the previous section, we have applied the proposed scheme of analysis to the data collected during 1992 by the radiometers devoted to the Olympus experiment. The instruments are calibrated periodically through the tip curve procedure. Data are collected with a sampling rate of 1 s. For the present work an average over 60 samples has been applied.
Fig. 3 shows the time series of the three measured temperatures for the 2 May 1992: that day was cloudy up to 2000 hrs and then the sky became clear. Correspondingly Figs. 4 a,b show the time series of the quantities R1 and R2.
The results confirm what expected: as one may see, during the period from 2000 to 2400 hrs R1 is definitely 0.4; the remaining part of the day was cloudy and the liquid water content along the path gave rise to a variability of R1 between 0.4 and 0.8. At the same time the R2 provides the same information but not as clearly. In fact, the transition between clear and cloudy sky is less evident being its variation towards the cloudy situation not marked (and partly masked by a very large noisy band, despite of filtering).

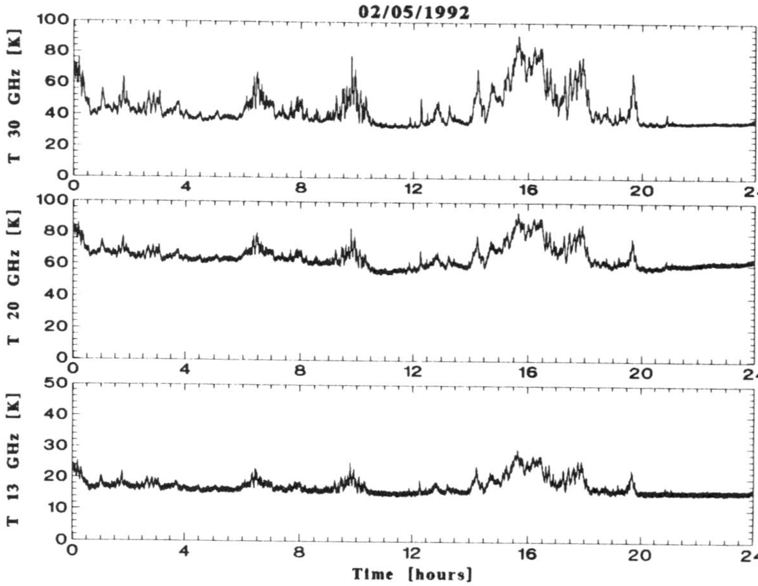

Figure 3 - Time series of radiometric brightness temperatures collected during 2 May 1992.

Another case is shown in Figs. 5, and 6a,b for the 30 May 1992 that was partly raining (the raingauge at the station measured a peak point rainfall intensity of 90 mm/h at about 1930 hrs). Again R1 clearly reveals the periods of different meteorological situations and also rain (from 1730 to 2130 hrs) is well identified.

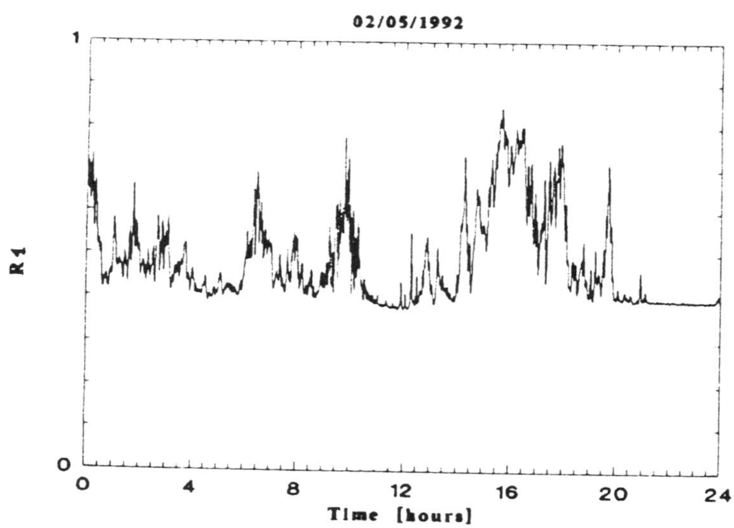

Figure 4a - R1 time series: 2 May 1992.

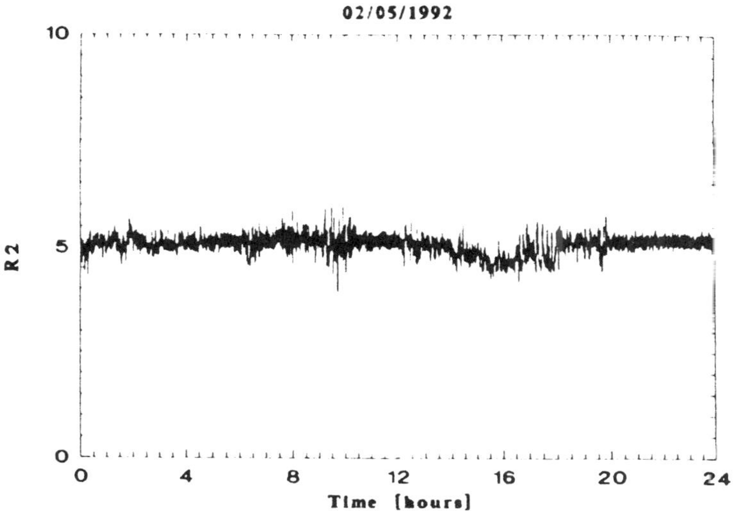

Figure 4b - R2 time series: 2 May 1992.

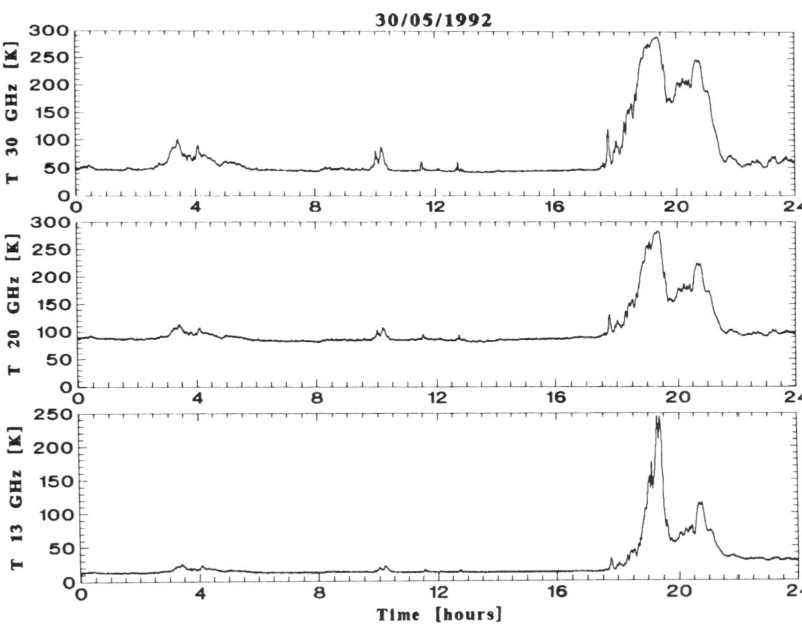

Figure 5 - Time series of radiometric brightness temperatures collected during 30 May 1992.

Once more, the time series of R2 let only guess the presence of clouds. On the contrary, the rainy time interval is really well marked with R2 changing from 5 to about 1.2 (and with the superposed noise almost disappeared). Moreover, the trace shows another "interesting" effect; differently from that of R1, the value of R2 remains low (2-2.5) also after the end of the rain event. The reason of this difference has to be attributed to a thin film of water which remained on the antenna dish after the passage of the storm over the station and not completely eliminated in few minutes by the blower as it happens usually. It gave rise to an enhancement of the measured temperature. This effect that occurs (usually for a short time) at almost all the instruments despite the accurate mechanical design is deeply marked by R2, because of the low absolute value of the 13 GHz brightness temperature in non rainy condition on which, an anomalous increase of only some Kelvin may represent a quite high percentage increment with obvious consequences on R2. The same malfunctioning would has less influence on R1.

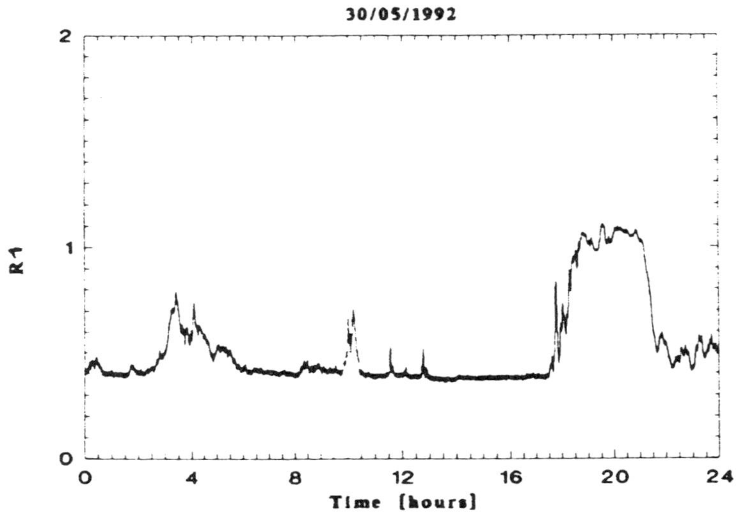

Figure 6a - R1 time series: 30 May 1992.

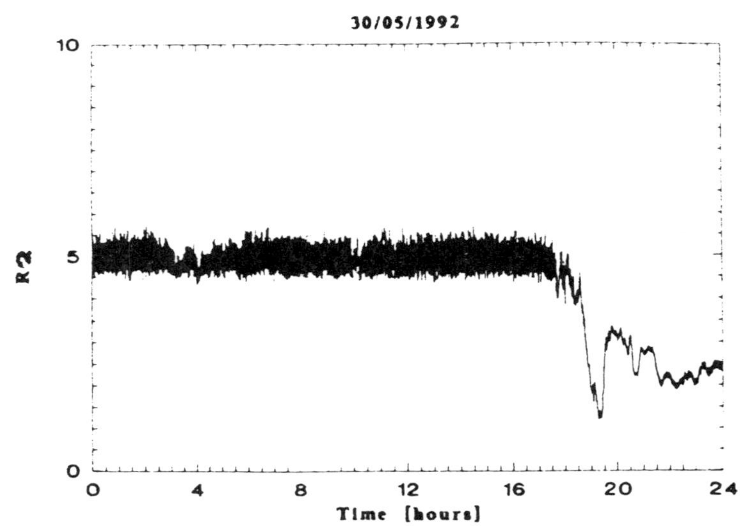

Figure 6b - R2 time series: 30 May 1992.

According to the indications of the above results an attempt was made to analyze statistically the quantity R1 which proved to be suitable to characterize the status of the atmosphere. The 1992 data base has been used to this aim after a check on the quality of the data. Fig. 7 shows the cumulative distribution function of R1 with two characteristic changes in the slope: one around R1= 0.4 and the other around R1= 1 that are just the threshold values pointed out in the previous section. Around the R1=1 value the curve shows a smoother behaviour because the transition between rainy and cloudy sky is less sharp than that between clear and cloudy sky.

If we assume 0.4 and 0.95 as the two threshold values we can deduce that for about 60% of the time the antenna observed a clear sky condition while it rained for about 2% of the time.

For reasons explained above, R2 cannot be used as well. Nevertheless, its use jointly with R1 let anticipate some useful applications (as for instance the check of the data quality). An example is reported in Fig. 8 where the scatterplot of R1 versus R2 is shown for the 30 May (see Figs. 5 and 6 a,b). The behaviour looks as one may expect except for the part of the plot circled. It refers in fact to the end of the day where R2 got wrong values because of the water on the antenna dish. A complete analysis of this kind of data presentation is beyond the scope of this paper and will be done in a future work.

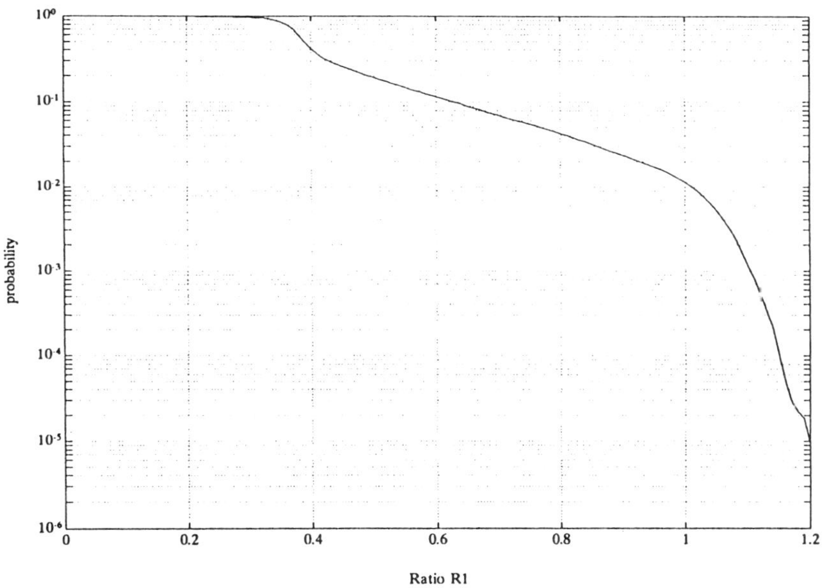

Figure 7 - Cumulative distribution of R1, 1992. Here the two slope changes show the transition between two different atmospheric situations.

CONCLUSIONS

It has been studied the effectiveness of the use of the ratios between the brightness temperatures measured by radiometers at different frequencies as a tool for automatically revealing the status of the atmosphere seen by the equipment carried out in terms of clear, cloudy or rainy sky. In particular, reference has been made to the 13, 23.8 and 31.65 GHz

measurements on experiment carried out at the experimental station of CSTS sited at Spino d'Adda in the framework of the Olympus propagation.
The ratio between the pair 31.65-23.8 GHz and to a less extent the one between the pair 31.65-13 GHz properly biased has shown to be adequate to this purpose and it has been evaluated for both of them the range of variation as a function of the sky condition.
The statistical analysis in terms of the cumulative distribution function of the ratios R1 (between 31.65 and 23.8 GHz) clearly allows the evaluation of the percentage time of cloudy and rainy sky.
A first insight to the behaviour of the scatterplots between the two ratios showed some very interesting aspects of checking the quality of the data.

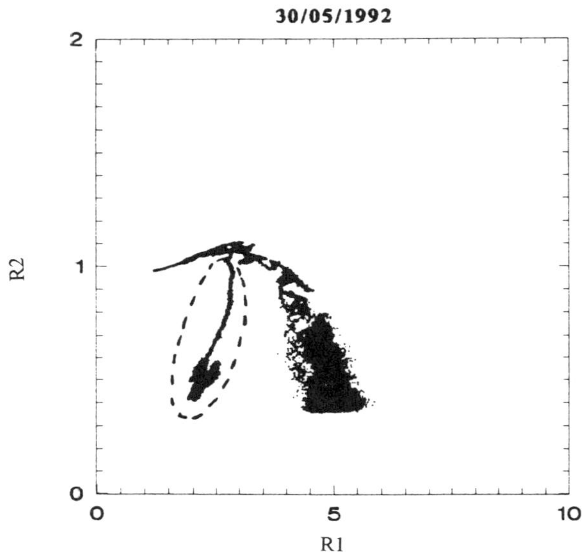

Figure 8 - Scatterplot between the two chosen ratios. Circled dots represent erroneous data (see text).

REFERENCES

1. Westwater E.,"The accuracy of water vapour and cloud liquid determination by dual frequency ground-based microwave radiometer", Radio Science, Vol. 13, No. 4, pp. 677 - 685, Jul.-Aug. 1978.
2. Schiavon G., D. Solimini, E. Westwater,"Performance analysis of a multifrequency radiometer for predicting atmospheric propagation parameters", Radio Science, Vol. 28, No. 1, pp 63-76, Jan.-Feb. 1993.
3. Barbaliscia F., M. Boumis,"The use of radiosonde data for atmospheric modeling", Proc. of Olympus Utilisation Conference, pp.635-639.20-22 April 1993, Sevilla (Spain).
4. Westwater E.R, J.B. Snider, M.J. Falls,"Ground-Based Radiometric Observations of Atmospheric Emission and Attenuation at 20.6,31.65,90.0 GHz:A Comparison of Measurements and Theory, IEEE Transaction on Antennas and Propagation, Vol.38,No.10, Oct.1990.
5. Wu Sien-Chong,"Optimum Frequencies of a Passive Microwave Radiometer for tropospheric Path-Lenght Correction", IEE Trans on ANtennas and propagation, Vol.27,No.2,pp 233-239, March 1979.
6. Liebe H.J.,"MPM - an atmospheric millimeter-wave propagation model", Int. J. Infrared Millimeter Waves, vol.10, pp.631-650. 1989
7. Bosisio A.V., C. Mallet, "A three frequency non linear algorithm for water vapour and liquid water content retrieval", Proc. of Progress on Electromagnetics Research (PIERS) 1994, 11-15 July, Noordwijk (NL)
8. Barbaliscia F., E. Fionda, P.G. Masullo, "Characterization of the status of the sky by means of radiometric observations at 20/30 GHz", Proc. of 1993 International Symposium on Radiopropagation (ISRP'93), 18-23 Aug.,Beijing, China.
9. Brussaard G., "Radiometry - A useful Prediction Tool?", ESA SP-1071, 1984.

SPATIAL FEATURES OF THE SKY BRIGHTNESS TEMPERATURE AT ROMA*

FRANCESCO BARBALISCIA, ERMANNO FIONDA, PIERGIORGIO MASULLO
Fondazione Ugo Bordoni, Viale Europa 190, 00144 Roma Italy
Tel: +39-6-5480-2110 - Fax: +39-6-5480-4401

Abstract In the paper some results are presented concerning the analysis of the spatial structure of the sky on the basis of radiometric measurements collected at two locations nearby Roma at 23.8 and 31.4 GHz. Attention is focused on the deviation from the cosecant law that is found to be statistically limited although polarised toward negative values. Several parameters are observed, such as elevation, site, azimuth, sky conditions, frequency and seasonal effects. Empirical fitting of the cosecant assumption is also offered as resulting from the measured data.

Key words: Radio-wave propagation, Radiometry, Atmosphere, Satellite links.

INTRODUCTION

Radiometric sensing of the atmosphere is of paramount relevance to both environment monitoring and telecommunication systems design. The space-time structure of minor atmospheric constituents, such as water vapour and clouds droplets, affects in fact the performance of those satellite communication services where the availability is radically lowered to reduce costs.

Radiometric measurements of the brightness temperature of the sky allows the derivation of the atmospheric composition, but, since the measurements are typically performed at zenith for environmental studies and along slant paths for telecommunications, great aid is offered by deriving the sky brightness temperature at a certain elevation from that measured at the zenith.

To that purpose it is commonly assumed that the sky is horizontally stratified, in such a way that the elevation angle dependence of the thermal emission results in a cosecant law. The validity of this assumption depends on the conditions of the sky: for clear sky, in the absence of clouds and with low water vapour content, the assumption is likely acceptable, while in principles it is no longer valid in the presence of clouds and with a high and irregular degree of humidity. On a long-term statistical basis, however, no difference should in principles occur among different points of the sky in terms of clouds coverage or water vapour concentration, unless in the case of particular microclimatic conditions.

To shed light on that aspect a correction term has been calculated for each scanning cycle, expressing the percent deviation of the elevation profile from the cosecant one and taking into account the type of the sky : clear, cloudy and covered. The cumulative distribution of this deviation index is analysed and correlated to sky conditions, elevation, site, azimuth and frequency in order to determine the influence of the various parameters on the accuracy of the cosecant assumption.

---

* Work carried out in the framework of the agreement between the Italian P.T. Administration and the Fondazione Ugo Bordoni.

Once the law governing the vertical scanning of the sky temperature is known, several applications can be correctly performed: radiometer tipping-curve calibration, beacon zero-level definition and, more in general, the conversion of the measurements from one path to another, typically for the feasibility of orbital diversity techniques.

THE EXPERIMENTAL SET-UP

A dual-channel radiometer (23.8/31.4 GHz), portable model manufactured by Radiometrics Co. (USA), of the noise injection type, with 5° beamwidth at 3 dB, sampling time of 2 minutes and accuracy around 2K, has been operating for 8 months at the PT labs at EUR, south side of Roma and for the subsequent 8 months at the Alenia Spazio facilities of Via Salaria, north side of Roma.

The measurements cycles were as follows : at EUR mainly at zenith, sometimes at 32° elevation towards the Olympus satellite, every clock hour a scanning was performed along 101° azimuth. At synoptic hours the azimuth was switched toward the Pratica di Mare airport (176°), where radiosonde is launched. The routine pointing at Salaria was toward the Italsat satellite, 41°and 180°, for zenith and azimuth, respectively, and the scanning were made at the same elevation angles as for EUR, again hourly, towards Pratica di Mare (220°).

The calibration of the instrument, performing however a self-calibrating cycle at each measure, was made by tipping-curve procedure in clear sky conditions. No significant interruption occurred in the acquisition period, the only problem being the condensation of moisture on the radome, in particular conditions of temperature and humidity. To avoid that a blower system is automatically switched on when the brightness temperature exceeds a fixed threshold.

THE DATA SET

Eight months of data were available for each site concerning hourly vertical scanning for four elevations (90°, 60°, 45°, 30°), two frequencies (23.8 and 31.4 GHz), two azimuth for Roma EUR (101° and 176°).

Data have been also divided according to the status of the sky classified on the basis of the values of T_b at 31.4 GHz: at 90° the sky is assumed clear when $T_{b31} < 20$ K, cloudy for 20 K $< T_{b31} < 40$ K and covered for $T_{b31} > 40$ K. For the other elevations the limiting values are corrected according to the cosecant law within 15% of tolerance. Doubt cases have been clarified or discarded on the basis of the sky observations available from the station logbook.

Pre-processing of the data was very accurate, leading to the exclusion of about 30% of the scanning cycles, mostly because of the suspicion of presence of moisture on the mylair radome of the radiometer. Daily log-book observations were largely used for analysing doubt events.

CRITERIA OF THE ANALYSIS

Data were processed calculating the deviation of the scan profiles from the cosecant law (horizontal atmospheric stratification). To transform that hypothesis in terms of brightness temperature two basic assumptions must be made: isothermal atmosphere and low values of attenuation (opacity), that correspond to:

$$T(z) = T \qquad \text{and} \qquad e^{-\tau} = 1 - \tau + \tau^2/2$$

Under those hypotheses the radiative transfer equation results as:

$$T_{bc}(\theta) = T_0 + [T_{bm}(90) - T_0)] \operatorname{cosec} \theta - 0.5\, \tau^2 (T - T_0) (\operatorname{cosec}^2 \theta - \operatorname{cosec} \theta) \qquad \textbf{(1)}$$

expressing the brightness temperature, $T_{bc}(\theta)$, at an elevation θ as a function of the one, $T_{bm}(90)$, measured at the zenith. T_0 is the cosmic background temperature, T the physical temperature and:

$$\tau = \ln \{(T - T_0)/[T - T_{bm}(90)]\}$$

is the attenuation, in nepers, along the path.

A deviation index is defined as:

$$CSER(\theta) = \{[T_{bm}(\theta) - T_{bc}(\theta)] / T_{bc}(\theta)\} * 100 \quad \textbf{(2)}$$

which expresses the percent deviation of the measured brightness temperature $T_{bm}(\theta)$ from that, $T_{bc}(\theta)$, calculated via the cosecant law. CSER can be used to convert zenith measurements to slant path ones and vice versa.

THE RESULTS

Influence of the sky

In order to have a first general indication the cumulative distributions of the average (all the elevations) cosecant deviation CSER is shown in Figure 1 as a function of the sky conditions at EUR, 23.8 GHz for the azimuth 101°. The term "all skies" indicates the entire pool of events, regardless of the sky conditions.

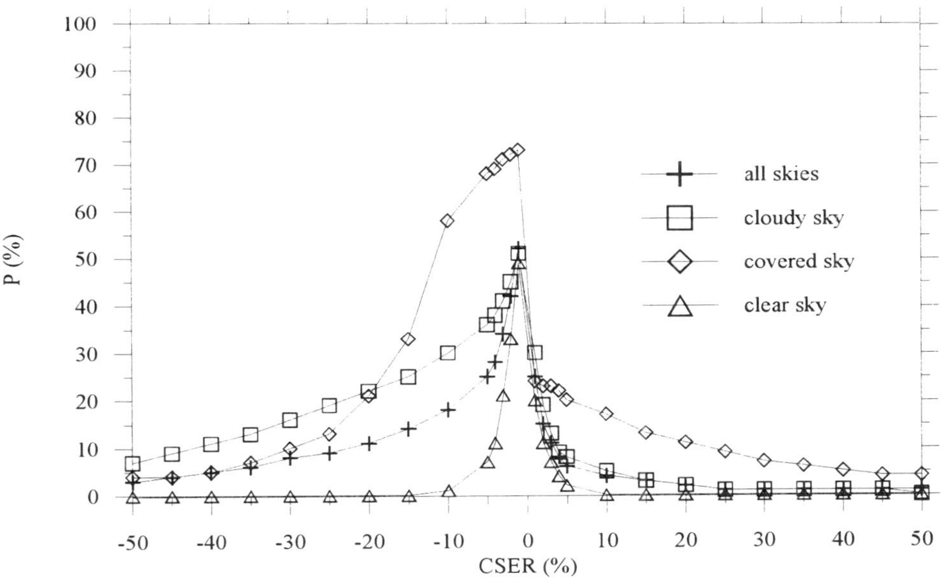

Figure 1. Cumulative distributions of the average deviation index. Roma-Eur, 23.8 GHz, azimuth=101°.

It can be noted that, although the atmosphere is not strictly cosecant, the deviation is very limited (within 10 %) for the majority of the events, but in the presence of clouds, when a few events show distortions up to 60 %.

Hence, while in clear air the cosecant approximation is acceptable together with isothermal atmosphere and law values of opacity hypothesis, depending of course on the required accuracy, in the general case (all skies) some events can give strong deviation.

The sign of the deviation

Figure 2 presents the scatterplot of the measured ratios of the brightness temperatures at elevation θ to the correspondent ones at the zenith versus the cosecant of the angles, as an average of the three elevations, for the entire pool of EUR data, at 23.8 GHz. As well as from the Figure 1, the polarisation of the CSER toward negative values can be noted, in particular in the presence of clouds. This effect might be attributed to the occurrence of moisture on the top of the radome that causes the zenith temperature to rise and the correspondent ratio to that along the path to lower. In order to clarify that behaviour several checks have been made bydedicated measurements carried on in various sky conditions, even with other higher precision radiometers where no radome was present, but the tendency to negative values was always found. No explanation of this behaviour appears evident. In terms of local climatology, however, the influence of vapour generation in urban areas, stronger with respect to the geography underneath the paths, where the vapour concentration is likely to be lower should be considered, hence causing values of T_b on slant paths to be lower than expected.

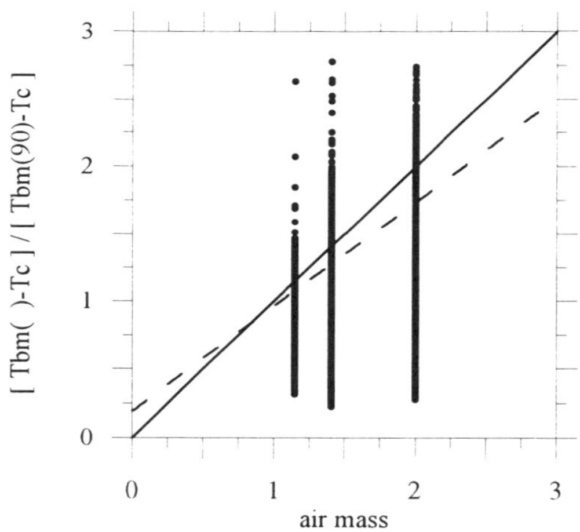

Figure 2. Scatterplot of measured ratios of T_b versus air mass. Roma-EUR, 23.8 GHz all skies.

For the direct conversion of the measurements from one elevation to another, the fitting equation can be used :

$$[\,T_{bm}(\theta) - T_0\,]/[\,T_{bm}(90) - T_0\,] = 0.78\ \mathrm{cosec}\,(\theta) + 0.19 \qquad \textbf{(3)}$$

Influence of the elevation

The cumulative distributions of the cosecant deviation are shown for three elevation angles, 60°, 45° and 30°, in Figure 3a for all the skies and in Figure 3b for clear air events. The deviation is greater for the lowest elevations, being the horizontal uniformity hypothesis weaker along the longest paths crossing denser atmosphere. As far as the behaviour with respect to the status of the sky is of concern, in clear sky the dependence on the elevation is very low and the cosecant approximation is valid.

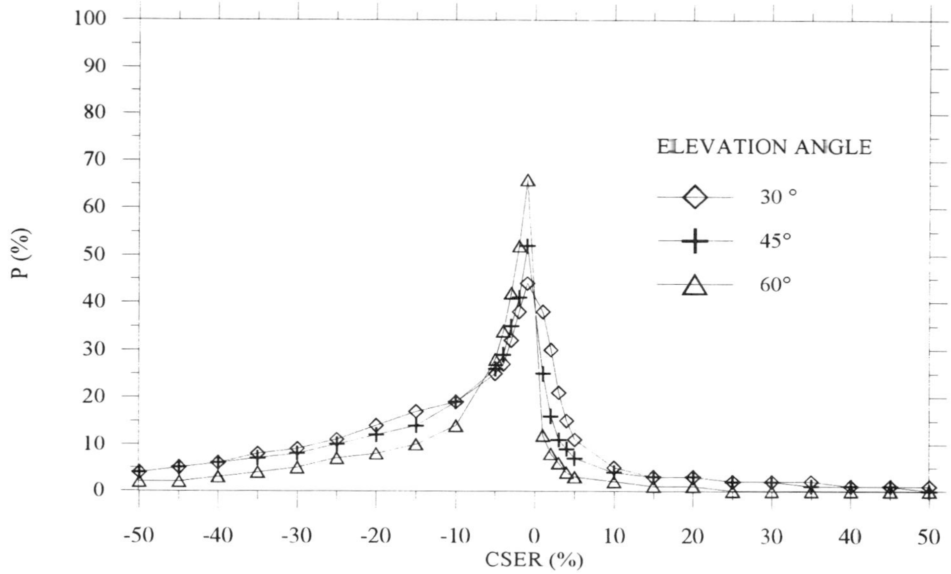

Figure 3a. Cumulative distributions of deviation index. Roma-EUR, 23.8 GHz, Azimuth=101°, all skies

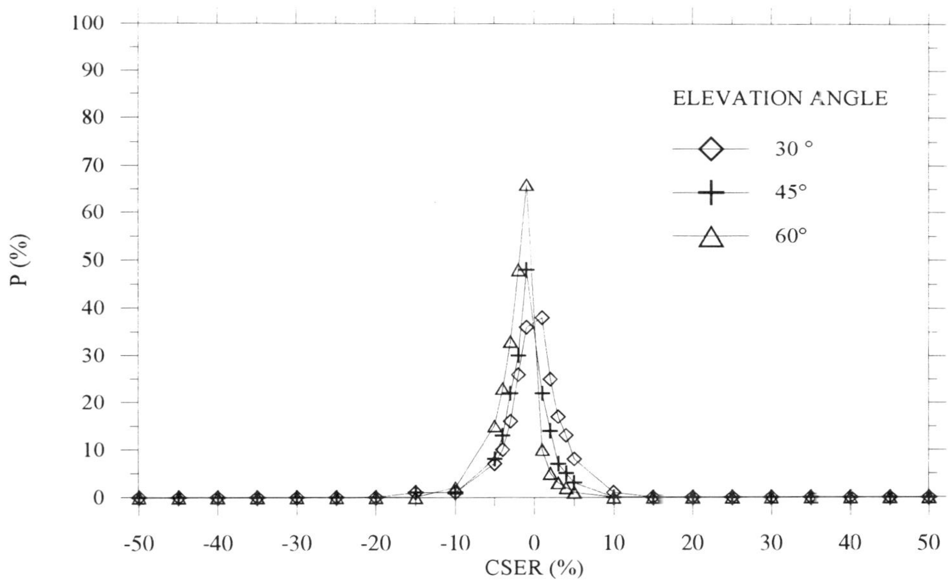

Figure 3b. Cumulative distributions of deviation index. Roma-EUR, 23.8 GHz, Azimuth=101°, clear sky.

Site and azimuth diversity

The distribution of the CSER is reported in Figure 4a for the two sites at 23.8 GHz, for the full pool of data, where no significant differences appear evident. It is noticeable that the deviation in the case of EUR is more balanced with respect to the sign. No difference occurs between the two azimuth at EUR : this would imply that orbital diversity countermeasures could be designed only on the basis of the elevation (cosecant) dependence (Fig. 4b).

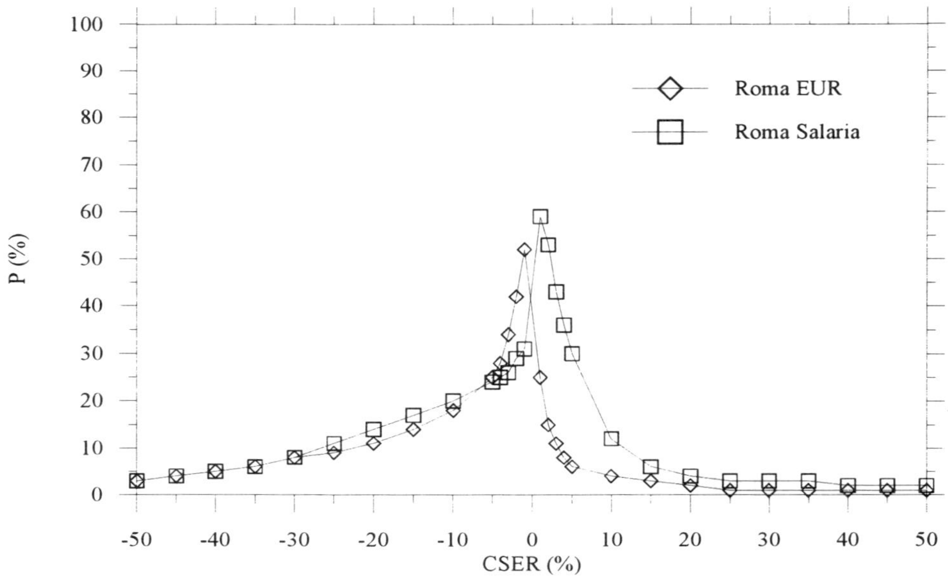

Figure 4a. Cumulative distributions of average deviation index, 23.8 GHz, all skies.

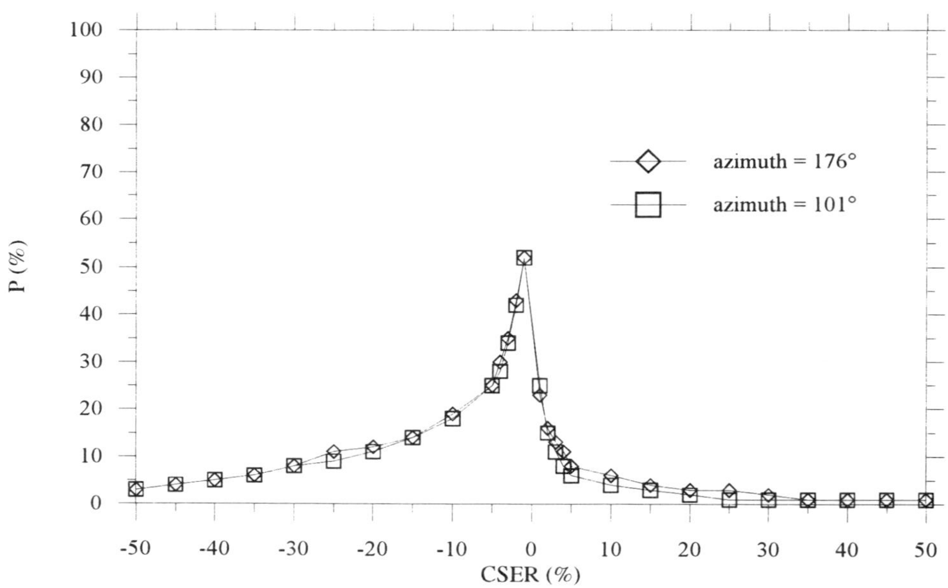

Figure 4b. Cumulative distributions of average deviation index. Roma-EUR, 23.8 GHz, all skies.

Influence of the frequency
The dependence of the deviation on the frequency, reported in Figure 5, does not appear to be significant in neither cloudy nor clear sky conditions.

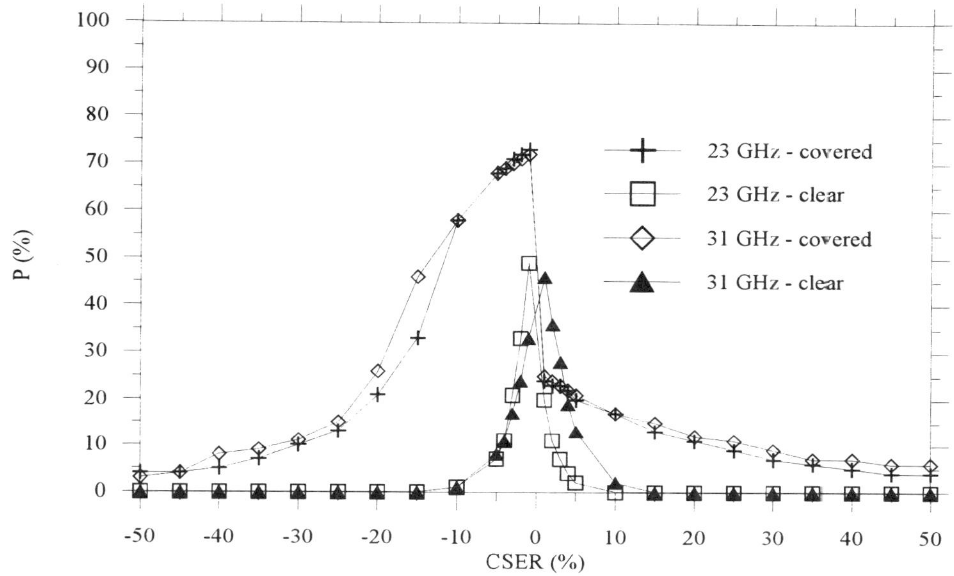

Figure 5. Cumulative distributions of average deviation index. Roma-EUR, 23.8 GHz, Azimuth 101°.

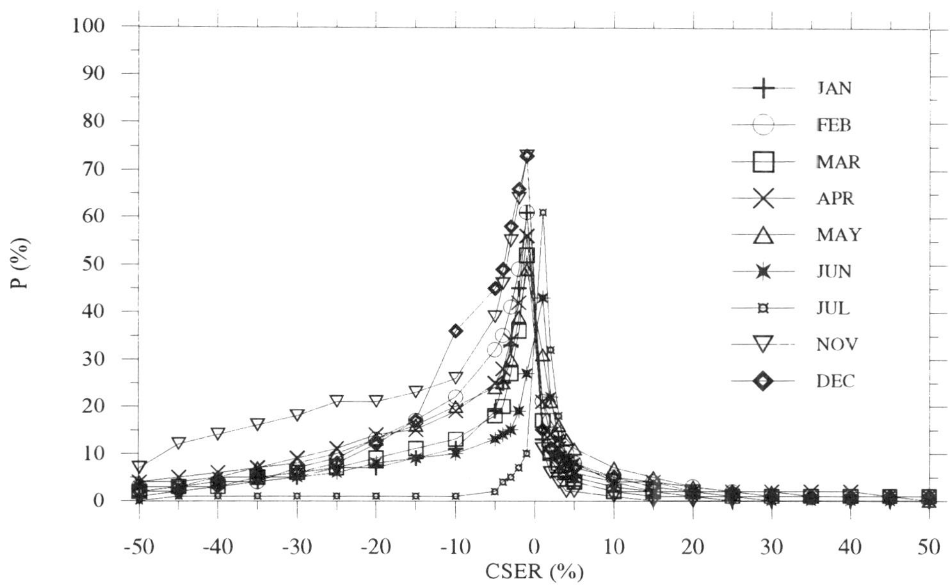

Figure 6. Monthly cumul. distributions of average deviation index. Roma-EUR, 23.8 GHz, Az. 101°,all skies.

The negative polarisation, occurring only in cloudy conditions, might be attributed to high values of attenuation that weakness the hypothesis assumed to derive the Equation (1) from the radiative transfer equation.

Seasonal effects
Both the magnitude of the deviation from the cosecant law and its negative trend are expected to depend on the season, since humidity and cloudiness vary markedly from month to month. Figure 6 confirms that the winter months form the worst month with values up to 30-40 % for the first decile, while clear sky season keeps within 10% for the major part of the time.

CONCLUDING REMARKS

The results are presented of the analysis concerning the spatial features of the brightness temperature for two sites nearby Roma. Under the assumption of isothermal condition and law values of opacity, the deviation from the cosecant hypothesis is calculated for eight months of hourly vertical scanning at four elevations and correlated with the status of the sky.
Results indicate that the sky is acceptably cosecant in clear conditions (within a 10% deviation), and that the distortion can be high in the presence of clouds: in general the cosecant hypothesis cannot be assumed for some 30% of the time.
It is also noted that the deviation is polarised toward negative values, resulting in an underestimation of the brightness temperature along inclined paths. This behaviour limits the applicability of the results for calibration purposes.
The deviation in the presence of clouds is, as expected, higher for lower elevations and depends on the season, being more marked during the winter months, when cloudiness is more frequent and longer lasting.
Empirical fitting is given for the direct conversion of measured data from zenith to slant paths and vice versa.
Although preliminary the results seem to give indications on that aspect. Further work is in progress with more sophisticate instruments capable of much faster sampling times. New data will be analysed considering temperature inhomogeneities and specific distributions of water vapour.

REFERENCES

1. Westwater E R & al 1990, Ground-based radiometric observations of atmospheric emission and attenuation at 20.6, 31.65 and 90.0 GHz: a comparison of measurements and theory, *IEEE Trans. on AP,* **10,** 1569-1580.
2. Matzler C, 1992, Ground based observations of atmospheric radiation at 5, 10, 21, 35 and 94 GHz, *Radio Science,* **3,** 403-415.
3. Gibbins C J and Cherry S M, 1989, The effect of spatial inhomogeneities on the elevation angle dependence of atmospheric thermal emission at millimetric wavelengths, *Radio Science,* **1,** 77-83.

Horizontal variability of water vapor and cloud liquid water as derived from space borne observations.

Stephen A. Tjemkes† and Marijke Visser
Institute for Marine and Atmospheric research Utrecht (IMAU)
The Netherlands

Abstract. Space borne observations by the High resolution Infra Red Sounder (HIRS) as archived in the International Satellite Cloud Climatology Project - TIROS-N Operational Vertical Sounder (ISSCP-TOVS) dataset and Special Sensor Microwave/Imager (SSM/I) are used to analyse spatial variability of temperature, moisture and cloud liquid water. Two regions are considered: the southern hemispheric storm track and the tropics. The observations are taken in July 1989. The specific humidity and temperature spectra derived from the HIRS follow a power law of the wavenumber. Two different values of the exponent in the power law are found. In the low wavenumber domain the exponent is approximately -3, and in the high wavenumber domain it is approximately -5/3. The former is thought to represent fluctuations induced by large scale air motions, while the latter represents fluctuations by small scale air motions. In the specific humidity spectrum the two regimes are best seen in the layers next to the surface, while for layers in the upper troposphere or lower stratosphere, the -3 power law regime is not visible. We attribute this to the decreasing zonal gradient of moisture with height. The spatial variance and the shape of the vertically integrated water vapor spectrum derived from the two sources mentioned above, are closely related. Finally the cloud liquid water periodogram derived from SSM/I follows theoretical distribution of the diffusion of a passive tracer in two-dimensional turbulence.

Key words: Spatial Variability, Hydrological Cycle.

INTRODUCTION

Model calculations in combination with observations provide a powerful tool to investigate the hydrological cycle. So far the focus has been on the quantification of the horizontal distribution of mean state. Spatial variability has received little attention. Yet, we believe that a quantification of the horizontal distribution of mean state alone, is not sufficient to understand the hydrological cycle as a whole. The atmosphere is in constant motion, resulting in variability of the various hydrological cycle components on different space and time scales. By studying the variability of these components one can not only gain insight in the hydrological cycle itself, but also of the dynamics of the atmosphere.

Spatial variability can effectively be observed from space borne platforms. Instruments mounted on these platforms are able to observe a relatively large spatial domain in a relatively short period. In this paper we present a study on the horizontal variability of atmospheric moisture derived from space borne observations by the High resolution InfraRed Sounder (HIRS) part of the Tiros Operational Vertical Sounder (TOVS) (*Smith et al.*, 1979) and the Special Sensor Microwave/Imager (SSM/I). These two radiometers operate at different part of the electromagnetic spectrum, and thus provide independent observations of the atmospheric moisture. A recent study by *Stephens and Jackson* (1993) shows that both types of sensors reproduce the bulk climatology of the vertical integrated water vapor content (V). The TOVS based V, tends to underestimate the

† Present Affiliation: EUMETSAT; Pallaswiesenstrasse 182; 6100 Darmstadt; Germany

SSM/I V by about 2 - 3 Wm^{-2}. This underestimation is the result of impossibility of the TOVS to sample the atmosphere column in presence of clouds. Besides the observations of atmospheric moisture we consider temperature observations by the HIRS, as a tracer of potential energy and vertically integrated cloud liquid water (L) observations by the SSM/I.

DATA: SOURCES AND ANALYSIS METHOD

TOVS

The TOVS data considered here are part of the International Satellite Cloud Climatology Project (ISCCP, *Schiffer and Rossow,* 1983) archives. They are derived from the infrared observations by the HIRS using a physical retrieval scheme implemented after 1987 (*Reale et al.,* 1989). This retrieval methodology follows the approach by *Smith and Woolf* (1979) with some additional tuning. The daily mean fields are on a 2.5°× 2.5° grid and contain information of temperature and specific humidity in 5 layers between the following pressure levels 1000, 800, 680, 560, 440 and 310 hPa. Data gaps due to persistent cloudiness exist in the fields.

SSMI

Observations over the global oceans by Special Sensor Microwave/Imager (SSM/I) were used to derive an alternative set of V. The SSM/I is a 7-channel passive microwave radiometer (*Hollinger et al.,* 1987). The radiometer observes the upwelling radiation at 19.35, 22.235, 37 and 85.5 Ghz. The observations at 22.235 GHz are made in the vertical polarization state only, in the other channels they are made in both horizontal and vertical state. The methodology to retrieve V and simultaneously L from these observations is based on depolarisation of the radiation, by the atmospheric constituents (mainly atmospheric moisture and liquid water) on its transfer from the surface to the satellite sensor. This method described by *Greenwald et al.* (1993), employes observations in 19.35 and 37 GHz. The daily mean horizontal mean distribution of V and L is on a 1°× 1° grid. Land areas are the prime source of data gaps in the fields.

Data Analysis

Two different regions are considered, the southern hemisphere storm track region between 38° – 58° S and the tropical region between 10°N - 5° S. The data are periodic in longitudinal direction which makes it suited for spherical harmonical analysis. Due to clouds (TOVS), the presence of continents (SSM/I) and other factors data gaps are present in the regular spaced grid. These gaps could be filled by means of interpolation or otherwise, but this could influence the power spectrum and therefor these gaps are not filled.

To characterize the spatial variability within the above mentioned regions, the modified periodogram analysis† developed by *Scargle* (1982) following *Lomb* (1976) is used. This analysis is equivalent to least-squares fitting of sinusoids to the data. Thus we assume the data fields are the result of a superposition of waves. In the case of one-dimensional data, *Scargle* (1982) proved that the modified periodogram equals the standard Fourier power spectrum as long as the data are evenly spaced.

We would like to calculate the periodogram in the two-dimensional (m,n) wavenumber space, with (m,n) the wavenumber in the longitundinal and meridional direction respectively, but this is not possible. The method of *Scargle* (1982) gives only a one-dimensional periodogram. Although extention to a two-dimensional dataset seems appropriate, this is not trivial. For now we use the method to calculate the periodogram for the n=0 mode only. Higher order modes are not considered. This is done by neglecting

† The terms *power spectrum* and *periodogram* are used interchangeably. Stricktly speaking a power spectrum is a theoretical quantity for continuous functions, the periodogram is merely an estimation of the power spectrum in case of a finite amount of discrete data (*Scargle,* 1982)

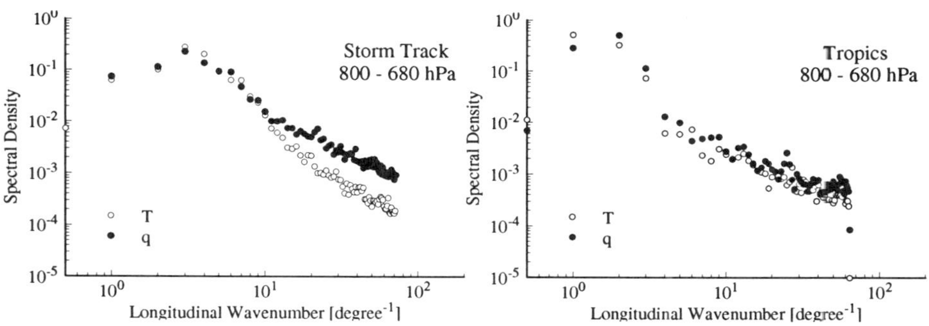

Figure $1^{a,b}$. Normalized spectral density of ISSCP-TOVS temperature (open circles) and specific humidity (filled circles) between 800 and 680 hPa. Observations are taken over the southern hemispheric storm track ($38° - 53°$S, Figure 1^a - left panel) and over the tropics ($10°N - 5°$S; Figure 1^b - right panel). Uncertainty in each datapoint is twice the size of the symbol.

the ϕ dependance of the data (See *Tjemkes and Visser*, 1994).

As mentioned above, the TOVS data are available on a $2.5° \times 2.5°$ grid (144 data points in logitudinal direction), the SSM/I data are available on a $1° \times 1°$ grid (360 data points in longitudinal direction). To reduce the variances in the periodogram, daily periodograms were calculated from the daily fields and averaged over 31 days to obtain a monthly mean periodogram. This averaging reduces the variance by a factor 31 (the standard deviation by $\sqrt{31}$).

RESULTS

Storm Track

In Figure 1 the temperature and specific humidity periodogram for the layer between 800 and 650 hPa, derived from ISCCP-TOVS for the southern hemisphere storm track region is presented. Shown is the one-dimensional periodogram for the n = 0 mode as explained above. Here we are primarily interested in the shape of the spectra, not in the absolute value. So each periodogram is normalized such that the surface under the curve equals unity. On the x-axis the longitudinal wavenumber (m) is given (unit: Degree^{-1}), $m = 1$ corresponds to a wave with a wavelength circumvent the globe. The conversion from wavenumbers to wavelength depends on latitude. Without further reference a constant latitude of 45°S is used for this conversion throughout the paper.

Maximum variance in the temperature spectrum is found around $m = 3$ which corresponds to a wavelength of roughly 9000 km. This wavelength is somewhat larger than the typical wavelength of synoptic disturbances generated by baroclinic instabilities which is given by the Rossby deformation radius which is approximately 2000 km.

In the temperature periodogram two different regimes can be identified: a low and a high wavenumber regime. In both these regimes the periodogram follows a power law, which appears as a straight line in the log-log plot. The exponent of the power law changes from a value of ≈ -3 in the low to a value of $\approx -5/3$ in the high wavenumber regime. The former is thought to be the result of an enstrophy cascade from a large scale source towards smaller scales (*Charney*, 1971). The latter is generated by an inverse energy cascade from a small scale source to larger scales (*Gage*, 1979; *Lilly*, 1983[a,b]; *Gage and Nastrom*, 1986[a]). The possible energy source is the breaking of internal waves or the collapse of three-dimensional turbulence generated by cumulus convection in the region under consideration (*Lilly*, 1983[b]).

Verification of these results is difficult to obtain. Independent aircraft observations analysed by *Nastrom and Gage* (1985) indicate the existence of a transition from a m^{-3}

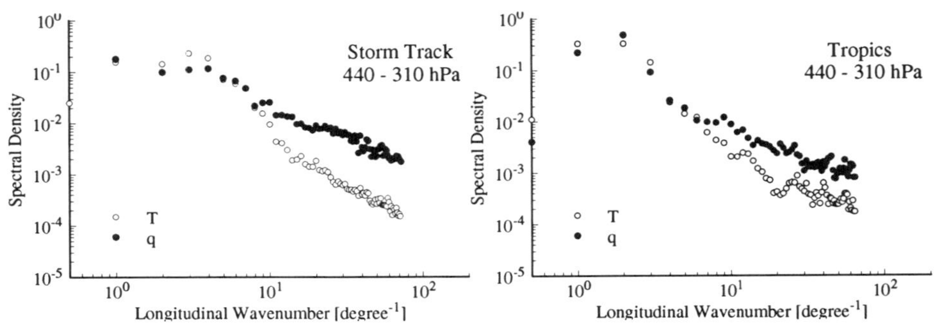

Figure $2^{a,b}$ As Figure $1^{a,b}$ but now for the layer between 440 - 310 hPa.

to a $m^{-5/3}$ power law in the temperature spectrum. These aircraft observations were taken between 9 and 14 km altitude, thus a direct comparison with the present results is not possible. It is likely that the small scale source responsible for the $m^{-5/3}$ power law, extends below the layers probed by the aircraft. Thus the reduced decay found in the ISCCP-TOVS temperature periodogram could be the result of a physical phenomena.

Inspection of Figure 1 shows that the shape of the specific humidity periodogram follows the temperature spectrum. The transition from the m^{-3} to the $m^{-5/3}$ occurs at a smaller wavenumber in the specific humidity spectrum than in the temperature spectrum. In the top three layers, the m^{-3} behaviour in specific humidity periodogram disappears. This can clearly be seen in Figure 2. As will be discussed below we attribute this to the decreasing horizontal gradient of specific humidity with increasing height.

The SSM/I observations can be used over cloudy as well as cloudfree pixels (with the exception of heavy precipitation cases *Greenwald et al.* 1993), the HIRS observations can only be used under clear sky conditions. This non-uniform sampling is the likely cause of the slight difference between the monthly mean V derived from TOVS and SSM/I observations as found by *Stephens and Jackson* (1993). A close agreement in the total variance between the ISCCP-TOVS and SSM/I V is found (*Tjemkes and Visser*, 1994).

The periodogram of cloud liquid water (L) and columnar water vapor (V) derived from SSM/I observations is shown in Figure 3. The V-spectrum is quite complicated. There appears to be a change from a m^{-3} to a $m^{-5/3}$ power law is around wavenumber 10, and a second transition back to the steep decay of the m^{-3} power law at wavenumber 30. An explanation of this second transition is difficult. It could be the result of the averaging process which is employed to generate a daily mean distribution. This process basically calculates the avereage of all observations within a single day, neglecting any diurnal cycle. As a consequence small scale variability will be reduced. This suggests that small scale variability can only be analysed from the observations without any gridding.

The L-spectrum differs considerably from the V-spectrum. For wavenumbers between 8 and 40 a $m^{-5/3}$ power law is consistent with the data, around wavenumber 50 a transition to a steeper decay is present. This steeper decay could be the result of the method to calculate the daily mean distribution. It is worth to notice that for wavenumbers between 5 and 30 a m^{-1} power law followed by a $m^{-5/3}$ power law can not be excluded. This is consistent with the theoretical spectrum of scalar diffusion in two-dimensional turbulence. This consistency seems odd given the strong coupling between dynamics and L. While the coupling between dynamics and moisture exists throughout the entire wavenumber domain, for L this coupling is only present during the formation and dissipation process which takes place at very specific wavenumbers and doesnot extend throughout the entire domain. The absence of a typical size of the moisture source might explain why the observed moisture spectra differ so much from the theoretical results.

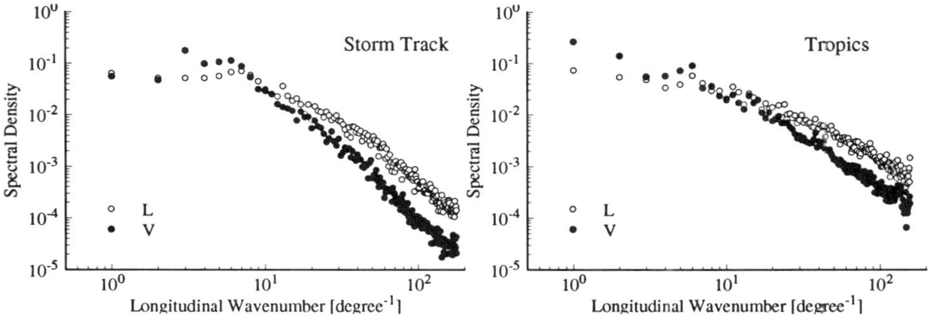

Figure 3^{a,b}. As Figure. 1^{a,b} but for V and L derived from SSM/I observations.

Tropics
In tropics the spectra have similar characteristics as in the storm track region. Yet, tropical dynamics differ considerably from dynamics in storm track. The similarity between the spectra suggests that spectral analysis might not be the appropriate tool for studying atmospheric dynamics. This might also have consequences for a comparison with general circulation models. Although to a large degree the tropical spectra resemble the midlatitude spectra, there are some subtle differences. Notably in the tropics the L-spectrum in the low wavenumber regime follows the theoretical result of the diffusion of a passive tracer in two-dimensional turbulence closer than in the southern hemispheric storm track.

DISCUSSION

For large parts of the wavenumber spectrum water vapor can be considered a passive scalar, which states that water vapor will trace the relative air motions. The diffusion of a passive tracer in two-dimensional turbulence has been studied by *Lesieur and Herring* (1985). They argue that the spectrum of the tracer has also two regimes. When the source is located in the enstrophy cascade the tracer spectrum follows a m^{-1} power law when located in the inverse energy cascade it follows a $m^{-5/3}$ power law. Surprisingly the spectrum cascade in both case will be forward, which means from larger scales towards smaller. The SSM/I V-spectrum nor the ISCCP-TOVS q-spectrum is consistent with these theoretical results. The SSM/I derived L spectra appears to be more consistent with these theoretical results. This result suggests that L is a passive tracer of atmospheric motions. The coupling between cloud water formation and dynamics is apparently restricted to a specific wavenumber. At present it is not clear why in the tropical region the agreement between theory and observations is better than in the southern hemisperic storm track region.

In contrast to L, the source of V doesnot have a distinct wavenumber. This could explain why the V-spectra differ from the theoretical results of the diffusion of a passive scalar in two-dimensional turbulence. Following *Gage and Nastrom* (1986^{a,b}), *Tjemkes and Visser* (1994) argue that in that case the moisture fluctuations are induced by air motions, and as such reflect the absolute gradient of the moisture field. While in the high wavenumber domain the horizontal moisture fluctuations are induced by vertical air motions, in the low wavenumber domain, the horizontal moisture fluctuations are induced by horizontal air motions. That is, consider a region where the gradient of a passive scalar in the latitudinal direction is much larger than in the longitudinal direction. Then horizontal air motions, will redistribute the passive scalar, from high latitudes to low latitudes and as such superimpose horizontal fluctuations upon the background gradient. The wavelength of these fluctuations will be closely associated with the wavelength of the source of the perturbations, at the wavenumbers considered here most likely a baroclinic instability. The strength of the fluctuations is thus proportional to the horizontal gradient.

From these arguments it can be shown that in the upper layers the synoptic scale air motions induce fluctuations in the specific humidity field, but that due to a small horizontal gradient, this signal is too weak to detect against a stronger signal produced by a small scale source (see *Tjemkes and Visser* 1994). This indicates that there is no dissipation of energy at wavenumbers in-between the meso and synoptic scale as hypothesised by *Gage and Nastrom* (1986[a]). This result is consistent with the model for two-dimensional turbulence discussed by *Lilly* (1989)

An interpretation of the tropical spectra, should include effects of gravity wave perturbation (*Gardner et al.*, 1993 and references therein). At this stage of our research we are studying that particular topic.

SUMMARY

Sensors mounted on space borne platforms observe a large spatial domain almost instantaneously, which makes them an attractive tool to study spatial variability. Here space borne observations by the HIRS (as archived in the ISCCP-TOVS dataset) and SSM/I are used to analyse spatial variability of temperature, water vapor and cloud liquid water. The southern hemispheric storm track and the tropics in July 1989 are considered. In an independent study *Stephens and Jackson* (1993) show that the columnar water vapor derived from the HIRS and SSM/I instruments differ by approximately $1-3\ \mathrm{kgm}^{-2}$. Here it is shown that the spatial variance and the shape of the columnar water vapor (V) spectrum derived from these two sources are related. Since the employed ISCCP-TOVS data have a somewhat lower spatial resolution than the employed SSM/I data, the SSM/I derived V-spectrum shows more detail in the high wavenumber domain.

In general we find the tempererature and specific humidity spectrum have two distinct regimes. In both cases it follows a power law of the wavenumber but two different values for exponent in the power law are found: ≈ -3 in the low wavenumber domain and $\approx -5/3$ in the high wavenumber domain. The former is thought to represent moisture fluctuations induced by large, while the latter is induced by small scale air motions. The shape of the temperature spectrum doesnot change with height. The specific humidity spectrum varies with height to the extend that for layers in the upper troposphere or lower stratosphere the low wavenumber regime tends to disapear. We attribute this to the decreasing zonal gradient of moisture with height. The large scale air motions will generate a m^{-3} shape in the moisture spectrum, but the amplitude of the variations is much smaller than the amplitude of the variations generated by a small scale source. This result indicates that there is no dissipation at wavenumbers in-between the meso and synoptic scale as hypothesised by *Gage and Nastrom* (1986[a]).

Besides the columnar water vapor power spectrum we used the SSM/I dataset to provide a cloud liquid water (L) power spectrum as well. Again two regimes can be distinguished in the L-spectrum but these are not related to the regimes found in the columnar water vapor spectrum. This is most likely associated with the typical size of the source of L and V.

The interpretation of the spectra collected over the tropical regions should include effects of gravity waves. That is a topic of future research.

Acknowledgments

Part of this study was performed when the first author was visiting the Department of Atmospheric Science of the Colorado State University. He thanks for the hospitality of and discussion with prof. G.L. Stephens. This work was partially supported by NOAA Grant #NA3RJ0202-Item 2 to CSU, by EEC grant EV5v-CT92-0131 and a grant from the Dutch foundation of Space Research (SRON) both to the University of Utrecht. The data employed in this study was provided by T. Greenwald (SSM/I) and D. Randall, I. Wittmeyer and prof. T. Vonder Haar (ISCCP-TOVS).

REFERENCES

1. Charney, J.G., 1971. Geostrophic turbulence, *J. Atmos. Sci.*, **28**, 1087 – 1095.
2. Gage, K.S., 1979. Evidence for a $k^{-5/3}$ law inertial range in mesoscale two-dimensional turbulence, *J. Atmos. Sci.*, **36**, 1950 – 1954.
3. Gage, K.S. and G.D. Nastrom, 1986[a]. Theoretical interpretation of atmospheric wavenumber spectra of wind and temperature observed by commercial aircraft during GASP, *J. Atmos. Sci.*, **43**, 729 – 740.
4. Gage, K.S. and G.D. Nastrom, 1986[b]. Spectrum of atmospheric vertical displacements and spectrum of conservative scalar passive additives due to quasi-horizontal atmospheric motions, *J. Geophys. Res.*, **91**, 13211 – 13216.
5. Gardner C.S., C.A. Hostetler and S.J. Franke, 1993. Gravity wave models for the horizontal wavenumber spectra of atmospheric velocity and density fluctuatuions. *J. Geophys. Res.*, **98**, 1035 – 1056.
6. Greenwald, T.J., G.L. Stephens, T.H. Vonder Haar and D.L. Jackson, 1993. A physical retrieval of cloud liquid water over the global oceans using Special Sensor Microwave/Imager (SSM/I) observations, *J. Geophys. Res.*, **98**, 18471–18488.
7. Hollinger J.P., R. Lo, G. Poe, R. Savage, and J. Pierce, 1987. Special Sensor Microwave/Imager user's guide, *NRL Memo rep.*, Nav. Res. Lab., Washington, D.C.
8. Lesieur, M. and J. Herring, 1985. Diffusion of a passive scalar in two-dimensional turbulence, *J. Fluid. Mech.*, **161**, 77 – 95.
9. Lilly, D.K., 1983[a]. Stratified turbulence and the mesoscale variability of the atmosphere, *J. Atmos. Sci.*, **40**, 749 – 761.
10. Lilly, D.K., 1983[b]. Mesoscale variability of the atmosphere, in Proceedings of the NATO-advanced study institute on *Mesoscale Meteorology Theories, Observations and Models*, D.K. Lilly and T. Gal-Chen (eds.), NATO-ASI-Series, D. Reidel Publ.
11. Lilly, D.K., 1989. Two-dimensional turbulence generated by energy sources at two scales, *J. Atmos. Sci.*, **46**, 2026 – 2030.
12. Lomb, N.R., 1976. Least-squares frequency analysis of unequally spaced data, *Astroph. Space Sci*, **39**, 447 – 462.
13. Manney, G.L. and J. Stanford, 1990. Wavenumber spectra from Temperature-Humidity Infrared Radiometer 6.7 μm water vapor data, *J. Geophys. Res.*, **95**, 909 – 913.
14. Nastrom, G.D. and K.S. Gage, 1985. A climatology of atmospheric wavenumber spectra of wind and temperature observed by commercial aircraft, *J. Atmos. Sci.*, **42**, 950 – 960.
15. Nastrom, G.D., W.H. Jasperson and K.S. Gage, 1986. Horizontal spectra of atmospheric tracers measured during the Global Atmospheric Sampling Program, *J. Geophys. Res.*, **91**, 13201 – 13209.
16. Reale, A.L., M.D. Goldberg and J.M. Daniels, 1989. Operational TOVS soundings using a physical approach, *12th Canadian Symposium on Remote Sensing, July 10-14, Vancouver BC*, 2653 – 2657.
17. Scargle, J.D., 1982. Studies in astronomical time series analysis II. Statistical aspects of spectral analysis of unevenly spaced data, *Ap. J.*, **263**, 835 – 853.
18. Schiffer, R.A. and W.B., Rossow, 1983. The International Satellite Cloud Climatology Project (ISCCP): The first project of the World Climate Research Program, *Bull. Amer. Meteor. Soc.*, **64**, 779 – 784.
19. Smith, W.L., H. Woolf, C. Hayden, D. Wark, and L. McMillin, 1979. The TIROS-N Operational Vertical Sounder, *Bull. Amer. Meteor. Soc*, **58**, 1177 – 1187.
20. Smith, W.L, and H.M. Woolf, 1979. The use of eigenvectors of statistical covariance matrices for interpreting satellite sounding radiometer observations, *J. Atmos. Sci.*, **33**, 1127 – 1140.
21. Stephens, G.L. and D.L. Jackson, 1993. A comparison of SSM/I and TOVS data over the global oceans, *submitted MAP.*
22. Tjemkes, S.A. and M. Visser, 1994. Horizontal variability of temperature, specific humitidy and cloud liquid water as derived from space borne observations, *submitted J. Geophys. Res*

Estimation of cloud liquid water contents from SSM/I and METEOSAT observations: application to the SOFIA-ASTEX experiment

C. PRIGENT[1] and G. SEZE[2]

[1] *Département de Radioastronomie Millimétrique, CNRS Observatoire de Meudon, 92190 Meudon, France.*

[2] *Laboratoire de Météorologie Dynamique, CNRS Ecole Polytechnique, 91128 Palaiseau, France.*

Abstract

A non-linear inversion of the SSM/I observations is shown to yield accurate estimates of the surface wind speed, the integrated water vapor and the liquid water content. The inversion scheme is based on the analysis of the radiative transfer equation, coupled to a realistic first estimation of the meteorological situation. The method is applied to the SSM/I data collected during the SOFIA-ASTEX campaign. The retrieved geophysical parameters are compared to the in situ measurements available during the observation campaign.

Keywords: satellite retrieval, radiative transfer, cloud liquid water.

INTRODUCTION

A number of statistical algorithms have been developed to retrieve the atmospheric integrated water vapor content as a combination of the SSM/I microwave observations (Alishouse et al., 1990; Petty and Katsaros, 1990, among others). These algorithms are easy to apply to global satellite data and have proved an accuracy suitable for meteorological applications. Recently, the SSM/I integrated water vapor estimates have been successfully used as inputs in the assimilation scheme of an operational weather forecast model (Filberti et al., 1994).

However, for the retrieval of cloud liquid water and rain, the current statistical algorithms often undergo deficiencies (Hollinger, 1991). They have difficulty to take into account the high variability of all the related parameters and they suffer from the lack of validation measurements. Today, new methods are proposed, based on more sophisticated radiative transfer models and associated to realistic cloud profiles (Kummerow et al., 1991; Smith et al., 1992, Hinton et al., 1992). But, so far, these works have been tested in tropical or continental areas, where the liquid water content is large.

In this study, we propose to analyze the sensitivity of the SSM/I responses to the liquid water content and to the cloud structure, using a radiative transfer model, in a region where the liquid water contents are not so high. A physical approach is adopted to retrieve the cloud liquid water content from SSM/I observations, based on an analysis of the radiative transfer equation, coupled to a weather forecast model analysis and to a cloud classification derived from infrared/optical images. The method is applied to data from the SOFIA-ASTEX experiment, in June 1992 in the Azores area (Weill et al., 1994; NASA report, 1992). During this well documented observation campaign, coincident SSM/I and METEOSAT images have been collected together with the European Center for Medium range Weather Forecast (ECMWF) analysis and with the associated in situ measurements.

A PHYSICALLY-BASED METHOD TO RETRIEVE LIQUID WATER CONTENTS

With a realistic first estimation of the atmospheric characteristics and the cloud structure in the meteorological area of concern, this scheme performs an inversion of the SSM/I brightness temperatures to retrieve simultaneously physically consistent estimates of the surface wind speed, the integrated water vapor content, the liquid water content and the rain rate. The sensitivity of the SSM/I channels with the other atmospheric parameters (sea surface temperature, humidity and temperature profiles, cloud altitude and thickness, among others) is too weak to retrieve these ancillary parameters. However, these parameters can be considered as a source of noise in the retrieval process of the geophysical quantities of interest (Petty and Katsaros, 1992). Introducing prior information on the atmospheric situation reduces this additive noise.

The approach we used is based on a non-linear inversion procedure, which aims at minimizing the difference between simulated and measured brightness temperatures, i.e. the solution x for the atmospheric state will minimize the following function:

$$\chi(x)^2 = (TB_{obs} - TB(x))^T . E^{-1}. (TB_{obs} - TB(x)) \qquad \textbf{(1)}$$

where x is a vector of length 4 representing the four atmospheric variables to be estimated; TB(x) is the vector of SSM/I brightness temperatures simulated with a forward radiative transfer model and corresponding to the atmospheric state x; TB_{obs} is the vector of the measured SSM/I brightness temperatures; E is a diagonal matrix the non-zero elements of which are the channel noise variances. A standard deviation of 2 K is adopted for the five lower channels, and 3 K for the 85.5 GHz channels, including the instrumental noise, unmodeled contributions to the radiative transfer, inhomogeneities in the scenes. For the nonlinear inversion problem, the method of newtonian iteration is applied and yields (Rodgers, 1976) :

$$x_{n+1} = x_n + (K(x_n)^T. E^{-1} . K(x_n))^{-1} (K(x_n)^T . E^{-1}. (TB_{obs} - TB(x_n))) \qquad \textbf{(2)}$$

with K(x) containing the partial derivatives of TB(x) with respect to the elements of x.

The matrix $S(x_n) = (K(x_n)^T . E^{-1} . K(x_n))^{-1}$ is an estimate of the error covariance for the step n.

To avoid contamination of the liquid water retrieval by unrealistic estimates of water vapor content and wind speed, when dealing with regions where the cloud liquid content is rather low, it is important to keep the integrated water vapor content and the surface wind speed as parameters in the inversion procedure.

This inversion process is a severe validation test for the radiative transfer model to be used which has to be accurate enough to give a good estimate not only of the brightness temperatures, but also to the first derivative of the brightness temperatures with respect to the geophysical parameters, for all the frequencies. Thus, special efforts have been brought to the analysis of the radiative transfer model including the gaseous absorption, the sea surface response, and the interaction of the signal with the water particles. The radiative transfer of polarized radiation is considered through a horizontally infinite plane parallel atmosphere, although we are aware that it can lead to errors when considering partial cloud coverage (Liu and Simmer, 1994; Petty, 1994; Roberti and Kummerow, 1994). The MPM 93 gaseous absorption model of Liebe et al., 1993 is used: it predicts a stronger water vapor absorption than the MPM 89 and on the average, agrees better with radiometric observations (Guillou et al., 1994; Snider, 1994; Han et al., 1994). For the sea surface emissivity, a geometric optics approach is adopted, the validity of which is discussed in Guillou et al., 1994. In case of rain or large particles, the radiative transfer is solved iteratively, using the successive orders of scattering. Up to 90 GHz and below 100 μm radius, cloud drops do not scatter the radiation significantly (i. e. the Mie theory reduces to the Rayleigh approximation) and as a consequence, in absence of rain or large particles, the scattering process is neglected.

For each SSM/I pixel, a first estimation of the atmospheric profile and surface parameters is derived from the operational weather forecast analysis provided by the European Centre for Medium-range Weather Forecast (ECMWF). The initial ECMWF profile is then altered to take into account the cloud information provided by METEOSAT images.

A statistical cloud classification method based on clustering on two-dimensional histograms is applied to the visible and infra-red window channels of the METEOSAT imagery (Seze and Desbois, 1987; Raffaeli and Seze, 1994). It provides a cloud classification with an horizontal resolution of 5 km. Assuming in a first approximation that the SSM/I observations are performed with an averaged gaussian field of view of 25 km, the METEOSAT cloud classification results are weighted within the averaged SSM/I field of view, and the most probable cloud characteristics are deduced for each SSM/I pixel, with the associated cloud coverage. Before, this treatment, one has to verify that the SSM/I and the METEOSAT images are well collocated, which is attempted by careful exam of the coastal response of all the instruments.

Then, within the inversion scheme, the initial atmospheric parameters (wind speed, integrated water vapor, integrated liquid water content and rain rate) are modified iteratively until satisfactory convergence is obtained. The water vapor density in each layer of the initial ECMWF profile is uniformly scaled, for the atmospheric profile to fit the integrated water vapor content imposed during the inversion process. Water vapor exceeding saturation are truncated and are compensated by an additive water vapor content in the adjacent layers. Within the cloud, the liquid water content is assumed constant with altitude. When existing, the rain layer is supposed to extend from the ground up to the 0° C isotherm, with a constant drop size distribution given by Marshall and Palmer (1948). In this study, the clouds are assumed to contain no ice: in the area of concern, even when cirrus clouds are present, they should not interfere significantly with the signal, except at 85GHz, in some convective areas. Usually, convergence is obtained in less than four iterations.

APPLICATION TO THE SOFIA-ASTEX EXPERIMENT

In the context of the SOFIA-ASTEX campaign, in the Azores area, in June 1992, the SSM/I data have been collected, together with the METEOSAT images and the ECMWF analysis. In this study, the discussion is limited to the SSM/I orbit of the 12 June 1992 at 1100GMT over the Azores, with the coincident METEOSAT image of 1100GMT and the ECMWF analysis of 1200GMT.

The meteorological surface analysis shows a high position near 33N, 30W, with a maximum surface pressure of 1032 mb, and the Azores are all within the influence of this strong high pressure. The METEOSAT visible image with the corresponding cloud classification are presented on figures 1 and 2. These images show a large area of stratocumulus in the north, dislocating toward the south in broken patchy stratocumulus. The collocated SSM/I images at 19 GHz (vertical polarization) and at 85 GHz (vertical polarization) are displayed on figure 3 and 4.

With the previously described method, the surface wind speed, the integrated water vapor and the cloud liquid content have been retrieved. Figures 5 and 6 present the integrated water vapor and liquid water retrieved fields.

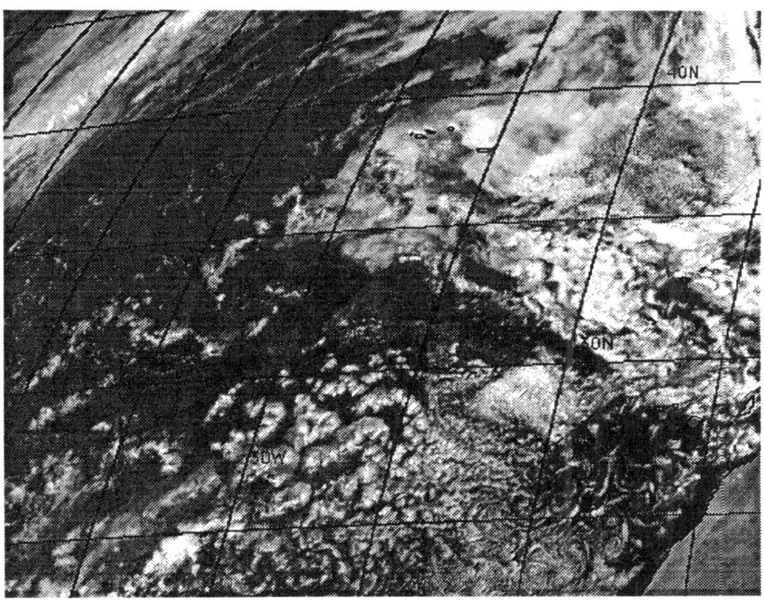

Figure 1. The METEOSAT visible image above the Azores, the 12 June 1992 at 1100GMT.

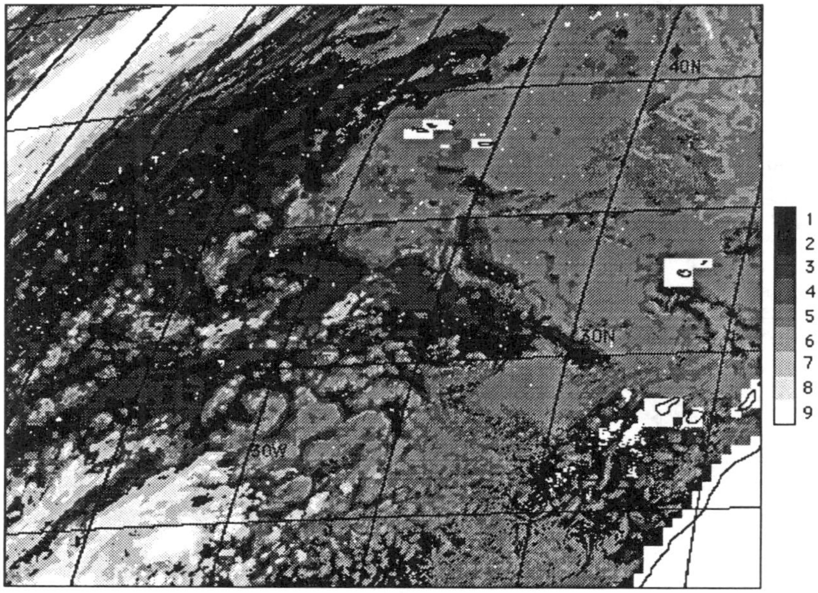

Figure 2. Cloud classification derived from METEOSAT, above the Azores, the 12 June 1992 at 1100GMT.

1 cloud edge	2 sea	3 thin cloud edge
4 low level cloud (partial coverage)	5 thick low level cloud	6 medium level cloud
7 cirrus	8 thick cirrus	9 cumulonimbus-nimbo-stratus

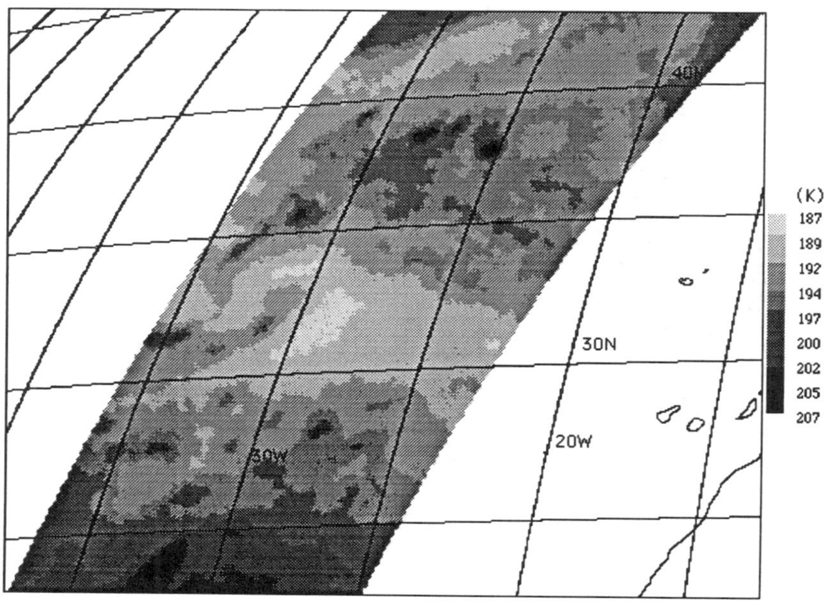

Figure 3. The SSM/I image at 19 GHz (vertical polarization), above the Azores, the 12 June 1992 at 1100GMT.

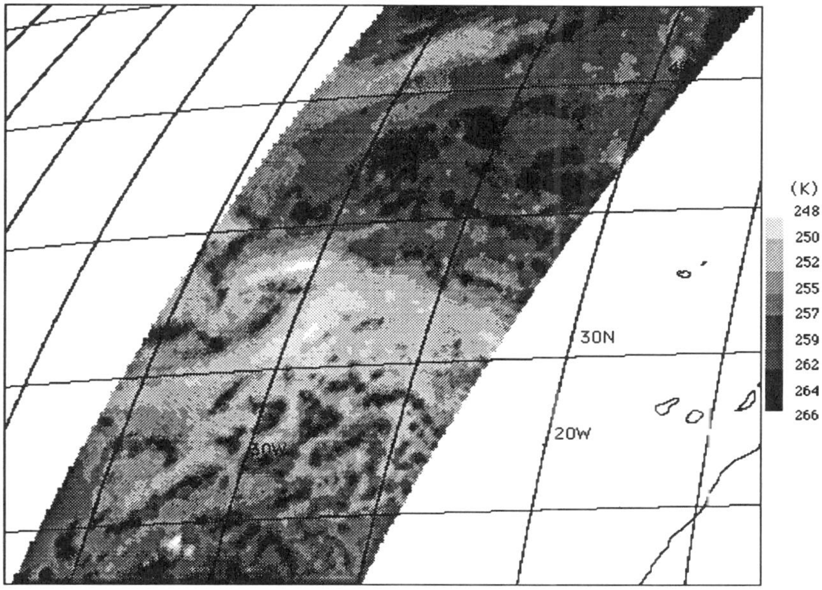

Figure 4. The SSM/I image at 85 GHz (vertical polarization), above the Azores, the 12 June 1992 at 1100GMT.

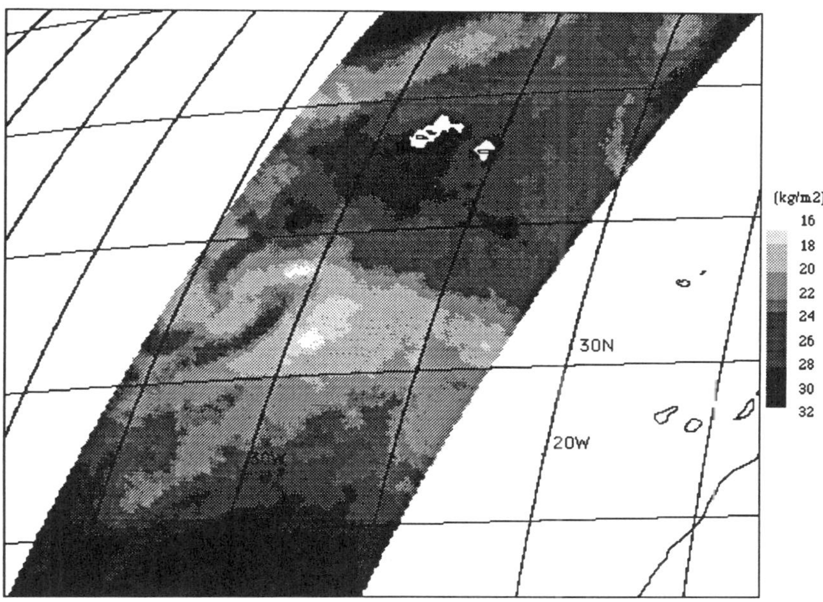

Figure 5. The integrated water vapour field (our method), above the Azores, the 12 June 1992 at 1100GMT.

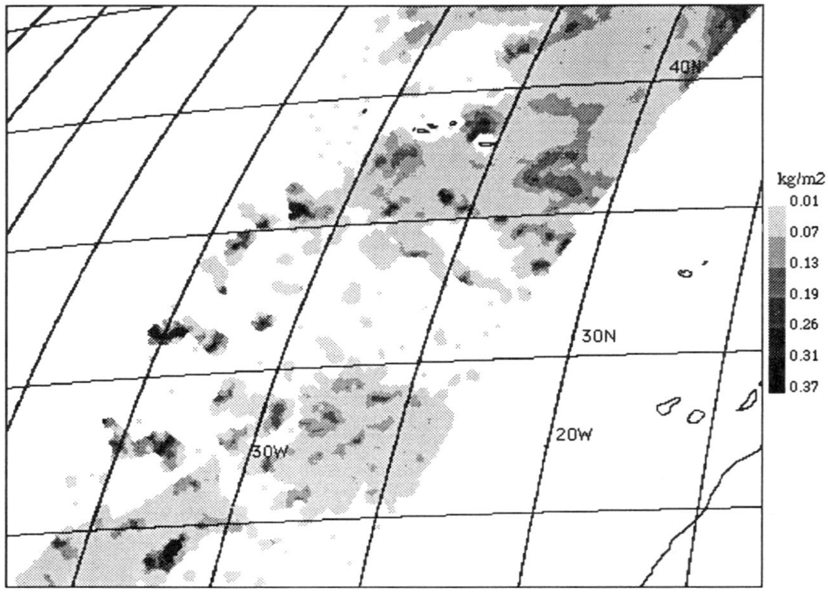

Figure 6. The integrated liquid water field (our method), above the Azores, the 12 June 1992 at 1100GMT.

The retrieved integrated water vapor contents and wind speeds are shown (Figure 7), compared to the ECMWF fields used to initialize the inversion process and also compared to the statistical algorithms of Alishouse et al. (1990) for the water vapor content and of Goodberlet et al. (1989) for the surface wind speed. With completely independent retrieval methods, the statistical SSM/I algorithms and our physically-based inversion yield comparable results.

In Table 1, the retrieved products are compared with in situ measurements coming from radiosoundings and from drifting buoys (see Weill et al., 1994 and the NASA report, 1992 for a description of the in situ measurements during the SOFIA-ASTEX campaign).

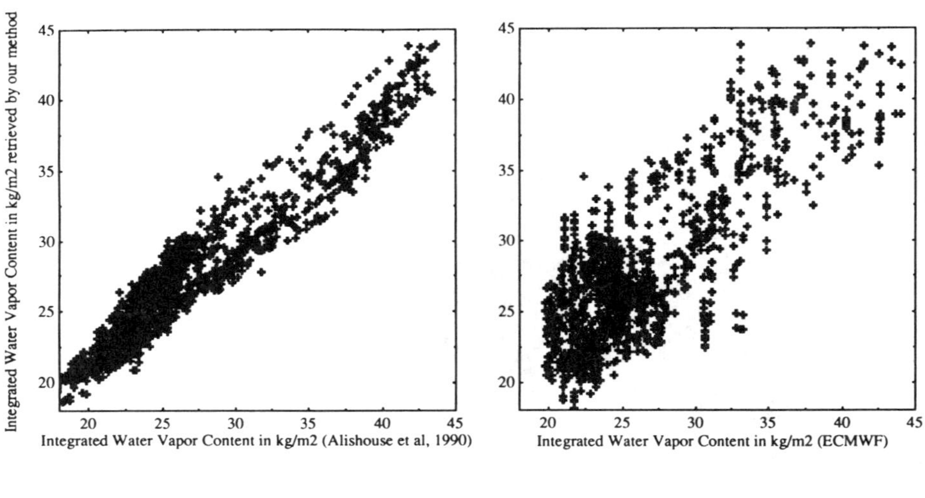

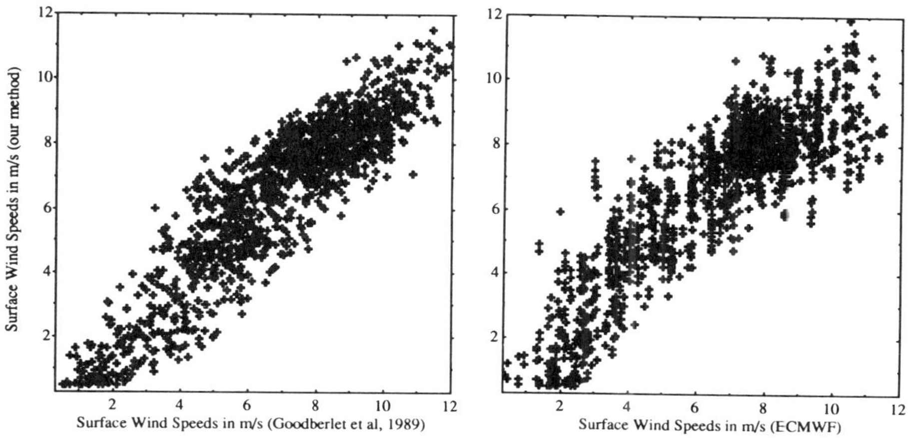

Figure 7. Comparisons between retrieved products

Table 1. Comparisons between our retrieved products and in situ measurements.

(a) Comparison of integrated water vapour contents in kg/m2

Radiosounding location	Lat	Lon	Radiosounding	Alishouse et al, 1990	Our rerieval	ECMWF
Oceanus	40.28N	24.00W	25.01	24.32	25.53	22.49
Santa Maria	36.99N	25.17W	26.14	24.57	27.01	23.60

(b) Comparison of surface wind speeds in m/s at 12.5 m altitude

Buoy number	Lat	Lon	Buoys	Goodberlet et al, 1989	Our retrieval (Cox and Munk)	ECMWF
'10121'	35.08N	21.80W	9.9	8.3	7.4	6.7
'15496'	35.11N	21.45W	9.3	8.7	7.8	7.3
'15498'	35.36N	21.60W	8.7	10.3	8.1	7.[illegible]
'15499'	34.86N	22.45W	9.9	8.0	7.7	6.0

For each SSM/I pixel, the matrix S (eq. 2) is calculated at the end of the iterative process and gives the error covariance of the retrieved products. Averaged on the SSM/I image we studied, the theoretical accuracy are of 1.7 m/s for the surface wind speed, of 1.6 kg/m2 for the integrated water vapor, and of 0.04 kg/m2 for the integrated liquid water content. The cost of χ^2 (eq. 1) at the convergence of the minimization is a useful parameter to examine the quality of the retrieval products. Figure 8 shows χ^2 with the corresponding discrepancy between the observations and the simulations for all the channels. For 88% of the processed pixels, the cost χ^2 is less than 7, i. e. for these pixels, the average discrepancy between the simulated and the measured brightness temperature is within the channel noise for the seven channels used in the retrieval. In homogeneous meteorological regions where the spatial correlation between the five lower frequencies and the 85 GHz is high enough, the inversion scheme performs well. However, when spatial heterogeneities are present, the inversion process has trouble to converge with the seven channels, mostly because of the diversity in horizontal resolution: all the channels do not see the same scene and have difficulty to fit on the same set of geophysical parameters (see the tail on the χ^2 histogram on Figure 8). A method similar to the approach of Farrar and Smith (1992) is to be tried, to take advantage of the spatial-over-sampling of the low frequencies in order to obtain equal resolution in all the channels.

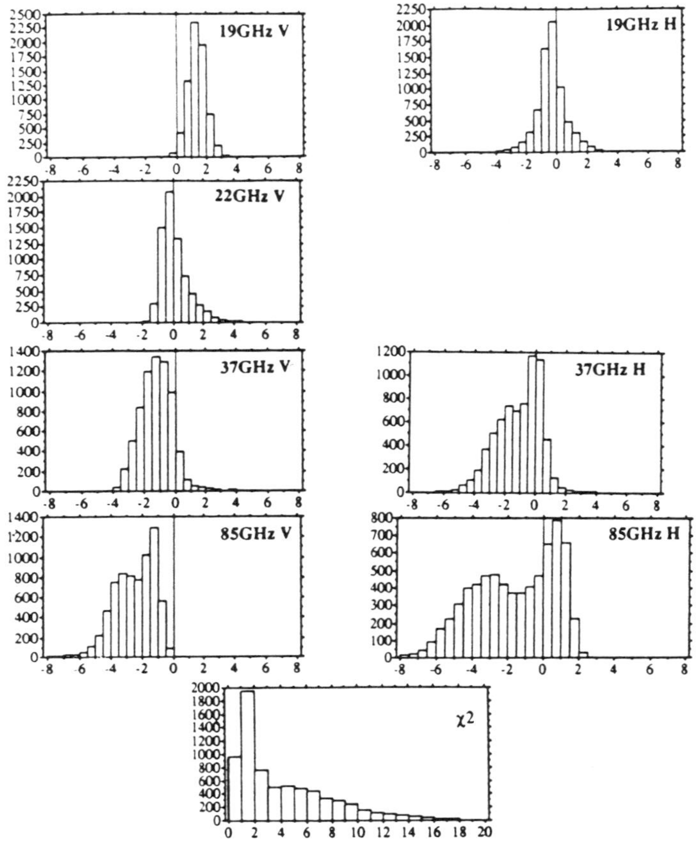

Figure 8. Differences between measurements and simulations (after inversion) for the SSM/I channels. The resulting χ^2 is also shown.

Other ASTEX situations will be soon processed, and compared with the liquid water contents estimated with other independent measurements, when available (microwave airborne estimates, microphysics airborne measurements...).

CONCLUSION AND PERSPECTIVES

A non-linear inversion of the SSM/I observations has been shown to yield simultaneously accurate estimates of the surface wind speed, the integrated water vapor and liquid water contents. The inversion scheme is based on the radiative transfer equation and associated to a realistic first estimation of the meteorological situation. This process is a severe validation test for the radiative transfer model: we have shown that we are able to retrieve consistent geophysical parameters with a radiative transfer model that agrees with the seven SSM/I channels. Now being confident in the radiative transfer accuracy, a parametrized version of the radiative transfer model is to be implemented, for numerical efficiency purposes. Although not present in this case study the inversion method has also been shown to yield realistic estimates of the rain rates for mid-latitudes oceanic situations (Prigent et al., 1994). The next step in the development of this scheme will consist in estimating the ice parameters within the clouds.
The numerical weather forecast models still lack a realistic representation of the cloud processes. Even when cloud water is explicitly introduced into the models, it is only derived from the other thermodynamics and water vapor variables and is never constrained by direct observations. Within the assimilation scheme of the operational weather prediction models, the trend today is to directly assimilate the satellite IR radiances rather than retrieved products,

provided that the radiative transfer associated with these radiances is accurate enough. We can reasonably foresee that the same tendency will prevail for satellite microwave brightness temperatures: this study lies within this prospect.
A cloud classification method, with coupled SSM/I and METEOSAT images, is now under investigation. For the METEOSAT pixels in coincidence with the center of a SSM/I pixel, the SSM/I brightness temperature are added as inputs in the cloud classification scheme. This method relies upon a good collocation of both satellite images, but is a preliminary step in the treatment of coincident IR, visible and microwave images, as will be soon available to the public from the SSM/I and Operational Linescan System (OLS) imagers on board the satellites of the Defense Meteorological Satellite Program (DMSP).

Acknoledgements: The SOFIA-ASTEX program is a cooperative work involving several groups. The authors are grateful to A. Weill (Centre d'étude pour l'Environnement Terrestre et Planétaire, CETP, France) for the access to the SOFIA-ASTEX data bank.

REFERENCES

Alishouse J. C., S. Snyder, J. Vongsathorn, R.R. Ferraro : "Determination of oceanic total precipitable water from the SSM/I", *IEEE Trans. Geosci. Remote Sensing*, 28, n° 5, pp. 811-816, 1990.

Farrar M. R., E. A. Smith : "Spatial resolution enhancement of terrestrial features, using deconvolved SSM/I brightness temperatures", *IEEE Trans. Geos. Remote Sensing*, vol. 30, n° 2, pp. 349-355, 1992.

Filberti M.A., Eymard L., Urban B. : "Assimilation of satellite precipitable water in a meteorological forecast model", *Monthly Weather Review*, in press, 1994.

Goodberlet M. A., C. T. Swift, J. Wilkerson : "Remote sensing of ocean surface winds with the Special Sensor Microwave / Imager", *J. Geo. Res.*, vol.94, pp.14547-14555, 1989.

Guillou C., Prigent C., English S. : "How to described the ocean roughened surface in microwave emissivity models?", *Specialist Meeting on Microwave Radiometry and Remote Sensing of the Environment*, Rome, Italy, 14-17 Feb, 1994.

Guillou C., Prigent C., English S., Brunel P. : "Gaseous absorption in atmospheric windows at millimeter wavelengths: comparison between models and measurements", *International Conference on Millimeter and Submillimeter Waves and Applications*, Jan. 10-14, San Diego, USA, 1994.

Hinton B., Olson W., Martin D., Auvine B. : "A passive microwave algorithm for tropical oceanic rainfall", *Journ. of Applied Meteorology*, vol. 31, pp.1379-1395, 1992

Hollinger J. P. : "DMSP Special Sensor Microwave / Imager. Calibration / Validation", Final report, Naval Research Laboratory, Washington D.C., vol. 2, 1991.

Han Y., Snider J., Westwater E., Melfi S., Ferrare R. : "Observations of water vapor by ground based microwave radiometers and raman lidar", *Specialist Meeting on Microwave Radiometry and Remote Sensing of the Environment*, Rome, Italy, 14-17 feb, 1994.

Kummerow C., I. M. Hakkarinen, H. F. Pierce, J. A. Weinman : "Determination of precipitation profiles from airborne passive microwave radiometric measurements", *J. Applied Meteor.*, vol. 8, pp.148-158, 1991.

Liebe H. J. : "MPM – an atmospheric millimeter-wave propagation model", *International Journal of Infrared and Millimeter Waves*, vol. 10, n° 6, pp.631-650, 1989

Liebe, H. J. : "Propagation modeling of moist air and suspended water/ice particles at frequencies below 1000GHz", *AGARD Specialist Meeting of the Electromagnetic Wave Propagation Panel*, Spain, 17-21 May 1993.

Liu Q., Simmer C. : "3-D radiation transfer effects of a raining cloud", *Specialist Meeting on Microwave Radiometry and Remote Sensing of the Environment*, Italy, 14-17 Feb 1994.

Marshall, J. S., W. McK. Palmer : "The distribution of raindrops with size", *J. Meteorol.*, vol. 5, pp. 165-166, 1948.

NASA report : "ASTEX operation plan", prepared by the Fire project and the ASTEX working group, March 1992.

Petty G. W. and K. B. Katsaros : "The response of the SSM/I to the marine environment. Part I: An analytic model for the atmospheric component of observed brightness temperatures", *Journ.. Atmospheric and Oceanic Technology*, vol. 9, pp.746-761, 1992.

Petty G. W., K. B. Katsaros : “New algorithms for the special sensor microwave imager", *Proc. of the 5th international conference on satellite meteorology and oceanography*, London, pp. 247-251, 3-7 Sept 1990.

Petty G. W. : "Physical retrievals of over-ocean rain rate from multichannel microwave imagery. Part 1: theoretical characteristics of normalized polarization and scattering indices", *Meteorology and Atmospheric Physics*, in press, 1994.

Prigent C., A. Sand, C. Klapisz, Y. Lemaitre : "Physical retrieval of liquid water contents in a north Atlantic cyclone using SSM/I data", *Quaterly Journal of the Royal Meteorological Society*, in press, 1994.

Roberti L., Kummerow C. : "A quantitative comparison between 3-D and plane parallel microwave radiative transfer codes applied to horizontally and vertically structured precipitating clouds", *Specialist Meeting on Microwave Radiometry and Remote Sensing of the Environment*, Rome, Italy, 14-17 Feb. 1994.

Rodgers C. D. : "Retrieval of atmospheric temperature and composition from remote measurements of thermal radiation", *Review of geophysics and space physics*, vol. 14, n° 4, pp. 609-624, 1976.

Raffaeli J. L., Seze G.: "Cloud type separation using local correlation between visible and IR satellite images", *8th Conference on Atmospheric Radiation*, pp.383-385, 23-28 Jan. 1994.

Seze G., M. Desbois : "Cloud cover analysis from satellite imagery using spatial and temporal characteristics of the data", *Journal of Climate and Applied Meteorology*, vol. 26, no.2, pp.287-303, 1987.

Smith E. A., A. Mugnaï, H. J. Cooper, G. J. Tripoli, X. Xiang : "Foundations for statistical-physical retrieval from passive microwave satellite measurements. Part I: brightness temperatures properties of a time-dependent cloud radiation model", *J. Applied Meteorol.*, vol. 32, pp.506-531, 1992.

Snider J. : "Observed and theoretical atmospheric emission at 20, 30 and 90GHz: recent results from land- and ocean-based locations", *Specialist Meeting on Microwave Radiometry and Remote Sensing of the Environment*, Rome, Italy, 14-17 Feb. 1994.

Weill A. et al. : "SOFIA 1992 experiment during ASTEX", submitted to the *Journal of Atmosphere and Ocean system.*, 1994.

An Improved Total Precipitable Water Algorithm For Cloudy Situations

R. FUHRHOP and E. RUPRECHT
Institut für Meereskunde an der Universität Kiel, Düsternbrooker Weg 20, 24105 Kiel, Germany

Abstract: Total water vapour content of an atmospheric column can be estimated with high accuracy today from microwave measurements of the Special Sensor Microwave / Imager on board of the satellites of the Defense Meteorological Satellite Program. This study investigates the retrieval errors by comparison of radiosonde observed and satellite estimated precipitable water. Two effects have been identified which are responsible for systematic estimation errors: temperature and cloud effects.
Different relationships between water vapour and air temperature in different climate regions lead to a systematic over- and underestimation of the total precipitable water. These systematic errors are eliminated by application of different algorithms for each climate region which is identified by its air temperature at the surface.
The effect of cloud liquid water on the retrieval of precipitable water is studied with the result that the effect could be quantified. An improved algorithm has been developed for cloudy situations by directly including the cloud effect into the algorithm.

Keywords: microwave radiometry, retrieval errors, SSM/I, total precipitable water

INTRODUCTION

Remote sensing of total water vapour over oceans from passive microwave measurements is today a common method. Especially for the Special Sensor Microwave / Imager (SSM/I) on board of the satellites of the Defense Meteorological Satellite Program (DMSP) various water vapour algorithms have successfully been developed, e.g. Schluessel and Emery [1] and Petty [2].

Our working group in Kiel also derived an algorithm to estimate the precipitable water (V) from SSM/I brightness temperatures. The algorithm is based on the brightness temperatures of the two vertically polarized channels at 22 and 37 GHz:

$$V = a_0 + a_1 \ln(T_0 - T_{22v}) + a_2 \ln(T_0 - T_{37v}) \;, \qquad (1)$$

where V is measured in kg/m^2, which is equivalent to columnar water measured in mm. T_0 is a reference temperature (=280 K). The coefficients have been determined as $a_0 = 131.95$, $a_1 = -39.50$ and $a_2 = 12.49$ (all in kg/m^2) by regression analysis with radiosonde observations over the Atlantic Ocean and corresponding simulated brightness temperatures.

In order to test the performance of the algorithm a comprehensive data set was collected, containing radiosonde ascents with coincident SSM/I measurements. The precipitable

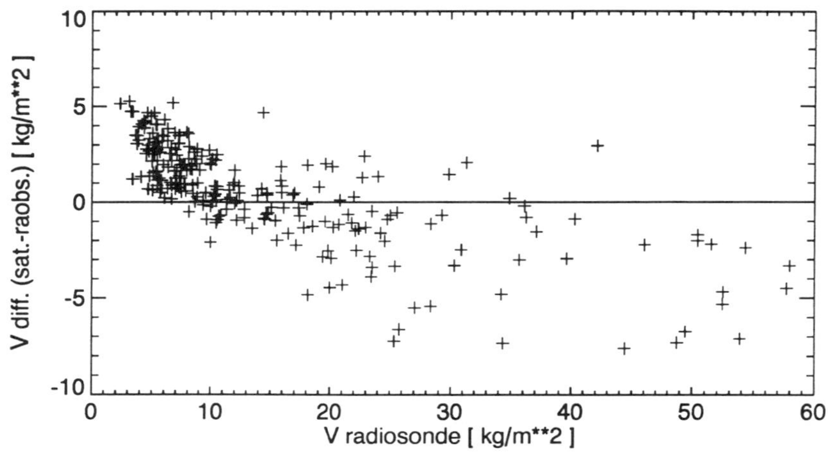

Figure 1. Total precipitable water differences (satellite minus radiosonde) for the global algorithm vs. radiosonde data.

water derived from the radiosonde profiles and the satellite estimations are compared in Figure 1. Here are plotted the differences between satellite estimation and radiosonde observation vs. the radiosonde observations. The striking feature is a systematic overestimation for low values and underestimation for high values of precipitable water. Such systematic differences are of course not satisfactory, but it seems to be a general problem. A similar pattern is found for example when applying the algorithm developed by Schluessel and Emery [1]. In this study we investigated the effects which are responsible for these estimation errors and we derived improved algorithms.

DATA AND RADIATIVE TRANSFER MODEL

Two independent data sets are used for this study. One is taken for the development of the precipitable water algorithms. It consists of about 1300 radiosonde ascents launched over the Atlantic Ocean during April to October 1979. The other data set used for verification contains about 250 radiosonde ascents with coincident SSM/I measurements also taken over the Atlantic Ocean but in 1989. These 250 observations have been selected from a larger data set of more than 500 soundings. They had to pass the following two conditions to minimize the effects of the different measurement systems with their different spatial resolution :

- the time difference between radiosonde launch and SSM/I overpass has to be less than ± 2 hours, in order to minimize errors due to changes in synoptic conditions,
- and the standard deviation of the 22 and 37 GHz vertically polarized brightness temperatures of the 9 nearest pixels at the location of the radiosonde ascent must be less than 1.5 K, to ensure horizontal homogeneity of precipitable water and cloud liquid water.

The above defined thresholds of ± 2 hours and 1.5 K standard deviation have been choosen to minimize the scatter but to retain a sufficient sample size.

The radiosonde profiles have been used as input for the radiative transfer model, which has been developed by Simmer [3]. Cloud water is introduced as modified adiabatic liquid water content calculated from the radiosonde profiles (see in Hargens [4]). The ocean surface roughness and foam coverage both depend on the wind speed. The surface roughness is parameterized with a modification of the mirror reflection by Wisler and Hollinger [5]. For low wind speeds (< 10m/s) the effect of foam is parameterized by the formulation of Stogryn [6], while for higher wind speeds the parameterization of Wilheit [7] is used. The application of two different foam parameterizations slightly reduces the differences between satellite and radiosonde estimated precipitable water.

ADJUSTMENT OF BRIGHTNESS TEMPERATURES

Since the precipitable water algorithm (1) was developed with simulated brightness temperatures calculated with our radiative transfer model, the model results had to be verified. This test was performed with the second data set excluding all cloudy cases. For the latter task the cloud liquid water was estimated from the measured brightness temperatures with an algorithm given in [4]. An integrated liquid water content of 0.03 kg/m^2 was assumed as a threshold to distinguish between cloud free and cloudy situations. This threshold was determined from a comparision of METEOSAT VIS/IR images and coincident microwave observations of SSM/I.

The simulated vs. the observed brightness temperatures of the five lower SSM/I channels for cloud free observations are shown in Figure 2. From these brightness temperatures

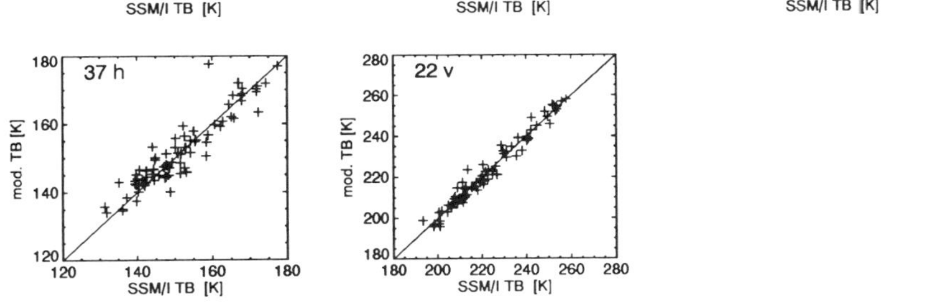

Figure 2. Simulated brightness temperatures (TB) calculated from radiosonde data vs. observed TB for the five lower SSM/I frequencies for cloud free situations.

Table 1. Adjustment coefficients for measured brightness temperatures (TB_{sat}) of the vertically polarized channels of the SSM/I ($TB_{ad} = b_0 + b_1 TB_{sat}$).

channel	b_0	b_1
19 GHz vertically polarized	-3.2	1.0
22 GHz vertically polarized	-6.73	1.0267
37 GHz vertically polarized	-16.04	0.9292

an adjustment was derived by linear regression for these channels, which are used in our algorithms. The coefficients are given in Table 1. For the 19 GHz channel a constant offset was found, while for the other two the adjustment depends on the observed brightness temperature itself. We adjust the observed brightness temperatures to the simulated, because long–term calibration errors in observed brightness temperatures can, if necessary, be corrected easily by recomputing the adjustment coefficients. If the simulated brightness temperatures would have been corrected, algorithms would have to be changed.

RESULTS

Since model effects are eliminated by the method just described, the systematic deviations revealed in the estimated precipitable water must have other causes. Those parameters are thought to be responsible for the systematic deviations, which affect the microwave radiation but are not explicitly considered in the algorithm. The following parameters or phenomena are investigated : (*i*) air temperature, (*ii*) clouds, and (*iii*) ocean roughness and foam. In this paper we focus on the first two, the latter is discussed in detail by Schrader (see this volume).

Temperature Effect

In order to investigate the temperature effect, we used the surface air temperature (T_a) as an indicator for the temperature of the whole tropospheric column. Figure 3 shows the simulated brightness temperatures for 22 GHz vertically polarized vs. the precipitable water in the range from 20 to 40 kg/m^2. Only cases without modelled clouds are given. Different air temperature intervals are marked with different symbols. It is obvious that for a constant precipitable water the brightness temperatures increase with increasing T_a. The differences in brightness temperatures can be larger than 10 K for the total possible range of air temperature. Such a large variability which produces errors in the retrieval of the total precipitable water can be reduced if surface air temperature is included in the algorithm.

The temperature effect is due to the fact that the microwave emission from the atmosphere is governed by both water vapour amount and tropospheric temperature. An increase of one or both leads to an increase in brightness temperarures. In general the water vapour amount and the tropospheric temperature are highly correlated, but the dependency is not constant and can vary from one climate zone to another, i.e. the vertical distribution of moisture depends in a wide sense on the temperature.

To correct for these errors we derived three algorithms for three temperature ranges : $T_a < 10$, $10 \leq T_a < 20$, $T_a \geq 20$ ^{0}C. The general structure of each of the algorithms is the same as for our global algorithm (1). For the temperature range below 10 ^{0}C we

Table 2. Coefficients of precipitable water algorithms (1) for each surface air temperature interval (T_A).

T_a (^{0}C)	a_0	a_{19v}	a_{22v}	a_{37v}
$T_a < 10$	84.6077	22.9640	-51.3013	10.6687
$10 \leq T_a < 20$	142.533	—	-45.0989	15.4418
$T_a \geq 20$	133.071	—	-38.1872	10.9215

could, however, improve the retrieval results by adding the brightness temperatures of the 19 GHz vertically polarized channel in the same form as those of the other channels. The coefficients for the algorithms are listed in Table 2.

The differences between the satellite estimated and radiosonde observed precipitable water are shown in Figure 4. The estimation errors, bias and rms–difference about the bias, are smaller than those of the global algorithm (Table 3). This is true for each T_a interval, but the error reduction is even more obvious if the global data set is analyzed and the three algorithms are applied according to T_a.

Cloud Effect

The presence of clouds leads to an increase in optical thickness of the atmospheric column. This has two effects on the microwave radiation : (*i*) increase of emission due to cloud droplets, and (*ii*) attenuation of the emission of the atmosphere below the cloud. For non–raining clouds the former effect dominates, thus the brightness temperatures increase. This increase depends on frequency. This fact allows the estimation of the cloud liquid water content. However, there is an ambiguity, the increase may result from either a larger amount of precipitable water or from liquid water. We have tested the ability of

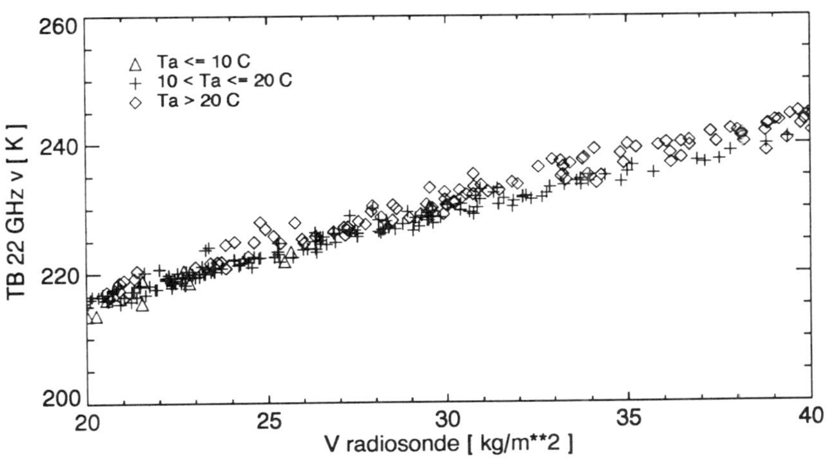

Figure 3. Simulated brightness temperatures at 22 GHz vertically polarized vs. total precipitable water. Surface air temperature intervals (T_A) are indicated by different symbols.

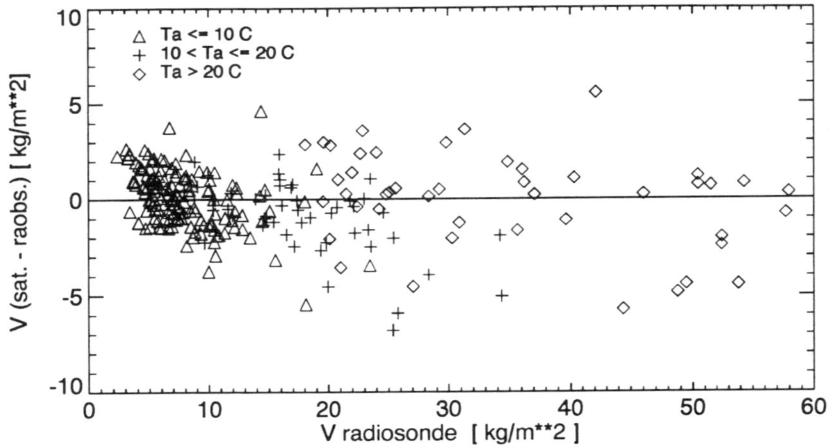

Figure 4. Total precipitable water differences (satellite minus radiosonde) vs. radiosonde data. Surface air temperature intervals (T_A) are indicated by different symbols.

our algorithm to resolve this ambiguity. We used the radiosonde data set for two runs of the radiative transfer model. First the brightness temperatures (TB_{cl}) are computed with the modelled cloud liquid water as input, in the second run cloud liquid water is set to zero (TB_0). All other parameters remain unchanged. The differences in the brightness temperatures are due to cloud water effects only.

The two different sets of brightness temperatures (TB_0. TB_{cl}) are applied to the precipitable water algorithms. Thus in the estimated precipitable water the cloud effects are also existent. We define the differences between the two estimated water vapour sets as the cloud effect, and show those in Figure 5 vs. the cloud liquid water content for the air temperature range from 10 to 20 ^{0}C. For the other two temperature intervals the results are very similar. It is obvious that with increasing cloud water the cloud effect increases. The cloud effect on the precipitable water can be larger than 1 kg/m^2 for this air tem-

Table 3. Bias and rms–difference of precipitable water (satellite minus radiosonde) in the three surface air temperature (T_a) intervals retrieved by the global algorithm (V global) and by the specific T_a algorithms (V T_a), respectively. The values in the last two columns (all T_a) and second row are calculated by applying the corresponding T_a –algorithms and then computing the statistics of the global data set.

	$T_a \leq 10$		$10 < T_a \leq 20$		$T_a > 20$		all T_a	
	bias	rms	bias	rms	bias	rms	bias	rms
V global	1.7	1.7	-1.2	2.3	-1.7	2.7	0.6	2.6
V T_a	0.1	1.5	-1.0	1.9	0.0	2.5	-0.1	1.8

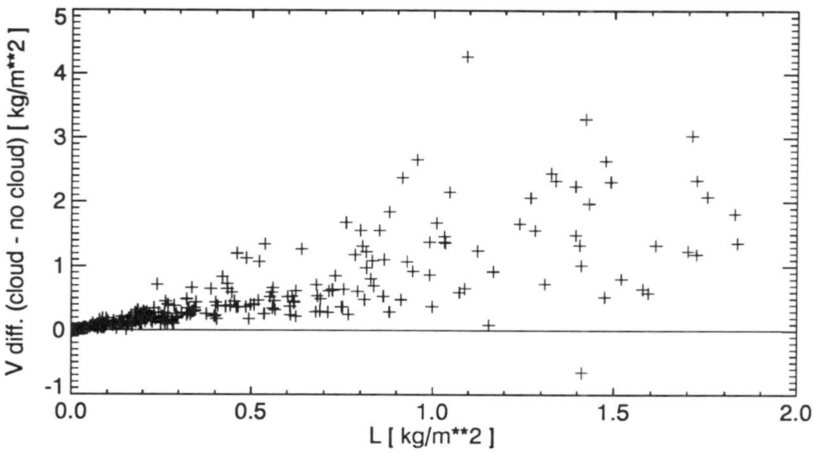

Figure 5. Cloud effect on total precipitable water retrieval from simulated brightness temperatures vs. cloud liquid water (L) for the surface air temperature interval from 10 to 20 ^{0}C.

perature interval. When the cloud water excceeds 0.5 kg/m^2 the cloud effect is more variable due to scattering effects of rain drops. The cloud effect depends strongly on the precipitable water algorithm. For our global algorithm (1) we found a cloud effect about two times larger than this one shown in Figure 5 based on the surface air temperature depending algorithms.

In order to eliminate the cloud effect we splitted the radiosonde data used for the algorithm development into two subsets, one for which the parameterization gives cloud water, the other which is cloud free. From each set a water vapour algorithm is derived for cloudy and cloud free situations, respectively. For the latter one we found an significant improvement with the new developed algorithm, the bias is reduced from -0.9 to -0.2 kg/m^2. For the cloudy situations, however, a significant improvement could not be achieved.

Therefore we took the cloud effect into account explicitly. The actual precipitable water is given as the sum of the estimated water vapour (WV) and an error term ε. With the modelled brightness temperatures of our two subsets we derived the following two equations for cloud free and cloudy situations, which are applied to TB_0 and TB_{cl}. respectively:

$$V = WV(TB_0) + \varepsilon_0 \quad (2)$$
$$V = WV(TB_{cl}) + \varepsilon_{all} = WV(TB_{cl}) + \varepsilon_0 + \varepsilon_{cloud} \quad (3)$$

with :

V	actual precipitable water
TB_0	brightness temperature without modelled cloud water
TB_{cl}	brightness temperature with modelled cloud water
WV	estimated water vapour
ε_0	water vapour error due to all effects except clouds
ε_{cloud}	water vapour error due to cloud effects
ε_{all}	total water vapour error due to all effects

We assume that we can split the error term ε_{all} into two terms, one includes all errors except those due to clouds (ε_0) and the other for cloud errors only (ε_{cloud}). Subtracting equation (3) from (2) yields

$$\varepsilon_{cloud} = WV(TB_{cl}) - WV(TB_0) \tag{4}$$

Thus ε_{cloud} is just the above defined cloud effect. As Figure 5 shows ε_{cloud} is always positive and increases with increasing cloud liquid water. That means, the estimated precipitable water in cloudy situations $WV(TB_{cl})$ is too large. The cloud effect ε_{cloud} has to be subtracted from the estimation. Since ε_{cloud} is (almost) linearly related to the cloud liquid water, it depends also on those brightness temperatures which are used in the precipitable water algorithm. Thus we can develop a new algorithm which includes this correction. The new algorithm is derived by a regression analysis based on the same brightness temperatures as before and $V - \varepsilon_{cloud}$. It follows that for given brightness temperatures the new algorithm gives precipitable water smaller by ε_{cloud} compared to the uncorrected algorithm.

New algorithms have been derived for all three T_a intervals. As a threshold for the cloud liquid water to distinguish between cloudy and cloud free situations we chose 0.05 kg/m^2, slightly higher than for the brightness temperature adjustment, because the cloud effect is small for very low amounts of cloud liquid water. Only for the T_a range 10 to 20 ^{0}C we found a significant improvement of the estimated water vapour: the bias is reduced and the rms–difference remains unchanged. Table 4 shows the error values, bias and rms–difference, for cloudy and cloud free situations for both types of algorithms, without cloud selection and with distinguishing between cloudy and cloud free observations.

CONCLUSIONS

This study shows that several effects produce systematic errors in the total precipitable water retrieved from SSM/I observations.
The temperature effect results from the fact that both water vapour and tropospheric temperature contribute to the microwave signal measured at satellite level. Because of the high correlation between both parameters, remote sensing of water vapour is possible with good accuracy, even without direct temperature knowledge. Since the functional dependence, however, of the total precipitable water and the tropospheric temperature can be different in different climate regions, systematic errors are introduced. To solve

Table 4. Bias and rms–difference in kg/m^2 of the estimated precipitable water for the surface air temperature interval from 10 to 20 ^{0}C for cloud free and cloudy situations. Values are for the algorithm without any selection (no sel.) and for the algorithms with separating into cloudy and cloud free situations (with cloud sel.).

	cloud free		cloudy	
	bias	rms	bias	rms
no sel.	-0.9	1.8	-1.2	2.1
with cloud sel.	-0.2	1.9	-1.0	2.2

this problem we derived different algorithms for different temperature ranges, which more or less correspond to different climate regions. In order to apply these algorithms temperature information is needed a priori. Since the T_a ranges are very wide this information should be easily available (e.g. weather forecast models or climate averages). Another approach may be to include satellite observations at frequencies of the O_2 lines between 50 and 60 GHz as a measure of temperature distribution.
Although the cloud effect is caused by physical reasons (ambiguity in increasing brightness temperatures by an increase in water vapour or by cloud liquid water), its amount depends strongly on the algorithm applied, i.e. the systematic error. For algorithms which are based on regression analysis as ours the quantitative cloud effect depends on the representativeness of the basic radiosonde data set. An increase in radiosonde data may reduce the cloud effect. However a residual effect still remains. unless the general form of the algorithm will be changed by including more indepent information.

Although the errors due to the ocean surface emission and reflection are not discussed in detail in this study, we assume that a further improvement of the total water vapour estimation from microwave measurements needs the 'correct' modelling of ocean roughness and foam effects to reduce the rms–difference.

REFERENCES

1. P. Schluessel and W. J. Emery, 1990, Atmospheric water vapour over oceans from SSM/I measurements, *Int. J. Remote Sens.*, **11**, 753 – 766.

2. G. W. Petty, 1990, *On the response of the Special Sensor Microwave / Imager to the marine environment – Implications for atmospheric parameter retrievals.* PhD thesis, University of Washington.

3. C. Simmer, 1994, *Satellitenfernerkundung hydrologischer Parameter der Atmosphäre mit Mikrowellen.* Hamburg: Verlag Dr. Kovač.

4. U. Hargens, C. Simmer, and E. Ruprecht, 1994, Remote sensing of cloud liquid water, *Meteor. and Atm. Physics.* Accepted for publication.

5. M. M. Wisler and J. P. Hollinger, November 1977, Estimation of marine environmental parameters using microwave radiometric remote sensing systems, Tech. Rep. NRL Memo. Rep. 3661, Naval Research Laboratory, Washington, D.C.

6. A. Stogryn, 1972, The emissivity of sea foam at microwave frequencies, *J. Geophys. Res.*, **77**, 1658 – 1666.

7. T. T. Wilheit, 1979, A model of the microwave emissivity of the ocean's surface as a function of wind speed, *IEEE Trans. Geosci. Electr.*, **GE – 17**, 244 – 249.

Spectral and Temperature Dependencies of the Millimeter and Centimeter Wave Absorption in Clouds

B. G. KUTUZA
Institute of Radio Engineering and Electronics of the Russian Academy of Sciences, Moscow, GSP-3, 103907, Russia.

Abstract Theoretical and experimental data on the temperature influence on the microwave absorption in clouds are presented. It is shown, that the water molecule relaxation time of the Debye formula determines the temperature dependence of microwave absorption in liquid clouds. It has been found, that the millimeter absorption spectrum departs essentially from the λ^{-2} wavelength dependence, which is widely used in the literature. The influence of large drops and of the temperature profile in clouds are examined. Data on the total microwave absorption in cumulus and stratiform clouds are presented.

Keywords: Microwave absorption in clouds, liquid drops, Debye formula parameters, temperature dependence.

INTRODUCTION

Millimeter and centimeter wave absorption in clouds is observed in cumulus and in the lower and middle layer of stratiform clouds. These clouds usually have liquid drops or mixed phase structure. The coefficient of cloud absorption is defined by dielectric permitivity of water as well as size, shape and temperature of drops, humidity, phase structure and other physical parameters. More detailed information about physical parameters of clouds is collected in monographs [1, 2, 3]. Liquid water dielectric properties in comparison with those of ice have exteme small values of imaginary part of complex dielectric penetration. Hence, the absorption coefficient of microwaves in cristal clouds is about two orders less, than in supercooled liquid drops, at the same temperature and water content. The water content of cristal clouds is smaller, than in liquid drops and is lower then 0.7 g/m^3 [4]. In practice often the absorption in cristal phase clouds can be negligible. The absorption in liquid drop clouds will be taken into account in this paper. It is assumed in modern literature, see [8] for example, that microwave absorption is inversely proportional to the wavelength square. This is correct at positive temperatures for frequencies that are less then 10 GHz. Author's theoretical and experimental investigations show strong temperature influence upon the microwave absorption spectrum [15, 17, 18, 19]. It is also shown, that the millimeter absorption spectrum in clouds essentially differs from a λ^{-2} dependence. Here λ is the wavelength (cm). Many results of author's investigations in this field were published in Russian scientific journals, which are poorly known among English language readers.

MICROWAVE ABSORPTION IN ISOTERMAL CLOUDS

It is known that the coefficient of cloud absorption for Rayleigh scattering

$$\frac{2\pi a}{\lambda} \cdot |m| \tag{1}$$

where a is radius of drops, m is refractivity factor, can be expressed in the following form [5]:

$$\gamma_c = \frac{0.6\pi}{\lambda} \cdot \mathrm{Im}(\frac{\epsilon_c - 1}{\epsilon_c + 2}) \cdot w \quad (\mathrm{Np/km}), \tag{2}$$

where ϵ_c is the complex permitivity of water and w is the liquid water content in clouds $(\mathrm{g/m^3})$.

The frequency dependence of the dielectric constant of liquid water is defined by the Debye formula [7]:

$$\epsilon_c = \epsilon_o + \frac{\epsilon_s - \epsilon_o}{1 + (\frac{\lambda_s}{\lambda})^2} - j\frac{\lambda_s}{\lambda}\frac{\epsilon_s - \epsilon_o}{1 + (\frac{\lambda_s}{\lambda})^2}, \tag{3}$$

where $\lambda_s = 2\pi c \tau_r \frac{\epsilon_s + 2}{\epsilon_o + 2}$; τ_r is the relaxation time of water; ϵ_s is static dielectric constant of water at frequencies $\tau_r \ll \frac{1}{\nu}$; ϵ_o is high-frequency (or optical) dielectrical constant of water at frequencies $\tau_r \gg \frac{1}{\nu}$.

Water dielectric properties depends upon temperature. The temperature changes from -50^oC to $+20^oC$ in clouds. Although liquid drop condition predominates in clouds for the temperature range from -17^oC to 0^oC, it is possible to have liquid drops up to -40^oC [9]. The Debye parameters $\epsilon_s, \epsilon_o, \tau_r$ of pure water are well known for temperatures higher than 0°C [8, 10, 11, 12, 13]. Detailed review of theoretical and experimental data is contained in monographs [6, 8].

The coefficient $\mathrm{Im}(\frac{\epsilon_c - 1}{\epsilon_c + 2})$ defines the temperature influence on the cloud absorption. It can be expressed as a function of the dimensionless parameter y in the following form:

$$K_\epsilon = \mathrm{Im}(\frac{\epsilon_c - 1}{\epsilon_c + 2}) = \frac{3(\epsilon_s - \epsilon_o)y}{(\epsilon_s + 2)^2 + (\epsilon_o + 2)^2 y^2}, \tag{4}$$

where y is $\frac{\lambda_s}{\lambda}$. It can be shown, that the water molecule relaxation time τ_r exerts the most influence on the cloudy absorption spectrum [14, 15]. This parameter has a strong dependence on temperature. Data extrapolation of τ_r, obtained by different authors above 0^oC, to negative temperatures shows considerable dispersion of τ_r, especially for temperatures $T < -10^oC$.

The method of determination of τ_r by means of simultaneous measurements of cloud absorption at several wavelength was proposed in [14, 15]. In this way τ_r can be determinated at negative temperatures, in contrast to laboratory experiments, where it is extremely difficult to obtain supercooled water, especially for $T < -5^oC$.

The ratio of absorption coefficients at two wavelengths (or total absorption in isotermal clouds) is written as:

$$N = \frac{\gamma_c(\lambda_1)}{\gamma_c(\lambda_2)} = \frac{{\lambda_2}^2 + 4\pi^2 c^2 \tau_r^2}{{\lambda_1}^2 + 4\pi^2 c^2 \tau_r^2}. \tag{5}$$

Hence, the relation N for two fixed wavelengths is a function of the Debye parameter only. Values of τ_r could be determined by measurements of one wavelength absorption according to (2) and (3), but this method requires measurements of cloud water content

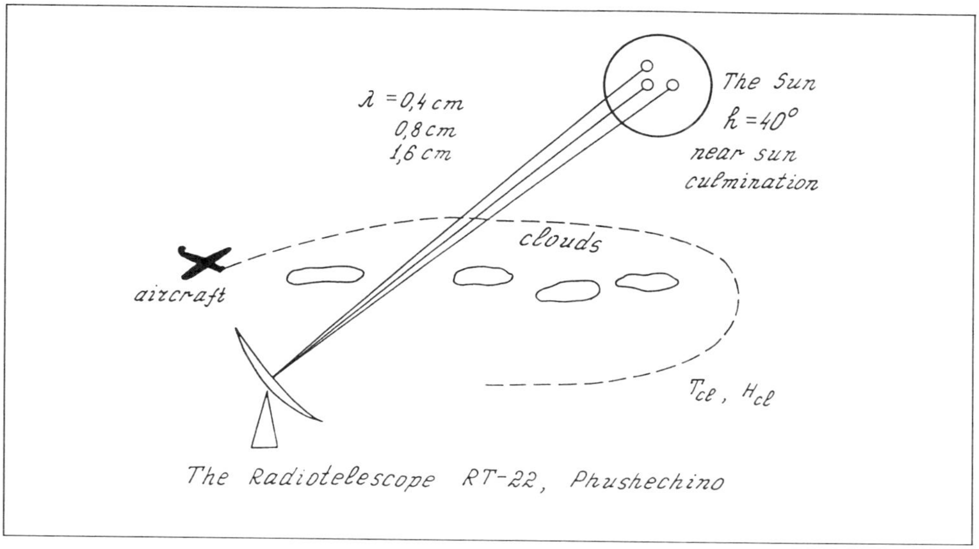

Figure 1. Scheme of experiment

in the direction of wave propagation. Instead, values of τ_r can be determined from the relation:

$$\tau_r = \frac{1}{2\pi c}\sqrt{\frac{\lambda_2{}^2 - N\lambda_1{}^2}{N-1}}. \tag{6}$$

Simultaneous observations of the Sun emisssion at the wavelengths of 0.4, 0.8 and 1.6 cm during cloudy weather conditions have been carried out at the 22-meter radiotelescope PhIAN (Pushino town, Moscow region). The data on τ_r have been obtained for the temperature range of -13^oC to 5^oC.

The scheme of experiment is presented in Figure 1. The measurements have been carried out in the beginning of September. Cumulus (Cu hum, Cu med) and stratuscumulus (Sc) clouds were observed in this time. The thickness of clouds was less than 1.5 km. Parameters of clouds were measured by means of a small aircraft (AN-2) in an area of several km from the RT-22 radiotelescope. Aircraft and microwave measurements were performed at the same time. The view angles at the wavelengths of 0.4, 0.8 and 1.6 cm were 1.9', 2.0', 3.5'. The beam axes at the 0.4 cm and 1.6 cm wavelengths were declined from the beam axis at 0.8 cm by 4'. The Sun radiation was observed during ±1.5 hours from the Sun culmination. The angle of Sun declination for this lattitude and season was about 40^0.

As results of this experiment data were obtained of relaxation time in sec, which agree with Eq. (7):

$$\tau_r = |\exp 9.8(\frac{273}{T_{cl} + 273} - 0.955)| \cdot 10^{-12}, \tag{7}$$

where T_{cl} is in ^oC.

Figure 2 shows the dependence of the K_ϵ coefficient on y. There are two curves for two values of water static dielectric constant ϵ_s, corresponding to $+20^oC$ and -10^oC . Curves 1 and 2 lie close to each other, which shows that the temperature change of ϵ_s is small.

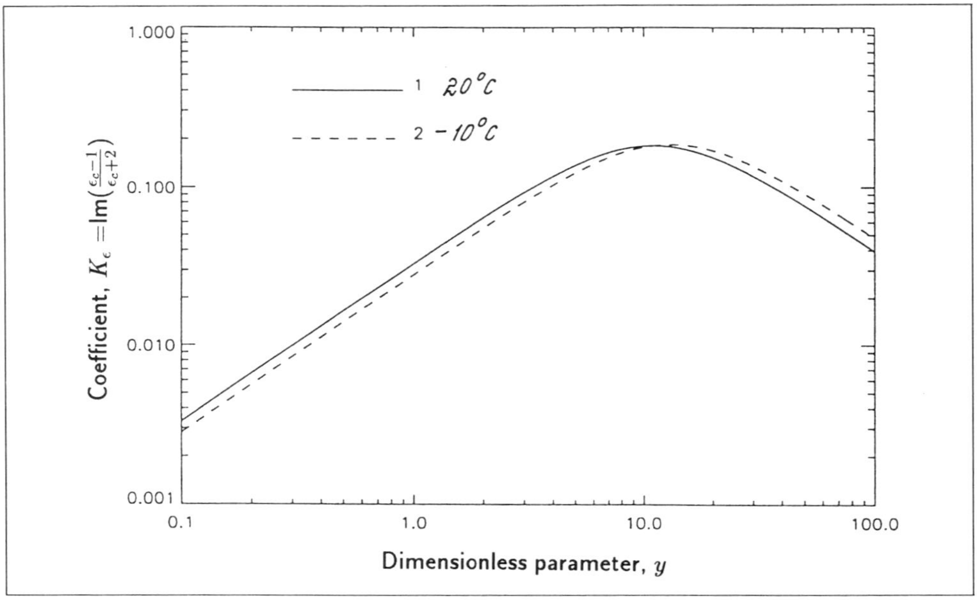

Figure 2. Dependence of the coefficient K_ϵ on the dimensionless parameter y

There are three different regions of the K_ϵ dependence on y. While values of y are less than 5, there is a direct proportional dependence between K_ϵ and y. Here the cloud absorption coefficient will be inversely proportional to the square of wavelength. This spectral dependence of cloud absorption manifests itself at longer wavelength ($\lambda > 2$ cm) and for positive temperatures (above 0^0). There is a comparitively strong dependence of the absorption coefficient upon temperature in this region .

In the next transition region, at interval of y values between 5 and 20, the curve slope also changes (see Figure 2). here, the cloud absorption spectrum deviates from the λ^{-2} dependence, and the temperature influence on the absorption coefficient change becomes lower. The maximum of K_ϵ occurs when $y = 11$ (curve 1) and when $y = 12.6$ (curve 2). Near this multiple maximum, the cloud absorption coefficient is inversely proportional to the wavelength and practically independent of temperature. The region of multiple maximum corresponds to the range of millimeter waves, after which the influence of temperature increases, but the sign of the gradient changes.

The third region ($y > 20$) corresponds to low temperatures and short wavelengths. Further increase of y leads to a dicrease of K_ϵ. In the limit this coefficient becomes inversely proportional to y. Hence, absorption becomes independent of wavelength.

One can show, that the wavelength and cloud temperature at the maximal point are connected by relation:

$$\lambda_m = 2\pi c \tau_r(T_{cl}). \tag{8}$$

Data of the characteristic wavelength of the cloud absorption are presented in Table 1.

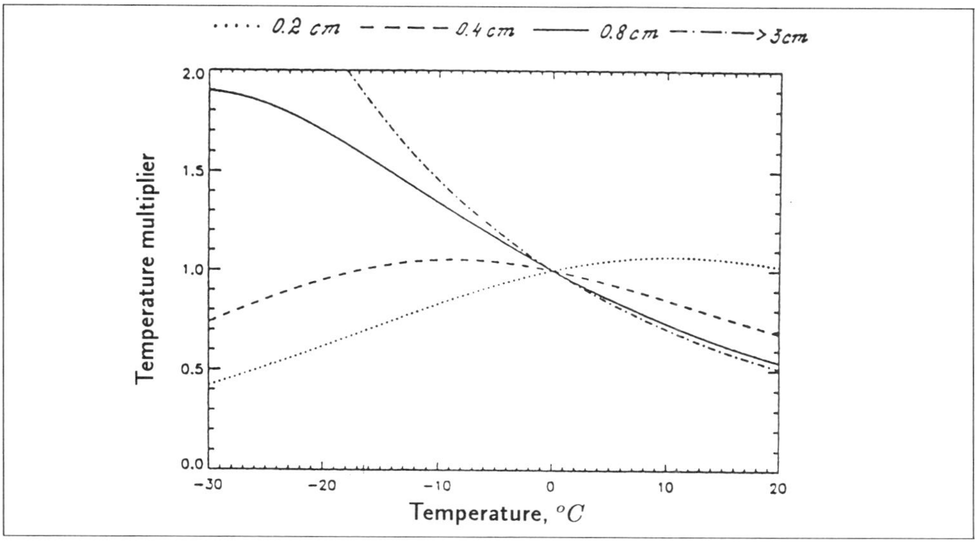

Figure 3. Relative absorption coefficient in clouds as function of temperature.

Table 1. Characteristic wavelengths of cloud absorption

$T_{cl}, ^oC$	-10	0	10	15	20	30
λ_m, mm	2.1	2.9	4.3	5.2	6.4	9.8

If the absorption coefficient of an isotermal cloud is known at wavelenth λ_m the absorption coefficient at another wavelength can be obtained by relation:

$$\gamma_c(\lambda) = \gamma_c(\lambda_m)\frac{2\lambda_m{}^2}{\lambda^2 + \lambda_m{}^2}. \tag{9}$$

Characteristics of microwave cloud absorption are shown in Figures 3 - 5.

Figure 3 presents the temperature dependence of cloud absorption, normalized for each one of the four wavelengths to the temperature $T_{cl} = O^oC$. When the temperature increases from -30^oC to 20^oC the coefficients of absorption at wavelengths of 0.8 cm and 3 cm fall monontonically, and , the temperature dependence at the wavelength of 3 cm is greater. For shorter wavelengths, the temperature dependence of absorption in liquid drop cloud is the same as for 3 cm wavelength. Millimeter wave absorption coefficient curves have maximums, which shift to the side of higher temperatures with decreasing wavelengths.

Temperature dependencies of millimeter wavelengths are given in Figure 4. Curves are plotted for 0.1, 0.2, 0.4 and 0.8 cm wavelengths. The curves diverge with increasing temperature. The temperature difference for $T_{cl} = -30^oC$ at wavelengths of 0.1 and 0.8 cm is 1.7 , and increases up to 13 times for $T_{cl} = 10^oC$. In this temperature interval, the value of γ_c changes more than 1.6 times at 0.4 cm wavelength and 6.4 times at 3 cm wavelength.

Figure 5 represents the cloud absorption spectrum at 20, 0 and -30 ^oC. To make comparison easier, we plot curves in logarithmic scale. It is also seen, that temperature makes a strong influence on the character of the absorption spectral dependence.

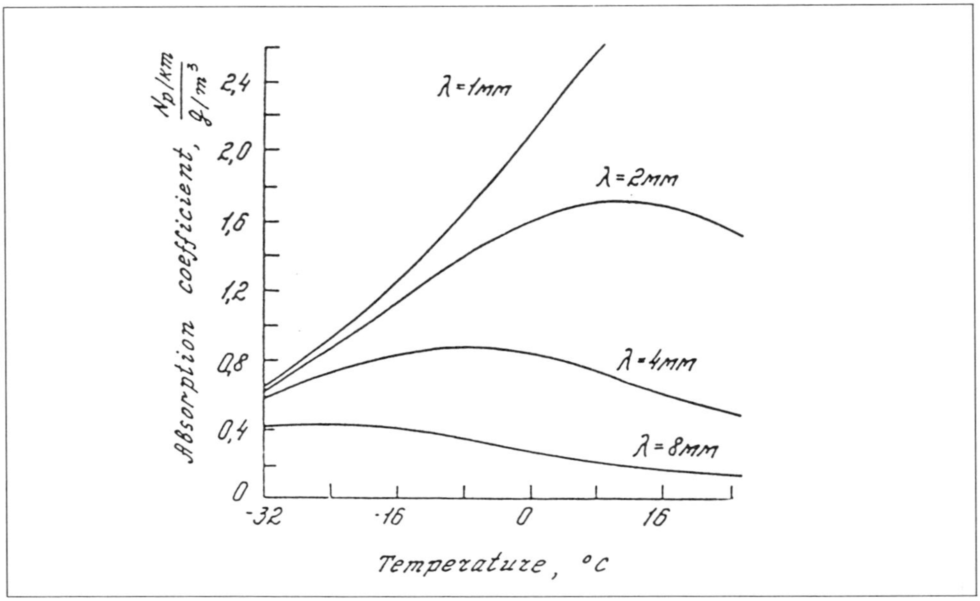

Figure 4. Temperature dependence of absorption coefficient at millimeter wavelengths.

INFLUENCE OF LARGE DROPS AND TEMPERATURE PROFILE

A real non-precipitating cloud can contain large drops with radius $a > 30$ microns, which do not satisfy the conditions of Rayleigh scattering. Inside of powerful clouds one can detect large temperature gradients which reaches $20-30^oC$ at bottom and top boundaries.

Calculation for absorption at larger drops requires the more precise formulas of Mie. This way we obtain more valuable drop absorption especially in a range of short millimeter waves in comparison with Rayleigh one. Aerological observations give average values of drops for layer formed clouds about 3.9-7.1 microns, and 3.2-4.5 microns for modal ones. The most large drops are contained in layer rain clouds (Ns) and smallest ones in layer (Sc). Calculations were run for stratiform (Ns) and cumuli (Cu cong) cloudiness at wavelength of 2 -10 mm with 0.25 microns step of integration. Results of calculation differences in percent between Mi and Rayleigh formulas are shown in Table 2.

Table 2. Differences in percent of absorption coefficients calculated by means of Rayleigh and Mie formulas.

$T_{cl}, ^oC$	$\lambda = 2$	$\lambda = 5$	$\lambda = 10$ mm
Cu,			
0	5.6	1.3	0.7
20	7.8	3.0	1.9
St,			
0	0.7	0.14	0.1
20	0.8	0.3	0.2

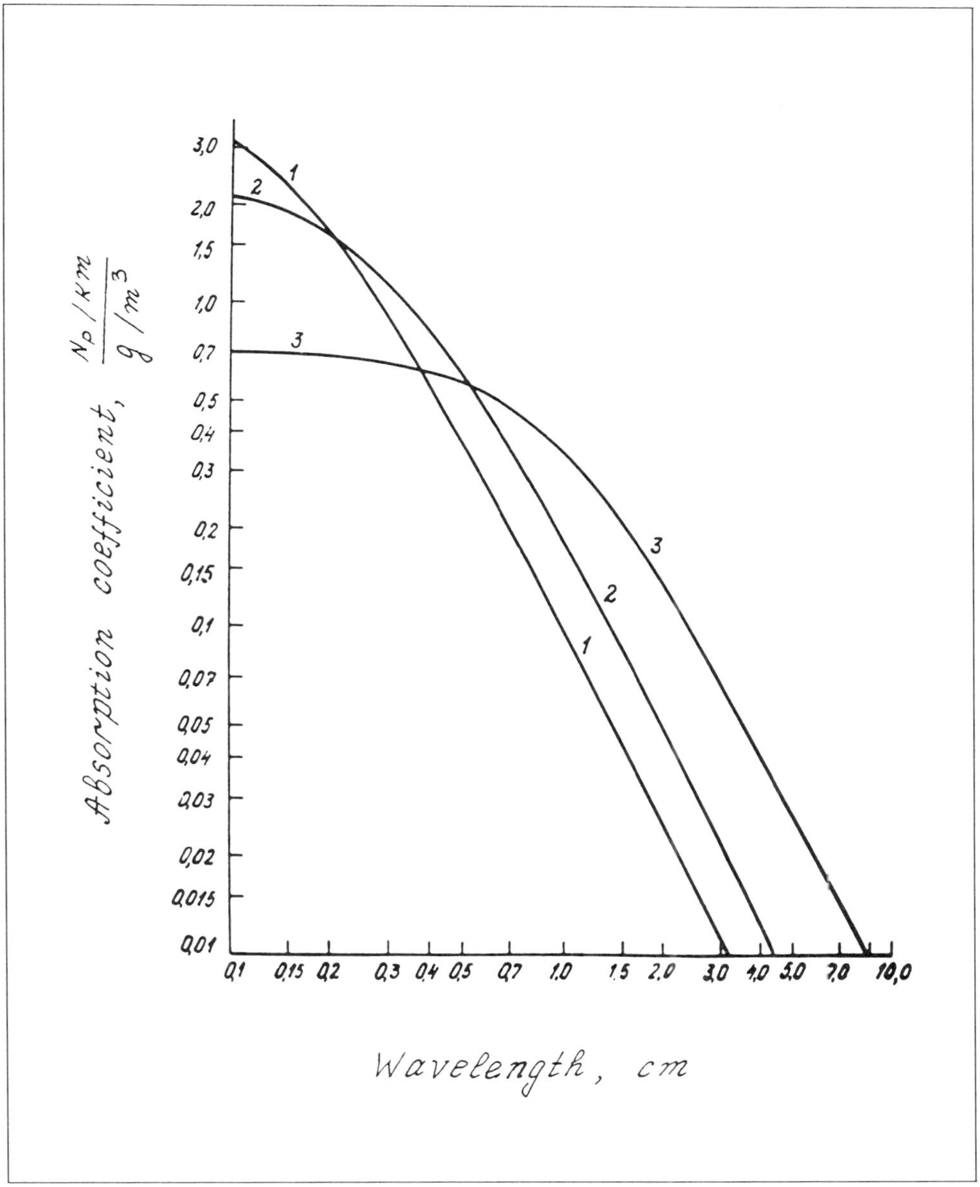

Figure 5. Microwave cloud absorption spectrums for three different temperatures. 1. 20^0 C, 2. 0^0 C and 3. -30^0 C.

The data presented in Table 2 show that at wavelengths longer than 10 mm the error cannot reach more than 2 percent. For Ns, St and cumulus clouds of fine weather the error calculations by Rayleigh scattering formula (2) do not exeed 1%.

Absorption in a non-isotermal cloud layer looks like:

$$\begin{aligned} \Gamma &= \int_{H_1}^{H_2} \gamma_c(h)dh \\ &= \int_{H_1}^{H_2} K_\epsilon[\lambda, T_{cl}(h)]w(h)dh \\ &= K_\epsilon[\lambda, T_{cl}(h_o)] \int_{H_1}^{H_2} w(h)dh, \end{aligned} \tag{10}$$

where H_1 and H_2 correspond to the lower and upper boundaries of the cloud layer.

The temperature $T(h_o)$ is an integral character. One can compare it with the temperature of a horizontal cloud layer at height h^*, which devides a cloud to top and bottom parts with equal values of liquid water content.

The deviation between $T(h_o)$ and average cloud layer temperature obviously is small enough. Even for cumulus clouds with a dephts of 4.5 km at the wavelength of 0.4 cm the maximum deviation between these temperatures reach 1.7 K. For a cloud thickness $\Delta H < 3$ km the absolute value of difference is $|T(h_o) - T(h^*)| < 1$K. Microwave cloud absorption data were obtained by two ways: Measurements of atmospheric emission and observations of Sun radiation. The second way allows to obtain considerable better accuracy in comparison with the first one at wavelengthes $\lambda < 2$ cm. But this case demands more strict limitations on the angles and time of observations.

The total RT 22 observation time of Sun emission at cloudy weather was about 150 hours. Measurements were run simultaneously at 0.4, 0.8 and 1.6 cm wavelengths in limits of 2 - 3 hours before or after Sun culmination. The data of the total microwave absorption in cumulus and stratiform clouds data are presented. This data were obtained by multichannel measurements of brightness temperature of the cloudy atmosphere at wavelengths of 0.4, 0.8, 1.6, and 3.3 cm. Table 3 represent values of average cloud total absorption in nepers, obtained by measurements of atmospheric emission.

Table 3. Average values of microwave absorption in clouds. Experimental data.

$T_{cl}, ^oC$	0-5	0-5	-10	5
clouds	Cu med	Cu cong	St, Sc	St, Sc
λ,cm				
0.4	0.4	1.1	0.13	0.2
0.8	0.14	0.4	0.05	0.06
1.6	0.04	0.12	0.015	0.019
3.3	0.009	0.026	0.0034	0.0036

The data were normalized to the equal values of temperature at the bottom cloud boundary and zenith angle. The experimental data of Table 3 confirm a strong variability of absorption in the given range of wavelengths. During the transition from 0.4 cm to 10 cm wave cloud absorption decrease by 300 - 400 times. This is inconsistent with the dependence of cloud absorption upon wavelength as λ^{-2}. The deviation of total absorption values from this kind of dependence substantially exceeds a measurement error.

REFERENCES

1. Khrgian, A.Kh., 1969, *The Physics of the Atmosphere.* Leningrad: Gidrometeoizdat.
2. Feigelson, E.M., 1984, *Radiation in a Cloudy Atmosphere.* Dodrecht/ Boston/ Lancaster: D.Reidel Publishing Company.
3. Mason, B.J., 1971, *The Physics of Clouds, second edition.* Oxford: Clarendon Press.
4. Minervin, V.E., 1965, Water reserves of clouds. *Trudy TsAO*, no.64, p.44.
5. Van de Hulst, H.C., 1957, *Light Scattering by Small Particles.* Wiley, New York.
6. V.I.Rosenberg, V.I., 1972, *Scattering and Attenuation of Electromagnetic Radiation of Atmospheric Particles.* Leningrad: Gidrometeoizdat.
7. Debye, P., 1929, *Polar Molecules.* New York: Dover.
8. Ulaby, F.T., R.K. Moore, and A.K. Fung, 1981, *Microwave Remote Sensing, Active and Passive 1.* Addison Wesley, Reading, Massachusetts.
9. Borovikov, A.M., and I.E. Demidova, *Trudy TsAO*, no.64, 66-72.
10. Saxton, J.A., 1946, The anomalous dispersion of water at very high radio frequencies. *Phys.Soc.and Roy.Meteorol.Soc.*, London, 292-306.
11. Saxton, J.A., 1952, Dielectric dispersion in pure polar liquids at very high radio frequencies, II, Relation of experimental results to theory. *Proc.Royal Soc.*, 213A, 473-492.
12. Lane, J.A., and J.A.Saxton, 1952, Dielectric dispersion in pure polar liquids at very high radio frequencies. *Proc. Royal Soc.*, 213, no.1114, 400-408.
13. Buchanan, T.J., 1952, Balance methods for the measurement of permiability of microwave region. *Proc.Inst.Elec.Eng.*, vol.99, no.57, 61-70.
14. Basharinov, A.E., and B.G.Kutuza, 1965, On the nature of the cloud layer of Venus (from radiometric observation at microwaves). *Radio Science*, vol.69D, no.12, 1580-1583.
15. A.E.Basharinov, and B.G.Kutuza, Investigation of emission and attenuation of a cloudy atmosphere at millimeter and centimeter wavelengths. *Trudy GGO*, no.222, pp.100-110, 1968.
16. Basharinov, A.E., and B.G.Kutuza, 1968, Investigation of radiation and absorption of cloudy atmosphere in microwave range. *Bull. Amer. Meteor. Soc.*, vol.49, no.5, p.2, 597-598.
17. Basharinov, A.E., and B.G.Kutuza, 1974, Determination of the temperature dependence of water molecular relaxation time in clouds and of the possibility of estimating the effective temperature of drop clouds from microwave radiometric measurements. *Izv. VUZov, Radiophys.*, vol.17, no.1, 52-57.

18. Akvilonova, A.B., and B.G.Kutuza, 1978, Microwave radiation of clouds. *Radiotekhnika i Elektronika*, v.23, no.9, pp.1792-1806, 1978. *Radio Eng. and Electron.Phys., (USA)*, no.23, 12-24.

19. Kutuza, B.G., 1986, Millimeter and centimeter wavelength absorption in clouds and its temperature dependence. *In the book: Electromagnetic waves in the atmosphere and space.* Moscow, Nauka, 180-192.

20. Voit, F.Y., and I.P. Mazin, 1972, The water content of cumulus clouds. *Izv. Acad. Nauk SSSR, FAO*, vol.8, no.11, 1166-1176.

2. RADIOMETRIC SENSING OF RAIN

OVERVIEW

- **Models of precipitation**

  Modelling the atmospheric brightness temperature at high microwave frequencies is complicated by the presence of rain, since scattering by the relatively large water and ice particles must be included beside absorption. In this case, the polarimetric characteristics of the emitted radiation may depend on the microphysics of the ice region of the precipitation. The assumption of a plane parallel layered troposphere simplifies the computations, but when the radiometric footprint approaches the dimensions of convective cells, as will be indeed the case for the highest frequency channel of the Multifrequency Imaging Microwave Radiometer (*MIMR*), a three-dimensional radiative transfer model must be used to account for the horizontal and vertical variability of precipitation.

- **Retrieval of rain**

  Through suitable precipitating cloud models, different kinds of clouds can be classified, rain intensity retrieved, and vertical profiles of cloud parameters estimated from multifrequency radiometric data. Dual-polarization measurements of brightness temperature also contain information on drop size distribution. Cooperative experiments involving the deployment of advanced research radiometers and multiparameter radars, as in the Convection and Precipitation/Electrification Experiment (*CaPE*), are starting to provide evidence of the feasibility of quantitatively characterizing the structure of precipitation cells.

Beside measurements taken in *ad hoc* experiments, data from radiometers aboard orbiting platforms are being extensively used to test and refine schemes and procedures for the retrieval of rain and liquid and ice water content in precipitating and nonprecipitating clouds. In particular, a number of algorithms have been developed for the *SSM/I* observations over both ocean and land. A considerable activity is now on the way for the systematic intercomparison of the performances of the different rain rate algorithms and for the assessment of their robustness both against changes in water vapor and cloud liquid and against variations of the non-raining fraction of footprint.

Effects of Hydrometeor Shape and Orientation Upon Passive Microwave Brightness Temperature Measurements

J. TURK[1] and J. VIVEKANANDAN[2]
[1]*Department of Electrical Engineering, Colorado State University, Fort Collins, CO 80523 USA*
[2]*National Center for Atmospheric Research, Boulder, CO 80307 USA*

Abstract. The upcoming NASA Tropical Rainfall Measuring Mission (TRMM) Microwave Imager (TMI) and the Multifrequency Imaging Microwave Radiometer (MIMR) planned for the Earth Observing System (EOS) will include an 85 GHz polarized passive microwave channel with a 5 km resolution, three times better than imagery currently available from the Special Sensor Microwave Imager (SSM/I). With this improvement, the resolution begins to approach the dimensions at which rainfall is generated in convective rainbands. At this frequency, the emission from the rain is strongly scattered by the ice aloft, and therefore it is possible that the emergent polarized brightness temperature T_B will contain not only information on the total ice water path, but additional information on the microphysics (size, shape, density) of the ice region. Microwave radiative transfer simulations for both ground- and space-based passive microwave applications are presented.

Keywords: polarization, radiometry, microwave radiative transfer, nonspherical particles.

INTRODUCTION

The 85 GHz passive microwave imagery from the Special Sensor Microwave Imager (SSM/I) polar orbiting satellite radiometers provides the finest spaceborne microwave resolution to date, 15 km at an altitude of near 860 km. The 85 GHz polarization capability of the SSM/I is especially useful for precipitation and ocean wind speed retrieval algorithms. Both the TMI and the MIMR will include an 85 GHz polarized channel and improve upon the resolution of the current Special Sensor Microwave Imager (SSM/I) by about a factor of three, as provided in Table 1. Currently, aircraft-based radiometric measurements from the NASA ER-2 aircraft enable fine-scale (1-3 km) microwave T_B measurements and reveal details of the cloud microphysical structure otherwise masked by coarser scale measurements. Vivekanandan et al. (1993) analyzed imagery from the NASA Advanced Microwave Precipitation Radiometer (AMPR), which operates at 10.7, 19.35, 37.1, and 85.5 GHz, against ground-based polarimetric radar data coincidentally gathered during the overflights [1]. However, fine-scale polarimetric T_B is not measured, as the AMPR operates a single receiver per channel with the feedhorn configured such that the polarization state skews from horizontal (H) to vertical (V) across the scan swath. Over precipitating clouds, strong 85 GHz ice scattering characteristics can produce T_B depressions below 100 K (with the fine-scale AMPR), and often below 200 K (with the coarser resolution SSM/I). Given the wide variability in ice and snow hydrometeor shapes, sizes, and densities, it is possible that polarized T_B measurements may also be capable of remotely sensing information on these same ice microphysical properties [2]. In this article we present several polarized microwave radiative transfer simulations for various nonspherical hydrometeor shapes, and discuss the utility of polarization for meteorological remote sensing applications.

HYDROMETEOR SHAPES

The nonspherical particles considered in this study are illustrated in Figure 1. These include oblate spheroidally shaped plates and raindrops, where the axis ratio a/b of the plates is fixed at 0.2, and rain follows the equilibrium axis ratio relationship. Graupel assumes a conical shaped top section connected to a half-oblate spheroidal bottom section. Although this approximates typical graupel shapes, we have not modeled other ice shapes such as hail, which is better modeled with spherical shapes due to their randomly tumbling nature in convective clouds. Exponential size distributions of the form $N(D) = N_0 \exp(-\Lambda D)$ are assumed, with plates, rain, and graupel having a fixed N_0 of 4000, 8000, and 13000 $mm^{-1}\ m^{-3}$, respectively. The slope Λ is calculated as to provide a desired rainfall rate R ($\Lambda = 4.1R^{-0.21}$) or mass content M ($\Lambda^4 = \pi\rho N_0/M$), where ρ is the particle density.

A polarized plane parallel microwave radiative transfer model using doubling and adding techniques was used to solve the case of a mulilayered atmosphere containing gaseous oxygen and water vapor, nonprecipitating cloud water, and nonspherical particles as described above [3]. The Extended Boundary Condition Method (EBCM) was used to compute the azimuthally averaged phase matrix for each combination of incident and scattered direction μ and μ', respectively, and the extinction matrix and emission vector for each scattered direction μ', where μ is the cosine of the zenith angle [4].

MODEL COMPUTATIONS

The case of upwelling radiation is discussed first, where the model consists of a 4 km layer of graupel overlying a 4 km layer of rain. The surface is either a specular, polarized ocean, or in the case of land, the emissivity is fixed at 0.9. It is well established that increasing amounts of rainfall or cloud water warm the net T_B and obscure the polarized ocean emission $T_B^V - T_B^H$ [5]. Past 35 GHz, scattering processes begin to produce a net decrease in the overall emergent T_B. It is instructive to quantify the net polarization difference that is contributed due to scattering processes within the rain and ice hydrometeors and not from the surface. This information could lead to possible separation of convective and stratiform regions of mesoscale convective systems, due to the different ice microphysical habits. Spencer et al. (1989) have noted 10 K polarization differences with the 85.5 GHz SSM/I channels over the stratiform anvils of a squall line, while simultaneously detecting little or no T_B differences over the convective storm core [5]. The upwelling polarization difference may also potentially provide information on cirrus cloud properties. Evans and Stephens performed a theoretical study to determine the utility of using T_B polarization differences between 85 and 340 GHz as a means to infer the total cirrus cloud ice water path [6].

Figure 2 illustrates T_B vs. θ ($\theta = \cos^{-1}\mu$, the angle from nadir) behavior for the case of a 4 km, 5 $mm\ hr^{-1}$ layer of oblate rain underlying a 4 km layer of conical graupel over ocean. The geometry appears circular when viewed at nadir; hence the polarization difference is zero in this case. At 20 GHz, no noticeable effect occurs compared to the rain layer-only case, while scattering-induced T_B depressions increasingly enhance the polarization difference $T_B^V - T_B^H$ away from nadir. Up to 20 K T_B difference is noted at 85 GHz for view angles between 45 and 60° from nadir. Although not shown, the oblate raindrops do not produce much polarization difference; they behave very similar to the case of equivalent-volume spheres.

Instrument	Operation	Frequency	Resolution	Polarization
SSM/I (Special Sensor Microwave Imager)	95° inclination 860 km altitude 1400 km swath	19.35	60 km	V,H
		22.235	40 km	V
		37.1	25 km	V,H
		85.5	15 km	V,H
TRMM-TMI (Tropical Rainfall Measuring Mission)	30° inclination 350 km altitude 760 km swath width (220 km radar) 1997 launch	10.7	40 km	V,H
		19.35	19 km	V,H
		21	18 km	V
		37.1	10 km	V,H
		85.5	5 km	V,H
MIMR (Multifrequency Imaging Microwave Radiometer)	EOS Polar Platform Late 1990s	6.8	60 km	V,H
		10.65	38 km	V,H
		18.7	22 km	V,H
		23.8	22 km	V
		36.5	12 km	V,H
		90.0	5 km	V,H

Table 1. Current and future spaceborne radiometer system parameters.

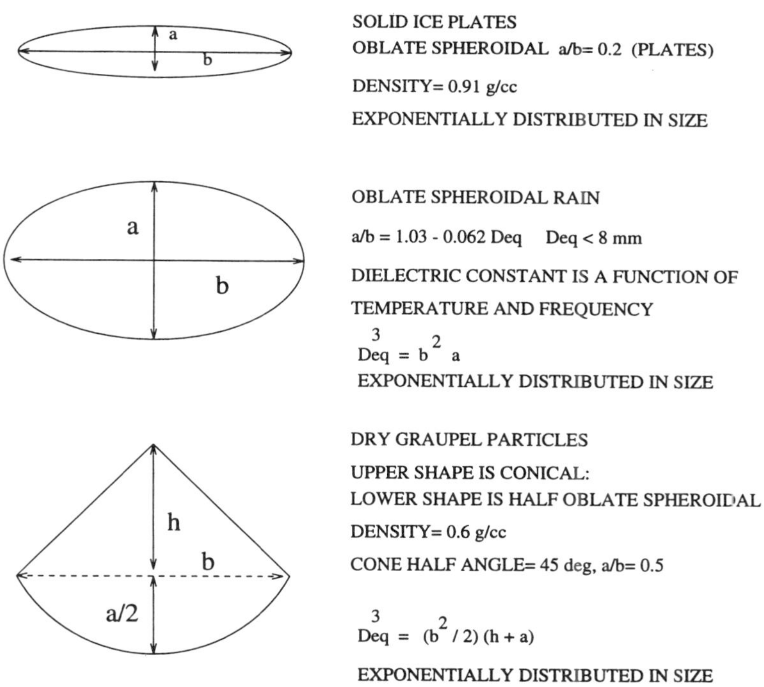

Figure 1. Shape parameters of nonspherical atmospheric hydrometeors. D_{eq} is defined as the diameter of an equivolume sphere.

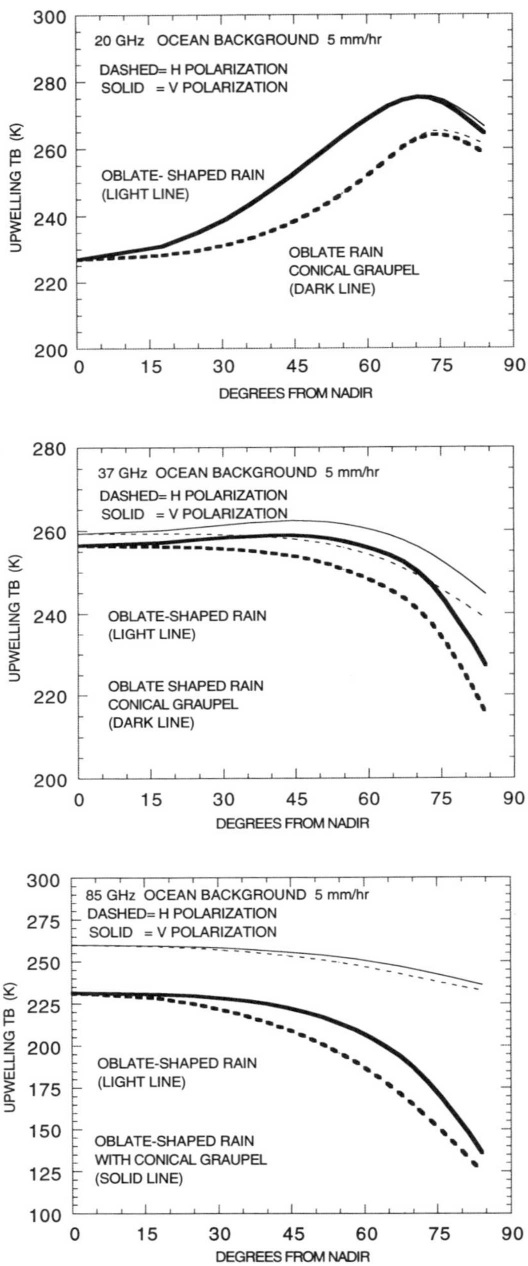

Figure 2. Upwelling T_B at 20, 37, and 85 GHz as a function of view angle away from nadir. Dark lines denotes a model consisting of a 4 km, 5 $mm\ hr^{-1}$ oblate spheroidally shaped rain layer beneath a 4 km layer of conical shaped graupel, and the light line denotes a model consisting of a 4 km oblate rain layer only.

Figures 3 and 4 consider the case of oblate rain and conical graupel over both land and ocean surfaces as a function of rainfall rate. The view angle is fixed at 48° from nadir. The surface becomes obscured at increasingly smaller rain rates as frequency increases. Similar polarization differences $T_B^V - T_B^H$ of $\approx$ 20 K occur at 37 and 85 GHz past 10 $mm\ hr^{-1}$ rain rate, due to scattering within the graupel layer. Since precipitating clouds typically have nonprecipitating cloud liquid spread out throughout the cloud, Figures 3 and 4 also depict the same model results with the addition of a cloud liquid profile of 0.2 $g\ m^{-3}$ from 3-7 km altitude. There is a general T_B warming which increases with frequency, but the net polarization difference appears unchanged.

For the case of the downwelling radiation, the radiometer is viewing against the radiometrically cold sky background. For the case of a purely absorbing atmosphere, only the profile of atmospheric gases and cloud water (and cosmic sky noise temperature) determines the net downwelling T_B. This assumption is valid up to several hundred GHz where the albedo of nonprecipitating cloud liquid water (less than 100 μm size) is very close to zero. While integrated water vapor and cloud liquid measurements using 20.6 and 31.65 GHz channels are well-developed, the addition of a 90 GHz channel improves the sensitivity for small amounts of cloud liquid, but the T_B saturates for integrated liquid amounts past 2 mm [7]. At 90 GHz, the effects of ice in the radiometer field of view may also influence the net T_B by scattering the atmospheric and surface emission into the radiometer field of view. This may occur with the presence of high altitude cirrus clouds [6]. Hazardous aircraft icing conditions often contain supercooled liquid water coexisting with snow crystals [7].

Figure 5 compares the net downwelling 85 GHz T_B with and without a 2 km layer of plates placed between 12-14 km altitude for ice mass contents of 0.05, 0.1, and 0.2 $g\ m^{-3}$. The net downwelling T_B warms with increasing ice mass content as the ice layer scatters the water vapor thermal emission downwards. Also, the direction of polarization difference $T_B^V - T_B^H$ is reversed compared to the upwelling case, since the horizontally polarized component is more strongly scattered and produces a warmer T_B. Up to -25 K is noted for the 0.1 $g\ m^{-3}$ situation for a view angles between 45 and 60° from nadir. Calculations at 20 GHz show that the same amount of ice is nearly transparent. Figure 6 compares the effect of plates against equivalent spheres. T_B^V appears to better approximate the spherical particle assumption than T_B^H. Figure 7 illustrates that the net polarization difference $T_B^V - T_B^H$ appears insensitive to the placement of ice layer within the atmosphere. However, Figure 7b considers the case of a layer of plates between 1-3 km with and without the presence of an equal amount of coexisting cloud liquid. In this case, the net albedo of the ice particles is reduced and the cloud liquid tends to mask the polarization effect while warming the overall T_B.

CONCLUSIONS

To determine the net effect of nonspherical ice particle scattering on polarized T_B measurements, several polarized microwave radiative transfer simulations were performed. Both satellite- and ground-based radiometric applications were considered. Modeling graupel particles with a conical shape produced up to 20 K of upwelling polarization difference $T_B^V - T_B^H$ for view angles near 45° from nadir. While ground-based T_B measurements near 85 GHz are quite responsive to increases in water vapor and cloud liquid, the presence of even small amounts of plate-shaped ice particles scatters the atmospheric emission back into the radiometer field of view, producing a polarized T_B warming over the ice-free sit-

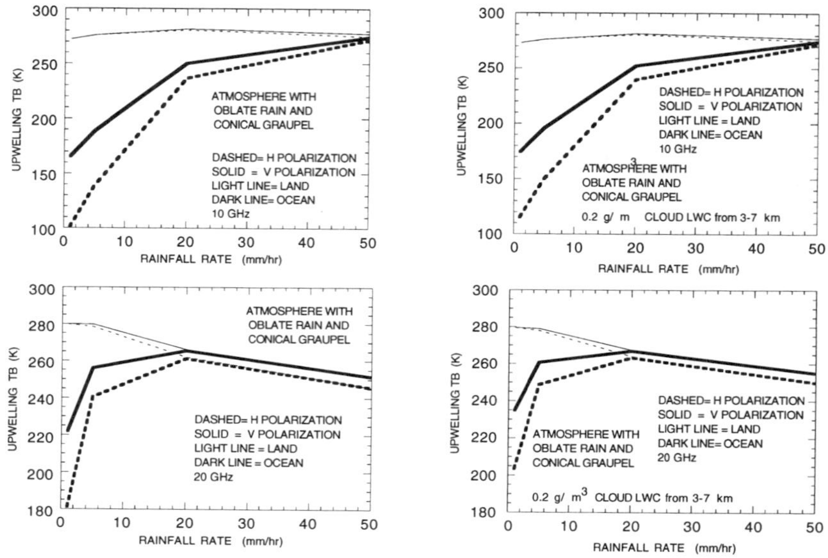

Figure 3. (Left) Upwelling T_B at 10 and 20 GHz as a function of rainfall rate for a view angle of 48° from nadir, and land and ocean backgrounds. The model consists of a 4 km layer of oblate spheroidal rain beneath a 4 km layer of conical shaped graupel. (Right) Same as left but with $0.2\ g\ m^{-3}$ of nonprecipitating cloud liquid water added between 3-7 km altitude.

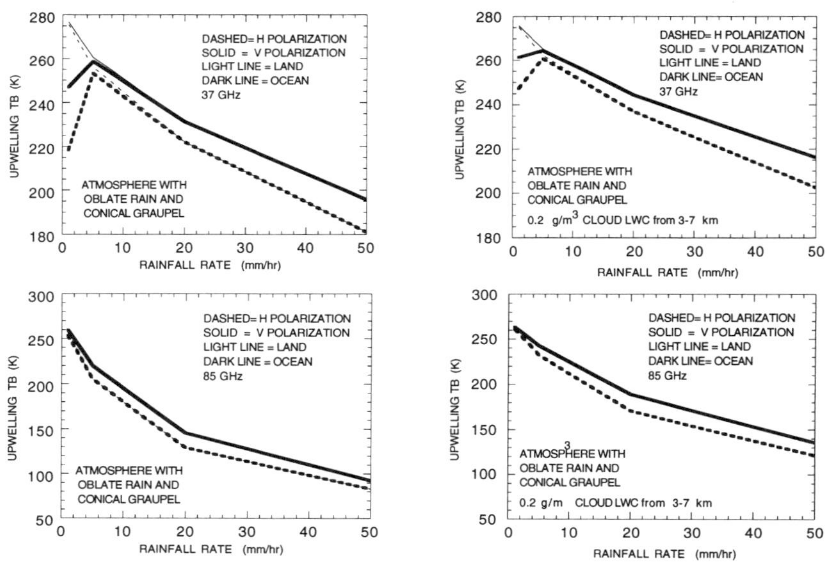

Figure 4. Same as Figure 3, except results are for 37 and 85 GHz.

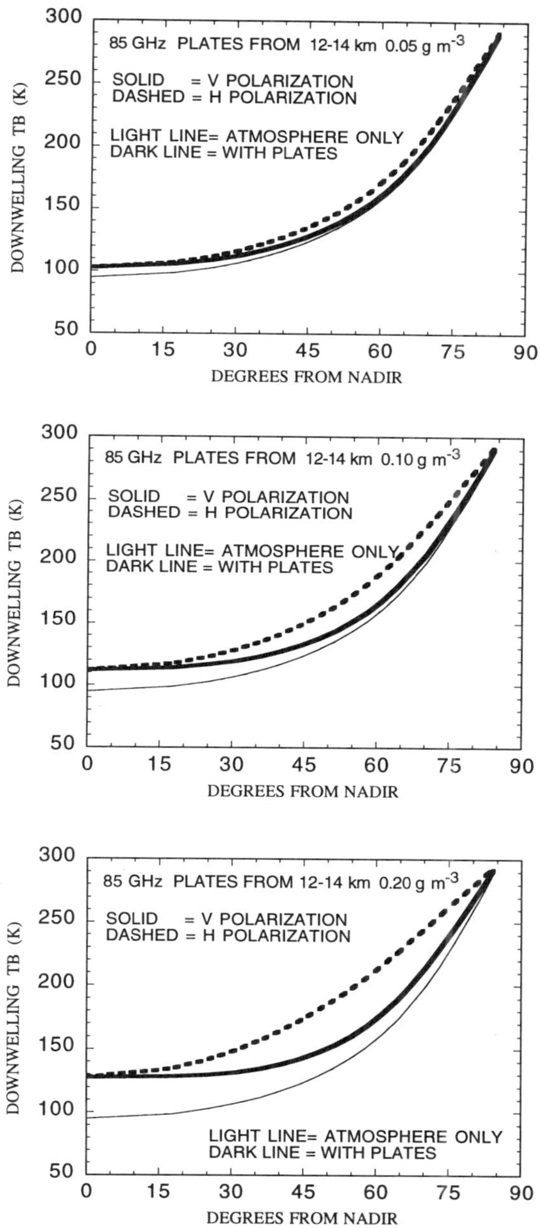

Figure 5. Downwelling 85 GHz T_B as a function of view angle away from nadir. The model consists of a 2 km layer of plate-shaped ice between 12-14 km, with ice water mass contents of 0.05, 0.1, and 0.2 $g\ m^{-3}$. Light line denotes the atmosphere only, dark line denotes the atmosphere and plates.

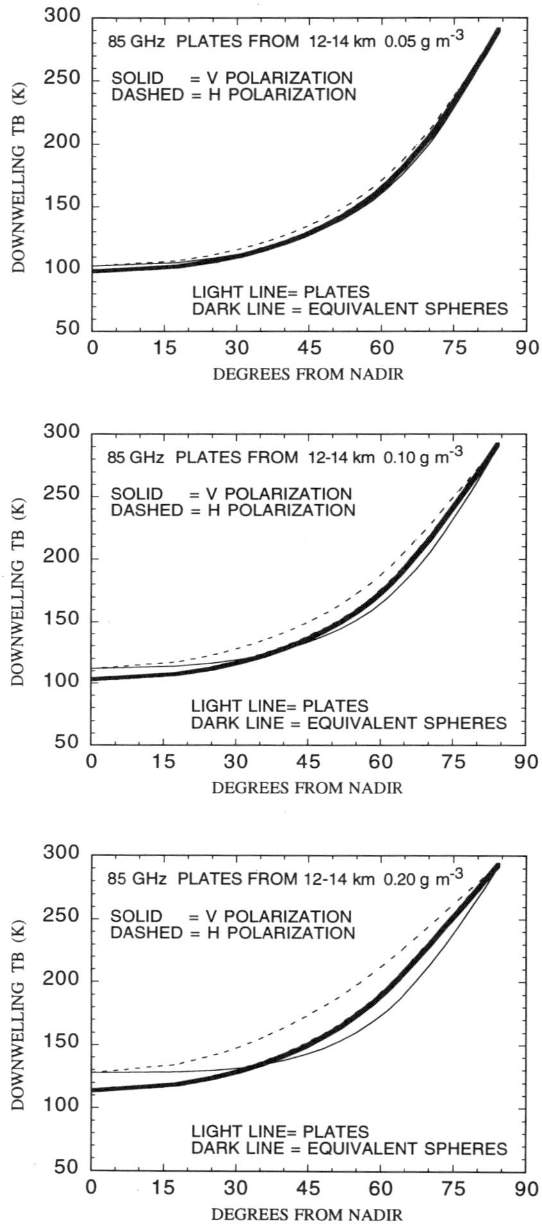

Figure 6. Downwelling 85 GHz T_B as a function of view angle away from nadir. The model consists of a 2 km layer of plate-shaped ice between 12-14 km, with ice water mass contents of 0.05, 0.1, and 0.2 $g\ m^{-3}$. Dark line denotes the case where the plates are represented by equivalent-volume spheres.

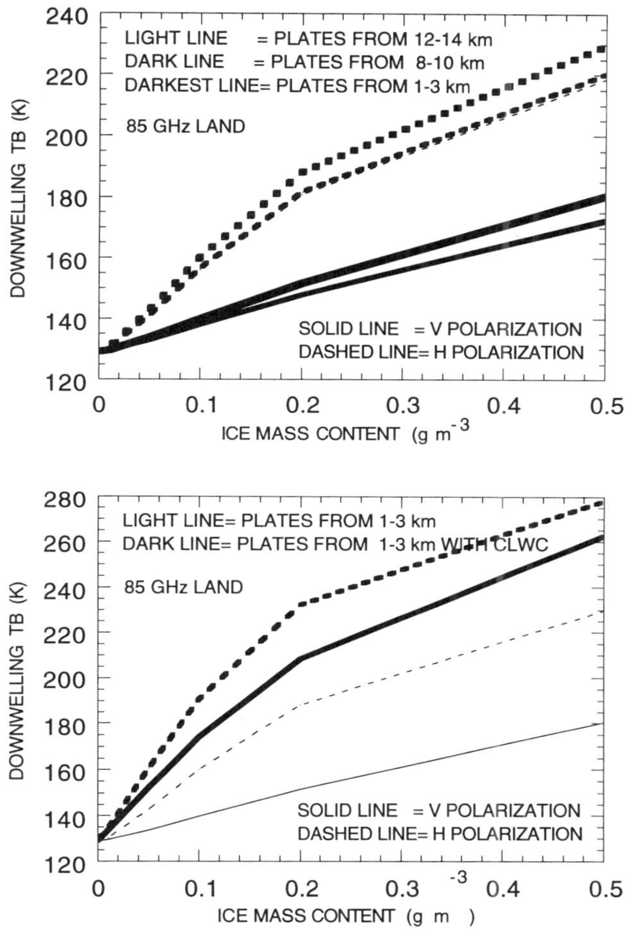

Figure 7. Downwelling 85 GHz T_B as a function of ice water mass content. Top figure compares the cases where the 2 km layer of plates is situated between 12-14, 8-10, and 1-3 km altitude. Bottom figure compares the 1-3 km altitude placement after adding an amount of cloud liquid equal to the ice mass content.

uation. Even for an ice water path as low as 0.1 $kg\ m^{-2}$, the T_B warming was near 10 K, with polarization differences $T_B^V - T_B^H$ of -10 K for a 45° radiometer elevation angle. This polarization difference further decreases with increasing ice water path. The sign of the polarization difference is reversed for the downwelling geometry, since the horizontally polarized component more strongly scatters the atmospheric emission back towards the radiometer. These results suggest the use of a dual polarized ground-based radiometer operating near 85 GHz in conjunction with lower unpolarized frequency channels near 20.6 and 31.65 GHz for detection of ice, in addition to integrated water vapor and cloud liquid. Measurements taken with the Advanced Microwave Precipitation Radiometer (AMPR) during the recent TOGA-COARE experiment in the South Pacific Ocean show considerably weaker 85 GHz ice scattering signatures as compared to earlier measurements from Florida. This demonstrates the impact of ice scattering in delineating both warm- and cold-rain precipitation processes, respectively. More exhaustive treatment must be given to modeling efforts before attempting to utilize polarized T_B measurements in a quantitative fashion.

Acknowledgements

The authors acknowledge support from NASA Earth Science and Applications Division grant NAG8-890. We thank Dr. Roy Spencer and Ms. Robbie Hood of NASA Marshall Space Flight Center for their support of AMPR measurements and modeling, and Dr. Frank Evans of the University of Colorado-Boulder, who developed the RT4 radiative transfer model.

References

[1] Vivekanandan, J., J. Turk, and V.N. Bringi, 1993: Comparisons of precipitation measurements by the Advanced Microwave Precipitation Radiometer (AMPR) and multiparameter radar. *IEEE Trans. Geoscience Rem. Sens.*, **31**, 860-870.

[2] Vivekanandan, J., J. Turk, and V.N. Bringi, 1991: Ice water path estimation and characterization using passive microwave radiometry. *J. Appl. Meteor.*, **30**, 1407-1421.

[3] Evans, K.F., and J. Vivekanandan, 1990: Multiparameter radar and microwave radiative transfer modeling of nonspherical atmospheric ice particles. *IEEE Trans. Geoscience Rem. Sens.*, **28**, 423-437.

[4] Barber, P., and C. Yeh, 1975: Scattering of electromagnetic waves by arbitrarily shaped dielectric bodies. *Appl. Optics*, **14**, 2864-2872.

[5] Spencer, R.W., H.M. Goodman, and R.E. Hood, 1989: Precipitation retrieval over land and ocean with the SSM/I: Identification and characteristics of the scattering signal. *J. Atmos. Ocean. Tech.*, **6**, 254-273.

[6] Evans, K.F., and G. Stephens, 1992: A theoretical basis for passive microwave remote sensing of cirrus. *Spec. Meeting on Microwave Radiometry and Remote Sensing Applcations*, E.R. Westwater, Editor, NOAA/ETL, Boulder, CO, 262-266.

[7] Westwater, E.R., 1993: Ground-based microwave remote sensing of meteorological variables. Chapter 4 of *Atmospheric Remote Sensing by Microwave Radiometry*, Michael A. Janssen, Editor, John Wiley, 145-213.

Influence of Cloud and Rainfall Characteristics on Brightness Temperatures of the Earth Measured by Satellite

B. G. KUTUZA[1], A. HORNBOSTEL[2] and A. SCHROTH[2]
[1]Institute of Radio Engineering and Electronics of the Russian Academy of Sciences, Moscow, GSP-3, 103907, Russia.
[2]German Aerospace Research Establishment (DLR), Institute for Radio Frequency Technology, D-82230 Oberpfaffenhofen, Germany.

Abstract The results of model calculations of the upwelling microwave radiances at the frequencies of 22.235, 37 and 75 GHz and data of temporal and spatial variations of the downwelling brightness temperature at the frequency 19.2 GHz are presented in this report. The influence of liquid water content in non-precipitating clouds and water vapor on the brightness temperature over ocean and land are considered. The microwave emission calculations were carried out for nadir viewing angle for rainfall rates from 1 mm/h to 100 mm/h. The variation intensities of the atmospheric emission were obtained for cumulus and stratiform precipitating clouds and for different temporal and spatial intervals.

Keywords: Upwelling brightness temperature, microwave radiometer, radiative transfer model, rain rate, liquid water content in clouds, temporal and spatial variations.

INTRODUCTION

Numerous methods of calculating the microwave radiance in rain have been developed over the past years based on radiative transfer equation. To examine the influence of separate parameters on the brightness temperature it is expedient to use approximate formulas. One of those approximate calculation methods is employed in this paper.

Brightness temperatures of the Earth have been observed by spaceborn microwave radiometers at wavelengths from 0.4 cm to 1.35 cm (frequency range 75.0-22.235 GHz) and beyond. To investigate the sensitivity to different atmospheric parameters, microwave radiance values were calculated using two methods. The first one is based on the solution of the radiative transfer equation in vector form for a plane-parallel homogeneous atmosphere using the method of successive orders of scattering and the Gauss quadrature technique [1, 2]. In the second method, approximate equations are used to calculate the atmospheric emission of a one-dimensional scattering medium [3]. According to [4], the accuracy of the radiance is better than 5K for frequencies below 40 GHz, when compared with the first method.

The input parameters to the model are surface reflectivity, temperature, rain rate, drop size distribution, height of the rain layer, physical temperature of rain and cloud layers, oxygen absorption, total water vapor mass in the atmosphere, and liquid water content in clouds.

Ground measurements of wind speed and the temporal variation of downwelling atmospheric emission are used to estimate the effect of spatial variations on the brightness temperatures. Such estimates are necessary for spaceborn radiometer applications, where the minimal footprint of the antenna is 5-20 km.

MICROWAVE EMISSION OF A PRECIPITATING CLOUDY ATMOSPHERE AS A ONE-DIMENSIONAL SCATTERING MEDIUM

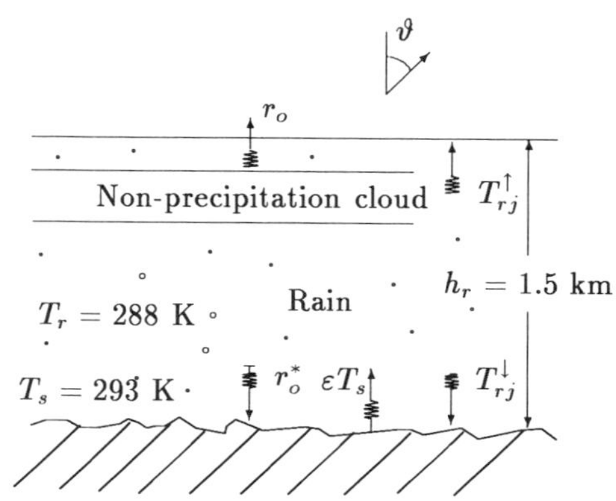

Figure 1. Model of a one-dimensional scattering atmosphere

The total atmospheric absorption without precipitation is the sum of the absorptions by oxygen, water vapor and clouds:

$$\tau_0(\nu) = \tau_{O_2}(\nu, T_o) + k_\rho(\nu, T_o)Q + k_w(\nu, T_{cl})W, \tag{1}$$

where k_ρ and k_w are the mass absorption coefficients for water vapor and clouds, with Q the total water vapor mass in the atmosphere (g/cm^2), W the integral liquid water content in clouds (kg/m^2). T_o is the atmospheric temperature near the surface, T_{cl} is the effective cloud temperature and $\tau_{O_2}(\nu, T_o)$ is the total absorption due to oxygen. The absorption properties of oxygen and water vapor are presented in [5, 6, 7].

The microwave attenuation due to rain can be approximately expressed as:

$$\tau_r(\nu, \vartheta) = k_{rj}(\nu, \vartheta, T_r, d_e)M, \tag{2}$$

where k_{rj} is the mass attenuation coefficient in the rain layer. The index $j = v, h$ denotes the polarization, ϑ is the incident angle, T_r is the effective temperature of the rain layer, d_e is the effective diameter of the rain drops, and M is the integral liquid water content.

Radiative transfer theory for a one-dimensional medium can be used to determine the transmittance q and reflection coefficient r_o of an isolated rain layer with finite thickness. This theory is presented in detail in [3]. According to it, the radiance transmitted through an one-dimensional medium is given by:

$$I_1 = qI_o = \frac{(1-r^2)e^{-k_1\tau_r}}{1-r^2e^{-2k_1\tau_r}}I_o, \tag{3}$$

where I_o is the initial radiance and r is the reflectivity of the infinitely optically thick one-dimensional medium. The reflectivity r is defined by the following equation:

$$r^2 - 2\frac{1-\xi\mu}{\xi(1-\mu)}r + 1 = 0, \tag{4}$$

where μ and $1-\mu$ are the probabilities of forward and back scattering, respectively; ξ is the ratio of the scattering and attenuation coefficients. The coefficient k_1 is determined by:

$$k_1 = \frac{\xi}{2}(1-\mu)\frac{1-r^2}{r}. \tag{5}$$

The reflectivity of the finite one-dimensional medium r_o is defined by the expression:

$$r_o = r\frac{1-e^{-2k_1\tau_r}}{1-r^2e^{-2k_1\tau_r}}. \tag{6}$$

The brightness temperature of the upwelling emission of the isolated rain layer for polarization $j = v, h$ is determined by the expression:

$$T_{rj}^{\uparrow} = (1-r_o)T_r(1-qe^{-\tau_o}), \tag{7}$$

where T_r is rain layer temperature and τ_o is the total absorption in oxygen, water vapor and small drop clouds. r_o is the rain layer reflection coefficient for observations from above. The factor $(1-r_o)$ denotes the transmitted part of emission, which is not reflected at the boundary (see Figure 1). Accordingly, for downwelling emission the following expression can be obtained:

$$\begin{aligned} T_{rj}^{\downarrow} &= (1-r_o^*)T_r(1-qe^{-\tau_o}) + T_s r_o^* \varepsilon_j \\ &+ r_o^* \varrho_j T_r(1-qe^{-\tau_o}), \end{aligned} \tag{8}$$

where r_o^* is the rain layer reflection coefficient for observations from below (see Figure 1). T_s is the surface temperature and ε_j and $\varrho_j = 1-\varepsilon_j$ are the emissivity and reflectivity of the ground surface, respectively.

Finally, the total upwelling brightness temperature, including ground emission, ground surface reflection and absorption by water vapour, oxygen and clouds can be presented in the following form ([4]):

$$\begin{aligned} T_{bj} &= T_{rj}^{\uparrow} + T_s\varepsilon_j q(1-r_o)e^{-\tau_o} \\ &+ T_{rj}^{\downarrow}\varrho_j(1-r_o)qe^{-\tau_o}. \end{aligned} \tag{9}$$

The accuracy of these formulas in comparison with digital radiative transfer equation calculation by the Monte-Carlo method [8, 4] and by the method of successive orders of scattering [2] is better than 5 K for frequencies below 40 GHz (see Figure 2 in the fourth section).

MODELS OF THE ATMOSPHERE - SURFACE SYSTEM

The theoretical model for calculation of the brightness temperature of a precipitating atmosphere proposed by T. Wilheit et al. was described in [9, 10]. This model was used to relate observed brightness temperatures to rain rates, including SSM/I data.

The magnitudes of brightness temperature T_{bj}, which are presented in this report, were obtained by using a model, which is based on Wilheit's model. It is conditionally called "Model A". In this model it is assumed that the surface temperature is 293 K, the effective rainfall temperature is 288 K, and the freezing level (rain-layer level) is 1.5 km (see Figure 1). Marshall-Palmer drop-size distribution is used for calculation of the attenuation and scattering coefficients. In contrast to Wilheit's model this model has the following free parameters: the reflectivity of the surface ϱ_j, the total water vapor mass in the atmosphere Q, the integral liquid water content in clouds W, and the effective temperature of clouds T_{cl}. A smooth surface (sea or land) is assumed, atmospheric inhomogenities are neglected. The assumed regions of variations of the model parameters are: 0-0.8 for ϱ, 0.5-5.0 g/cm^2 for Q, 0.05-5.0 kg/m^2 for W, and -18-+10°C for T_{cl}. The following average values of the parameters are assumed: $Q = 2.0$ g/cm^2, $W = 0.5$ kg/m^2, and $T_{cl} = 0^o$C.

RESULTS OF UPWELLING MICROWAVE EMISSION CALCULATIONS

The results of model calculations of the brightness temperature T_{bj} at three frequencies 22.235, 37.5 and 75 GHz (wavelengths 0.4, 0.8 and 1.35 cm) are presented in this section. The T_{bj} magnitudes are computed with formulas (3) - (9) by using the Model A, which was described in the last section. The values of attenuation and scattering coefficients of spheroidal oblate rain drops, which are required to calculate the quantities τ_r and ξ in these equations, were obtained by the Waterman T-matrix method [2].

Figures 2 and 3 show computer-simulated dependencies of brightness temperature T_{bj} on rainfall rate. The curves correspond to two observation conditions over ocean and land ($\varepsilon_j = 0.96$) at the 0° nadir angle. At this angle the brightness temperature is the same for both polarizations. With increasing rain rate the brightness temperature over sea at the frequencies of 22.235 and 37.5 GHz increases first, reaching a maximum, and then decreases. The T_{bj} values over land and over sea at 75 GHz decrease monotonically with increasing rain rate, which can be explained by the change of the rain layer albedo. The obtained curves agree basically with the data [11, 9]. Unlike the calculations in [12], the higher T_{bj} obtained in our calculation for $R > 10$ mm/h are explained by the lack of an ice layer in the Model A.

Figure 3 shows a comparision of the one-dimensional model (model A) and the more complex three-dimensional radiative transfer model, described in [2]. The attenuation and scattering coefficients are the same in both models. The maximum deviation is about 5 K.

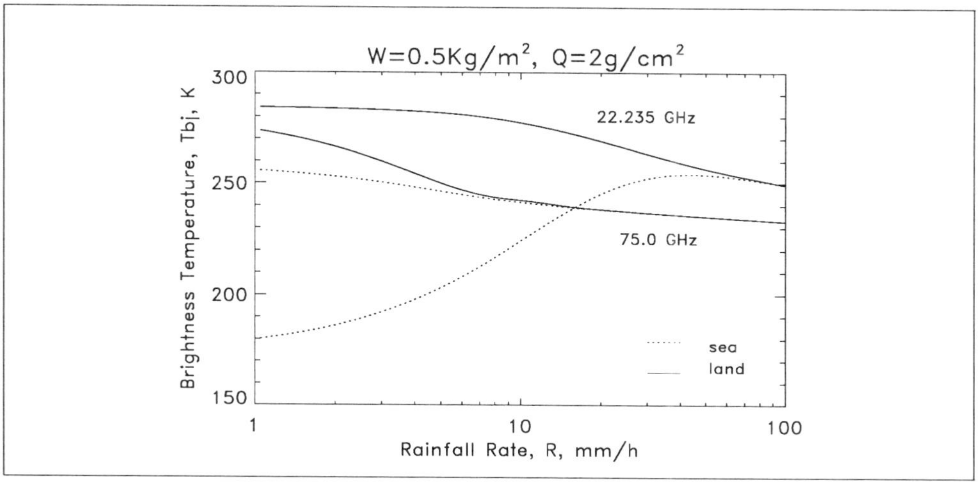

Figure 2. The upwelling brightness temperature at the wavelengths 0.4 and 1.35 cm as a function of rainfall rate over land (solid lines), and over ocean (dashed lines) at the nadir view angle $\vartheta = 0^o$.

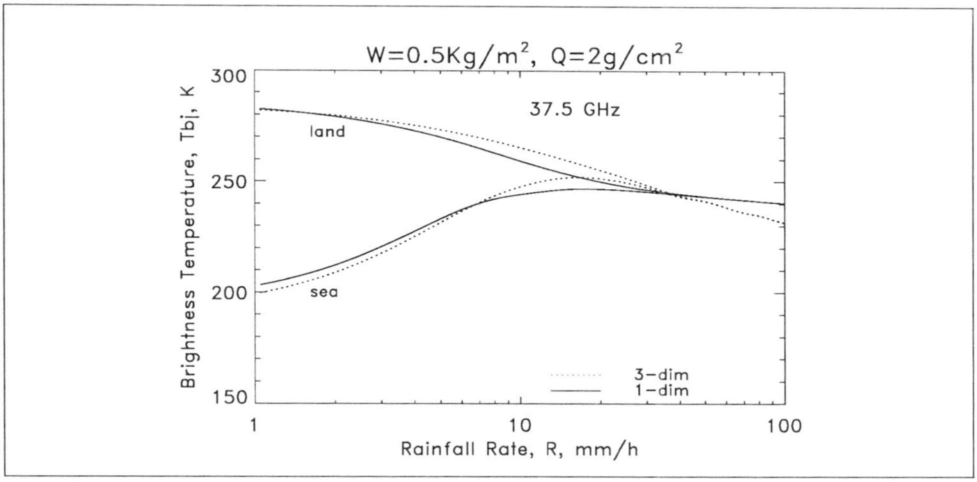

Figure 3. Comparison of the upwelling brightness temperature at the wavelength 0.8 cm as function of rainfall rate over land and over ocean at the nadir view angle $\vartheta = 0^o$ obtained with the one-dimensional model (Model A) and the three-dimensional digital model.

The T_{bj} dependencies on water vapor are shown in Figure 4. As expected, the influence of water vapor on the brightness temperature is strongest at 1.35 cm. The gradient $\frac{\partial T_{bj}}{\partial Q}$ is 15 K/g/cm^2 at this wavelength for $R = 1$ mm/h and for $W = 0.5$ kg/m^2. The gradient at 0.4 cm is 6 times less than at the 1.35 cm wavelength, though the partial absorption coefficient in water vapor k_ρ at 0.4 cm is more than at the 1.35 cm wavelength.

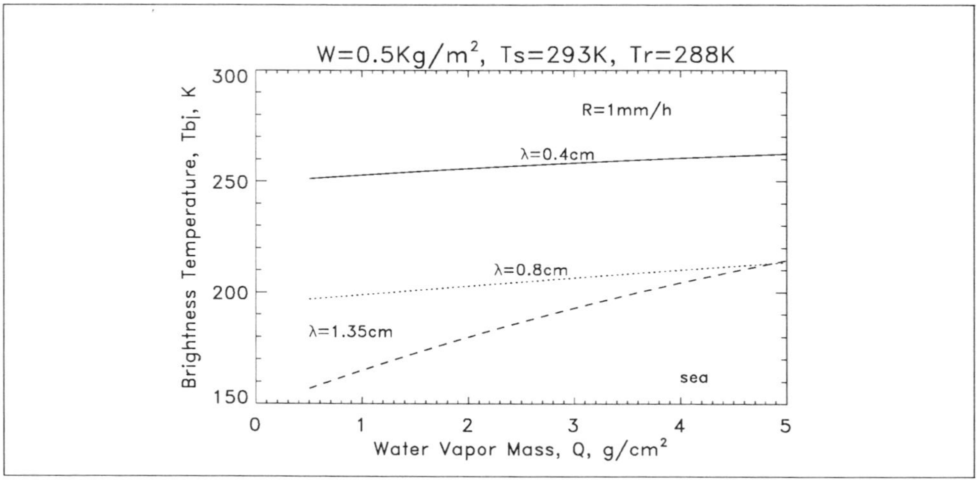

Figure 4. Calculated brightness temperature over ocean at wavelengths of 0.4, 0.8 and 1.35 cm at nadir as a function of the total water vapor for a rainfall rate of $R = 1$ mm/h. The assumed liquid water content in non-precipitating clouds is $W = 0.5$ kg/m^2.

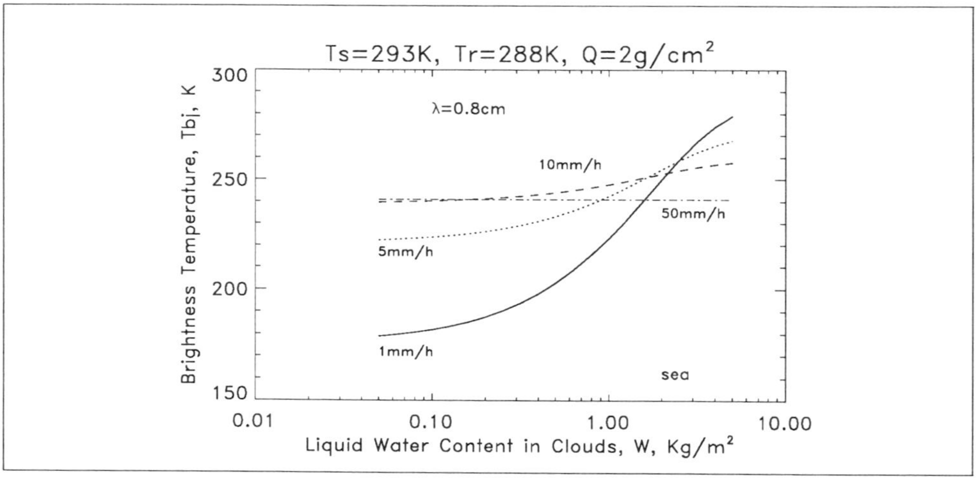

Figure 5. The upwelling brightness temperature over ocean at the wavelength 0.8 cm and nadir view angle as a function of the liquid water content in non-precipitating clouds. The assumed total water vapor mass in the atmosphere is $Q = 2$ g/cm^2.

The sensitivity of the brightness temperature to the parameters of a non-precipitating cloud layer are given in the next figures. The effects of the integral liquid water content in clouds W on the upwelling radiance T_{bj} are illustrated in Figures 5 and 6. The curves for the 0.4 cm wavelength (see Figure 6) show that small liquid drops influence T_{bj} only for $R = 1$ mm/h, and for $W < 2$ kg/m^2. This influence is strong at the other two wavelengths (see Figures 5 and 6 for $R = 5$ mm/h). The gradients $\frac{\partial T_{bj}}{\partial W}$ reach 57 K/kg/m^2 at the 0.8 cm wavelength and 23 K/kg/m^2 at the 1.35 cm wavelength for $R = 1$ mm/h and for small values of W. This result agrees with experimental data, where the large scattering

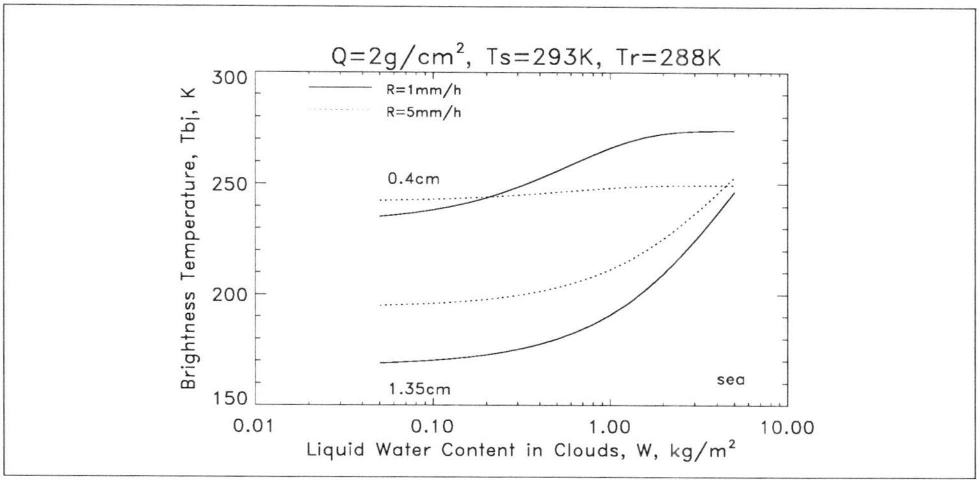

Figure 6. Same as Figure 5 but at the wavelengths 0.4 cm and 1.35 cm.

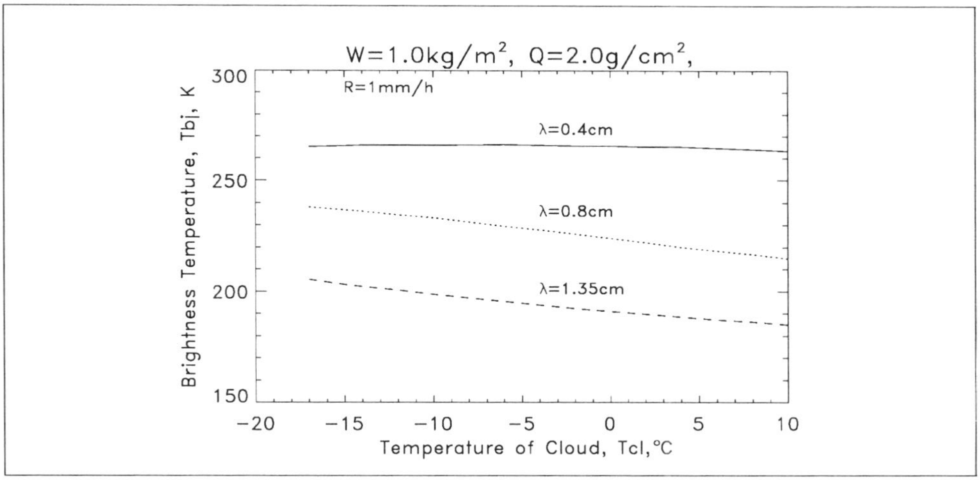

Figure 7. Upwelling brightness temperature over ocean at wavelengths of 0.4, 0.8 and 1.35 cm at nadir as a function of the non-precipitating cloud temperature for a rainfall rate $R = 1$ mm/h. The assumed liquid water content in clouds is $W = 1$ kg/m^2.

of the radiance value for the same magnitude of the rain rate is observed; see for example [10]. The variations of the brightness temperature, which are observed by ground based microwave radiometers, can reach $100 - 150$ K at wavelengths of 0.8 and 1.35 cm due to variations in the cloud liquid water content W.

Figure 7 shows the brightness temperature as a function of the effective cloud temperature. These dependences are observed only at wavelengths of 0.8 and 1.35 cm. The gradient $\frac{\partial T_{bj}}{\partial T_{cl}}$ is approximatly the same at these wavelengths; it is equal to ~ 0.8 K/K for $R = 1$ mm/h and for $W = 1$ kg/m^2.

TEMPORAL AND SPATIAL CHARACTERISTICS OF ATMOSPHERIC BRIGHTNESS TEMPERATURE DURING RAINFALL

To study the effect of footprint averaging by the satellite antenna, an investigation of brightness temperature variations of cloudy atmospheres have been made. Temporal and spatial variations of atmospheric emission are caused by variations of the meteorological parameters such as liquid water content, drop size distribution, phase structure, humidity, etc. These variations result from the turbulence and convection in the troposphere. The spatial resolution (footprint) of current and future microwave radiometers (MIMR, TRMM, IKAR), range between 5-20 km. The brightness temperature of a rainfall is highly variable within this footprint [10, 13].

For a description of the temporal or spatial field of the atmospheric emission variations the so-called structure function D_{T_b} is used [14]:

$$D_{T_b} = \overline{[\xi(t) - \xi(t + t^*)]^2} \tag{10}$$

where $\xi(t) = T_b(t) - \overline{T_b(t)}$ is the brightness temperature at time t and $t^* = t_1 - t_2$ is the time difference (temporal interval). The square root of the structure function is a measure of brightness temperature variation ΔT_b.

Spatial and temporal variations of atmospheric emission can be investigated using ground microwave radiometers by applying the Taylor "freezing" turbulence hypothesis. This hypothesis assumes that the relation between the spatial interval l^* and the temporal interval t^* is given by:

$$l^* = v_h \cdot t^*. \tag{11}$$

The velocity v_h is the horizontal component of the wind velocity, which is responsible for the transport of turbulence.

Studies concerning the variation of atmospheric emission were performed at the Institute for Radio Frequency Technology, DLR. The investigation used a microwave polarimetric radiometer at frequency 19.2 GHz (wavelength of 1.56 cm), which was built for the Olympus satellite propagation experiment (OPEX). Simultaneous measurements of rain parameters were obtained by the C-band weather radar, the distrometer and meteorological ground based instruments. These T_b measurements have been performed automatically during several years at the elevation angle 27° with integration time 1 sec.

A program for processing T_b data was developed to obtain the statistical characteristics of brightness temperature variations. The program outputs the values of the brightness temperatures at the H polarization $TB1H$ and at the V polarization $TB1V$, their difference $TDIFF$, and the rainfall rate R in time and space (using Eq. (11) and the wind speed measurements). Also, the structure function of these parameters $\Delta TB1H$, $\Delta TB1V$, $\Delta TDIFF$, ΔR can be determined. The temporal interval of data averaging is equal 10 sec.

The results of this data processing for two days with different weather conditions are shown in Figures 8 - 12. Heavy rain (convective precipitation) with rain rates up to 24 mm/h was observed in the first case on September, 9, 1993. Temporal variations of the downwelling brightness temperature and of the rainfall rate are presented in Figures 8 and 9, respectively.

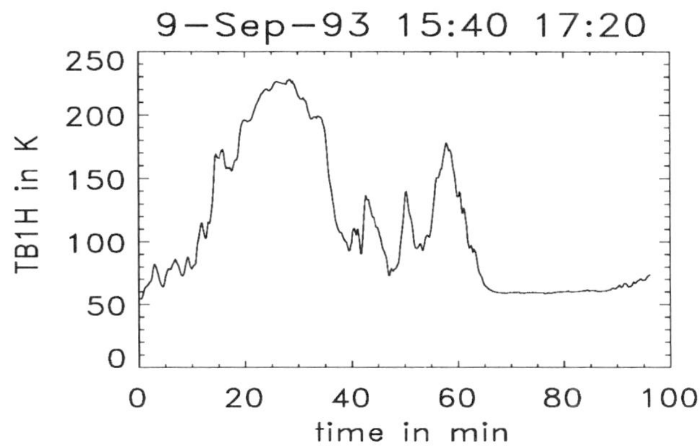

Figure 8. Temporal variations of the downwelling microwave brightness temperature at horizontal polarization, $TB1H$, at 19.2 GHz. Measurements on September, 9, 1993, convective precipitation.

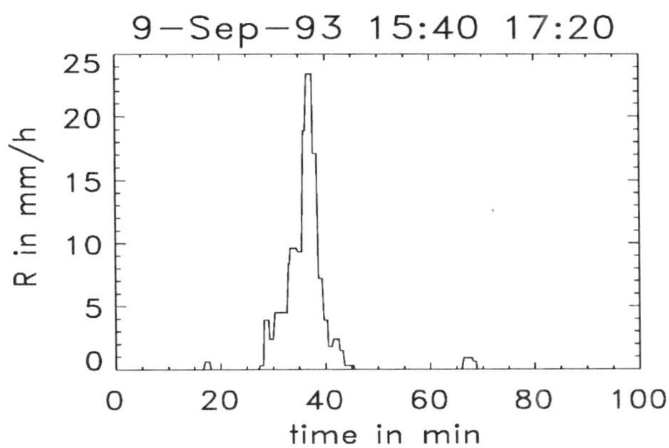

Figure 9. Temporal dependencies of the rainfall rate R. Measurements on September, 9, 1993, convective precipitation.

Using the wind speed data, measured simultaneously with $TB1H$, the temporal intervals can be converted into spatial intervals. Spatial variations of the brightness temperature structure function $\Delta TB1H$ are illustrated in Figure 10.

This figure shows the square root of the structure function Eq. (10) for the convective event of September, 9, 1993 and for a mainly stratiform event with rainfall rates up to 12 mm/h, which was observed on October, 12, 1993. The observation time in the second case was 10 hours; that is 6 times more than in the convective case. It can be seen, that the gradient of the curves decreases monotonically with increasing spatial interval.

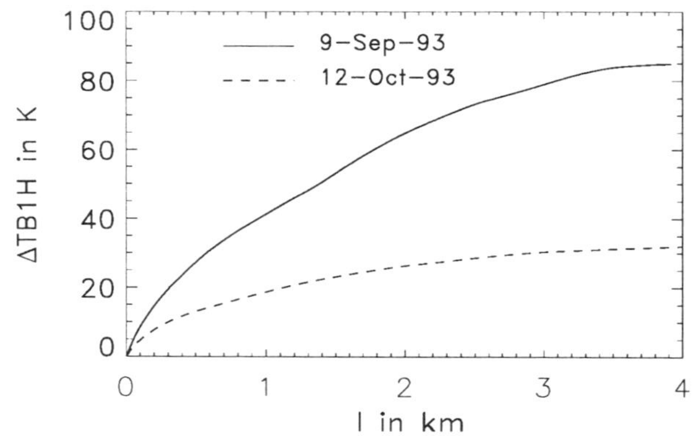

Figure 10. Spatial dependencies of the variation intensities (square root of the structural function) of the microwave radiances $\Delta TB1H$ at horizontal polarization for convective rain measured on September, 9, 1993 (solid line) and stratiform rain measured on October, 12, 1993 (dashed line).

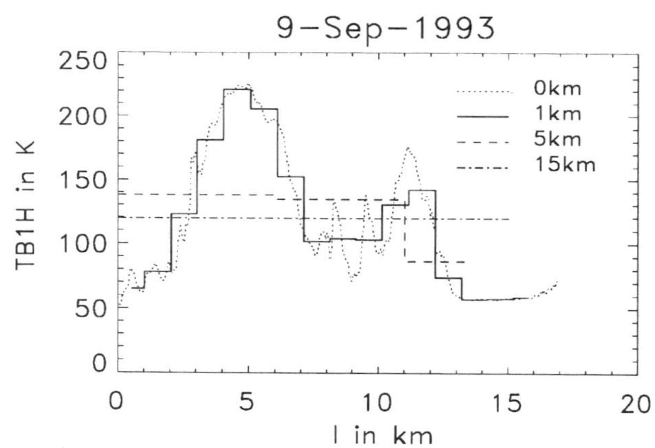

Figure 11. Spatial dependence of the brightness temperature $TB1H$ for different spatial averaging.

There are differences between the magnitude of the brightness temperature variation measured during stratiform and convective clouds. The variation intensities and their gradient $\frac{\partial TB1H}{\partial l^*}$ are 2.5 times more for convective precipitation than for stratiform rain.

The influence of spatial averaging on the measured brightness temperature and rainfall rate are presented in Figures 11 and 12. The shape of the curves change dramatically for large spatial averaging. The average with a spatial interval of 1 km follows well the original curves (dotted line) but averaging with scales of 5 km and higher do not sufficiently represent the unaveraged curves.

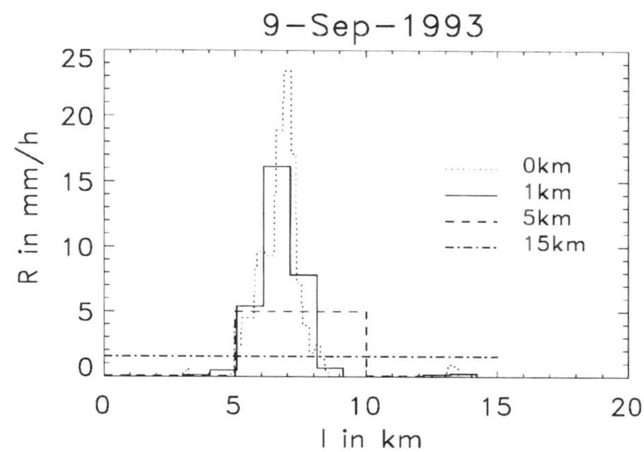

Figure 12. Spatial dependence of the Rain rate R for the different spatial averaging.

CONCLUSION

The simulation of upwelling microwave emission of a precipitating atmosphere shows that the brightness temperature depends beside the rainfall rate essentially on the following parameters: Total water vapour mass in the atmosphere, liquid water content and temperature of non-precipitating clouds. The integral parameters of the cloud layer - liquid water content and effective cloud temperature - have a strong influence on brightness temperature at millimeter and short centimeter waves for rain rates less than 5 mm/h. Water vapor, non-precipitating clouds and rain cause some spectral peculiarities of atmospheric emission. Therefore multi-channel measurements at different wavelengths and polarizations are necessary to study rain from space. The most applicable spectral regions for this purpose are 75-100 GHz, 30-40 GHz, 21-23 GHz, and 13-19 GHz.

Ground measurements of the temporal and spatial variations of downwelling atmospheric emission were performed at a frequency of 19.2 GHz. These experimental data indicate a strong influence of spatial averaging on the brightness temperature for spatial intervals up to 15 km, which correspond to the typical footprint sizes of satellite microwave radiometers.

REFERENCES

1. Liou, K.-N., *An Introduction to Atmospheric Radiation*, Int. Geophysics Series, vol.26, Academic Press 1980.
2. Hornbostel, A., A. Schroth, and B.G. Kutuza, 1994, Polarimetric measurements and model calculations of downwelling rain brightness temperatures. *Proceedings of μrad'94*, Rome.
3. Sobolev, V.V., 1956, *Radiative Transfer in Atmospheres of Stars and Planets.* Moscow: GTI.
4. Kutuza, B.G., and M.T. Smirnov, 1991, Signature of rain microwave emission. *Proceedings of IGARSS'91*, vol.1, 191-194.

5. Ulaby, F.T., R.K. Moore, and A.K. Fung, 1981, *Microwave Remote Sensing, Active and Passive*, vol.1, Addison-Wesley, Reading, Massachusetts.

6. Basharinov, A.E., A.S. Gurvich, and S.T. Yegorov, 1974, *Radiation of the Earth as a Planet*, Moscow: Nauka.

7. Barrett, A.H. and A.V. Chung, 1962, A method for the determination of high-altitude abundance from ground-based microwave observations. *J. Geophys. Res.*, vol.67, 4259-4266.

8. Smirnov, M.T., 1984, Modelling the rain microwave thermal emission by Monte-Carlo method. *Izv. Acad. Nauk SSSR, FAO*, vol.20, no.9, 820-826.

9. Wilheit, T.T., A.T.C. Chang, M.S.V. Rao, E.B. Rodgers, and J.S. Theon, 1977, A satellite technique for quantitatively mapping rainfall rates over the oceans. *J. Appl. Meteor.*, vol.16, 551-560.

10. Wilheit, T.T., 1986, Some comments on passive microwave measurement of rain. *Bull. Amer. Meteor. Soc.*, vol.66, no.10, 1226-1232.

11. Savage, R.C., and J.A. Weinman, 1975, Preliminary calculations of the upwelling radiance from rain clouds at 37.0 and 19.35 GHz. *Bull. Amer. Meteor. Soc.*, vol. 56, 1272-1274.

12. Wu, R., and J.A. Weinman, 1984, Microwave radiances from precipitating clouds containing aspherical ice, combined phase, and liquid hydrometeors. *J. Geophys. Res.*, vol.89, no.D5, 7170-7178.

13. Gagarin, S.P., and B.G. Kutuza, 1983, Influence of sea roughness and atmospheric inhomogeneities on microwave radiation of the atmosphere-ocean system. *IEEE J. Ocean. Eng.*, vol. OE-8, no.2, 62-70.

14. Tatarskii, V.I., 1961, *Wave Propagation in Turbulent Medium*, R.A. Silverman, Trans., McGraw-Hill Book Co., New York.

A quantitative comparison between 3-D and plane parallel microwave radiative transfer codes applied to horizontally and vertically structured precipitating clouds.

LAURA ROBERTI[1] and CHRISTIAN KUMMEROW[2]

[1] *Department of Electronic, Politecnico di Torino, Corso Duca degli Abruzzi, 24, 10129 Torino, Italy*

[2] *Laboratory for Atmospheres, NASA/Goddard Space Flight Center, Greenbelt, MD 20771,*

Abstract This study specifically investigates the differences between three-dimensional radiative transfer results and those obtained by plane parallel approximations in the microwave spectrum, making use of realistic three-dimensional distributions of cloud constituents. A backward Monte Carlo method is used to calculate the true radiances emerging from the cloud. Analyses between this model and plane parallel approximations reveal that plane parallel approximations introduce two distinct types of errors: physical and geometric. These sources of error are analysed both for nadir and a 50 degrees polar viewing angle at frequencies of 10.7, 19.35, 37.0 and 85.6 GHz. The errors are related to the spatial inhomogeneity of the cloud.

Keywords: Microwave, 3-D radiative transfer, clouds

INTRODUCTION

The detection and measurement of clouds and precipitation on a global scale is of great importance for a wide range of meteorological problems. While microwave rainfall retrievals offer a direct relationship between brightness temperature and precipitation, accurate simulation and interpretation of the brightness temperatures are necessary before estimates of precipitation are possible. To meet this demand a number of plane parallel radiative transfer codes have been used and described in the literature. However 3-D radiative transfer models must be used to account for the horizontal and vertical variability of rain over scale lengths smaller than the radiometric footprint.

The development of three dimensional radiative transfer techniques has been led largely by the solar and infrared community and used to study the effect of broken cloud fields upon radiation. Early studies [1,2] used Monte Carlo methods while analytical methods such as [3,4] used two stream approximations. Microwave studies date back to Weinman and Davies [5], who first recognized the need for 3-D radiative transfer schemes to properly simulate precipitating clouds. While their three-dimensional results showed significant deviations from plane parallel calculations, limited computing capabilities restricted their study to single layer cuboidal clouds. Later studies [6] expanded the single layer cuboidal cloud to include liquid and ice layers, but the realism that could be achieved with such models was limited. Cloud measurements have not generally been available with the detail necessary to justify 3-D radiative transfer calculations. Recent advances in dynamical cloud modelling [7, 8], however, have provided a great deal of fine scale cloud hydrometeor information. Such information may now be used to study the effects of arbitrary horizontal distributions. The complexity of 3-D models, the steady advances in computational capabilities together with the fact that absorbtion is fairly strong at these frequencies, have made Monte Carlo calculations a viable option in the microwave regime. In the Monte Carlo method the photons are followed as they propagate through the medium. This allows a simple understanding of their interactions with the medium itself and the computation of quantities like the mean number of scattering events or weighting functions for a 3-D geometry. Problems such as the definition of physically realistic boundary conditions disappear once a model for the background plane parallel atmosphere is defined. Photons are allowed to cross the sides of the cloud and can propagate in the clear atmosphere and vice versa.

2. The Cloud Model

All the radiative transfer computations in this paper refer to realistic cloud profiles obtained from the cloud microphysical-dynamical model in [7]. This model supplies a complete atmospheric profile, including temperature, relative humidity, cloud liquid water, cloud ice, precipitable liquid and two classes of precipitable ice (snow and graupel). The cloud domain consists of 64x64 pixels. To account for the horizontal variability, each grid point was considered as a subcloud having 1.5 km in the horizontal and 19 km in the vertical dimension. Each subcloud has 25 vertical layers. The particular simulation considered in this study involves a fast moving squall line over a tropical ocean. The associated rainfall intensities produced at the surface are shown in Figure1. Both liquid as well as ice are assumed to follow a Marshall-Palmer drop size distributions. All scattering parameters were calculated from Mie theory. To simplify the Monte Carlo calculations the phase function was replaced everywhere by the appropriate Henyey-Greenstein phase function having the calculated asymmetry factor.

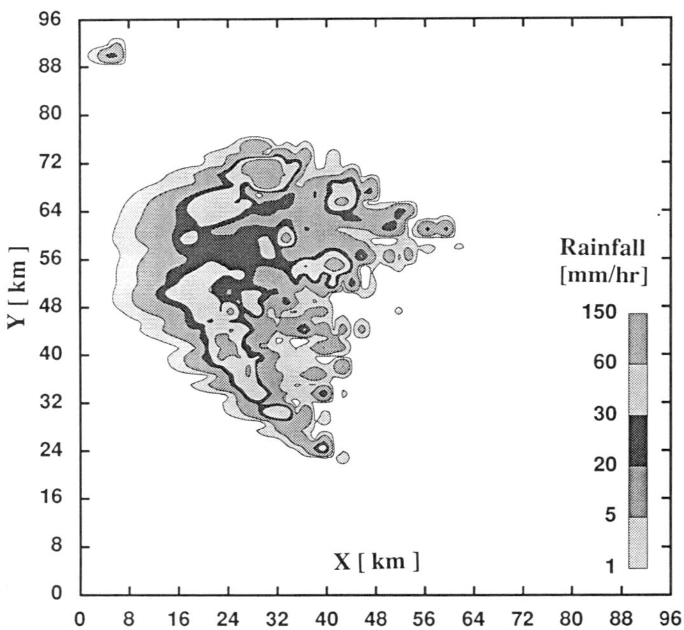

Figure 1. Surface rainfall map [mm/h] produced by the three-dimensional cloud dynamical model.

The cloud is represented as a 3-D grid in which the photons can travel in every direction and cross the external boundaries. In every cell of the grid the temperature decreases linearly from the bottom to the top, while the extinction coefficient, the single scattering albedo and the phase function are assumed constant. The radiation incident to the cloud top is considered the cosmic background with a temperature equal to 2.7K, since the atmospheric profile above the cloud top was not specified in the cloud model [7]. This can be considered a good approximation since the interactions of the photons with the medium outside the raining cloud are almost negligible. The lower boundary at the surface is specified to be either specular or Lambertian. The atmosphere surrounding the cloud is modelled as plane parallel, considering the atmospheric profile relative to one of the subcloud in the non raining portion of the cloud.

3. Three-dimensional Monte Carlo method

The main purpose of this study is to compute radiances that might be seen by airborne or satellite borne radiometers. The natural choice is therefore a backward Monte Carlo scheme since it considers only photons which ultimately escape the cloud in a specified direction [9,10]. In this scheme the photons are started at the point at which the brightness temperature (T_B) is computed, in the direction opposite to that in which they would physically propagate inside the medium. Each photon is then traced backwards through the medium following probabilistic interaction laws which are sampled by a selection of numbers from a quasi random sequence. The interactions can be due to scattering and/or absorption. When the photon is absorbed it is re-emitted at the point of absorption x with T_B equal to the physical temperature of the medium at that point $T(x)$. This assumption can be made because in the microwave regime at typical tropospheric temperatures, the Planck function is virtually a linear function of temperature and can therefore be replaced by $T(x)$. If a scattering event was prescribed, a new direction is determined by the phase function. The Monte Carlo code can account for atmospheric emission, surface emission, cosmic background radiation, although almost negligible, and multiple scattering. For further information the reader is referred to [11]. The cloud top is considered as a matrix of (N_x by N_y) subcloud tops. For each subcloud, a set of Np computational points at which radiances emerge is determined. In the present calculations the distance between each computational point is 0.3 km, for a total of 25 points for each subcloud. Four thousand (N=4000) photons are started at each computational point for a total of 10E5 photons for each cloud. In this code no weight is attached to the photons, i.e. no biasing involved. After a photon is started, it's trajectory is followed as it diffuses through the medium [12,13] The T_B for each subcloud is computed as

$$T_B = \frac{1}{N}\frac{1}{N_P}\sum_{np=1}^{N_P}\sum_{n=1}^{N} T_B(n,np) \qquad (1)$$

From a computational point of view the code is very efficient because the contribution from each photon is calculated independently. The final result of the computation is a (N_x by N_y) map of brightness temperatures.

4. Plane Parallel Modelling

Three variations of a plane parallel model are considered to understand and quantify its accuracy relative to a 3-D radiative transfer algorithm. The first is the classical one in which the TB is computed for each subcloud considered as plane parallel. A variation from this method may be defined when angles other than nadir are involved. In this variation, all

subclouds are taken into consideration at the same time. The path of a given ray is then traced trough the 3-dimensional structure and a new plane parallel cloud is constructed based upon the layers intercepted by the ray in question. This variation better represents the geometry of the problem. It will be referred to as the modified plane parallel approach. The last variation is applicable to larger footprints and consists of averaging hydrometeors before the radiative transfer calculations are performed. For footprint sizes of ~25km corresponding to current satellite sensors, this method introduced errors one order of magnitude higher than the other two methods. Averaging of hydrometeors was therefore not pursued any further in this study. An Eddington radiative transfer code has been used for the plane parallel computations [14].
In general, the plane parallel model suffers from two problems that must be analysed separately: the first is physical, the second is geometric. The main physical problem is that the radiation is "trapped" in each subcloud considered as plane parallel. The main consequence is that there can be non-physical abrupt variations of TB from subcloud to subcloud. At low frequency the warm radiation emitted inside the cloud cannot exit the sides of the cloud and, consequently, the plane parallel results are slightly warmer. At high frequency, when there is scattering in the higher layers, the average temperature is lower for the same reason. The geometric problem is due to approximations which plane parallel models must make to accommodate edges. This problem is very difficult to model, since it depends strongly on the frequency, on the cloud constituents and on the viewing angle.

5. Quantitative Intercomparisons

The characteristics of the results depend on frequency, but it is possible to distinguish three basic behaviours at low, medium and high frequencies. These are well represented at 10.7, 37.0 and 85.6 GHz respectively. Quantitative errors are also computed for 19.35 GHz because it is the lowest frequency currently observed by the SSM/I instrument. The radiation is considered horizontally polarized with emissivities, for a sea surface, given in Table 1.

Table 1. Sea surface emissivities values

Freq [GHz]			
10.70	19.35	37.0	85.5
0.265	0.281	0.318	0.411

Qualitative comparisons are given through difference maps between the 3-D and plane parallel results. Quantitative errors are tabulated using the mean deviation Δ given by:

$$\Delta = \frac{1}{N_{S.C.}} \sum_{i=1}^{N_{S.C.}} \left| T_B(M.C.) - T_B(P.P.) \right| \quad \textbf{(2)}$$

where M.C. refers to Monte Carlo and P.P. to plane parallel methods. The bias, β is obtained using the same expression but without the absolute value.
Referring to Figure 1, the rainfall can be seen to cover approximately 20% of the image. The T_Bs of the background are fairly uniform and the error introduced by the plane parallel approximation is negligible. To make comparisons useful, it is therefore necessary to define the area over which comparisons are made. Two areas are defined: the raining cloud itself, and the area beyond the rain itself which is radiometrically affected by the rainfall. This area, referred to as the "radiometrically affected area", is defined operationally as the area whose brightness temperature differs from the background temperature by more than 5K. Monte Carlo T_B maps and differences between Monte Carlo results and plane parallel results for a specular surface and nadir viewing angle are shown in Figure 2. The azimuthal angle is always zero. Quantitative comparisons between Monte Carlo results and plane parallel results for

nadir and 50 degrees polar viewing angle are given respectively in Tables 2 and 3. The modified plane parallel approach is employed for off-nadir comparisons. The comparisons are made over the rainy area for nadir viewing. For 50 polar viewing angles, the T_B pattern is shifted and comparisons are made over the radiometrically affected area. Equivalent results using the classical plane parallel scheme are not presented because the shift in the T_B patterns at off-nadir viewing angles produce very large deviations which have no physical meaning on a pixel by pixel basis. However, results for both the modified plane parallel as well as the shifted classical approach described above for 25 km footprints are presented in Table 4. These areas are commensurate with current satellite footprints and are large enough to eliminate deviations due to small misalignments of the T_B patterns. The overall shift in the classical plane parallel approach is accounted for by minimizing the error with respect to horizontal shifts in the brightness temperature patterns. The shifts introduced in such a manner roughly corresponds to $\Psi\tan(\vartheta)$, where Ψ is the mean height of the weighting function defined in [15] and ϑ is the viewing angle.

Table 2. Mean deviations, maximum deviations and biases computed over the raining cloud between 3-D and plane-parallel results for nadir-viewing angle and specular surface.

Freq. [GHz]	View. Ang.	Δ Mean Dev.	β Bias	Max Dev.
10.70	Nadir	1.65	-0.73	-12.61
19.35	"	2.56	-1.43	-15.38
37.00	"	5.62	+0.57	+24.35
85.60	"	10.03	+8.77	+47.81

Table 3. Mean deviations, maximum deviations and biases computed over the radiometrically affected area between 3-D and the modified plane-parallel approach. Surface is specular. 50 degrees polar viewing angle.

Freq. [GHz]	View. Ang.	Δ Mean Dev.	β Bias	Max Dev.
10.70	50 Deg.	2.88	-0.83	-20.86
19.35	"	3.26	-0.76	-21.85
37.00	"	5.80	+1.97	+47.10
85.60	"	13.26	+9.54	+80.92

Table 4. Mean deviations, maximum deviations and biases computed over the radiometrically affected area between 3-D and both the classical and the modified plane parallel approach. The deviations are averaged to 25km resolution. Surface is Lambertian and the viewing angle is 50 degrees.

Classical Plane Parallel				
Freq. [GHz]	View. Ang.	Δ Mean Dev.	β Bias	Max Dev.
10.70	50 Deg.	7.29	+7.29	+12.71
19.35	"	9.47	+8.35	+19.24
37.00	"	15.79	+7.57	+37.57
85.60	"	8.64	+8.64	+13.36
Modified Plane Parallel				
Freq [GHz]	View. Ang.	Δ Mean Dev.	β Bias	Max Dev.
10.70	50 Deg.	1.27	0.00	+1.53
19.35	"	1.28	-0.18	-2.21
37.00	"	2.18	+2.08	+4.36
85.60	"	8.17	+8.34	+13.22

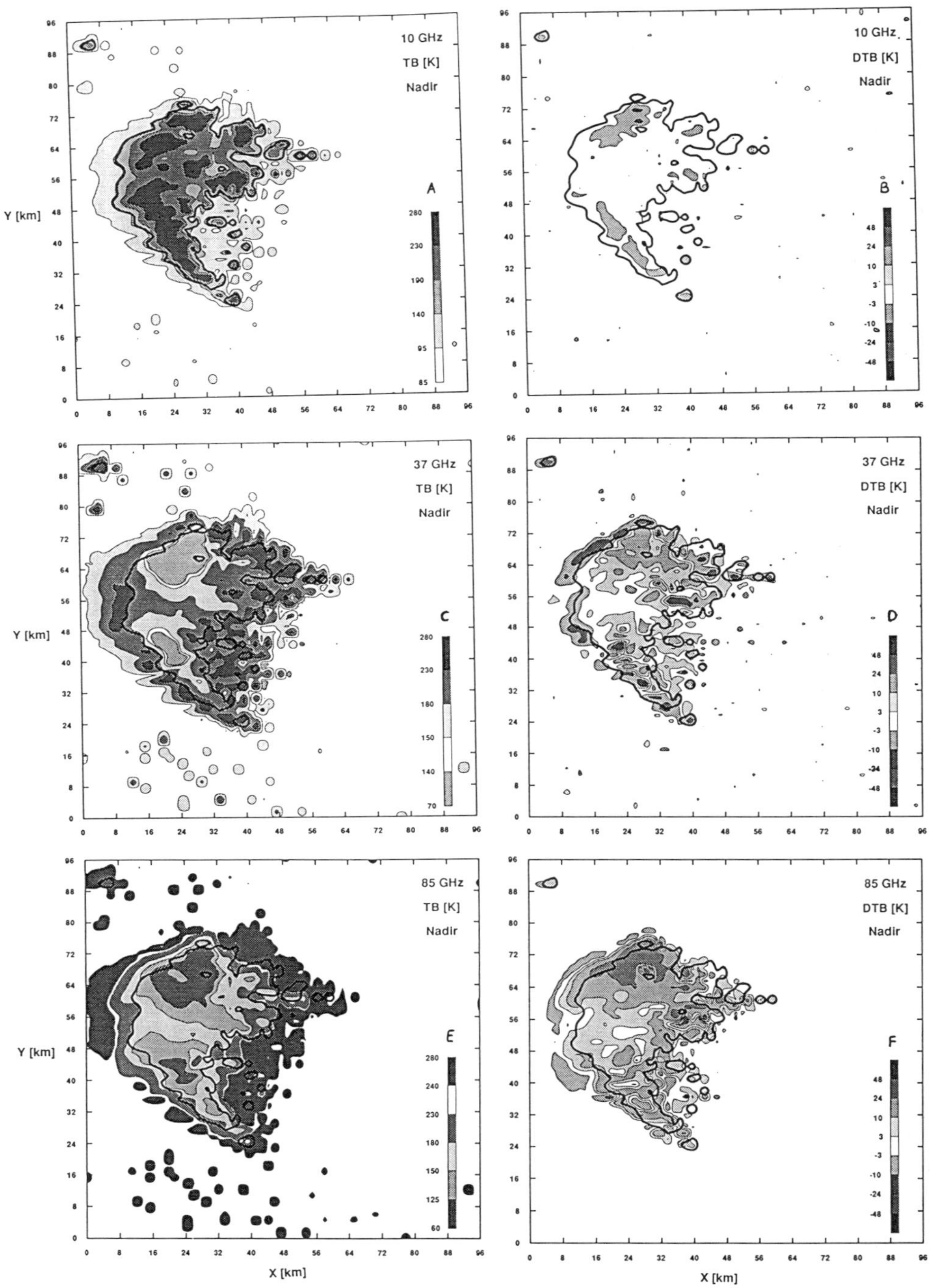

Figure 2. Brightness temperature maps and difference maps between the Monte Carlo and the plane parallel solution for nadir viewing angle at 10.7 GHz(A, B), 37.0 GHz (C, D) and 85.6 GHz (E, F). The 10 mm/hr rain profile is superimposed.

Low Frequency - At 10 (Figure 2A) and 19 GHz, the cloud is characterized by emission signatures, especially in the regions of heavy rainfall. It is consequently warmer than the atmospheric background. If a plane parallel approximation is used, the warm radiation emitted inside each subcloud cannot exit the sides of the subcloud itself. The plane parallel results are consequently warmer in the heaviest raining portions of the cloud (Figure 2B). Quantitative comparisons are given in Table 2. The plane parallel modelling appears a good approximation to 3-dimensional calculations.
For off-nadir observations, geometric effects are inevitably present and different problems arise since plane parallel radiative transfer cannot properly deal with radiation at cloud edges. One of these problems consists in the above mentioned geometric shift of the cloud radiometric footprint with respect to the real cloud position. This effect can be partially taken into account if the modified plane parallel approach is used. Overall, the same arguments made for nadir observations hold true for the comparison on a pixel by pixel basis (Table 3) between 3-D and plane parallel results obtained with the modified approach. Nevertheless, for off-nadir viewing, the radiometric effect of the cloud is extended by the propagation of the radiation from the sides of the cloud and the resulting errors are generally higher than for nadir viewing results. Maximum deviations are almost doubled at all frequencies.

Medium Frequency - At 37 GHz (Figure 2C), the T_B relative to the raining cloud are warmer or colder than the background depending whether emission or scattering are the dominant events. At this frequency scattering is due to heavy rainfall rates and to the associated ice presence in the higher layers of the cloud. If a plane parallel model is used, the radiation emitted in the lower layers of a subcloud is reflected back to the lower layer by the ice canopy and is thus prevented from exiting the top of the cloud. The resulting radiative budget is obviously altered and the plane parallel results are generally colder than 3-D (Figure 2D). For low rainfall rates, emission is still relevant and plane parallel T_Bs are warmer as is the case with lower frequencies. For the particular simulation, scattering effects in the core of the storm dominate over emission at the edges and the resulting bias is positive (Table 2 and 3).

High frequency - At 85 GHz (Figure 2E) the raining cloud is characterized by intense scattering and appears colder than the background. The warmer T_Bs that surround the raining cloud are due to emission by water vapour and cloud liquid water in those regions. The effects of emission are extremely reduced inside the raining cloud and the plane parallel results are generally colder (Figure 2F). Both the mean deviation and the bias are significant and cannot be neglected in this case. In general, it was found that the errors introduced by the plane parallel approximation are mostly due to atmospheric scattering so that the errors increase with the frequency (Table 2 and 3).

Comparisons were also carried out with the classical plane parallel approach. Because the geometric effects prevail with this method, even if the results are shifted to minimize the error as described above, comparisons were made only after the results were averaged to 25 km areas. Table 4 summarizes these discrepancies for both the classical and the modified plane parallel results for lambertian surface and 50 degrees polar viewing angle. As can be seen, the discrepancies with the classical plane parallel approach are still very large. The fact that all biases seem positive (i.e. plane parallel is higher than 3-D) is rooted in the fact that the plane parallel model, in dealing with the edges of the cloud, tends to exaggerate the radiometric impact of these edges. The modified plane parallel results are consistently better than those obtained with the classical approach.

6. Errors related to cloud inhomogeneity.

Errors introduced when plane parallel approximations are used to describe cloud structures are a function of the actual inhomogeneity of clouds themselves. To explore this point further, comparisons between 3-D and plane parallel calculations have been made as a function of a cloud inhomogeneity parameter defined using the extinction coefficient. The extinction coefficient, in fact, determines the average mean free path of photons inside a cloud. In this study, the average extinction coefficient is determined for each subcloud by averaging over the 25 vertical layers. To measure the inhomogeneity of the resulting 2-D extinction coefficients map, a window of 5 x 5 pixels surrounding each 1.5 km subcloud is considered. The inhomogeneity parameter is then calculated for each subcloud according to

$$\Gamma = \frac{1}{24} \sum_{i,j} \left| K_{ext}(i,j) - < K_{ext} > \right| \quad \textbf{(3)}$$

where the average extinction $<k_{ext}>$ is computed over the 5 by 5 subcloud domain. The inhomogeneity defined in this fashion may now be related to brightness temperature errors introduced by plane parallel transfer schemes. Figure 3 shows the correlation between inhomogeneities as defined above and the mean departure of the modified plane parallel results from actual 3-D values. Because inhomogeneities as well as T_B differences depend upon neighbouring subclouds, the results shown in Figure 3 correspond to averages over 4 by 4 subclouds (i.e. 6x6 km). The correlations can be seen to be quite high for all the frequencies. The good correlations achieved in Figure 3 lend hope to the concept that errors introduced by plane parallel radiative transfer methods may be corrected empirically if the inhomogeneity of the radiative field is known. Ufortunately the inhomogeneity parameter defined in eq.(3) cannot be computed directly from radiometer measurements, so research has to be done to improve the definition inhomogeneity itself.

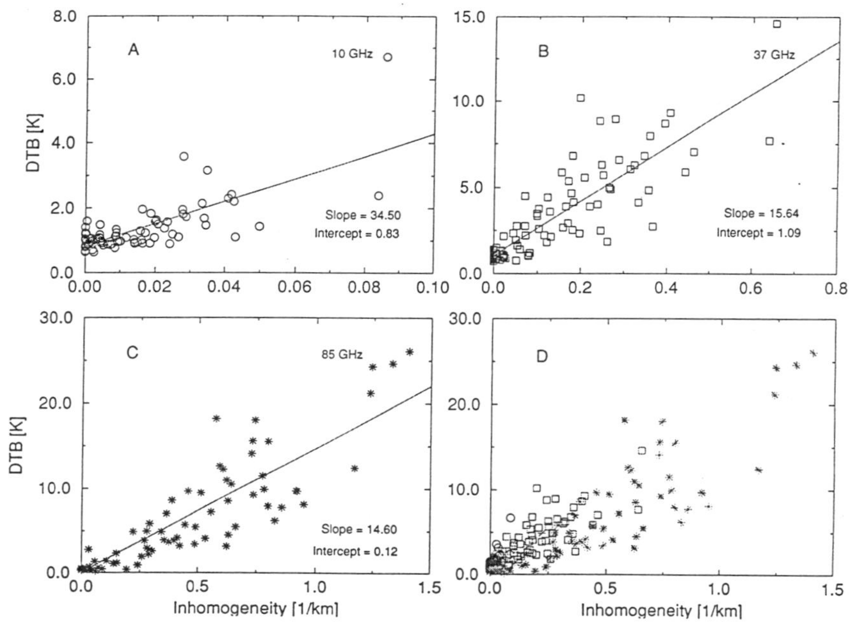

Figure 3. Correlation plots between subclouds inhomogeneities and brightness temperature differences averaged over an area of 6km x 6km at 10.7 GHz(A), 37.0 GHz (B), 85.6 GHz (C). In (D) the 3 correlation plots are shown together.

7. Conclusions.

Three different methods for the computation of the microwave brightness temperature relative to an horizontally and vertically structured cloud have been compared. The methods are: 3-D Monte Carlo and two different variations of plane parallel modelling For very complex raining cloud structures, the Monte Carlo solution turns out to be not only simple, but also computationally efficient. The classical plane parallel method gives relatively good results only in the case of specular surface and nadir viewing angle. When the emerging radiation crosses more than one subcloud, it was found that the modified plane parallel method gave far superior results. This method partially accommodates for the geometrical problems due to surface scattering and oblique viewing angle, but still cannot take into account the effects due to atmospheric scattering. For low frequencies (10 and 19 GHz), this study found the modified plane parallel method to be a good alternative to a 3-D code. A first approach to the problem of parameterizing the effects of 3-D structure upon radiation was explored. Good correlations were found between radiative errors and inhomogeneity parameters defined using the extinction coefficient of the medium.

Acknowledgements
Laura Roberti has been financially supported by the Fondazione SANPAOLO di Torino as part of the "Global Climate and Environment" project. Christian Kummerow is supported by NASA headquarters as part of the pre-TRMM algorithm development effort. Special thanks are due Bill Ridgeway for helpful advise and discussions on the radiative models.

REFERENCES

1 Kargin, B. A., L. D. Krasnokutskaya and E. M. Feigelson, 1972: Reflection and absorption of solar radiant energy by cloud layers. Izv. Acad. Sci. USSR Atmos. Oceanic Phys., Eng. Transl., 8, 287-293.
2 McKee, T. B., and S. K. Cox, 1974: Scattering of visible radiation by finite clouds. J. Atmos. Sci., 31, 1885
3 Davies, R.,1978: The effect of finite geometry on the three-dimensional transfer of solar irradiance in clouds. J. Atmos. Sci., 35, 1712-1724.
4 Harshavardan, J. A. Weinman and R. Davies, 1980: Transport of infrared radiation on cuboidal clouds. J. Atmos. Sci., 38, 2500-2513.
5 Weinman, J. A. and R. Davies, 1978: Thermal microwave radiances from horizontally finite clouds of hydrometeors. J. Geophys. Res., 83, 3099-3107.
6 Kummerow, C. and J. A. Weinman, 1988: Determining microwave brightness temperatures from horizontally finite, vertically structured clouds. J. Geophys. Res., 93, D3720-3728.
7 Tao, W.-K., J. Simpson, and S.-T. Soong, 1986: Statistical properties of a cloud ensemble: A numerical study. J. Atmos. Sci., 44, 3175-3187.
8 Tripoli G. J., 1992: A nonhydrostatic model designed to simulate scale interactions. Mon. Wea. Rev., 120, 1342-1359.
9 Nelson, H.F,1991:Backward Monte Carlo Modeling for Rocket Plume Base Heating, J Thermoph., 6, 556-558.
10 O'Brien, D. M., 1992: Accelerated Quasi Monte Carlo Integration of the Radiative Transfer Equation, J. Quant. Spectrosc. Radiat. Transfer., 8, 41-59.
11 Roberti,L, J. Haferman and C. Kummerow, 1993. Microwave radiative transfer through horizontally inhomogeneous precipitating clouds, submitted to J. Geophys. Res.
12 Cashwell, E. D. and C. J. Everett, 1967: A Manual of the Monte Carlo Method for Random Walk Problems, Pergamon Press, New York.
13 House, L. L. and L. Avery, 1969 : The Monte Carlo Technique Applied to Radiative Transfer. J. Quant. Spectrosc. Radiat. Transfer., 9, 1579-1591.
14 Kummerow, C., 1993: On the accuracy of the Eddington Approximation for Radiative Transfer in the Microwave Frequencies. J. Geophys. Res., 98, 2757-2765.
15 Wu, R., and J. A. Weinman, 1984: Microwave radiances from precipitating clouds containing aspherical ice, combined phase, and liquid hydrometeors. J. Geophys. Res., 89, 7170-7178.

3 -D Radiation Transfer Effects of A Raining Cloud

QUANHUA LIU and CLEMENS SIMMER

Institut für Meereskunde, University of Kiel

Abstract

A 3-dimensional transfer model for polarized radiation is developed using the Monte Carlo method. A combination of the backward and forward Monte Carlo methods (backfor method) allows the polarization dependent scattering of photons. A comparison of the results of a 3-D model with calculations from those of a horizontally homogeneous cloud (1-D model) explains and quantifies two distinct radiation effects of the 3-dimensional structure of a raining cloud. First, upwelling radiation reaching the satellite is reduced due to leakage of radiation out of side walls of the cloud. Second, over water surfaces the reflection of radiation from the cloud leads to an increase of upwelling radiation for the cloud free areas in the vicinity of the clouds. Both effects compete with each other and can lead to higher or lower 3-D brightness temperatures, depending on the surface reflectivity, surface temperature, the horizontal width and microphysics of the clouds, and on the coverage of the cloud within the field of view. Calculations at 37 GHz for a vertically homogeneous cloud extending from the surface up to 4.4 km and consisting only of rain drops give the following results: Assuming a fully cloud-covered field of view the 3-D brightness temperature shows a similar behaviour with increasing rain rate (RR) as the horizontally infinite cloud, namely an increase at low rain rates due to emission effects and then a decrease due to saturation and scattering effects. For small horizontal dimensions, however, the 3-D brightness temperatures can drop even lower followed by another increase. Assuming a cloud with the same size as the field of view of the radiometer the 3-D brightness temperature at nadir is always lower than the 1-D brightness temperature with differences ranging up to 20 K (1 km by 1 km, 10 mm/h). Contrary results are found at a viewing angle of 53° for clouds with large horizontal scales. For the 37 GHz channel of SSM/I the plane parallel slab model tends to underestimate the rainfall rate for small rainfall rates (< 1 mm/h), overestimates for 1 ~ 5 mm/h, but again underestimates the rainfall rates for larger rainfall rates. Results for 19 GHz show similar behaviour.

1. Introduction

Rainfall is an important parameter in the atmosphere which is very variable in time and space. It is, however, difficult to measure particularly over the ocean. Rainfall is related to the amount of latent heat released to the atmosphere, a main energy source for many processes. Meteorological satellites provide a way to estimate the rainfall globally. Visible and infrared (VIS/IR) satellite observations have been applied to estimate rainfall from space [1], [2], [3]. Turpeinen et al. [4] demonstrated the possibility to calculate rainfall rates in certain climatological regions with the geostationary meteorological satellite METEOSAT, which provides a high temporal resolution. However, in the visible and infrared spectral range, rain rates must be inferred indirectly, since clouds are almost opaque. The indirect method depends on climate zones and cloud category. In the microwave region, e.g. at 37 GHz, clouds and rain are semi-transparent. In this part of the spectrum, water surfaces appear cold and provide a uniform background for identifying rain clouds. But a shortcoming is encountered when applying the satellite-borne microwave radiometers to the rainfall estimation. The field of view (FOV) of SSM/I at 37 GHz is about 30 km, which is much larger than the scale of the rain cells, which have diameters of only a few kilometers. Therefore large inhomogeneities can occur within a microwave pixel.

In order to solve the radiative transfer problem for finite rain clouds, several authors have studied the differences between 3-D and 1-D radiative transfer models. Kummerow and Weinman [5] demonstrated that the one dimensional plane-parallel slab model tends to underestimate the rainfall rates at a viewing angle of 50 degree. Haferman et al. [6] found opposite results at nadir, where the 1-D model overestimates rainfall rates after saturation. Our results explain both arguments.

In the following sections we discuss different radiative transfer models and the differences between 3-D and 1-D models, and the validation of our 3-D radiative transfer model.

2. Radiative Transfer Model

Various radiative transfer models for horizontally inhomogeneous atmospheric layers (3-D model) have been developed. Weinman and Davis [7] developed an Eigenvalue Method while Kummerow and Weinman [5] extended the Eigenvalue Method to vertically inhomogeneous cases. McKee and Cox [8] used the Monte Carlo method (MCM) to study the scattering of visible radiation by finite clouds. Stephens [9] provided a general 3-D radiative transfer model based on the Fourier Transform method. A 3-D Discrete Ordinates model is described by Lovejoy et al. [10] and Girens [11]. Recently, Haferman et al. [6] have studied 3-D raining clouds using this method. The main problems in using 3-D radiative transfer models is the limit of present computer capacity and the treatment of boundaries.

Generally, there are two alternative Monte Carlo methods; one is the backward method and the other is the forward method. Using the forward method, the thermal sources within the atmosphere-earth system are simulated by emitting many photons isotropically from small cubic cells in all directions. Therefore, the forward method requires a large amount of photons, which is very time consuming, and its uncertainty is about 2 K for non polarized brightness temperatures [7]. The backward method retraces the photons from the receiver (e. g. satellite radiometer) to their source, where they have been emitted. This method is very time efficient because only photons are considered which actually reach the receiver. The backward method has also been applied to the microwave spectral region [12]. But the method can not treat the polarizations of photons during scattering events within clouds. We combine backward and forward methods. The backward method is used to determine those photons and their energy which contribute to the signal at radiometer. The energy of the photons is then dependent on the temperature at the point of emission. The photons emitted from the surface reaching a radiometer have high energy because of the high surface temperature. However, the photons scattered by hydrometeors reaching the radiometer from outer space having little energy. Then the forward method is used for these photons to determine their polarization state at the radiometer caused by the scattering events. This combination is almost as fast as a pure backward method, but treats the polarization within clouds correctly. We call it the backfor method. For our MCM 3 random numbers R_i are required:

$$R_1 = \exp(-\delta\ ) = \exp(-\int_0^s \beta\ (x,y,z)dl) \qquad (1)$$

where δ is the optical depth, β is the extinction coefficient, and s is the distance along the photon path.

The scattering and absorbing status is determined by

$$R_2 - \varpi(x,y,z) = \begin{cases} \leq 0, scattered \\ > 0, absorbed \end{cases} \qquad (2)$$

where ϖ the local single scattering albedo.

The scattering angle Θ of each photon is computed from

$$R_3 = 2\ \pi \int_0^{\Theta} P(\theta)\, d\cos(\theta) \qquad (3)$$

where P(θ) is the phase function for total intensity.

We assume that each local source emits two photons at the same time; one is vertically polarized and the other is horizontally polarized. Energies of the two photons after k scattering processes is determined by forward calculation using

$$\begin{bmatrix} I_v \\ I_h \end{bmatrix} = S_1\ S_2 \ldots S_k \begin{bmatrix} I_{vo} \\ I_{ho} \end{bmatrix} \qquad (4)$$

$$I_{v0} = I_{h0} = I_0$$

where I_0 is the energy of the local source, I_v and I_h are the energies in vertical and horizontal polarization after the k scattering events, respectively. The matrix S_i, which represents the i -th scattering process, is calculated from the rotated phase matrix (the subscript i is dropped here)

$$S = \frac{2}{P_{11}+P_{12}+P_{21}+P_{22}} \begin{bmatrix} P_{11} & P_{12} \\ P_{21} & P_{22} \end{bmatrix} \qquad (5)$$

The rotated phase matrix elements P_{ij} are calculated from:

$$L(i_2 - \pi\ )\ \bar{P}(\Theta)\ L(i_1\ ) = \begin{bmatrix} P_{11} & P_{12} & P_{13} & P_{14} \\ P_{21} & P_{22} & P_{23} & P_{24} \\ P_{31} & P_{32} & P_{33} & P_{34} \\ P_{41} & P_{42} & P_{43} & P_{44} \end{bmatrix} \qquad (6)$$

where L is the rotation matrix. The rotation angles i_1 and i_2 depend on the incoming and outgoing directions. $\bar{P}$ is the scattering matrix, which depends only on the scattering angle and the optical properties of the medium. For details about the rotation matrix we refer to the book of Tsang et al. [13]. In our calculations, only the four elements in the upper left corner of the rotated phase matrix are required because the photons in our MCM are linearly polarized. The random numbers R_1, R_2, and R_3, which represent stochastic processes of photons, are generated by computer.

The boundaries are defined by a rectangular volume with horizontal extentions of 800 km by 800 km and 40 km in the vertical direction. The cloud is placed in the center area of the volume. Radiative flux exchange through the side boundaries cause an error of less than 0.2 K. The space above 40 km is considered vacuum while the lower boundary is modeled as an ocean surface according to Wisler and Hollinger [14]. Our 3-D MCM can be used for inhomogeneous clouds in both vertical and horizontal dimensions. To simplify the discussion, the raining clouds in the present calculations are considered horizontally and vertically homogeneous.

3. Intercomparison of the Radiative Transfer Model

In order to verify our model intercomparisons are made with other models. At first, our 1-D model based on the Matrix Operator Method (MOM) [15],

[16], was compared against the iterative method of Weinman and Guetter [17]. The parameterizations of the single scattering albedo ϖ and optical depth δ with rainfall rates R are taken from Weinman and Guetter [17]. The top of the raining cloud is at a height of 4.57 km and a temperature of 258 K. The base temperature is 288 K. Table 1 shows that the results from both methods are almost the same. There are small differences for the rainfall rate of 32 mm/h. This is probably due to the coarse vertical resolution used in the iterative method. A second comparison was performed between our MCM and MOM for a horizontally infinite cloud to check whether our MCM can correctly treat the polarization processes within the cloud. Both results agree well (see table 2). The maximum difference is 0.2 K. This implies that our MCM can correctly treat the vertical and horizontal polarizations.

A third comparison was performed with the model of Weinman and Davis [7] for a finite cubic cloud. The horizontal and vertical scales are determined by the optical depth in x, y, and z direction. The temperature of the cloud is 275 K. Outside of the cubic cloud is a vacuum with no fluxes from outside into the cloud. Results calculated from the MCM and the analytical method of Weinman and Davis [7] agree well although the differences in few cases are up to 4 K (see table 3). For the horizontally homogeneous cloud the results calculated from our MCM show good agreement with those calculated from the 1-D MOM. We therefore conclude that our 3-D Monte Carlo model yield reasonable results when used to investigate 3-D effects in radiative transfer.

Table 1. Brightness temperature at 37 GHz emerging from a rain cloud with a height of 4.57 km located over calm water. Ite.: Iterative method; MOM: Matrix Operator Method; θ viewing zenith angle.

R(mm/h)	1	4	8	32
ϖ	0.20	0.27	0.33	0.40
δ	0.37	1.33	2.59	10.2
θ = 76				
Ite. Tv	253.8	240.0	231.3	223.6
Ite. Th	235.8	237.0	228.5	220.2
MOM Tv	253.8	240.0	231.3	220.8
MOM Th	235.8	237.0	228.4	217.3
θ = 48.6				
Ite. Tv	230.7	251.7	245.1	235.4
Ite. Th	200.6	247.3	242.9	232.4
MOM Tv	230.6	251.7	245.2	233.8
MOM Th	200.5	247.4	242.9	231.2
θ = 16				
Ite. Tv	203.7	248.4	248.7	238.6
Ite. Th	196.0	246.8	248.1	237.9
MOM Tv	203.6	248.4	248.8	237.2
MOM Th	195.9	246.8	248.2	236.6

Table 2. Comparison of brightness temperatures calculated with MCM (first row) and MOM (second row) for different rainfall rates R and viewing angle θ.

R mm/h	θ =26 horiz.	θ =26 verti.	θ =49 horiz.	θ =49 verti.
1				
	196.9	207.1	201.9	232.8
	196.9	207.1	201.9	232.7
2				
	227.9	233.5	234.8	248.5
	227.9	233.5	234.6	248.4
4				
	247.4	249.5	247.8	252.3
	247.3	249.5	247.7	252.4
8				
	247.9	249.0	243.0	245.6
	247.9	248.8	243.0	245.5
16				
	241.4	242.5	236.3	239.0
	241.4	242.3	236.2	238.9
32				
	236.2	237.4	231.3	234.5
	236.3	237.2	231.4	234.3

Table 3. Mean brightness temperature at $\theta = 50°$ for an isothermal cloud computed by the Monte Carlo technique and the analytical method of Weinman and Davis [7]. The value in parentheses is the uncertainty of the MCM given by Weinman and Davis [7]. The cloud is characterized by a asymmetry factor g = 0. constant temperature T = 275 K. and a single scattering albedo $\varpi = 0.6$. The environment around the cloud is assumed vacuum.

δz	$\delta x = \delta y$	0.3	3	10	∞
0.3	A	21.6 (0.2)	44.0 (0.6)	48.7 (0.7)	52.3 (0.3)
	B	21.7	42.6	46.3	48.1
	C	22.0	44.5	49.0	51.3
	D				51.4
1	A	31.1 (0.5)	94.6 (1.4)	119 (2.0)	136.7 (0.7)
	B	31.3	92.5	116	129.7
	C	31.9	96.0	121.1	136.6
	D				136.9
3	A	35.9 (0.6)	131 (2.0)	169 (3.0)	210 (1.0)
	B	36.0	130	171	208
	C	36.8	132.9	173.3	210.9
	D				211.2
10	A	37.9 (0.6)	158 (2.0)	189 (3.0)	224.5 (1.5)
	B	38.0	159	191	219
	C	38.9	162.6	191.9	221.9
	D				222.1

A: MCM of Weinman and Davis [7]
B: analytical method from Weinman and Davis [7]
C: MCM. our model
D: MOM for the horizontal homogeneous cloud
$\delta x = \delta y$: optical depth in horizontal directions x and y. respectively.
δz: optical depth in vertical direction z.

4. Results

For our calculations, the rain cloud and its environment are described as follows: cloud height 4.4 km, homogeneous distribution of rain water in vertical and horizontal directions, top temperature To = 273.1 K, surface temperature Ts = 290 K; the absorption coefficients of the atmospheric gases are taken from Liebe [18]. The raining cloud consists of water spheres with Marshall-Palmer size distribution. The sea surface reflectances are calculated from an empirical model from Wisler and Hollinger [14] with a surface wind speed of 5 m/s. Results will be shown for 37 GHz and 19.35 GHz at nadir direction and a viewing angle of 53°. The rain cloud is in the center of the model area, surrounded by a clear atmosphere with the same atmospheric temperature and humidity profiles as within the cloud. The whole model area has the size of 800 km by 800 km in the horizontal plane, which is large enough to neglect effects from horizontal boundaries (see table 4).

Table 4. Effects of the size of the cloud free environment on the brightness temperature of a rain cloud (1 km x 1 km horizontal size and a height of 4.4 km). The environment is a clear sky with the same atmospheric profiles as in the cloud. The rain rate is 5 mm/h.

environment around cloud (km)	$\theta = 0$
0	210.2
1	214.8
10	228.8
50	233.8
200	234.6
500	234.7

The different brightness temperatures in Table 4 result from radiation lost or gained from the cloud side walls to the environment.

The side effect (exchange of radiation between cloud and clear sky environment) depends on the:

(1) ratio of area of the side walls to the area of the top and bottom of the cloud
(2) optical properties of the cloud
(3) environmental conditions around the cloud

Figure 1a shows the typical Tb-R relationship for 37 GHz: The brightness temperature first increases with increasing rain rate due to the increase of the emissivity of the cloud; it reaches a maximum and then decreases due to the increase of scattering and the decrease of the cloud temperature with height. Smaller Tb for the smaller clouds (1 km by 1 km, 5 km by 5 km) are due to the side effects, which become smaller when the rain rate increases. For the 1 x 1 km cloud, the side effects overcompensate the scattering and temperature effects. Thus, Tb increases with larger rain rates. For 19.35 GHz similar results are obtained (see Fig. 1b), but the saturation point shifts to the rain rate of 20 mm/h.

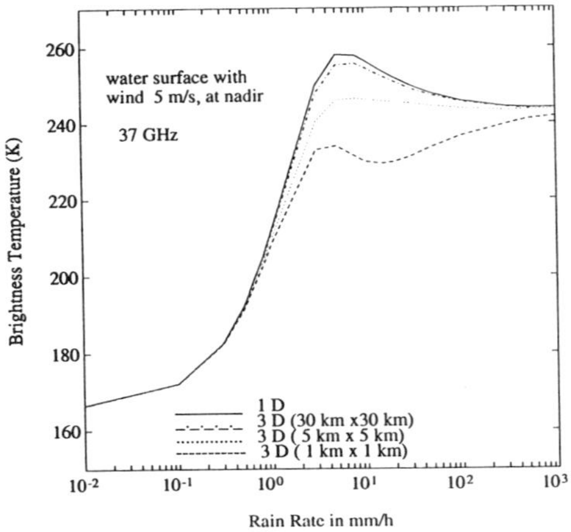

Figure 1a. Variation of the upward brightness temperature at nadir as a function of rain rate for different horizontal scales of the raining cloud.

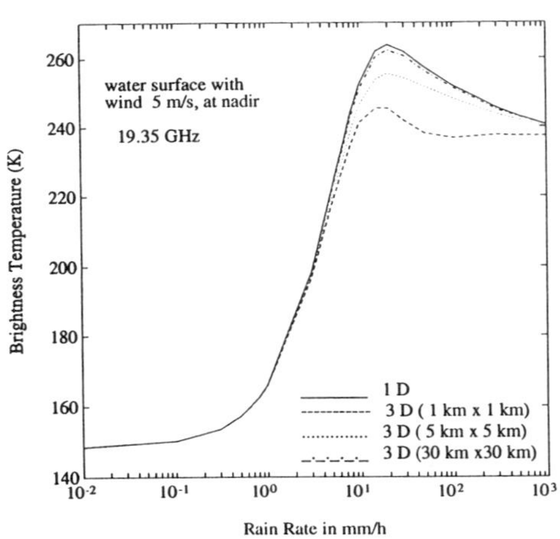

Figure 1b. Same as Fig. 1a, but for 19.35 GHz.

In order to study the effect of cloud coverage within the field of view, clouds of different size are placed in the centre of the pixel surrounded by cloud-free area. For the 1-D model a linear relationship is evident. As shown in Figure 2a, the relationship between Tb at 37 GHz and cloud cover for the 3-D model is non-linear. This is due to the side effects. An interesting feature is the over- and underestimation of Tb of the 1-D model for larger and smaller clouds, respectively. The underestimation is due to two effects: first, the high energy from the cloud which leaves the side walls is partly reflected at the surface and can still reach the radiometer. Second, the radiometer views the lower (warm) part of the cloud. These two effects overcompensate the side effects for the larger clouds, at least for larger zenith angles. For 19.35 GHz, rain clouds with 5 mm/h are semi-transparent. The side effect for 19.35 GHz is larger than that for 37 GHz (Fig. 2b).

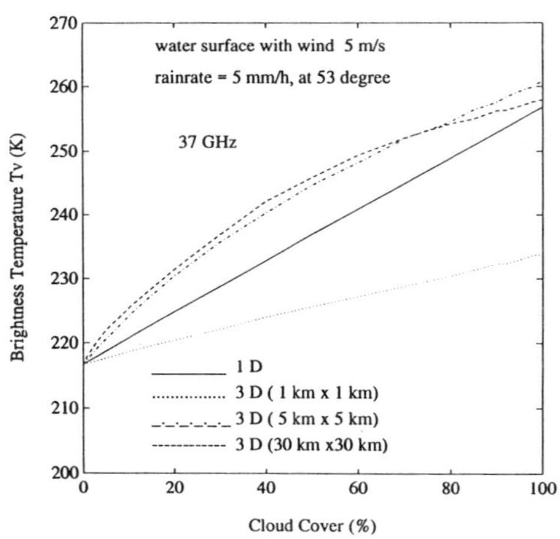

Figure 2a. Variations of the 37 GHz vertical polarized brightness temperature with cloud coverage at the viewing angle = 53° .

To illustrate these effects we performed a numerical experiment with an airplane flying along the center line of a rain cloud of 10 km x 10 km horizontal dimensions and a

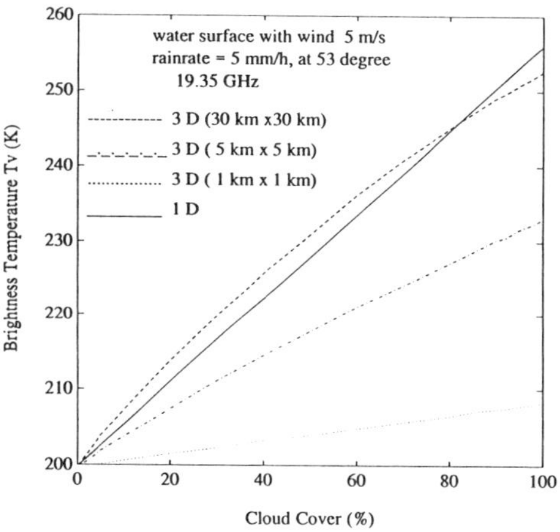

Figure 2b. Same as Fig. 2a. but for 19.35 GHz.

height of 4.4 km. The plane flies from west to east (left to right in Fig. 3a) and the radiometer looks forward at a zenith angle of 53°. The FOV of the radiometer is 1 km x 1 km. The radiometer first receives a signal from the cloud when viewing the area still about 6 km away from the cloud edge when the radiometer views the reflection of the warm rain cloud. The signal reaches its maximum at the edge of the rain cloud because the radiometer now observes the lower warmer part of the cloud where the temperature is high. Then the brightness temperature decreases monotonically to the clear sky value about 6 km behind the cloud edge when the radiometer viewing beam does not cross the cloud anymore. For 19.35 GHz the signal increases from the left side to about center of the rain cloud and then decreases from the center to the right side. At the left side of edge brightness temperature for 19.35 GHz is lower than the brightness temperature for 37 GHz. This is due to the fact that the cloud of 5 mm/h is more transparent for 19.35.

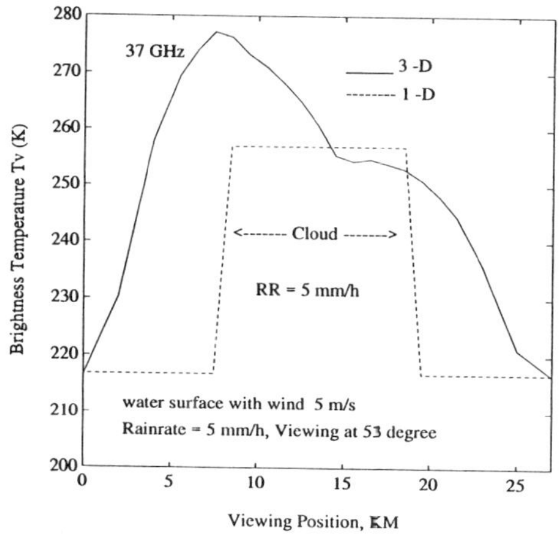

Figure 3a. Calculated vertically polarized brightness temperature with respect to the viewing positions at the ocean surface. The zenith angle is 53°. The rainfall rate is 5 mm/h.

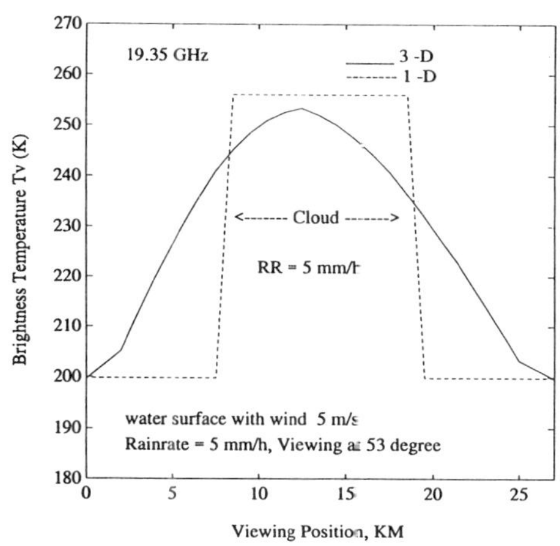

Figure 3b. Same as Fig. 3. but for 19.35 GHz.

5. Conclusion

Brightness temperatures at 37 GHz and 19.35 GHz are calculated from 3-D and 1-D models at a viewing angle of 53°. Differences can approach 50 K for clear sky regions in the neighbourhood of a rain cloud. Rainfall errors of 1~5 mm/h can occur. At nadir, the brightness temperature calculated from the 3-D model is lower than that from the 1-D model because of the energy escaping from the sides. The difference depends on frequency, viewing angle, and the vertical distribution of optical properties of the cloud. For a finite rain cloud with a rainfall rate of 10 mm/h and a horizontal scale of 5 km by 5 km, the calculated brightness temperature from the 1-D model at 37 GHz is larger than that from the 3-D model at nadir, but smaller at the viewing angle of 53°. We can conclude:

(1) Non-linear relationships exist between the brightness temperatures and cloud coverage.

(2) At non-nadir angles there are dramatical increases of Tb for clear sky areas in the neighbourhood of rain clouds due to the fact that the high energy from the cloud which leaves the side wall is partly reflected at the surface and can reach the radiometer in the clear area.

(3) For the 37 GHz channel of SSM/I, the plane parallel slab model tends to underestimate the rainfall for small rates (< 1 mm/h), overestimate the rainfall for rates 1 ~ 5 mm/h, and underestimate the rainfall for larger rates.

REFERENCES

1. Kilonsky. B. J. and C. S. Ramage. 1975. A technique for estimating tropical open ocean rainfall from satellite observations. J. Appl. Meteor.. 15. 972-975.

2. Arking. P. N.. and B. N. Meisner. 1987. The relationship between large-scale convective rainfall and cold cloud over the western hemisphere during 1982-84. Mon. Wea. Rev.. 115. 51-74.

3. Adler. R. F. and A. J. Negri. 1988. A satellite infrared technique to estimate tropical convective and stratiform rainfall. J. Appl. Meteor.. 27. 30-51.

4. Turpeinen. O. M.. and A Abidi. and W. Belhouane. 1987. Determination of rainfall with the ESOC precipitation index. Mon. Wea. Rew.. 115. 2699-2706.

5. Kummerow. C.. and J. A. Weinman. 1988. Determining Microwave Brightness Temperatures from Precipitating Horizontally Finite and Vertically Structured Clouds. J. Geophy. Res.. 93. D4. 3720-3728.

6. Haferman. J. L.. W. F. Krajewski. T. F. Smith. and Sanchez A.. 1993. Radiative Transfer for a Three-Dimensional Raining Cloud. Applied Optics. 32. 2795-2802.

7. Weinman. J. A.. and R. Davis. 1978. Thermal Microwave Radiances from Horizontally Finite Clouds of Hydrometeors. J. Geophy. Res.. 83. C6. 3099-3107.

8. McKee. T. B. and Stephen K. Cox. 1974. Scattering of Visible Radiation by Finite Clouds. J. Atmosph. Sci.. 31. 1885-1892.

9. Stephens. G. L.. 1988. Radiative Transfer through Arbitarily Shaped Optical Media. J. Atmosph. Sci.. 45. 1818-1848.

10. Lovejoy. S. et. al.. 1990. Discrete Angle Radiative Transfer. 1. Scaling and Similarity. Universality and Diffusion. J. Geophy. Res.. 95. 11699-11715.

11. Girens. K. M.. 1993. A Fast Six-Flux Radiative Transfer Method for Application in Finite Cloud Models. Contributions to Atmospheric Physics. 66. 73-87.

12. Roberti. L. and C. Kummerow. 1994. A quantitative comparison between 3-D and plane parallel microwave radiative transfer codes applied to horizontally and vertically structured precipitating clouds. In Specialist Meeting on Microwave Radiometry and Remote Sensing of the Environment. Feb. 14-17. 1994. Rome. Italy

13. Tsang. L.. J. A. Kong. and R. T. Shin. 1985. Theory of Microwave Remote Sensing. John Viley & Sons. Inc. Canada. 613 pp.

14. Wisler. M. M.. and J. P. Hollinger. 1977. Estimation of marine environmental parameters using microwave radiometric remote sensing systems. N. R. L. Memo. Rept. 3661.

15. Liu. Q.. 1990. An analytical solution of transmission and reflection operators for homogeneous atmospheres. Contributions to Atmospherical Physics, 63. 128. 128-133.

16. Liu. Q.. C. Simmer. and E. Ruprecht. 1991. A general analytical expression of the radiation source function for emitting and scattering media within the matrix operator method. Contributions to Atmospheric Physics. 64. 73-82.

17. Weinman. J. A. and P. J. Guetter. 1977. Determination of rainfall distributions from microwave radiation measured by the Nimbus 6 ESMR. J. Appl. Meteorol.. 16. 437-442.

18. Liebe. H. J.. 1985. An updated model for millimeter wave propagation in moist air. Radio Science. 20. 1069-1089.

A Microwave Radiometry Characterization of Precipitating Clouds

P. BASILI[1], P. CIOTTI[2], G. D'AURIA[3], F.S. MARZANO[3] and N. PIERDICCA[3]

[1] *Institute of Electronics, University of Perugia, Santa Lucia, Perugia, Italy*
[2] *Department of Electric Engineering, University of L'Aquila, Poggio di Roio, L'Aquila, Italy*
[3] *Department of Electronic Engineering, University of Rome "La Sapienza", via Eudossiana 18, Rome, Italy*

Abstract. Precipitating clouds are classified by means of both the equivalent water contents of four hydrometeor species and their upward brightness temperature at four frequencies (SSM/I channels). The merits of the two classification are discussed.
Five convective clouds characterized by the brightness temperature they produce are generated according to a Gaussian multivariate distribution. The microphysics of the generated data set agrees with a space-time evolving model assumed for a storm (UW-NMS). The relatively high correlation between some hydrometeor contents makes it possible to estimate profiles of cloud parameters using only four radiometric measurements. The accuracy of these retrievals is discussed when a multiple multivariate regression technique is used.

Key words: Microwave radiometry, Clouds, Retrieval algorithms

INTRODUCTION

One of the problems in investigating the characterization of clouds from space using microwave radiometers is the scarce in situ data on clouds. This makes it almost impossible to test the satellite data. Even if we limit ourselves to only one cloud parameter, the rain rate, the surface rain gauges available are sparsely sampled and indeed doubtfully reliable. On the other hand the use of other sensors such as radar's, IR radiometers, and ground based microwave radiometers may add useful indirect information about the cloud structure, but for their use one must assume statistical relations from among the constituents of the cloud. Moreover, these relations are also needed to obtain a statistical inversion of radiometric data and we must therefore assume this information from suitable cloud models. Models which have been used by various authors range from a fully random generation of profiles of the constituent hydrometeors of the cloud with few physical constraints [1], [2], [3], to those which use a simulated reconstruction of a selected precipitation event, fully based on the microphysics of the cloud [4], [5].
Numerous techniques have been developed to retrieve the rain intensity at the surface, but the determination of profiles of other hydrometeors is also important. This knowledge is particularly important for characterizing different types of clouds and for many applications.
In this paper, both the geophysical data set obtained from a microphysical space-time evolving cloud model and the corresponding simulated data set of brightness temperatures are analyzed and classified using clustering algorithms. The classes obtained from the data set of simulated brightness temperatures show the various kinds of clouds which can be singled out by microwave signature. Finally, different regression algorithms are applied to retrieve cloud parameters and their accuracy is determined. It is worth noting that no attempt was made to include other sources of uncertainty, such as surface emissivity variations and antenna beam

filling effects in the simulations. However, instrumental noise was included (1 K).

GEOPHYSICAL AND RADIATIVE DATA SET

The geophysical cloud data used in this study is obtained from a three-dimensional time-dependent cloud mesoscale model (UW-NMS) [6]. It is based on the results of a 4-hour simulation of an intense summer hailstorm which occurred in Alabama, and takes into consideration the microphysical relations between four different species of hydrometeors: precipitating liquid water (rain drops), non-precipitating liquid water (cloud droplets), precipitating ice (graupel particles) and non-precipitating ice (ice crystals and aggregates). Water vapor, temperature profiles and the dimensional distribution of particles are also included in the model. The hydrometeor profiles are sampled every minute and the horizontal and vertical resolutions are respectively 1 km and about 0.5 km. This model was further modified by horizontally averaging the profiles over 10 km to account for the antenna beam effects. The resulting data set is composed of 768 cloud structures sampled in the horizontal plane and in time. For each cloud structure, four profiles of each hydrometeor species are expressed in Equivalent Water Content (EWC's), and the profiles relative to pressure, temperature and specific humidity are given. Each structure represents different features of the storm during its lifetime. We assume that each structure represents the entire population of a continental precipitation random process.
In order to obtain the radiometric signatures of these cloud structures the radiative transfer equation must be solved. The solution is calculated for the four frequencies of the Special Sensor Microwave Imager (SSM/I): 19.3, 22.2, 37.0, 85.5 GHz, so that comparisons can be made with available measurements. The radiative transfer model assumes a plane parallel geometry. All particles are supposed to be spherical and the Mie theory is used to calculate the single scattering parameters. Extinction coefficients are computed by averaging over the size distribution of each particle. The effects of temperature and gaseous absorption are taken into account using Liebe's model [7]. The Heyney-Greenstein approximation is used for the phase function of each scatterer.

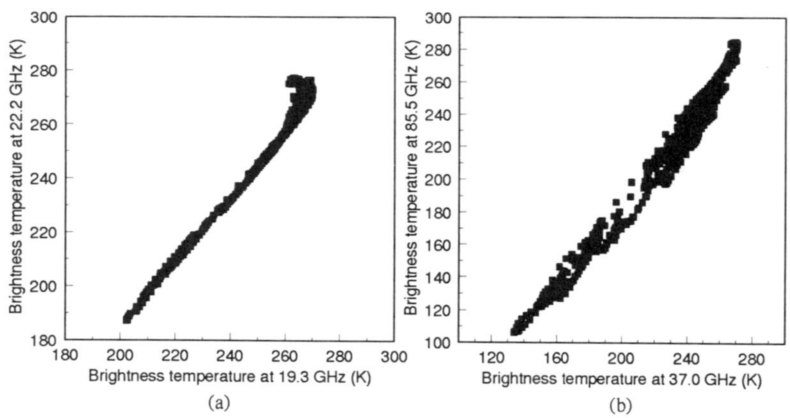

Figure 1. a) brightness temperatures (T_B) at 22.2 GHz vs. T_B's at 19.3 GHz for the microphysical data set (768 points); b) the same as in Figure 1a, but for T_B's at 37.0 GHz vs. 85.5 GHz.

The non-polarized radiative transfer equation has been solved numerically using the discrete-ordinate method [8], applied to 41 atmospheric layers as corresponding to the vertical resolution of the original microphysical cloud model. The numerical solution solves a set of linear differential equations corresponding to a finite number of elevation angles θ, while the azimuthal dependence is ignored (Fourier expansion of zero order). The Earth's surface is modeled as Lambertiana with an emissivity of 0.85 and at a temperature of 300 K. The result of the radiative transfer solution is a random vector **t**, whose elements are the upwelling brightness temperatures at the four frequencies, for each cloud structure. In Figure 1 (a,b) scatter plots of the brightness temperatures (T_B's) for two sets of frequencies, 19.3-22.2 GHz and 37.0-85.5 GHz, are shown. The dispersion in these plots is due to the e.m. interaction mechanisms with the particles whose difference at the two frequencies is determined by the intrinsic variability of the geophysical data set.

CLASSIFICATION AND LAYERING PROCEDURE

The microphysical time evolving model used covers a large variety of cloud structures from the initial stage to the mature and declining stage of a storm. It is interesting to group these structures in order to see if they can represent diverse typologies of clouds. A classification algorithm has been implemented based on two different points of view: geophysical and radiometric. In both cases the classification is made by using a disjoint clustering analysis based on the Euclidean distance [9]. In the first case the classification is made by considering either the profiles of the EWC of each hydrometeor species or their columnar EWC's (CEWC's). The differences between EWC and CEWC classifications are found non essential. In the second case the cloud structures are classified on the basis of the radiometer measurements for the cloud, that is, according to the four simulated brightness temperatures. We consider five possible classes as a compromise between a sufficient number of cloud structures in each class and an efficient stratification of the data set for further statistical analysis.

Table 1. Confusion matrix between different kinds of classification criteria applied to the 7-layer cloud data set with 768 points.

		Classification on CEWC.					
		Class 1	Class 2	Class 3	Class 4	Class 5	total
Classification on brightness temperatures	Class 1	256					256
	Class 2	116	17	62			195
	Class 3	5	62	101	1		169
	Class 4			20	35	24	79
	Class 5				22	47	69
	total	377	79	183	58	71	768

To appreciate the effects of the different criteria of classification, Table 1 shows the confusion matrix relative to classifications based on CEWC's and T_B's. The Table indicates the number of individuals (over a total of 768) which belong at the same time to classes of both classifications. We can note that classes of both classifications (with some exceptions; i.e., class 2) represent fairly well the same kinds of cloud structures. That is, for the chosen set of frequencies (SSM/I channels) and within the reported limitations, radiometric measurements are able to characterize fairly well the physical structure of the observed clouds. However the CEWC's classification is more suitable to understand the physical nature of the classes and therefore to compare them to particular situations which are present in an evolving storm (the model from which we have started the analysis). In fact, referring to the CEWC's classification it is possible to see that some of these classes are directly connected to specific part of the storm, during its space-time evolution [4]. Specifically, class 2 represents the rear anvil of the storm, while the first cell is mainly in class 3, and the main core is in class 4 and 5. Class 1 groups diverse parts of the storm, principally connected with its contours (flanks) and partially

with the first cell. Class 1 is therefore less characteristic of a single cloud feature [4].
In the following we only report the results obtained by classifying according to the simulated brightness temperatures for the different clouds, since this classification copes with the actual measurements and possible retrieval schemes.
In Figures 2 and 3, for example, separate scatter plots of T_B's are shown for two different sets of frequencies, for each of the five classes. They represent a segmentation of plot of Figure 1.

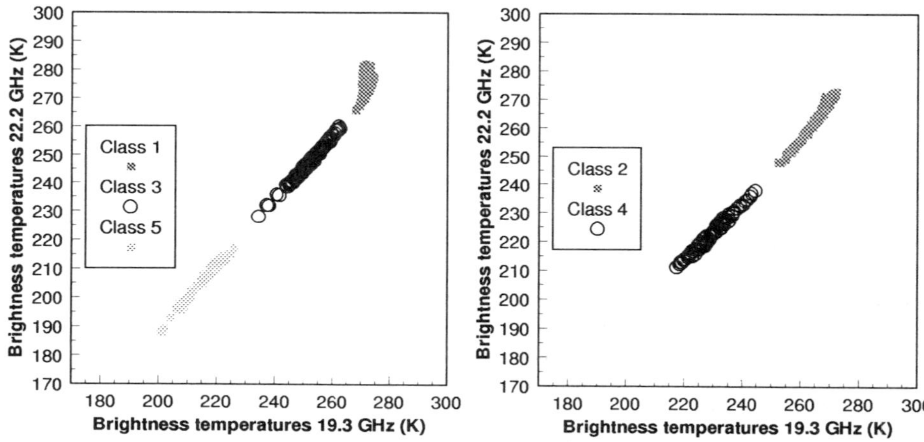

Figure 2. T_B's at 22.2 GHz vs. T_B's at 19.3 GHz for the classified (5 classes) and reduced (7 layers) data set (768 points).

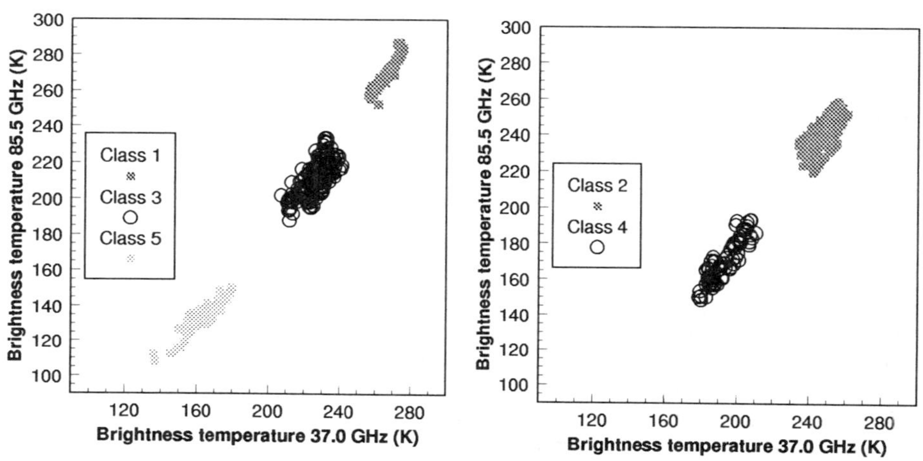

Figure 3. T_B's at 85.5 GHz vs. T_B's at 37 GHz for the classified (5 classes) and reduced (7 layers) data set (768 points).

Figures 4 and 5 plot CEWC values of relevant hydrometeor species, for classes based on TB's classification. In Figure 4 the CEWC of cloud liquid is plotted against the CEWC of rain drop, while Figure 5 shows the CEWC of ice crystals plotted against the CEWC of graupel particles. According to the results reported in the confusion matrix of Table 1, classes in these planes show less defined contours and an evident partial overlapping. A slighter overlapping of the graupel values is noted since these particles affect to a greater extent the T_B's. A classification based on CEWC (not shown) would exhibit more defined contours in comparison with Figures 4 and 5.

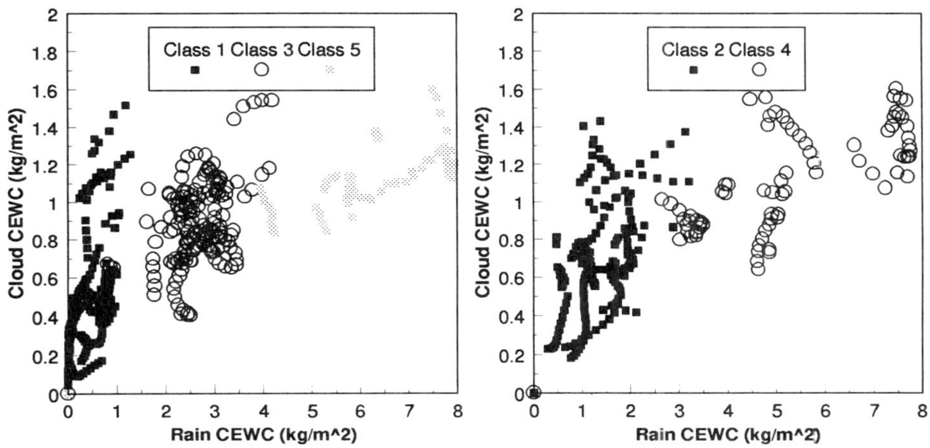

Figure 4. Columnar Equivalent Water Content (CEWC) of cloud droplets vs. CEWC of raindrops for the microphysically modeled data set (768 points), when clustered in 5 classes on the base of the T_B's at 19.3, 22.2, 37.0 and 85.5 GHz.

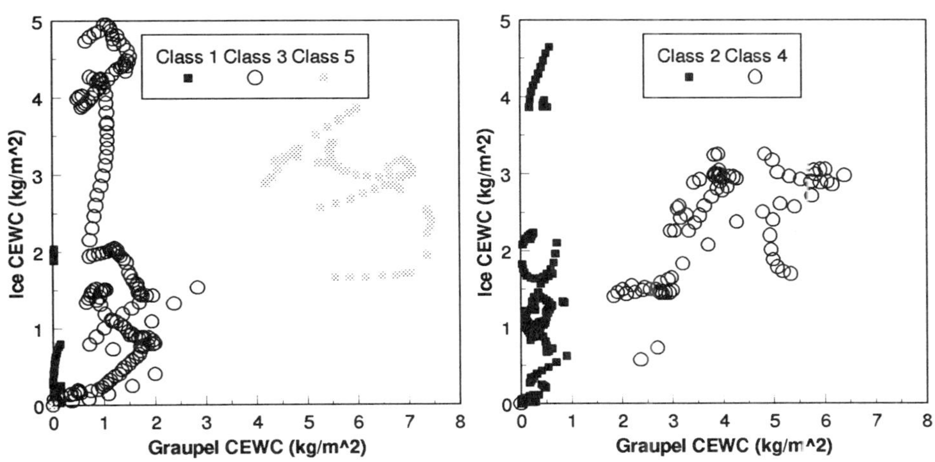

Figure 5. Columnar Equivalent Water Content (CEWC) of ice crystals vs. CEWC of groupel particles for the microphysically modeled data set (768 points), when clustered in 5 classes on the base of the T_B's at 19.3, 22.2, 37.0 and 85.5 GHz.

All the classes obtained from the diverse classifications seem to indicate clouds of convective origin. This is revealed by the presence of liquid water well beyond the freezing level and by the fairly high percentage of ice of different kinds in various layers of the cloud. Figure 6 shows profiles of the four hydrometeor EWC's averaged over gross layers of atmosphere for classes 1 and 4; these classes are characterized, respectively, by a low and a high presence of graupel particles.

Although the five classes can represent only some clouds typologies, the statistical information contained in them can become valuable in the process of generating other kinds of clouds. In fact the microphysical relationships among the hydrometeor profiles of different species appear to be statistically valid for a larger typology of clouds.

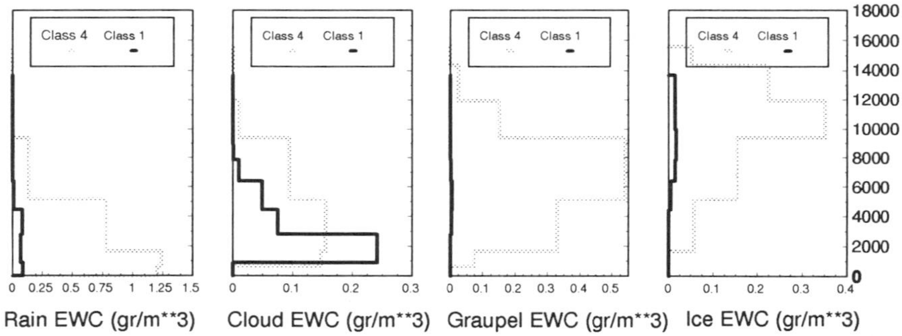

Figure 6. Mean vertical profiles of each hydrometeor density for two classes: class 1 with a small graupel content and class 4 with large graupel content.

In order to investigate this, the number of variables describing a cloud structure is reduced by simplifying the cloud geometry of each class to seven horizontal layers, where the EWC values of the four hydrometeors are assumed to be equal to their vertical mean. Within each class the altitude limits of each layer have been averaged and likewise the profiles of pressure, temperature and specific humidity. Subsequently, we have calculated for each class the correlation matrix of the cloud variables: four hydrometeor EWC's for the seven layers (28 variables). Referring to class 2 Figure 7 shows plots of the correlation between rain drops (in layer 1) and graupel particles (in layer 4) and the other particles as a function of height. Here

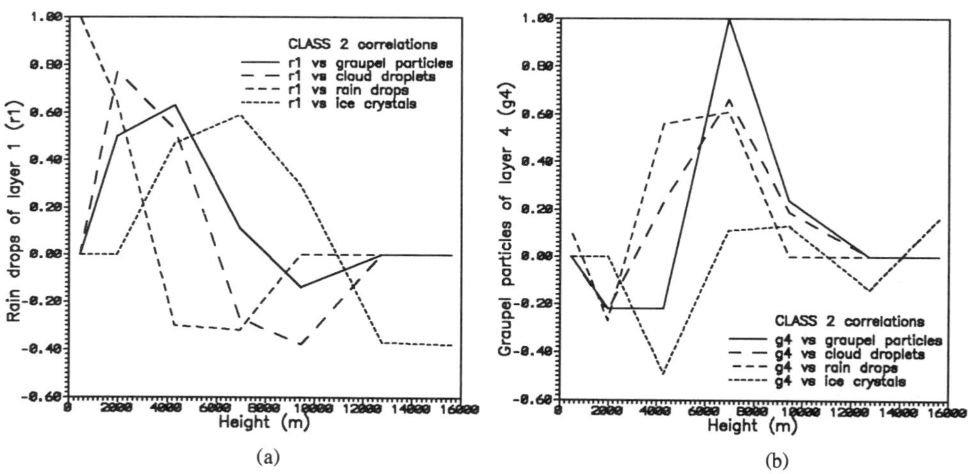

Figure 7. a) correlation coefficients of rain drops at the first layer with the hydrometeor EWC's in each layer vs. mean layer altitude. b) the same as in Figure 5a, but for graupel particles at the fourth layer.

we can appreciate the positive or negative correlation that indeed exists, for instance, between rain in the first layer (counting from the bottom) and the graupel particles just around the freezing level (around 4000 m). Also, we observe the correlation between graupel well above the freezing layer and the cloud in that region, and the ice at the higher levels and so on.

GENERATION OF CLOUD DATA SETS

We have seen from Table 1 that the population of some classes is rather scarce. It is necessary to enlarge the population of a given data set in order to make the data set more meaningful from a statistical point of view. Statistical inversions of cloud parameters will then be more reliable. Furthermore, modifications of a given data set are necessary to cope with other kinds of clouds which may be observed from actual radiometric measurements. We have tackled the problem by producing new cloud data sets for each class using a random Gaussian generation of variables having the same mean vector, correlation matrix and variances as the class. This ensures a microphysically consistent data set.

As already noted, for a seven layer cloud model we need no more than 28 random variables (plus the associated rain rate, if required) to describe the cloud structure. The data set is therefore defined by a matrix **G**, whose rows are made of the 28 standardized EWC's (i.e. the ratio between the centered values of EWC's and their standard deviation) of the four hydrometeors and the number of columns is n, the number of individuals of each class. Associated to the matrix **G** are the vectors $<g>$ and σ^2 which respectively represent the means and the variances of each variable. The correlation matrix $\mathbf{R_g}$ is, therefore, given as:

$$\mathbf{R_g} = (n-1)^{-1}\, \mathbf{G}\, \mathbf{G}^T \quad (1)$$

We have investigated the use of these new data sets obtained using a Gaussian multivariate generation. First of all, we have reproduced the previous classes of clouds, increasing to one thousand the number of individuals in each class. This is obtained by keeping the same matrix **G**, and the same $<g>$ and σ^2, pertaining to each class. From the radiative transfer model, the random vector **t** of the four brightness temperatures has then been obtained. These results are shown in Figure 8, where the scatter plots of T_B's for two frequencies (37.0 and 85.5 GHz) are given. By superimposing the two scatter plots, the signature of a more general cloud can be obtained, and compared with that of Figure 1, relative to the original model.

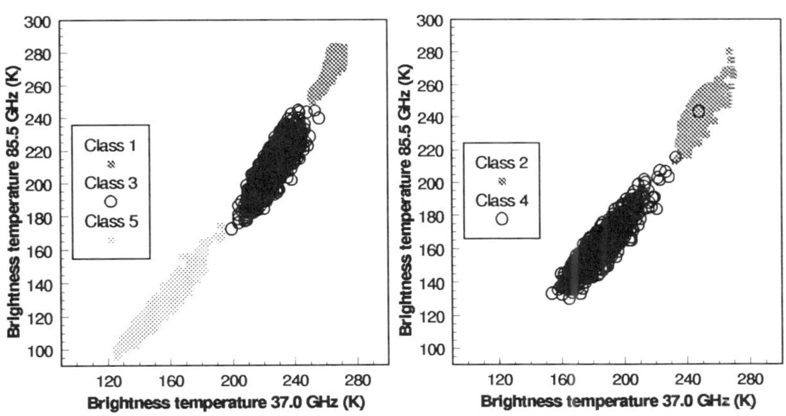

Figure 8. T_B's at 85.5 GHz vs. T_B's 37.0 GHz for the statistically generated data set, classified into 5 classes (1000 points for each class).

However, this procedure can also be applied to generate other classes of clouds, by keeping matrix **Rg** as a constant and by varying <g> and σ^2. For instance, by increasing the values of σ^2 a wider range of points is obtained, whereas, when increasing <g> by a scale factor, points of the scatter plot move towards lower brightness temperatures. Other alterations of <g> elements can be introduced to cope with actual radiometric measurements of clouds.

STATISTICAL INVERSION AND CONCLUSIONS

A regression technique has been applied to each enlarged class of clouds to show the potential of retrieving the random vector **g** (the EWC's of hydrometeors in each layer) from a vector **t** of brightness temperatures. This is accomplished for each class by computing the standard deviation of residual terms using multivariate multiple linear regression, which may be written in matrix form:

$$\mathbf{g} = \mathbf{B}\,\mathbf{t} + \mathbf{e} \tag{2}$$

where **B** is the matrix of the weights (regression coefficients) and **e** is a random vector of the residual terms. Subject to the minimization of the average of ($\mathbf{e}^T\,\mathbf{e}$), we get:

$$\mathbf{B} = \mathbf{G}\,\mathbf{T}^T\,(\mathbf{T}\,\mathbf{T}^T)^{-1} \tag{3}$$

where **T** is the matrix of random vector **t** standardized about the mean.

Results of this regression analysis are given by computing the ratio between the standard deviation of the residual terms and the standard deviation calculated from the data set. Figure 9 shows the ratios obtained for each EWC of the four hydrometeors at each layer and for the five classes of clouds. This ratios give an appraisal of the efficiency of the regression procedure for inferring the required parameters. Figure 9 also shows the large differences in error found when estimating the diverse cloud parameters. For example, ice crystals result to have large errors, while graupel particles are retrieved more accurately. The resulting errors are still optimistic since the residual terms are computed by applying Eq. (3) to the same data set used for estimating the regression coefficients.

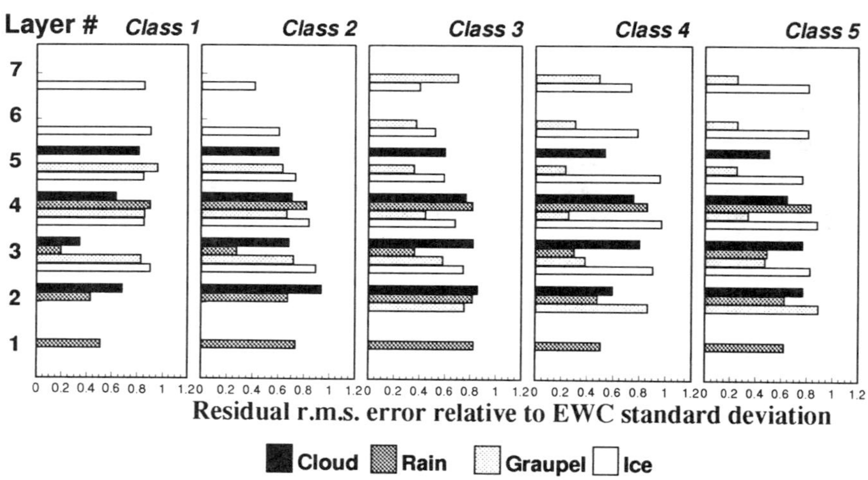

Figure 9. Residual r.m.s. errors of the Equivalent Water Contents within the layer for each hydrometeor, of the multiple linear regression. The r.m.s. errors are normalized to EWC standard deviation of the data set.

Furthermore, when processing radiometric measurements, a discrimination analysis should be made in advance to select the classes and, therefore, the involved errors should also be considered. Finally, errors originated from the uncertainties in the Earth's surface emissivity and antenna beam filling effects have not been considered.

Acknowledgments

This work has been partially supported by Italian Space Agency (ASI) and Ministry of University and Research (MURST). We are grateful to Dr. A. Mugnai (CNR/IFA) for providing us the UW-NMS data and for his collaboration.

REFERENCES

1 Wilheit, T.T., 1986. Some comments on passive microwave measurement of rain, *Bull. Am. Meteor. Soc.*, **67**, **10**, 1226-1332.

2 Kummerow, C., R.A. Mack, and I.M. Hakkarinen, 1989. A self-consistency approach to improve microwave rainfall rate estimation from space, *J. Appl. Meteor.*, **28**, 869-884.

3 Basili, P., P. Ciotti, G. d'Auria, F.S. Marzano, N. Pierdicca, and A. Mugnai, 1992. A simulation study for retrieving rainfall from space-borne microwave radiometers, Proc. of Specialist Meeting on Microwave Radiometry and Remote Sensing Appl., Boulder (CO), 251-256.

4 Smith, E.A., A. Mugnai, H.J. Cooper, G.J. Tripoli, and X. Xiang, 1992. Foundations for statistical-physical precipitation retrieval from passive microwave satellite measurements. Part I: brightness temperature properties of time-dependent could-radiation model, *J. Appl. Meteor.*, **31**, **6**, 506-531.

5 Basili, P., P. Ciotti, G. d'Auria, F.S. Marzano, N. Pierdicca, and A. Mugnai, 1992. Cloud microphysical model application to multivariate analysis of satellite microwave radiometric data, Proc. of Specialist Meeting on Microwave Radiometry and Remote Sensing Appl., Boulder (CO), 245-250.

6 Tripoli, G.J., 1992. A nonhydrostatic model designed to simulate scale interaction", *Mon. Wea. Rev.*, **120**, 1342-1359.

7 Liebe, H., 1985. An update model for millimeter wave propagation in moist air, *Radio Sci.*, **20**, **6**, 1069-1089.

8 Stamnes, K., S.C. Tsay, W. Wiscombe and K. Jayaweera, 1988. A numerically stable algorithm for Discrete-Ordinate-Method Radiative Transfer in Multiple Scattering and Emitting Layered Media, *Appl. Opt.*, **27**, 2502-2509.

9 Richards J. A., 1986. Remote Sensing Digital Image Analysis: An Introduction, Springer-Verlag, Berlin, 173-204.

Polarimetric Measurements and Model Calculations of Downwelling Rain Brightness Temperatures

A. HORNBOSTEL[1], A. SCHROTH[1] and B. G. KUTUZA[2]

[1]German Aerospace Research Establishment (DLR), Institute for Radio Frequency Technology, D-82230 Oberpfaffenhofen, Germany.

[2]Institute of Radio Engineering and Electronics of the Russian Academy of Sciences, Moscow, GSP-3, 103907, Russia.

Abstract Dual polarization measurements of sky brightness temperature are obtained for a convective and a stratiform rain event. Concurrent ground based measurements were also obtained of rain rate, distrometer data and other meteorological parameters like ambient temperature and wind speed. The paper analyzes the correlation of brightness temperature and differential brightness temperature with rain rate and drop size distribution. The measurements are also compared with radiative transfer model calculations. These model calculations are used to develope an algorithm to derive an effective (path integrated) two-parameter drop size distribution from polarimetric radiometer measurements. Comparisons with drop size distributions obtained by distrometer and polarimetric radar measurements show good agreement.

Keywords: Rain rate, drop size distribution, radiometry, radiative transfer, polarization, brightness temperature.

INTRODUCTION

In the framework of the Olympus propagtion experiment (OPEX), ground based radiometer measurements at 19 GHz have been carried out together with 20 GHz beacon measurements along the Olympus Earth-satellite path with an elevation of 27.6 degrees. Additionaly, the rain rate, distrometer data and other meteorological parameters like ambient temperature and wind speed were measured at the ground station [1, 2]. Due to the high correlation of sky brightness temperature with rain rate and path attenuation, it is possible to determine both parameters by radiometer measurements. However, the relationship of brightness temperature with rain rate and attenuation is non-linear, and for high values of both the radiometer measurement is saturated. Also, contributions by atmospheric gases, clouds and ground surface to the measured brightness temperatures have to be taken into account. It is also recognized that the brightness temperature and attenuation are path integrated values, but the rain rate and other ground measurements are measured at the surface.

Beacon receiver and radiometer of DLR form an integrated system [1, 2]. As a special feature, it is possible to measure the sky temperature in two linear orthogonal polarizations, if no beacon signal is received. Such dual polarization measurements have been carried out continuously since March 1993. Oblate rain drops, having their axis of symmetry aligned in the vertical direction, produce the largest difference between horizontal and vertical linear polarization measurements. Model calculations and measurements [2, 3, 4, 5] show that the maximum value of the differential temperature (difference between horizontal and vertical polarization) at 19 GHz is about 15 K. Therefore, for such measurements, the radiometer has to be calibrated carefully.

For a plane-parallel homogeneous rain slab of known thickness, the drop size distribution has a strong influence on the differential temperature. It is shown that it is possible to derive a two-parameter exponential drop size distribution by measurements of the brightness temperature and differential brightness temperature for a precipitating atmosphere. Furthermore the rain rate dependends on the drop size distribution, and can also be derived from the measurements.

RADIOMETER SPECIFICATIONS AND CALIBRATION

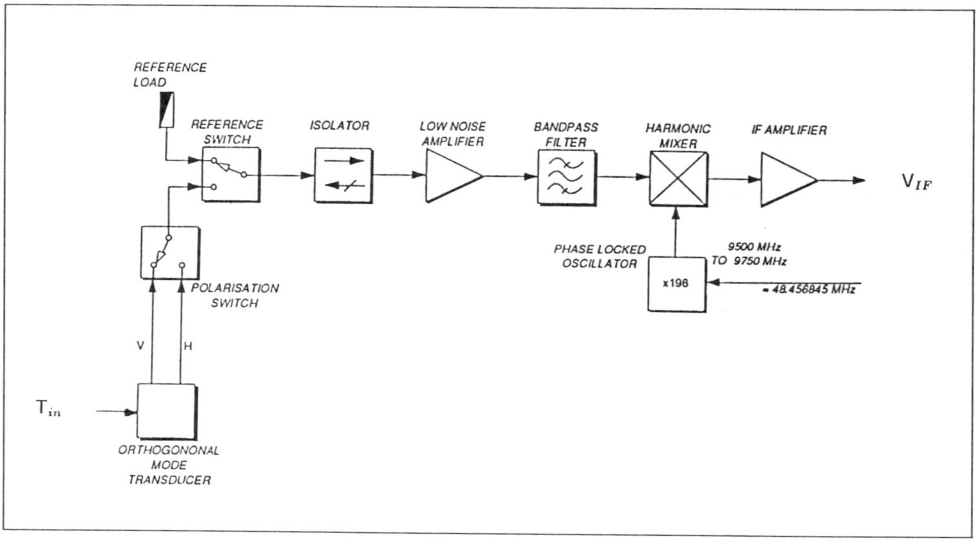

Figure 1. Front end of the polarimetric DLR radiometer.

The block diagram in Figure 1 shows the front end of the polarimetric Dicke radiometer. The reference switch and polarization switch are ferrite circulators, having switching frequencies of 933 and 1866 Hz, respectively. The center frequency is tunable from 19 to 19.5 GHz and the bandwidth is selectable between 25, 50, 100 or 200 MHz. The reference load temperature T_{ref} of 323 $\pm$ 0.2 K is controlled permanently. After low-pass filtering and further amplification stages, including AGC, the IF-Voltage V_{IF} is detected by a coherent detector. Measurements demonstrated an excellent linearity between the detector output voltage V_{out} and the input temperature T_{in}.

In the normal radiometer mode of operation, T_{in} is the antenna temperature T_a, which still has to be corrected for feed losses and sidelobe contributions to yield the sky temperature T_{sky}. The antenna is a Gregorian antenna with a beamwidth of 1.1 degree. Sidelobe supression and cross polarization discrimination are both better than 32 dB. The antenna pattern integration factor is ≈ 0.99 and the pattern diagrams show good axial symmetry with respect to the axis of the main beam.

Tests showed, that rain drops on the radome of the antenna horn can cause severe measurement errors of more than 10 K in the differential temperature. It was found that the removal of drops by blowing warm air on the radome was not sufficient in all cases. Therefore, a small roof, which prevents the drops from falling on the radome, was installed above the antenna horn in such way that it does not disturb the antenna pattern.

The slope and the offset temperature of the linear radiometer transfer curve must be derived by a two point calibration procedure for each frequency, polarization, and bandwidth of operation. This is done by using a cold load calibration with liquid nitrogen $T_{nit} = 77.4$ K and a second load containing ice-water $T_{ice} = 273.2$ K. During the calibration procedure, the receiver front end has to be removed from the antenna and connected to the cold load unit with different waveguide components. Due to losses inside this connecting network and mismatch of the different components, the receiver measures an *effective* temperature, which is different from the original temperature delivered by the cold load. The variation of this effective temperature is 15 K in the range 19-19.5 GHz in the case of liquid nitrogen. Also, there is a polarization dependence due to different connections of the load to the front end for both polarizations [2].

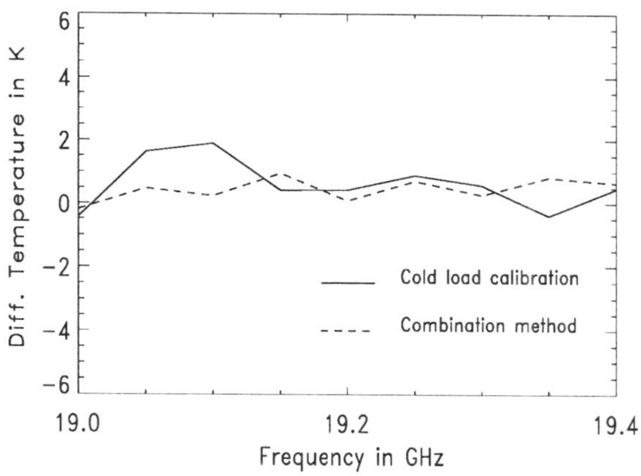

Figure 2. Measured differential temperatures of clear sky after calibration with two different methods: 1. Cold load calibration with liquid nitrogen and ice-water. 2. Combination of effective ice-water temperatures and antenna temperatures.

The accuracy of the calibration was checked by clear sky measurements. Under clear sky conditions, differences between the sky brightness temperatures measured along the horizontal and vertical polarization, the differential temperature, should be zero and the frequency dependence of the total brightness temperature from 19 to 19.5 GHz should be small. The results of the measurements are shown in Figure 2 and display the calibration error. The error bounds are between -1 and 2.0 K. Further improvement was achieved by use of the effective cold load temperatures of ice-water and replacement of the liquid nitrogen cold load temperatures by averaged clear sky antenna temperatures. With this method a calibration error of less than 1 K was achieved for the differential temperature. Clear sky measurements with different elevation angles (tip curves) showed that the absolute temperature error is below 2 K.

MODEL SIMULATIONS

Radiative Transfer Model

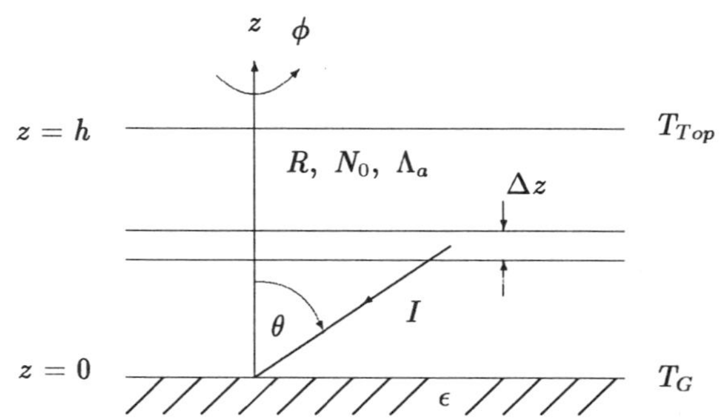

Figure 3. Modelling of a rain slab as a plane-parallel atmosphere with azimuth symmetry.

Figure 3 represent the model of a homogeneous rain slab of height h above the ground surface with linear temperature profile. Mathematically, this model is described by the vector equation of radiative transfer for a plane-parallel atmosphere with azimuth symmetry:

$$\mu \frac{d}{dz}[I(z,\mu)] = -[\Lambda(\mu)][I(z,\mu)] + [\Lambda(\mu)] \int_{-1}^{1} [P(\mu,\mu')][I(z,\mu')]d\mu' + [J_\epsilon(z,\mu)]. \tag{1}$$

$[I]$ is the intensity vector with two elements I_1 and I_2, which are the intensities in vertical and horizontal linear polarization. These are the first two elements of the modified Stokes-vector [6], which are independent of the components U and V in this case. Also, $\mu = cos\theta$, where θ is the zenith angle.

The function P ist the 2×2 phase-matrix for the azimuth symmetry with elements:

$$p_{ij}(\mu,\mu') = \frac{1}{\Lambda_{ii}} \int_0^{2\pi} |\rho f_{ij}(\mu,\mu',\phi-\phi')|^2 d\phi'; \quad i,j = 1,2, \tag{2}$$

where $f_{ij}(\mu,\mu',\phi-\phi')$ are the bistatic scattering coefficients of spheroidal oblate rain drops with Pruppacher-Pitter axial ratio, which are computed with the Waterman T-matrix method. The operator ρ denotes an integration with a two parameter exponential drop size distribution:

$$\rho f_{ij} = \int_{D_e} N(D_e)\, f_{ij}(D_e) dD_e. \tag{3}$$

$$N(D_e) = N_0 e^{-3.67 D_e/D_0}, \tag{4}$$

where D_e is the equivolumetric drop diameter.

The quantity $[\Lambda]$ in Eq. (1) is the 2×2 extinction matrix of the medium with elements:

$$\begin{aligned} \Lambda_{ii}(\mu) &= -\frac{4\pi j}{k} \rho f_{ij}(\mu, \mu' = \mu; 0) + \Lambda_a \\ \Lambda_{ij}(\mu) &= 0; \qquad i, j = 1, 2, \quad i \neq j, \end{aligned} \tag{5}$$

where Λ_a is the atmospheric absorption due to water vapour and oxygen. $[J_\epsilon]$ is the emission vector, describing emission by rain and atmospheric gases, which is calculated with the Rayleigh-Jeans aproximation and absorption coefficients, which are the difference between extinction and scattering cross section for each polarization.

After introduction of a source function

$$\begin{aligned} J_i(z, \mu) &= \int_{-1}^{1} p_{ii}(\mu, \mu') I_i(z, \mu') d\mu' \\ &+ \int_{-1}^{1} p_{ij}(\mu, \mu') I_j(z, \mu') d\mu' \\ &+ \frac{\epsilon_i(\mu)}{\Lambda_{ii}(\mu)}; \qquad i, j = 1, 2; \quad j \neq i. \end{aligned} \tag{6}$$

and definition of the optical depth for an inclined path through the slab with angel θ

$$\tau_{\mu,i}(z) = \int_0^z \frac{1}{|\mu|} \Lambda_{ii}(z', \mu) dz', \tag{7}$$

for both components, a simple first order differential equation is derived:

$$\frac{d}{d\tau_{\mu,i}} I_i(\tau_{\mu,i}) = -I_i(\tau_{\mu,i}) + J_i(\tau_{\mu,i}). \tag{8}$$

The formal solution of this equation is:

$$\begin{aligned} I_i^+(\tau_{\mu,i}) &= \int_0^{\tau_{\mu,i}} J_i(\tau'_{\mu,i}) e^{(\tau'_{\mu,i} - \tau_{\mu,i})} d\tau'_{\mu,i} \\ &= J_i(\tau_{\mu,i}) * e^{-\tau_{\mu,i}} \end{aligned} \tag{9}$$

for upwelling radiation ($\mu > 0$), and

$$\begin{aligned} I_i^-(\tau_{\mu,i}) &= \int_{\tau_{\mu,i}(h)}^{\tau_{\mu,i}} J_i(\tau'_{\mu,i}) e^{(\tau_{\mu,i} - \tau'_{\mu,i})} d\tau'_{\mu,i} \\ &= J_i(\tau_{\mu,i}) * e^{-\tau_{\mu,i}} \end{aligned} \tag{10}$$

for downwelling radiation ($\mu < 0$). The $*$ denotes a convolution integral.

A numerical solution is obtained using the method of successive orders of scattering [7, 8] in connection with the Gauss quadrature technique. The integral with μ' in Eq. (7) is replaced by a sum of 16 discrete angles μ_k multiplied with the Gauss-Legendre weight coefficients. For computation of the convolution integrals in Eq. (9) and Eq. (10) the slab is devided in a number of thin layers. The number of the layers is varied from 50 to 1000 in dependence of the rain rate. The intensity and source function components are expanded in a sum of nth order scattering contributions, which are computed recursively. In obtaining the results of the next section, multiple scattering up to 4th order is included.

Computed Results

Figures 4 - 6 show the simulated results for the rain slab shown in Figure 3 with height $h = 2.5$ km, surface temperature $T_G = 293$ K, $T_{Top} = 273$ K, ground emissivity $\epsilon = 0.96$, absorption by water vapour and oxygen $\Lambda_a = 0.035$ Np/km and rain rates $R = 1 - 100$ mm/h.

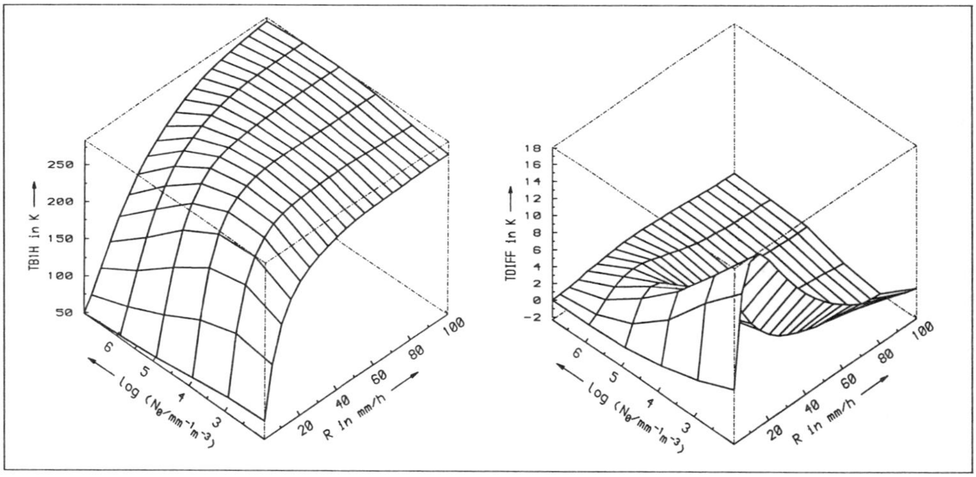

Figure 4. Dependence of the downwelling brightness temperature in horizontal polarization $TB1H$ and the differential temperature $TDIFF$ on rain rate R and particle number density N_0. Elevation 27.3^0.

Figure 4 show the dependence of the downwelling brightness temperature in horizontal polarization $TB1H$ and the differential brightness temperatures $TDIFF$ at 19 GHz on rain rate R and the drop size distribution parameter N_0 of Eq. (4). The path elevation $90^0 - \theta$ is 27.3^0. Rain rate and drop size distribution are directly related. Therefore, for constant N_0 the second parameter D_0 is calculated from the rain rate. The influence of N_0 on the brightness temperature TB1H is weak. For rain rates higher than 40 mm/h the radiometer measurements approach saturation values. On the other hand the drop size distribution has a large effect on the differential temperature. High values of N_0 correspond to a large number of small nearly spherical drops, while small values of N_0 contain less but large oblate drops which have different scattering properties in the vertical and horizontal polarization. The maximum differential temperatures appear for rain rates in the range 10-25 mm/h. The decrease at high rain rates is due to the saturation effect.

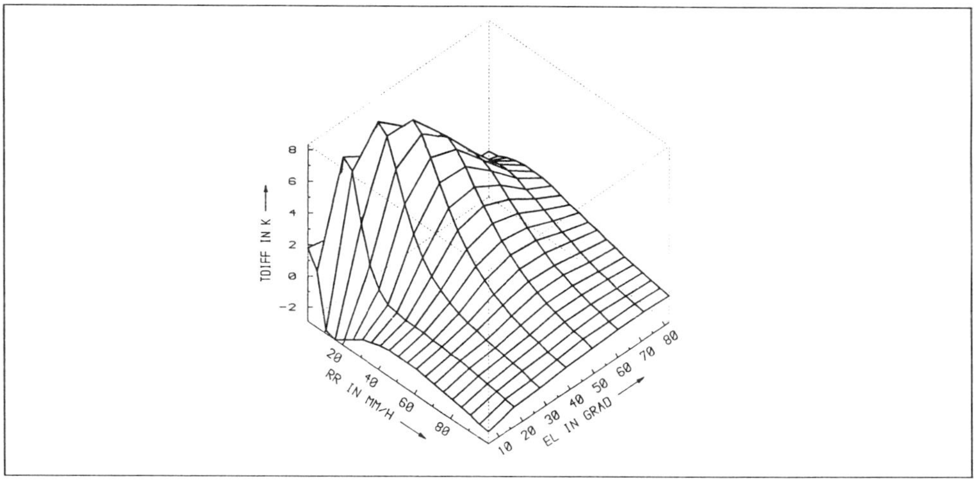

Figure 5. Differential temperature $TDIFF$ vs. rain rate R and elevation EL. $N_0 = 8000$ m^{-3}mm^{-1}.

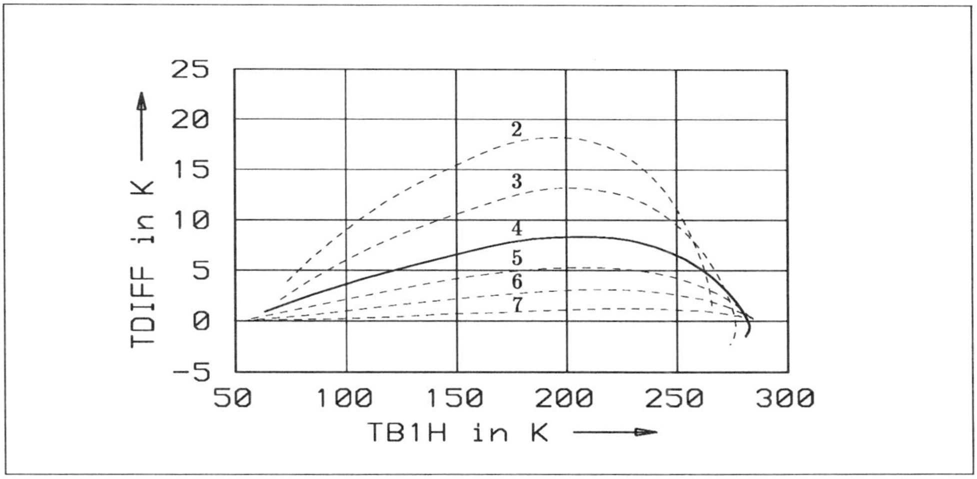

Figure 6. Differential temperature $TDIFF$ vs. brightness temperature $TB1H$. Curve parameter: n with $N_0 = 0.8 \cdot 10^n$ m^{-3}mm^{-1}. Elevation 27.3^0.

Figure 5 also shows the differential temperature, but for constant $N_0 = 8000$ mm^{-1}m^{-3} (Marshall-Palmer distribution) and different elevation angles. The maximum values appear again at rain rates of 10-25 mm/h for elevations between 25^0 and 35^0. This may be explained as follows: For horizontally aligned oblate rain drops the polarization difference with respect to the scattering coefficients is largest for an elevation of 0^0 and vanishes for 90^0 elevation. However, for low elevation angles the path length through the atmosphere is so long, that the radiometer is saturated already for low rain rates, i. e. the differential temperature is low and and may even change its sign. An elevation of abcut 30^0 results in the best compromise between the two opposing effects.

Figure 6 shows the differential temperature versus the horizontal brightness temperature.

For each curve the parameter N_0 is constant and the parameter D_0 grows corresponding to rain rates from 1 to 100 mm/h. As shown in the next section, these curves allow the retrieval of N_0 and D_0 from polarimetric radiometer measurements.

MEASUREMENTS

Convective Event 17th June 1993

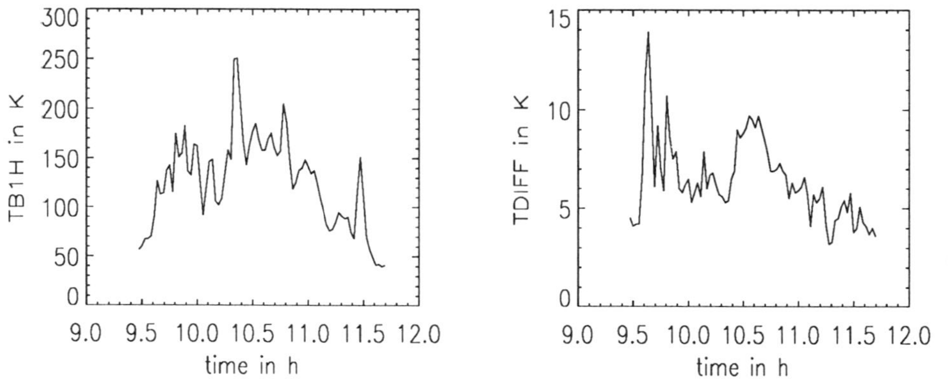

Figure 7. Brightness temperature for horizontal polarization $TB1H$ and differential temperature $TDIFF$ vs. time measured on 17th June 1993.

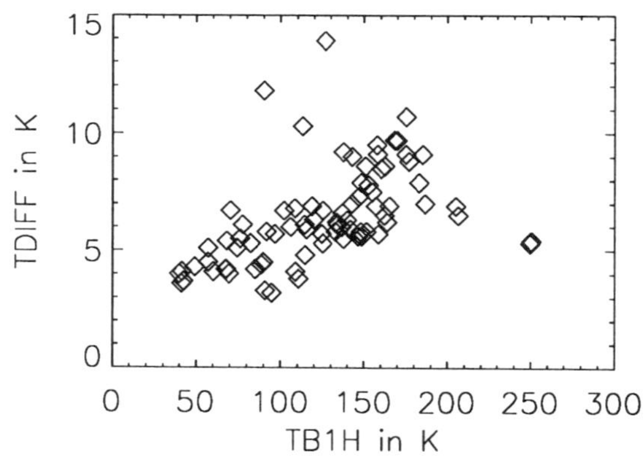

Figure 8. Scatterplot of differential temperature $TDIFF$ and brightness temperature $TB1H$; 17th June 1993.

Figure 7 shows the measurements of $TB1H$ and $TDIFF$ during a convective event on 17th June 1993. The scatterplots of both values in Figure 8 show a clear correlation, but also some spread. The distribution of the points lies between the simulated curves 2 and 5 of Figure 6. Figure 9 shows the correlation of $TB1H$ and $TDIFF$ with the rain rates measured by a rain gauge. Although there is some correlation, the spread is large for low rain rates. This spread cannot be explained only by the drop size distribution. Other reasons include path inhomogeneities and time delay effects between the path integrated radiometer and point rain fall measurements. Also, the influence of cloud emission, which is attenuated by the rain layer for high rain rates, must be accounted for.

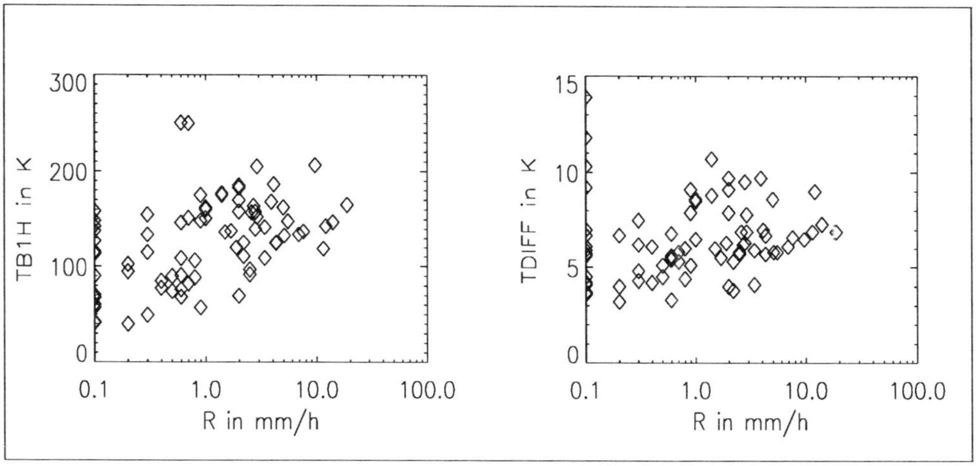

Figure 9. Brightness temperature $TB1H$ and differential temperature $TDIFF$ vs. R; 17th June 1993.

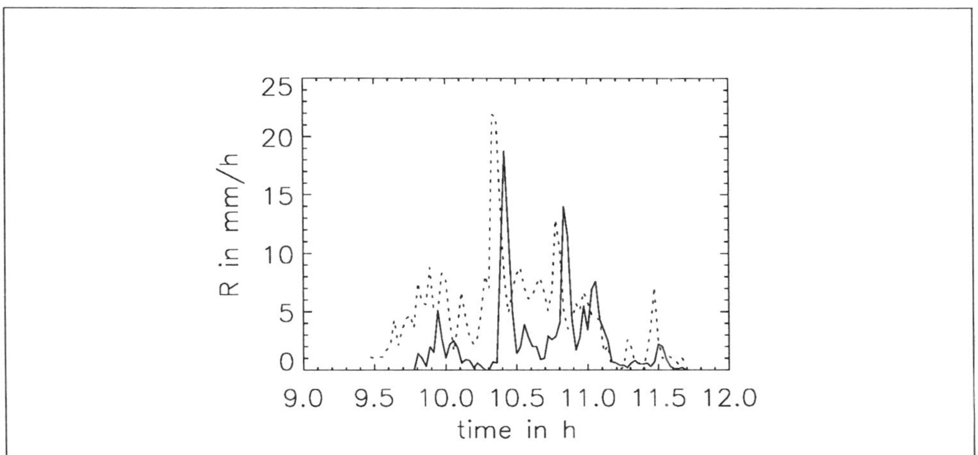

Figure 10. Rain rate R vs. time on 17th June 1993. Solid line: Rain gauge. Dashed line: Retrieval from radiometer measurements.

In Figure 10 the rain rates measured by the rain gauge are compared with rain rates computed by inversion of the theoretical relations underlying Figure 6. Briefly, the inversion algorithm looks for the two points on the curves in Figure 6, which are closest to the measured values of $TB1H$ and $TDIFF$, and does a linear interpolation between the corresponding values of N_0 and D_0. The values of N_0 and D_0 (dashed line) are compared with values derived from distrometer data (solid line) in Figure 11. With the exception of some spikes the agreement is good, although here again path integrated (effective) values are compared with point measurements.

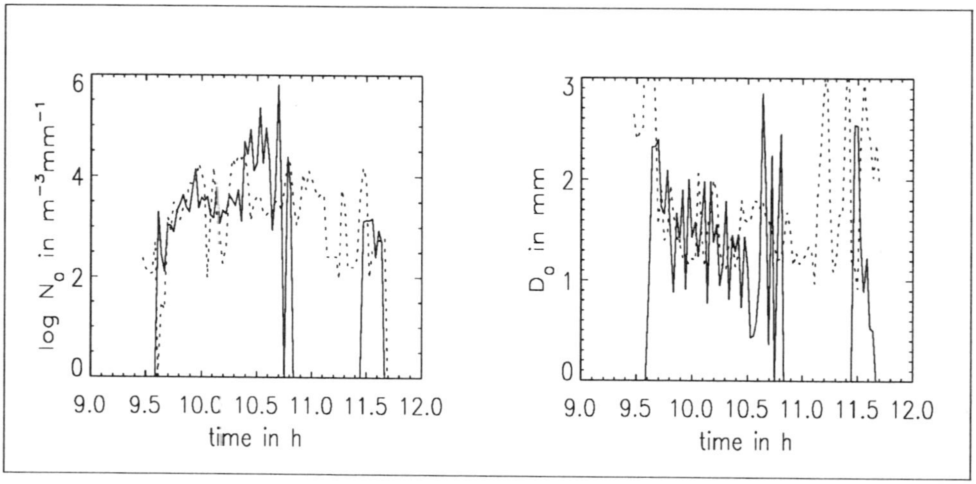

Figure 11. Particle number density N_0 and volume median drop diameter D_0 vs. time on 17th June 1993. Solid line: Distrometer. Dashed line: Retrieval from radiometer measurements.

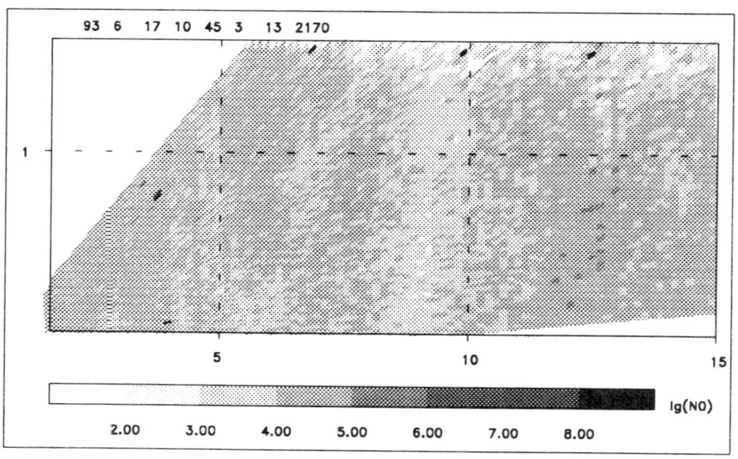

Figure 12. Spatial distribution of the particle number density N_0 on 17th June 1993 derived from a radar RHI-scan. x-axis: Distance from radar in km. y-axis: Height in km.

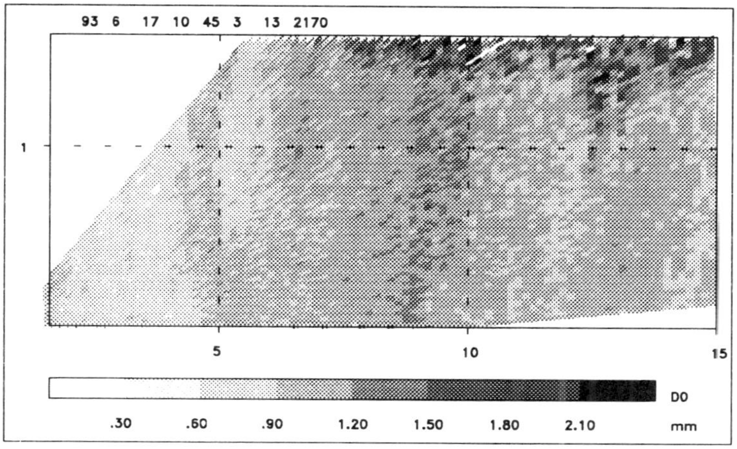

Figure 13. Spatial distribution of the volume median drop diameter D_0 on 17th June 1993 derived from a radar RHI-scan. x-axis: Distance from radar in km. y-axis: Height in km.

The spatial distributions of N_0 and D_0 within the rain layer are presented in Figures 12 and 13. N_0 and D_0 are derived here from polarimetric reflectivity values of a radar RHI-scan along the path direction with the $Z - Z_{DR}$ technique [9]. The values of the drop size parameters are in the same range as in Figure 11. However, the distribution of D_0 is more inhomogeneous than N_0. Also the results in Figure 11 show higher fluctuation and less good agreement between the radiometer and distrometer values for D_0 compared to N_0.

Stratiform event 17th *October 1993*

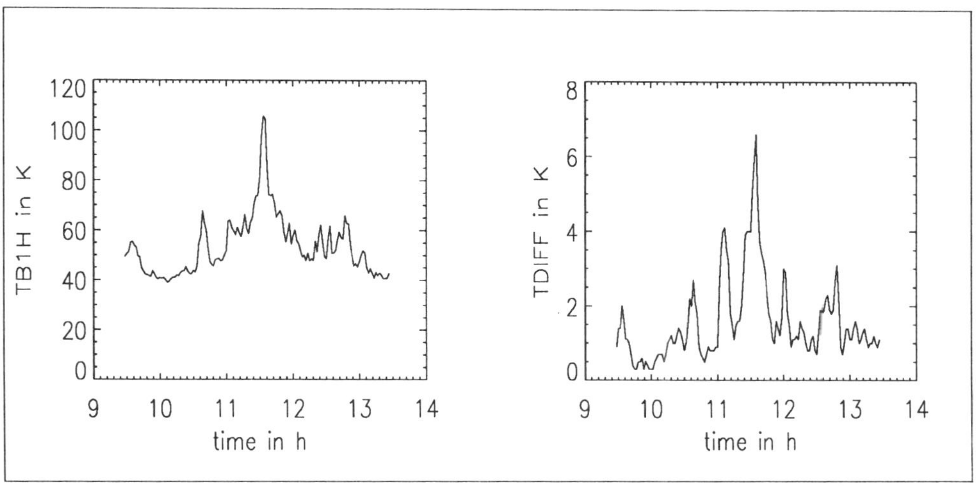

Figure 14. Brightness temperature for horizontal polarization $TB1H$ and differential temperature $TDIFF$ vs. time on 17th Oct. 1993.

The horizontal polarized and differential brightness temperature for a stratiform rain event are presented in Figure 14. Here the shape of the two curves agree well (compare Figure 7). This high correlation can be seen also in Figure 15. However, the correlation with the rain rate is not as good. The reasons are again the influence of clouds on the brightness temperature for low rain rates and time delay effects between rain gauge and radiometer values. In Figure 17 the drop size parameters N_0 and D_0 derived from the radiometer and distrometer measurements are compared and are in good agreement.

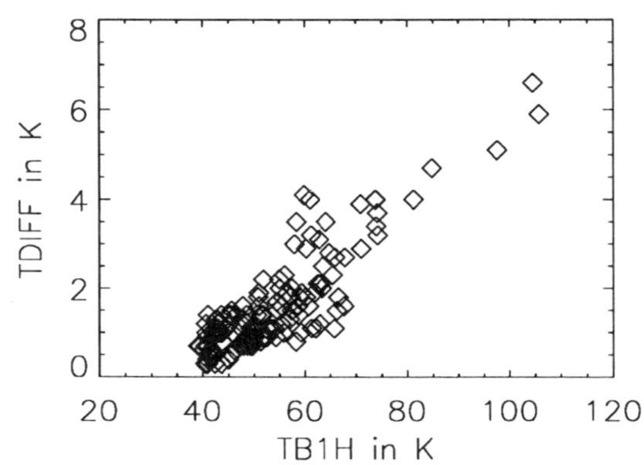

Figure 15. Scatterplot of differential temperature $TDIFF$ and brightness temperature $TB1H$; 17th Oct. 1993.

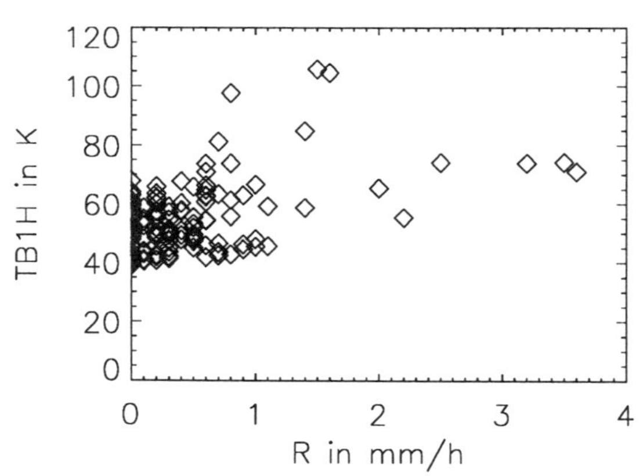

Figure 16. Brightness temperature $TB1H$ vs. R; 17th Oct. 1993.

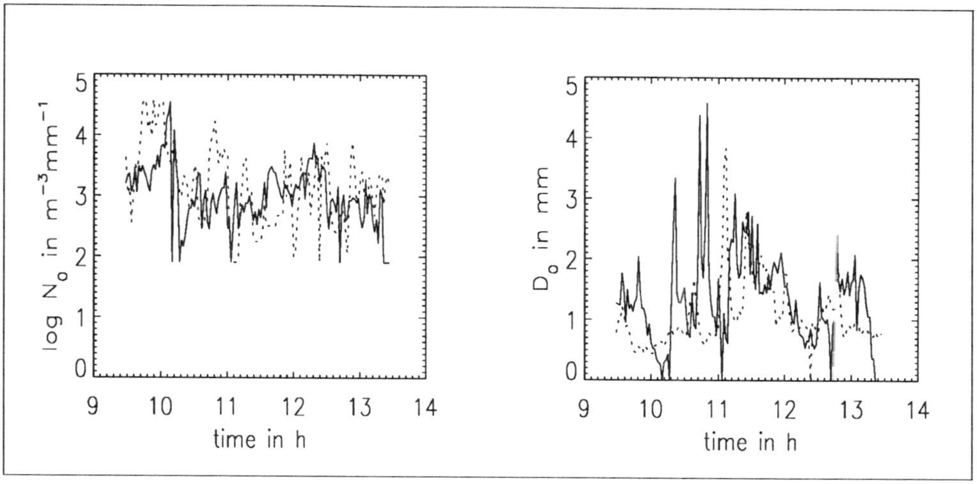

Figure 17. Particle number density N_0 and volume mean drop diameter D_0 vs. time on 17th Oct. 1993. Solid line: Distrometer. Dashed line: Retrieval from radiometer measurements.

STATISTICS OF RAIN RATE AND BRIGHTNESS TEMPERATURE

In Figure 18 an equal probability distribution of brightness temperature and rain rate is shown, which is derived from monthly cumulative distributions for the period May-August 1992. In this statistical approach the correlation is better than for the single events, because the time delay effects between the two measurement values are averaged out. Simulations with the model described in the third section for a 2.5 km thick atmosphere with two different drop size distributions are included for comparison.

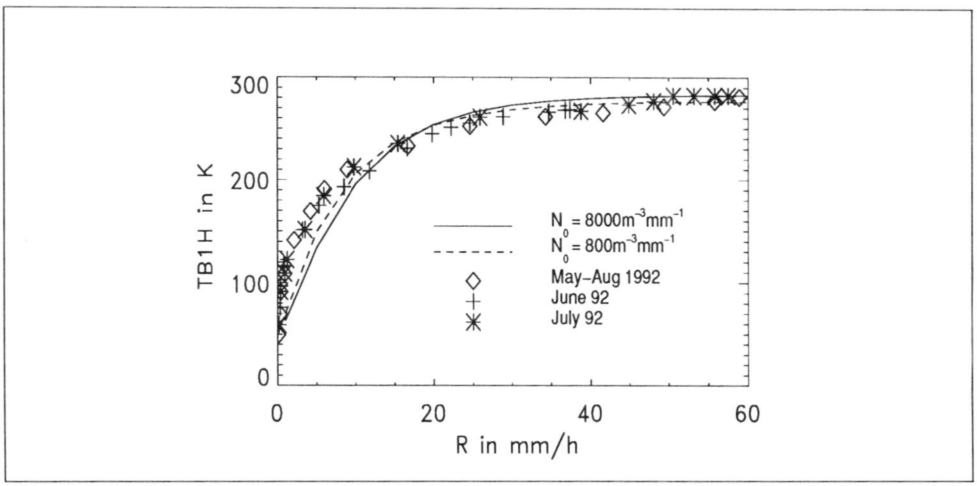

Figure 18. Equal probability distribution of brightness temperature and rain rate for the period May-Aug 92. Path elevation 27.6^0. Lines: Model simulations.

SUMMARY

Model simulations and measurements show differences up to 15 K between the brightness temperatures for horizontal and vertical polarization measured from the ground at 19 GHz. The dropsize distribution has a major influence on the differential temperature. Simulations using a plane-parallel homogeneous atmosphere show that it is possible to derive an effective two parameter drop size distribution from polarimetric radiometer measurements. The radiometer derived values are in good agreement with distrometer and radar measurements. Analysis of stratiform and convective events show that the rain rate estimation using ground based radiometers is quite accurate. The use of polarimetric information may even improve the quality.

Acknowlegements

We express our gratitude to Dr. J. Vivekanandan from NCAR and Dr. J. Turk from Colorado State University, for providing us with the Waterman T-matrix code and further programs, their assistance in the calculation of the bistatic scattering coefficients and the useful discussion.

REFERENCES

1. Hornbostel, A., and A. Schroth, 1993. Statistical and Event Based Analysis of 20 GHz Attenuation on the Earth-Olympus Satellite Path. *Proc. of ICAP'93 Edinburgh*, Conf. Publication No. 370, part 1, pp. 131-134.
2. Hornbostel, A., and A. Schroth, 1994. Propagation Research on the 20 GHz Earth-Olympus Satellite Path. To be published in *Journal of Electromagnetic Waves and Applications* soon.
3. Brussard, G., 1985. *Radiometry, a useful prediction tool ?* . ESA SP-1071.
4. Kutuza, B. G., and M. T. Smirnov, 1991. Signatures of Rain Microwave Emission, *Proc. of IGARSS'91 Finland*, IEEE Cat.No. 91CH2971-0, Vol. 1, P. 193-194.
5. Universite Catholique de Louvain, 1976. *Mesures radiometriques a deux polaristions a 11 GHz*, ESA CR(P)-821.
6. Ishimaru, A., 1978. *Wave Propagation and Scattering in Random Media*, Vol. 1, Academic Press.
7. Liou, K.-N., 1980. *An Introduction to Atmospheric Radiation*, Int. Geophysics Series, Vol. 26, Academic Press.
8. Lenoble, J., 1985. *Radiative Transfer in Scattering and Absorbing Atmospheres: Standard Computational Procedures*, A. Deepak Publishing, Hampton, Virginia.
9. Schnabl, G., 1988. *Tropospheric Influences on Satellite-Earth Paths and Their Determination by Means of Radar Measurements*, ESA-TT-1143, translation of DFVLR-FB 88-53.

Precipitation profile retrieval from airborne microwave radiometers: a case study over ocean during CaPE

F.S. MARZANO[1], A. MUGNAI[2], N. PIERDICCA[1], E.A. SMITH[3], J.TURK[4], and J. VIVEKANANDAN[5]

[1]*Dipartimento di Ingegneria Elettronica, Università "La Sapienza", Roma, Italy*
[2]*Istituto di Fisica dell'Atmosfera, Consiglio Nazionale delle Ricerche, Frascati, Italy*
[3]*Department of Meteorology, Florida State University, Tallahassee, Florida, USA*
[4]*Department of Electrical Engineering, Colorado State University, Fort Collins, Colorado, USA*
[5]*National Center for Atmospheric Research, Atmosperic Research Program, Boulder, Colorado, USA*

Abstract During the Convection and Precipitation/Electrification Experiment (CaPE) in the summer of 1991 in Florida (USA), several precipitating storms were contemporarily observed by the airborne Advanced Microwave Precipitation Radiometer (AMPR), having four channels at 10.7, 19.35, 37.1, and 85.5 GHz, and by the ground-based NCAR CP-2 multiparameter radar at 3 GHz and 10 GHz frequency bands. In this paper, we compare the estimated hydrometeor contents and rainfall rates derived from the AMPR radiometer and from the CP-2 radar, for a precipitating storm over ocean that occurred near Cape Canaveral on August 12, 1991. The retrieval scheme applied to the multifrequency AMPR data uses a physically-constrained cloud-radiation dataset and a maximum-likelihood discriminant analysis for estimating the columnar ice and water contents and the gross vertical structure of the precipitating cloud; rainfall rates are then derived by a suitable fallout model. Results are discussed and interpreted only for along-track nadir-looking data, because of AMPR scan geometry and polarization changing for off-nadir observations. A fairly good agreement between the radar and radiometer derived retrievals is often achieved within the intense core cell, considering the different model assumptions within the used retrieval algorithms and the diverse geophysical information content from the two types of measurements. The effect of raindrop size distribution on radar retrieval is analyzed in order to explain radar-radiometer discrepancies within the decaying stage of the observed storm.

Key words: Microwave remote sensing, Airborne radiometry, Radar meteorology, Precipitation retrieval.

INTRODUCTION

After the launch of the Defense Meteorological Satellite Program (DMSP) in 1987 carrying the new generation of multifrequency microwave radiometers, i.e. the Special Sensor Microwave/Imager (SSM/I), many aircraft-based scientific campaigns have been carried out in order to better understand the interaction of microwave radiation at SSM/I frequencies with precipitating clouds. Recently, the Convection and Precipitation/Electrification Experiment (CaPE) in the summer of 1991 has been devoted to precipitation studies. During CaPE, the Advanced Microwave Precipitation Radiometer (AMPR) [1], having three of the four SSM/I frequencies (i.e., 19.35, 37.1, and 85.5 GHz) plus an additional frequency at 10.7 GHz, was flown in Florida aboard the NASA aircraft ER-2 over several precipitating storms that were contemporarily scanned by the National Center for Atmospheric Research multiparameter CP-2 radar settled in Cape Canaveral.

In the last years, many authors have realized that passive microwave observations may be exploited to retrieve cloud and precipitation profiles, i.e. the vertical distributions of the equivalent water contents (EWC's) of the various hydrometeors in different phases [2]. The physical foundations of passive microwave precipitation profile retrieval stem from the fact that the upwelling brightness temperatures (T_B's) respond, in a frequency-dependent fashion, to the vertical distribution of the various hydrometeors in addition to atmospheric gases and surface characteristics [3], [4]. In particular, the upwelling T_B's at frequencies greater than about 19 GHz are more suited to retrieve hydrometeor profiles rather than to estimate directly the surface rain rate, since they are almost insensitive to the microphysical structure of the lower rain layers. Surface rainfall rate can be derived from the precipitation profile by means of rain fallout models. The use of additional lower frequencies, such as 10.7 GHz, that penetrate all the way down to the surface, may help the precipitation retrieval [5], [6].

A way to develop a passive microwave profile retrieval algorithm is to resort to precipitating cloud models. In this paper, we first describe a new profile algorithm that uses the microphysical outputs from a three-dimensional time-dependent cloud/mesoscale model. Then, we show the intercomparison of precipitation retrievals from the airborne multifrequency AMPR and from the ground-based dual polarization CP-2 radar, for a CaPE case-study consisting of an intense precipitating storm over ocean near Cape Canaveral, Florida, on August 12, 1991, that has been already analyzed [7]. The radiometer-based retrieval algorithm uses a maximum likelihood discriminant method that selects the solution set of hydrometeor profiles from a statistically generated cloud-radiation dataset satisfying the physical consistency of the cloud/mesoscale model. The dual polarization radar-based retrieval algorithm utilizes range-resolved backscatter parameters (i.e., the reflectivity and the differential reflectivity) in an iterative way to derive vertical profiles of hydrometeor drop-size distributions and equivalent water contents.

RADIOMETER AND RADAR PRECIPITATION RETRIEVAL

We now describe the basic features of the techniques we have utilized to extract precipitation parameters from AMPR and CP-2 microwave measurements. A detailed description of the multiparameter radar and of the multifrequecy radiometer algorithms can be found in [7] and [8], respectively.

Multifrequency-radiometer algorithm

The passive microwave retrieval algorithm is based on the use of cloud-radiation models and of multivariate statistical analysis. As compared to other methods inferring the vertical structure of the precipitating cloud, it adopts a maximum likelihood criterion to identify the actual profile; moreover it exhibits fairly good performance in terms of processing time since the algorithm itself is not iterative and the time consuming task of solving the radiative transfer equation is accomplished only once, when generating the cloud-radiation dataset. Its main features and steps are schematically illustrated as follows.

Modeling phase: statistical generation of the cloud-radiation dataset. The three-dimensional time-dependent cloud/mesoscale model named University of Wisconsin - Nonhydrostatic Modeling System (UW-NMS) [9], is used for generating a microphysical dataset, explicitly presenting four species of hydrometeors: cloud droplets, rain drops, graupel particles (hereafter, "ice particles" when comparing with radar results), and ice crystal/aggregate particles. Then, the number of cloud layers is reduced in an automatic way to at most seven homogeneous layers in order to simplify the radiative transfer calculations and to reduce the redundancy of the cloud microphysical structures with respect to the information carried by the standard microwave frequencies currently used. The original dataset is then extended by means of a pseudo-random multidimensional generator, which is based on the classification of the microphysical dataset into various cloud typologies and on the use of the covariance matrix of precipitating cloud parameters, thus retaining the physical and statistical features of the primary cloud model. Finally, by applying a plane-parallel radiative transfer scheme to the cloud dataset, as described in [10], a cloud-radiation database (consisting of five thousand cloud structures and the associated T_B's) is generated. This database is considered as a random sample of space-borne microwave radiometer observations of precipitation.

Inversion phase: cloud and precipitation profile retrieval. The maximum likelihood estimation can be expressed rigorously by matrix notation. If the set of hydrometeor vertical profiles is identified by a vector **g**, consisting of equivalent water contents of each hydrometeor category, and the multi-frequency T_B measurement is expressed by vector **t**, the most probable cloud structure within the cloud-radiation dataset is found by maximizing the following discriminant function with respect to **g**:

$$d(\mathbf{t},\mathbf{g})=\ln[p(\mathbf{t}|\mathbf{g})] + \ln[p(\mathbf{g})] - \ln[p(\mathbf{t})] \qquad (1)$$

where $p(\mathbf{t}|\mathbf{g})$ is the conditional probability density function (pdf) of **t** conditioned to **g**, and $p(\mathbf{g})$ and $p(\mathbf{t})$ are the pdf's of **g** and **t**, respectively. Eq. (1) has been derived by means of the Bayes theorem that allows one to express the posterior pdf $p(\mathbf{g}|\mathbf{t})$ as a function of the prior pdf $p(\mathbf{t}|\mathbf{g})$. Assuming a Gaussian multidimensional joint-distribution of vector **g**, the covariance matrix $\mathbf{C_g}$ and the mean vector **m** are sufficient to describe the statistics of **g**. Moreover, if **t'** is the T_B vector associated with **g** by radiative transfer computations, the errors (**t-t'**) are assumed uncorrelated at the different frequency bands with the same variance σ_t^2. Thus, considering $p(\mathbf{t})$ as a common term, Eq. (1) becomes:

$$d(\mathbf{t},\mathbf{g}) = (-1/\sigma_t^2)\,(\mathbf{t} - \mathbf{t}')^T\,(\mathbf{t} - \mathbf{t}') - \ln[\det(\mathbf{C_g})] - (\mathbf{g} - \mathbf{m})^T\,\mathbf{C_g}^{-1}\,(\mathbf{g} - \mathbf{m}) \qquad (2)$$

where $\det(\mathbf{C_g})$ and $\mathbf{C_g}^{-1}$ indicate the determinant and the inverse matrix of Cg, respectively, and "T" represents the transpose operator. Once the vector **g** has been selected, the surface rainfall rate is computed from the retrieved hydrometeor profiles by means of a fallout model consistent with that used in the UW-NMS model. Note that Eq. (2) can be applied to the whole dataset as well to a classified dataset, by inserting the class probability into the equation itself and by calculating **m** and $\mathbf{C_g}$ for each class. Noteworthy, considering that a number of profiles could be statistically indistinguishable since the discriminant function could not differ significantly, the average between such indistinguishable profiles can be computed together with the r.m.s. of the differences so that a measure of the retrieval accuracy can be associated to the final solution. The performed estimation is, of course, valid under the assumption of horizontal homogeneity of the precipitating cloud within the radiometer antenna beam.

Multiparameter-radar algorithm

In general, the radar microphysical inversion process can be stated as:

$$[Z_H, Z_{DR}] \Rightarrow [N_o, m, \Lambda, D_{max}, \rho, f_{water}] \qquad (3)$$

where, Z_H and Z_{DR} are the measured reflectivity at horizontal polarization and differential reflectivity, respectively; N_o, m, and Λ are parameters defining the hydrometeor size distribution, having used a modified gamma drop size distribution (DSD) for all hydrometeors; D_{max} is the hydrometeor maximum diameter, ρ is the hydrometeor mass density, and f_{water} denotes the included volume fraction of water.
Within the retrieval, the values of Z_H and Z_{DR} are used to assign each point as rain, ice, or a mixture of rain and ice. A Z_H threshold of 10 dBZ is applied, below which no data are used. The ice region extends from the height where Z_{DR} remains near 0 dBZ up to the 10 dBZ cloud top, while the rain region extends from the lowest observed level (approximately 1 Km) up to 4 km. In some regions characterized by high Z_H values, the rain may extend only up to 3 km height when a mixture of rain and ice (probably hail) reduces the Z_{DR} below that expected for the associated Z_H assuming pure rain. In between the rain and ice regions, lies a band of positive Z_{DR} that is assumed to contain wet ice composed of a ice/water mixture.
It should be mentioned that the occurrence of wet ice is not taken into account in the cloud-radiation database used for radiometric inversion and therefore, when comparing radar and radiometer derived retrievals, we have to associate the radar-derived mixture to either graupel particles or rain drops, or both. Considering that the radar response of wet ice hydrometeors is best comparable to water drops, we have merged the radar-derived mixture with the radar-derived rain before doing comparisons.

MICROWAVE OBSERVATIONS OF A STORM OVER OCEAN

CaPE was carried out from July to August 1991 in Florida as a joint effort of NASA, universities and research centers. One of the aim of the experimental campaign was to provide a complete dataset relative to precipitation events, consisting of high spatial resolution T_B measurements from airborne AMPR and contemporary ground-truth data from ground-based multiparameter CP-2 radar.
The four-channel AMPR is a scanning radiometer mounted on board of the NASA ER-2 aircraft flying at an altitude of about 20 Km. Scanning is performed in the cross-track direction through a plate-reflector, making an off-nadir angle of ±45° during which the received linear polarization state changes from fully horizontal at +45° to fully vertical at -45°. The nadir ground resolutions at 10.7, 19.35, 37.1, and 85.5 GHz are 2.8 Km, 2.8 Km, 1.5 Km, and 0.6 Km, respectively. The operational configuration of AMPR allows contiguous cross-track and along-track sampling at 85.5 GHz in nominal flight conditions. The CP-2 radar operates in a linear polarization basis at both S (3 GHz) and X (10 GHz) bands, measuring both the copolar and cross-polar reflectivities with a range resolution between 200 and 300 meters. Its maximum range is 100-150 Km, while the antenna beam is 0.93° and the antenna scan rate is about 6° sec^{-1} in azimuth. The CP-2 system uses a coherent receiver at S band (allowing the measurement of fully polarimetric features) and incoherent receiver at X band.

AMPR and CP-2 measurements during CaPE

On August 12, 1991, the ER-2 aircraft flew in central Florida near Cape Canaveral. Close to the Atlantic Ocean coastline, a moderately intense precipitating storm was imaged between 21:52 GMT and 21:55 GMT, 100 km south of the CP-2 radar. A detailed description of the event and of the AMPR and CP-2 observations has been given elsewhere [7].
AMPR T_B data are characterized by the variation of linear polarization along the scan from fully horizontal to fully vertical, and, in the selected case, by the transition from land to sea. Thus, in order to avoid the radiometric and geometric distortion introduced by the AMPR

scanning geometry and the polarization effects for off-nadir observations, we shall consider only the nadir-looking track over ocean (i.e., from pixel 25 to 60). Figure 1 shows the AMPR T_B's for the nadir track. In this figure, all T_B's have been reduced to the ground resolution of the two lower-frequency channels, i.e., to 2.8 Km.

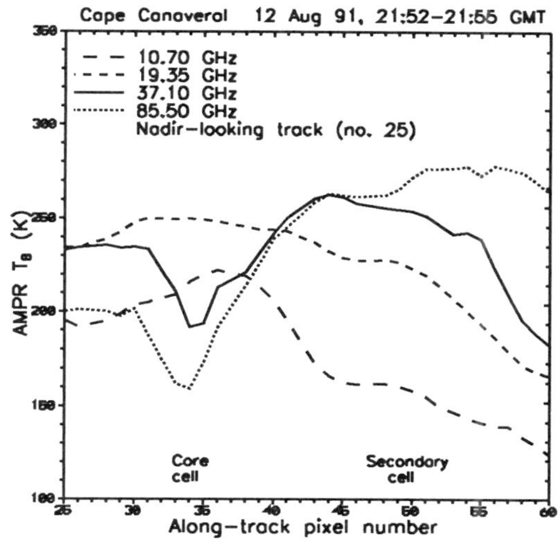

Figure 1. AMPR T_B's at 10.7, 19.35, 37.1, and 85.5 GHz versus along-track nadir-looking pixel number for August 12, 1991 storm. The core cell is between pixels 25-42 and the secondary cell between pixels 42-60.

The storm is well discernible by the two higher frequencies (37 and 85 GHz) that show a large T_B depression around pixel 34 (core cell). The T_B behaviour shows that the storm is gradually decaying from the core cell up to pixel 60, which exhibits T_B values typical of a rain-free ocean surface.

Figure 2 shows the cross-sections of the S-band horizontal reflectivity Z_H and differential reflectivity Z_{DR} corresponding to the AMPR nadir-looking track. From the Z_H cross-sections, the two-cells structure of the observed storm is apparent; the one in the center is active and intense (core cell), while the one to the right is smaller and less intense (secondary cell). It is worth noting that the AMPR radiometric signature for the secondary cell is not as strong as that appearing from CP-2 measurements (see Figure 1). This particular feature and the information apparent from Z_{DR} cross-section will be discussed in the following section.

Comparison with the simulated brightness temperature dataset

Since we have trained our retrieval algorithm on a simulated cloud-radiation dataset, it is necessary to analyze the adequateness of the dataset itself to represent the AMPR microwave appearance of the considered event.

To this end, we consider the two scatterplots of Figure 3 that shows the simulated T_B's from the statistically-generated cloud-radiation dataset together with the AMPR-measured T_B's for the nadir track; in particular, left panel shows 19 GHz versus 10 GHz, while the right panel shows 85 GHz versus 37 GHz. In general, the simulated T_B's agree with the measurements at the two higher frequencies, while some differences between measured and simulated T_B's are present at the two lower frequencies. However, the largest differences (up to 20 K) are observed for low T_B's values for clear and cloudy sky conditions. This fact mainly depend on the properties of the ocean surface wave spectrum and foam coverage, which are unknown.

Noteworthy, the domain of the generated cloud-radiation dataset fits the high resolution AMPR measurements cold-rain events quite well. A similar result was found in another AMPR rainfall case study over land [6].

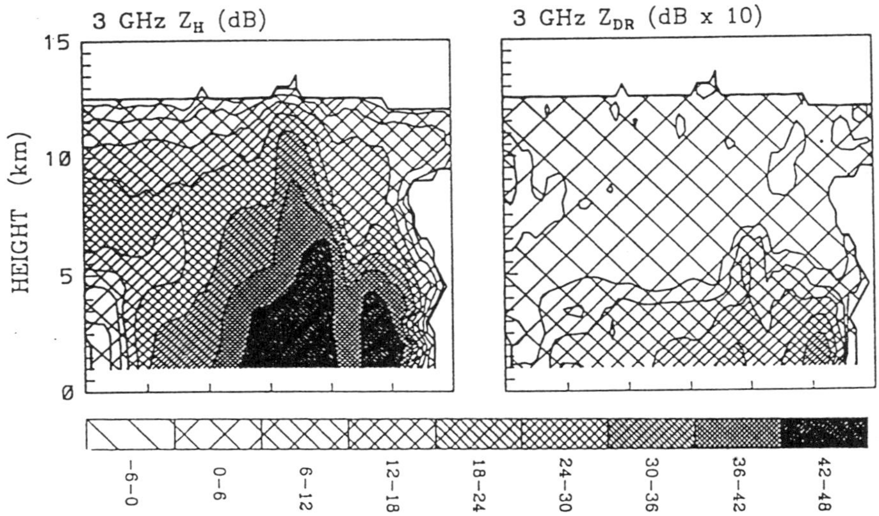

Figure 2. Vertical cross sections of CP-2 radar horizontal and differential reflectivities (Z_H and Z_{DR}) at 3 GHz, corresponding to AMPR nadir-looking track of Figure 1. Right and left panels are for Z_{DR} and Z_H, respectively. The high Z_{DR} values within the secondary cell imply melting processes with large ellipsoidal raindrops.

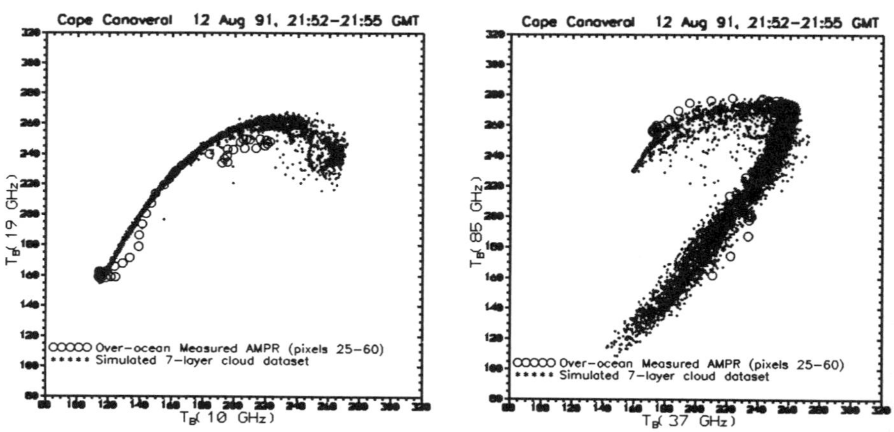

Figure 3. Comparison between the AMPR-measured nadir-looking T_B's and the T_B's from the simulated cloud-radiation dataset. Left panel presents 19.35 GHz vs 10.7 GHz; right panel presents 85.5 GHz vs 37.1 GHz. Only AMPR data collected over-ocean are plotted.

COMPARISON OF AMPR AND CP-2 MICROWAVE SENSOR RETRIEVALS

In the following, we will first discuss the intercomparison between the radiometer-derived and radar-derived columnar EWC's and rainfall rate estimates. Then, we will show an example of retrieved precipitating cloud profiles and explain some discrepancies between the AMPR and CP-2 retrievals.

Estimates of columnar contents, rain rates and precipitation profiles

Figure 4 shows the intercomparison between the radar-derived and radiometer-derived rain and ice columnar EWC's for the nadir-looking track 25. For the radar, we also show results for the contiguous tracks 24 and 26. The left panel shows that the radiometer generally tends to overestimate (with respect to the radar) the ice columnar EWC in the core cell, with differences up to 1 Kg/m^2. On the contrary, in the secondary cell (around pixel 50) the AMPR derived ice EWC is practically zero, while the CP-2 measures ice EWC up to 0.4 Kg/m^2. The rain columnar EWC retrievals, shown in the right panel, exhibit an opposite behavior with respect to ice, i.e., the AMPR derived rain content is generally less than that measured by the CP-2. This is particularly evident in the secondary cell, where the AMPR result is up to 2 Kg/m^2 lower than the CP-2 result; in the core cell, on the other hand, the AMPR columnar rain falls within the radar retrieval variability.

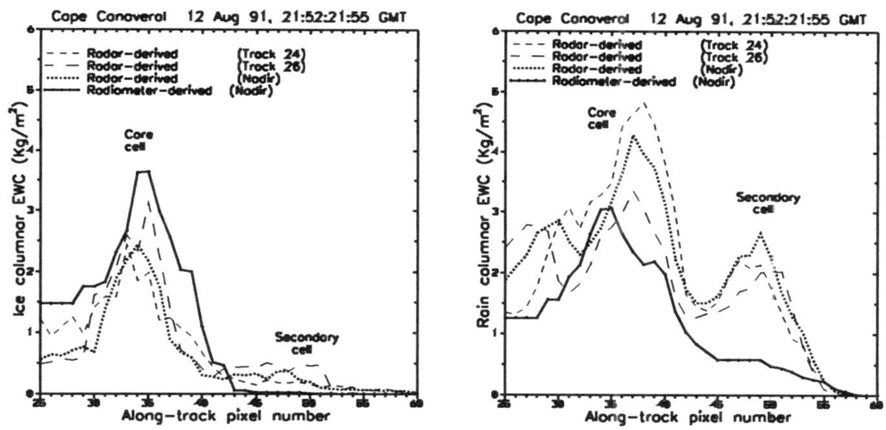

Figure 4. Comparison between ice (left panel) and rain (right panel) Columnar Equivalent Water Contents (EWC's) estimated from AMPR and CP-2 measurements for the nadir-looking track 25. For the radar, results for adjacent tracks 24 and 26 are also shown.

In the core cell, the two retrievals are generally in fairly good agreement. In the secondary cell, on the other hand, there are significant retrieval differences, that are strongly related to the diverse spectral signatures of measured T_B's and Z_H's which have been discussed in a previous section (these differences will be examined in detail in the next sub-section). Also, the radiometer retrieval for ice is better than for rain simply because large ice particles are directly sensed by a space-borne radiometer, while the accuracy of the rain retrieval depends on our knowledge of the internal microphysics of the storm. The overestimation of columnar ice is partly due to the fact that the UW-NMS model uses a size distribution that emphasizes

the large ice particles [6].
Figure 5 compares the AMPR estimated surface rainrates for the nadir-looking track, with the rainfall rates derived from CP-2 radar at 1 Km height. For each pixel, the dashed curves indicate the accuracy of the AMPR estimate in terms of root mean square error. The surface rainrates are usually less than 15 mm/h within the storm core. The AMPR estimated rainfall rate at the surface is generally lower than that estimated by the radar at 1 Km height. Again, within the core, the AMPR and CP-2 estimates compare better than in the secondary cell where differences may be as large as 10 mm/h. This behavior is related to what has been already observed for Figure 4.

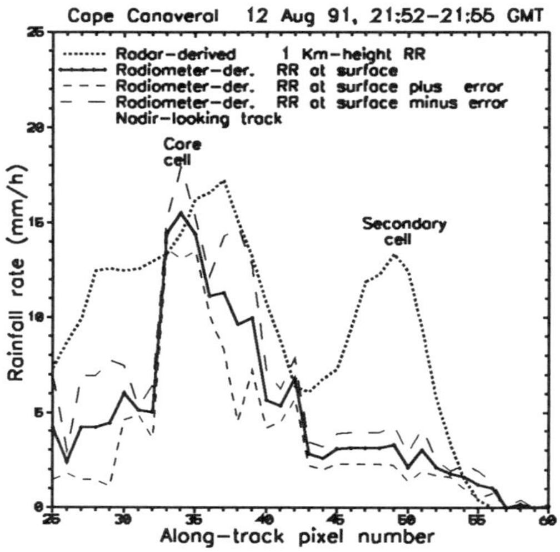

Figure 5. Comparison between AMPR-derived and CP-2-derived rainfall rates for the nadir-looking track. Rainrates refer to the surface and to 1 km height for the AMPR and for the CP-2, respectively. For each pixel, the dashed and long-dashed lines indicate the uncertainty range of the AMPR estimates in terms of root mean square error.

As an example of precipitation profile retrieval, Figure 6 shows the comparison of the rain and ice EWC profiles, estimated by the radar and radiometer algorithms, for pixel 34 in the core cell (ice profile in the left panel, rain profile in the right panel). At this pixel, the AMPR-derived columnar rain is fairly comparable to that measured by the radar (3 and 2.7 kg/m^2, respectively); on the other hand, the columnar ice is about 1.2 kg/m^2 larger (3.6 and 2.4 kg/m^2, respectively). Correspondingly, the AMPR-estimated rainrate is almost the same as the radar estimate (about 15 mm/h). In spite of that, the AMPR and CP-2 profile retrievals look quite different. Note, however, that the vertical resolution of the radar is constant (of the order of 0.5 Km) while that of the AMPR retrieval depends on the stratification of the cloud-radiation dataset, which generates layers of variable thickness from 0.5 up to about 7 km. Moreover, the AMPR retrieval shows a mixed layer of both rain and ice particles that is not analogous to the mixture layer observed by the radar for the reasons explained previously. Finally, the fact that the AMPR-derived ice layer extends to significantly lower altitudes than the mixture layer is related to the different assumptions of the UW-NMS model and the radar algorithm.

Effect of raindrop size distribution parametrization

Unlike the core, the secondary cell in Figure 2 shows moderately high Z_H's (up to 45 dBZ) but relatively high positive Z_{DR}'s (up to 3.5 dBZ). Such reflectivity features have been observed by multiparameter radars in the case of decaying convective clouds, characterized by substantial lack of ice with associated weak air motion. The large differential reflectivities near the cloud base is due to large ellipsoidal raindrops whose growth occurs "during continuous collection of cloud water as the particle melts and descends in moderate updrafts" [11].

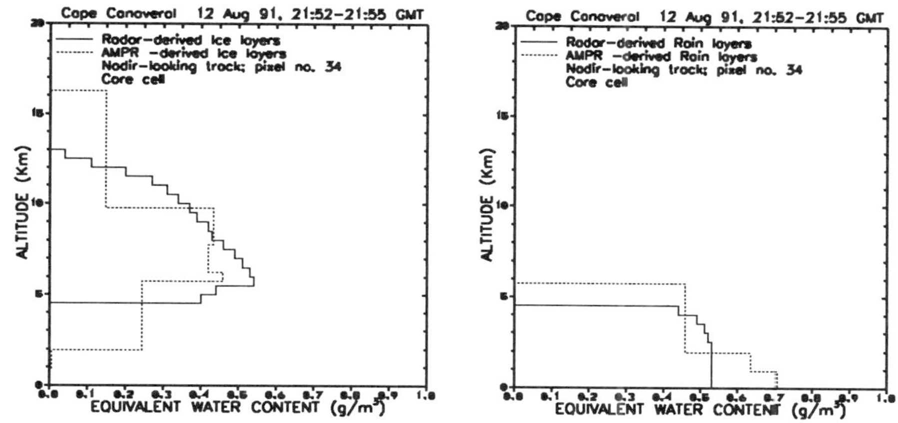

Figure 6. Comparison between the AMPR-derived and the CP-2-derived hydrometeor vertical profiles for nadir-looking pixel 34 (ice profile in the left panel, rain profile in the right panel). The solid line represents the radar retrieval and the dashed line the radiometer retrieval.

The modified Gamma drop size distribution, used within the radar retrieval scheme, does not contain an adequate presence of large raindrops for reproducing such warm-rain conditions (by contrast, the cloud model distributions used within the radiometer-based scheme emphasize the number of large drops). Thus, in order to check if an enhanced concentration of large raindrops can explain the observed radiometer-radar discrepancies in the secondary cell, we have replaced the modified gamma DSD. For the sake of simplicity, we have used a monodisperse DSD (i.e., all particles have the same size) whose resultant reflectivity is the same as that measured by the CP-2 radar. Note that as the drop radius increases a decreasing EWC is needed to produce the same reflectivity (on the other hand, larger drop radii imply higher extinction and, thus, lower T_B's at 10 GHz).

Figure 7 shows the effect on rain retrieval within the secondary cell (pixels 40 through 60) after assuming two mono-disperse raindrop size distributions with drop diameters of 2 mm and 4 mm, respectively; for comparison, the columnar rain derived from the AMPR and that derived from the radar using the modified gamma DSD are also shown. It is evident that a significant presence of large raindrops in the lower cloud layers enhances the agreement between the AMPR and CP-2 rain estimates.

CONCLUSION

The problem of retrieving precipitation from airborne microwave radiometers has been tackled by applying a maximum likelihood algorithm based on a cloud-radiation database statistically generated from a cloud mesoscale simulation. The case study consists of a convective storm over ocean in the sub-tropical region, which was observed by the airborne AMPR radiometer and the ground-based CP-2 radar during CaPE. The high spatial resolution of all AMPR channels (less than 3 Km) is able to match the fine scale of rainfall events better than current spaceborne microwave radiometers, such as SSM/I. AMPR estimates of rain and ice columnar EWC's and vertical profiles, as well as surface rainrates, have been compared with corresponding CP-2 retrievals. Fairly good agreement has been found within the core cell, while significant radar-radiometer discrepancies have been found within the secondary cell. The latter fact, however, may be explained assuming an enhanced presence of large raindrops in the lower portion of the decaying storm, even though uncertainties remain very large.

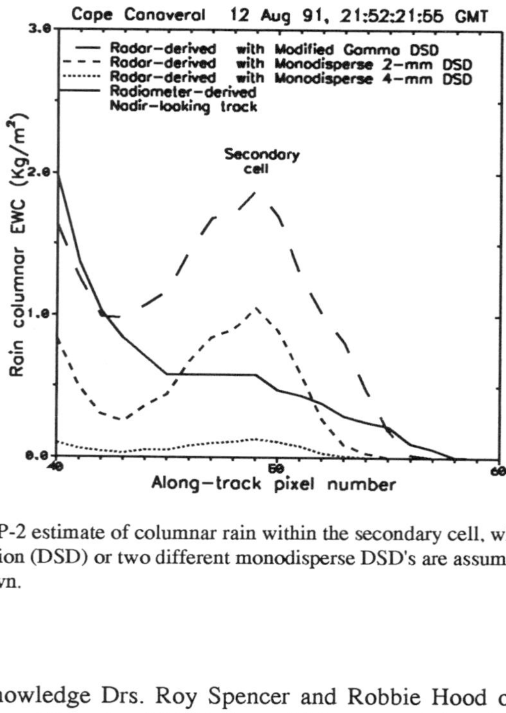

Figure 7. Effect on CP-2 estimate of columnar rain within the secondary cell, when either a modified gamma rain drop size distribution (DSD) or two different monodisperse DSD's are assumed. For comparison, AMPR estimates are also shown.

Acknowledgments

We gratefully acknowledge Drs. Roy Spencer and Robbie Hood of NASA/Marshall Space Flight Center for providing the AMPR data. This research has been supported by Gruppo Nazionale per la Difesa dalle Catastrofi Idrogeologiche and by the Italian Space Agency, by Commission of the European Communities (contract EV5V-CT92-0167) within research program Climatology and Natural Hazards of the EC ENVIRONMENT Programme 1991-1994, and by National Aeronautics and Space Administration grant NAGW-991.

REFERENCES

1. Galliano, J.A., R.H. Platt, R.E. Hood, R.W. Spencer, V.L. Griffin, M.W. James, and F.J. LaFontaine, 1992. Advanced Microwave Precipitation Radiometer observations during the Convection and Precipitation/Electrification Experiment, Proceedings of the Specialist Meeting on Microwave Radiometry and Remote Sensing Applications, (Westwater, E.W., Ed.), National Oceanic and Atmospheric Administration, Boulder, CO, 428-432.
2. Smith, E.A., C. Kummerow, and A. Mugnai, 1994. The emergence of inversion-type profile algorithms

for estimation of precipitation from satellite passive microwave measurements, *Remote Sens. Rev.*, in press.

3. Smith, E.A., A. Mugnai, H.J. Cooper, G.J. Tripoli, and X. Xiang, 1992. Foundations for statistical-physical precipitation retrieval from passive microwave satellite measurements. Part I: Brightness-temperature properties of a time-dependent cloud-radiation model, *J. Appl. Meteor.*, **31**, **6**, 506-531.
4. Mugnai, A., E.A. Smith, and G.J. Tripoli, 1993. Foundations for statistical-physical precipitation retrieval from passive microwave satellite measurements. Part II: Emission source and generalized weighting function properties of a time- dependent cloud-radiation model, *J. Appl. Meteor.*, **31**, **6**, 506-531.
5. Smith, E.A., X. Xiang, A. Mugnai, R.E. Hood, and R.W. Spencer, 1994. Behavior of an inversion-based precipitation retrieval algorithm with high resolution AMPR measurements including a low frequency 10.7 GHz channel, *J. Atmos. Oceanic Tech.*, in press.
6. Marzano, F.S., A. Mugnai, J. Turk, J. Vivekanandan, E.A. Smith, and X. Xiang, 1994. Active and passive microwave remote sensing of precipitating storms during CaPE. Part I: Intercomparison of AMPR and CP-2 retrievals, *Meteorol. Atmos. Phys.*, in press.
7. Turk, J., J. Vivekanandan, F.S. Marzano, R.E. Hood, R.W. Spencer, and F.J. LaFontaine, 1994. Active and passive microwave remote sensing of precipitating storms during CaPE. Part I: Advanced Microwave Precipitation Radiometer and polarimetric radar measurements and models, *Meteorol. Atmos. Phys.*, in press.
8. Basili, P., P. Ciotti, G. d'Auria, F.S. Marzano, A. Mugnai, and N. Pierdicca, 1994. Precipitation retrieval from spaceborne microwave radiometers based on maximum likelihood discriminant analysis, *IEEE Trans. Geosci. Remote Sens.*, submitted.
9. Tripoli, G.J., 1992. A nonhydrostatic model designed to simulate scale interaction, *Mon. Wea. Rev.*, **120**, 1342-1359.
10. Basili, P., P. Ciotti, G. d'Auria, F.S. Marzano, and N. Pierdicca, 1992. A Microwave Radiometry Characterization of Precipitatin Clouds, Proceeding of the Specialist Meeting on Microwave Radiometry and Remote Sensing Applications, Roma 14-17 February 1994. This issue.
11. Bringi, V.N., D.A. Burrows, and S.M. Menon, 1991. Multiparameter radar and aircraft study of raindrop spectral evolution in warm-rain based clouds, *J. Appl. Meteor.*, **30**, 853-880.

Sensitivity of SSM/I Rain Rate Algorithms to Variations in Water Vapor and Cloud Liquid Water - A Modeling Study

BARBARA A. BURNS and FRIEDRICH FLENDER
Institute of Environmental Physics
University of Bremen
D-28334 Bremen, Germany
Tel: +49 421 218 2584
FAX: +49 421 218 4555
email: burns@diana.physik.uni-bremen.de

Abstract Model-generated brightness temperatures for the SSM/I instrument are used to evaluate the sensitivity of a number of algorithms currently used to retrieve rain intensity. The effects of changes in cloud liquid water and water vapor profiles over both land and ocean are considered. The radiative transfer calculations assume a multilayered plane parallel atmosphere with the vertical structure specified in terms of the average relative humidity, cloud liquid water content, and precipitation rate of liquid and frozen hydrometeors in each layer. The algorithms evaluated include model-based algorithms, linear or log regression algorithms, and algorithms utilizing the so-called scattering index. All algorithms employ the 85 GHz channel alone or in combination with the lower frequency channels. The results show that a 30% variation in the cloud water content above the surface rain layer changes the retrieved rain rate by up to 3 mm/hr over the ocean, and up to 2 mm/hr over land, assuming the rain cloud fills the radiometer footprint (a beam-filling event). The algorithms most sensitive to this variation are those which rely heavily on the 19 and 37 GHz channels. Algorithm sensitivity to variations in relative humidity is very low, less than 0.5 mm/hr, again assuming a beam-filled situation. The effect of partial beam filling on the algorithm sensitivity is most important for ocean algorithms relying mainly on the 85 GHz channel.

Keywords: Passive Microwave Radiometers, Rain Retrieval

INTRODUCTION

Satellite Passive Microwave Imagers (PMI) can provide estimates of rain rate, cloud liquid water, and water vapor on spatial and temporal scales compatible with regional weather prediction and global climate models. A large number of algorithms have been proposed to estimate rain rate from data obtained by satellite microwave imagers. These algorithms were initially developed for frequencies less than 22 GHz where the absorption and emission from rain drops dominates the microwave radiation from raining clouds (e.g. [1]). Such algorithms are limited to retrievals over ocean and for rain rates up to about 20 mm/hr. Also the retrieval results are typically biased due to the combined effect of the non-linear relationship between rain rate and brightness temperature and the large footprint size relative to the typical rain cell. Recent observations at high resolution (30 km at 37 GHz, 12 km at 85 GHz) from the DMSP series of Special Sensor Microwave Imagers (SSM/I) reduces the effect of non-linear beam averaging. The higher frequencies also enable measurements of rain rate over land [2]. At these high frequencies the brightness temperature contains contributions from both emission from cloud droplets and rain drops and scattering from rain drops and ice particles. As such, the interpretation of the high frequency measurements in terms of the surface rain layer is less direct (e.g. [3]).

Present algorithms exploit both the emission and scattering characteristics of the higher frequency channels along with their complementarity with the lower frequency channels. Most algorithms use relationships between rain rate (RR) and brightness temperatures (Tb) based on physical models or results of radiative transfer calculations (e.g. [4]), whereas others use mostly empirical Tb-RR relationships (e.g. [5]). Algorithm coefficients are typically derived from regression against radiosonde or weather radar data (e.g. [6]).

The relative accuracy of these algorithms and their geographic limitations depend on their sensitivity to changes in surface conditions and atmospheric conditions other than rain rate. The sensitivity of the higher frequency channels (37 and 85 GHz) to variations in water vapor and cloud liquid water is attested to by their use in retrieval algorithms for these quantities. The 37 GHz channels is used by [7], for example, to retrieve precipitable water (water vapor) from SSM/I brightness temperature data, and the best-fit four channel algorithm for cloud liquid water from [8] includes both 37 and 85 GHz channels.

The role of cloud liquid water in satellite passive microwave measurements has been examined by [3,9] using an evolving cloud model superimposed on various rain layers. Their results showed that although cloud water tends to obscure the Tb-RR relationships, its effect is highly frequency dependent. Considering only the emission contribution at 19 and 22 GHz, the brightness temperature increase due to rain is always larger than that from cloud water. This is also true at 37 GHz for light and medium rainfall, but for heavy rain, the rain and cloud liquid water have almost equal importance. At 90 GHz, for all but light rain cases, the cloud liquid water emission is either equivalent to or totally masks the rain emission. However, in these cases over land, both emission contributions are small compared to the scattering by ice and rain drops at this frequency. Thus the ambiguity between rain and cloud water signals can be reduced by combining the lower and higher frequency channels in rain retrieval algorithms.

A multichannel combination (19 and 22 GHz) was suggested by [10] to reduce the effect of water vapor variability on rain retrievals, which can amount to 2 mm/hr for a single frequency (19 GHz) algorithm. However they also indicated that errors in the minimum detectable rain rate depends on the contribution resulting from non-precipitating clouds.

The effects of variable atmospheric conditions and fractional coverage on rain retrievals were also modeled by [6]. They found that both factors reduce the correlation between brightness temperature and rain rate, especially at the higher frequencies. Using a multivariate analysis of radiosonde and model-generated Tb data they showed that very low rain rates would be overestimated even with multichannel algorithms due to the similar effect of vapor, cloud droplets, and rain drops on the radiometer signal.

In this study we use model-generated brightness temperature data to evaluate the sensitivity of a number of algorithms currently used with SSM/I data to retrieve rain rate. We consider a range of cloud water and water vapor profiles over both land and ocean, as well as the effects of variable surface emissivity and fractional cloud coverage. In the next section we briefly describe the radiative transfer model and, in more detail, the cloud models used to generate the brightness temperature "data". These data are then examined in section 3 to determine how variations in the amount and vertical distribution of liquid water and water vapor (relative humidity) affect the brightness temperatures of clear air, nonprecipitating and precipitating clouds at SSM/I frequencies. The precipitation algorithms examined in this study are briefly described in section 4 and their response to the variabilities in the model data are presented in section 5.

DESCRIPTION OF MODELS

Radiative Transfer Model

Brightness temperatures were obtained with a radiative transfer model developed by Kummerow and Liberti (see [11]). The radiative transfer model takes account of the absorption and scattering by hydrometeors as well as the cloud temperature structure and ice/water mixing ratio variability. Model radiances are calculated for a multilayered plane parallel atmosphere using Eddington's second approximation to the equation of radiative transfer. Accuracies of 2-3 K in brightness temperature (not including errors due to finite cloud effects) are expected with this model.

The model includes the effects of single and multiple scattering from precipitating hydrometeors with the Mie calculation of the extinction coefficients being carried out assuming the Marshall-Palmer drop size distribution for the liquid phase. In this study the distribution given by [12] is used for the solid phase. Drops of cloud liquid water are assumed to be too small to produce scattering at microwave frequencies so that only Rayleigh absorption is accounted for. Absorption coefficients for water vapor and atmospheric oxygen are also included in all levels; these are the sole contributors in levels not having cloud as well as in the clear sky case.

The model assumes horizontal homogeneity and extends from the surface (z=0) to a maximum height of 15 km. The vertical structure of the layered medium (cloud) must be specified in terms of the average relative humidity (rh), cloud liquid water content (CWC), and precipitation rate of liquid and frozen hydrometeors in each layer. The height and temperature at each layer interface as well as the surface emissivity must also be specified.

Surface Models

Both ocean and land surfaces were considered. For ocean, a calm surface with frequency and temperature dependent emissivities was assumed. Wind roughening was found to have little effect on the brightness temperature in cloudy regions, although the effect of surface variations is significant in clear areas. This variability in surface emissivity will be important for fractional cloud-cover cases (see [6]) but was not examined here.

Five different land surfaces have been considered: wet soil, rocky soil, short grass, bare soil, and a mixture of bare and vegetated soils. The emissivities were taken from [13] and are given in Table 1.

Table 1.

Land Surface Emissivities

Channel	Wet Soil	Rocky Soil	Short Grass	Bare Soil	Mixed Surface
6.6 H	0.73	0.80	0.93	0.75	0.84
6.6 V	0.87	0.90	0.95	0.89	0.92
10.7 H	0.81	0.86	0.95	0.84	0.90
10.7 V	0.89	0.92	0.96	0.90	0.93
19 H	0.87	0.89	0.94	0.90	0.92
19 V	0.90	0.93	0.94	0.92	0.93
22 H	0.87	0.89	0.94	0.90	0.92
22 V	0.90	0.93	0.94	0.92	0.93
37 H	0.90	0.91	0.95	0.91	0.93
37 V	0.91	0.94	0.94	0.92	0.93
85 H	0.93	0.91	0.94	0.93	0.94
85 V	0.94	0.94	0.95	0.94	0.95

Cloud/Atmosphere Models

Standard cloud and atmosphere models are used in the radiative transfer model. Five general situations have been considered: clear air, stratiform nonprecipitating and precipitating clouds, and convective nonprecipitating and precipitating clouds. Information on the heights of the cloud bottom, the cloud top, and the boundaries of intermediate layers, as well as the total water content in each layer, was taken from Table 5-7 "Properties of Standard Cloud Models" in [14]. Total water content includes the equivalent water contents of liquid and frozen, precipitating and nonprecipitating hydrometeors. The drop size distribution of the hydrometeors in each layer is also given in terms of the mode radius and shape parameters of the modified gamma distribution from [15].

The atmospheric temperature below the cloud, above the cloud (but below the tropopause) and in the clear air case is assumed to decrease with height from the given surface temperature (290 K for ocean and for land) at a constant lapse rate of G_c = -6.56 K/km. In the tropopause (10-15 km) the temperature is assumed constant. Within a cloud layer the saturated adiabatic lapse rate is assumed.

The relative humidity in the clear air cases and outside the clouds is calculated from the pressure profile used in the radiative transfer program, the temperature profile described above, and the specific humidity profile according to [16]. The specific humidity is assumed to decrease exponentially with height from the surface value, calculated from the given value of rh_o, to a constant value above the tropopause. In cloud layers the relative humidity is set to 100%. The nominal value assumed for rh_o is 50%.

The total water content (TWC) given in the standard cloud models must be partitioned between cloud droplets (CWC) and rain drops (RWC). Model results presented in [9] show that both the drop size distribution and total water content changes not only as the cloud evolves but also varies with height in the cloud. They also indicate rain drop size hydrometeors even in cases when no rain layer is present (at least in the convective cases they modelled). Their results are in generally good agreement with the convective cloud models assumed here. Since the drop size distribution at each level is given for these models, the water contents of small and large drops can be calculated. Assuming an upper cut-off of 100µm for cloud droplets, this distribution is integrated over the small and large drop regimes to obtain cloud and rain water contents, respectively. The resulting CWC and RWC profiles show distinctly different structures for convective and stratiform cases, the latter having very low RWC above the surface rain layer.

The rain water content is further partitioned between liquid and frozen hydrometeors depending on the temperature structure in the cloud. RWC rather than TWC is used since in the radiative transfer model all ice particles are taken to be > 100µm in diameter and precipitating. Following [17] the ice water content (IWC) is given by

$$
\begin{aligned}
&z < z(0^{o}C), && \mathrm{IWC} = 0 \\
&z(0^{o}C) < z < z(-35^{o}C), && \mathrm{IWC} = \mathrm{RWC} \cdot \frac{z - z(0^{o}C)}{z(-35^{o}C) - z(0^{o}C)} \qquad (1) \\
&z > z(-35^{o}C), && \mathrm{IWC} = \mathrm{RWC}
\end{aligned}
$$

As the main parameter of interest in the radiative transfer scheme, rain rate rather than water content must be specified for each layer. RR is related to RWC through the drop size distribution N(D):

$$RWC = \int_{D_{min}} \frac{4\pi}{3}\left(\frac{D}{2}\right)^3 N(D)dD \quad (2)$$

where ρ_L is the density of liquid water (1.0 g/cm^3), D is the drop diameter and N(D) is typically observed to have an exponential dependence on RR

$$N(D) = N_o \exp(-\Lambda \cdot D), \qquad \Lambda = \beta\, RR^{-\alpha} \quad (3)$$

For stratiform rain the values of N_o, β and α are those of the Marshall-Palmer distribution (N_o=8.0x10^{-6} (m^{-4}); Λ=4100 RR$^{-0.21}$ (m^{-1})); for convective rain the values given by [18] as typical of thunderstorms (N_o=1.4x10^{-5} (m^{-4}) ; Λ=3000 RR$^{-0.21}$ (m^{-1})) are used.

The precipitation rate for ice particles (IR) is determined analogously with ρ_I = 0.91 g/cm^3 replacing ρ_L and using a size distribution similar to (3) with N_o = 2.5 x 10^6 IR$^{-0.94}$ (m^{-4}) and Λ = 2290IR$^{-0.45}$ (m^{-1}) [12].

BRIGHTNESS TEMPERATURE RESULTS

Examples of the model-generated brightness temperatures for five atmospheric conditions are shown in Figure 1 for a calm ocean surface and a bare soil surface. In non-raining cases the frequency behavior generally follows the spectral characteristics of the surface emissivity. Over ocean the brightness temperatures increase with increasing total water content as expected when emission dominates the signal, whereas over land they exhibit relatively little change with water content. The results for light stratiform rain indicate that at the 19, 22 and 37 GHz channels emission still dominates (Tb(with rain) > Tb(no-rain)) but at 85 GHz scattering by rain drops reduces the signal. Also as the optical thickness of the clouds increases self emission comes from increasingly higher and colder cloud layers (see [3]).

For convective rain Figure 1 displays a decrease in brightness temperature at all frequencies greater than 19 GHz due to scattering from precipitating hydrometeors. At 85 GHz the brightness temperatures over ocean is lower than that of the fair weather cumulus by at least 25 K, and that of the cumulus congestus cloud (not shown) by about 5 K. Figure 1 (top) shows that this creates ambiguities between convective rain events and non-precipitating water clouds at 37 and 85 GHz. Over land (Figure 1, bottom) the depression at 85 GHz below the brightness temperatures of no-rain clouds demonstrates the well recognized utility of this channel for identifying rain over land.

The reduction in brightness temperature due to scattering is not as great as that predicted by other radiative transfer calculations (e.g. [17, 19]). However this can be attributed to the differences in the cloud descriptions, especially the comparatively large total water contents used here, and to the linear partitioning of water content between ice and rain drops in the mixed layer (see also [6]).

The effect of surface variability on brightness temperatures over land is shown in Figure 2 for horizontal and vertical polarizations. The range is bounded by wet soil (bottom) and short grass (top) with bare soil falling midway between (see Table 1). In general the variation seen is greater for horizontal than for vertical polarization. At 22 GHz and below, the variability in Tb produced by changes in surface emissivity is greater than that due to changing cloud properties (i.e. water content). This leads to ambiguities between clear, cloudy, and raining conditions at these frequencies. At 37 GHz ambiguities arise from the variability of the clear air signature and are greatest at horizontal polarization. At 85 GHz the surface emissivity variations have little if any effect on the modeled brightness temperatures of rain clouds. In the following analyses the emissivity of bare soil is used to represent the average behavior.

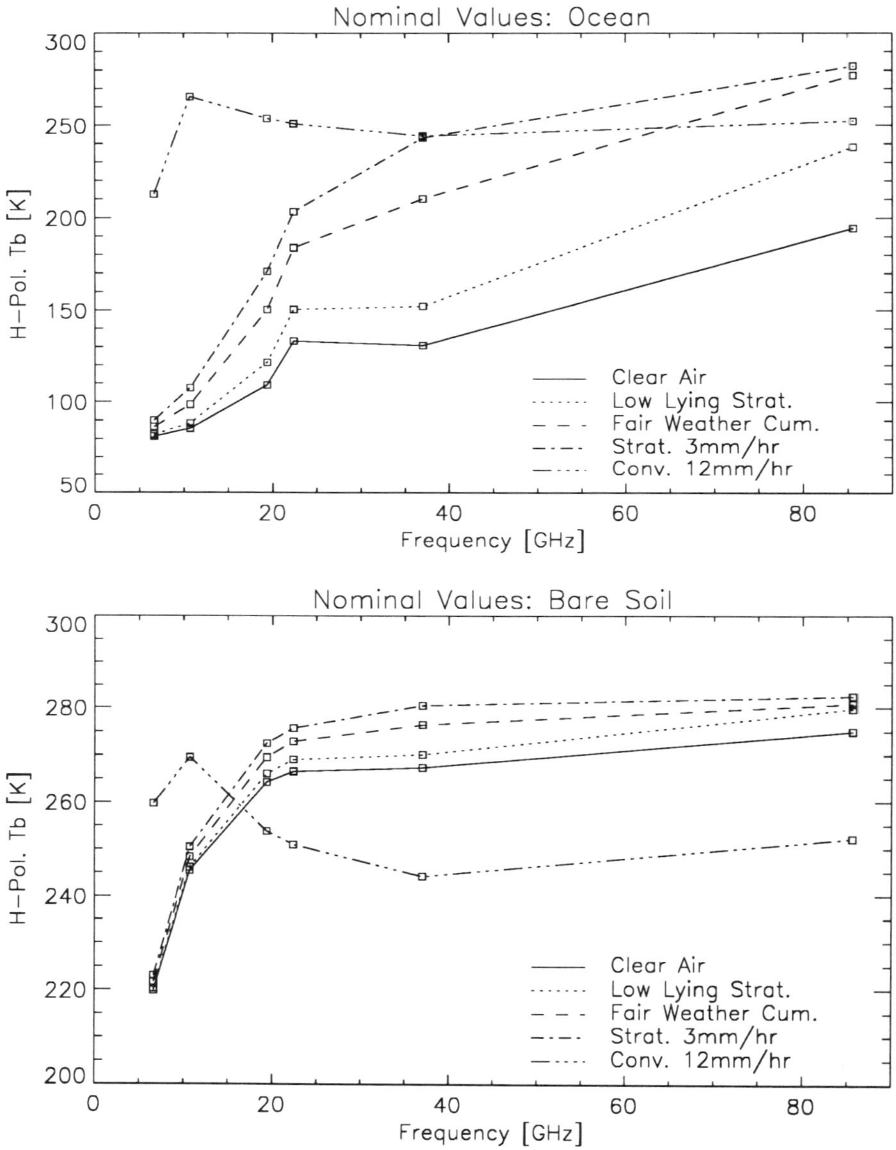

Figure 1. Model-generated brightness temperatures for five atmospheric conditions with an ocean, and with a bare soil background.

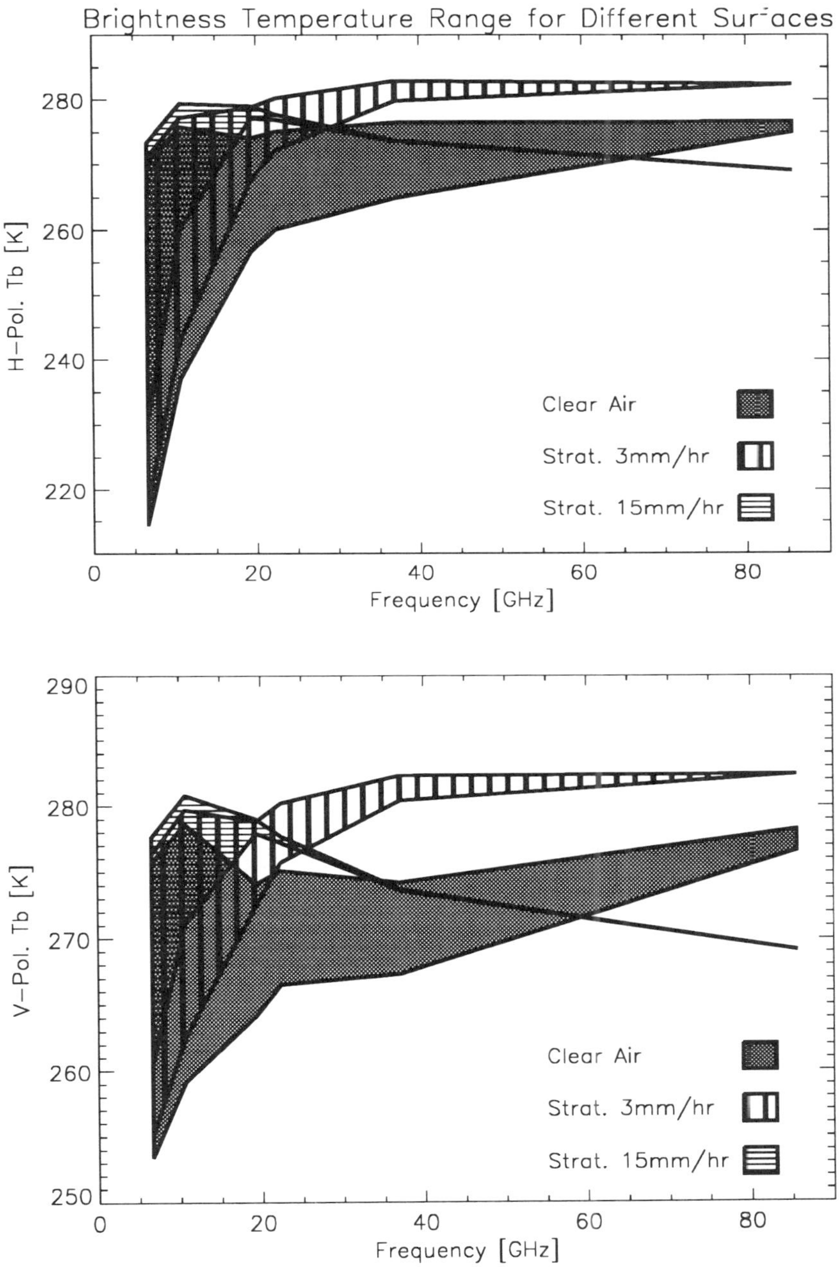

Figure 2. Range of brightness temperatures produced by variable land surface emissivity as a function of frequency and atmosphere/cloud model. (top) horizontal polarization, (bottom) vertical polarization.

The effect of changing atmospheric conditions on brightness temperature is shown in Figures 3 and 4 for ocean and bare soil surfaces, respectively. Surface relative humidity was varied from its nominal value of 50% to 25% and 75% (i.e. $\Delta rh_0 = \pm 25\%$) to effect a change in the water vapor profile. Cloud liquid water variability was produced by varying the total water content in each cloud layer (excluding surface rain layers) by 30% of the given values (i.e. CTWC = TWC $\pm$ 0.30·TWC). For clarity in the figures only Tb variations due to decreases in CTWC and increases in rh_0 are shown. Changes of opposite sign result mostly in symmetrical changes in Tb.

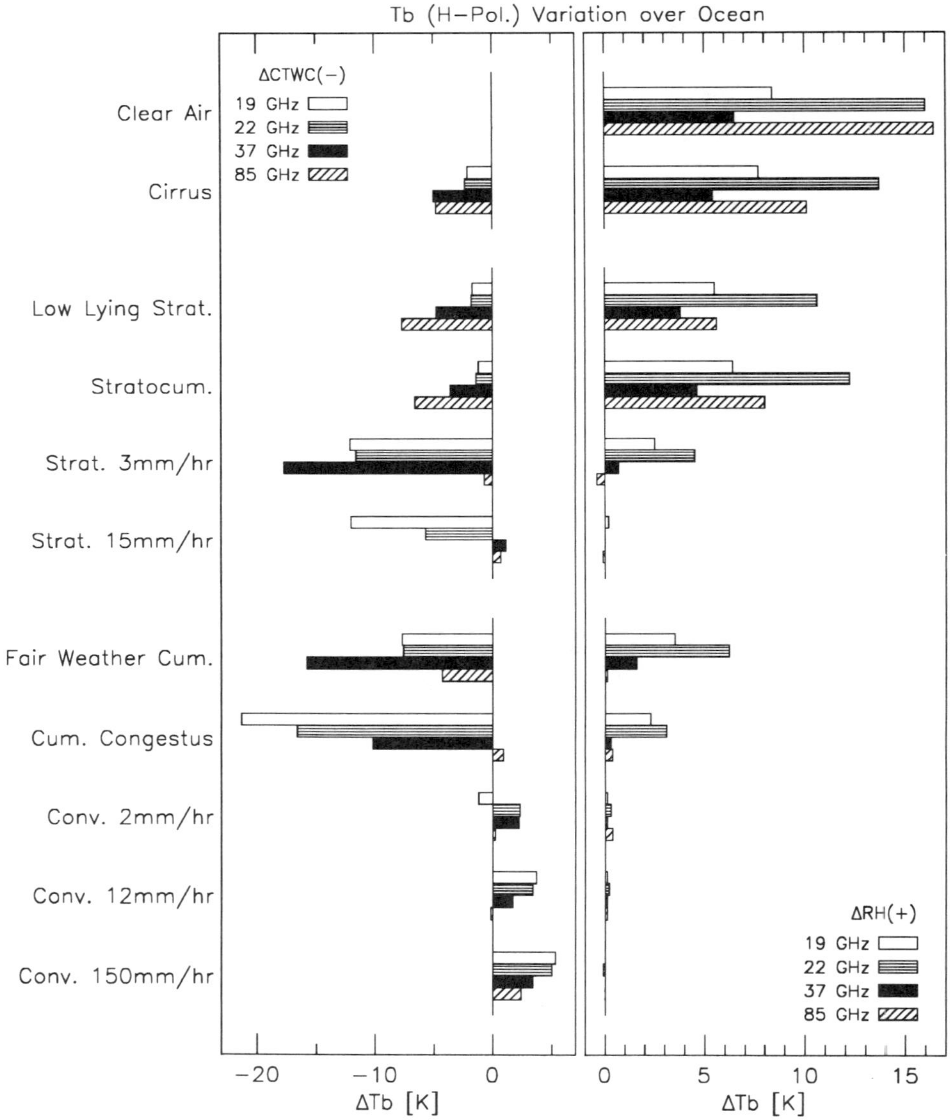

Figure 3. Change in brightness temperatures over ocean for (left) a 30% decrease in cloud water content and (right) an increase of $\pm$ 25% in surface relative humidity.

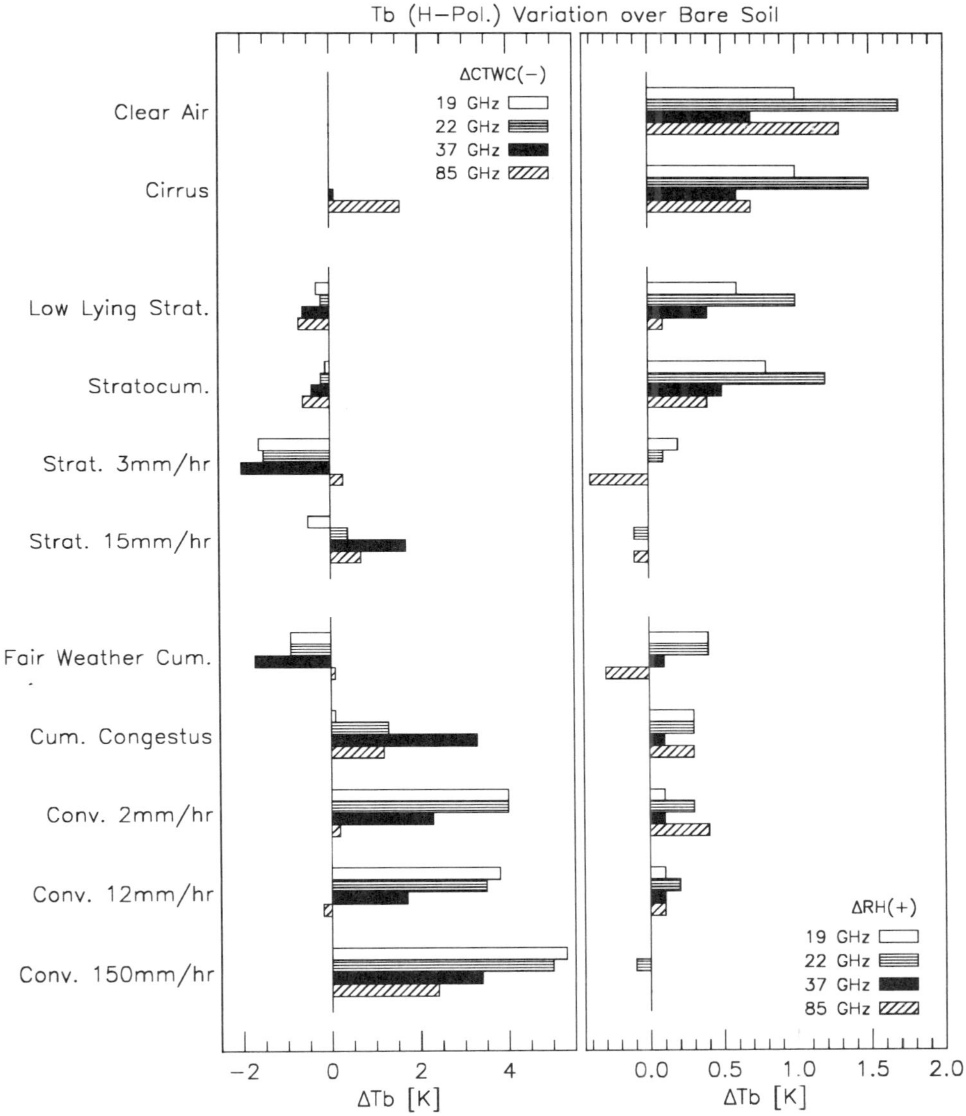

Figure 4. Change in brightness temperatures over land for (left) a 30% decrease in cloud water content and (right) an increase of ± 25% in surface relative humidity.

The magnitude of the Tb change due to the change in CTWC (ΔCTWC) reflects the relative amount of water in each cloud, at least when considering nonprecipitating and precipitating clouds separately, since the ΔCTWC is a percent of the nominal value. As seen in Figure 3, for non-raining clouds over ocean, decreases in CTWC generally result in decreased emission and therefore a decrease in Tb at all frequencies. The exception is in the cumulus congestus cloud at 85 GHz which shows a positive but small change. Tb behavior in precipitating clouds differs with rain type. For stratiform clouds, where most of the CTWC is in the form of cloud droplets, the 19 and 22 GHz channels respond in the same way as in the non-raining clouds to ΔCTWC. At the lower rain rate the 37 GHz Tb also decreases with decreasing water content (ΔTb = -18 K), but at the higher rate ΔTb at both 37 and 85 GHz is very small (< 2K) and possibly increases due to decreased scattering (fewer rain drops). For the convective rain cases all four channels generally show small (< 5K) increases with decreased CTWC since most of the water in these models is in the form of rain drops.

Figure 4 shows that over land the changes in Tb for clouds with the lowest TWC's are similar in nature to those seen for the ocean but smaller (< 2K) because in these cases the signal is dominated by the surface emission. The non-raining convective clouds and stratiform rain clouds represent transition cases where first the higher frequencies and then the lower ones become optically thick to surface emission. For these transition cases changes in CTWC result in only small changes in Tb; only the change at 37 GHz is really significant. The response of convective clouds over land is similar in both sign and magnitude as over ocean because the surface emission contributes little to the signal. For these cases the significant changes in Tb are at 19 and 22 GHz.

Comparison of Figures 3 and 4 shows that the Tb changes produced by varying rh_0 are greater over ocean than over land. For clear air the ΔTb over ocean is more than 15K at 85 GHz but less than 2K at the same frequency over land. Over ocean Δrh is important at both 22 and 85 GHz for thin cloud and clear air, but only at 22 GHz (and possibly 19 GHz) for thicker and raining clouds. For raining clouds over land, the change at 85 GHz can be as large or larger than that at 22 GHz.

SSM/I RAIN RATE ALGORITHMS

Having presented the effect of varying cloud water content and relative humidity on the calculated brightness temperatures, we now wish to apply various algorithms to these data and examine the effect of these changes on the rain rates retrieved. The algorithms applied here were taken from those used and fully documented in the AIP-2 [20] and/or PIP-1 [21] analyses and selected to represent different concepts or types of Tb-RR relationships. Also only algorithms making use of the 85 GHz channel alone or in combination with the lower frequency channels were chosen. Although the algorithms using iterative application of radiative transfer models (e.g. [11]) represent the most physical retrieval methods, statistical regression-based algorithms were chosen here for their ease of implementation.

The four ocean algorithms evaluated here are defined with the following expressions:

adler: $RR = 120.1 - 0.478 \cdot Tb85H$

bauer:
$$\log(RR) = 17.45 - 0.1761E11 \cdot (Tb19V)^{-4} - 0.057 \cdot Tb22V + 0.1404E6 \cdot (Tb19V)^{-2} - 0.1188E\text{-}6 \cdot (Tb19V - Tb19H)^4 + 0.1944E\text{-}3 \cdot (Tb22V - Tb85V)^2$$

ferriday: $RR = (Tb19V + Tb37V - Tb85V - Tb85H + 50) / 10$

ferraro: $SI = -182.7 + 0.75 \cdot Tb19V - 2.543 \cdot Tb22V$
$+ 0.00543 \cdot (Tb22V)^2 - Tb85V$
$RR = -1.05 + 0.149 \cdot SI$

Note that the name indicated next to each algorithm is the reference name used in this study and is not an author citation. The land algorithms are defined as follows:

adler: $RR = 59.9 - 0.239 \cdot Tb85H$

ferriday: $RR = (Tb19V + Tb22V - Tb37V - Tb85V) / 7$

ferraro: $SI = 438.5 - 0.46 \cdot Tb19V - 1.735 \cdot Tb22V$
$+ 0.00589 \cdot (Tb22V)^2 - Tb85V$
$RR = -2.71 + 0.362 \cdot SI$

smith: $RR = 125.5 - 0.455 \cdot Tb19 + 0.108 \cdot (Tb19-Tb85)$

Most algorithms produced reasonable rain rates in a relative sense (that is comparing cloud models) although the absolute values were not necessarily comparable with the model rain rates or between algorithms. This is primarily due to the determination of algorithm coefficients with regressions against data or model results of which the models used here are not representative. This limitation also meant that no screening schemes could be evaluated. Except where the algorithm was also based on the radiative transfer model used here ("ferriday"), most rain cases would have been screened out. The reason for this was generally the relatively small depression at 85 GHz mentioned in the previous section. Our study was therefore restricted to looking at only the changes in rain rates retrieved.

ALGORITHM SENSITIVITIES

The sensitivity of the various algorithms to changes in relative humidity and cloud liquid water was evaluated assuming first that rain clouds fill the radiometer beam and then that the beam is only partially filled by rain with either non-raining clouds or clear air as the other component. The rain rate variations (ΔRR) over ocean for beam-filling events are presented in Figure 5. Variations greater than 1 mm/hr are observed due to $\Delta CTWC$. In the stratiform rain models, the "ferriday" algorithm exhibits a 1.5 mm/hr variation for the 3 mm/hr model which results from this algorithm's use of the 37 GHz channel. The largest variation in the convective models is approximately 3 mm/hr with the "bauer" algorithm which is explained by the large response of the 19 GHz channel to $\Delta CTWC$ (see Figure 3). Variations in rain rate retrieved with the same algorithms due to Δrh_o are also shown in Figure 5. Here the ΔRR's are much smaller (< 0.5 mm/hr) as expected from the smaller ΔTb's (Figure 3). The response to Δrh_o is largest for the "ferraro" algorithm which uses Tb^2(22 GHz) in the scattering index.

Rain rate variations for the land algorithms are shown in Figure 6. Changes in CTWC produce ΔRR > 1 mm/hr for convective rain models. The "ferraro" algorithm retrieval variation of approximately 1 mm/hr is again due to the relatively large response at 19 and 22 GHz to CTWC changes in convective clouds. This is also the cause of the retrieval variation of up to 2 mm/hr with the "smith" algorithm that relies on the 19 - 85 GHz relationship. The retrieval variations due to Δrh_O over land are very small (0.1 mm/hr or less) for all algorithms.

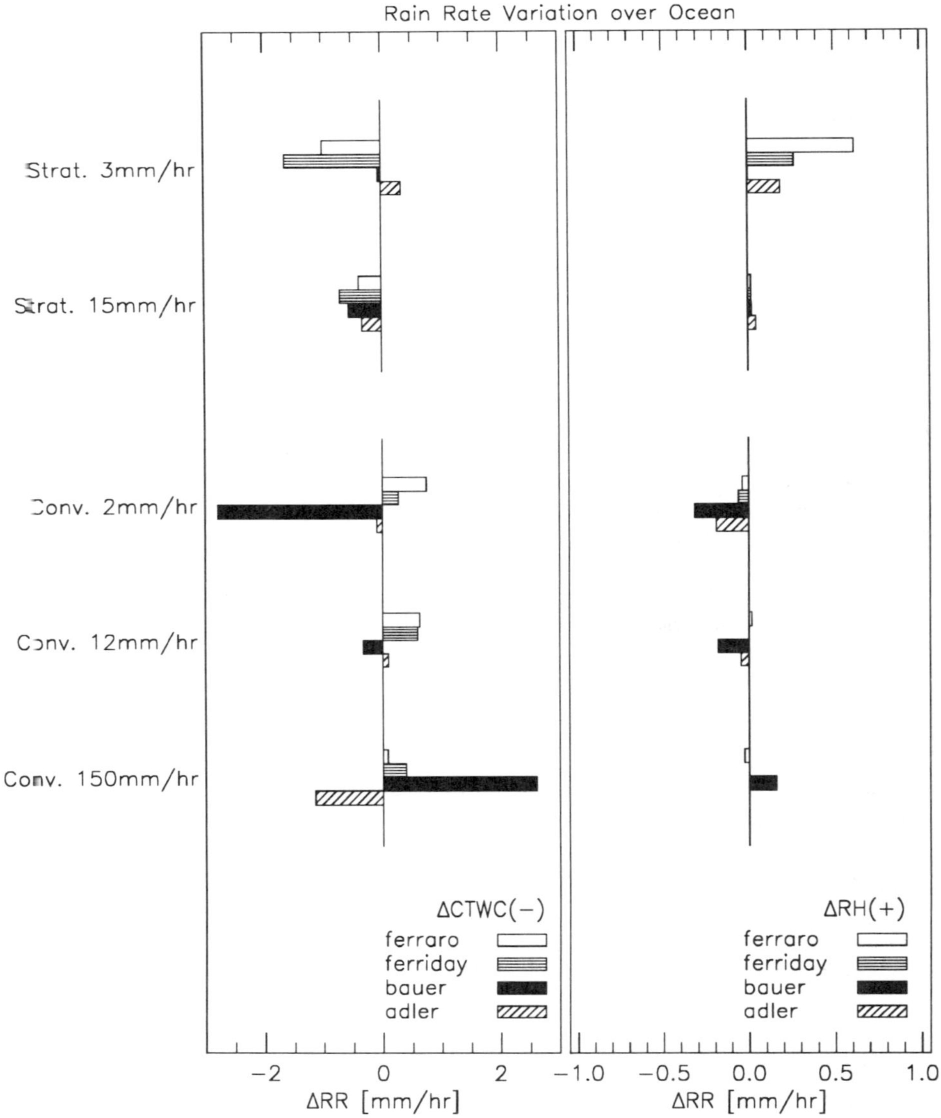

Figure 5. Algorithm sensitivity to (left) cloud water and (right) relative humidity changes for beam-filled rain events over ocean.

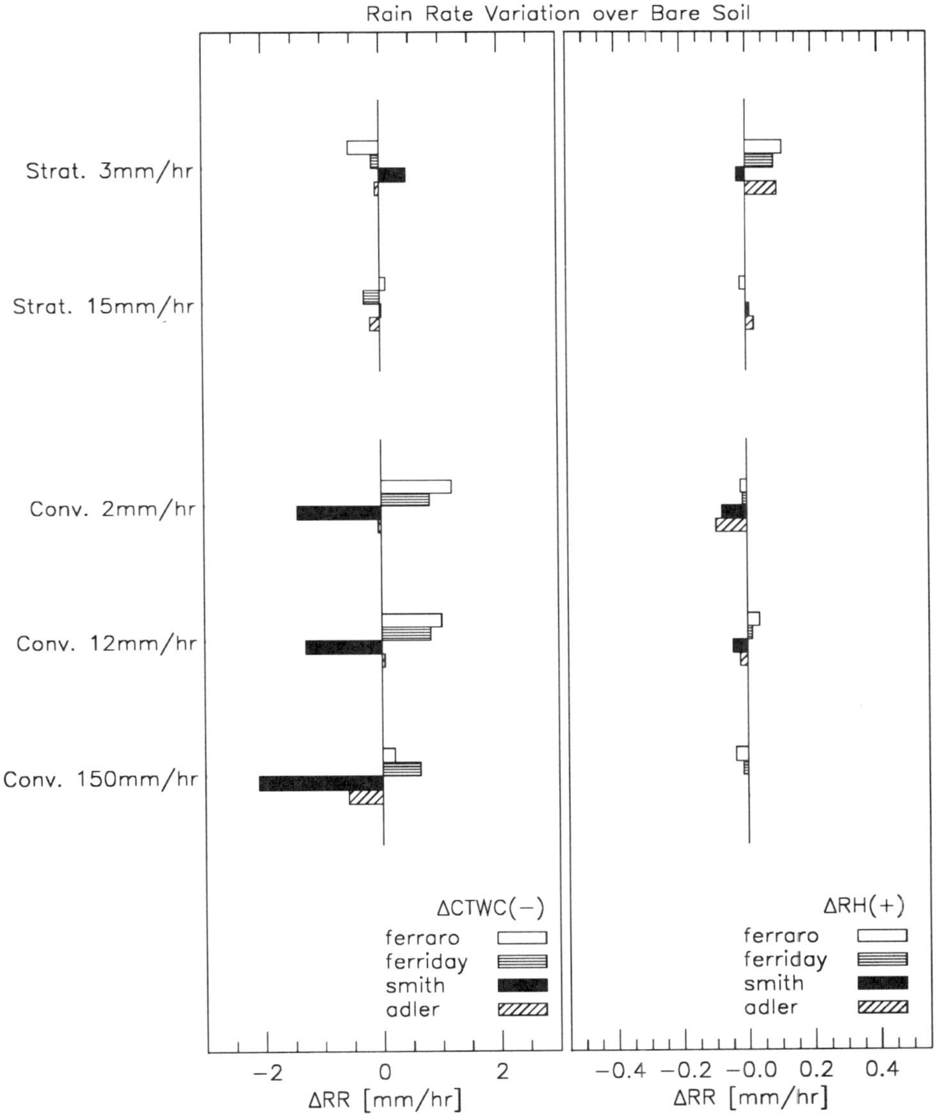

Figure 6. Algorithm sensitivity to (left) cloud water and (right) relative humidity changes for beam-filled rain events over land (bare soil).

To investigate how these results change when the radiometer beam is not filled with a rain cloud, the brightness temperatures of rain/no-rain cloud and rain/clear air combinations were calculated and the algorithms applied to these mixed signature data. The rain/no-rain cloud combination tests the effect of changing cloud water content, whereas the rain/clear air combinations tests the changes in relative humidity.

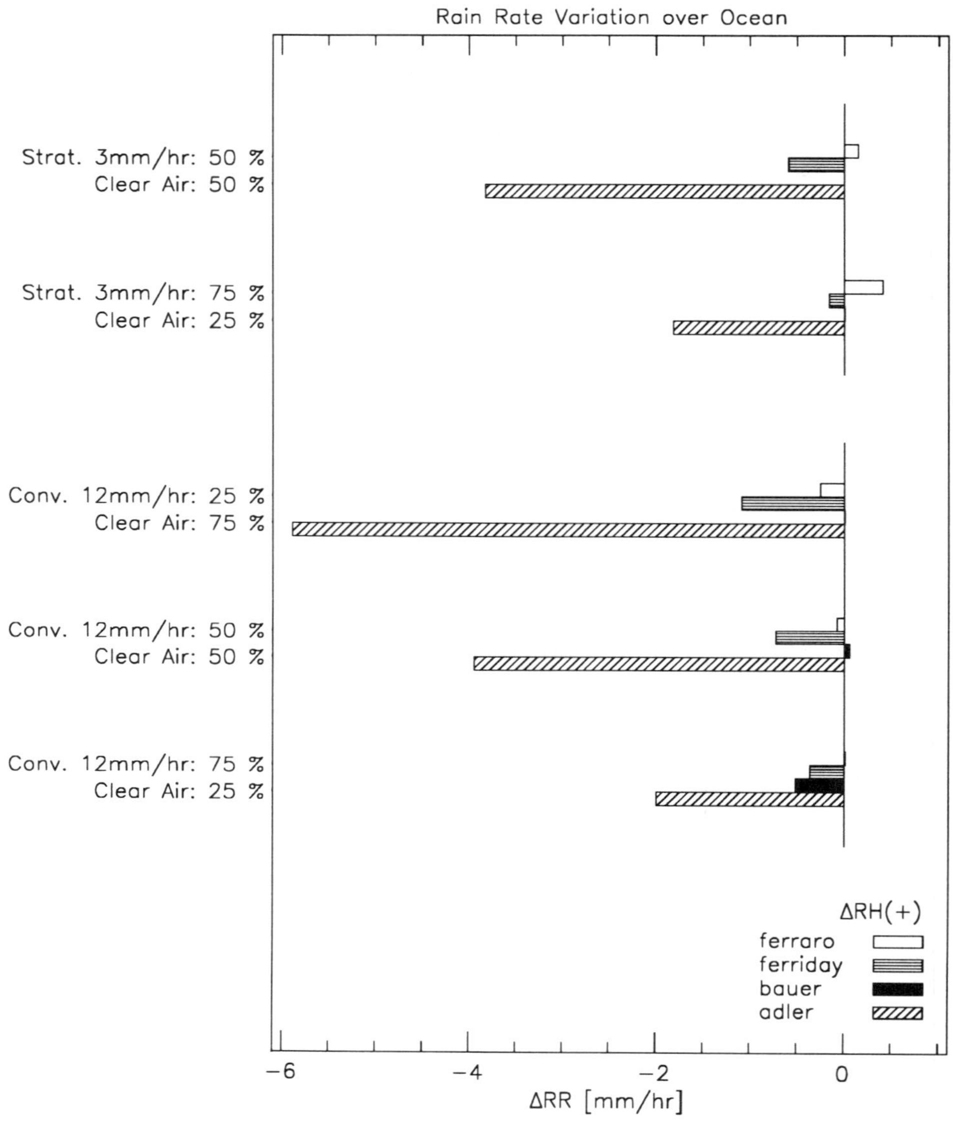

Figure 7. Algorithm sensitivity to relative humidity changes for partially beam-filled situations (rain/clear air) over ocean.

Over ocean the rain rate retrieval variation due to ΔCTWC is reduced for all but the "adler" algorithm. This can be explained by referring to Figure 3 which shows that at the lower frequencies the ΔTb due to ΔCTWC is smaller for nonprecipitating than for precipitating cloud for the stratiform models, and of opposite signs for the convective models. At 85 GHz, however, the ΔTb's are generally larger for no-rain clouds with the result that the "adler" algorithm shows a rain retrieval variation of 2 mm/hr compared to 0.3 mm/hr for the beam-filled case (stratiform model). Over land all algorithms show reduced variation with partial beam filling mainly because of the effect of compensating ΔTb seen in the convective models over ocean.

A rain/clear air situation leads to increased rain retrieval variability due to Δrh_0 for the "ferriday" and "adler" ocean algorithms (Figure 7). This can be attributed to the importance of the 85 GHz channel in these algorithms. The ΔRR for the "adler" algorithm which relies on 85 GHz alone increases from about 0.2 mm/hr in the beam-filled case to over 5 mm/hr. On the other hand, for algorithms using differences of 85 GHz and lower frequency channels ("ferraro") retrieval variations are reduced with partial beam filling. Over land algorithms relying heavily on 85 GHz show very slight increases (to 0.3 mm/hr) in retrieval variability.

SUMMARY

A radiative transfer model has been used to simulate brightness temperatures at the SSM/I frequencies for clear air, nonprecipitating and precipitating clouds and compute the variations due to changes in water vapor and cloud liquid water. A number of rain rate algorithms developed for use with SSM/I data have been applied to these "data" to evaluate the effect of these atmospheric variations on the retrievals.

The results show that a 30% change in the water content in cloud layers above the surface rain layer can lead to changes in the retrieved rain rate of up to 3 mm/hr for a beam-filling rain event over the ocean, and up to 2 mm/hr over land. The most sensitive algorithms are those relying heavily on 37 GHz when applied to thin (stratiform) rain clouds and those utilizing mainly 19 GHz with thick (convective) rain clouds. Algorithm sensitivity to variations in relative humidity (water vapor) in a beam-filled situation is very low, less than 0.5 mm/hr over ocean and 0.1 mm/hr over land.

The effect of partial beam filling on the algorithm sensitivity is most important for ocean algorithms relying mainly on the 85 GHz scattering signal. In this case a situation of combined rain/no-rain stratiform clouds or convective rain cells surrounded by clear air can lead to high variability in the retrieved rain rate due to variability of cloud water or water vapor in the non-raining portions of the footprint. Increases from 0.2 mm/hr for the beam-filled case to over 5 mm/hr with partial beam filling (rain/clear air) are seen for a change in surface relative humidity of 25%. This effect is much less important for algorithms combining 85 GHz with lower frequency channels.

Acknowledgements

We wish to thank J. P. V. Poiares Baptista for his help and guidance in this study and C. Kummerow for supplying the radiation transfer model code. We are also grateful to E. A. Smith and A. Mugnai for enlightening discussions, and to an anonymous reviewer for useful comments on the manuscript. This study was funded by ESA/ESTEC Contract No. 10347/93/NL/NB(SC).

REFERENCES

1. Wilheit, T.T., A.T.C. Chang, M.S.V. Rao, E.B. Rodgers, and J.S. Theon, 1977, A satellite technique for quantitatively mapping rainfall rates over the ocean, *J. Appl. Meteor.*, **16**, 551-650.

2. Spencer, R.W., H.M. Goodman, and R.E. Hood, 1989, Precipitation retrieval over land and ocean with the SSM/I: Identification and characteristics of the scattering signal, *J. Atmos. Oceanic Technol.*, **6,** 254-273.
3. Smith, E. A. and A. Mugnai, 1988, Radiative transfer to space through a percipitating cloud at multiple microwave frequencies. Part II: Results and analysis, *J. Appl. Meteor.*, **27**, 1074-1091.
4. Smith, E.A., A. Mugnai, H.J. Cooper, G.J. Tripoli, and X. Xiang, 1992, Foundations for statistical-physical precipitation retrieval from passive microwave satellite measurements. Part I: Brightness-temperature properties of a time-dependent cloud-radiation model, *J. Appl. Meteor.*, **31**, 506-531.
5. Grody, N. C., 1991, Classification of snow cover and precipitation using the Special Sensor Microwave Imager, *J. Geophys. Res.*, **96**, 7423-7435.
6. Bauer, P. and P. Schluessel, 1993, Rainfall, total water, ice water, and water vapor over sea from polarized microwave simulations and Special Sensor Microwave/Imager data, *J. Geophys. Res.*, **98,** 20737-20759.
7. Liu, W.T., W. Tang, and F.J. Wentz, 1992, Precipitable water and surface humidity over global oceans from Special Sensor Microwave Imager and European Center for Medium Range Weather Forecasts, *J. Geophs. Res.*, **97**, 2251-2264.
8. Alishouse, J.C., J.B. Snider, E.R. Westwater, C.T. Swift, C.S. Ruf, S.A. Snyder, J. Vongsathorn, and R.R. Ferraro, 1990, Determination of cloud liquid water content using the SSM/I, *IEEE Trans. Geosci.Remote Sensing,* **GE-28**, 817-822.
9. Mugnai, A. and E. A. Smith, 1988, Radiative transfer to space through a percipitating cloud at multiple microwave frequencies. Part II: Model description, *J. Appl. Meteor.*, **27**, 1056-1073.
10. Wilheit, T.T., A.T.C. Chang, and L.S. Chiu, 1991, Retrieval of monthly rainfall indices from microwave radiometric measurements using probability distribution functions, *J. Atmos. Oceanic Technol.*, **8,** 118-136.
11. Kummerow, C. and L. Giglio, A passive microwave technique for estimating rainfall and vertical structure information from space. Part I: Algorithm description, 1994, *J. Appl. Meteor.,* **28**, 869-884.
12. Sekon, R.S., and R.C. Srivaastava, 1971, Doppler radar observations of drop-size distributions in a thunderstorm, *J. Atmos. Sci.*, **28**, 983-994.
13. Mätzler, C., 1990. Algorithms for retrieving snowpack properties from a Multichannel Imaging Microwave Radiometer (MIMR), in *Study of Microwave Interaction with the Earth's Surface.* Report, ESA Study Contract No. 8447/89/NL/PB/(SC).
14. Colwell, R.N. (editor-in-chief), 1983. *Manual of Remote Sensing,* 2nd. Edition, Falls Church, VA: American Society of Photogrametry, 196-197.
15. Deirmendjian, D., 1964, Scattering and polarization properties of water clouds and hazes in the visible and infrared, *Appl. Optics,* **3,** 187.
16. Rogers, R.R. and M.K. Yau, 1989. *A Short Course in Cloud Physics*, Third Edition, Pergamon Press, New York.
17. Wu, R. and J. A. Weinman, 1984, Microwave radiances from precipitating clouds containing aspherical ice, combined phase, and liquid hydrometeors, *J. Geophys. Res.*, **89**, 7170-7178.
18. Meneghini, R. and T. Kozu, 1990. *Spaceborne Weather Radar*, Boston, MA: Artech House.
19. Liu, G. and J.A. Curry, 1992, Retrieval of precipitation from satellite microwave measurement using both emission and scattering, *J. Geophys. Res.*, **97**, 9959-9974.
20. Allam, R., G. Holpin, P. Jackson, and G.-L. Liberti, 1993. *Second Algorithm Intercomparison Project AIP-2 Pre-Workshop Report,* U.K. Meteorological Office.
21. Kniveton, D. R., C. Kidd, E. C. Barrett, B. Motta, F. LaFontaine, M. Smith, and M. Goodman, 1994, WetNet PIP-1 Results Version 2, Marshall Space Flight Center, Huntsville, AL, U.S.A.

Retrieval of liquid and ice water content in atmosphere using Special Sensor Microwave Imager (SSM/I)

Fuzhong Weng and Norman C. Grody

NOAA/Satellite Research Laboratory, Camp Springs, MD

Abstract The Special Sensor Microwave Imager (SSM/I) is a microwave radiometer having dual polarized channels at 19.35, 37 and 85.5 GHz and a vertically polarized channel at 22.23 GHz. The measurement data at these frequencies are applied for retrieving the liquid and ice water content in precipitating and non-precipitating clouds. The algorithms for the cloud liquid water retrieval are calibrated with the ground-based measurements. Three separate algorithms which are accurate to measure the liquid water from different cloud systems are composited to measure the large dynamic range of liquid water. For the ice water retrieval, the algorithm is derived using data sets from radiative transfer simulations. Retrieved cloud liquid and ice water distributions are compared with satellite visible and infrared images and show that the algorithms are capable for detecting all clouds except low level very thin cloud and high level cirrus cloud.

Key words: microwave, retrieval, cloud liquid, cloud ice.

INTRODUCTION

Global distribution of the cloud water is important for the climate and climate change issues. For example, a temperature change in the earth climate due to an increase of CO_2 amount may alter cloud liquid water content and therefore cause a strong feedback[1].There are many other potential applications of the liquid water content such as validations of cumulus parameterization and initialization for the numerical weather prediction models[2]. Cloud liquid water represents one of the earliest parameters derived using satellite microwave radiometers[3,4,5,6,7]. However, these studies used frequencies less than 37 GHz and were therefore mainly applicable to clouds having liquid water greater than 0.2 mm. Smaller amounts of water found in low stratus clouds are best detected using frequencies higher than 37 GHz. Petty[8] pointed out the importance of microwave measurements at 85.5 GHz to derive the liquid water for stratus and stratocumulus clouds formed in the marine oceanic atmosphere.

This study first develops a procedure for combining individual algorithms, each optimized for measuring different ranges of liquid water. A composite algorithm utilizes all SSM/I frequencies to measure the liquid water for stratiform clouds as well as highly convective systems. Retrieved cloud liquid is compared with ground-based microwave radiometer measurements. Then, an algorithm is proposed for cloud ice water retrieval. Finally, derived cloud ice and liquid water distributions are compared against visible and infrared satellite imagery for the cloud analysis.

CHARACTERISTICS OF SSM/I DATA

SSM/I instruments aboard the satellites of the Defense Meteorological Satellite Program (DMSP) include four frequency measurements at 19.35, 22.23, 37, and 85.5 GHz with both vertical and horizontal polarization except at 22.23 GHz which has only vertical polarization. To illustrate characteristics of these channel measurements, four brightness temperature data sets of clear, cloud, rain-only and rain-ice are generated from the radiative transfer model[9] which includes scattering and polarization effects. Each of these data sets corresponds to an ensemble of atmospheric situations according to their name. Namely, the clear sky data set represents upwelling brightness temperatures in the clear sky condition. The cloud data set corresponds to brightness temperatures in the rain-free cloud condition. The rain-only data set includes brightness temperatures in the rain-bearing clouds which have cloud tops at the freezing level. The rain-ice data set includes the brightness temperatures in the clouds which have raindrops below the freezing level and ice particles above that level. Detailed atmospheric and surface parameters used in the simulation (mean and standard deviation) are shown in Table 1.

Table 1. Atmospheric and surface parameters used for generating radiative transfer data sets. Variables, V, Ts, WVP, CWP, RWP and IWP represent surface wind, surface temperature, water vapor path, cloud water path, rain water path, and ice water path, respectively

Variables	V (m/s)	Ts (K)	WVP(mm)	CWP(mm)	RWP(mm)	IWP (mm)
Mean	4.85	291.6	28.73	0.19	2.00	0.57
Std. Dev.	2.87	7.0	16.73	0.13	1.63	0.34

Figures 1a-1d show the brightness temperatures over a water surface at the SSM/I frequencies of 19.35, 22.23, 37 and 85.5 GHz for vertical polarization. The different brightness temperatures are denoted as TV19, TV22, TV37 and TV85 and are plotted against the liquid water path (LWP) which is a vertically-integrated liquid water content. The LWP is the sum of the cloud liquid water and rain water paths, where the points numbered 1, 2, 3 and 4 correspond to the data for clear, cloud, rain-only and rain-ice data sets. Note that the rain-free clouds (2) have LWP much less than 1.0 mm, while rain-bearing clouds (3, 4) have LWP values extending to as high as 6 mm (also see Table 1).

For clear conditions (LWP=0), the brightness temperature exhibits a large variation due to atmospheric water vapor. This variation is maximum for TV22 since the frequency is at the center of the 22.23 GHz water vapor line. Increases in liquid water produce an increase in brightness temperature for all channels, followed by a saturation effect. Saturation occurs for the highest LWP (~ 3.0 mm) at the lowest frequency (19.35 GHz). At this frequency, the brightness temperature approaches the temperature of the cloud tops, which is near freezing. The highest frequency channel (85.5 GHz) saturates at the lowest LWP (~ 0.3 mm). Higher values of LWP at this frequency (also 37 GHz) result in a decrease in brightness temperature due to scattering by the large liquid and ice hydrometeors in the rain clouds. Horizontally polarized brightness temperatures, THs are much lower than the corresponding TVs for the clear data set due to the lower sea surface emissivity (figure not shown). This causes the THs to increase more rapidly than the TVs as LWP increases from 0 to 1.0 mm. The THs therefore appear to be more sensitive to lower LWP than the TVs. However, the variations in sea surface emissivity due to wind roughening and water vapor result in a larger variation in the THs than for TVs.

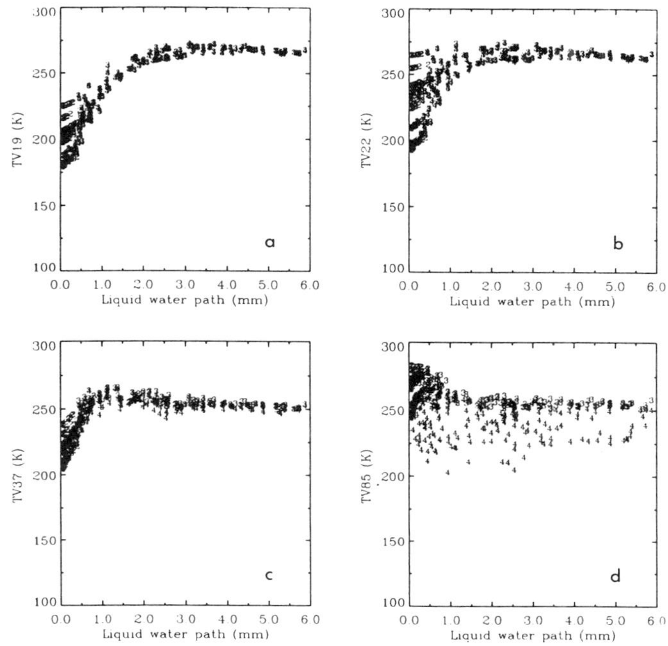

Figure 1. Simulated vertically polarized brightness temperatures at a 53° zenith angle against the liquid water path for (a) 19.35 GHz; (b) 22.23 GHz; (c) 37 GHz; and (d) 85.5 GHz. The numbers 1, 2, 3 and 4 represent data set name, i.e. clear, cloud, rain-only and rain-ice, respectively.

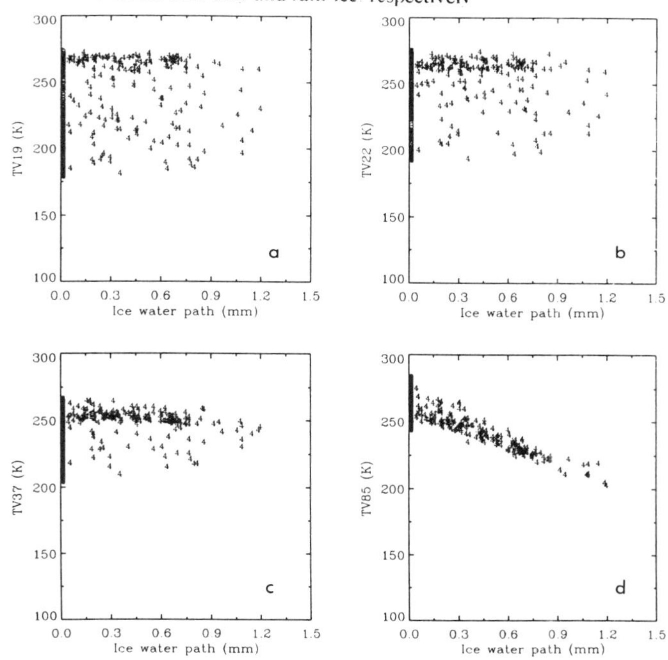

Figure 2. As explained in Figure 1, except for brightness temperatures against the ice water path.

Figures 2a-2d plot the TVs against the ice water path (IWP). Notice that brightness temperatures display a large variation for 19.35 and 22.23 GHz due to the liquid water. At IWP=0, such variation becomes much larger because of the clear, cloud and rain data sets which do not include any ice phase hydrometeors. The brightness temperature at 85.5 GHz shows a strong correlation with the IWP due to the scattering effects from the ice particles and provides useful information for the retrieval of the ice water content. The characteristics of the relationships between TH and IWP are similar to those in Figure 2, except the large variations at the low frequencies and at low IWP values (figures not shown here).

ALGORITHM FOR LIQUID WATER RETRIEVAL

As shown before, brightness temperatures are affected by many other factors such as sea surface wind and atmospheric water vapor in addition to the cloud liquid water. Brightness temperatures at multiple frequencies can be used to eliminate these effects for the cloud liquid water retrieval. To derive an algorithm for this retrieval, we first approximate the radiative transfer equation at microwave frequency as two separate differential equations of TV and TH

$$\mu \frac{dTV(\tau,\mu)}{d\tau} = TV(\tau,\mu) - (1-\omega)T(\tau) - \omega t_{vms}(\tau,\mu), \tag{1}$$

$$\mu \frac{dTH(\tau,\mu)}{d\tau} = TH(\tau,\mu) - (1-\omega)T(\tau) - \omega t_{hms}(\tau,\mu), \tag{2}$$

where T is the atmospheric temperature; μ the cosine of zenith angle; ω single scattering albedo; τ the optical thickness; t_{vms} and t_{hms} the contributions from the multiple scattering process. Alternatively, we may formally express the solution of Eqs. (1) and (2) as

$$TV(\tau,\mu) = T_{vu} + \zeta[\varepsilon_v T_s + (1-\varepsilon_v)T_{vd}] + \omega T_{vms}, \tag{3}$$

$$TH(\tau,\mu) = T_{hu} + \zeta[\varepsilon_h T_s + (1-\varepsilon_h)T_{hd}] + \omega T_{hms}, \tag{4}$$

where ε_v and ε_h are surface emissivities for vertical and horizontal polarization, respectively; T_{vu} and T_{hu} the upwelling radiation at vertical and horizontal polarization; T_{vd} and T_{hd} the downwelling radiation at vertical and horizontal polarization; ζ the atmospheric transmittance, and T_{vms} and T_{hms} the contributions from the multiple scattering process.

As a first approximation, the scattering terms in both equations can be neglected. If the atmosphere is assumed isothermal, previous two equations can be written as

$$TV(\tau_1,\mu) = T_s[1 - \zeta_o^2 \zeta_1^2 (1-\varepsilon_v)], \tag{5}$$

$$TH(\tau_1,\mu) = T_s[1 - \zeta_o^2 \zeta_1^2 (1-\varepsilon_h)], \tag{6}$$

where ζ_o is the transmittance function of oxygen which is nearly constant; ζ_l the transmittance function of the cloud liquid and water vapor and may be approximated as

$$\zeta_l = \exp\{-[c_1(f)WVP + c_2(f)LWP]/\mu\}, \tag{7}$$

where WVP is atmospheric water vapor path, c_1 and c_2 are frequency-dependent parameters.

It is seen that brightness temperature is a function of sea surface temperature, surface emissivity, water vapor and cloud liquid water paths. Under a certain circumstance where variations of brightness temperatures are mainly due to the water vapor and liquid water, two channel measurements may be enough for the retrieval of the LWP [10], i.e.,

$$LWP = a_0[\ln(290 - TB_1) - a_1 - a_2 \ln(290 - TB_2)] \quad \mathbf{(8)}$$

where TB_2 is the brightness temperature at the moisture absorption band; TB_1 the brightness temperature sensitive to the cloud liquid water; a_0, a_1 and a_2 are the coefficients which can be statistically determined using some training data sets. The coefficient, a_0 is referred as a scaling factor in this study.

Different data sets may produce distinct coefficients in Eq. (8). In another word, the algorithm is highly data-set dependent. For instance, the scaling factor using clear and cloud data sets can be much larger than that using clear, cloud and rain-only data sets. This is shown in Table 2 and Table 3 for comparisons of low frequency algorithms. In both tables, the subscript of the LWP represents a predicted LWP using corresponding two channel data. For example, the LWP_{19V} represents a predicted LWP using TV19 and TV22. The parameters r and β describe a statistical correlation coefficient and an explained variance, respectively. The LWP_{max} in the tables illustrates the maximum LWP in the simulated data sets. Notice that a_0 for LWP_{19V} based on the clear and cloud data sets is about twice as large as that using clear, cloud and rain-only data sets, whereas a_1 and a_2 do not differ significantly.

Table 2. Derived coefficients in Eq. (8) using *clear* and *cloud* data sets which are generated from the radiative transfer simulations. The correlation coefficient, r and the ratio of the predicted variance to the variance in the data set, β are also shown in the table.

LWP	TB_1	TB_2	LWP_{max} (mm)	a_0	a_1	a_2	r	β (%)
LWP_{19V}	TV19	TV22	0.3	-5.19	2.91	0.39	0.76	58
LWP_{19H}	TH19	TV22	0.3	-2.21	3.58	0.36	0.59	34
LWP_{37V}	TV37	TV22	0.3	-1.52	2.82	0.35	0.94	88
LWP_{37H}	TH37	TV22	0.3	-1.23	3.51	0.33	0.88	77
LWP_{85V}	TV85	TV22	0.3	-0.53	-1.89	1.25	0.74	55
LWP_{85H}	TH85	TV22	0.3	-0.42	-1.70	1.39	0.86	74

Table 3. Derived coefficients in Eq. (8) using *clear, cloud* and *rain-only* data sets which are generated from the radiative transfer simulations. Notice that the coefficients for the algorithms at 85.5 GHz are the same as those in Table 2 because only clear and cloud data sets are used.

LWP	TB_1	TB_2	LWP_{max} (mm)	a_0	a_1	a_2	r	β (%)
LWP_{19V}	TV19	TV22	3.0	-2.70	2.96	0.39	0.99	98
LWP_{19H}	TH19	TV22	3.0	-1.87	3.79	0.32	0.99	98
LWP_{37V}	TV37	TV22	1.0	-1.15	2.89	0.34	0.97	95
LWP_{37H}	TH37	TV22	1.0	-0.76	3.71	0.30	0.94	89
LWP_{85V}	TV85	TV22	0.3	-0.53	-1.89	1.25	0.74	55
LWP_{85H}	TH85	TV22	0.3	-0.42	-1.70	1.39	0.86	74

Notice that the LWP_{max} in Table 2 is 0.3 mm. This constrains derived algorithms in a pure emission region. Thus, the algorithms can be only applicable for the non-raining clouds. Such limitation makes the algorithms impractical for the SSM/I data because the SSM/I sensor often observes both raining and non-raining clouds. Therefore, uses of the algorithms in Table 2 for raining clouds may introduce a large uncertainty to the LWP. For example, the algorithms of the LWP_{19V} and LWP_{19H} in Table 2 may produce twice as large as the actual LWP because of the large scaling factor. For a satisfactory retrieval of the LWP in the convective clouds, the algorithms should be developed with rain-bearing cloud data sets. In Table 3, the LWP_{max} are extended to 3.0 mm for LWP_{19V} and LWP_{19H} and to 0.8 mm for LWP_{37V} and LWP_{37H} in order to take into account of possible scattering effects of large rain drops. Thus, the algorithms of the LWP_{19V} and LWP_{19H} can be used for retrieval of high amount of LWP, whereas, the algorithms of the LWP_{37V} and LWP_{37H} can be used for retrieval of a moderate amount of LWP.

To find precise conditions which individual algorithms can be applied to, we analyze each algorithm in Table 3 using the four simulated data sets. Figures 3a-3f show various predicted LWP indices against the LWP. Apparently, the LWP_{19V} and LWP_{19H} estimate the LWP in the range from 0 to 3.0 mm very well. To measure LWP larger than 3.0 mm, the microwave information at frequency lower than 19.35 GHz may be required. Apparently, the LWP_{37V} and LWP_{37H} (see Figures 3c and 3d)l estimate the LWP in the range from 0 to 0.8 mm very well and include less noises in comparison to the LWP_{19V} and LWP_{19H}. They saturate at 0.8 mm. The LWP_{85V} and LWP_{85H} (see Figures 3e and 3f) may only estimate the LWP in the range from 0 to 0.3 mm and become largely negative for the LWP greater than 0.3 mm.

It is important to note that the algorithms of LWP_{37V} and LWP_{37H} in Table 3 still poorly retrieve low LWP especially for the clouds forming in the moisture-deficit climate regimes, although the correlation coefficients are relatively high for all climate regimes. This is illustrated in Figures 4a-4f. Shown data points are a subset of the simulated clear and cloud data sets with a water vapor path less than 30 mm. Notice that the algorithms using vertically polarized data produce less noise (left portion) than the algorithms using horizontally polarized data (right portion). This is because the vertically polarized brightness temperatures are affected less by the surface effects. It is also seen that the algorithm of the LWP_{85V} shows the smallest noise. It is highly recommended that the retrieval of the cloud liquid in the dry climate regime should be based on the high frequency algorithms such as LWP_{85V} and LWP_{85H}.

To use advantages of individual algorithms in terms of the dynamic range and the sensitivity to the low LWP, we composite three algorithms in Table 3 to a single algorithm as followed:

$$LWP = \begin{cases} LWP_{19V} & \text{if} \quad LWP_{19V} \geq 0.6\text{mm} \\ LWP_{37V} & \text{if} \quad LWP_{37V} \geq 0.2\text{mm} \quad \text{or} \quad WVP \geq 30\text{mm} \\ LWP_{85H} & \text{otherwise} \end{cases} \qquad (9)$$

ALGORITHM FOR ICE WATER RETRIEVAL

Single and multiple scattering and possible thermal emission processes by ice particles strongly interact with the microwave radiation. The radiation from ice layer aloft is coupled with that from low level liquid. To identify the scattering due to the atmospheric ice, Grody [11] defines a scattering index as

$$SI = F - TV85, \qquad (10)$$

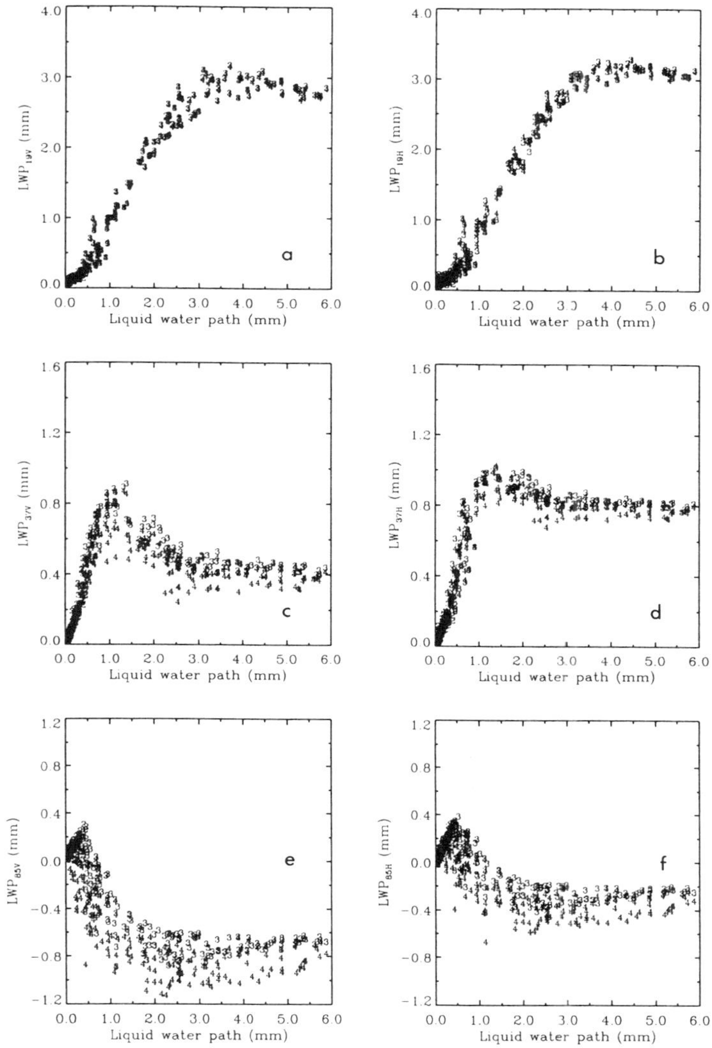

Figure 3. Cloud liquid water path predicted using various two channel algorithms against the truth liquid water path. (a) LWP_{19V} is the LWP using TV19 and TV22; (b) LWP_{19H} is the LWP using TH19 and TV22; (c) LWP_{37V} is the LWP using TV37 and TV22; (d) LWP_{37H} is the LWP using TH37 and TV22; (e) LWP_{85V} is the LWP using TV85 and TV22; and (f) LWP_{85H} is the LWP using TH85 and TV22.

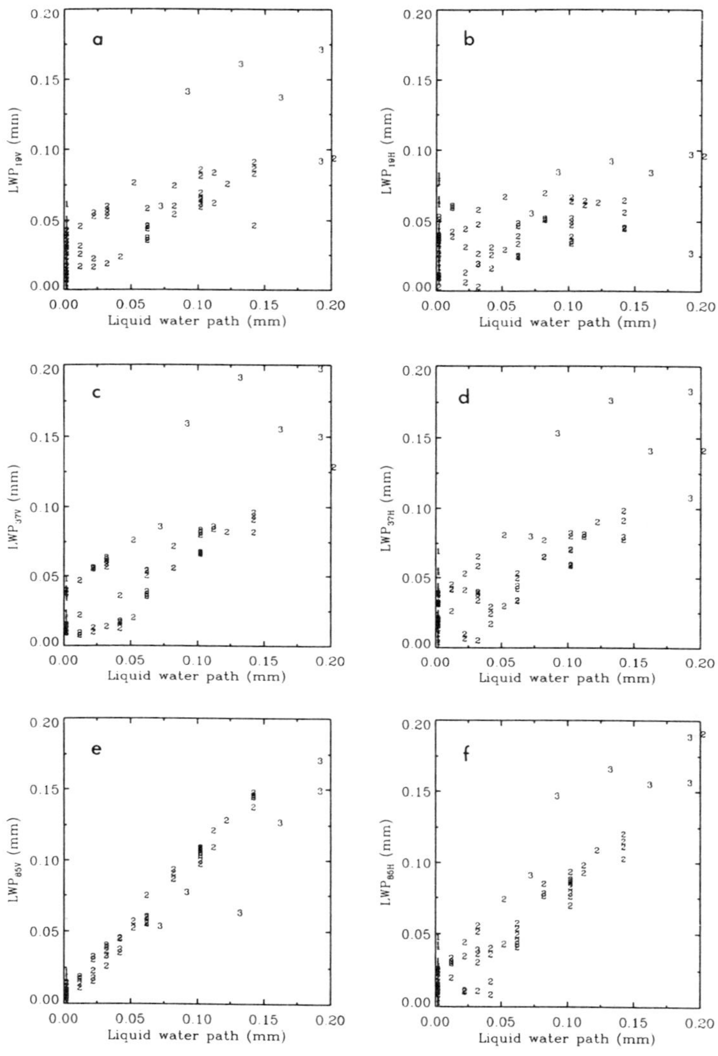

Figure 4. As in Fig. 3, except for the clouds forming in the moisture-deficient atmosphere. Shown data are a subset of the four simulated data sets with the water vapor path less than 30 mm. The LWP is typically less than 0.2 mm.

where the TV85 is the brightness temperature at 85.5 GHz measured over the atmospheric top, F is a function which estimates the brightness temperature at 85.5 GHz due to the atmosphere without ice. In fact, the scattering index reflects the brightness temperature depression due to the scattering effects of the atmospheric hydrometeors. The function, F is related to observed TV19 and TV22 with

$$F = A + B \times TV19 + C \times TV22 + D \times TV22^2. \tag{11}$$

where the coefficients A, B , C and D can be derived using two distinct ways. If the clear and cloud data sets are used, the derived SI represents the temperature depression at 85.5 GHz due to the scattering of the large rain drops as well as ice particles. If the clear, cloud data and rain-only data sets are used, the SI reflects the depression due to the ice only. Table 4 lists the coefficients using various combinations of the data sets.

Table 4. Derived coefficients for the scattering indices in Eq. (11) using data sets from radiative transfer simulations. The values in the first row are derived using clear and cloud data sets, whereas those in the second row are derived using clear, cloud and rain-only data sets.

A	B	C	D	Data sets used
-364.0	1.80	3.01	-0.00792	clear, cloud
74.3	-0.68	1.99	-0.00239	clear,cloud, rain-only

Figures 5a-5b plot the SIs against the IWP over the ocean using the coefficients in Table 4. It is very interesting to see that the SI based on the clear and cloud data sets (Figure 5a) are much larger than the SI based on the clear, cloud and rain-only data sets (Figure 5b) and is not well correlated with the IWP. This is because the SI in Figure 5a includes scattering effects from both ice particles and large raindrops. The SI based on the clear, cloud and rain-only data sets (see Figure 5b) is strongly correlated with the IWP. The variation for the SI in Figure 6b at IWP=0 is in the range from -10 K to 10 K and is associated with the clear, cloud and rain-only data sets. In this range, the scattering due to the ice can not be separated from the non-ice conditions. A threshold of IWP is needed to identify the significant amount of the ice. The IWP (with a unit of mm) is related to the SI as followed

$$IWP = 0.021 \times SI, \tag{12}$$

where the SI is the scattering index based on the clear cloud and rain-only data sets.

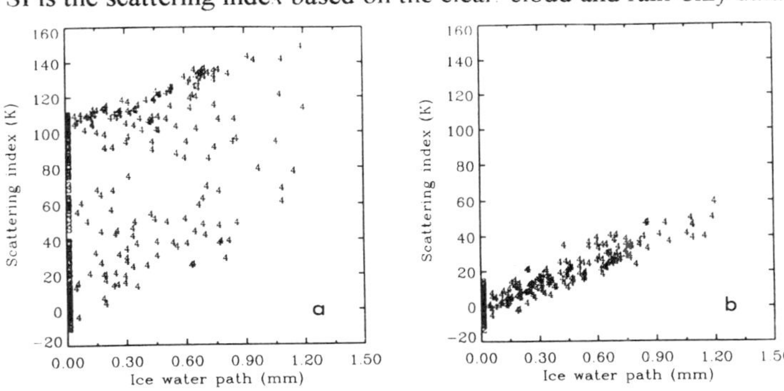

Figure 5. Scattering indices against the ice water path for (a) using *clear* and *cloud* data sets, and (b) using *clear, cloud* and *rain-only* data sets.

CALIBRATION AND VALIDATION

It is well known that brightness temperatures calculated from radiative transfer models represent point measurements and are distinct from satellite antenna temperatures, which measure an integrated effect in a satellite footprint. Therefore, the algorithm based on the radiative transfer model is subject to the beam filling problem. It is very important to calibrate theoretical algorithms in terms of satellite measurements and other ground truth data. An ideal method to calibrate and validate the algorithm is to directly compare the retrieved cloud liquid water path with observed one. However, this calibration can not be easily made because the cloud liquid water content is not a routinely-measured product. Therefore, this study develops a few indirect methods to do our algorithm calibration and verification.

Verification of algorithm coefficients using clear-sky SSM I data

If the LWP is set a zero, Eq. (8) becomes as a relationship between the TB_1 and TB_2, i.e.,

$$\ln(290 - TB_1) = a_1 + a_2 \ln(290 - TB_2), \quad \textbf{(13)}$$

In another word, the logarithmic functions of brightness temperatures at two channels are linearly correlated. The SSM/I data in the clear sky conditions can be used to compute the coefficients, a_1 and a_2. Uses of the coefficients from the SSM/I data in the retrieval algorithm may increase the accuracy for the cloud water retrieval because impacts of other parameters such as surface wind speed and atmospheric lapse rates, possible beam filling effect are properly taken into account.

To derive a SSM/I data set in clear conditions, we use visible imagery from geostationary satellite platforms for the classification. This is because clear and cloud areas on the visible imagery usually exhibit a large contrast and are easily separated. For this study, the GOES-7, Meteosat3 and GMS-3 visible data with a 16 km resolution are remaped to the SSM/I data coordinate. A count value less than 50 on the remaped images is used as a threshold for the clear and cloud separation. SSM/I data during four seasons over different climate regimes are used to generate a comprehensive training data set which includes sufficient climatological information. Figures 6a-6f show logarithmic functions of SSM/I channels of TV19, TH19, TV37, TH37, TV85 and TH85 against predicted values using a logarithmic function of TV22. The 1:1 line shown as solid represents a perfect prediction. It is seen that the TV22 predicts the TV19 and TV37 better than the TH19 and TH37. This may be because a horizontally polarized emissivity at microwave frequencies lower than 37 GHz is much lower than a vertically polarized one and therefore is more affected by the surface wind speed. However, the TV22 predicted TH85 better than the TV85.

Table 5 shows a_1 and a_2 derived using the SSM/I data sets. For a comparison, the values based on the simulated data sets are also shown. It is surprisingly seen that the values of a_1 and a_2 from the simulated data sets agree with those from the SSM/I measurement data very well. This implies that our simulated data sets include sufficient climatological information for the cloud liquid water retrieval.

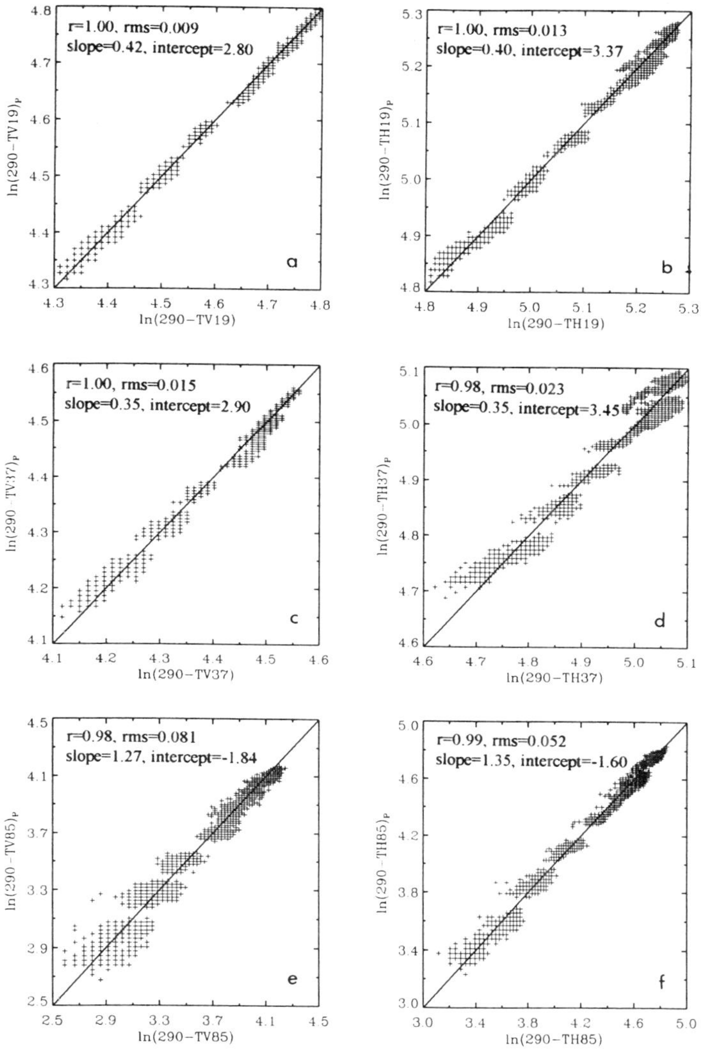

Figure 6. Predicted brightness temperatures at 19.35, 37 and 85.5 GHz against observed data using the observed brightness temperature at 22.23 GHz as a predictor. The correlation coefficient (r); the root mean square error (rms); slope (a_2) and intercept (a_1) in Eq. (13) are also shown.

Table 5. A comparison between the model generated algorithms and observation generated algorithms for the cloud liquid water retrieval. The values in the RTM column are derived from the radiative transfer modeling data sets and the values in the OBS column from the observational data sets during the Atlantic Stratocumulus Transition Experiment (ASTEX).

LWP	a_0		a_1		a_2	
	RTM	OBS	RTM	OBS	RTM	OBS
LWP_{19V}	-2.70	-3.20	2.96	2.80	0.39	0.42
LWP_{37V}	-1.15	-1.66	2.89	2.90	0.34	0.35
LWP_{85H}	-0.42	-0.44	-1.79	-1.60	1.39	1.35

Verification of scaling factor using ground-based microwave radiometer data

Previous method discusses how to derive the coefficients, a_1 and a_2 using actual measurement data sets. It is also critical to calibrate the scaling coefficient, a_0 in the retrieval algorithm. Such a calibration requires highly accurate measurements of cloud liquid water content from other remote sensing techniques. As a fairly robust technique, ground-based microwave radiometers operating at 20.6 and 31.65 have been used for the cloud liquid water retrieval for a long time. Unlike satellite instruments which receive radiation from both atmosphere and surface, the ground-based radiometers receive radiation coming mostly from the atmosphere rather than from the cosmic background which is very colder (3 K) and stable. As a result, the ground-based method can retrieve the LWP with a relative error less than 15% under various conditions[12]. For the present study, cloud liquid water from the ground-based radiometer data collected during the Atlantic Stratocumulus Transition Experiments (ASTEX) are provided to us by J. B. Snider in the NOAA Wave Propagation Laboratory (WPL) at Boulder, Colorado. The WPL radiometer instruments actually measure downwelling microwave radiation at three microwave frequencies of 20.6, 31.65, and 90 GHz, respectively and record the data with 2 minute interval.

A direct comparison between two data sets is not usually recommended because the satellite retrieved LWP represents a value over an area, whereas the ground-based radiometer retrieved LWP represents a value for a single point. It is common to average the LWPs from the ground-based method during a time period in order to compensate for the spatial average of the LWP inherently performed by the satellite instrument. The time period for averaging depends on the weather systems and should be different from time to time and from area to area, rather than a fixed value (e.g. 1 hour) used by Alishouse et al.[6]. At present, four different averaging time (ΔT) are tested in order to obtain a best matchup.

Figures 7a-7d show the LWP retrieved from SSM/I data (LWP_s) against the ground-based radiometer retrieved LWP (LWP_g) with ΔT of 3, 6, 12 and 60 minutes, respectively. Apparently, a ΔT of 60 minutes produces a best correlation between the LWP_s and LWP_g with a highest correlation coefficient (0.85) and a highest explained variance (82%).

The LWP retrieved from the ground-based method is also used to verify the scaling factor. In doing so, the logarithmic values of the SSM/I data in Eq. (8) and the retrieved LWP with the ΔT of 60 minutes are statistically regressed to derive the scaling factor a_0 which is referred as observed values (see Table 5 in the column "OBS"). For the LWP_{85V}, a_0 from the simulated data sets agrees with the observed value very well. However, for the LWP_{19V} and LWP_{37V},

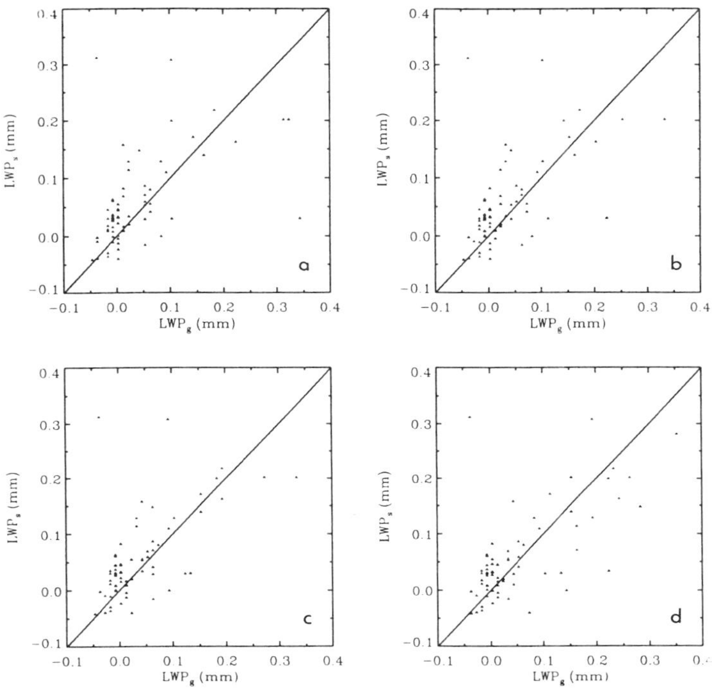

Figure 7. SSM/I retrieved LWP against the measurements from the ground-based microwave radiometer. The time interval, ΔT for averaging the ground-based radiometer data is (a) 3 min, (b) 6 min, (c) 12 min, and (d) 60 min.

the difference between the simulated and observed a_0 is about 0.5. The reason for this disagreement is because the dynamic range of the LWP during the ASTEX is smaller than that in the simulated data sets and is not sufficient for validating a_0 in LWP_{19V} and LWP_{37V}. Observational data sets from the ground-based method in the tropics are highly desired for such validation.

A CASE STUDY

Cloud liquid and ice water distributions retrieved from the algorithms may be compared with the visible and infrared (IR) imagery from the geostationary platforms for the cloud detection analysis. Figure 8 shows retrieved cloud ice and liquid water distributions (unit of mm) for a cloud system over North Atlantic ocean. Remaped visible and IR data from Meteosat3 show an elongated cloud system with a most convective part (marked as A) in the southern portion where the IR temperature is below 235 K. A large area (B) on the IR image with a temperature range from 235 and 250 K is quite inhomogeneous on the visible image and is presumably associated with cirrus clouds. Cirrus clouds also appear over a large area on the upper left portion (C).

A comparison analysis among retrieved IWP, LWP, visible and IR data may help identify various clouds and illustrate important cloud microphysical processes. Notice that the retrieved IWP distribution is primarily associated with the brightest areas on the visible image. Lack of ice water on area B of the visible imagery illustrates the clouds are mainly thin cirrus, within which the ice particles are too small to be detected by SSM/I sensor at 85.5 GHz (3.5 mm wavelength). The LWP distribution for a threshold of 0.02 mm reflects the major cloud system on the visible and IR image well. For instance, high LWP over area A indicates most likely raining-clouds which include large size raindrops (millimeter size). A narrow band of a zero LWP corresponds to area B on the visible image again indicates the cirrus clouds. The retrieved LWP also reflects the low level cloud area on the visible data very well. It is very interesting to see a high LWP area on the upper left corner (marked as D) corresponds to low reflectance on the visible data and warm temperature (above 273 K) on the IR data. Therefore, the clouds over the area are probably associated with warm rain processes.

SUMMARY AND CONCLUSIONS

This study develops the new algorithms for retrieving cloud liquid and ice water path from the SSM/I data. The algorithms are based on the analyses of the brightness temperature data sets simulated from radiative transfer model. Our analyses show that brightness temperatures at 19.35 and 22.23 GHz are capable for retrieving high amount of liquid in the convective clouds, whereas the brightness temperatures at 85.5 and 22.23 GHz can be used to retrieve low amount of liquid in the stratus or stratocumulus clouds. Our analyses also show that the LWP can be expressed as a function of two brightness temperatures with three coefficients (a_0, a_1 and a_2), which can be derived using either simulated data sets or observed SSM/I data. Derived values from both data sets agree very well. Different combinations of two channel at SSM/I frequencies can be used to generate the algorithms. However, this study uses only three of these algorithms which are accurate for retrieving liquid in different cloud systems. Retrieved values from individual algorithms are analyzed and composited to the liquid water

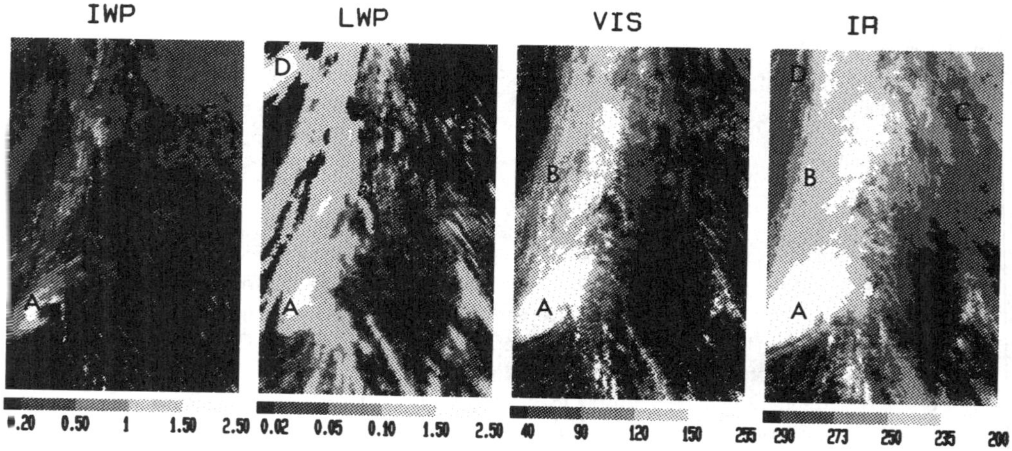

Figure 8. Cloud ice and liquid water path distributions derived from SSM/I using present algorithms against visible and infrared data from Meteosat3 which are remaped to the SSM/I data coordinate. Shown is a frontal system over North Atlantic ocean at 1300 UTC on September 20, 1993.

field in the clouds. The cloud liquid water derived from SSM/I data agrees with ground-based microwave radiometer measurements. For the ice water retrieval, the algorithm is derived using the simulated data sets. At the present time, no calibration and validation are performed because of the lack of observation data sets. A composite analysis of retrieved cloud liquid, ice water paths, visible and infrared image shows that the area with a high amount of ice is often associated with the brightest clouds on the visible image and the coldest temperature on the infrared image and therefore implies intense convective clouds. The area with a high amount of liquid water and without apparent ice is often related to very warm infrared temperature (higher than 273 K) and means a significant warm rain process in the cloud system. A comparison between the liquid water and visible images also identifies cirrus areas which can not be detected from individual data alone.

Acknowledgments This research was supported by National Oceanic and Atmospheric Administration under the climate change contract: NOAA-91055.

REFERENCES

1. Stephens, G. L., and T. J. Greenwald, 1991, Observations of the Earth's radiation budget in relation to atmospheric hydrology, II. Cloud effects and cloud feedback, *J. Geophys. Res.*, **96**, 15325-15340.
2. Kristjánsson, J. E., 1991, Initialization of cloud water in a weather prediction model, *Preprints of the Ninth Conference on Numerical Prediction*, Dever. CO. American Meteorological Society, 823-824.
3. Grody, N. C., 1976, Remote sensing of atmospheric water content from satellite using microwave radiometry, *IEEE Trans. Antennas Propag.*, **AP 24**, 155-162.
4. Staelin, D. H., K. F. Kunzi, R. L. Pettyjohn, R. K. L. Poon, and R. W. Wilcox, 1976, Remote sensing of atmospheric water vapor and liquid water with the Nimbus 5 microwave spectrometer, *J. Appl. Meteorol.*,**15**, 1204-1214.
5. Takeda, T., and G. Liu, 1987, Estimation of atmospheric water amount by Nimbus 7 SMMR data: A new method and its application to the western north-Pacific region, *J. Meteorol. Soc. Jpn*, **65**, 931-94 6.
6. Alishouse, J. C., J. B. Snider, E. R. Westwater, C. T. Swift, C. S. Ruf, S. A. Snyder, J. Vongsathorn, and R. R. Ferraro, 1990, Determination of cloud liquid water content using the SSM/I, IEEE *Trans. Geosci. Re mote Sensing*., **28,** 817-822.
7. Greenwald, T. J., G. L. Stephens, and T. H. Vonder Haar, 1993, A physical Retrieval of cloud liquid water over the global oceans using special sensor microwave/imager (SSM/I) observations, *J. Geophys. Res.*, **98**, 18471-18488.
8. Petty, G. W., and K. B. Katsaros, 1993, The response of the SSM/I to the marine environment. Part I: an analytic model for the atmospheric component of the observed brightness temperatures, *J Atmos. and Oceanic Tech.*, **9,** 746-761.
9. Weng, F., 1992, A multi-layer discrete-ordinate method for vector radiative transfer in a vertically-inhomogeneous, emitting and scattering atmosphere - I. Theory, *J. Quant. Spectrosc. Radiat. Transfer,* **47,** 19-33.
10. Grody , N. C., and R. R. Ferraro, 1992, A comparison of passive microwave rainfall retrieval methods, *Proc. Sixth Conference on Meteorology and Oceanography,* AMS, Atlanta, GA, 60-65.
11. Grody, N. C., 1991, Classification of snow cover and precipitation using Special Sensor Microwave Imager (SSM/I), *J. Geophys. Res.*, **96**, 7423-7435, 1991.
12. Westwater, E. R., 1978, The accuracy of water vapor and cloud liquid determination by dual-frequency ground-based microwave radiometry. *Radio Sci.,* **13**, 677-685.

Review of the SSM/I-based algorithms submitted for the GPCP-AIP/2

G.L.LIBERTI
LMD-CNRS, Ecole Polytechnique, F-91128 Palaiseau Cedex, France

Abstract

A short review of the 19 SSM/I-based algorithms submitted for the second Algorithm Intercomparison Project of the Global Precipitation Climatology Project (GPCP-AIP/2) is presented. Algorithms are grouped on the basis of different characteristics adopted in their development (physical basis, calibration, satellite information used, structure). Summary tables are included. Apart from its value as a review, the present work can offer a useful tool for the interpretation of the AIP/2 results as well as for the development/improvement of algorithms.

Key words: GPCP-AIP/2, Precipitation, SSM/I.

INTRODUCTION

As part of the Global Precipitation Climatology Project (GPCP) a series of Algorithm Intercomparison Project's (AIP) was undertaken, in order to intercompare the performances of different satellite-based precipitation estimation techniques. This paper reviews the algorithms using SSM/I data, submitted for the GPCP-AIP/2.

The project and the data sets are described in [1]. Table 1 lists the participants, their affiliation and principal references for the submitted algorithms. The name associated with each algorithm is consistent with the nomenclature used in [1]. In the following, the symbol reported in the first column will be used to characterize each algorithm (e.g. $\mathcal{P}$ = NEGRI).

Among the 19 different algorithms submitted, some of the algorithms represent a different versions of the same tecnique. In particular:

- $\mathcal{K}$ and $\mathcal{L}$ differ, respectively, whether the 85 GHz is used or not.
- $\mathcal{B}$ and $\mathcal{C}$ differ whether the statistical properties of a data set defined spatially ($\mathcal{B}$) or temporally ($\mathcal{C}$) are used.
- $\mathcal{G}$ and $\mathcal{H}$ differ, respectively, whether 37 GHz or 19 GHz measurements are used;
- $\mathcal{I}$ and $\mathcal{J}$ differ whether a unique algorithm is used for all surfaces type ($\mathcal{J}$) or two different algorithms are used over land and over ocean ($\mathcal{I}$).

Only 3 algorithms [1]($\mathcal{G}, \mathcal{H}, \mathcal{L}$) submitted estimates for cases where both 85 GHz channels were not available from the F8 satellite instrument. This is symptomatic of the importance given, by almost all the participants, to the 85 GHz channels data (see discussion below).

A detailed description of each algorithm can be found in [1], as well as in the references reported in the last column of Tab.1. In the following, the algorithms are analysed and grouped, taking into account the following characteristics:

[1]Although $\mathcal{E}$ does not employ 85 GHz measurements, estimates for the F8 selected cases were not submitted.

- Physical mechanism on which the precipitation estimation is based.
- Type of satellite information used.
- Calibration of the algorithm.
- Algorithm structure.

The performances of each algorithm (i.e. CPU-time, error etc.) are not treated in this review, principally for lack of documentation.
Finally, the results of the intercomparison are not included since the results are not yet available in their final form at the present time.

Table 1. GPCP-AIP/2 Participants Algorithms name, affiliation and principal *R*eferences.

#	ALGORITHM NAME	AFFILIATION	WORKING GROUP	R
$\mathcal{A}$	Barrett & Kidd	Remote Sensing Unit Dept. of Geography, Univ. of Bristol, UK	E.C.Barrett, C.Kidd	2
$\mathcal{B}-\mathcal{C}$	Barrett 1/2		E.C.Barrett, D.Kniveton	2
$\mathcal{D}$	Barrett & Todd		E.C.Barrett, M.Todd	2
$\mathcal{E}$	Berg	Univ. of Colorado, Colorado Center for Astrodynamics Research, USA	W.Berg	3
$\mathcal{F}$	Cuddapah	NASA-GSFC, Laboratory for Atmospheres, USA	C.Prabhakara J.J.Nucciarone, G.Dalu	4
$\mathcal{G}-\mathcal{H}$	Ferraro 1/2 (Q_{37}/Q_{19})	NOAA/NESDIS, Satellite Research Laboratory, USA	R.Ferraro, N. Grody	5
$\mathcal{I}-\mathcal{J}$	Ferraro 3/4 ($SI_{G/LW}$)	NOAA/NESDIS, Satellite Research Laboratory, USA	R.Ferraro, N. Grody	5
$\mathcal{K}-\mathcal{L}$	Ferraro NRL1/2	Naval Research Laboratories, USA	W.Olson, J.Hollinger	6
$\mathcal{M}$	Ferriday	Univ. of Colorado Colorado Center for Astrodynamics Research, USA	J.Ferriday	1
$\mathcal{N}$	Kummerow	NASA-GSFC, Laboratory for Atmospheres, USA	C.Kummerow, L.Giglio	7
$\mathcal{O}$	Marzano	Univ. of Rome, Dept. of Electronics, Italy	F.S.Marzano, P.Basili, P.Ciotti, G.d'Auria, A.Mugnai, N.Pierdicca	8
$\mathcal{P}$	Negri	NASA-GSFC, Laboratory for Atmospheres, USA	R.Adler, A.Negri	9
$\mathcal{Q}$	Olson	Univ. of Wisconsin, USA	W.Olson	1
$\mathcal{R}$	Petty	Purdue Univ., Dept. of Earth and Atmospheric Sciences , USA	G.W.Petty	10
$\mathcal{S}$	Schlüssel	Universität Hamburg, Meteorologisches Institut, Germany	P.Schlüssel, P.Bauer	11

PHYSICAL MECHANISMS.

It is well known, from radiative transfer theory as well as from observations, that precipitation within the Field Of View (FOV) of a satellite/air-borne microwave radiometer affects the measured radiances through different physical mechanisms, commonly referred to as follows:

- EMISSION: emission from cloud water droplets, particularly over low emissivity ocean background;
- POLARIZATION: depolarization due to precipitation and cloud particles, particularly over polarized ocean background;
- SCATTERING: extinction due to the scattering mostly from precipitation sized ice particles.

Table 2 indicates the physical mechanism(s) applyed, within each algorithm, for the rain estimation (NB: rain detection is not considered here).
Over all the mechanism principally considered is the scattering while the emission is least used. However, over water, polarization and emission signatures of precipitation are highly correlated at low frequencies, so that one or the other signal can be used to detect precipitation. At high frequencies, such correlation is reduced.
To estimate the precipitation the majority of the algorithms adopts a combination of the signatures except for the following:

Table 2. GPCP-AIP/2 SSM/I Algorithms Summary

#	PHYS[1,2]			SURF[3]			INFORMATION[4]				CAL[5]		SSM/I CHANNELS USED[6,2]							EXTERNAL DATA[7]	RANGE[8,2]	RES[9,2]	
	S	E	P	W	C	L	P	S	T	M	M	D	1	2	3	4	5	6	7		mm/hr	A	B
$\mathcal{A}$	S		P	W	C	L		y		Y		D	y	y	y			Y	Y	No	0.005 ÷ 600		B
$\mathcal{B}$	Ŝ		P̌	w		l		y		Y		D	y	y	y	Y	Y̌	Ŷ		LSM	0̌.05/0̂.25÷ 64	A	
$\mathcal{C}$	Ŝ		P̌	w		l			y	Y		D	y	y	y	Y	Y̌	Ŷ		LSM,lat,lon	0̌.05/0̂.01÷ 64	A	
$\mathcal{D}$	Ŝ		P̌	w	c	l		y		Y		D	y	y	y	Y̌		Y	Y̌	No	0.1÷∞	Â	B̌
$\mathcal{E}$	S	E	P	W						Y	M	D		Y	Y	Y	Y			LSM,lat,month	0.01 ÷∞	A	
$\mathcal{F}$	S	E		W				y		Y		D		Y			Y		Y	LSM	0 ÷∞	A	
$\mathcal{G}$	S	E		W						Y	M	D	Y		Y	Y				LSM	0.84÷∞	A	
$\mathcal{H}$		E		W						Y	M	D	Y		Y					LSM	0.81÷∞	A	
$\mathcal{I}$	S			W		L				Y		D	Y	y	Y			Y		LSM	1.24 ÷ 40	A	
$\mathcal{J}$	S			w		l				Y		D	Y	y	Y			Y		LSM	0̌.44/0̂.91÷ *40*	A	
$\mathcal{K}$	S	E	P	w		l				Y		D	Y̌	y	Y̌	Y̌	y	Y	Y	LSM	0÷ *26.0*	A	
$\mathcal{L}$	S	E	P̌	w		l				Y		D	Y	y	y	Y	Y̌			LSM	0÷ *26.0*	A	
$\mathcal{M}$	S	E	P̌	w	c	l				Y	M		Y		Y	Ŷ	y	Ŷ	Y	LSM	1÷ 40	A	
$\mathcal{N}$	S	E	P	W				Y		Y	M		Y	Y	Y	Y	Y	Y	Y	LSM,Z_f	0.2÷25	A	
$\mathcal{O}$	S	E		W	C	L				Y	M		Y	Y	Y	Y	Y	Y	Y	No	0 ÷∞		B
$\mathcal{P}$	S			W	C	L	Y	y		y	M		y		y		y	y	Y	No	1.0 ÷ *55*		B
$\mathcal{Q}$	S		P	W						Y	M		Y	Y	Y	Y	Y	Y	Y	LSM	0.05÷∞	A	
$\mathcal{R}$	S		P	W				Y		Y	M		Y	Y	Y	Y	Y	Y	Y	LSM	0 ÷ 50		B
$\mathcal{S}$	S	E	P	W						Y	M		Y	Y	Y		y	Y	y	LSM	0.3 ÷ *31.7*	A	

[1]: Physical mechanism used: Scattering, Emission, Polarisation.

[2]: ˆ land surface only, ˇ water surface only

[3]: Surface of validity of the algorithm: Land Water Coastline. If in small letter, the algorithm is different according to the surface type.

[4]: Imagery information used: Single Point, Spatial, Temporal, Multispectral. If in small letter the use is partial (see Text).

[5]: Calibration Type Model, Data (see Text)

[6]: (1) 19V; (2) 19H; (3) 22V; (4) 37V; (5) 37H; (6) 85V; (7) 85H. Channels used only for rain detection are indicated with y

[7]: LSM: Land Sea Mask, z_f: freezing level.

[8]: If in *italics* the maximum is not given with the algorithm but is derived from the maximum rainrate available within the calibration data set

[9]: Final product resolution A or B scan.

- $\mathcal{B}, \mathcal{C}, \mathcal{D}$ over land and $\mathcal{I}, \mathcal{J}, \mathcal{P}$ purely scattering based;
- $\mathcal{B}, \mathcal{C}, \mathcal{D}$ over water purely polarization based;
- $\mathcal{H}$ over water purely emission based.

Different physical mechanisms are applied depending on the surface. Table 2 indicates the surface of validity for each algorithm and whether data are treated differently, inside the algorithm, depending from the surface type. In general, microwave radiometry over water surfaces allows the use of all physical mechanisms, while for land surfaces the scattering mechanism is most efficient. Consequently, while all the algorithms were able to retrieve precipitation over water, 11 [2] algorithms retrieved precipitation over land and only 5 over coastlines. The difficulty in estimating precipitation over coastlines, where the FOV includes both water and land surfaces, is due to its similar behaviour to precipitation, particularly when using an emission or polarization-based method. Moreover, the uncertainties in the navigation are such that any treatment of the data based on the exact quantification of each surface type within the FOV is not possible. It is interesting to notice the specific treatment of coastline adopted in $\mathcal{P}$, to discriminate coastlines from precipitation based on the correlation between measurements at 37 and 85 GHz over a box centered on the analysed pixel (see [9]).

Table 2 also shows the channels employed by each algorithm and whether they are used for rain estimation or not. The channel mostly used (18 algorithms), considering both rain detection and estimation is the 22V channel, while, if only the use for rain estimation is considered it is the 85V (13). It must also be taken into account that: $\mathcal{L}$ is an algorithm specifically designed for 85V being not available and $\mathcal{P}$ uses the 85H because of the absence of 85V within the original calibration data set.

In general, horizontal polarization channels are less used because they are more sensitive to surface variables than the corresponding vertical ones. In contrast, however, $\mathcal{F}$ is based only on horizontal channel measurements.

Due to the non-unique relationship between single channel Tb and precipitation, the channels are generally combined (see Tab.6) and often, such combinations are defined as *indices*.

In particular 9 algorithms applied, at least for the detection of precipitating areas, the concept of scattering index (NB: although in $\mathcal{Q}$ and $\mathcal{R}$ a scattering index is also defined, it will be treated as a polarization index (see below)). Such an index, first introduced by [13], is based on the hypothesis that the presence of precipitation sized ice within the cloud is related to the presence of precipitation and on the fact that precipitation sized ice does not affect low frequencies ($f \leq 37$ GHz) through extinction by scattering as much as the high frequency ones. The scattering indices generally have the form of a difference between a function of low frequencies Tb's and an high frequency one. Table 3 shows the scattering indices used by the algorithms.

Among the algorithms using polarization indices two different types can be defined:
- indices using polarization differences at low frequencies ($\mathcal{B}, \mathcal{C}, \mathcal{Q}$ and $\mathcal{R}$).
- indices using polarization differences at 85 GHz ($\mathcal{A}, \mathcal{D}, \mathcal{O}, \mathcal{Q}$ and $\mathcal{R}$). Such indices, first introduced by [12], combine the scattering from precipitation sized ice particles, and the polarization information to eliminate the ambiguity in cloud-free or non-precipitating clouds over ocean.

Emission indices are employed only by $\mathcal{F}$. In particular, to classify different types of precipitation, an index is defined based on the analysis of tri-dimensional (T^H_{19}, T^H_{37} and T^H_{85}) distribution.

[2] Although $\mathcal{N}$ is able to retrieve precipitation over all surfaces, only results over water surface were submitted for the GPCP-AIP/2 due to lack of confidence in the results obtained over land.

Finally, several algorithms employ SSM/I derived geophysical parameters, such as cloud liquid water content, for both, precipitation detection and estimation. The equations used are reported in Table 3.

Table 2. SSM/I derived indices.

INDEX	NAME	USER
SCATTERING INDICES		
$SI_{85} = [450.2 - 0.506 \cdot T_{19}^V - 1.874 \cdot T_{22}^V + 0.00637 \cdot T_{22}^{V^2}] - T_{85}^V$	Scattering index (Grody (1991))	$\mathcal{A}, \mathcal{B}, \mathcal{C}, \mathcal{D}, \mathcal{O}, \mathcal{P}$
$AFD = a \cdot T_{37}^V - T_{85}^V - b$	Adjusted Frequency Difference	$\mathcal{B}, \mathcal{C}$
$\mathrm{FTFA} = [a \cdot T_{37}^V + b] - T_{85}^V$	Fine-Tuned Frequency Algorithm	$\mathcal{D}$
$SI_{37} = (62.18 + 0.773 \cdot Ta_{19}^V) - Ta_{37}^V$	Scattering Correction Term	$\mathcal{G}$
$SI_G = [256.17 - 0.375 \cdot Ta_{19}^V - 0.200 \cdot Ta_{22}^V + 0.0237 \cdot (Ta_{22}^V)^2] - Ta_{85}^V$	Global Scattering Index	$\mathcal{I}$
$SI_{LW}^L = [438.5 - 0.46 \cdot Ta_{19}^V - 1.735 \cdot Ta_{22}^V + 0.00589 \cdot (Ta_{22}^V)^2] - Ta_{85}^V$	Land Scattering Index	$\mathcal{J}$
$SI_{LW}^W = [-182.7 + 0.75 \cdot Ta_{19}^V - 2.543 \cdot Ta_{22}^V + 0.00543 \cdot (Ta_{22}^V)^2] - Ta_{85}^V$	Water Scattering Index	$\mathcal{J}$
POLARIZATION INDICES		
$T_{37}^V - T_{37}^H$	Polarisation Difference	$\mathcal{B}, \mathcal{C}$
$P_f = \frac{T_f^V - T_f^H}{T_f^{V\,cl} - Tb_f^{H\,cl}}$ $[f = 19, 37]$	Normalized Polarization Differences	$\mathcal{Q}, \mathcal{R}$
$PCT = 1.818 \cdot T_{85}^V - 0.818 \cdot T_{85}^H$	Polarisation Corrected Temperature	$\mathcal{O}$
$PCT^{Kidd} = (1 + t) \cdot T_{85}^V - t \cdot T_{85}^H$	Polarization Corrected Temperature	$\mathcal{A}$
$OSI = (a + b \cdot T_{85}^H) - (T_{85}^V - T_{85}^H)$	Oceanic Scattering Index	$\mathcal{D}$
$S_{85} = [P_{85} \cdot T_{85}^{V\,cl} + (1 - P_{85}) \cdot 273] - T_{85}^V$	Scattering Index	$\mathcal{Q}, \mathcal{R}$
EMISSION INDICES		
$\frac{T_{19}^H - T_{19}^{H*}}{T_{37}^H - T_{37}^{H*}}$, $(T_{37}^H - T_{37}^{H*})$		$\mathcal{F}$
CLOUD LIQUID WATER		
$Q_{37} = -1.679 \cdot [ln(290 - Ta_{37}^V - SK) - 2.777 - 0.3851 \cdot ln(290 - Ta_{22}^V)]$	Atmospheric Liquid Water	$\mathcal{G}$
$Q_{19} = -6.723 \cdot [ln(290 - Ta_{19}^V) - 2.830 - 4.135 \cdot ln(290 - Ta_{22}^V)]$	Atmospheric Liquid Water	$\mathcal{H}$
$L = -0.323 \cdot log P_{85}$	Atmospheric Liquid Water	$\mathcal{R}$
$TWP = 10 \cdot 10^{(0.8124 - 6.799 \cdot 10^4 (T_{37}^H)^{-2} - 8.733 \cdot 10^{-4} (T_{19}^V - T_{22}^V)^2 + 1.734 \cdot 10^{-2} (T_{22}^V - T_{37}^H) - 7.193 \cdot 10^{-5} (T_{37}^H - T_{85}^H)^2)}$	Integrated Water Path	$\mathcal{S}$

SATELLITE INFORMATION.

In general, satellite imagery allows the use of different kinds of information as:
- ***single point:*** the parameter is a function of a measurement at a single point and at a single time;
- ***spatial:*** the parameter at a the given point is also a function of measurements at different points;
- ***temporal:*** the parameter at a given time is also a function of measurements at different times;
- ***multispectral:*** the parameter at a given location and time is a function of the measurements at different wavelenghts.

Table 2 shows the information used by each algorithm.
Multispectral information is used by all the submitted algorithms. Although $\mathcal{P}$ derives the rainrate using a single channel (T_{85}^H), it still applies a multispectral rain detection scheme.
Spatial information is used by the following:
- $\mathcal{P}$ for the coastline discrimination test (see above);
- $\mathcal{N}$ for the discrimination of precipitation type and also for the postprocessing screening of low consistency estimates;
- $\mathcal{R}$ where the algorithm searches for consistency between estimates at different resolutions.

$\mathcal{A}, \mathcal{B}, \mathcal{D}$ and $\mathcal{F}$ use, instead, the statistical properties of a spatially defined set of data.
Temporal information is only used by $\mathcal{C}$ to obtain a temporal series of measurements at a given site, but without taking into account the time sequence order.
Although the limited use of ***temporal*** and ***spatial*** information may derive by the fact that

the sampling characteristics of polar orbiting SSM/I are insufficient to properly monitor the life cycle of precipitating systems, even in geostationary VIS/IR-based precipitation algorithms a limited use of such informations is observed.
External information such as surface types, the freezing level, latitude and month, are also indicated in Table 2 when needed by a particular algorithm.

CALIBRATION.

Microwave radiometry provides a more direct relationship between observed radiance and precipitation than the radiometry in the VIS/IR spectrum. As a consequence it is common to find algorithms based on modelization of the radiative transfer within the cloud-atmosphere system (so called *physical* algorithms) in addition to the empirically derived algorithms.
The classification of algorithms as *physical* or *empirical* can be misleading, because it assumes the absence of any physical reasoning employed within an *empirical* algorithm and viceversa. Moreover, assuming the availability of the required data, *physical* algorithms can be generally transformed into an *empirical* one and viceversa.
A different classification criterion is proposed: based on the calibration method used for the rain estimation (the rain detection scheme being, generally, independently developped and calibrated). Algorithms are divided therefore into: *data*-based or *model*-based according to whether precipitation measurements or radiative transfer simulations where used to calibrate the algorithm, respectively.
Table 2 indicates the type of calibration used for each algorithm. Within the reviewed set of algorithms the two kinds of algorithms are approximately equally represented. $\mathcal{E}, \mathcal{G}, \mathcal{H}$ report a *mixed* calibration (e.g. a *model*-based result adjusted with ground truth data). Except for $\mathcal{E}$, which uses rain-gauge derived climatology, the rest of the *model*-based algorithms were calibrated with radar data; the radars being located whithin different climatological regions.
In general, the lack of a common calibration data set results in a source of differences, among the final precipitation estimates, independent from the scheme and physical approach of the algorithm itself.
A positive consequence of the recent efforts in producing common comprehensive and documented data sets, such as the AIP's ones, is the fact that 5 algorithms included radar data from the GPCP-AIP/1 within the calibration data set.
Model-based algorithms are no longer independent from the calibration data set used. Calibration-derived differences among algorithms may be generated, from both, the radiative transfer model and the input used.
In general, the modeling of radiative transfer in the microwave spectrum, for precipitation applications must include the modelization of: the SSM/I radiometer, the cold space emission, the surface emission-reflection, the atmospheric gas absorbtion/emission and the radiative processes within the cloud.
The model input must include: surface temperature and radiative characteristics, atmospheric profile and cloud vertical structure. Particularly important for the radiative transfer within a precipitating cloud is the solution adopted for the treatment of the multiple scattering as well as the cloud input model adopted; both are reported in Table 4 for the *model*-based algorithms.
Model-based algorithms may also be grouped, according with the use done of the radiative transfer model, into *calibration-only* and *in-line*, respectively, whether radiative transfer simulations are performed only once in order to produce a precipitation-Tb's data set to be used, subsequently to calibrate the algorithm, or they are performed also during the

retrieval phase ($\mathcal{N}$ and $\mathcal{O}$ research the input conditions, for the radiative transfer model, such to reproduce, as close as possible, the observations).
The different use of radiative transfer may be a key factor when selecting the radiative transfer model and relative input: an *in-line* use requiring a faster model and subsequently a simpler input than a *calibration-only* use. While for *data* based algorithms it is theoretically possible to use the same *in-line* approach, this has not yet been exploited.

Table 4. GPCP-AIP/2 Algorithms radiative transfer summary.

#	MULTIPLE SCATTERING SCHEME	CLOUD INPUT
$\mathcal{E}$	Variational-Iterative	Radiosounding and climatology derived cloud profiles
$\mathcal{G},\mathcal{H}$	Multiple scattering not included	Radiosounding derived cloud profiles
$\mathcal{M}$	Two-stream Delta-Eddington Approximation	Conceptual Cloud Profile
$\mathcal{N}$	Two-stream Delta-Eddington Approximation	Conceptual Cloud Profile
$\mathcal{O}$	Delta-M scaling scheme	Microphysical Cloud Model
$\mathcal{P}$	Variational-Iterative	Microphysical Cloud Model
$\mathcal{Q}$	Two-stream Delta-Eddington Approximation	Microphysical Cloud Model
$\mathcal{R}$	Inverse Monte-Carlo	Conceptual Cloud Profile
$\mathcal{S}$	Matrix Operator	Radiosounding derived cloud profiles

ALGORITHM STRUCTURE.

A generic MW precipitation algorithm may be schematized as a sequence of four different phases: *Preprocessing, Rain detection, Rain estimation* and *Post-processing.* Each phases it is, in principle, independent from the others and affect the final performance of the algorithm (e.g. a good algorithm estimation scheme may perform poorly because of the rain detection scheme employed).

PREPROCESSING.

Within this phase the operations needed to prepare the raw data to the application of the precipitation detection/estimation scheme are performed. The preprocessing of the algorithms reviewed include one or more of the following processes.
- Conversion of antenna temperature to brightness temperature. All the algorithms, except $\mathcal{G},\mathcal{H},\mathcal{I},\mathcal{J}$, convert antenna temperature into brightness temperature according with the scheme proposed in [14]. While the use of brightness temperatures is necessary if, in any phase of the algorithm development, brightness temperature from radiative transfer models are used, it is not indispensable for the *data*-based algorithms. - Remapping. Algorithms making use of a reference grid (e.g. $\mathcal{A}$ and $\mathcal{B}$ collocate SSM/I data into the UK National Grid in order to compute the necessary statistics) require the SSM/I data to be remapped into such grid.
- Ingestion of external variables. Algorithms using external data need to ingest such data taking into account the location in time and space of the SSM/I measurement.
- Surface discrimination. As discussed previously, the physical approach implied to detect and retrieve the rainfall rate is, generally dependent on the type of surface observed. It is therefore necessary, for most of the algorithms, to know, for each data point, the surface type. As observed in [7] while it is possible to classify the surface type from the data, such procedure is generally more time consuming than the ingestion of an external mask. SSM/I data are given with a surface flag index, that is often used, otherwise external mask are used. The use of different masks as well as the different criteria adopted to eliminate coastal pixels, results into a different size of the area for which the rain estimation is given and the consequent difficulty in interpreting the intercomparison statistics.
- Screening of 'low quality' data. 'Low quality' data are flagged throughout tests searching, generally, for physically meaningless Tb's. (e.g. $\mathcal{K}$ and $\mathcal{L}$ rejects the data having a

polarization difference lower than a given negative value).
- Deconvolution. The spatial resolution (FOV) of the SSM/I data is a function of the frequency. Particularly algorithms based on radiative transfer simulations require the observations to be made at the same resolution (i.e. each channel observing the same scene). Several methods have been developped for reducing as well as enhancing the data resolution, depending on the desired one.
- Variables definition. In this phase the indices (see Tab.3) or variables derived from SSM/I brightness temperatures, that will be used to detect and/or estimate the precipitation, are calculated.

RAIN DETECTION.

The rain detection consists in determining, within the data, the area of probable precipitation. Table 5 lists the tests adopted.
The majority of algorithms (17) include tests based on scattering signature both, in the form of scattering indices thresholds (a general agreement on the value of 10 K is found) or differences between low and high frequencies thresholds. In particular, the Grody's classification test [13] is adopted by 7 algorithms.
Polarization and emission tests differ not only on the test but also on the threshold required (e.g. for the difference T_{37}^{V}- T_{37}^{H} the threshold, over ocean varies from 30 to 40 K).
$\mathcal{G}$, $\mathcal{H}$, $\mathcal{R}$ and $\mathcal{S}$ require the atmospheric liquid water content to be greater than a certain value (0.1 to 0.5 Kg/m^2). It is interesting to notice that the lower one, suggested by $\mathcal{S}$, is actually a threshold for the total (i.e. cloud + rain water and ice) atmospheric water path.
Finally it must be noticed that test thresholds can be classified as dynamical or fixed: whether their value is scene dependent or not (e.g. T_{37}^{H*} adopted from $\mathcal{F}$ is computed from the statistical properties of the scene).

RAIN ESTIMATION.

The rain estimation consists in computing appropriate values of rain-rate for each point previously classified as raining. Table 6 provides the equations used by the algorithms to estimate the precipitation.
Only 4 algorithms $\mathcal{E}$, $\mathcal{M}$, $\mathcal{P}$ use a linear functional form ($\mathcal{N}$ uses linear for rain intensity intervals). An exponential form is generally preferred.
Finally several algorithms include by-products (i.e. geophysical parameters different for the precipitation at the ground) such as liquid water content, water vapor content or precipitation type. In particular, $\mathcal{O}$ and $\mathcal{N}$ give, at the end of the retrieval a complete vertical structure of the atmosphere-cloud system, such information being particularly useful for vertical latent heat profiles.
The resolution of the final estimates differs from algorithm to algorithm and is reported in Table 2.

POST PROCESSING.

In this phase the algorithm derived rainrate field is analysed. $\mathcal{K}$ and $\mathcal{L}$ eliminate meaningless negative rainrate, while $\mathcal{S}$ due to low confidence in low rainrate estimates, do not consider measurements below 0.3 mm/hr. Similar tests, eliminating values of rainrate out of an expected range, may also be included in other algorithms but were not mentioned in the submitted algorithm description.

Table 5. Rain detection tests summary. Rainrate is estimated if the conditions given in the table are satisfied. If not specified the conditions are AND.

<table>
<tr><th>#</th><th>LAND</th><th>WATER</th></tr>
<tr><td rowspan="2">$\mathcal{A}$</td><td>Grody (1991) snow test</td><td></td></tr>
<tr><td colspan="2">$PCT^{Kidd} < PCT^{Kidd}_{threshold} - 4$ K</td></tr>
<tr><td>$\mathcal{B}-\mathcal{C}$</td><td>Grody (1991) snow test
$a \cdot T^V_{37} - b\, T^V_{85} > 0$</td><td>Cavaleri et al. (1984) ice mask
$T^V_{37} - T^H_{37} > 0$</td></tr>
<tr><td>$\mathcal{D}$</td><td>Grody (1991) snow test
$a + b \cdot T^V_{37} - T^V_{85} > -3$ K</td><td>$OSI > OSI_{threshold}$</td></tr>
<tr><td>$\mathcal{E}$</td><td></td><td>$T^H_{19} > 190$ K
$T^V_{37} - T^H_{37} < 30$ K</td></tr>
<tr><td>$\mathcal{F}$</td><td></td><td>$T^H_{37} > T^{H*}_{37}$</td></tr>
<tr><td>$\mathcal{G}-\mathcal{H}$</td><td></td><td>$Q > 0.40$ Kg/m^2</td></tr>
<tr><td>$\mathcal{I}-\mathcal{J}$</td><td colspan="2">$SI_{G/LW} > 10$ K
Grody (1991) snow test</td></tr>
<tr><td>$\mathcal{K}-\mathcal{L}$</td><td>$T^V_{22} - T^V_{19} \leq 4K$
$T^V_{37} > 268$ K
$T^V_{19} + T^V_{37} - T^H_{19} - T^H_{37} \begin{cases} \leq 8K & T^V_{85} - T^V_{37} < -1K \\ > 8K & \begin{cases} T^V_{85} - T^V_{37} < -5K \\ T^V_{37} - T^V_{19} < -3K \\ T^H_{85} - T^H_{37} < -3K \end{cases} \end{cases}$</td><td>$T^V_{37} - 3.638 \cdot T^H_{37} < -432.48$ K</td></tr>
<tr><td>$\mathcal{M}$</td><td>$T^V_{19} > 250$ K
$T^V_{37} - T^H_{37} < 7$ K
$T^V_{19} - T^V_{85} > 20$ K</td><td>$T^V_{19} > 230$ K</td></tr>
<tr><td>$\mathcal{N}$</td><td></td><td>$T^V_{37} - T^H_{37} < 40$ K
non-precipitating cloud test</td></tr>
<tr><td rowspan="2">$\mathcal{O}$</td><td colspan="2">$SI_{85} > 10$ K
$PCT < 250$ K</td></tr>
<tr><td>$PC_1 < 50K$
$PC_2 > -40K$</td><td></td></tr>
<tr><td>$\mathcal{P}$</td><td colspan="2">$T^H_{85} < 247$ K
$T^H_{37} > 185$ K
$T^H_{85} - T^H_{37} \leq 0$
$r(T^H_{37}, T^H_{85}) < 0$ OR $T^H_{85} - T^H_{37} \leq -10$ K
Grody (1991) snow test</td></tr>
<tr><td>$\mathcal{Q}$</td><td></td><td>$P_{37} \leq 0.8$</td></tr>
<tr><td>$\mathcal{R}$</td><td></td><td>$L > 0.5$ Kg/m^2
$T^V_{85} - T^H_{85} < 6$ K
$S_{85} > 10$ K</td></tr>
<tr><td>$\mathcal{S}$</td><td></td><td>$TWP > 0.1$ kg/m^2</td></tr>
</table>

Table 6. Rain estimation summary.

<table>
<tr><th></th><th>LAND</th><th>WATER</th></tr>
<tr><td>$\mathcal{A}$</td><td colspan="2">$0.048 \cdot (PCT^{kidd} - PCT^{kidd}_{threshold}) + 0.00808 \cdot (PCT^{kidd} - PCT^{kidd}_{threshold})^2$</td></tr>
<tr><td>$\mathcal{B}-\mathcal{C}$</td><td>$c \cdot AFD + d \cdot AFD^2 \quad c, d > 0$</td><td>$e \cdot (T^V_{37} - T^H_{37}) + f \cdot (T^V_{37} - T^H_{37})^2 \quad e, f > 0$</td></tr>
<tr><td>$\mathcal{D}$</td><td>$exp[c + d \cdot FTFA] \quad c, d > 0$</td><td>$exp[c + d \cdot (OSI - OSI_{threshold})] \quad c, d > 0$</td></tr>
<tr><td>$\mathcal{E}$</td><td></td><td>$4.0 \cdot RR^* \cdot \sum_{i=0}^{7} C_{i,m} \cdot (L_m)^i$
$RR^* = 210.28 + 0.1217 \cdot T^H_{19} - 0.7829 \cdot T^V_{22} - 0.1830 \cdot T^V_{37} + 0.0998 \cdot T^H_{37}$</td></tr>
<tr><td>$\mathcal{F}$</td><td></td><td>$exp(A) \cdot exp(B) - 1$
$A = \frac{T^H_{19} - T^{H*}_{19}}{T^H_{37} - T^{H*}_{37}} \cdot [(.018 + .03 \cdot (.147 \cdot (T^H_{37\,min} - 126))^{0.8}) \cdot (T^H_{37} - T^{H*}_{37})]^{1.3}$
$B = \begin{cases} 0 & \text{if } T^H_{85} > 260K \\ (260 - T^H_{85})/40 & \text{if } T^H_{85} < 260K \text{ and } T^H_{37} > (T^{H*}_{37} + 40 \\ 2.5 & \text{if } T^H_{85} < 180K \text{ and } T^H_{37} > (T^{H*}_{37} + 40 \end{cases}$
$T^{H*}_{37} = T^H_{37\,min} + 15$ K</td></tr>
<tr><td>$\mathcal{G}$</td><td></td><td>$-0.17 + 0.31414 \cdot Q_{37} + 5.501 \cdot Q^2_{37}$</td></tr>
<tr><td>$\mathcal{H}$</td><td></td><td>$-0.25 + 2.65760 \cdot Q_{19}$</td></tr>
<tr><td>$\mathcal{I}$</td><td></td><td>$-1.65 + 0.289 \cdot SI_G$</td></tr>
<tr><td>$\mathcal{J}$</td><td>$-2.71 + 0.362 \cdot SI^L_{LW}$</td><td>$-1.05 + 0.149 \cdot SI^W_{LW}$</td></tr>
<tr><td>$\mathcal{K}$</td><td>$exp(3.29716 - 0.01290 \cdot T^V_{85} + 0.00877 \cdot T^H_{85}) - 8$</td><td>$exp(3.06231 - 0.0056036 \cdot T^V_{85} + 0.0029478 \cdot T^H_{85} - 0.0018119 \cdot T^V_{37} - 0.00750 \cdot T^V_{22} + 0.0097550 \cdot T^V_{19}) - 8$</td></tr>
<tr><td>$\mathcal{L}$</td><td>$exp(-17.76849 - 0.09612 \cdot T^V_{37} + 0.15678 \cdot T^V_{19}) - 1$</td><td>$exp(5.10196 - 0.05378 \cdot T^V_{37} + 0.02766 \cdot T^H_{37} + 0.01373 \cdot T^V_{19}) - 2$</td></tr>
<tr><td>$\mathcal{M}$</td><td>$0.143 \cdot (T^V_{19} + T^V_{22} - T^V_{37} - T^V_{85})$</td><td>$0.100 \cdot (T^V_{19} + T^V_{22} - T^V_{85} - T^H_{85})$</td></tr>
<tr><td>$\mathcal{N}$</td><td></td><td>$C^M_{0i} + C^M_{1i} \cdot T^V_{19} + C^M_{2i} \cdot T^H_{19} + C^M_{3i} \cdot T^V_{22} + C^M_{4i} \cdot T^V_{37} + C^M_{5i} \cdot T^H_{37} + C^M_{6i} \cdot T^V_{85} + C^M_{7i} \cdot T^H_{85}$</td></tr>
<tr><td>$\mathcal{O}$</td><td colspan="2">$A^m_0 + A^m_1 \cdot PC1^m + A^m_2 \cdot (PC1^m)^2$</td></tr>
<tr><td>$\mathcal{P}$</td><td></td><td>$59.9 - 0.239 \cdot T^H_{85}$</td></tr>
<tr><td>$\mathcal{Q}$</td><td></td><td>$a_0 - a_1 \cdot ln(P_{37}) + a_2 \cdot S_{85} + a_3 \cdot [ln(P_{37})]^2 - a_4 \cdot ln(P_{37}) \cdot S_{85} + a_5 \cdot S^2_{85}$</td></tr>
<tr><td>$\mathcal{R}$</td><td></td><td>$S_{85}/4.19$ or $[-\frac{1}{a \cdot Z_f} \cdot [\frac{\cos\theta \cdot ln(P)}{2} + k_{e,l} \cdot L]]^{1/b}$</td></tr>
<tr><td>$\mathcal{S}$</td><td></td><td>$10 ** [17.45 - 1.761 \cdot 10^{10} T^{V\,-4}_{19} - 5.700 \cdot 10^{-2} T^V_{22} + 1.404 \cdot 10^5 T^{H\,-2}_{19} - 1.188 \cdot 10^{-7} (T^V_{19} - T^H_{19})^4 + 1.944 \cdot 10^{-4} (T^V_{22} - T^V_{85})^2]$</td></tr>
</table>

$\mathcal{E}$ modifies the rainrate field as obtained from the Hughes D-Matrix algorithm [15] according to a correction function of the month and of the latitude.
$\mathcal{N}$ substitutes rainrate pixels showing low consistency between simulated and measured Tb's with the median rainfall value of nine pixels corresponding to itself and the eight surrounding pixels.

AKNOWLEDGMENTS

The author wishes to thank all the GPCP-AIP/2 participants for their collaboration in providing information on their algorithms and Dr. Isabelle Jobard, at LMD-CNRS, for the helpful discussions and comments. The work was started when the author was at the UK Met. Office under a EUMETSAT contract and continued during his permanence, as a NERC research fellow at the Dept. of Meteorology of the University of Reading, UK.

REFERENCES

1. Allam, R., G.Holpin, P.Jackson and G.L.Liberti, 1993. Second Algorithm Intercomparison Project of the Global Climatology Project. Pre-Workshop Report. UK Met. Office. pp. 200.
2. Kidd, C. and E.C.Barret, 1990: The use of Passive Microwave Imagery in Rainfall Monitoring. Remote Sens. Rev. 4 (2), 415-420.
3. Berg, W.K., 1993. Estimation and analysis of climate-scale rainfall over the tropical Pacific. Ph.D. Dissertation, Colorado Center for Astrodynamic Research, Univ. of Colorado, Boulder, Colorado, 202 pp.
4. Prabhakara, C., J.J.Nucciarone and G.Dalu, 1992. Convective and Stratiform Rain: Multichannel Microwave Sensing over Oceans. To be published in Met. and Atm. Physics.
5. Grody, N.C and R.R.Ferraro, 1992. A comparison of passive microwave rainfall retrieval methods. 6th AMS Conference on Satellite Meteorology and Oceanography, Atlanta, GA, Jan. 1992. 60-65.
6. Hollinger, J., 1991. DMSP SSM/I calibration/validation. Final report part II. Naval Research Laboratories, Washington DC, 257 pp.
7. Kummerow, C. and L. Giglio, 1994: A Passive Microwave Technique for Estimating Rainfall and Vertical Structure Information from Space. Part I: Algorithm Description. J. of Appl. Meteor., 33, 1, 3-18.
8. Basili P., P.Ciotti, G.d'Auria, F.S.Marzano, N.Pierdicca and A.Mugnai, 1994. A precipitation retrieval algorithm by space-borne multispectral radiometry based on cloud radiation models and multivariate techniques. In preparation.
9. Adler R.F, A.J.Negri, P.R.Keehn and I.M.Hakkarinen, 1992. Estimation of monthly rainfall over Japan and surrounding waters from a combination of low-orbit microwave and geosynchronous data. J. of Appl. Meteor. 32, 2, 335-356.
10. Petty, G.W., 1993a: Physical Retrieval of Over-Ocean Rain Rate from Multichannel Microwave Imagery. Part II: Algorithm Implementation. Submitted to Meteo. and Atmos. Physics.
11. Bauer, P. and P.Schlüssel, 1993. Rainfall, Total Water, Ice Water and Water Vapour over Sea from polarized microwave simulations and SSM/I data. J. of Geophys. Res. 98, 20737-20760.
12. Spencer, R.W., H.M.Goodman and R.E.Hood, 1989. Precipitation retrieval over land and ocean with SSM/I: identification and characteristics of the scattering signal. J. Atm. Ocean. Tech. 6, 254-273.
13. Grody, N.C., 1991. Classification of snow cover and precipitation using the SSM/I. J. of Geophys. Res., 96, 7423-7435.
14. Wentz F.J., 1991. User's Manual SSM/I Antenna Temperature Tapes. Revision 1. RSS Technical Report 120191. 70 pp.
15. Hollinger, J.P., R.Lo, G.Poe, R.Savage and J.Pierce, 1987. Special Sensor Microwave/Imager Users' Guide. Naval Research Laboratory. Washington D.C.. 120 pp.

3. RADIOMETRIC SENSING OF THE EARTH'S SURFACE

OVERVIEW

- **Snow and ice**

  Over the ocean the *SSM/I* channels can routinely provide meteorological data, such as rainfall, cloud liquid and water vapor, whereas over the land the identification and separation of precipitation from snow cover are difficult and require suitable decision schemes, based essentially on the different spectral characteristics of emission. In the future, the availability of *MIMR* observations and of appropriate inversion algorithms should open new fields of applications, for which quantitative data and adequate ground resolution are crucial. In particular, it is expected that this sensor will allow operational monitoring of snow coverage, depth and water equivalent. The availability of channels in the higher microwave frequency range (85 GHz for *SSM/I* and 89 for *MIMR*) makes these sensors also suitable to produce maps of sea ice, which are of primary importance in ship navigation.

- **Ocean surface**

  The wind speed, if on one side contaminates the radiometric data over the ocean because of the induced variable roughness of the sea surface, on the other is an important meteorological parameter by itself. The Advanced Microwave Sounding Unit (*AMSU*) is provided with window channels with the additional potential of wind speed retrieval. The attainment of the needed accuracy relies upon the availability of adequate emissivity models of the rough sea surface, which, anyway, are needed for the retrieval of atmospheric parameters to properly take into account the surface contribution to the radiometric signal. The required validity over an extended range of frequencies, and the inclusion of realistic features such as foam cover, make the development of such models rather challenging.

- **Crops and forest**
  The radiometric characterization of the land surface covered with vegetation is even more intricate, because of the complexity of the interacting medium both on the small scale of the individual element and on the large scale of the many interacting elements. Nevertheless, significant results are being obtained by the microwave emission simulation models worked out for agricultural crops and for forests. Through these simulations the feasibility of detecting forest fires in their early stages is also explored.

  SSM/I data too appear to contain information on land characteristics, due to the sensitivity of spectral and polarization features to land cover types. The coarse ground resolution of present spaceborne systems limits their use to large spatial scales. However, the reduced footprint size of *MIMR* will help promoting the exploitation of satellite radiometric data in land applications also for regions with relatively high spatial variations of land features, as many European and Mediterranean countries.

Identification of snowcover and precipitation using the Special Sensor Microwave Imager (SSM/I)

Norman C. Grody[1], Ralph R. Ferraro[1], and Alan N. Basist[2]

[1]. NOAA/NESDIS/ORA/SRL, Washington D.C.
[2]. NOAA/NWS/NMC/CAC, Washington D.C.

Abstract The Special Sensor Microwave Imager (SSM/I) was developed as part of the Defense Meteorological Satellite Program (DMSP). It was first launched in November 1987 aboard the F-8 polar orbiting satellite and later placed aboard the F-10 and F-11 satellites. The SSM/I contains dual polarized channels at 19.35, 37.0 and 85.5 GHz and a vertically polarized channel at 22.23 GHz to measure various atmospheric and surface parameters. One of the most difficult problems involves the identification and separation of precipitation from snowcover. Separate decision trees were developed to measure snowcover and precipitation over land and ocean. The decision trees contain various steps to identify these parameters and filter out features that can lead to misclassification (e.g., deserts, arid land and sea ice). The paper describes the decision trees and the various filters needed to remove false signatures.

Key words: precipitation, snowcover, SSM/I

INTRODUCTION

Satellite instruments are particularly useful for measuring the global distribution of snowcover and precipitation. Since the launch of the first polar orbiting satellite in 1966, experienced analysts have produced weekly composite snowcover maps by manually inspecting visible images to identify cloud free regions [1]. This weekly product is useful for climate analysis, although it does not meet the needs of weather forecast models which require daily measurements [2]. For this application a more objective, automated technique must be developed to identify snowcover. While clouds are a hindrance when viewing the surface, the cloud top temperatures inferred from infrared measurements have been used to identify precipitation. The objective technique relies on a statistical relationship between the hourly rainfall over a given area ($\approx$2500 km^2) and the population of infrared cloud top measurements below a 235 K threshold [3] obtained using observations from the GOES satellites. However, errors result from cirrus clouds or other non-precipitating clouds having cold cloud tops.

The cloud problems associated with visible and infrared techniques are significantly reduced using microwave measurements. Microwave radiometers aboard satellites have been used to identify precipitation and observe the surface under cloud covered conditions [4].

Unfortunately these measurements were previously only available for limited time periods using sensors aboard research satellites, such as the Scanning Multichannel Microwave Radiometer (SMMR) on Nimbus 7. However, beginning in 1987, the Air Force has launched three Special Sensor Microwave Imager's (SSM/I) into polar orbit aboard the F-8, F-10 and F-11 satellites. Data from this sensor is routinely processed by the National Oceanic Atmospheric Administration (NOAA) as well as by the Navy and Air Force.

The SSM/I contains dual polarized channels at 19, 37 and 85 GHz and a vertically polarized channel at 22 GHz [5]. These channels are particularly useful for measuring rainfall and snowcover over land. Over oceans the SSM/I provides measurements of rainfall, cloud liquid water, water vapor, sea surface winds and sea ice concentration. Although many of these parameters were obtained using earlier satellite instruments (e.g., SMMR), the SSM/I is the first sensor to contain an 85 GHz channel, which significantly increases the sensitivity to identify precipitation and snowcover. This paper describes the algorithms developed at NOAA to identify precipitation and snowcover, and contains improvements to the original algorithm [6].

SCATTERING SIGNATURES OVER LAND

Since the launch of the Nimbus-E Microwave Spectrometer (NEMS) on Nimbus-5, it has been recognized that the ice particles within snow are strong scatterers of microwave radiation [7]. More recently, SMMR measurements have shown that the large ice produced as part of the precipitation process also scatters microwave radiation [8]. Less intense scattering results from the liquid rain drops themselves. In addition to the scattering by ice and water, studies using SSM/I data have shown that sandy deserts and dry bare soils can also scatter microwave radiation [9]. As discussed later, these sparsely vegetated dry surfaces can be misclassified as rain or snowcover if not accounted for.

Volume scattering results in a decreased brightness temperature for increasing frequency. The physical basis of this scattering signature is best illustrated by referring to Figure 1, which shows a simplified diagram of the radiation transfer for snowcover and precipitation.

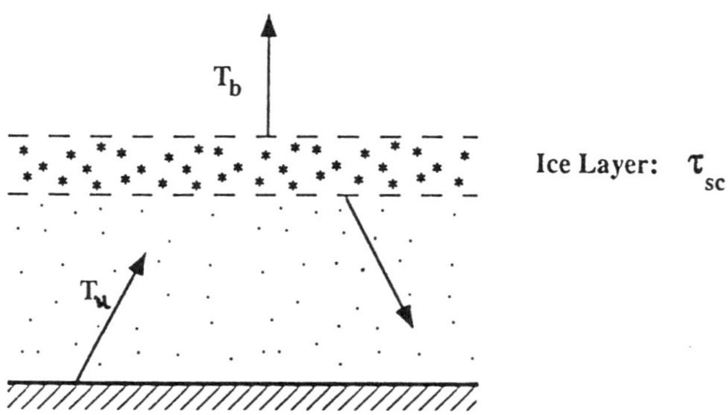

Figure 1. Schematic of different radiation contributions and variables contained in the scattering model.

In both cases, T_u represents the upwelling brightness temperature emitted below the scattering ice layer and T_b is the radiation emerging from the top of the layer. Ice particles produce little absorption or emission at microwave frequencies so that T_b can be approximated as

$$Tb\ (\upsilon) = \tau_{sc}(\upsilon)\ Tu,$$

where τ_{sc} is the transmittance of the scattering layer at frequency, υ. A more exact model of precipitation would include an emission contribution due to liquid water drops within the ice layer.

The transmittance defines the percent of incident radiation, T_u, that is not scattered by the ice particles. Neglecting multiple scattering, the transmittance of a uniform layer of particles having thickness L is

$$\tau_{sc}(\upsilon) = \mathrm{Exp}[-L\int_0^\infty \pi a^2 Q_e(a/\lambda)n(a)da],$$

where the integral contains the density of ice particles, n(a), having radius a, and the scattering cross section, $\pi a^2 Q_e$. The frequency dependence of transmittance (and brightness temperature) is defined by the efficiency factor, Q_e, which depends on the ratio of particle radius to radiation wavelength λ. Considering spherical particles, Figure 2 shows a plot of Q_e as a function of particle diameter for the SSM/I frequencies of 19, 37 and 85 GHz. For light rain or newly formed snow the particles are generally less than 2 mm so that Rayleigh scattering applies (i.e., $Q_e \propto \upsilon^4$) and the cross-section at 85 GHz is about 27 times greater than at 37 GHz. In the case of convective rain or refrozen snow the ice particles are much larger. This results in smaller differences in both Q_e as well as the brightness temperatures of the two channels (see Figure 2).

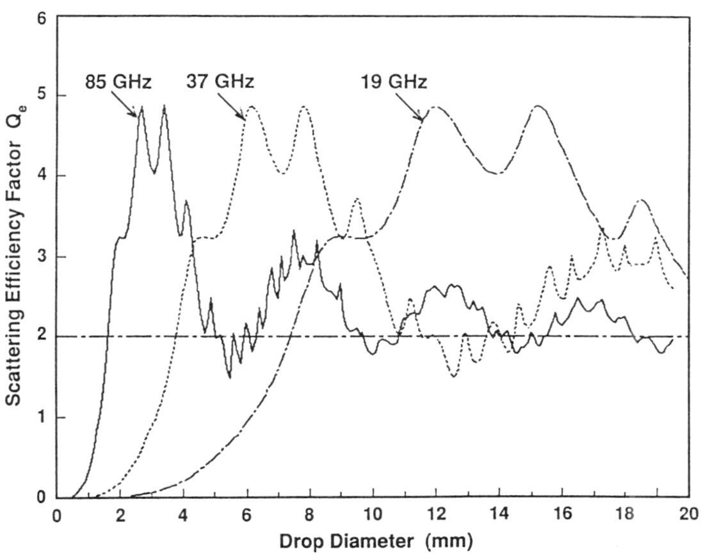

Figure 2. Scattering efficiency for spherical ice particles as a function of diameter.

Models such as that discussed above provide insight on the physical mechanisms affecting the channel measurements and are in general agreement with actual data. Examples of actual SSM/I measurements are shown in Figure 3 for rain, snow and deserts. Both vertical and horizontal polarization measurements are shown, although the largest scattering signature occurs for vertical polarization. The light and heavy rain data were obtained over Texas during November 1987, while the snowcover data is from Canada and the United States during February 1993. Figure 3 also shows the measurements obtained in the cold Gobi desert and hot Sahara desert during February 1993. These sandy deserts display a decrease in the vertical polarized brightness temperature for increasing frequency, which is weaker than the scattering signatures shown for rain and snow. The smallest scattering signatures occur for sparsely vegetated dry soils, which are not shown in the Figure but are discussed later.

The different spectral characteristics illustrated in Figure 3 enable the SSM/I to identify snowcover and precipitation with minimal ambiguity. As discussed next, a decision tree approach is used to identify and filter out various features that can lead to misclassification. The filters are based on the physical knowledge of microwave characteristics as well as empirical observations. Figures 4, 5 and 6 show the decision trees developed for identifying snowcover, precipitation over land and ocean and are discussed in detail below. It should be noted that the original SSM/I measurements, called antenna temperatures, are used throughout this paper. Brightness temperatures contain a correction to account for the fractional amount of cold-space radiation collected by the antenna when viewing earth. The corrections at 19, 22, 37 and 85 GHz are about 7, 6, 4 and 3 K, and must be added to the antenna temperatures when using brightness temperatures.

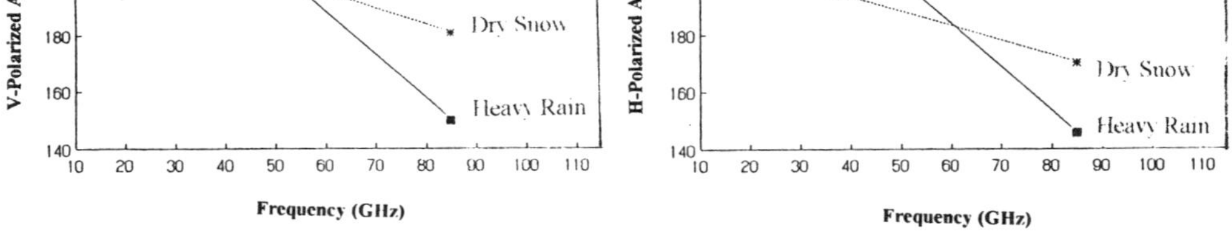

Figure 3. Examples of SSM/I measurements for precipitation, snowcover and deserts. Both vertical (left) and horizontal antenna temperatures are shown.

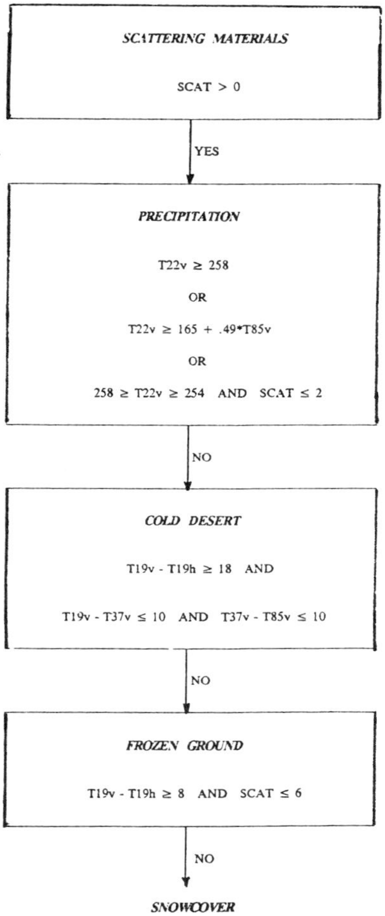

Figure 4. Decision tree to identify snowcover.

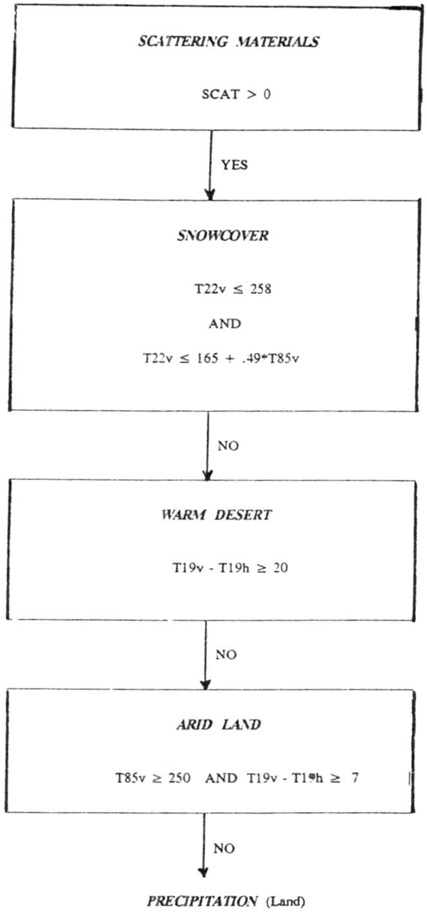

Figure 5. Decision tree to identify precipitation over land.

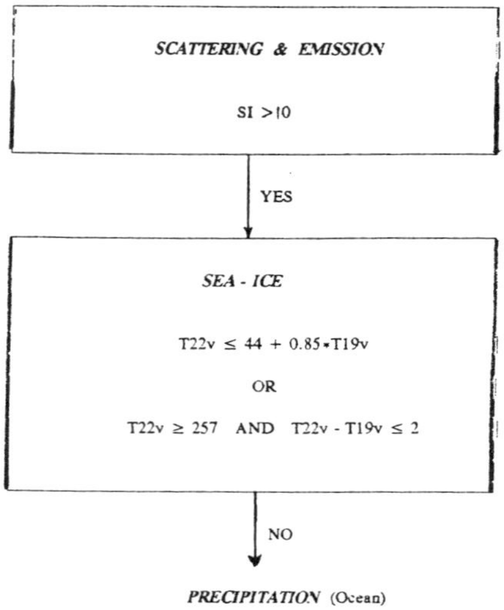

Figure 6. Decision tree to identify precipitation over oceans.

SNOWCOVER

As discussed above, snowcover produces a positive difference between the low and high frequency SSM/I channels. Although this scattering signature identifies snow, it also represents precipitation, sandy deserts and cold bare ground. The latter two surfaces were not accounted for earlier [6] but are included in the improved decision trees developed here.

The first step in the identification of snowcover is given by

$$T22v - T85v > 0 \quad \textbf{(1a)}$$

OR

$$T19v - T37v > 0 \quad \textbf{(1b)}$$

where T19v, T22v, T37v and T85v denotes the vertically polarized antenna temperatures at the four respective frequencies. In general, (1a) is sufficient to identify snowcover, although (1b) provides an additional check when there is considerable cloud liquid water. The relationships given by (1a) and (1b) are used throughout the paper with different thresholds and shall be denoted as SCAT > N, when N is the threshold. As shown in Figure 4, the constraint SCAT >0 represents the first of many steps.

Precipitation Filter

The density and size of ice particles found in snow are generally larger than those in precipitating systems. Snowcover therefore produces larger scattering effects at low frequencies than precipitation. Furthermore, even in the absence of low frequency scattering, the low frequency measurements are coldest for snowcovered surfaces. Based on these factors, precipitation is identified when the 22 GHz measurements[1] are greater than a given threshold, i.e.,

$$T22v > 257. \qquad \textbf{(2a)}$$

The 257 K threshold represents average conditions and can deviate by ±3 K depending on the surface and atmospheric conditions. For example, melting snow can increase T22v to about 260 K (see Figure 3) while cold rain can reduce T22v to 254 K. To minimize the problems associated with melting snow only the early morning satellite observations are used. Also, cold rain is approximately filtered using the relationships;

$$T22v \geq 254 \ \text{ AND } \ SCAT \leq 2, \qquad \textbf{(2b)}$$

where these conditions are based on the smaller scattering signatures generally observed for cold rain than for snowcover. In the future, water vapor sounding channels around 183 GHz from the SSM/T2 sensor may be used to improve the separation between surface features and precipitation.

Although (2a) and (2b) are the primary conditions to filter precipitation, a third relationship is needed to account for deep convective rain. Convective rain contains large ice particles so that scattering occurs at 22 GHz as well as at higher frequencies (see Figure 3). This reduces T22v below the 257 K threshold, and occurs when T85v is less than about 188 K. As such, the following condition must also be included to filter out precipitation[2];

$$T22v \geq 165 + .49 \times T85v, \qquad \textbf{(2c)}$$

where (2a), (2b) and (2c) define the second step in the decision tree of Figure 4. As an important side benefit, (2a) also filters out warm deserts, although cold deserts are not removed.

Cold Deserts

The cold deserts found in outer Mongolia, Tibet and central Iran scatter microwave radiation and must be accounted for since T22v is less than the 257 K threshold. These regions are relatively flat so that they produce a large polarization difference at 19 GHz (see Figure 3). Unfortunately, snowcover can occasionally produce similar scattering and polarization signatures. After considering various channel combinations, the following set of conditions were found to best filter cold deserts without significantly removing any snowcover;

$$T19v - T19h \geq 18 \quad \text{AND} \qquad \textbf{(3a)}$$

$$T19v - T37v \leq 10 \quad \text{AND} \quad T37v - T85v \leq 10 \qquad \textbf{(3b)}$$

These conditions are used as the third step in the decision tree of Figure 4.

---

[1] The 22 GHz channel was chosen (rather than 19 GHz) since water vapor absorption reduces the channels sensitivity to surface emissivity variations for moist atmospheres. As such, its measurements are generally greater than 257 K, even for light rain over wet surfaces.

[2] In the earlier study [6], the 85 GHz measurements were convolved to the 22 GHz resolution of 55 km, whereas the full resolution measurements are used here. The higher resolution (15 km) results in larger peak scattering values at 85 GHz, which is reflected by the larger intercept in (2c).

Frozen Ground

In addition to the scattering observed for cold deserts, SSM/I measurements can produce a small scattering signature during cold periods over vegetated areas of high elevations. Such areas include North Africa, Western Bolivia, New Mexico and the Great Basin of the United States. These dry soils are usually snow-free during their corresponding cold season but produce scattering values between 1 and 6 K due to the thermal gradient within the soil as well as its scattering property [10]. Since these frozen surfaces often occur adjacent to areas having actual snowcover, they can be misinterpreted as light snowcover if not accounted for.

As in the case of cold deserts, a compromising set of conditions is sought to filter frozen ground surfaces without removing any significant amounts of snowcover. Compared to snowcover, frozen ground generally produces a lower scattering signature with a higher polarization difference at 19 GHz, i.e.,

$$\text{SCAT} \leq 6 \quad \text{AND} \quad \text{T19v - T19h} \geq 8, \qquad \textbf{(4)}$$

where these relationships define the last step of the decision tree in Figure 4.

PRECIPITATION OVER LAND

Over low emissivity oceans, precipitation is detected by combining its high frequency scattering signature with a low frequency emission signature, and will be discussed in the next section. However, over high emissivity land their is very little contrast between the emission by rain drops and the underlying surface. This prohibits the use of emission techniques, so that precipitation is only detected by its scattering signature. Scattering due to precipitation is always larger at 85 GHz than at 37 GHz so that only (1a) is used as the first step in the decision tree of Figure 5.

Snowcover Filter

The second step of the decision tree filters out snowcover measurements using (2a) and (2c). Since condition (2a) identifies precipitation only when T22v exceeds 257 K, the scattering due to cold deserts and frozen ground are also filtered. However, an additional filter must be used to screen out the scattering signatures obtained from warm deserts and arid land.

Warm Deserts

Sandy deserts, such as that found over North Africa and Saudi Arabia, have polarization differences exceeding 20 K. This value is larger than that obtained for precipitation so that the polarization difference alone serves as an effective filter [6], i.e.,

$$\text{T19v - T19h} \geq 20, \qquad \textbf{(5)}$$

where this equation defines the third step in the decision tree. Note that polarization information alone was insufficient to separate cold deserts from snowcover since snow can be highly polarized (see Equations (3a) and (3b)).

Arid land

As discussed previously, the scattering due to dry bare soil must by filtered out when identifying snowcover or precipitation. For precipitation it is only necessary to identify and remove the surfaces where T22v is greater than 257 K, i.e., arid land. A prominent arid area is found in the Sahel region of Africa during the winter months. This region is adjacent to the Sahara desert to the north and is used for agriculture, although there is very little rain or vegetation present during the winter season.

It is found that the frozen ground filter given by (4a) and (4b) must be modified to separate precipitation from arid land. In particular, a larger scattering threshold of 8 K is needed in (4a) to account for the larger scattering found for arid land. An alternate filter that has been tested more extensively is given by

$$T85v > 250 \quad \text{AND} \quad T19v - T19h > 7 \qquad \textbf{(6)}$$

and is used as the last step in the decision tree of Figure 5. In addition to filtering out measurements in the Sahel region, it also removes the arid land found in southern Africa, Australia and adjacent to the Andes mountains.

PRECIPITATION OVER OCEANS

Although (1a) adequately detects rain over high emissivity land surfaces, it is not sufficient over oceans. The low emissivity ocean surface reduces the low frequencies measurements so that a strong scattering effect is needed at 85 GHz to obtain a positive difference between T22v and T85v. As such, only the heaviest (convective) rain events are observable using (1a). A more sensitive indicator of rain is required which includes both the scattering and emission signatures of rain over ocean. Such an indicator, called a scattering index, was developed earlier [6]. However, the index was derived using data over land and oceans and therefore was not as effective over oceans. An improved index is obtained here using only oceanic measurements.

The scattering index, SI, is obtained by correlating T85v with the lower frequency channels (T19v, T22v) for rain-free conditions over oceans. It is given by

$$SI = F - T85v \qquad \textbf{(7a)}$$

$$F = 0.75 \times T19v + (2.543 - 0.00543 \times T22v) \times T22v - 182.7 \qquad \textbf{(7b)}$$

where F provides an estimate of T85v with a standard error of 3.5 K. The function F is obtained using regression analysis on a rain-free data set, which contains variations in water vapor, wind speed, sea surface temperature and cloud liquid water. The data set contains cloud liquid water less than 0.3 mm, which is considered the maximum value for non-precipitating clouds, and is obtained using the SSM/I measurements [11].

Precipitation is identified using a 10 K threshold of the SI, where this value is about 3 standard deviations of the error in F. The scattering index increases beyond the threshold due scattering and/or emission by precipitation. This is due to the scattering effects by ice particles which decreases T85v while the low frequency emission due to liquid hydrometers increases F.

Sea-Ice Filter

The condition SI > 10 is used to identify the precipitation over ice-free oceans. However, the emissivity of sea-ice is similar to land at low frequencies. Also, multiyear ice scatters at high frequencies. The result is that sea-ice increases the scattering index similar to precipitation and therefore must be filtered.

To filter sea-ice from all but the heaviest precipitation, we apply the condition [6],

$$T22v \leq 44 + 0.85 \times T19v. \qquad \textbf{(8a)}$$

This relationship accounts for all types of sea-ice, including partially filled ice within the 19 GHz field of view. A second condition is used to identify and separate sea-ice from heavy precipitation, and is given by

$$T22v \geq 257 \quad \text{AND} \quad T22v - T19v \leq 2. \qquad \textbf{(8b)}$$

As shown in Figure 6, equations (7) and (8) complete the decision tree for identifying precipitation over oceans.

CONCLUSIONS

This paper describes the decision trees developed at NOAA to identify snowcover and precipitation using SSM/I measurements. In the case of snowcover and precipitation over land, the scattering signature at 85 GHz provides the first step in both decision trees (see Figures 4 and 5). This is followed by a series of filters to remove other features such as deserts. frozen ground, and arid land. Over oceans, a scattering index is used to identify the scattering as well as emission signatures due to precipitation. This defines the first step in the decision tree of Figure 6. A second step is used to filter out sea-ice.

The objectively derived precipitation and snowcover obtained from the SSM/I has great application in an operational mode. This is particularly true for forecast models, which require daily measurements. Also, the monthly products are applicable to hydrologic studies pertaining to climate and global change studies. These applications require highly stable sensors, and objectively derived precipitation and snowcover, for detecting small climatic changes.

The SSM/I measurements from the F-10 and F-11 DMSP satellites are currently used to produce daily snowcover and precipitation fields on a $1/3^{o} \times 1/3^{o}$ latitude-longitude grid using the decision tree algorithms. The two satellites offer up to four observations/day at high latitudes although the sampling is no more than two observations/day in the mid-latitudes and tropics. For climate analysis, the daily snowcover maps are combined to obtain the percent of snowcover for a given month. Similarly, the daily precipitation maps of rainfall are combined to obtain the monthly average rainfall. As discussed in (12), radar data is used to pre-calibrate the daily scattering parameters (SCAT and SI) into rain rates.

During the past year the Climate Analysis Center (CAC) of NOAA began an extensive evaluation of the SSM/I experimental products by comparing the results with the operational snowcover [1] and rainfall products [3] derived using visible and infrared sensors. Beginning in January 1994, the inter comparisons between the SSM/I and operational products have been published in the *Climate Diagnostic Bulletin.*

As an example, Figure 7 displays the percent of snowcover in the northern hemisphere for March, 1994 based on the SSM/I (right) and the operational analysis (left). The variation in snowcover is indicated by the gray scale at the bottom of the Figure. Differences between the two results is attributed to the fact that the SSM/I product uses daily observations, while the operational analysis averages four weekly products to obtain the monthly snowcover. Smallest differences between the two analyses are noted where snowcover was observed for the entire month. However, large differences occur where snowcover was less persistent during the month, such as over the northern tier of the United States, southwestern and southeastern Canada, the Alps, Tibet, and many mountain ranges around the hemisphere.

Figure 8 shows the rainfall for March, 1994 derived from the SSM/I (top) and operational analysis based on the GOES Precipitation Index (GPI) method. The rainfall variations obtained from both techniques are displayed in gray shades, contoured with 100 mm intervals. Differences between the two rainfall estimates are due to the different signatures observed from the microwave and infrared sensors (see introduction). Also, the sampling is quite different for the GPI (four observations/day) and SSM/I. (generally less than two observations/day). Furthermore, the GPI was calibrated using data over oceans (GATE) and thus, likely provides better estimates over oceanic than land areas. The SSM/I technique was calibrated against radar and station data world-wide.

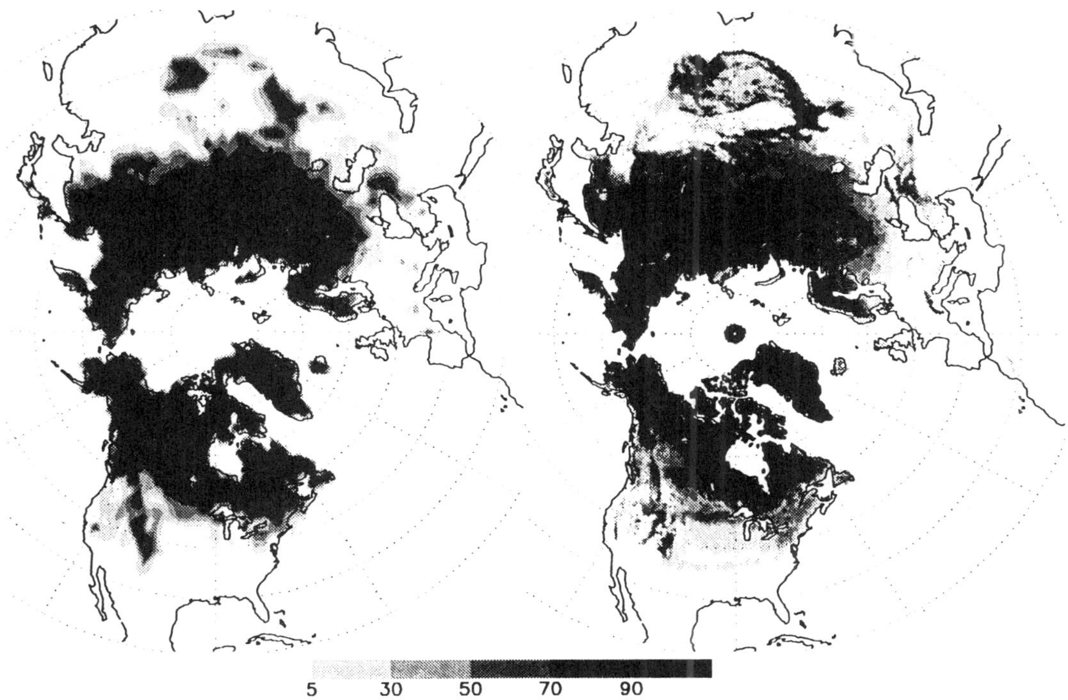

Figure 7. Monthly average snowcover for March 1994 obtained using the SSM/I (Right) and NOAA operational analysis (Left). In the operational analysis a given area is identified as being snow covered for the month if at least 50% of the once-weekly analyses indicates snowcover in that area. The SSM/I technique indicates the percent of days during the month in which snowcover was observed at a given location.

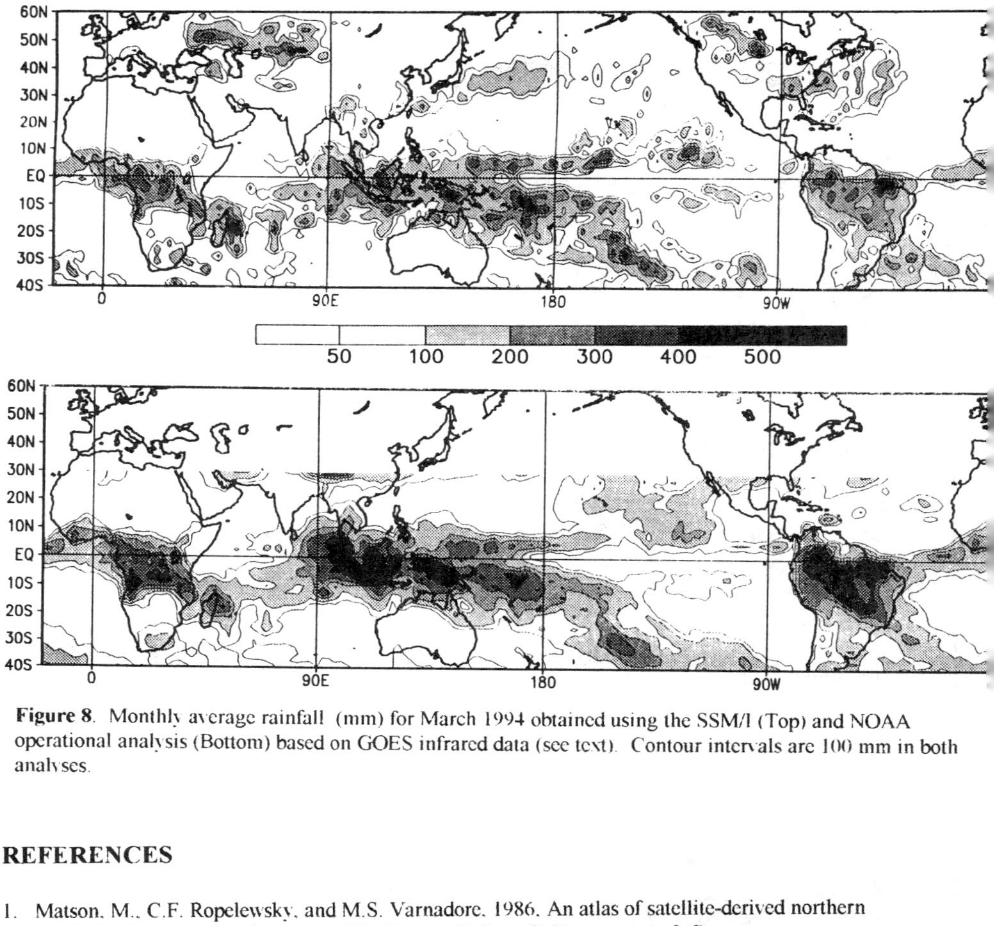

Figure 8. Monthly average rainfall (mm) for March 1994 obtained using the SSM/I (Top) and NOAA operational analysis (Bottom) based on GOES infrared data (see text). Contour intervals are 100 mm in both analyses.

REFERENCES

1. Matson, M., C.F. Ropelewsky, and M.S. Varnadore, 1986, An atlas of satellite-derived northern hemispheric snow cover frequency, Washington D.C., U.S. Department of Commerce.
2. Peterson, R.A. and J.E. Hoke, 1989, The effect of snow cover on the Regional Analysis and Forecast System (RAFS) low-level forecasts, *Wea. and Forecasting*, **4**, 253-257.
3. Arkin, P., and B. N. Meisner, 1987, The relationship between large scale convective rainfall and cold cloud cover over the western hemisphere during 1982-84, *Mon. Wea. Rev.*, **8**, 51-74.
4. Ferraro, R.R, N.C. Grody, and J.A. Kogut, 1986, Classification of geophysical parameters using passive microwave satellite measurements, *IEEE Trans. Geosci. Remote Sensing*, **GE-24**, 1008-1013.
5. Hollinger, J.R., B. Lo, G. Poe, R. Savage and J. Pierce, 1987, Special Sensor Microwave Imager User's Guide, 120 pp., Naval Research Laboratory, Washington, D.C.
6. Grody, N.C., 1991, Classification of snowcover and precipitation using the special sensor microwave imager, *J. Geophys. Res.*, **96**, 7423-7435.
7. Kunzi, K.F., A.D. Fisher, D.H. Staelin, and J.W. Waters, 1976, Snow and ice surfaces measured by the Nimbus 5 microwave spectrometer. *J. Geophys. Res.*, **81**, 4965-4980.
8. Spencer, R.W., W.S. Olson, W. Rongzhang, D. Martin, J.A. Weinman and D.A. Santek, 1983, Heavy thunderstorms observed over land by the Nimbus-7 scanning multichannel radiometer, *J. Clim. Appl. Meteorol.*, **22**, 1041-1046.
9. Neale, M.U., M. J. McFarland and Kai Chang, 1990, Land-Surface Classification Using Microwave Brightness Temperatures from the Special Sensor Microwave/Imager, *IEEE Trans. Geosci. Remote Sensing*, **28**, 829-838.
10. England, A.W., 1990, Radiobrightness of diurnally heated, frozen soil, *IEEE Trans. Geosci. Remote Sensing*, **28**, 464-474.
11. Grody, N.C and R.F. Ferraro, 1992, A comparison of passive microwave rainfall retrieval methods, *Proc. Sixth Conference on Meteorology and Oceanography*, AMS, Atlanta, GA, 60-65.
12. Ferraro, R.R and N.C. Grody, 1992, The calibration of an SSM/I scattering index for rain retrievals using RADAP-II and Amedas radar data, *Proc. Sixth Conference on Meteorology and Oceanography*, AMS, Atlanta, GA, 290-294.

Multi-Temporal Aspects in Snow Retrieval using Spaceborne Radiometers

J. NOLL[1], J.P.V. POIARES BAPTISTA[1], M. BORGEAUD[1], and A. ROGNES[2]
[1]*ESA ESTEC-XEP, NL 2200 AG Noordwijk*
[2]*ESA ESRIN-DPE/OD, I 00044 Frascati*

Abstract
ESA's Multi-frequency Imaging Microwave Radiometer (MIMR) is briefly described, focusing on instrument features to meet the user requirements for the retrieval of geophysical parameters. Snow retrieval is chosen as an example, in land applications, of models developed for the retrieval of geophysical parameters from MIMR. Direct models for the brightness temperature prediction of a snow covered terrain and associated inversion algorithms are explained. Taking into account atmospheric information, multitemporal data analysis for the Norwegian test-site 'Hardangervidda' is performed. The retrieved snow parameter is qualitatively compared to AVHRR images and validated against observed ground truth.

Keywords: MIMR, SSM/I, snow retrieval, microwave passive remote sensing

1 INTRODUCTION

The increasing concern for the environment has promoted the use of using remote sensing of the Earth's surface and atmosphere. Microwave measurements from space can provide monitoring on a global, day and night all weather basis.

Since the first Cosmos satellite (U.S.S.R.) was launched in 1968, the complexity of microwave spaceborne radiometers has increased and several multi-frequency, multi-polarization radiometers are currently in operation in space. The Special Sensor Microwave/Imager (SSM/I) first launched in 1987 on the American DMSP-F08 satellite is such a type of advanced radiometer, operating at 19, 22, 37 and 85 GHz with both horizontal and vertical polarizations. In Europe, an even more advanced radiometer (Multi-frequency Imaging Microwave Radiometer – MIMR) is under development and planned to fly on the European METOP-1 and -2 and the American EOS-PM platform.

With the first launch foreseen around 2000, MIMR will offer increased capabilities for atmosphere, ocean, ice and land monitoring. This paper deals with the objectives of this instrument and the user requirements. Multitemporal monitoring of snow is chosen as one example of a geophysical product to be generated from MIMR data. Direct models for emissivity simulation and inverse algorithms for snow height and snow water content retrieval are presented.

2 SPACE-BORNE RADIOMETERS

Observation of the Earth in the last decade has shown, that local events may often have an impact on a worldwide scale (e.g. Pinatubo eruption → greenhouse effect). Due to this global interaction, researchers in climatology or meteorology do not concentrate on one unique parameter, but cover a whole set of observables. For the atmosphere the basic required parameters are water vapour (WV), liquid water (LW) and precipitation.

Space-borne radiometers are of high importance for parameter retrieval as they offer day and night global all weather observation. Starting with the Cosmos 243 satellite in 1968, microwave radiometers have been carried aboard several spacecrafts (e.g. NEMS, SMMR, SAMIR, SSM/I, SSM/T).

2.1 Multi-channel radiometers

To fulfill multiple requirements, multi-channel radiometers have been designed and launched, one of them being the SMMR (Scanning Multi-channel Microwave Radiometer) on Nimbus-7 and SEASAT in 1978. Covering the frequencies 6.6, 10.7, 18, 21 and 37 GHz, the SMMR was able to monitor sea surface temperature (SST) and wind speed (WS) together with the atmospheric parameters WV, LW and rain rate.

At present, the SSM/I is collection passive microwave data from the Earth. Since the first instrument was launched in 1987 on the DMSP-F08 platform continuous global data has been collected. The multi-channel SSM/I covers with it's 19.3, 22.2, 37.0 and 85.5 GHz channels a wide spectrum of geophysical parameters, especially for ocean, sea ice and atmosphere applications [1]. The development of land applications has been poor so far, as the resolution of 70×45 km for 19 GHz and 16×14 km for 85 GHz is not very high.

2.2 MIMR performance

MIMR, developed under the responsibility of ESA, is an instrument with the same principle of operation as the SSM/I. It will cover all the frequencies of the SMMR and SSM/I (Tab. 1).

Table 1. MIMR system performance (f $\mapsto$ frequency, pix $\mapsto$ pixel, samp. $\mapsto$ sampling interval, sens. $\mapsto$ radiometric sensitivity, stab. $\mapsto$ radiometric stability, acc. $\mapsto$ radiometric accuracy)

f [GHz]	6.8	10.65	18.7	23.8	36.5	90
pix. [Km]	60	38	22	20	11.6	4.9
samp. [Km]	15.5	15.5	15.5	15.5	7.75	3.87
sens. [°K]	0.2	0.4	0.5	0.5	0.5	0.7
stab. [°K]	0.2	0.4	0.5	0.5	0.5	0.7
acc. [°K]	1	1	1.5	1.5	1.5	1.5

Table 2. Channel utilization for MIMR

Product	6.8	10.65	18.7	23.8	36.5
Atmosphere					
Water vapor			x	x	x
Cloud liquid water				x	x
Precip. over ocean				x	x
Ocean					
Temperature	x	x	x	x	
Wind speed	x	x	x	x	
Sea-ice					
Concentration			x		x
Type (age)			x		x
Edge location				x	x
Land					
Soil moisture	x	x	x		
Snow water equiv.			x		x

Due to the increase of the antenna size of 1.6 m compared to 0.65 m for the SSM/I the spatial resolution of the corresponding channels is improved by about the factor three for MIMR. The radiometric accuracy is, despite the antenna size, kept very similar to the SSM/I. With a sampling distance of 15.5 km for the 6.8 to 23.8 GHz channels and even smaller for the 36.5 and 90 GHz channels (Tab. 1), the use of MIMR data will open new fields of applications.

2.3 *Geophysical parameter retrieval*

The usefulness of the retrieved geophysical data depends on the choice of instrument parameters such as frequency, spatial resolution and radiometric accuracy. Tab. 2 gives an overview of the most important geophysical parameters for each application area and the related frequencies used in the retrieval algorithm.

The system parameters of Tab. 1 have been selected in accordance with user requirements and feasible instrument parameters [2]. Even though detailed studies have to be undertaken to establish direct models and inverse algorithms, it is expected that MIMR will provide the geophysical parameters mentioned in Tab. 3.

Table 3. Environmental products as expected from MIMR data (Sampling interval 15.5 km, WV $\mapsto$ water vapour, CLW $\mapsto$ cloud liquid water, precip $\mapsto$ precipitation over ocean, cloud $\mapsto$ cloud amount, SST $\mapsto$ sea surface temperature, WS $\mapsto$ wind speed, ice cov. $\mapsto$ ice coverage, m_v $\mapsto$ surface moisture, W_{equ} $\mapsto$ snow water equivalent, type $\mapsto$ surface type)

Parameter	Range	Quant. levels	Absolute accuracy
WV	0...1 Kg/m^2	0.05 Kg/m^2	±0.1 Kg/m^2
CLW	0...60 Kg/m^2	0.1 Kg/m^2	±0.2 Kg/m^2
precip	0...25 mm/h	0, 5, 10, 15 20, $\leq$ 25 mm/h	±5mm/h
Cloud	0...100 %	1 %	±20 %
SST	0...30 °C	0.2 °C	±1 °C
WS	3...25 m/s	1 m/s	±2 m/s
Ice			
- cov.	0...100 %	5 %	±10 %
- age	First year multi-year		
- edge			±2 Km
m_v	0...60 %	1 %	
W_{equ}	0...30 cm	1 cm	±3 cm
type	12 types		

3 SURFACE - ATMOSPHERE MODELS

The link between the measured brightness temperature T_B and the geophysical parameters is established by models and algorithms. Direct models, either empirical, semi-empirical or physical, perform the calculation of the brightness temperature as a function of geophysical parameters, frequency and polarization. Inversion algorithms, mainly based on linear combinations of different instrument channels, retrieve geophysical parameters from the satellite measured T_B.

In order to model the MIMR brightness temperature, European scientists have under ESA supervision put together an emission simulation and inversion tool [3, 4].

3.1 *Emission simulation*

In this section a description of the emission simulation as part of the MIMR-software is briefly described [3, 4]. Emissivity models are implemented to simulate the surface

emissivity ϵ_s from [3]
- ocean surface using Pandey's model,
- snow-covered land using HUT's model,
- vegetation covered land using Kerr's model, and
- different ice types using constant emissivity values.

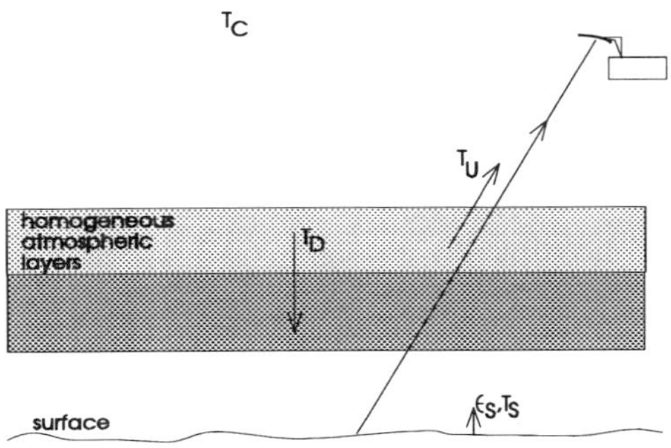

Figure 1. Simulation model of a non scattering atmosphere

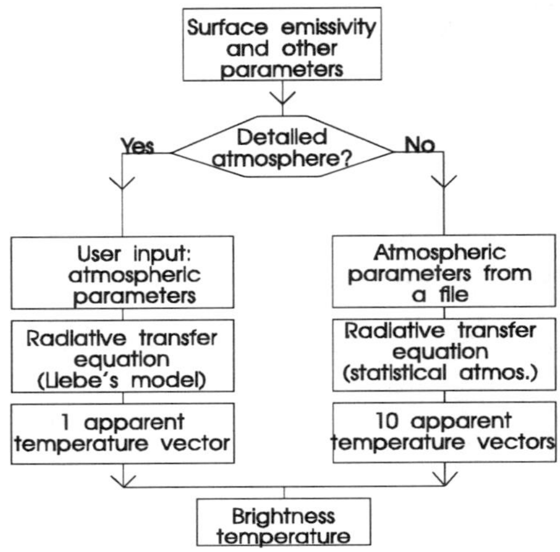

Figure 2. Modelling with statistical or detailed atmosphere

Knowing the emissivity of the surface the apparent temperature observed from space (Fig. 1) is calculated using

$$\begin{aligned} T_B &= T_U(1 - t_{trans}) + \epsilon_S t_{trans} T_S + (T_D(1 - t_{trans}) + \\ &\quad + t_{trans} T_C)(1 - \epsilon_S) t_{trans} \end{aligned} \quad \textbf{(1)}$$

where ϵ_s is the surface emissivity, t_{trans} is the transmissivity of the atmosphere, T_U and T_D are the atmospheric apparent radiative temperatures for up-welling and down-welling radiation. T_C and T_S are respectively cosmic and surface temperatures.

Two options are implemented to consider the influence of the atmosphere (Fig. 2). For a detailed atmosphere specific input parameters are requested. Liebe's MPM model is then used for calculating the upwelling and downwelling contributions of the atmosphere [5].

If no a priori information about the atmosphere is available, the space observed brightness temperature is simulated by using statistical transmissivity values. Using

$$T_{U,D} = \alpha_{U,D} T_s (1 - t_{trans}), \quad \textbf{(2)}$$

the up- and downwelling temperatures T_U, T_D are based on the atmospheric profile factors $\alpha_{U,D}$ and the transmissivity t_{trans}. For a full explanation of this approach see Pullianen et. al. [4]. Equation (2) is an approximation that has shown very good results for climates similar to Norway's.

3.2 Inversion algorithm

Inversion for retrieving geophysical parameters are implemented following the two different approaches shown in Fig. 3:

(A) inversion algorithms as (linear) combinations of MIMR channels or

(B) search for an inverse solution by using a model with input parameters, which represents the actual measurements.

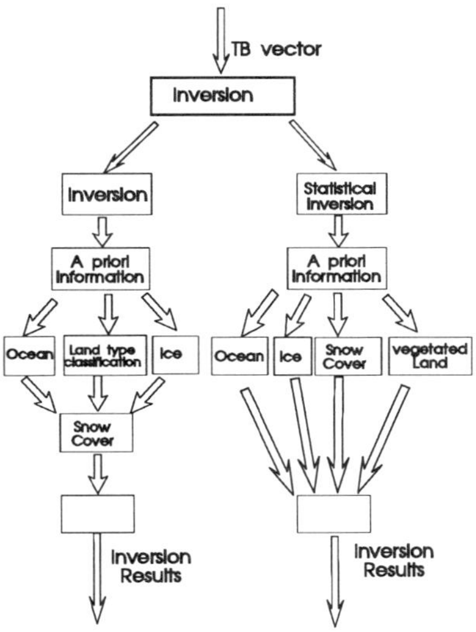

Figure 3. Inversion procedure

Figure 4. Test site Hardangervidda in southern Norway

Procedure (A) is based on inversion algorithms described in the literature [4].

Procedure (B) uses a maximum likelihood (ML) inverse solver for fitting the model with the results of multichannel measurements [4]. In this case the atmospheric contribution to the overall emissivity is reduced to one single integral parameter in order to simplify the application of the ML algorithm.

Analysis of both the direct models and inversion algorithms is performed in the next section for properties of land covered snow.

4 SNOW RETRIEVAL

With a high resolution of less than 5 km for the 90 GHz channel, it is expected that MIMR will open operational applications in snow and ice monitoring. In addition to the parameters listed in Tab. 3 snow parameters to be retrieved are snow depth and snow coverage.

4.1 Test site: Hardangervidda (N)

The test site selected for snow retrieval is the Hardangervidda in southern Norway (Fig. 4). The area is a plateau at an altitude of 1200 - 1400 m with some tops up to 1600 m. The basin of this plateau is controlled by dams (e.g. Koldal, Soldal, Sanga, Gausta in the south and Tyssedal, Hol, Nes and Nore in the north) and used for hydro electric power. Knowledge of the amount of water in the snow in the basin is of vital interest for the

Norwegian state power board (Statkraft). So far this snow information is gathered by in situ measurements and estimations from AVHRR images. Meteorological information is provided by the Norwegian meteorological service (DNMI).

4.2 *Snow emissivity including atmosphere*

Simulation of the emissivity by the snow covered land surface is done following the model described in section 3.1. Snow data measured by Statkraft on 11th April 1988 at the station Dyratjönn in the north of the Hardangervidda are used as inputs for the emissivity model. Meteorological data from local observations at the same time are used to describe the atmosphere. The input parameters as used in this emission simulation are summarized in Tab. 4.

Table 4. Geophysical parameters used for the emission simulation, station Dyratjönn, 11 Apr 1988

Ground emissivity	from [3]
Ground temperature	$-1°C$
Grain size	0.8 mm
Snow density	364 Kg/m^3
Snow depth	147 cm
Snow water equivalent	535 mm
Forest coverage	0%
Atmosphere	using [5]
– pressure	1013 mbar
– temperature	$-6.5°C$
– humidity	64%
– standard atmosphere	southern Finnland from [4, 6]

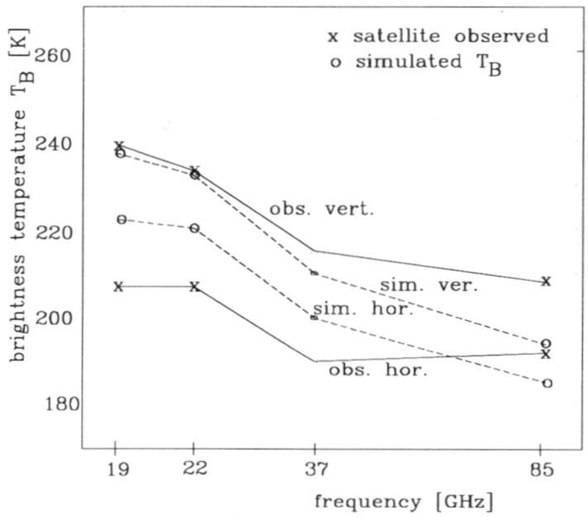

Figure 5. Comparison between space observed ('x') and ground based simulated ('o') brightness temperature T_B. Parameters in Tab. 4 were used

The results of the simulation are shown in Fig. 5 and compared with SSM/I observations. It can be noted that the observed brightness temperature polarization difference for the 19 GHz channel (PD_{19}) is in the order of 30 to 35°K, whereas the simulated temperatures have a PD_{19} of less than 20°K. Second, the gradient for the spectral difference of the 37 and 85 GHz channels ($SD_{37/85}$) has a negative value for the emission simulation which is slightly higher than the observed vertical gradient and opposite to the positive gradient for the observed horizontal T_B.

A more detailed analysis was carried out to quantify the contributions from the snow parameters and those related to the atmosphere.

4.2.1 Statistical atmosphere. The influence of the atmosphere on the measured T_B is simulated by statistical values of the atmospheric transmissivities as described in section 3.1. These transmissivity values, calculated from data for southern Finland, apply relatively well for northern latitudes (50 - 70°) [6, 4].

The emission simulation shows, that atmospheric effects are less then 5°K for the 19 GHz channels. The PD_{19} remains in the order of less then 20°K, leading to the conclusion that either the surface parameter of the snow or the implemented snow emissivity models are

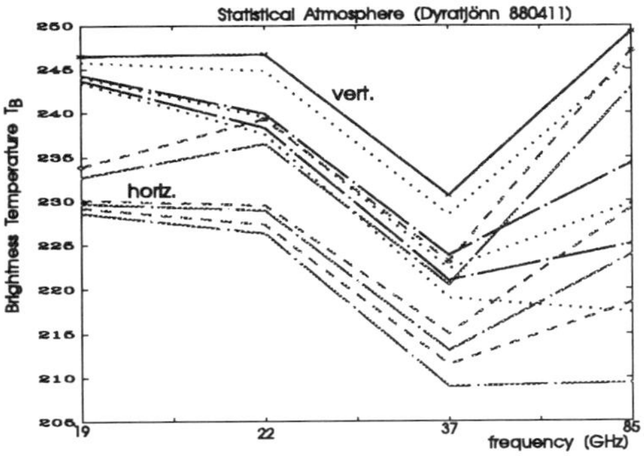

Figure 6. Emission simulation with a statistical atmosphere (typical transmissivities)

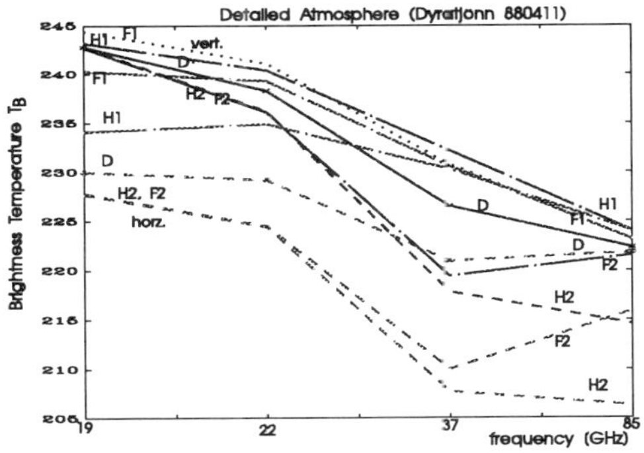

Figure 7. Detailed atmosphere information taken from the stations listed in tab. 5

inadequate. As expected, the impact of the atmosphere on the 85 GHz channel data is higher with up to 40°K and different slopes of the $SD_{37/85}$.

4.2.2 Detailed atmosphere. Meteorological data from February till April 1988 measured by the DNMI station on the Hardangervidda (Tab. 5) are used as input parameters for a detailed atmosphere.

Table 5. Detailed atmosphere information from stations Dyratjönn (D), Haukeliseter (H1, H2), and Finse (F1, F2). Cloud types are stratus (2), stratocumulus (3) and cumulus (7) clouds.

station	D	H1	H2	F1	F2
pressure [mbar]			1013		
max. cloud thickness [km]	1	2	0.2	4	1
cloud type	7	7	2	7	3
rain rate [mm/h]	0	0	0	5	0
air temp.[°C]	-1	2.5	-7.7	0.8	-11
humidity [%]	64	86	64	99	77

The simulated brightness temperature as shown in Fig. 7 leads to the following conclusions.

If no precipitation occurs, the 19 GHz channel is only slightly affected (<5°K). As mentioned for the statistical atmosphere, the simulated PD_{19} is significantly smaller than the observed PD_{19}. For the gradient of the $SD_{37/85}$ no general trend is noticeable. However, both negative gradients as for the observed vertical $SD_{37/85}$ and positive gradients for the observed horizontal $SD_{37/85}$ are represented within the simulation. A more detailed analysis is required to evaluate the influence of the input parameters.

4.2.3 Sensitivity analysis. A first step towards a detailed analysis is a sensitivity study. For the frequencies foreseen for MIMR, this sensitivity analysis was carried out by Pulliainen et al. [3].

The results listed in Tab. 6 underline the outcome of the previously performed simulations. The atmosphere plays a significant (strong) role only for the 90 GHz channel and is

Table 6. Sensitivity analysis of surface and atmosphere parameters for MIMR frequencies, with W = weak (<5K), M = moderate (5K < ... < 25K), S = strong (> 25K) influence

MIMR analysis					
freq. [GHz]	6.8	10.6	18.7	36.5	90.0
ϵ_S (0.88...0.96)	M	M	M	M	W
T_S (-20...0°C)	M	M	M	M	W
T_{snow} (-20...0°C)	W	W	W	M	M
snow W_{equ} (15mm)	W	W	W	W	W
grain size (0.4...1.2mm)	M	M	S	S	M
snow dens. (0.1...0.5g/cm^3)	M	M	S	S	M
surf. rough. (0.1...0.4mm)	W	W	M	M	S
atmosphere (10...90%)	W	W	M	M	S

less dominant (moderate) for the 19 and 37 GHz channels. Main contributions to these lower frequency channels are snow parameters such as grain size and snow density. The simulation points out that inversion algorithms, based on the 19 and 37 GHz channels tend to be very sensitive to these snow parameters.

4.3 Inversion algorithm

Three major snow parameters are required for this type of application: snow extent, snow height, and snow water equivalent. Inversion algorithms based on linear combinations of the observed channels are briefly reviewed and applied to SSM/I observed T_B.

4.3.1 Snow detection algorithms. Amongst all snow parameters dry versus wet snow shows the highest spectral contrast. As a consequence retrieval algorithms have been developed for dry and wet snow surfaces.

In this paper the relationships from Mätzler are used for dry snow coverage and from Pulliainen et. al. for wet snow coverage [3].

For **dry snow**, mainly the spectral and polarization differences for the 19 and 37 GHz channels are used. The footprint is assumed to be completely covered by snow, if at least one of the following relations is true

$$T37V - T19V < -5\text{K}, \quad (3)$$
$$(T37V - T19V) * (T19V - T19H) < -60\text{K}^2, \quad (4)$$
$$(T37V - T19V) * (T37V - T37H) < -50\text{K}^2, \quad (5)$$
$$(T37V - T19V) * (T90V - T90H) < -30\text{K}^2. \quad (6)$$

Even though the algorithms presented are developed for MIMR frequencies, validation is performed with SSM/I data. Due to missing 85 GHz data in 1988 eq. 6 is left out of the inversion performed here.

Mapping of **wet snow** is more difficult, as slightly wet soils, wet snow and open water have similar emissivities. The relations proposed by Pulliainen et. al. [3]

$$T37V > 263\text{K}, \quad (7)$$
$$|T37V - T19V| < 3\text{K}, \quad (8)$$

$$12K < T19V - T19H \;<\; 33\text{K}, \tag{9}$$
$$|T19V - T19H - 1.4 * (T37V - T37H)| \;<\; 3\text{K}, \tag{10}$$

have all to be true in order to exclude false detections.

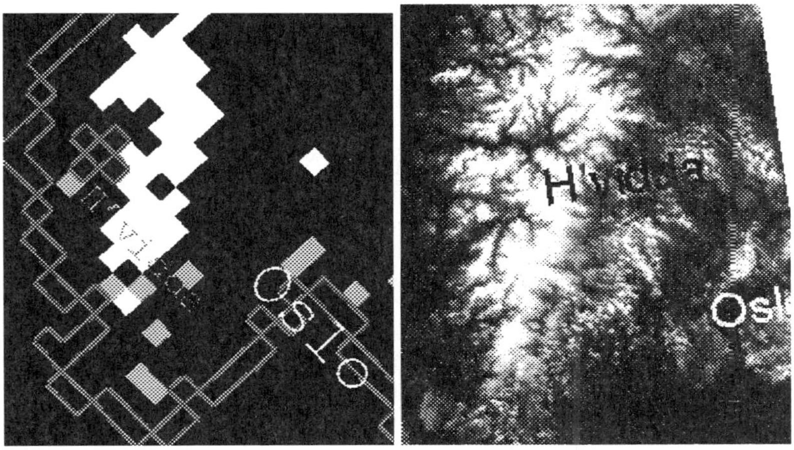

Figure 8. Comparison of SSM/I retrieved snow coverage (left) and AVHRR image (right) from 13th June 1990. On the SSM/I image, dry snow is indicated by the white pixels, wet snow by gray pixels, the coast by the double lines.

The algorithms are applied for multi-temporal SSM/I data of the Hardangervidda (Fig. 8). The test site can be identified by the outlined cross, which indicates the Sognefjorden (see Fig. 4). Both AVHRR image and SSM/I snow cover image are oriented north to be comparable to Fig. 4. The images show a quite similar snow coverage, indicating the possibility of retrieval from space of snow extent. It has to be mentioned however that the spatial resolution is rather unsatisfactory. From the multi-temporal data analysis (see Fig. 9) it can be concluded that the lower bound of eq. 9 and the higher bound of eq. 10 depend on the time of the year and general meteorological conditions and from site to site. The same threshold cannot be applied throughout the whole winter season, as false identification of either wet soil or open water will occur.

4.3.2 Snow height and snow water content. For snow height s_d and snow water content W_{equ} two approaches, the Chang/Künzi algorithm and the Aschbacher algorithm, are applied to SSM/I data of the test site [7, 3]:

Chang/Künzi

$$s_d = 1.59(T19H - T37V)\ [cm] \tag{11}$$
$$W_{equ} = -2.6\frac{\left(\frac{T37H-T19H}{37-19} - 0.085\right)}{0.036}[mm] \tag{12}$$

and the **S**pectral and **P**olarization **D**ifference algorithm proposed by Aschbacher [3]

$$SPD = (T19V - T37V) + (T19V - T19H)$$
$$s_d = 0.68 \cdot SPD - 0.67\ [cm] \tag{13}$$
$$W_{equ} = 2.20 \cdot SPD - 7.11\ [mm]. \tag{14}$$

The comparison between both algorithms leads to approximately 1.6 times higher values for the Künzi W_{equ} algorithm compared to the SPD algorithm. Snow height is overestimated by both algorithms by about a factor of two.

Table 7. Comparison between retrieved snow water content from SSM/I data using eq. 14 and ground truth observed by Statkraft

Station	Snow type	Date	W_{equ} simu.	W_{equ} meas.
Larsbudalen	DRY	2 Mar 88	502 mm	293 mm
Juklevatn	DRY	6 Apr 88	550 mm	918 mm
Stigstudalen	DRY	7 Apr 88	538 mm	435 mm
Dyratjönn	DRY	11 Apr 88	477 mm	535 mm

The comparison between ground observed and spaceborne radiometer retrieved snow properties shows that the results are in the same order of magnitude. One reason for the differences is what is being compared. It is obvious that a comparison between a 25×25 km pixel (SSM/I) with a single ground data point is not appropriate. Second, dry snow tends to be blown away from exposed areas and accumulates in protected zones. The advantage of passive microwave observations lies in the multi-temporal capabilities, which are explained in the following section.

4.3.3 Multi-temporal snow monitoring. The advantage of satellite remote sensing with passive microwaves is the weather independent day and night capability of monitoring. In order to demonstrate this capability a multi-temporal analysis for snow coverage (winter 87/88) and snow height (winter 88/89) is performed.

Figure 9 shows a snow extent chart similar to Fig. 8. No rotation into the north direction is performed. For that winter almost no snow is present in October, then snow coverage increases up to mid January. Melting occurs in the beginning of February, followed by an almost complete snow cover in March.

Snow height detection is performed for the test site Dyratjönn. The winter 88/89 is chosen because of a dramatic run-off (melting) at the beginning of March. In addition to the snow height, Fig. 10 contains two more pieces of information. The solid line in the bottom indicates the existence of snow ('—'). The crosses ('+') above this line indicate a lack of data. 'No snow' detection is quite often caused by missing data, except for the periods in the beginning of March and mid April. For these both periods a run-off was reported.

The quantitative retrieval of snow height seams to be insufficient with the existing retrieval algorithms. It can be noted however that changes in snow conditions can be identified. These changes are indicated by rapid variations and represent dates of melting or refreezing (Jan/Feb and April/May).

5 RECOMMENDATIONS

It is shown in this paper that passive space-borne microwave remote sensing of snow properties is possible. Even though the order of magnitude is correctly retrieved for snow

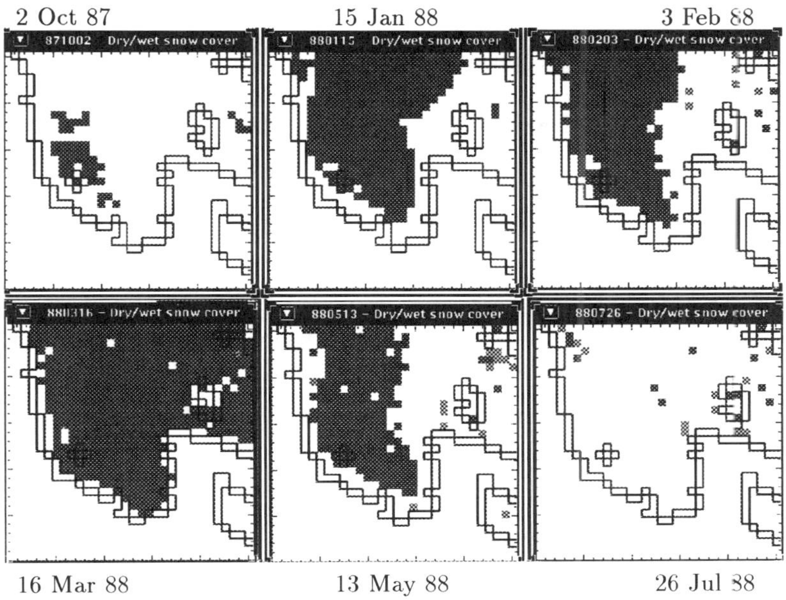

Figure 9. SSM/I retrieved snow cover in winter 1987/88 in southern Norway

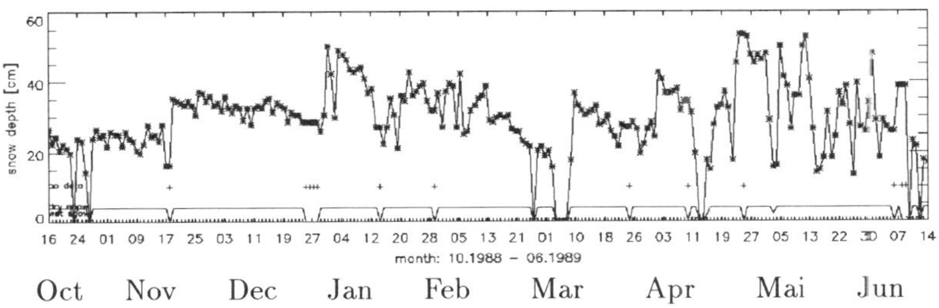

Figure 10. Snow height in Dyratjönn in Winter 1988/89 as obtained by the SPD approach

height and snow water content, the differences to ground observed data are in the order of a factor 2-3. The difficulty in the validation of the retrieved snow properties are mainly due to the comparison of 25×25 km SSM/I data with ground truth at selected points. The major limitation here is the footprint size, which for land applications should be in the order of 10-15 km.

The superiority of passive microwaves is the all weather, day and night remote data retrieval. This advantage has to be further evaluated in models describing the change of snow properties rather than performing singular event calculations.

The simulations have shown also, that atmospheric parameters affect significantly the measured emissivity, thus combined models for atmosphere and surface parameter retrieval are required.

Acknowledgments

The authors would like to thank Frank Wentz at the University of Colorado (World Data Center A), the National Snow and Ice Data Center in Boulder and Gerhard Triebnig at ESA–ESRIN for providing us with SSM/I data. Special thanks applies to the scientists who were involved in ESA contracts during the last years and developed various direct and inverse algorithms. The authors wish also to thank Yvon Menard at ESA–ESTEC for comments on future radiometer missions.

References

1. Hollinger, J.P., Peirce, J.L., and Poe, G.A., 1990, SSM/I Instrument Evaluation, *IEEE Trans. Geo. Rem. Sens.*, **28**, 781–790

2. ESA, 1990, The Multi-frequency Imaging Microwave Radiometer, *ESA SP-1138*

3. Pulliainen, J., Hallikainen, M., Somersalo, E., Kärnä, J.-P., Jääskeläinen, V., Hyyppä, J., Talvela, J., Luntama, J.-P., and Manninen, T., 1990, Study of Microwave Interaction with the Earth's Surface, *final report ESA ESTEC contract 8447/89/NL/PB*

4. Pulliainen, J., Kärnä, J.-P., and Hallikainen, M., 1993, Development of Geophysical Algorithms for MIMR, *IEEE Trans. Geo. Rem. Sens.*, **31**, 268-277

5. Liebe, H., 1989, MPM - an atmospheric millimeter-wave propagation model, *Int. J. Infrared and Millimeter Waves*, **10**, 631-650

6. Salonen, E., Karhu, S., Jokela, P, Zhang, W., Uppala, S., Aulamo, H., Sarkkula, S., and Poiares Baptista J.P.V., 1992, Modelling and Calculation of Atmospheric Attenuation for Low-Fade Margin Satellite Communications, *ESA Journal*, **16**, 3, 299-317

7. Künzi, K., Patil, S., and Rott, H., 1982, Snow-cover parameters retrieval from Nimbus-7 Scanning Multichannel Microwave Radiometer (SMMR) data, *IEEE Trans. Geo. Rem. Sens.*, **20**, 452-467

Retrieval of snow and sea ice parameters from SSM/I data

JOCHEN GRANDELL, LAURI KURVONEN and MARTTI HALLIKAINEN
Helsinki University of Technology, Laboratory of Space Technology
Otakaari 5A, 02150 Espoo, Finland
tel: +358-0-451 2170
fax:+358-0-460 224
email: Jochen.Grandell@hut.fi

Abtract: Special Sensor Microwave Imager (SSM/I) data was used to verify existing inversion algorithms for retrieval of snow and sea ice parameters, and to develop algorithms using higher frequency channels (85 GHz). The test sites were located in Finland (snow studies) and in the Baltic Sea (sea ice studies). For snow studies, the linear correlation values between satellite-derived and ground-measured snow water equivalent values using the SPD-algorithm (Spectral and Polarization Difference) exceeded 0.80. A higher frequency algorithm was developed and tested, and proved to be useful with equal or better correlation for snow water equivalent values < 50 mm. For sea ice studies, the developed 85 GHz algorithm gave slightly better correlation values than the NASA-TEAM algorithm, used as a reference. All results are preliminary, and should be verified with more higher frequency data.

Key words: Inversion, snow water equivalent, sea ice concentration, SSM/I

INTRODUCTION

The aim of the study is to provide tools for the analysis of multi-frequency multi-polarization spaceborne radiometer data in the retrieval of snow and sea ice characteristics. The emphasis was laid on studying the possible advantages of using higher frequency channels (up to 90 GHz). The most advanced spaceborne radiometers currently in operation (Special Sensor Microwave Imager SSM/I) or under consideration (Multi-Frequency Imaging Microwave Radiometer MIMR), operate in a frequency range up to 90 GHz. Table 1 shows the main characteristics of the SSM/I and the MIMR instuments [1], [2].

Table 1. SSM/I and MIMR main characteristics.

SSM/I			MIMR		
Frequency (GHz)	**Pixel size (km x km)**	**Polarization (Vert./Hor.)**	**Frequency (GHz)**	**Pixel size (km x km)**	**Polarization (Vert./Hor.)**
19.35	69 x 43	V , H	6.8	60 x 60	V , H
22.235	60 x 40	V	10.65	38 x 38	V , H
37	37 x 29	V , H	18.7	22 x 22	V , H
85.5	15 x 13	V , H	23.8	20 x 20	V , H
			36.5	11.6 x 11.6	V , H
			89	4.9 x 4.9	V , H

The study was considered to have two major parts, first data acquisition and second the implementation and verification of algorithms for the extended MIMR frequency range. For sea ice studies, the feasibility of SSM/I data for sea ice concentration inversion is examined. Snow cover and snow water equivalent with respect to the underlying vegetation are the main subjects of land studies.

DATA ACQUISITION

Satellite data

Satellite data from the DMSP satellite SSM/I radiometer was acquired from the National Snow and Ice Data Center, Colorado, USA for a time period from middle-September 1990 to the end of May 1991.

Test sites and ground truth information

For the retrieval studies, a set of test sites had to be carefully chosen. For sea ice studies, practically all of the Baltic sea north of Aland was chosen as a test site. This area was then divided into individual test sites of 1.0^0 lon x 0.5^0 lat (55 km x 55 km), leaving out coastal areas and areas including islands. Each of these main test sites were later divided into four subareas of 28 km x 28 km as it corresponds approximately to the 37 GHz footprint. Figure 1 shows the Baltic sea with the chosen test sites.

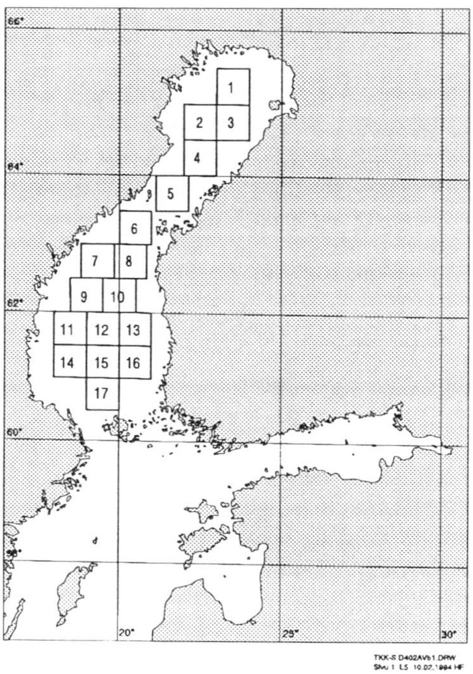

Figure 1. Chosen test sites for sea ice studies in the Baltic sea.

For snow studies, the test sites were chosen from all parts of Finland, with an emphasis put on areas used during previous studies with SMMR data [3]. The sizes of the sites were chosen for practical reasons to be the same for both sea ice and snow studies (55 km x 55 km and subareas of 28 km x 28 km). Figure 2 shows a map of Finland with the chosen snow test sites.

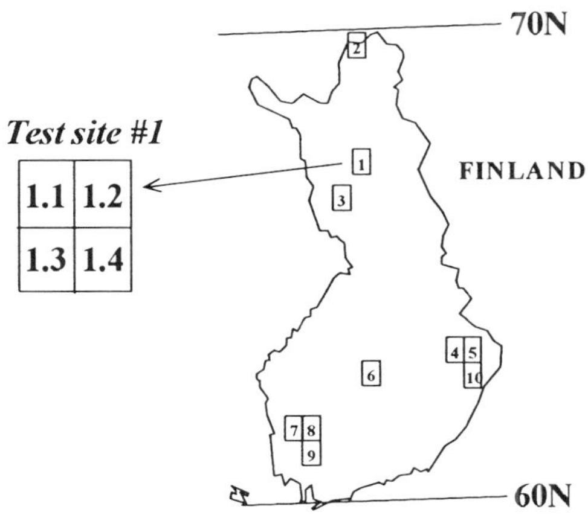

Figure 2. Chosen test sites for snow studies.

Ground truth information were provided by several institutions, including Finnish Institute of Marine Research for ice maps containing concentration information, Finnish Meteorological Institute for weather information (temperature, precipitation) and Hydrological Office for snow maps with snow water equivalent information. All data was in paper format, and had to be digitized in order to make computerized analysis possible. After digitizing, the maps were transformed to geodetic coordinates and rasterized with a 0.25^0 x 0.25^0 raster size. These rasterized pictures were then converted to ASCII-format, as later analysis was not to be performed as an image analysis.

RETRIEVAL OF SNOW PARAMETERS FROM SSM/I DATA

Review of existing inversion algorithms

A study of existing inversion algorithms for snow retrieval has been done previously for example by [4], and was used as a main reference. The algorithms have been collected and widely examined during this before mentioned study, although not tested with real space-borne data. Pulliainen et al. suggest that the SPD-approach is the most promising for snow studies, and these algorithms were selected as the basis of this snow study. Another previous study with an emphasis on the influence of vegetation for inversion of snow cover is presented in [3], but vegetation maps were not used during this study. Thus, the main subject of this chapter is the evaluation of the SPD-algorithm, which is still not widely tested and verified.

The SPD (Spectral and Polarization Difference) algorithms are based on a combination of SMMR channels:

$$SPD = [T_B(18V) - T_B(37V)] + [T_B(18V) - T_B(18H)] \quad \textbf{(1)}$$

The following equations for snow depth (h_s) and snow water equivalent (W_{eq}) are obtained from the regression analysis based on a study by Aschbacher [5]:

$$h_s = 0.68 * SPD - 0.67[cm] \quad \text{for all data} \quad \textbf{(2)}$$
$$h_s = 0.72 * SPD - 1.24[cm] \quad \text{if } T_{\max} < 0 \quad \textbf{(3)}$$
$$W_{eq} = 2.20 * SPD - 7.11[mm] \quad \text{for all data} \quad \textbf{(4)}$$
$$W_{eq} = 2.02 * SPD - 7.42[mm] \quad \text{if } T_{\max} < 0 \quad \textbf{(5)}$$

In Equations (2) to (5), $T_{\max}$ is the maximum daily air temperature. The snow water equivalent calculations were first made for each test site, although the most southern areas (areas 7,8,9 in Figure 2) had very little snow during the whole interval of study because of the relatively mild winter 1990-1991. These areas had to be excluded from the final results. Obtained linear correlation between SPD-derived snow water equivalent values and ground truth are presented in Table 2. An example of the calculated snow water equivalent with respect to ground truth from the digitized snow maps is presented in Figure 3. It should be kept in mind that the results in Table 2 and Figure 3 are obtained by Equation (1), Equation (2) to (5) employ SMMR frequencies , 18 and 37.6 GHz, which differ somewhat from the SSM/I frequencies. This may cause slight differences.

Table 2. Linear correlations between SPD-derived snow water equivalents and ground truth data.

Test site number	Linear correlation coefficient	W_{eq}-values measured (mm)	Number of data points
1.1	0.862	10 - 125	130
1.2	0.878	10 - 120	139
1.3	0.867	7 - 125	137
1.4	0.856	5 - 120	143
3.1	0.856	10 - 130	134
3.2	0.863	10 - 130	143
3.4	0.816	10 - 130	137
4.1	0.804	5 - 140	134
4.2	0.791	5 - 140	141
4.3	0.727	5 - 120	147
4.4	0.726	7 - 140	146
6.1	0.715	5 - 80	141
6.2	0.635	5 - 100	132
6.3	0.656	5 - 80	130
6.4	0.609	5 - 80	135

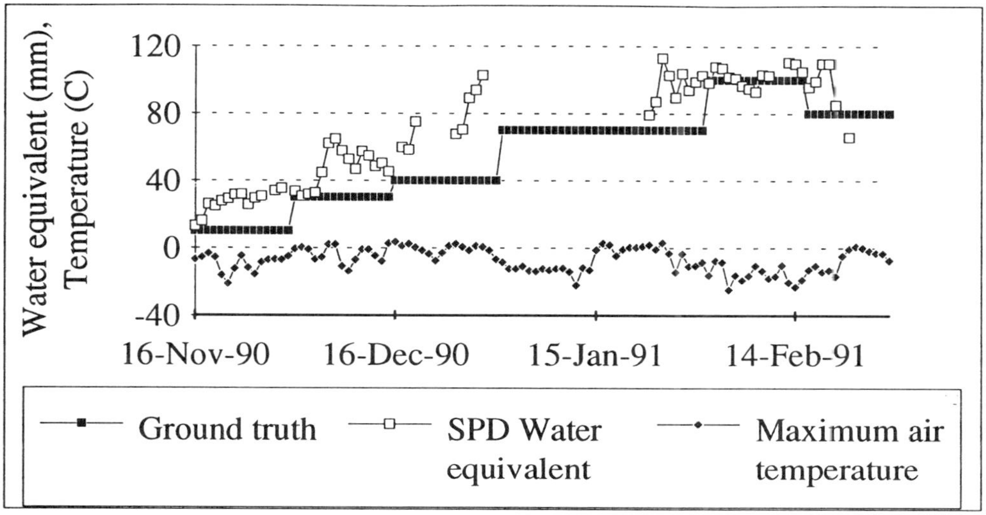

Figure 3. SPD-derived snow water equivalent vs. corresponding ground truth data and maximum air temperature for test site 1.2 (Figure 2).

The results show that the linear correlations for the northern test areas reach values from 0.80 to 0.88. This agrees with the results of Aschbacher. For southern areas the correlation values drop to values from 0.60 to 0.70. One reason for this can be derived from the fact that the possibility for a large relative error in the used ground truth data is higher when the measured snow water equivalents are small.

New inversion algorithms developed during the study

New possible snow inversion algorithms using higher frequency channels were searched using the methods described by Hallikainen and Jolma, [3]. This includes plotting several possible ΔT_B -timeseries with a reference to ground truth in order to visually observe the correlation. A reference function of $\Delta T_B(W_{eq} = 0)$ [3] was not used, as it wasn't considered necessary for this study. In the study of Hallikainen and Jolma [3], both the vertical and horizontal channels of 37 GHz were found to provide information concerning the snow water equivalent. Thus, it was expected that the ΔT_B function including a 37 GHz channel would give the most satisfactory results. An 85 GHz channel was included in order to enhance the spatial resolution, despite the atmospheric effects. These were not compensated by the algorithm, as measurements with a high probability of atmosphere related errors were removed from further processing. The method is described in more detail when sea ice inversion is discussed. Figure 4 shows the examined ΔT_B functions and the corresponding ground truth. Function number 4 is the previously discussed SPD-algorithm for snow water equivalent.

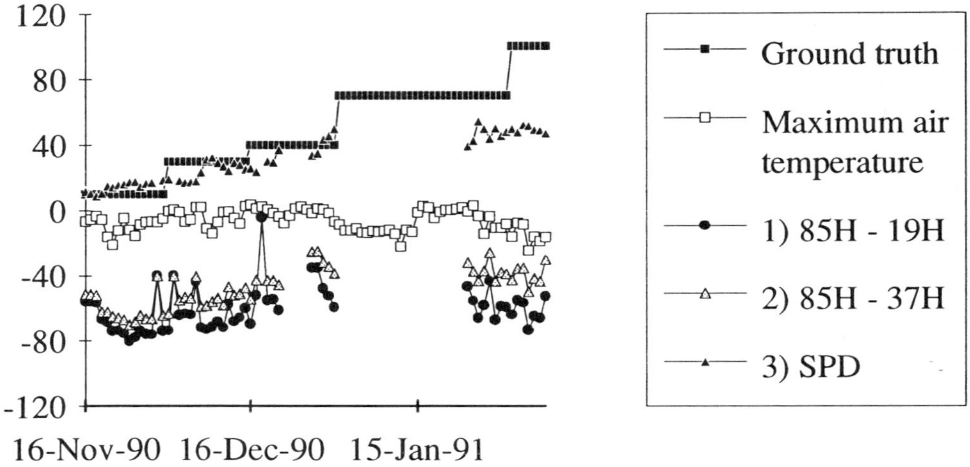

Figure 4. Examined ΔT_B functions and ground truth test site 1.2. Functions 1 and 2 (numbering refers to the legend) are shifted down in order to make the Figure more readable.

The coefficients of the most promising function $T_B(85H) - T_B(37H)$ (function 2 in Figure 4) were determined by a simple regression analysis. As a matter of fact, function 1 could have been used as well, but both functions were not selected for simplicity. The final algorithm is an equation of the determined coefficients and the ΔT_B function. The algorithm for northern areas can be written as

$$W_{eq} = 1.1 * [T_B(85H) - T_B(37H)] + 35. \quad \text{[mm]} \tag{6}$$

During a previous study [3], the difference between the measured water equivalent and the ground truth has been found to be the best measure of feasibility of an examined algorithm. This method was used during this study as well. The results of this examination using the previously determined coefficients are in Table 3.

The results suggest that higher frequency channels are only suitable for retrieval of snow water equivalents less then 40-50 mm which is, of course, expected when the weaker penetration of the 85 GHz channels is taken into account. This is demonstrated in Table 3, which include the average values for two cases, (a) all data for which $T_{max} < 0$, and (b) all data for which $T_{max} < 0$ and $W_{eq} \leq 50$ mm. It is obvious from the results in Table 3 that different coefficients are needed for southern and northern areas, as the difference in the measured snow water equivalent and ground truth grows for southern test sites (test sites 4 and 6 in Figure 2).

Table 3. Average differences (mm) between satellite-derived and ground based water equivalent values for northern test sites using Equation (6).

Test Site	Vegetation	Average W_{eq} difference, $T_{max} < 0$	Average W_{eq} difference, $T_{max} < 0$ and $W_{eq} < 50$ mm	Number of data points $T_{max} < 0$	Number of data points $T_{max} < 0$ and $W_{eq} < 50$ mm
1.1	mostly tundra and clear cut areas, sparse forest	-27.5	-13.9	434	326
1.2	- " -	-20.6	-9.3	449	335
1.3	- " -	-20.8	-9.3	451	339
1.4	- " -	-18.1	-4.9	468	351
3.1	- " -	-20.9	-6.5	456	342
3.2	- " -	-21.7	-6.1	452	339
3.3	- " -	-19.1	0.6	452	341
3.4	- " -	-17.1	-2.5	451	338
4.1	pine-dominant forest, farmland and bogs	-17.4	3.1	445	334
4.2	- " -	-17.7	4.5	430	322
4.3	- " -	-9.7	7.9	462	347
4.4	- " -	-16.9	7.1	457	340
6.1	mostly pine-dominant forest and farmland	2.4	12.9	463	410
6.2	- " -	0.8	14.6	444	392
6.3	- " -	No value	No value	-	-
6.4	- " -	5.9	16.5	441	390

RETRIEVAL OF SEA ICE PARAMETERS FROM SSM/I DATA

Review of existing inversion algorithms

For this study, a suitable inversion algorithm developed mainly for the SSM/I instrument had to be found. An algorithm which is later referred to as the NASATEAM algorithm [6] met these demands and was implemented in the software and included as a reference algorithm for any new algorithms developed during the study. No suitable algorithm using the higher frequency channel was found, although an interesting possibility is provided by Svendsen et al., [7].

The NASATEAM algorithm uses three SSM/I channels (19H- and 19V-polarization, 37V -polarization) from which radiances are gridded into 50 x 50 km SSM/I grid cells. This cell size corresponds approximately to the footprint of the SSM/I 19 GHz channel [1]. Two independent variables, the polarization ratio PR and gradient GR ratio are used in the algorithm. These variables are defined by:

$$PR = \frac{T_B(19V)-T_B(19H)}{T_B(19V)+T_B(19H)} \tag{7}$$

$$GR = \frac{T_B(37V) - T_B(19V)}{T_B(37V) + T_B(19V)}. \tag{8}$$

Using these two variables and coefficients derived from algorithm tie points (tie points are chosen from the data for each measured ice type and open water) we get the equations for first year C_F and multiyear C_M concentrations:

$$C_F = \frac{(a_0 + a_1 PR + a_2 GR + a_3 PR \bullet GR)}{D} \tag{9}$$

$$C_M = \frac{(b_0 + b_1 PR + b_2 GR + b_3 PR \bullet GR)}{D} \tag{10}$$

where

$$D = c_0 + c_1 PR + c_2 GR + c_3 PR \bullet GR. \tag{11}$$

The total ice concentration is the sum of the first year and multiyear ice concentrations:

$$C_T = C_F + C_M. \tag{12}$$

The coefficients in Equations (9)-(11) are functions of nine brightness temperatures, and are set individually for the desired measurements area. The coefficients recently presented in literature are

meant to be used in either arctic or antarctic regions, and therefore new coefficients had to be derived for the Baltic sea. No multiyear ice can be found in the Baltic sea, and corresponding tie points were kept as they are chosen for arctic areas [6]. The tie points chosen are presented in Table 4. When the effect of using Baltic tie points instead of arctic values was observed, we found that arctic tie points gave 10-20% too small concentration values with respect to ground truth. Thus, when further analysis is made, the process of choosing tie points should be performed with care.

Table 4. NASATEAM algorithm Baltic tie points.

SSM/I Channel	Open Water	First-Year Ice	Multiyear Ice
19.35 (H)	114.37	225.04	203.91
19.35 (V)	184.22	251.76	223.19
37.0 (V)	207.63	241.65	186.31

The NASATEAM algorithm includes a weather filter in order to eliminate unreliable ice concentrations due to weather related effects (atmospheric moisture). These measurements are flagged as missing for future use. This weather filter also justifies the disregarding of atmospheric effects at 85 GHz, since measurements for all SSM/I channels are removed from further analysis. The same approach was used for all measurements (both sea ice and snow studies). The filter excludes all data from measurement days for which:

$$GR \geq 0.05 . \tag{13}$$

The filter value 0.05 is open to changes [6], but the effect of this was not considered during our study.

If the calculated concentration values are less than -20% or greater than 120% the calculation is assumed to be incorrect and these measurements are flagged as missing as well. If the computed ice concentration is between 0 and 20% the pixel is set to 0%, and if the computed ice concentration is between 100 and 120%, the pixel is set to 100%. These corrections were also kept as guidelines for calculating ice concentrations with any new inversion algorithms developed during this study.

Ice concentrations were calculated separately for two time periods, the first period covering the total ice season i.e. beginning of November - end of May, and the second period covering the time of operation for the 85H channel of the SSM/I instrument. This period is significantly shorther, since the 85H channel malfunctioned shortly after the beginning of February. Concentrations were calculated for this shorter period, because any new inversion algorithms developed during this study were to take advantage of the 85H channel and any calculations with these algorithms could not be reached further than February. Figure 5 shows an example of the calculated NASATEAM ice concentration for the test site 1.2 (Figure 1) with corresponding ground truth data from the digitized ice maps.

Linear correlation coefficients between the calculated ice concentrations and the gathered ground truth information for both time periods were calculated simultaneously, and the results are presented in Table 5. From the results it can be seen that the NASATEAM algorithm using Baltic tie points gives linear correlation values over 0.80 for the best measurement areas. This can be considered good, although the short time period must be taken into account. The use of the shorth time period gives significantly lower correlation values, due to the overestimated concentration values when open water is present, which is of course underlined for early 1991 measurements. When observed visually, it can be seen that there is a lack of lower concentration measurements. Therefore any conclusions of the feasibility of the algorithm using these linear correlation values are mostly based on either 0% or 90-100% situations.

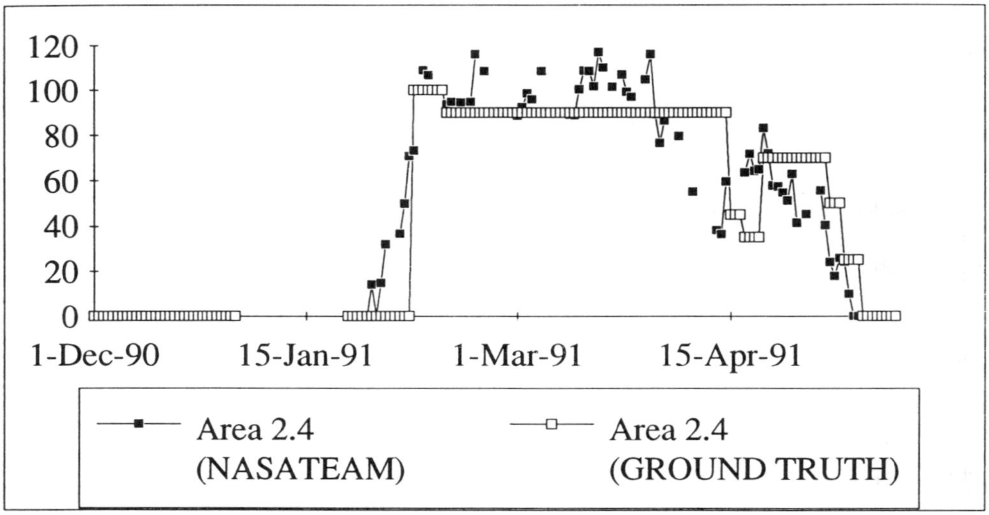

Figure 5. An example of the calculated NASATEAM sea ice concentration with respect to ground truth information from the digitized ice maps. The example is for test site 2.4.

Table 5. Linear correlation coefficients between NASATEAM ice concentrations and existing ground truth information for long and short time periods (see text). The short time period reaches from 16 November 1990 to 8 February 1991 and the long period from 16 November 1990 to 30 May 1991.

Test site number	Linear correlation (long period)	Number of data points (long period)	Linear correlation (short period)	Number of data points (short period)
1.2	0.875	302	0.871	115
1.3	0.836	317	0.611	111
1.4	0.892	307	0.853	111
2.1	0.872	308	0.463	121
2.2	0.881	302	0.436	113
2.3	0.865	314	0.524	124
2.4	0.921	287	0.576	105
3.1	0.898	293	0.806	102
3.2	0.859	299	0.702	107
3.3	0.929	303	0.839	105
3.4	0.901	307	0.871	110
4.1	0.941	307	0.871	112
4.2	0.884	322	0.536	120
4.3	0.925	300	0.462	111
4.4	0.777	297	0.651	112

New inversion algoritms developed during the study

The aim of this study was to provide tools for sea ice inversion using higher frequency channels. As no algorithms suitable for this study were found, new algorithms had to be developed and tested. Problems arose with the failing of the 85H-channel, since the 85V-channel had malfunctioned already earlier. Thus, the available 85 GHz data was truncated to one channel (85H) and a short period, reaching only to February. This lack of satellite data did not seriously harm the snow studies described in the previous chapter, since ground truth data was not available for March-April 1991. More higher frequency data is needed in order to verify the results obtained during this study.

While developing new algorithms, an emphasis was put on studying the difference of brightness temperatures of water and ice. The idea is known from snow inversion methods, and from the study mentioned earlier [7]. This was also nearly the only possible approach with the lack of valid, especially multi-polarized, higher frequency data. However, it proved to be useful, although simple as it utilizes data from only one channel (85H). Of course, using only one channel which has a weak penetration. makes it impossible to get any information on the structure of the measured ice. Thus, multiyear and first year ice can not be distinguished. The algorithm is mostly useful for determining the border of the ice floe and the surrounding water area, as it compares the radiances of sea ice and water.

In this approach, in addition to the sea ice test site of interest, an area containing only water is needed. This area should be as near the test site as possible, since this minimizes the chances of differences in the atmospheric conditions. For a relatively small area like the northern Baltic Sea, these differences can be neglected if the reference area is located reasonably close to the test site (< 400 - 500km). The basic formula for the sea ice concentration using this T_B difference can be expressed as:

$$C_T = a_0(T_{B,ice} - T_{B,water}) - b_0, \tag{14}$$

where

C_T	= total sea ice concentration,
a_0, b_0	= coefficients which are determined empirically,
$T_{B,ice}$	= brightness temperature values from the measurement area,
$T_{B,water}$	= brightness temperature values from the reference area.

The same correction procedures are used for this algorithm as for the NASATEAM algorithm, i.e values 100-120% are marked as 100%, and values >120% are flagged as missing. The algorithm may give values a great deal less than zero (because of the coefficients) when water is measured, and these values are all marked as 0% concentration.

The shorter time period explained previously when snow inversion was discussed, was used for inversion calculations with this experimental algorithm. Linear correlation to ground truth data was calculated using a stepwize sweeping procedure, setting the coefficients with the correlation values as a reference. Because of the correction procedure included in the algorithm, this sweeping method cannot be used as the only measure for setting the coefficients, as it will give wrong values when the coefficients are big enough. Also, it must be remembered that without the correction procedures, the b_0 term would not have any effect on the correlation coefficient, according to definition of correlation. An example of these inversion calculations are shown in Figure 6, with NASATEAM inversion results and the ice map ground truth information as a reference. An example of the results from the linear correlation analysis is shown in Table 6.

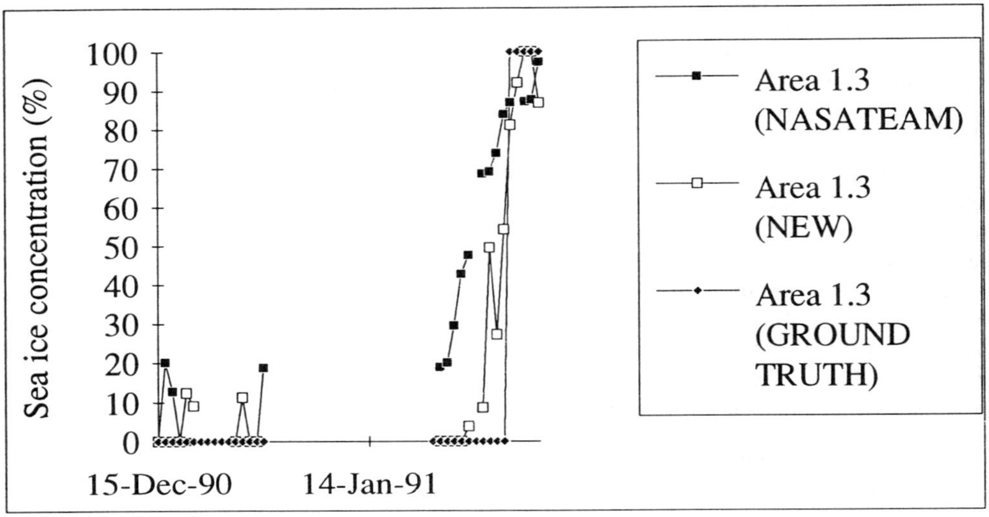

Figure 6. An example of the inversion results with Equation (14), with NASATEAM results and ground truth as a reference.

Table 6. Results of the linear correlation calculation of the developed sea ice inversion algorithm for a period of 16 November 1990 to 8 February 1991. Coefficients are $a_0 = 2.0$ and $b_0 = 20$.

Test site number	Linear correlation coefficient	Number of data points
1.2	0.834	456
1.3	0.764	450
1.4	0.902	452
2.1	0.518	476
2.2	0.469	477
2.3	0.605	478
2.4	0.596	463
3.1	0.818	463
3.4	0.955	448
4.1	0.797	465
4.2	0.537	473
4.3	0.597	458
4.4	0.798	463

The results suggest that higher frequency channels can be succesfully used for retrieval of the total sea ice concentrations. The obtained correlation values should, however, be considered as approximate only, since the 85 GHz data could be used only for a very limited time period. It must be noted that the number of data points is higher for the 85 GHz channel than for lower frequencies. This is due to the SSM/I scan geometry [8].

CONCLUSIONS

For ice studies, the tested NASATEAM algorithm gave acceptable results. For some test sites linear correlations exceeded 0.80. A new algorithm using only one higher frequency channel was also tested, with a reference to the multichannel NASATEAM algorithm. The results suggest that higher correlation values can be reached, but no final conclusions should be made as more higher frequency data is needed to verify the results.
For snow studies, the SDP-approach was used as a reference. These algorithms gave linear correlations up to 0.80-0.90 during the study. Algorithms using data from one higher frequency channel were tested, and no significant change in the correlation values could be found. These new algorithms, however, seem to saturate for relatively low snow water equivalent values (40-50 mm), which was expected. More data is needed to verify these results as well.

REFERENCES

1. Hollinger J, Peirce J, Poe G: SSM/I Instrument Evaluation, *IEEE Trans. Geoscience and Remote Sensing*, Vol. 28, No. 5, pp. 781-790, 1990.
2. Menard Y, Reynold M: The design of the ESA Multiband Imaging Microwave Radiometer MIMR, IEEE 1991 Int. Geosci. Remote Sensing Symposium (IGARSS'91), pp. 2359-2363, 1991.
3. Hallikainen M, Jolma P: Comparison of algorithms for retrieval of snow water equivalent from Nimbus-7 SMMR data in Finland, *IEEE Trans. Geoscience and Remote Sensing*, Vol. 30, No.1, pp. 124-131, 1992.
4. Pulliainen J, Hallikainen M, Somersalo E, Kärnä J-P, Jääskeläinen V, Hyyppä J, Talvela J, Luntama J-P: Study of microwave interaction with the earth's surface, ESTEC contract 8447/89/NL/PB(SC), Helsinki,1990.
5. Aschbacher J: Land surface studies and atmospheric effects by satellite microwave radiometry, PhD dissertation, University of Innsbruck, 1989.
6. Cavalieri D, Crawford J, Drinkwater M, Eppler D, Farmer L, Jentz R, Wackerman C: Aircraft active and passive microwave validation of the sea ice concentration from the Defence Meteorological Satellite Program Special Sensor Microwave Imager, *Journal of Geophysical res.*, Vol 96, No. C12, pp. 21989-22008, 1991.
7. Svendsen E, Mätzler C, Grenfell T: A model for retrieving total sea ice concentration from a spaceborne dual-polarized passive microwave instrument operating near 90 GHz, *International J. of Remote Sensing*, Vol 8, No 10, pp. 1479-1487, 1987.
8. Wentz F: User's Manual: SSM/I Antenna Temperature Tapes, Remote Sensing Systems, Inc., Santa Rosa, CA. RSS Technival Report 032588, 1988.

Airborne microwave radiometer measurements of Baltic sea ice

L.KURVONEN and M.HALLIKAINEN

Helsinki University of Technology, Laboratory of Space Technology
Otakaari 5 A, 02150 Espoo, Finland
tel: 385-0-451 2378
fax: 385-0-460 224
email: kurvonen@ava.hut.fi

Abstract: helicopter-borne multifrequency radiometer (24, 34, 48 and 94 GHz, vertical polarization) was used to investigate the behaviour of the brightness temperature of different sea ice types in the Gulf of Bothnia. The measurements were made in dry and wet conditions (air temperature above and below 0 °C). The angle of incidence was 45° in all measurements. The aim of the investigation is to determine: (a) frequency dependency of the brightness temperature of different ice types, (b) the capability of the multifrequency radiometer to classify ice types, and (c) the optimum measurement frequencies for mapping of sea ice.

Key words: Brightness temperature of sea ice, sea ice classification

INTRODUCTION

The northern Baltic Sea (Gulf of Bothnia) is annually covered by ice due to its low salinity (3 *o/oo*) and climatological location. The freezing starts usually in late October and the ice cover reaches the maximum extent in March. The maximum ice thickness is usually less than one meter. Typically, the sea is free of ice by the end of May. Finland and Sweden have important harbours on the coast of the northern Baltic, thus routes to harbours are kept open by ice breakers. The ice situation is regularly published in ice charts which present the ice extent, types and thickness. The information on ice drifting, ridging and open water is of primary importance in ship navigation. In Finland the Ice Service produces the ice charts from the information that ice breakers, weather stations and satellite-sensors provide. The area of interest is often (60 -70 % of the time) covered by clouds and the day-time is very short in winter. Thus, it is desirable that microwave images are also used in ice mapping [1].

In the winter of 1991-92 Finnish Institute of Marine Research (FIMR) organized two sea ice measurement campaigns of the Baltic Experiment for ERS-1 (BEERS), which is a part of the Programme for International Polar Oceans Research (PIPOR). The research cruises (February 7 - 14, 1992 and March 11 - 22, 1992) in the northern Baltic Sea, provided the Laboratory of Space Technology from Helsinki University of Technology (HUT) an opportunity to measure different sea ice types with an airborne radiometer and a scatterometer [2, 3].

INSTRUMENTATION

In 1991 a helicopter-borne multifrequency microwave radiometer was completed in Laboratory of Space Technology. It was designed for airborne measurements of snow, ice and forest. The radiometer operates at 24, 34, 48 and 94 GHz and its measurement polarization is vertical. The 24, 34 and 48 GHz receivers were constructed in 1990 and they operate in the Dicke mode. The 94 GHz receiver, which operates in the total power mode, was added to the system in 1991. The main technical characteristics of the system are presented in Table 1. The radiometer system is controlled by a microcomputer which also stores the data. The four receivers and a video camera are placed on a support structure below a helicopter. The angle of incidence can be set from 0 to 45° off nadir [4, 5].

Table 1. Main characteristics of the HUT radiometer system.

Centre frequency (GHz)	24	34	48	94
RF bandwidth (MHz)	650	650	650	2000
3 dB beamwidth (deg)	5.4	4.2	4.3	3
Noise temperature (K)	1500	1100	2400	1700
Integration time (s)	0.2	0.2	0.2	0.4
Sensitivity (K)	1.6	1.2	2.4	3.5

GROUND TRUTH

FIMR measured the ice and water parameters. The test site consisted of compact pack ice, ice ridges and new ice on both research cruises. The main results of the ice measurements are presented in Table 2 [2, 3].

On both research cruises FIMR measured the properties of snow: depth, wetness, density, and temperature. On the second cruise Helsinki University also studied physical properties of snow, including snow structure. The mean values of the snow parameters relevant to radiometer measurements are presented in Table 3. Snow existed only on compact pack ice [2, 3].

Table 2. Main characteristics of sea ice and water in the campaigns [2,3].

Ice type/ water	Mean thickness (cm)		Mean salinity (*o/oo*)	
	Feb.	March	Feb.	March
New ice	2	2	0.63	0.61
Compact pack ice	51	69	0.44	0.28
Water	-	-	3.3	3.2

Table 3. Measured snow parameters for 13 February, 13 March and 17 March [2,3].

Date	Snow type	Thickness (cm)		Wetness (% wgt.)	Density (kg/m^3)	Grain size (mm)	Temp. (°C)	
		mean	std			min - max	air	snow
13 Feb.	dry	6.5	2.4	-	-	-	-10	-4.4
13 Mar.	wet	5.8	1.9	3.8	0.3	0.5 - 3	0.5	0
17 Mar.	refrozen	9.8	1.1	-	0.2	1 - 6	-2	-1.2

RADIOMETER MEASUREMENTS

The dates for radiometer measurements were selected according to weather and snow conditions because these factors have a great influence on the brightness temperature. Thus, the measurement flights were made in three different snow conditions: (a) dry new snow, (b) wet snow and (c) refrozen snow covering compact pack ice (Table 3). The measured brightness temperatures were associated with ice types, based on videotapes. Only representative measurement values were included in the data analysis. The main categories were:

- New ice: The thickness of new ice is typically few centimetres and its surface is smooth and snow free. New ice is common in early winter; later, it is found in leads.
- Compact pack ice: Compact pack ice is formed from crushed level or new ice. The crushed ice field has been packed and it is refrozen. This ice type is common in open areas where waves and wind can crush existing level or new ice.
- Close pack ice: Close pack ice is formed from crushed ice and slush, but it is not solid. This ice type exists typically between solid ice field and open water. It absorbs waves that keep it fractured and wet.
- Open water: Open water was measured to determine the capability of microwave radiometry to distinguish water from various ice types.

Dry ice and snow cover

Figure 1 shows the mean brightness temperature versus frequency of the measured ice types for 13 February. The measured new ice was a couple of days old and it was free of snow. The compact pack ice was covered by dry snow which had fallen during previous days. The mean thickness of snow was 6.5 cm on the test field, but beside ice blocks there was even 20 cm of snow. The air temperature was -10 °C. The measured close pack ice was wet and it formed a zone between the compact pack ice field and open water. In the case of new ice, close pack ice and open water, the brightness temperature increases with increasing frequency. At 34 and 48 GHz the behaviour of compact pack ice is similar to that of other ice types, but beyond that the dry snow causes the brightness temperature to decrease. The grain size of the snow was around 1-2 mm, but volume scattering was mainly sensed at 94 GHz due to the small snow depth.

Wet ice and snow cover

On 13 March, new ice did not exist due to warm weather and high winds. The compact pack ice was partially covered with wet snow. The snow temperature was 0 °C and the air temperature was +0.5 °C. The measured close pack ice formed a zone between the compact pack ice field and open water. Figure 2 shows the mean brightness temperature versus frequency of the measured ice types for 13 March. The effects of warm weather and wet snow are obvious: the mean brightness temperatures of the ice types are very close to each other and also close to their physical temperatures.

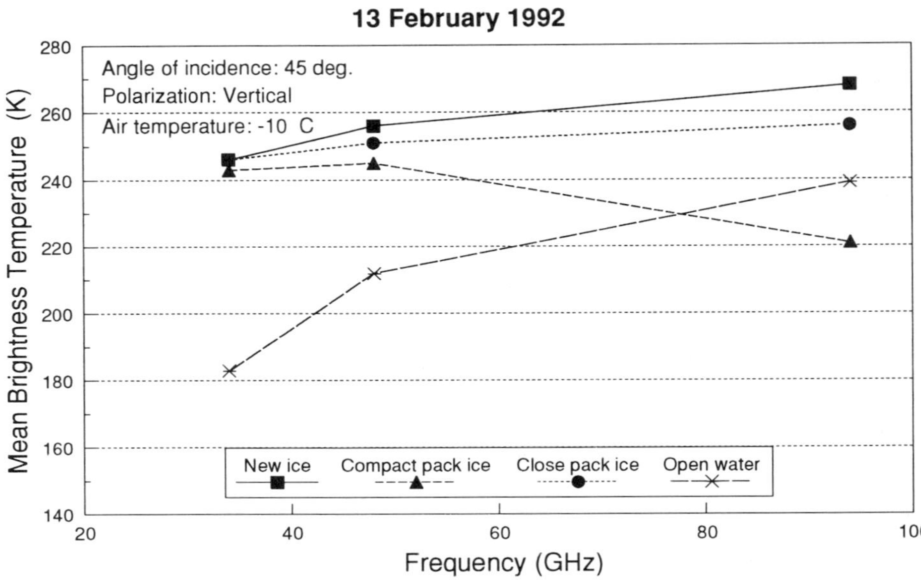

Figure 1. Mean brightness temperatures of measured ice types at 34, 48 and 94 GHz on 13 February 1992. Only compact pack ice was covered by snow.

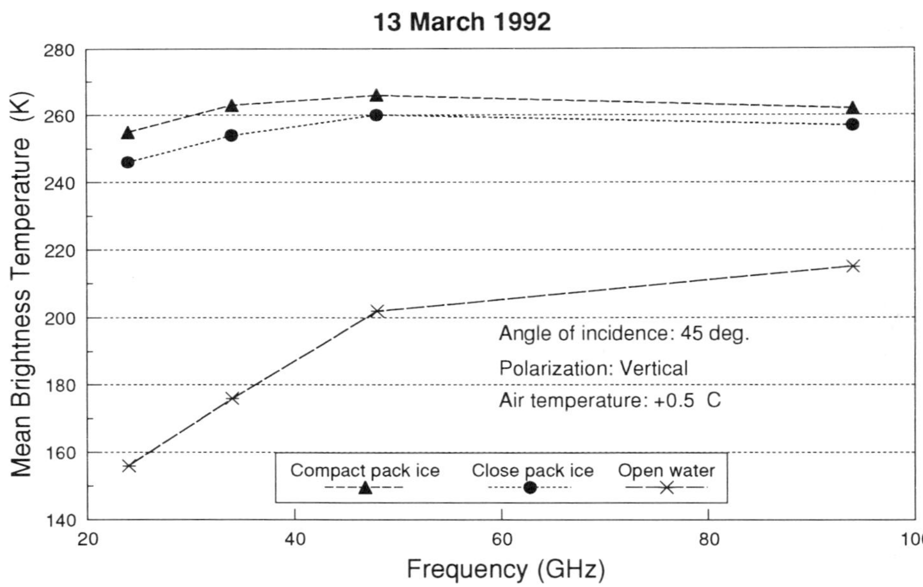

Figure 2. Mean brightness temperatures of measured ice types at 24, 34, 48 and 94 GHz on 13 March 1992.

Dry ice and refrozen snow cover

On 17 March the measured ice types were new ice, compact pack ice and close pack ice. The new ice had formed after 14 March. Due to the cold night the new ice was covered by frost which was about 1 cm thick. Frost crystals had an average size of 5 - 6 mm. The compact pack ice was covered by refrozen snow with a mean depth of 10 cm in the test field. On the top of the snow cover there was also a frost layer similar to that on the new ice. The snow temperature was -1.2 °C and the air temperature was -2 °C. The measured close pack ice was wet and it formed a zone between the compact pack ice field and open water as in the previous measurements. Figure 3 shows the mean brightness temperature versus frequency of the measured ice types for 17 March. The behaviour of the mean brightness temperatures is similar to that in Figure 1, except in the case of new ice. Frost on new ice decreases the brightness temperature at 94 GHz. Due to presence of refrozen snow on compact pack ice its brightness temperature decreases with increasing frequency. On 17 March the snow grain size was bigger than on 13 February; thus, the effect of refrozen snow is also remarkable at lower frequencies.

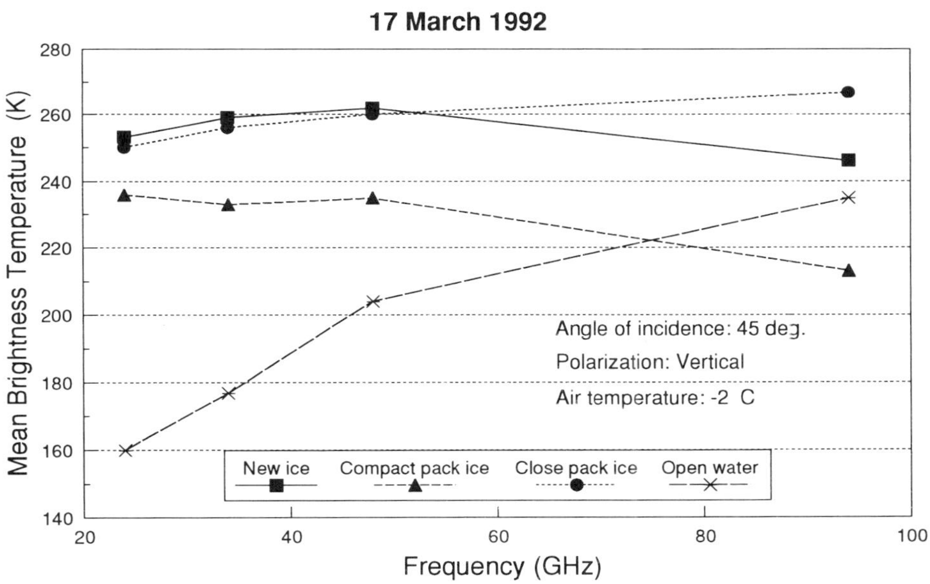

Figure 3. Mean brightness temperatures of measured ice types at 24, 34, 48 and 94 GHz on 17 March 1992.

CLASSIFICATION

The aim of this classification experiment is to estimate the upper limit of radiometric discrimination of various ice types with a multichannel radiometer. Ice classification was done separately for each snow (weather) condition, because a snow type has a critical effect to the measurement in the used frequency range. This procedure also guarantees that possible calibration variation between different measurement flights cannot interfere the classification. The normal distributions of the classes are the posterior information and all data points have been used for the calculation of the distributions.

The applied classification is based on the maximum likelihood rule [6]. The probability of a measurement value is calculated for all classification classes. The data point is assigned to a class, where the data point reaches the maximum probability. Classification has been processed with all available channel combinations (the 24 GHz channel was not available on 13 February).

Dry ice and snow cover

Figure 4 shows the mean accuracy of the classification, processed from the measurements acquired on 13 February. In each classification the mean accuracy is the average of the accuracies in the classes. If only one of the channels is used in classification, the classification capability of the ice classes increases as the frequency increases. This result agrees well with the behaviour of the mean brightness temperatures shown in Figure 1. With the 34 and 48 GHz channels open water is classified with an accuracy of 100 %. The lowest accuracies are for new ice at 34 GHz (20%) and for compact pack ice at 48 GHz (49%). At 94 GHz open water is distinguished with an accuracy of 91% and the lowest accuracy (63%) is for close pack ice.

The best result is achieved with a combination of the lowest (34 GHz) and the highest frequency (94 GHz). This combination gives the highest accuracy (93 %), but the difference is small (1 %) to the other combinations where the 94 GHz channel is involved. Figure 5 shows the result of the classification which gives the highest mean accuracy. The biggest error is for close pack ice: 22 % of the measurements have been assigned to the new ice category. In the other classes the errors are smaller than 5 %.

Wet ice and snow cover

The classification, which deals with wet snow/ice condition, is processed from the measurements acquired on 13 March. Due to weather, new ice did not exist. Thus only three classes (open water, compact pack ice and close pack ice) are used in the classification. Figure 6 shows the mean accuracy of the classification. Only the lowest and the highest accuracies are shown from two- and three-channel combinations. The mean accuracy is about 80 - 90 % with all the possible channel combinations. In the wet snow/ice condition the brightness temperatures for compact pack ice and close pack ice are rather independent of frequency (Figure 2). The 94 GHz channel gives the lowest accuracy (78%) because of the largest variation in the brightness temperature values. This may be partially caused by the 94 GHz receiver, which has a lower sensitivity than the other receivers. The variation of the values is the lowest at 48 GHz, thus the highest accuracy (91%) is achieved with the 48 GHz channel. All the combinations, which have several channels, have approximately the same accuracy (89 - 91%). Figure 7 shows the result of the classification which gives the highest accuracy. The biggest error is for close pack ice: 19 % of the measurements have been assigned to compact pack ice.

Dry ice and refrozen snow cover

Figure 8 shows the mean accuracy of the classifications which are processed from the measurements acquired on 17 March. The used classes are: open water, new ice, compact pack ice and close pack ice. In classifications with one channel the discrimination of the ice classes increases with increasing frequency. At 94 GHz open water, compact pack ice and close pack ice are recognized with an accuracy of over 90%. The frost cover on new ice increases the deviation of the measurements; thus, only 74 % accuracy is achieved for new ice at 94 GHz.

The best results are achieved when the lowest (24 or 34 GHz) and the highest frequencies (94 GHz) are used together in the classification. Then the mean accuracy of the classification is 95 - 97 %. Figure 9 shows the result of the classification which gives the highest accuracy (97%). The biggest error is for close pack ice where 6 % of the measurements have be classified as new ice. Discrimination of close pack ice and new ice is easier for the refrozen situation than for the dry snow situation as Figures 5 and 9 show. For the refrozen situation frost on new ice is only distinguished at 94 GHz, which separates new ice from the other classes in this case (see Figures 1 and 3).

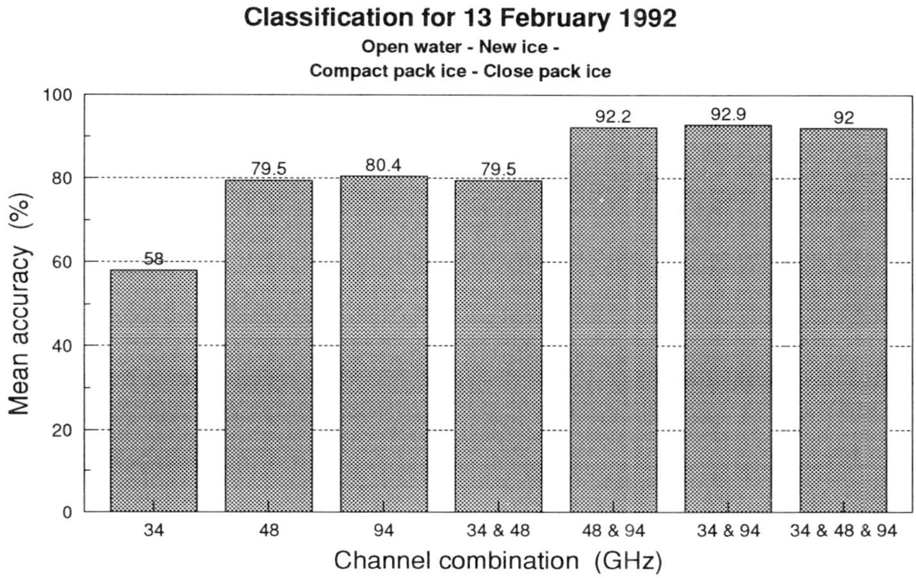

Figure 4. Mean accuracy of classification with different channel combinations based on data obtained on 13 February (dry snow and ice).

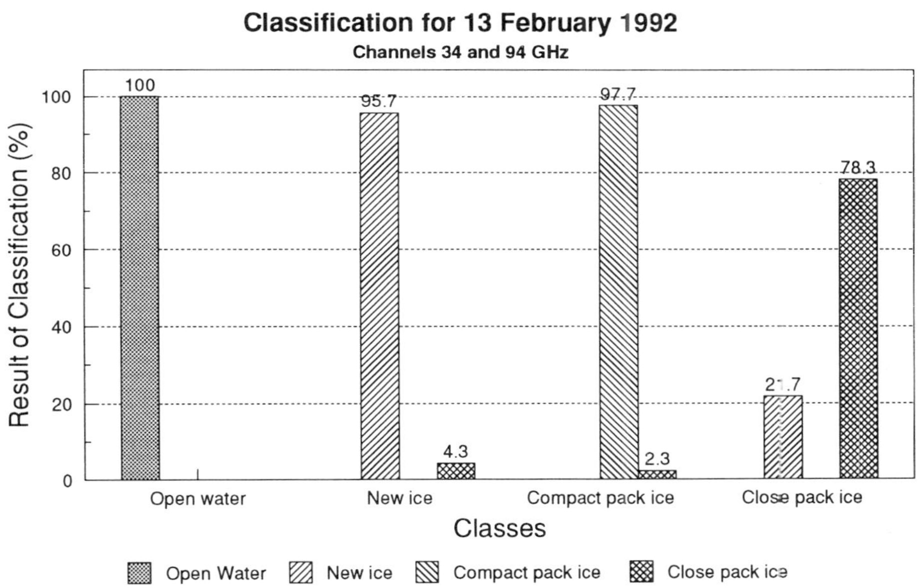

Figure 5. Result of classification with 34 and 94 GHz channels (dry snow and ice).

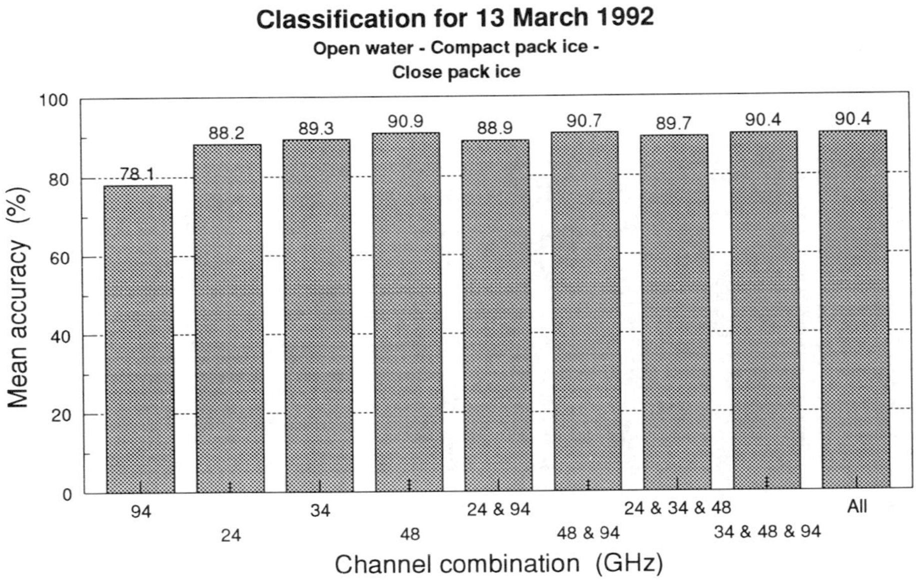

Figure 6. Mean accuracy of classification with different channel combinations based on data obtained on 13 March (wet snow and ice).

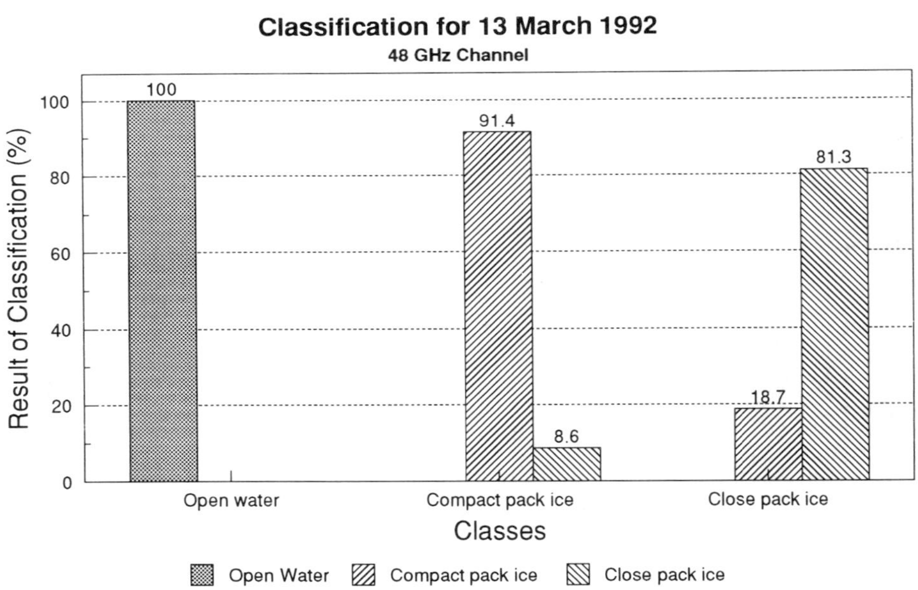

Figure 7. Result of classification based on 48 GHz data (wet snow and ice).

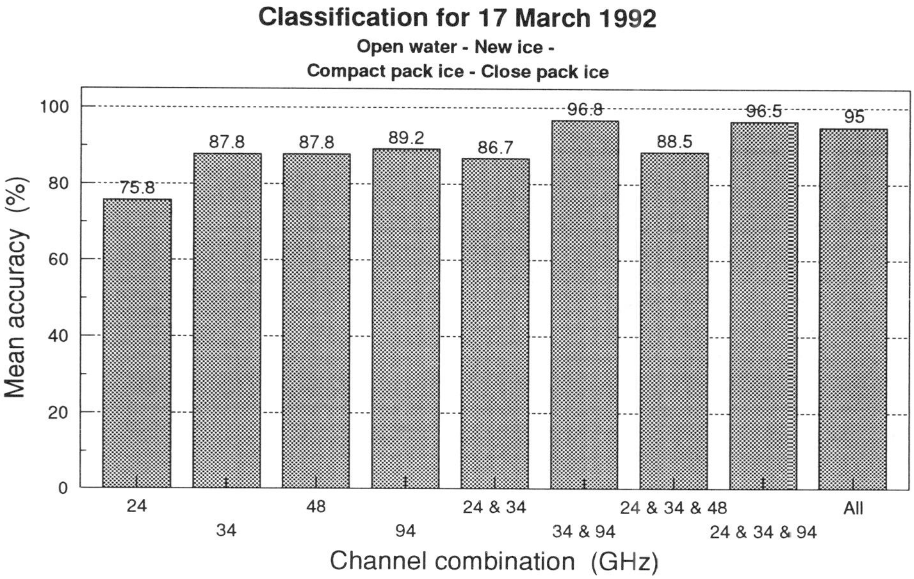

Figure 8. Mean accuracy of classification with different channel combinations based on data obtained on 17 March (refrozen snow and dry ice).

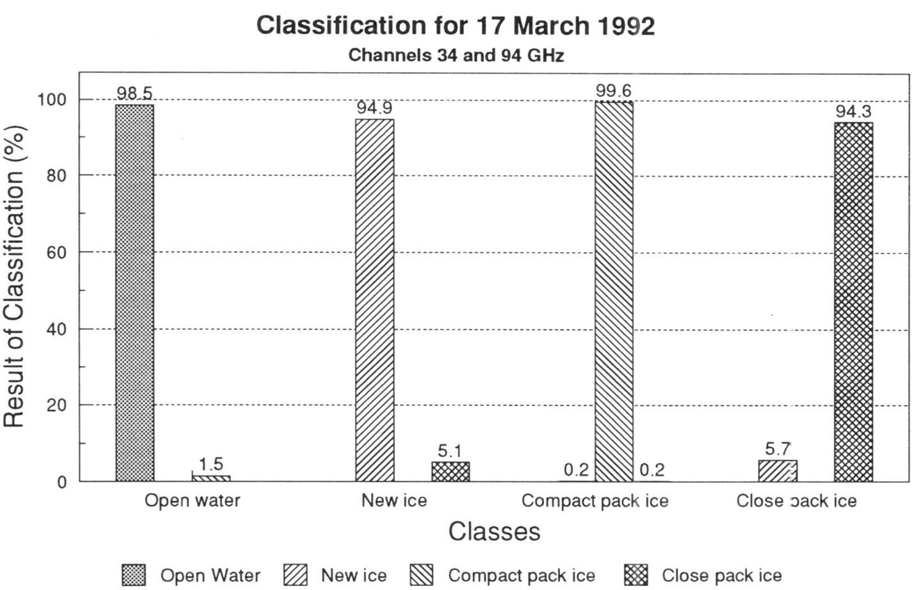

Figure 9. Result of classification with 34 and 94 GHz channels (refrozen snow and dry ice).

CONCLUSIONS

The measurements suggest that a high-frequency receiver (90 GHz) should be combined with a lower-frequency (35 GHz) receiver in ice type classification. The high frequency is very sensitive to dry and refrozen snow, hence it is very useful in detection of old ice (compact pack ice and fast ice) which is often covered by snow. The lower frequency is needed for detection of open water and new ice. The high frequency may cause problems because the brightness temperature of old ice can be equal to that of open water, if there is a suitable amount of snow on top of old ice. In that case the high frequency channel cannot detect leads in old ice. The lower frequency is also needed to detect new ice if it is covered by snow or frost. According to this experiment the optimum frequencies for ice mapping are 35 and 90 GHz in the studied range. Frequencies lower that 35 GHz provide good ice-water discrimination capability and are less sensitive to variation of atmosphere conditions. However, their main disadvantage is poor spatial resolution especially in spaceborne use.

The classification results are promising: open water can be detected clearly in all cases and the ice classes are discriminated with high accuracy. The overall accuracy of the classification is over 90 % and even in wet weather an accuracy of 90% has been achieved. The classification results show that an airborne imaging radiometer would be useful for ice mapping in the Northern Baltic Sea. An airborne radiometer could also provide the spatial resolution needed for ice mapping.

REFERENCES

1. SMHI, FIMR, *Climatological ice atlas for the Baltic Sea, Kattegat, Skagerack and Lake Vänern*, Swedish Meteorological and Hydrological Institute, Finnish Institute of Marine Research, 220 p., 1982.
2. Kosloff, P., Lensu M., Leppäranta, M., Pihkala, P., Similä, M., *Beers 92 R/V Aranda Base Date Report II: The Measurements*, Finnish Institute of Marine Research, Internal report 1993(5), 136 p., 1993.
3. Lensu, M., *Profilometer Measurements of Ice Ridges in the Bay of Bothnia*, Finnish Institute of Marine Research, Internal report 1993(6), 197 p., 1993.
4. Panula-Ontto, E., *Airborne radiometer system for remote sensing* (in Finnish), M.Sc. Thesis, Helsinki University of Technology, 64 p., 1990.
5. Kurvonen, L., *94 GHz airborne radiometer for remote sensing* (in Finnish), M.Sc. Thesis, Helsinki University of Technology, 60 p., 1991.
6. Therrien, C., *Decision estimation and classification, An introduction to pattern recognition and related topics*, John Wiley & Sons, 1989.

VALIDATION OF LIQUID WATER PATH AND SURFACE WINDSPEED RETRIEVALS FROM AN AIRBORNE MICROWAVE RADIOMETER

S.J. ENGLISH, D.C. JONES and R.W. SAUNDERS

Remote Sensing Instrumentation
U.K. Meteorological Office
Building Y70, D.R.A.
Farnborough, Hants GU14 6TD
U.K.
tel:+44-252-395726 *fax:+44-252-515523*
e-mail: senglish@email.meto.govt.uk

Abstract. Measurements of microwave brightness temperature at 89 and 157 GHz have been obtained from the Microwave Airborne Radiometer Scanning System (MARSS) on the U.K. Meteorological Office's C–130 aircraft. The radiometer has flown over 130 sorties of between 3 and 10 hours duration collecting data over open sea, sea ice and land for a wide variety of weather conditions. The effort is directed towards an improved modelling capability for the window channels of the high frequency component of the Advanced Microwave Sounding Unit (AMSU–B). A non–linear retrieval scheme has been applied to the observations to retrieve cloud liquid water paths for comparison with *in situ* measurements. Early results show that the liquid water path can be retrieved to an accuracy of $50gm^{-2}$ using downward views with a bias towards overestimating cloud water content. The error bars on the observations would imply an error of this order so errors in the radiative transfer model are not significantly affecting the retrieval.

Keywords: Microwave, radiometer, cloud, liquid water, retrieval, windspeed

INTRODUCTION

The Advanced Microwave Sounding Unit (AMSU) comprises a temperature sounder (AMSU-A) and a humidity sounder (AMSU-B). In addition to the sounding channels there are window channels at 23.4, 31, 50.3 GHz on AMSU–A and 89 and 150 GHz on AMSU–B. The purpose of the window channels is to identify fields of view contaminated by precipitation and, in the absence of very dense or precipitating clouds, to retrieve cloud and surface parameters.

Measurements of microwave brightness temperature[1] at 89 and 157 GHz have been obtained from the Microwave Airborne Radiometer Scanning System (MARSS) on the U.K. Meteorological Office's C–130 aircraft. The radiometer has flown over 130 sorties of between 3 and 10 hours duration collecting data over open sea, sea ice and land for a wide variety of weather conditions. This data has been used both to validate currently available models of surface emissivity and cloud absorption and also to test retrieval schemes for surface windspeed and cloud liquid water path (LWP) using the two single polarization window channels. The C–130 provides a unique platform for measuring the *in situ* cloud microphysical data providing a detailed comparison not available from satellite data. The retrieved products are compared with the *in situ* measurements made by the C–130's instrumentation.

AIRCRAFT INSTRUMENTATION

The MARSS radiometer has been described in past work[2,3]. In brief it is a two channel radiometer operating at 89 and 157 GHz measuring a single polarization which rotates with view angle from nominal horizontal at nadir. Recently the 157 GHz channel has had a polarizer added which rotates the vector through about 45°. Nine upward views, nine downward views and two calibration targets are observed during each scanning cycle which takes just under 3s. The nine views are nominally at intervals of 10° from –40° to +40° viewing along track. The calibration uses a hot target (at 334K) and a cold target which is allowed to remain at ambient temperature (230-290K).

The C–130's instrumentation is able to measure all parameters of importance to the radiative environment for comparison with the observed brightness temperatures. Windspeed is calculated using an Inertial Navigation System (INS) calibrated using the Global Positioning System (GPS). The surface windspeed can then be calculated using a boundary layer model. Cloud liquid water content (CLWC) is measured by a Johnson–Williams hot wire probe and also by various dropsize counters. The LWP is found by integrating the CLWC through an aircraft profile through cloud.

THE RADIATIVE TRANSFER MODEL

The gaseous absorption model was validated in clear air conditions using upward views. The observations have been compared with various empirical models [3] and the best agreement was given by the model of Liebe[4] except for very moist and very dry atmospheres. Liebe's model is used for all gaseous absorption in this paper.

The microwave emissivity of the surface depends on surface type, frequency and viewing geometry. A geometric sea surface emissivity model using a Gaussian distribution of facets slopes[1] is used to represent emission from the wind–roughened ocean. Only one scale of roughness is considered such that the reflection from each facet can be written in terms of Fresnel reflection coefficients[6]. A low frequency dielectric formula for saline water is adopted[7]. The Gaussian facet slope distribution depends on the surface stress vector which can be related to a windspeed at a standard height (e.g. 12.5m) At an incidence angle of 53° the emissivity depends more strongly on windspeed for horizontal than for vertical polarization. It has been shown[8] that it is possible to retrieve the surface windspeed and the surface temperature from the dual polarized SSM/I (Special Sensor Microwave/Imager). However for AMSU only a single polarization is measured and this is a mix of vertical and horizontal polarization depending on view angle. For a single polarization radiometer there is less information to retrieve windspeed and it is unlikely that these retrievals will be as useful as those from a dual polarized radiometer. However it is still necessary to correctly parameterise the surface. Wentz[9] has also shown that the bidirectional nature of the sea slope roughness gives a weak wind direction dependence. This effect is neglected in this paper.

For liquid clouds the scattering coefficient is negligible so the total extinction coefficient $K_{ex} \approx K_a$ (the Rayleigh absorption coefficient). The extinction can then be calculated using the Rayleigh approximation. At frequencies above 100 GHz the accuracy of the Rayleigh approximation can be in error by several percent for large droplets. In this work however we consider only cases of stratocumulus clouds where the Rayleigh expression remains good. We can calculate the dielectric constant using a single Debye relation[10] or a double-Debye relation[11], the latter being more appropriate for high frequency calculations.

In this paper we compare Grody's[10] algorithm to the observations. Grody's algorithm for cloud absorption fits Ray's[12] dielectric formula for the Rayleigh approximation such that:

$$K_{ex} = \frac{0.0241\nu^2 F(T)}{\nu^2 + F(T)^2} C \quad \text{where} \quad F(T) = 160\exp[7.2(1 - 287/T)] \tag{1}$$

where C is the cloud liquid water content in millimetres ($1mm = 1kgm^{-2} = 1000\text{gm}^{-2} = 0.1gcm^{-2}$), T is temperature in K, p pressure in hPa and ν is frequency in GHz. The relaxation frequency, $F(T)$, is between 100 and 140 GHz at temperatures of 270–285 K. As a result the temperature dependence of the optical depth is different at 89 and 157 GHz.

The radiative transfer equation is solved for a non-scattering atmosphere using microwave brightness temperature[1] as the radiative variable. The atmosphere is divided into 40 levels. Each level is determined such that all significant layers of the atmospheric profile are represented.

THE RETRIEVAL METHOD

Newtonian iteration is used for the retrieval scheme. In general an optimum solution can be found using this method for weakly non–linear cases[13]. The *n+1*th guess can be calculated from the *n*th guess and the background, $\mathbf{x_0}$.

$$\mathbf{\hat{x}_{n+1}} - \mathbf{x_0} = \mathbf{W_n}(\mathbf{y} - \mathbf{y_n} - \mathbf{K_n}(\mathbf{x_0} - \mathbf{x_n})) \tag{2}$$

where $\mathbf{W_n} = \mathbf{SK_n^T}(\mathbf{K_n SK_n^T} + \mathbf{E} + \mathbf{F})^{-1}$.
$\mathbf{K_n} = \mathbf{dy_n}/\mathbf{d\hat{x}_n}$
$\mathbf{\hat{x}_{n+1}}$ is the solution on the *n+1*th iteration
$\mathbf{\hat{x}_n}$ is the solution on the *n*th iteration
$\mathbf{x_0}$ is the *a priori* information vector
$\mathbf{S}$ is the error covariance matrix of the background field
$\mathbf{y}$ is the observed brightness temperature vector
$\mathbf{y_n}$ is the calculated brightness temperature vector
$\mathbf{E}$ is the error covariance matrix of the observations
$\mathbf{F}$ is the error covariance matrix of the forward model
If the model is linear then the solution can be found in one step. For weakly non linear cases the derivative matrix, $\mathbf{K}$, has to be recalculated at each step in the iteration. The value of $\mathbf{\hat{x}_{n+1}} - \mathbf{\hat{x}_n}$ is a suitable convergence criterion.

The covariance of the solution is given by Rodgers (1976) for the converged solution as

$$\mathbf{S_{n=\infty}} = (\mathbf{S} + \mathbf{K^T}(\mathbf{E} + \mathbf{F})^{-1}\mathbf{K})^{-1} \tag{3}$$

where $\mathbf{K}$ is evaluated for the final value of $\mathbf{S}$. If it is known we are observing very thin and non–precipitating clouds then if the first guess cloud amount is set to $0gm^{-2}$ with an *a priori* uncertainty of $50gm^{-2}$. Using equation 3 we can show, by assuming typical values for $\mathbf{E}$ ($\mathbf{F}$ is set to zero), that the uncertainty in the solution is about 15–25gm^{-2}, depending on the water vapour content of the atmosphere. For a more general scheme, applicable to thicker clouds, the error on the background of zero will be larger, for example $200gm^{-2}$, as we have less information. In this case the error on the solution is 30-50gm^{-2}. For marine stratocumulus the typical liquid water content can range from 10 to $200gm^{-2}$ although deeper and often drizzling stratocumulus with liquid water paths of 500 to $1000gm^{-2}$ are also observed. It is common to find embedded cumulus in a stratocumulus sheet and also scud cumulus below the stratocumulus base. The liquid water content in these various forms of cumulus can easily exceed $500gm^{-2}$. For deep clouds the MARSS channels start to saturate (i.e. they become sensitive to the cloud height rather than its liquid water content). If the initial uncertainty is set to $1000gm^{-2}$ then the uncertainty in the retrieved LWP increases from $50gm^{-2}$ for LWP less than $100gm^{-2}$ to $225gm^{-2}$ at LWP=$1000gm^{-2}$ for zenith views (equation 3). For nadir views the information loss is greater, ranging from $150gm^{-2}$ for LWP less than $100gm^{-2}$ to $850gm^{-2}$ at LWP=$1000gm^{-2}$. This provides a criterion to assess the retrievals. We expect the random error to be larger for deeper clouds. It also shows that an ideal LWP for validation is between 50 and $250gm^{-2}$.

The MARSS radiometer only has two window channels and is therefore limited in its ability to retrieve many variables. For this experiment the cloud liquid water path is the only free parameter for upward views and the first guess is taken as being zero cloud with an error variance of $250gm^{-2}$ (i.e. we assume thin stratocumulus is being observed). The water vapour and temperature profiles measured by the aircraft are used. The cloud is distributed between a cloud top and cloud base. The choice of cloud top and base can be made carefully using observations from the profile through cloud. In the absence of a valid cloud top or base a useful retrieval is still possible for runs above or below cloud if all the cloud is assumed to be in the lowest 3 km of the atmosphere. For example flight A209 (run 3.1) gave a retrieved liquid water path of $227gm^{-2}$. The cloud top was 955mb and the cloud base 997mb. The aircraft flew below all cloud. If a cloud top of 700mb and a cloud base of 1000mb was assumed the retrieval gives $255gm^{-2}$, an overestimate of 12%. The overestimate arises from the colder emission signal from the misplaced cloud, particularly on the 157 GHz channel. In practice it is always likely that a reasonable cloud top will be available from other instrumentation.

For downward views the cloud liquid water in the path from the aircraft to the surface and the surface windspeed are the free parameters. The dependence on wind direction is neglected as the wave slope distribution is taken to be Gaussian. The sea surface temperature measured by the C–130's Barnes PRT–4 infrared radiometer whilst flying at a height of $30m$ is used and is a constant in the retrieval scheme. Non–diagonal terms in the covariance matrix are neglected and the diagonal terms are set to $250gm^{-2}$ for cloud and $5ms^{-1}$ for windspeed. The cloud liquid water path along the reflected paths is constrained to be equal to the sum of the cloud viewed above the aircraft and that viewed between the aircraft and the surface. The background nadir cloud is also taken to be zero and a background windspeed is used equal to that measured at the base of the first aircraft profile.

These constraints are appropriate for thin stratocumulus cloud and for a reasonable background windspeed. The amount of information available for windspeed retrievals using these two channels is very small. However by allowing windspeed to vary in the retrieval it should be possible to distinguish between changes in windspeed and changes in cloud. The scheme is however primarily designed for the retrieval of a realistic cloud liquid water path.

CASE STUDIES

It can be seen from the discussion about the retrieval method that the optimum type of cloud for verifying the retrieval scheme and the cloud model is homogeneous stratocumulus of liquid water content 50–$250gm^{-2}$. Some inhomogeneous cases may still be useful if there is a very good match between the C–130 flight path above or below the cloud and the location of the profile through the cloud. Alternatively the variability of the measured cloud during a run in cloud can be taken as representative of the variability through the whole depth of cloud and the results compared.

The case studies used are mostly taken from the Atlantic Stratocumulus Transition Experiment (ASTEX) based in the Azores during June 1992. The MARSS was operated on fourteen of the C–130's sixteen sorties and collected about 100 hours of data flying below, in, and above the stratocumulus. The experiment was primarily designed at studying the breakup of stratocumulus on the edge of an extensive stratocumulus sheet during the diurnal cycle. The MARSS had two roles. Firstly to be able to observe the variability of the stratocumulus at night in the transition region. Secondly, and the main purpose, was to measure the radiative properties of stratocumulus in the more homogeneous cloud. In addition examples of homogeneous stratocumulus from other C–130 flights have been analysed, notably around the British Isles, near Ascension Island and in Finland. This has provided examples of cloud with varying temperature, altitude and dropsize distribution.

In each case the aircraft flew one or two profiles from the surface to the cloud top before starting a stack of straight and level runs at various altitudes. There would normally be at least one run below cloud, one in cloud and one above cloud in each stack. However not all runs will match equally well to the original profile and only runs with a very good or excellent match (given the advection of the cloud in the wind) are considered.

(a) ASTEX Flight A203
The C-130 flew a stack in stratocumulus varying in thickness between 50 and 250gm^{-2}. The cloud consisted of two main bands of stratocumulus and cumulus below the main base. Three runs (1.3, 1.4 and 1.5) matched the profile well. Run 1.3, at 500m, was below the stratocumulus base but above scud cumulus. Runs 1.4 and 1.5 were at 1200m and 1350m respectively and were in the stratocumulus layer.

(b) ASTEX Flight A206
Two runs (2.1 and 4.5) above thin stratocumulus (44-66gm^{-2}) had a moderate match to the aircraft profiles. The cloud structure was multileveled with scud cumulus beneath the stratocumulus layers. The cloud structure was observed to vary in thickness but both runs and the profile were through the denser stratocumulus.

(c) ASTEX Flight A207
Flight A207 had thin stratocumulus, LWP between 40 and 70gm^{-2}. Two runs (2.1 and 3.1) were almost coincident with the profile. Run 3.1 was at 1650m, above the cloud, and run 2.1 was at 30m and below all cloud. The profile through the cloud showed a typical stratocumulus profile (water content 57gm^{-2}) with some scud below (water content 14gm^{-2}).

(d) ASTEX Flight A208
The stratocumulus sheet for A208 was very thin (25-35gm^{-2}) but was very homogeneous except for scattered embedded cumulus. The match between profile and runs was excellent. The location of the runs and profiles for A208 are given in Fig. 1 and the cloud profiles in Fig. 2. The cloud shows a typical stratocumulus structure.

(e) ASTEX Flight A209
The cloud liquid water content for A209 was over 200gm^{-2} and very homogeneous, producing near ideal conditions for validation of the retrieval scheme. There were two good runs giving similar conditions (3.1, 3.2). The stratocumulus formed a typical profile with cloud density reaching 1gm^{-3} at the cloud top and falling linearly to the cloud base.

(f) FATE Flight A146
The measured profiles for A146 gave only 118gm^{-2} of cloud. However the run near the cloud top gave cloud density of 0.8–1.2gm^{-3} which, combined with a cloud depth of 600m, suggests liquid water contents of about 300gm^{-2}. This is an example where the run in cloud probably gives a more realistic cloud water content than the single vertical profile through the cloud. The liquid water content at the top of the profile was 0.6gm^{-3}. Three runs (2.1, 2.2 and 3.1) are used.

(g) SAAMEX Flight H986
Flight H986 was in the region of the Gulf of Bothnia in March 1990 and the stratocumulus was associated with a frontal system with very deep cloud (liquid water content about 700gm^{-2}). The cloud was very homogeneous giving a good validation case study.

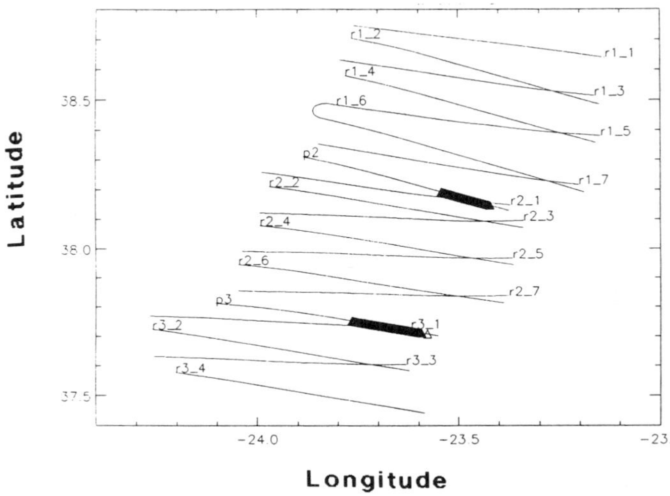

Figure 1: Location of aircraft runs for flight A208. Thick lines denote profiles through cloud.

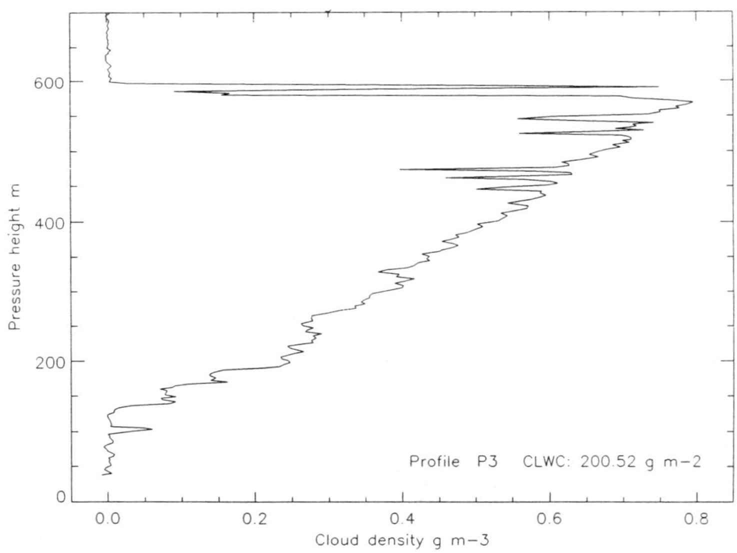

Figure 2: Cloud liquid water path (LWP) for Profile P3 (A208)

RESULTS

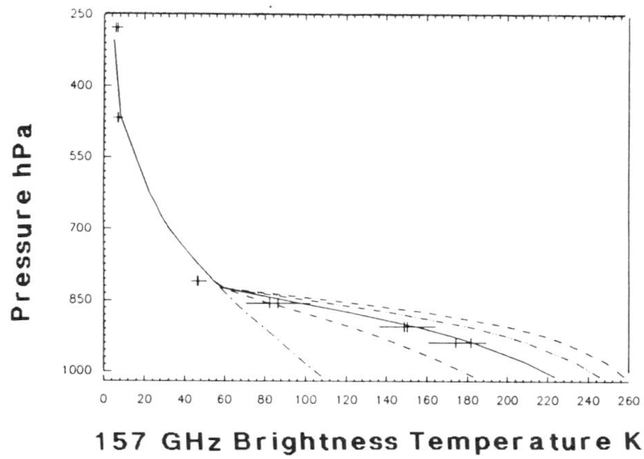

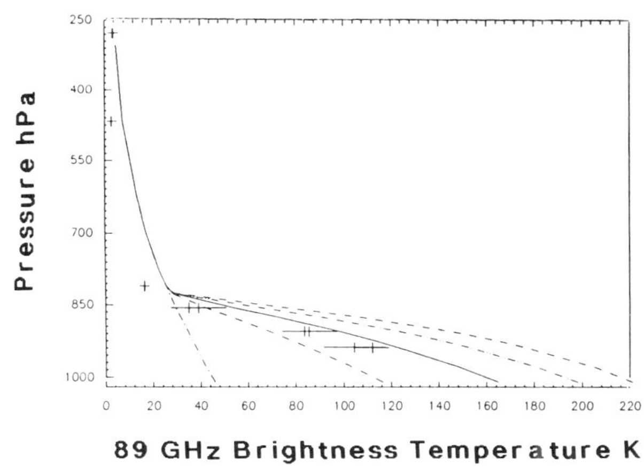

Figure 3: Brightness temperature for flight H986 at 89 GHz (above) and 157 GHz (below)

(a) Cloud model validation

Fig. 3 shows brightness temperature plotted as a function of altitude for flight H986. The continuous line is the brightness temperature predicted by Liebe's gaseous absorption model and Grody's cloud model and the crosses represent the observations. The four dashed lines represent the predicted brightness temperature profile with no cloud, 50%, 150% and 200% of observed cloud from left to right. The agreement between observed and calculated brightness temperature is very good at 157 GHz but shows a model overestimate at 89 GHz. The 'error bars' on the observations represent the standard deviation along each run. The mean temperature of the cloud was about -5°C and some ice crystals were present. The effect of these small ice crystals is neglected. The model may underestimate the cloud absorption in the presence of very large droplets. However comparison of extinction efficiency from a Rayleigh calculation and a Mie calculation shows this to be a very small error even for very maritime airmasses where many large drops exist. A more significant problem in the presence of large drops is the inability of the hot wire probe to efficiently detect the water content of drops with radii in excess of 30μm. For the flights presented here this is not however a significant source of error.

(b) Cloud retrievals using upward views
The mean values and standard deviations for the zenith cloud retrievals are given in column 3 of table 1. Only four flights: A207, A208, A209 (all ASTEX) and A146 (FATE), gave a sufficiently good colocation of the cloud profile and the below cloud straight and level run. The cloud was fairly homogeneous in all cases and generally at a temperature of between 7 and 10°C.

Table 1. MARSS LWP retrieval using upward views.

Flight and Run		JW LWP gm^{-2}	MARSS LWP (Z) gm^{-2}
A207	R3.3	57	17±7
A208	R3.3	32	28±12
A208	R3.5	32	40±18
A208	R3.6	32	24±12
A209	R2.1	98	202±39
A209	R3.1	200	227±30
A209	R3.2	200	268±39
A146	R1.0	118	216±36

Table 2. MARSS clear air windspeed retrieval using downward views.

Flight and Run		Windspeed ms^{-1}				Bias ms^{-1}	
		Ship	Aircraft	MARSS 12°	MARSS 45°	MARSS 12°	MARSS 45°
A012	R3.2	2.5	3.7	12.3	8.8	9.8	6.3
A013	R3.2	6.6	11.4	16.8	14.2	10.2	7.6
A014	R2.2	5.9	7.0	14.6	11.6	8.7	5.7

The agreement in Table 1 is broadly acceptable. Thin stratocumulus (LWP=17-40gm^{-2}) is retrieved for flights A207 and A208 and thicker stratocumulus (LWP=202-268gm^{-2}) for flights A209 and A146. Poor agreement is specifically observed for run 3.3 of A207, run 2.1 of A209 and run 1.0 of A146. In the case of A207 and A146 the colocation was not perfect and there is evidence to suggest the profile was not representative of the run. The overestimate for A209 run 2.1 is more surprising as the colocation is good and the cloud appeared homogeneous over a wide area. This appears to be the worst example of a general tendency for the retrieval to overestimate the cloud LWP at high LWP.

(c) Windspeed retrievals in clear air
For this purpose windspeeds measured by the aircraft and extrapolated to the surface are used as a first guess for cases where ship winds are available. In clear air conditions the apparent temperature of the surface is related to the windspeed such that a 1ms^{-1} change in windspeed changes the brightness temperature by about 0.6K. For the MARSS radiometer, which has a sensitivity of about 0.3K and typical error bars of 1 to 2K, we can potentially retrieve windspeeds to an accuracy of just over 2.5ms^{-1} (assuming an *a priori* uncertainty in the first guess of 5ms^{-1}), neglecting errors in the sea surface emissivity model.

The MARSS results[14] have implied that a geometric sea surface emissivity model underestimates the surface brightness temperature by 2-4K at near nadir angles and 0-2K at a nadir angle of 45°. This corresponds to a bias in windspeed of 4-8ms^{-1} near nadir and 2-4ms^{-1} at 45° from nadir. It should be noted that because the bias reduces with view angle no significant bias would occur at the SSM/I incidence angle of 53°. Examples are given in Table 2 of windspeed retrievals in clear air conditions at view angles of 12° and 45°. The bias with respect to ship winds is also indicated. All three examples are taken from flights from the Mediterranean Aircraft and Ship Transmission Experiment (MASTEX) over the seas around Crete during June 1990.

The observed bias between retrieved windspeed and the ship windspeeds is close to that expected from the known bias in the surface emissivity model. It can also be observed that the aircraft winds are generally in acceptable agreement with the shipwinds. For flights A012 and A014 the bias is similar and the agreement between the aircraft and ship winds was good. For flight A013 the ship windspeed was only $6.6ms^{-1}$ compared to the aircraft observation of $11.4ms^{-1}$ and a slightly larger bias was observed. The estimate of sea state from the aircraft supported the higher windspeed measured by the MARSS and inferred from the aircraft navigation aids. In conclusion the MARSS windspeeds, in clear air conditions, can give extra information but a very accurate surface emission model is required to eliminate biases. At lower frequency it has been shown that two scale roughness models can reduce the model bias. At higher frequency it is evident that similar biases occur. However it has yet to be proved that a two scale roughness model is required and English[15] provided evidence that simple adjustment to the dielectric model can improve the agreement.

(d) Windspeed and cloud liquid water retrievals using downward views

The retrieval of LWP for nadir (downward) views is of particular relevance to satellite data. Examples are taken from five ASTEX flights (A203, A206, A207, A208, A209) and one FATE flight (A146). The results are shown in Table 3. The retrieved LWP above the aircraft is given in the LWP below the aircraft in column 4 and the total LWP in column 5. For A146 and A209 a substantial overestimate in the cloud retrieval (in column 5 of Table 3) compared to the measured profile (in column 2) is observed. A smaller retrieval overestimate is observed for A206 with much improved agreement for A203 and A207.

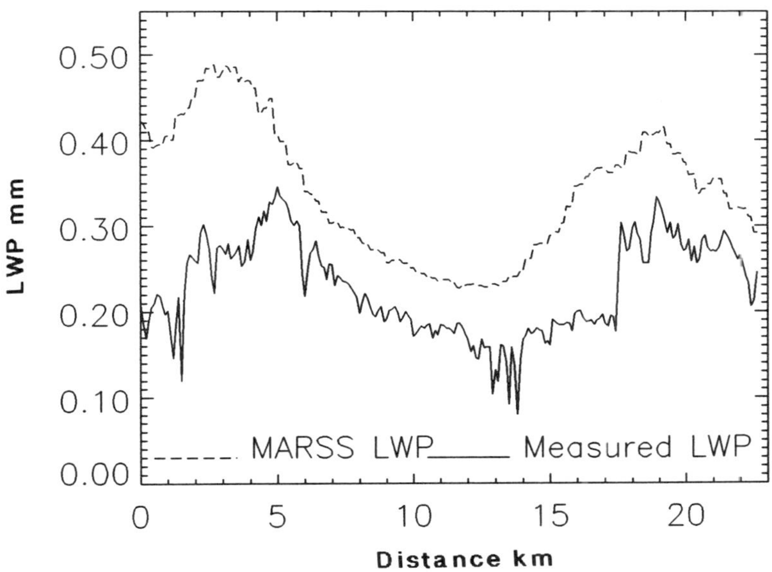

Figure 4: A146: LWP inferred from aircraft and retrieved from MARSS

However the windspeed retrievals do not show the bias observed for the clear air flights. The RMS error of the first guess windspeed was $2.0ms^{-1}$ with a standard deviation of

$1.5ms^{-1}$ whereas the MARSS RMS was $2.1ms^{-1}$ with a standard deviation of $1.6ms^{-1}$. The MARSS showed a bias of $1.5ms^{-1}$ and the first guess a bias of $1.3ms^{-1}$. The windspeed retrieval through cloud therefore provided no extra information. The model is still underestimating the surface brightness but the observed signal is being misinterpreted as cloud giving a cloud overestimate. The 2-4°C excess can give a $30\text{-}60gm^{-2}$ overestimate in the cloud retrieval. However the observed cloud retrieval is in excess of this and is consistent with the overestimate observed for large LWP for upward views.

Flight A146 showed a large difference between the aircraft profile of $118gm^{-2}$ and the retrieval of $300\text{-}400gm^{-2}$. However if the total LWP is calculated from the CLWC at the top of the cloud by $LWP = 0.5 \times CLWC \times \Delta z$ where Δz is the cloud depth then the LWP measured by the MARSS can be compared with the LWP inferred from the CLWC at cloud top. Fig. 4 shows this comparison for A146 run 3.1. The retrieval still overestimates but this is within the error bars discussed in section 4 and the variation along the run is captured well by the retrieval. Both the retrieved and the inferred LWP are between 200 and $500gm^{-2}$ in the nadir direction. The large retrieval error shows that a very good first guess is required (e.g. from AMSU–A) to use AMSU–B data at full resolution through thick stratocumulus (LWP greater than $200gm^{-2}$).

Table 3. MARSS LWP and windspeed retrieval

Flight and Run		Cloud LWPgm^{-2}				Wind Speed ms^{-1}		
		JW LWP	MARSS ↑	MARSS ↓	MARSS (Total)	MARSS	Aircraft	1st guess
A203	R1.3	206	93±77*	118±68	211±141	5.4±0.4	5.8±0.5	6.1
A203	R1.4	206	76±58*	163±96	240±142	6.2±0.4	5.8±0.5	6.1
A203	R1.5	206	55±42*	114±55	169±88	6.4±0.2	5.8±0.5	6.1
A206	R2.2	66	11±2*	85±5	96±7	7.7±0.1	5.5±0.5	7.7
A206	R4.5	44	16±5*	85±8	101±10	7.7±0.1	4.2±0.4	7.7
A207	R3.1	57	0±0	48±6	48±6	6.1±0.3	1.7±0.4	5.1
A208	R3.7	32	0±0	73±11	73±11	4.1±0.7	4.2±0.7	2.8
A209	R2.6	200	0±0	365±46	365±46	7.3±0.1	6.7±0.4	8.3
A209	R2.7	200	0±0	360±39	360±39	7.4±0.1	6.7±0.4	8.3
A146	R2.1	118	0±0	306±47	306±47	5.1±0.1	N/A	5.0
A146	R2.2	118	0±0	316±72	316±72	5.1±0.0	N/A	5.0
A146	R3.1	118	73±19*	317±80	390±93	5.2±0.0	N/A	5.0

Note: * LWP above aircraft for runs in cloud.

CONCLUSIONS

The MARSS channels have been shown to be capable of retrieving a realistic cloud LWP assuming that a reasonable temperature and humidity profile is available and some prior knowledge about the height of the cloud and the cloud type is available. It has been shown how a bias in the sea surface emissivity model can give a bias in windspeed retrievals in clear air. In cloudy conditions the bias in the surface emissivity model gives a bias in the cloud retrieval. This is unlikely to be a problem for a dual polarized radiometer. For AMSU–B, however, this places a high accuracy requirement on the surface emissivity model if cloud and water vapour are to be accurately retrieved.

The agreement between observed and retrieved LWP is usually within the retrieval error given by equation 3. The results confirm that the model used is adequate for thin stratocumulus and the LWP can be retrieved to an accuracy of $25gm^{-2}$ for zenith views and $50gm^{-2}$ for nadir views. However at high LWP there is an overestimation of the retrieved product for both nadir and zenith views. This overestimate can however only be tentatively accepted at present due to the difficulty of constructing a controlled experiment and possible errors in other aspects of the radiative environment (e.g. the gaseous absorption and the surface emissivity model). For deep clouds with large droplets the Johnson–Williams hot wire probe may not collect the liquid water very efficiently and hence underestimate the LWP. Also for large drops the Rayleigh expression begins to break down which would lead to a retrieval overestimate. Either would be consistent with the observations at high LWP so at present these results are still inconclusive.

Not enough data has been analysed to identify any trend of the error with temperature or cloud type. However the results have been shown to be only weakly sensitive to inaccuracy in the cloud height. For thin stratocumulus it is, therefore, of limited importance where the cloud is placed in the profiles.

ACKNOWLEDGEMENTS

The effort of the Royal Air Force air and ground crews and all the staff of the Meteorological Research Flight is gratefully acknowledged. The MARSS radiometer was jointly developed by the Laboratoire Meteorologie Dynamique de Paris and the staff of the remote sensing branch of the UK Meteorological Office. The authors thank their many colleagues from these establishments.

REFERENCES

1. Stogryn A., 1975. A note on brightness temperature at millimeter wavelengths, *IEEE. transactions on Geoscience Electronics*, **GE-13(2)**, 81-84
2. Jones, D. C., 1991. Microwave Airborne Radiometer Scanning System: Calibration and initial performance assessment., *Met O(RSI) Branch Memorandum No 3.* (Available from National Meteorological Library, Bracknell).
3. English S.J., Guillou C., Prigent C., Jones D.C., 1994. Aircraft measurements of water vapour continuum absorption at millimetre wavelengths, *Q. J. Royal Met. Soc.*, **120**, 603–625.
4. Liebe H. J., 1989a. MPM — An Atmospheric Millimeter Wave Propogation Model., *Int. J. Infrared and Millimeter Waves*, **10(6)**, 631–650.
5. Cox C. and Munk W., 1955. Some Problems in Optical Oceanography, *J. Marine Res.*, **14(1)**, 63-78.
6. Wilheit T.T., 1979. A Model for the Microwave Emissivity of the Ocean's Surface as a Function of Wind Speed., *IEEE Trans. Geosci. Electron.*, **GE-17**, 244-249.
7. Klein L.A. and C. T. Swift, 1977. An Improved Model for the Dielectric Constant of Sea Water at Microwave Frequencies., *IEEE J. Ocean. Eng.*, **OE–2**, 104-111.
8. Wentz F. J., L. A. Mattox and S. Peteherych, 1986. New Algorithms For Microwave Measurement of Ocean Winds: Applications to SEASAT and the Special Sensor Microwave/Imager., *J. Geo. Res.*, **91(C2)**, 2289–2307.
9. Wentz F. J., 1993. The effect of wind direction on MIMR geophysical retrievals., *Proceedings of PIERS, July 12–16 1993*, 766.
10. Grody N. C., 1993. Remote sensing of the atmosphere from satellites using microwave radiometry., in *Atmospheric Remote Sensing by Microwave Radiometry*, M. A. Janssen, ed., John Wiley and Sons, 259–334.
11. Liebe H. J., 1989b. Millimeter Wave Attenuation and Delay Rates Due to Fog and Cloud Conditions., *IEEE Trans. Antennas and Prop.*, **37(12)**, 1617–1623.
12. Ray P., 1972. Broadband Complex Refractive Indices of Ice and Water., *Appl. Optics*, **11**, 1836-1844.
13. Rodgers C., 1976. Retrieval of Atmospheric Temperature and Composition From Remote Measurements of Thermal Radiation., *Revs. Geo. Space Phys.*, **14(4)**, 609–624.
14. Guillou C., Prigent C., English S. J., 1994. How to describe the ocean roughened surface in microwave emissivity models?, *Microwave Radiometry and Remote Sensing of the Environment, Rome, Italy (ibid).*
15. English S. J., 1991. Remote Sensing of Meteorological Parameters by Microwave Radiometry, D. Phil Thesis, University of Oxford, UK (Available from the Met.O(RSI), Y70, DRA Farnborough, UK).

How to describe the ocean roughened surface in microwave emissivity models ?

C. GUILLOU[1], C. PRIGENT[1], and S.J. ENGLISH[2]

[1]*Département de Radioastronomie Millimétrique, CNRS*
Observatoire de Meudon, 92190 Meudon, France.
[2]*U.K.Meteorological Office*
Remote Sensing Instrumentation, Y70, DRA Farnborough, UK.

Abstract
This paper studies the influence of the sea surface description, especially the wave height and slope spectrum, on the radiometric and radar signatures of the ocean, from single and two-scale scattering models, in comparison with observations. A geometric optics model is shown to provide satisfactorily results, when coupled with the Cox and Munk's wave slope description, on the basis of a direct comparison with airborne radiometric measurements at 89 and 157GHz, and of an inversion process using SSM/I data collected during the SOFIA-ASTEX campaign. Sea surface slope variances inferred from 5GHz radar measurements are also found to be in close agreement with the Cox and Munk's predicted values.

Key words: emissivity models, sea wave spectrum, wind satellite retrieval

INTRODUCTION

The emission and scattering properties of the ocean surface, at microwave frequencies, are highly sensitive to the surface roughness and to the foam coverage induced by the wind, offering the possibility of remotely sensing the ocean from either passive or active techniques. If mesoscale and global scale maps of surface wind speeds can be retrieved from microwave imagers (SMMR, SSM/I...), active sensors, like Seasat or ERS-1 scatterometers, are necessary to provide the wind direction. Other oceanic variables, namely wave heights and slopes, directional gravity wave spectra, can also be inferred from radar measurements (altimeter and SAR onboard ERS-1 for instance). From another point of view, a good estimation of the sea surface emissivity is also required to properly describe the surface contribution of the radiometric signal, in case of atmospheric retrievals: integrated water vapor and liquid water contents [1], as well as temperature and humidity profiles from the current and future sounders SSM/T, AMSU, MHS [2], [3].
Empirical emissivity models have been widely used for operational purposes. Based on statistical regressions on previous measurements, in specific experimental conditions (most of them were established from SMMR data), their main advantage is to be fast. But they are limited to the low frequency range, within 1-40GHz, and strongly depend on the instrumental calibration. Moreover, their extrapolation to other frequencies and scanning geometries is not always valid. For instance, these models have been unsuccessfully used to construct statistical inversion algorithms for recovering surface wind speeds from SSM/I measurements [4]. For these reasons, a physical approach is desired to describe the interaction process of electromagnetic and oceanic waves.
In this paper, after a description of the ocean surface modeling and a brief historical record of the sea surface scattering models, we will focus on the influence of the sea surface description on the radiometric and radar signatures of the ocean. For different sea surface spectra as input, simulations will be compared with observations, from both passive and active sensors, including airborne radiometric measurements at 89 and 157GHz, SSM/I images from 19 to

85GHz, and 5GHz radar airborne observations.

SEA SURFACE DESCRIPTION

The sea surface is characterized by many scales of roughness, which form a continuous spectrum. The large scales are governed by the gravity force, while the smallest scales (capillary waves) are driven by the surface tension. The statistical properties of the surface are required in the electromagnetic calculations, through a slope probability density function (pdf), and a directional wave height and slope spectrum. The fairly gaussian slope pdf, as described by the well-known sun glitter measurements of Cox and Munk [5], is usually adopted in the scattering models. On the other hand, different slope variances and surface displacement spectra are proposed in the literature (Cox and Munk [5], Pierson and Stacy [6], Bjerkaas and Riedel [7], Phillips [8]...), offering a large disparity (figure 1), that can lead to confusing discrepencies between the results. These spectra are generally applicable to fully-developed seas with unlimited fetch, so that the sea state only depends on the surface wind speed, or on the friction velocity, both being related in a one way if neutral atmospheric conditions are assumed. However, swell conditions as well as unstable stratification at the air-sea interface have been shown to modify the mean square slope of the surface [9]. If the gravity wave range is relatively well-known, there is still a lack of knowledge concerning the capillary range [10]. Most of the measurements are conducted in tank conditions, having limited fetches, and their extrapolation to ocean conditions is not at all guaranteed. In particular, the small structures in wave tank and in open sea are not necessary identical. In addition, another misunderstood aspect of the sea surface characteristics is the angular behavior of the roughness spectrum, especially the upwind-downwind anisotropy [11].

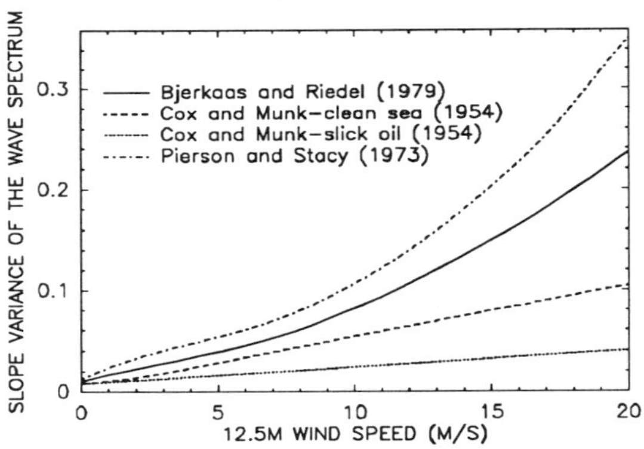

Figure 1. Sea slope variance as a function of the surface wind speed for various wave slope spectra.

The foam coverage induced by the wind is, to our knowledge, never taken into account in radar applications. However, it has an apparent effect on the ocean radiometric signal, by increasing the brightness temperature for both polarizations. The fractional coverage is usually described by a simple power-law function of the wind speed, and the foam emissivity is often taken to be 1. But, this is probably an oversimplification of the reality. Other parameters like the thermal stability of the lower atmosphere may have an impact on the whitecap coverage [12]. Measurements of the foam emissivity from an airborne experiment at 19 and 37GHz [13] confirmed a frequency dependance of the foam emissivity, and were shown to provide rather good agreement with the empirical frequency, incidence angle, and polarization dependent expression given by Stogryn [14].

SCATTERING MODELS

At low frequency, the simple scale scattering models (Stogryn [15], Wilheit [16]), based on the geometric optics assumption, have been shown to underestimate the classical measurements of

Hollinger [17] at 1.4, 8.3 and 19.3GHz, at near-nadir views, and especially at the lowest frequency. Better agreement has been found from the use of two-scale models superimposing small irregularities to the large undulation (Wu and Fung [18], Wentz [19], Guissard and Sobieski [20]). The development of such two-scale models to reproduce the sea surface apparent temperature is to be related to the radar backscattering modeling, where the presence of different mechanisms according to the observation angle is evident: while the specular reflection is the most important contribution of the backscattered signal at low incidence, the Bragg diffraction of the microwaves by the ripples becomes preponderant at incidences greater than about 30°.

The first encountered problem is the choice of the wave spectrum $W(K, \phi)$. Each author adopts a different one. The second problem is the definition of the splitting frequency K_c between the large and the small scales, as a function of the observation wavelength and of the surface wind speed. Once these quantities are well defined, the roughness parameters (rms slope of the large scale, rms height of the small scale) can be retrieved:

Given the expression of the roughness spectrum as $W(K, \phi) = K^{-1} S(K) F(K, \phi)$, one defines the slope variance of the large undulation by

$$m^2 = \int_0^{K_c} K^2 S(K) dK$$

and the rms height of the ripples, σ_1, from:

$$\sigma_1^{\ 2} = \int_0^{\infty} S(K) dK$$

At higher millimeter frequencies, the contribution of the small irregularities is expected to become less important, if not negligible. But, following the surface description adopted by each author, the results can significantly differ from one to the other. For instance, if one uses the Wu and Fung's parameters, the diffraction by the small scales should still have an impact on the brightness temperature up to 90GHz, while it doesn't seem to be the case from Wentz developments (figure 2). A single scale geometric optics approach may be valid at millimeter frequencies. Such a model has been used to derive the simulations presented on figure 3, based on Wilheit's [16] work, with the following specifications: in our calculations, the scattering component of the brightness temperature results from the integration of the sky radiation over the upper hemisphere, while Wilheit's formulation only takes into account the downwelling radiation coming from the specular direction. The gaussian approximation of the slope pdf defined by Cox and Munk is adopted, with slope variances calculated either from the Cox and Munk relations for a clean sea, from Bjerkaas and Riedel [7] or from Pierson and Stacy [6] empirical spectra.The dielectric constant of the saline water is computed from Klein and Swift [21]. Shadowing and multiple reflection effects are taken into account. A foam coverage, given by Wu [22], is also included and the foam emissivity is assumed to be 1. When this model is

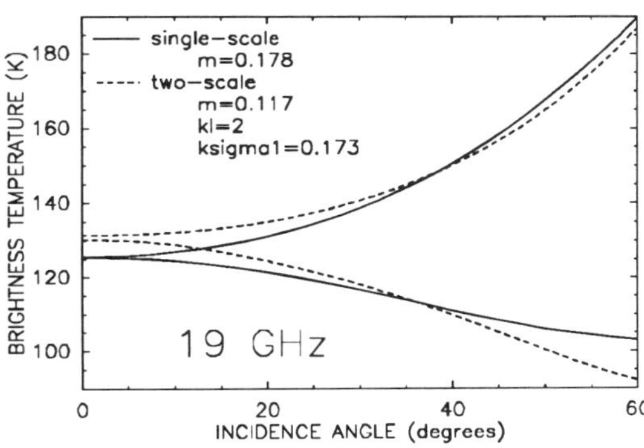

(a)

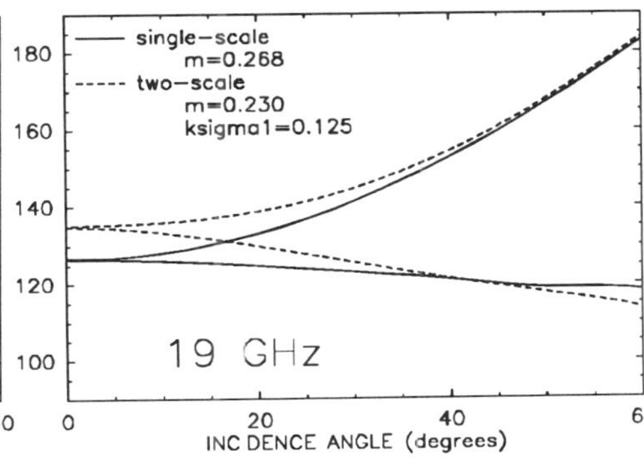

(d)

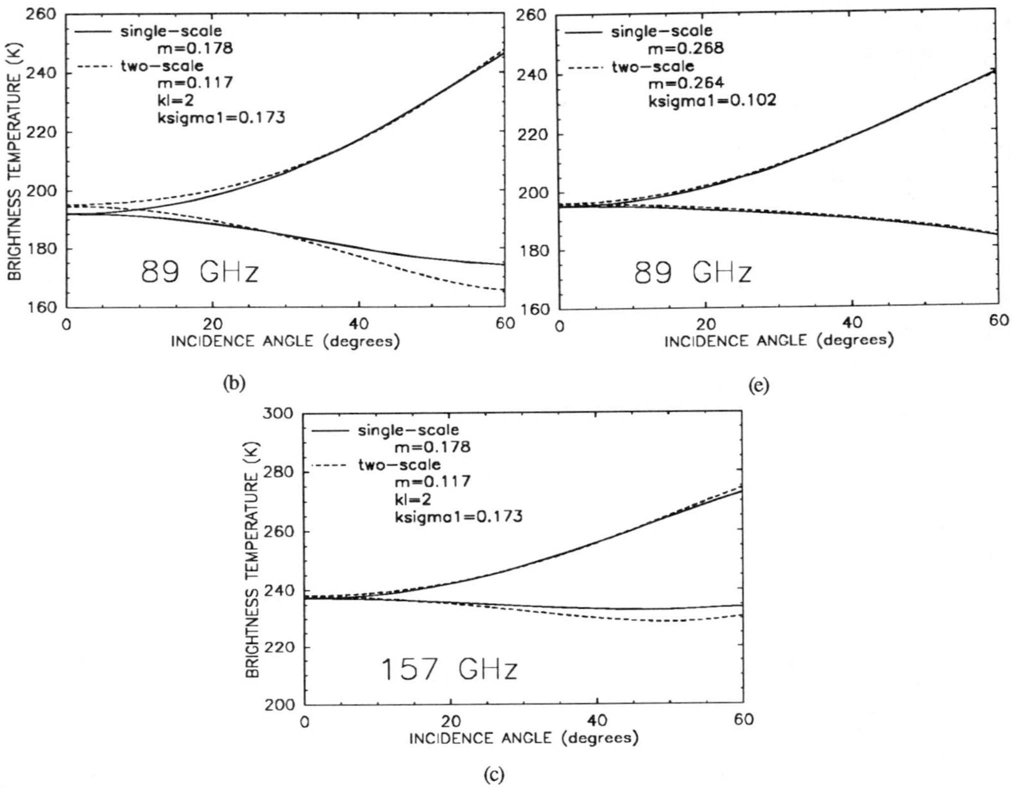

Figure 2. Comparison between single and two-scale approaches (a,b,c: Wu and Fung (1972)/Cox and Munk (1954); d,e: Wentz (1975)/Pierson and Stacy (1973))
Simulations using a US standard atmosphere. SST=290K, W=12m/s, no foam

supplied by Pierson and Stacy's spectrum, which provides a slope variance about twice higher than the variance given by Cox and Munk, this yields the same effect, in the radiometric response, as doubling the wind speed in the simulation. Such a difference in the polarized signals at high incidence should be apparent in retrieval schemes, from SSM/I data for instance.

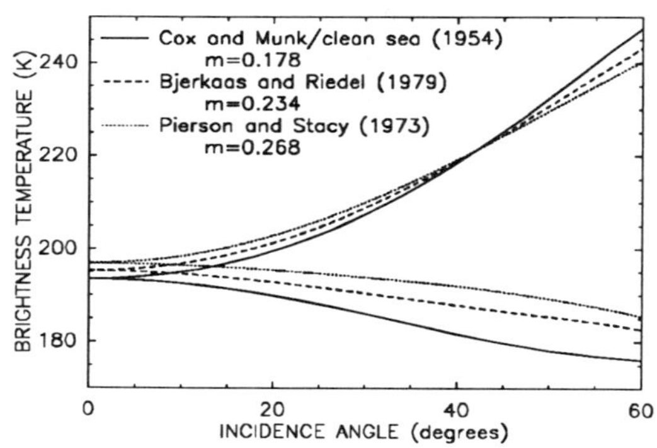

Figure 3. Influence of the slope variance on a single scale geometric optics model at 89GHz.
US standard atmosphere. SST=290K, W=12m/s

COMPARISON WITH MEASUREMENTS

Wind speed retrievals from SSM/I data

The SOFIA-ASTEX experiment, that took place in the Azores in June 1992, provided a good opportunity to test the validity of a radiative transfer model from an inversion procedure on SSM/I measurements, in a well documented situation. For the present purpose, SSM/I data from the orbit of the 12th June at 1100GMT have been collected.

The complete radiative transfer model is described in Prigent et al. [23]. For the sea surface contribution, the single scale geometric optics model, previously described, is adopted. The inversion process is based on an iterative algorithm, which finds the maximum probability solution to a non-linear problem (for more details, see Prigent and Seze [24]).

This method has been applied to the seven SSM/I channels to retrieve 10m surface wind speeds (figures 4, 5). In comparison with the ECMWF (European Center for Medium range Weather Forecast) winds, with a series of drifting buoy winds (see table 1), or even with estimates from the Goodberlet algorithms [25], a broadly good agreement is obtained, when using the Cox and Munk wave slope description, within the error bar (±2m/s) generally accepted for wind speed retrievals. As expected, retrievals based on the Bjerkaas and Riedel wave slope spectrum are found to slightly underestimate the ground truth.

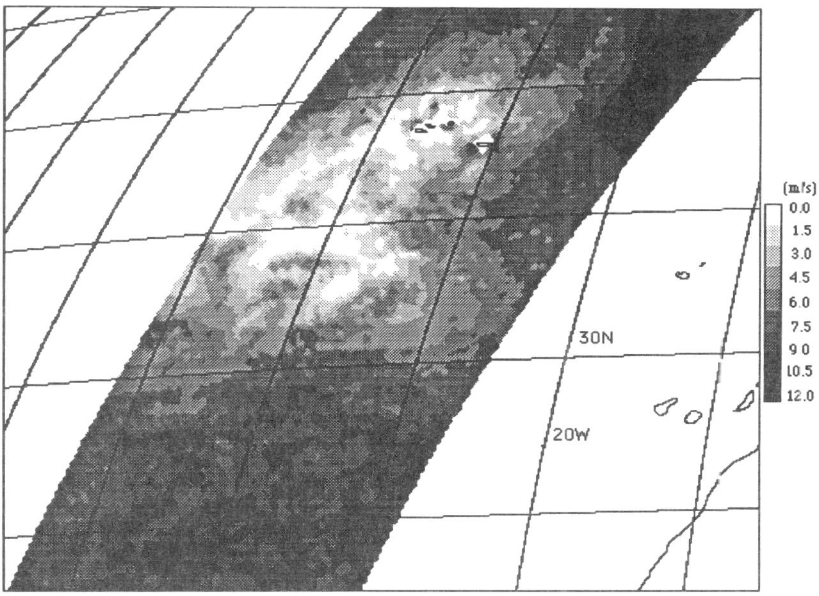

Figure 4. Wind speed fields (our retrieval) from SSM/I data, above the Azores, the 12 th June 1992, at 1100 GMT.

Table 1. Comparison of surface wind speeds in m/s at 12.5m altitude

	Lat	Lon	Buoys	Goodberlet et al, 1989	Our retrieval (Cox and Munk)	ECMWF	Our tretrieval (Bjerkaas, Riedel)
'10121'	35.08N	21.80W	9.9	8.3	7.4	6.7	5.7
'15496'	35.11N	21.45W	9.3	8.7	7.8	7.3	5.8
'15498'	35.36N	21.60W	8.7	10.3	8.1	7.1	6.3
'15499'	34.86N	22.45W	9.9	8.0	7.7	6.0	6.0

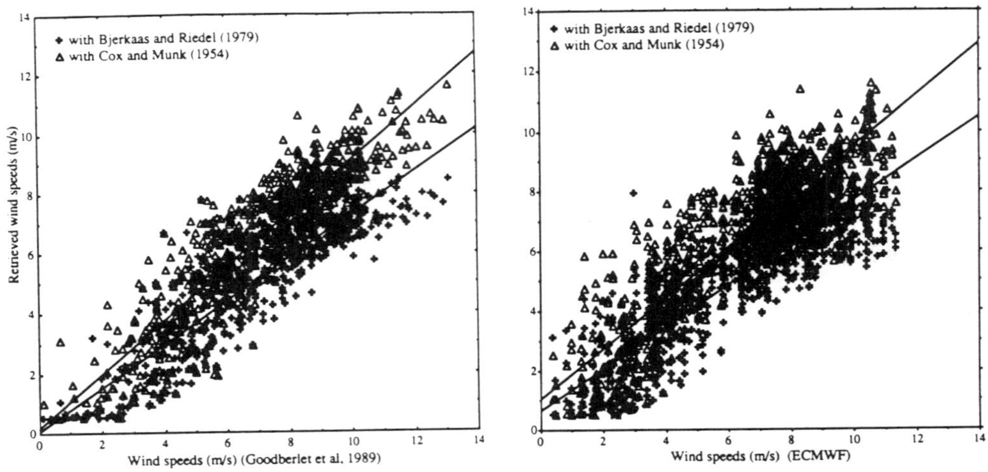

Figure 5. Our retrieved surface wind speeds from SSM/I data. Two different sea slope descriptions are used as input in the direct radiative transfer model for comparison.

Airborne radiometric measurements at 89 and 157GHz

In the context of the future humidity sounder AMSU-B in the water vapor line centered at 183GHz, the LMD (France) and the UK Meteorological Office have developed the Microwave Airborne Radiometer Scanning System, operating at 89 and 157GHz, frequencies close to the AMSU-B window channels. Fitted to the port side of the C-130 of the UKMO, and scanning along the aircraft track both upwards and downwards between +40° and -40°, the MARSS has flown in several international observation campaigns above the sea, since 1990. Brightness temperatures from a number of these experiments (SAAMEX, MASTEX, FATE, RENE) and from a few UK flights, covering a wide range of surface wind speeds (0 to 20m/s) and temperatures (275 to 300K), are analyzed here in comparison with radiative transfer simulations using the geometric optics model previously described, and an empirical emissivity model, based on Hollinger (1970) and Stogryn [14] measurements, as derived by Francis et al. [26]. Each measurement, given within ± 2K, is a combination of horizontal and vertical polarizations, depending on the scanning mirror position with respect to the sea surface. The results are presented here as the deviations between the measurement and the simulation (figure 6). They clearly indicate, first, that the empirical model adapted by Francis et al. to the SMMR applications cannot be extrapolated to the millimeter frequency domain, and secondly, that the use of Pierson and Stacy's [6] spectrum in the single scale model leads to a significant overestimation of the data, while most of the differences lie within the error bar of the data when using the Cox and Munk slope variances.

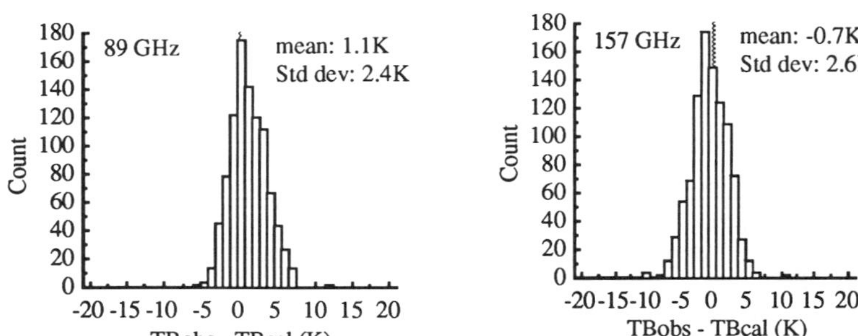

(a) Geometric Optics model - Cox and Munk (1954)

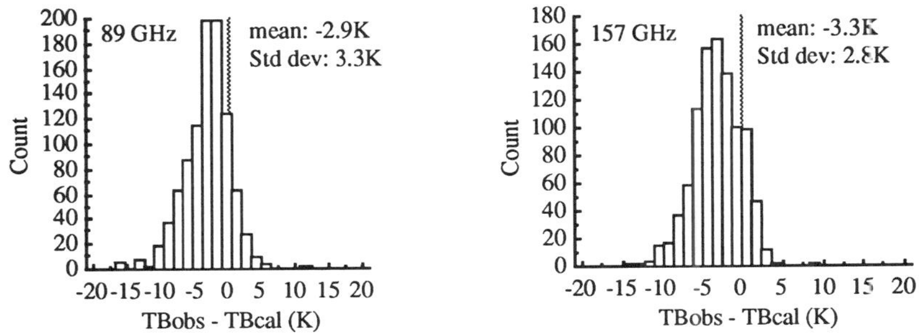

(b) Geometric Optics model - Pierson and Stacy (1973)

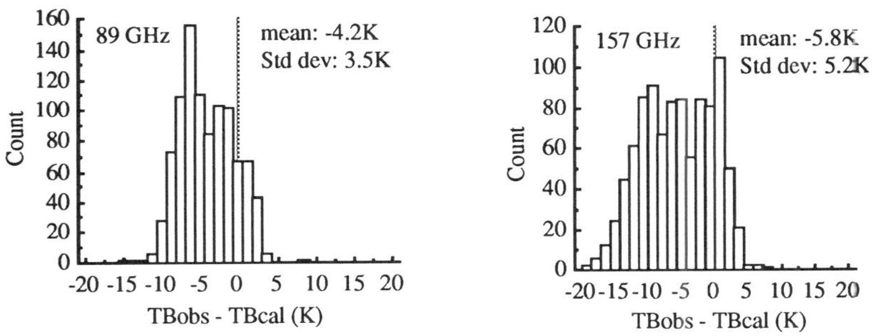

(c) Francis et al. (1983) empirical model

Figure 6. MARSS data at 89 and 157GHz.
Differences between observed and simulated brightness temperatures, with their associated means and standard deviations.

Slope variance estimates from radar measurements
Airborne radar measurements at 5.35GHz and at low incidence angle (around 14°) were also collected during a rehearsal campaign preparing for the RENE experiment, over the Haltenbanken area, off the coasts of Norway in 1990, from the RESSAC instrument (Hauser et al. [27]). A rotation system of the antenna around the vertical axis allowed to retrieve the azimuthal information of the signal. From measurements of the backscattering cross-sections against the viewing angles ranging from 7 to 21°, sea surface slope variances have been inferred (figure 7). Backscattering near the vertical incidence being dominated by the specular reflection mechanism, these slope variances cannot be representative of the whole wave spectrum, but only of the gravity wave domain. The results are consistent with other data sets (Jackson et al. [28]), in that the radar inferred slope variances are generally bounded by the Cox and Munk optical measurements for a clean sea and for an oil slick surface.

CONCLUSIONS AND PERSPECTIVES

An emissivity model based on the geometric optics approach has been shown to agree satisfactorily with the measurements when coupled with the Cox and Munk's description of the roughened surface, where the classical empirical emissivity models failed. In the prospect of new millimeter-wave radiometers (SSM/T, MHS, AMSU with frequencies from 20 to 200GHz), and in the context of simultaneous active and passive measurements of the ocean surface (ERS-1, Topex/Poseïdon), it is worth deriving a consistent approach that can be representative of both passive and active observations, covering a large frequency range and acceptable for diverse scanning geometries. New radiometric measurements at low frequency,

in well documented meteorological situations, are still required to analyze the validity limits of the geometric optics approach. As far as our knowledge of the ocean surface characteristics is limited, recourse to more complex scattering models (Integral Equation Models for instance), to describe the interaction process of electromagnetic and oceanic waves, is somewhat illusory.
From the early 1995, the DEIMOS radiometer, with channels at 23 and 50GHz, will be mounted on the C-130 of the UKMO, in addition to the MARSS instrument, to extend the range of observing frequencies. Airborne experiments are also planned for June 1994 above the Mediterranean sea, from the PORTOS (CNES) radiometer at frequencies between 5 and 90GHz, onboard the ARAT, to investigate the sea surface response, and notably its azimuthal dependance with wind direction.
Given the paucity of data and their limitation to 10GHz since Klein and Swift's [19] work, new laboratory measurements of the dielectric constant of the saline water, at higher frequencies (up to 110GHz), will be soon undertaken in a french laboratory (PIOM, Bordeaux, France), with a specific investigation in temperature (0 to 25°C).

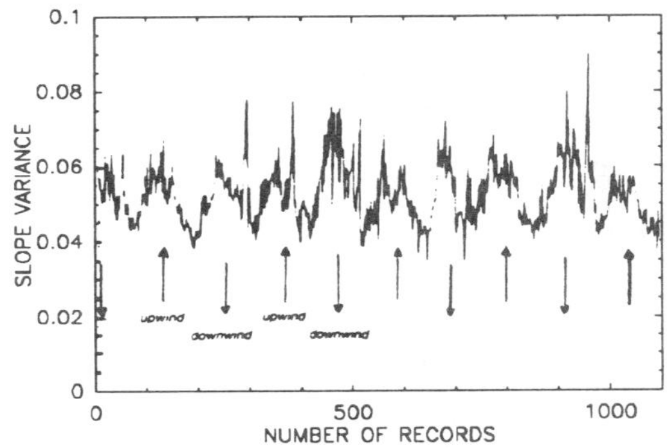

Figure 7. Slope variance retrievals in the gravity-wave range from airborne radar measurements: Time variation of the sea slope variance v inferred from the RESSAC scatterometer data, at 5.35 GHz below 20°. For an observed surface wind speed of 11m/s, the clean sea model of Cox and Munk yields $v=2\sigma_c^2=0.0482$ and $v=2\sigma_u^2=0.0696$ in the crosswind and upwind directions respectively, with an anisotropy $\sigma_c^2/\sigma_u^2=0.69$, comparable with the one observed (0.66) (from Hauser et al. 1992).

Acknowledgements
The SOFIA-ASTEX program is a cooperative work involving several groups. The authors are grateful to A. Weill (Centre d'étude pour l'Environnement Terrestre et Planétaire, CETP, France) for the access to the SOFIA-ASTEX data bank.

REFERENCES

1 Gagarin S. P. and B. G. Kutuza, Influence of sea roughness and atmospheric inhomogeneities on microwave radiation of the atmosphere-ocean system, *IEEE Journ. of Ocean. Engineen.*, vol. OE-8, No. 2, April 1983.
2 Kakar R. K., Retrieval of clear sky moisture profiles using the 183 GHz water vapor line, *J. Clim. Appl. Meteor.*, vol. 22, pp. 1282-1289, july 1983.
3 Wang J. R. and L. A. Chang, Retrieveal of water vapor profiles from microwave radiometric measurements near 90 and 183GHz, *Journ. Applied Meteoreology*, vol. 29, pp. 1005-1013, oct. 1990.
4 Schluessel P. and H. Luthardt, Surface wind speed over the north sea from Special Sensor Microwave/Imager Observations, *Jour. Geophysical Res.*, vol.96, no.c3, pp.4845-4853, march, 1991.
5 Cox C. and W. Munk, Measurements of the roughness of the sea surface from photographs of the sun's glitter, *Journal of the Optical Society of America*, vol. 44, n° 11, pp. 838-850, nov. 1954.
6 Pierson W. J. and R. A. Stacy, The elevation, slope, and curvature spectra of a wind roughened sea, NASA Contract. Rep., CR-2247, 1973.
7 Bjerkaas A. W. and F. P. Riedel, Proposed model for the elevation spectrum of a wind-roughened sea surface, *Tech. Memo. JHU/APL TG 1328*, 31 pp., Appl. Phys. Lab., Johns Hopkins Univ., Laurel, Md., 1979.

8 Phillips O. M., Spectral and statistical properties of the equilibrium range in wind-generated gravity waves, *J. Fluid Mech.*, vol. 156, pp. 505-531, 1985.

9 Hwang P. A. and O. H. Shemdin, The dependence of sea surface slope on atmospheric stability and swell conditions, *JGR*, vol. 93, No. C11, p 13,903-13,912, November 15, 1988.

10 Jähne B. and K. S. Riemer, Two-dimensional wave number spectra of small-scale water surface waves, *JGR*, vol. 95, No. C7, p 11,531-11,546, July 15, 1990.

11 Guissard A., C. Baufays, and P. Sobieski, Sea surface description requirements for electromagnetic scattering calculations, *Journal of Geophysical Research*, vol. 91, n° c2, 2477-2492, February 15, 1986.

12 Monahan E. C. and I. G. O'Muircheartaigh, Whitecaps and the passive remote sensing of the ocean surface, Review article, *Int. Remote Sensing*, vol. 7, n° 5, 627-642, 1986

13 Smith P. M., The emissivity of sea foam at 19 and 37GHz, *IEEE Trans. on Geos. and Remote Sensing*, vol. 26, n° 5, September 1988.

14 Stogryn A., The emissivity of sea foam at microwave frequencies, *J. Geophysical Research*, vol 77, pp 1658-1666, 1972.

15 Stogryn A., The apparent temperature of the sea at microwave frequencies, *IEEE Trans. Antennas Propag.*, vol. AP-15, pp. 278-286, 1967.

16 Wilheit T. Jr., A model for the microwave emissivity of the ocean's surface as a function of wind speed, *IEEE Trans. Geosci. Electron.*, vol. GE-17, n° 4, oct. 1979.

17 Hollinger J. P., Passive microwave measurements of sea surface roughness, *IEEE Transactions on Geoscience Electronics*, vol. GE-9, n° 3, pp. 165-169, juil. 1971.

18 Wu S. T. and A. K. Fung, A noncoherent model for microwave emissions and backscattering from the sea surface, *J. Geophys. Res.*, vol. 77, pp. 5917-5929, 1972.

19 Wentz F. J., "A two-scale scattering model for foam-free sea microwave brightness temperatures, *J. Geophys. Res.*, vol. 80, pp. 3441-3446, 1975.

20 Guissard A. and P. Sobieski, An approximate model for the microwave brightness temperature of the sea, *Int. J. Remote Sensing*, vol. 8, n° 11, pp. 1607-1627, 1987.

21 Klein L. A. and C. T. Swift , An improved model for the dielectric constant of sea water at microwave frequencies, *IEEE Trans. Antennas Propag.*, vol. AP-25, pp. 104-111, 1977.

22 Wu S. T., Oceanic whitecaps and sea state, *J. Phys. Oceano.*, vol. 9, pp. 1064-1068, 1979.

23 Prigent C., A. Sand, C. Klapisz, Y. Lemaitre, Physical retrieval of liquid water contents in a north Atlantic cyclone using SSM/I data, *Quaterly Journal of the Royal Meteorological Society*, in press, 1994.

24 Prigent C. and G. Seze, Estimation of cloud liquid water contents from SSM/I and Meteosat observations: Application to the Sofia-Astex experiment, *Specialist Meeting on Microwave Radiometry and Remote Sensing of the Environment*, Rome, Italy, 14-17 Feb, 1994.

25 Goodberlet M. A., C. T. Swift, J. Wilkerson, Remote sensing of ocean surface winds with the Special Sensor Microwave/Imager, *JGR*, vol. 94, pp 14,547-14,555, 1989.

26 Francis C. R., D. P. Thomas, and E. P. L. Windsor, The evaluation of SMMR retrieval algorithms, in Satellite Microwave Remote Sensing, edited by T. D. Allen, pp. 481-498, Ellis Horwood, Chichester, England, 1983.

27 Hauser D., G. Caudal, G. J. Rijckenberg, D. Vidal-Madjar, G. Laurent, P. Lancelin, RESSAC: A new airborne FM/CW radar ocean wave spectrometer, *IEEE Trans. on Geosci. and Rem. Sens.*, vol.30, n° 5, September 1992.

28 Jackson F. C., W. T. Walton and D. E. Hines, B. A. Walter, C. Y. Peng, Sea surface mean square slope from Ku-band backscatter data, *JGR*, vol. 97, No. C7, p 11,411-11,427, July 15, 1992.

On the Use of Different Ocean Surface Models in Radiative Transfer Calculation

MEENO SCHRADER and QUANHUA LIU

Institut fuer Meereskunde, Abt. Meteorologie D-24105 Kiel, Germany

Abstract

Surface reflectivity in radiative transfer models is often treated quite simple. In order to improve the retrieval of geophysical parameters more accurate models are necessary. A fast two scale model based on the KIRCHHOFF method is presented. Its results show good agreement with measurements.
Different foam models are compared. It is proposed to separate foam into its different constitutents: whitecaps, foamstreaks and spray. Their emissivities vary from 0.5 - 1.0 and have to be calculated separately. The application of the two scale model in a retrieval scheme for wind algorithms leads to the best results compared with commonly used parametrizations. We obtain an accuracy of 2.48 m/s and an explained variance of 74.9 %. The best values obtainable with the dataset used would be 2.19 m/s and 80.4 %.

Keywords: microwave radiometry, two scale model, foam, SSM/I, wind

Introduction

The retrieval of geophysical parameters from microwave radiometric measurements requires exact calculation of the contributions of the surface to the radiative transfer through the atmosphere. If a radiometer is directed from an aircraft or satellite towards the system earth - atmosphere, a main contribution of the signal received by the radiometer comes from the surface. Even over the ocean with low emissivity this contribution can reach up to 70 % of the total signal depending on frequency and atmospheric state. To obtain information about windspeed and -direction, surface tension, SST or energy fluxes it is very important to calculate the electromagnetical interaction (scattering and emission) in a most realistic manner. The objective of this paper is to look into the effect of different surface models on retrieval schemes.

Surface models

To calculate the reflectivity of a wind roughened ocean surface three effects must be considered (Fig.1):

1. the rough surface with all roughness scales from large gravity waves to small capillary waves
2. the foam appearances on the water surface
3. the degree of foam coverage

There are in general two possibilities to model the wind roughened foam-free surface:
1. Mirror reflection is assumed and the roughness contribution is parametrized in an empirical way. This model concept will be called MIRPAR in this study. The empirical methods are fast to calculate, simple to handle and quite accurate. Their disadvantage is the restriction in frequency, view angle and surface parameter interaction.
2. In a more physical way the reflectivity of the longer waves is calculated by the well known KIRCHHOFF - method of stationary phase approximation by angular integration of the scattered energy. The small scale contribution of the ripples can be taken into account by the small pertubation method. This is a so-called two scale model. The physical methods have a large range of validity and the inherent possibility to consider complex interactions. Those models need a lot of computer time and are impractible for retrieval schemes yet.

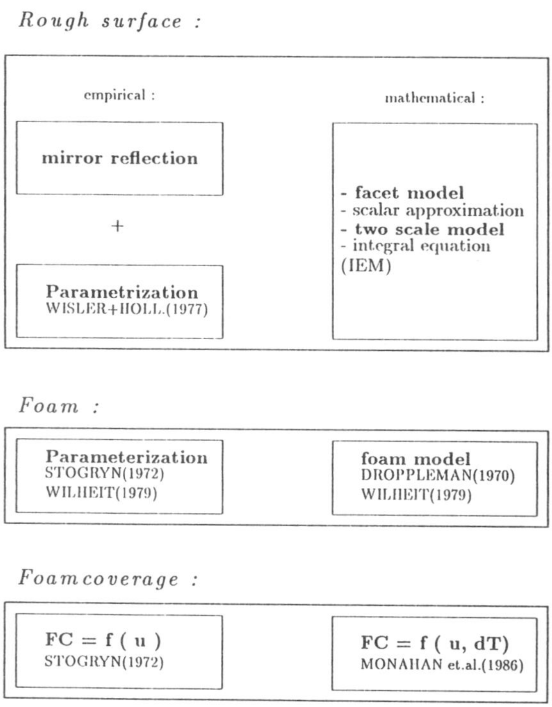

FIGURE 1. Surface models used in this study for comparison (bold printed)

Foam can be considered by using measurements of its reflectivity (e.g. [1], [2]) or their calculations by a foam model (e.g.[3], [4]). The latter gives the opportunity to take into account the microphysical parameters of foam in a direct manner. The effect of foam on the brightness temperature of the surface depends on the foamcoverage which has been measured as a function of the windspeed, the water temperature and the stability of the atmospheric boundary layer.

Comparison of the two scale model 2SCIFM with measurements

At our institute a two scale model for the surface, called 2SCIFM in the following, has recently been incorporated in our radiative transfer model.
The scattered electrical field of zeroth order is calculated by the KIRCHHOFF - model with stationary phase approximation formulated in ULABY et.al.[4]. This model gives the contribution to the reflectivity of the surface caused by the longer waves with curvature larger than the electromagnetical wavelength. It treats the surface as distribution of plane facets. The scattered energy is perceived as mirror reflection at individual facets weighted by their slope probability distribution. This is assumed to be gaussian and anisotropic in the upwind and crosswind direction. Due to its high sensitivity in the vertical component of the phase vector the probability function shows a sharp decrease. Inaccuracies arise in the case of backscattering ($\Theta_i = \Theta_s, \phi_i - \phi_s = 0$), where the contribution has to treat as a delta function. To approach this contribution by the probability function the inaccuracies could be minimized if the steps of integration are choosed very small $\Delta\phi \longrightarrow 0$. To calculate the reflectivity this leads to an inacceptable long computer time. Otherwise the error leads to an overestimated reflectivity which gives a negative emissivity at high incident angles ($\Theta_s > 75^o$).
To overcome this numerical inaccuracy a normalization procedure takes place in our facet model. This assumes the reciprocity in the scattering coefficient

$$\gamma^o(\Theta_s, \Theta_i)\cos(\Theta_i) = \gamma^o(\Theta_i, \Theta_s)\cos(\Theta_s) \tag{1}$$

which leads to a symmetrical scattering matrix in the incident and scattered angle. It has to be ensured that the contribution by the probability function P is

$$P\sin(\Theta_s)d\Theta_s d\phi_s \leq 1 \tag{2}$$

so that

$$\int P\sin(\Theta_s)d\Theta_s d\phi_s = 1 \tag{3}$$

This is achieved if the probability function is integrated over each incident angle. The normalized probability function is obtained by

$$P_{norm} = \frac{P(\Theta_i, \Theta_s, \phi_s)}{\int_0^{\pi/2}\int_0^{2\pi} P\sin(\Theta_s)d\Theta_s d\phi_s} \tag{4}$$

By the normalization procedure the azimuthal resolution can be reduced to 100 angles. This leads to reasonable values for the bistatic scattering coefficients especially at high incident angles and saves a large amount of computer time.
The scattered energy of higher orders is parameterized as function of the windspeed and zenith angles. This parametrization is derived by an adjustment of the KIRCHHOFF - model to foam-free measurements of HOLLINGER at nadir [5]. Due to HOLLINGER the measurements smaller than 15 degree view angle are influenced by the tower construction and have been excluded. The nadir values are obtained by extrapolating the measured values at higher view angles. Due to this parametrization in the second scale the model contains all residual effects of a rough surface except those caused by foam.

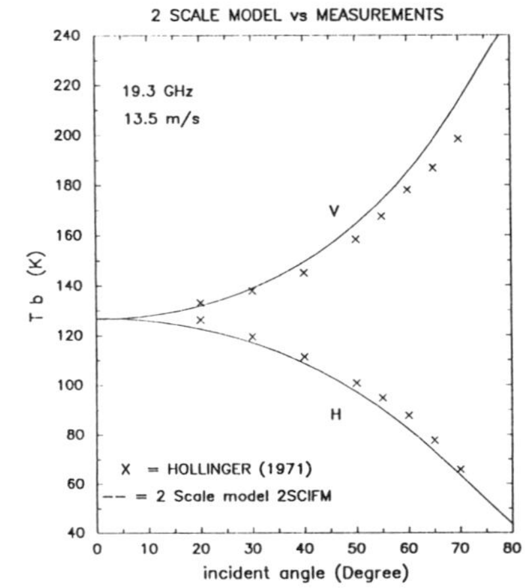

FIGURE 2. Comparison of the two scale model 2SCIFM of Kiel with measurements at 19.3 GHz for a wind speed of 13.5 m/s.

Figure 2 compares the brightness temperatures calculated with 2SCIFM with measurements. As expected there is a quite good agreement. However, the model tends to overestimate the vertically polarized brightness temperatures for incident angles larger than 35 degrees and to underestimate the horizontal polarization over a wide range of incident angles. But these differencies are still in the range of error of 5 - 10 percent given by HOLLINGER . A comparison of the 2SCIFM - model with the two scale model of [6] shows a very good agreement for the incident angles smaller than 45 degree in both polarizations. But there is a small over- and underestimation in vertical and horizontal polarization, respectively. Their results differ in maximum by 5 Kelvin in horizontal polarization at 60 degree and 14 Kelvin in vertical polarization at 70 degree. This can be explained by the normalization, which leads to a higher emissivity at large incident angles.

Foam

Due to its high emissivity (typically 0.7 - 1.0) an important contribution to the microwave signal is caused by foam. With foam in general we mean different features arising on the surface due to an increasing windspeed namely foamstreaks, whitecaps and spray. Their structures are different in general, but similar to a foamlayer and bubble accumulation, respectively. Their emission characteristics are strongly controlled by the thickness of the foam layer, the viewangle, the ratio of water and air, the frequency, the bubble size distribution and other constitutents.

Up to now these phenomena are treated in literature as only one physical quantity called foam or white water, although the emissivity of a foamlayer and of bubble rafts on water are not the same [7]. Tank measurements of WILLIAMS [8] show an increase of the emissivity of a foam layer over water from 0.7 to 1.0 with an increase of the foam thickness from 1 mm to 7 mm.

This is in qualitative agreement with radiometric tank measurements of EDGERTON et.al. and active measurements of ROOTH and WILLIAMS carried out at different view angles given by STOGRYN [1]. The most accurate spectral radiometric measurements of foamlayers are carried out by [9]. Their results show high emissivities of 0.965 and 0.977 for the horizontal and vertical polarization, respectively (Figure 3). The foam model of DROPPLEMAN (i.e. [3]) assumes foam streaks as a porous dielectric layer with a dielectric constant of foam ϵ_f

$$\epsilon_f = \epsilon_w \left[1 - \frac{3v_a(\epsilon_w - 1)}{(2\epsilon_w + 1) + v_a(\epsilon_w - 1)} \right] \tag{5}$$

ϵ_w is the complex dielectric constant of water, v_a is the air volume fraction in foam. This is assumed to be 0.0045 to achieve the best agreement with measurements.The foam layer is treated as a scattering medium regarding wave propagation as an incoherent process. The reflectivity of the foam is calculated as function of the SST, frequency, incident angle, height of the foamlayer and the ratio of water and air of the foam volume by a formulation of ULABY et.al.[4].

A validation of the model results is very difficult due to the few existing measurements. Figure 3 shows a comparison at 37 GHz, where most of the experimental data exist in literature. There is a good agreement in the case of vertical polarization (small symbols) for a thickness of the foam layer d = 13.4 mm. This value is calculated by a parametrization of [10]. For the horizontal polarization good agreement only exist with the measurements of BORDONSKIY [9]. A decrease in the thickness of the foam layer leads to a decrease in the emissivity and an increase in the difference of the polarization. This behaviour is represented by the foam model with decreasing thickness of the foam layer, but no data exist at high view angles.

POE [1] made active measurements on a foam layer with d = 5 mm in thickness at higher view angles. But only cruede assumptions lead to the emissivity derived from a single scattering coefficient.In particular at the horizontal polarization there is a high sensitivity of stratified foam in emissivity due to its structure. Another occurence of white water are the white caps. They have to be regarded as a mixture of bubbles on the water layer, in the air (spray) and foam streaks. This assumes a mixture in the emissivity characteristics of these components. Measurements of WILLIAMS [8] of the emissivity of bubble rafts over water show values in the range of 0.5 to 0.6 for bubble diameters of 1 to 2 mm. This bubble size is typical for whitecaps [11]. From there a lower emissivity has to be expected for white caps than for foam streaks. Radiometric measurements of breaking waves in the surf zone of a beach [12], from a tower [5] at different view angles and from an aircraft [13] show this behavior. Additionally, they show a decrease in brightness temperature with an increase in view angle. The interpretation of this view angle dependency has to be done carefully: 1.The geometry of the surface is disturbed by a breaking event.2.With increasing view angle the contribution to the radiometer signal is less determined by the signal of the white caps only. The measured brightness temperatures are influenced by contributions of foam in general as described above and the foamless water surface with alower emissivity.

STOGRYN [1] parametrized these foam effects based on the tower and surf measurements given above and aircraft observations. Independant aircraft measurements of [13] confirm this parametrization, but they show also a high variability of the emissivity of white caps (see the bars in Figure 3, left). His empirical model leads to lower values in emissivity than the foam model of DROPPLEMAN predicts. There is a decrease in emissivity with increasing view angle for the reasons discussed before. The foam effect parametrization of STOGRYN, called FOPASTO in the following, is used in this study. It is a function of the incident angle, frequency and SST. The validity is restricted to the range from 13.4

GHz to 40 GHz and from 0 degree to 70 degree in incident angle.

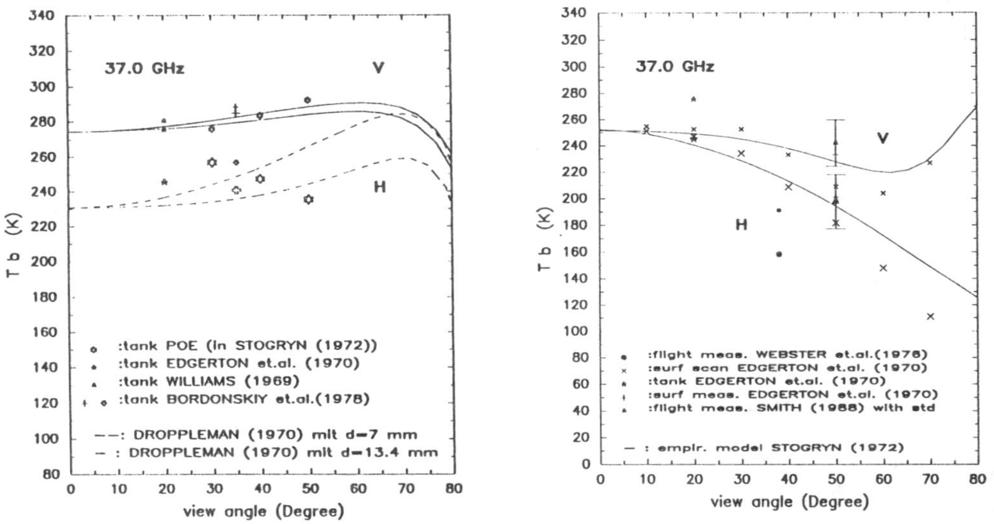

FIGURE 3. Comparison of the brightness temperatures out of the foam model of DROPPLEMAN(1970) with measurements at 37 GHz and SST = 295 K. The curves are obtained for a foam layer of d=7 mm and d=13.4 mm.Higher values in symbols and curves are for vertical, lower for horizontal polarization(**left**).
Brightness temperatures measured by different authors and calculated by the empirical model of STOGRYN at 37 GHz and SST=295 K. The model is derived from some of these measurements. The bars show the 95 %-confidence interval of the measurements of SMITH(**right**).

For comparison the foam model of DROPPLEMAN [3] is considered and is called FOMDROP in the following. FOPASTO and FOMDROP are also used in the retrieval scheme to derive wind algorithms in the last chapter.
In the study of different models calculating the reflectivity of foam a third approach is considered which are common in the use of radiative transfer models: a parametrization derived by WILHEIT [2] estimates a factor reducing the reflectivity of the rough surface due to foam. It depends on frequency, wind speed and the onset of foam development.
The impact of the foam features to the measured brightness temperature depends on the degree of foamcoverage. Usually foam coverage is estimated as a function of the windspeed alone [1] or, as a function of windspeed and stability of the boundary layer [14]. However, there are different assumptions when foam starts to occur. WILHEIT [2] assumes a threshold windspeed of 7 m/s. Recent investigations of MONAHAN et.al.[14] showed that foam build-up starts already at windspeeds of 3 - 4 m/s. They also postulate a dependence on the sea surface temperature and the stratification of the boundary layer.
For comparison the three models FOPASTO, FOMDROP and the parametrization of WILHEIT are used in the surface model 2SCIFM. Their contributions are regarded separately in Fig.4. A comparison of FOPASTO and FOMDROP is shown for 19 GHz and 37 GHz. There are only very small differences up to 40 degree incident angle in both polarizations. At higher angles the differences increase. At 37 GHz the differences increase especially at higher angles. They are larger for the horizontal than for the vertical polarization. At 53.0 degree the differences are only 0.5 K in the vertical and 1.0 K in the horizontal polarization.

The differences range from 1 to 3 degree Kelvin for all incident angles between the model of WILHEIT and FOMDROP and FOPASTO. The former leading to lower emissivities. These differences decrease to roughly 0.5 K, if the degree of foam coverage is reduced by assuming stable atmospheric conditions. This reduces the development of foam but has no influence on the parametrization of WILHEIT. The differences can be explained by the high foam onset threshold proclaimed by WILHEIT leading to an underestimation of the foam coverage.

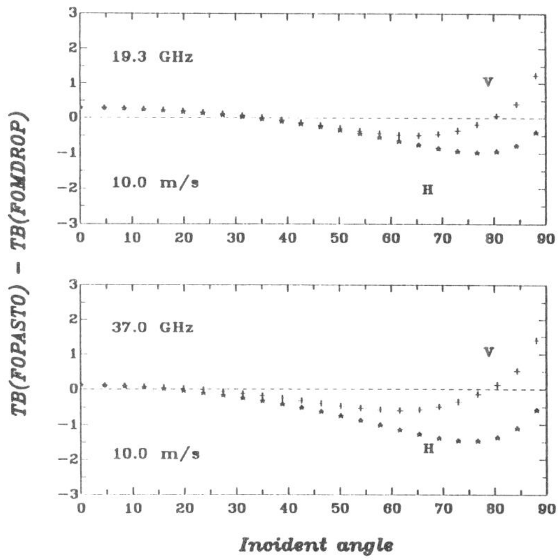

FIGURE 4. Differences in brightness temperature between the foam model of DROPPLEMAN and the empirical model of STOGRYN at 19.3 GHz (top) and 37.0 GHz (below) at a moderate windspeed of 10 m/s and neutral stability condition. This leads to a foamcoverage of 3 %. SST = 288 K, salinity = 35 % .

Concerning the development of foam there are many other dependencies as SST, duration of wind, fetch, salinity, dissolved organic materials and/or the presence of an organic monolayer on the surface.
No parametrized relationship between these quantities and the foam coverage is known to the authors. Much more work has to be done to arrive at more realistic values for the development of foam.

Derivation of wind algorithms

The detection of geophysical parameters of the ocean surface i.e. wind speed by satellite measurements requires a retrieval scheme. To derive algorithms for geophysical parameters we used multipie linear regression applied to simulated radiation temperatures. For the simulation we used the radiative transfer model of SIMMER [15] (see also FUHRHOP,R. this issue). The dataset used consists of 250 radiosonde ascents over the Atlantic Ocean from the year 1989. Wind algorithms were derived for the five lower frequencies of the SSM/I (Special Sensor Microwave Imager) for different combinations of the surface models as explained above. The satellite estimated windspeed is compared with in-situ measurements. Their time difference is less than 2 hours.

Different combinations, linear and nonlinear, for the SSM/I channels are applied to develop an optimum algorithm. To exclude rain and its masking influence to the brightness temperature only atmospheric profiles with a total liquid water content LWP below 0.5 kg/m^2 are taken into consideration. The method to calculate the LWP from radio ascents and the threshold in LWP for rain is described by HARGENS et.al.[16].
Best results are obtained with a nonlinear five channel algorithm

$$\begin{aligned} V &= 17.926\,ln(280 - TB_{19v}) + 0.53773\,TB_{19h} - 0.22837 TB_{22v} \\ &\quad +40.093\,ln(280 - TB_{37v}) + 0.18666 TB_{37h} - 286.956 \end{aligned} \tag{6}$$

with V in m/s.

Comparison of wind algorithms

According to SIMMER [15] only up to 82.5 % of the variance of surface wind can be explained by the microwave signal between 5 and 40 GHz. This leads to a maximum accuracy of 1.5 m/s. Figure 5 shows the comparison of satellite derived wind speed developed by the use of the surface model MIRPAR (left) and by the use of 2SCIFM (right) against measurements.

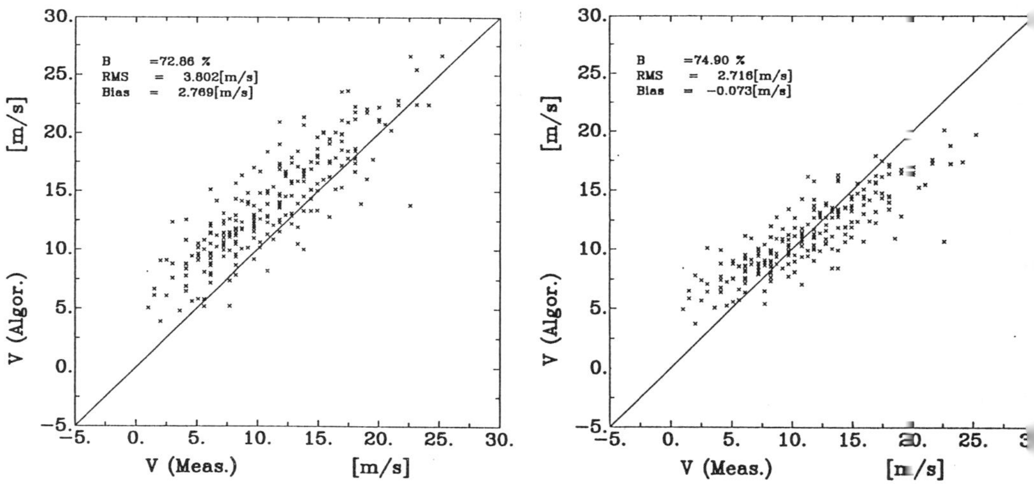

Figure 5. Satellite derived wind speed with an algorithm calculated with MIRPAR as surface model against measured values (left) and with an algorithm derived by 2SCIFM as surface model (right).

For the MIRPAR - algorithm there is a moderate scatter over the whole spectrum of observed windspeeds ranging from 3 m/s to 25 m/s in this dataset. There is an overestimation by the satellite leading to a bias of 2.7 m/s. 72.9 % of the variance can be explained by this algorithm.
With the 2SCIFM - algorithm the scatter is reduced. 74.9 % of the variance can be explained by this algorithm and there is almost no bias. But lower windspeeds up to 7 m/s are overestimated and windspeeds higher than 15 m/s are underestimated. Table 2 shows the best results of this study.

The best values obtainable with the dataset used are shown in the left column. The regression error is in the range of 0.9 m/s. The collocation and measurement error is 2.0 m/s[17]. The summation of these errors gives a rms of 2.19 m/s. The variance of the windfield of 24.5 m^2/s^2 leads to 80.4 % for the variance which could be explained in maximum by the use of this dataset.

Table 2. Statistical results for the comparison of different wind algorithms with measurements.

	best possible this dataset	2 scale + foam model	mirror + rough par. + foam model	2 scale + foam par.
rel. expl. variance in %	80.4	74.9	72.8	74.8
rms (m/s)	2.19	2.48	2.58	2.48
bias (m/s)	0.00	-0.07	3.27	-0.07

The next two columns show the results for the algorithm obtained with 2SCIFM and MIRPAR. Both retrievals are calculated with the foam model FOMDROP. The two scale algorithm can explain more of the variance than the others. The rms error is a little smaller for the two scale retrieval, but this value assumes that there is no bias. The remotely sensed wind speed has already an accuracy close to the best possible estimate. But improvements due to the over- / underestimation have to be established, yet.
The fourth column shows the results obtained with an algorithm using FOPASTO as foam model. There is almost no difference between the results with FOPASTO and FOMDROP, but the result tends to be better with FOMDROP as foam model.

Conclusions

Different surface models for the application in radiative transfer calculations are presented and compared in this study. The two scale model used is a fast and quite accurate surface model for the retrieval of geophysical parameters. Improvements are neccessary to remove the behavior of over- and underestimation at different windspeeds.
The calculation of the reflectivity of a foamlayer by a foam model of DROPPLEMAN leads to quite different results compared to the empirical model of STOGRYN. Incorporated in radiative transfer calculations the results in brightness temperature are more similar, but different by a few Kelvin depending on the wind speed. This is due to the small share of foamcovered water in the total area. Although small these differences can lead to high discrepancies in the retrieval of the precipitable water content, liquid water path and other parameters (e.g. $\Delta T_B = 1K \Rightarrow \Delta PWC = 1kg/m^2$ with PWC = precipitable water content).
In order to improve the contribution of foam different models have to be added for the reflectivity of foamstreaks, whitecaps and spray.
The best results for the detection of wind speed are obtained if the reflectivity of the surface is calculated with the two scale model combined with the foam model of DROPPLEMAN. A better consideration of the spectral features of the waves, the impact of ocean - atmosphere interaction, a more sophisticated distinction of the different forms of foam and their impact to the reflectivity and a more extensive consideration of the

processes which lead to the development of foam has to be taken into account. This promises higher accuracies in the modelled brightness temperatures and a more accurate estimate of the remotely sensed geophysical parameters.

References

[1] A. STOGRYN, 1972, The emissivity of sea foam at microwave frequencies, *Journal of Geophysical Research*, **77**, no. 9, 1658–1666.

[2] T. WILHEIT, Jr., 1979, A model for the microwave emissivity of the ocean's surface as a function of wind speed, *IEEE Trans. Geosc. Electronics*, **GE-17**, 244–249.

[3] J. DROPPLEMAN, 1970, Apparent microwave emissivity of sea foam, *Journal of Geophysical Research*, **75**, no. 3, 696–698.

[4] F. ULABY, R. MOORE, and A. FUNG, 1983, Microwave remote sensing, active and passive, volume i:fundamentals and radiometry, *Addison Wesley Publ. Comp., London*, 232ff.

[5] J. HOLLINGER, 1970, Passive microwave measurements of the sea surface, *Journal of Geophysical Research*, **75**, no. 27, 5209–5213.

[6] F. WENTZ, 1975, A two-scale scattering model for foam-free sea microwave brightness temperatures, *Journal of Geophysical Research*, **80**, no. 24, 3441–3446.

[7] G. WILLIAMS, Jr., August 1969, Microwave radiometry of the ocean and the possibility of marine wind velocity determination from satellite observations, *Journal of Geophysical Research*, **74**, 4591–4594.

[8] G. WILLIAMS, Jr., 1971, Microwave emissivity measurements of bubbles and foam, *IEEE Trans. Geosc. Electronics*, **GE-9**, no. 4, 221–224.

[9] G. BORDONSKIY, I. VASIL'KOVA, V. VESELOV, N. VORSIN, Y. A. MILITSKIY, V. MIROVSKIY, V. NIKITIN, V. RAYZER, K. Yu.B., Y. SHARKOV, and V. ETKIN, 1978, Spectral characteristics of the emissivity of foam formations, *Izvestiya, Atmospheric and Oceanic Physics*, **14**, no. 6, 464–469.

[10] Y. MIYAKE and T. ABE, 1948, A study of the foaming of sea water, *Journal of Marine Research*, **7**, no. 2, 67–73.

[11] R. BORTKOVSKI, 1987, Air-sea exchange of heat and moisture during storms, *D.Reidel Publ.Comp.*, 31.

[12] A. EDGERTON, G. TREXLER, A. POE, S. STOGRYN, J. SAKAMOTO, J. JENKINS, D. MEEKS, and F. SOLTIS, 1970, Passive microwave measurements of snow, soils, and oceanographic phenomena, *Tech. Rep. 6 SD 9016-16*, **Aerojet-General Corp.,El Monte, Calif.**

[13] P. SMITH, September 1988, The emissivity of sea foam at 19 and 37 ghz, *IEEE Trans. Geosc. and Remote Sensing*, **26**.

[14] E. MONAHAN and I. O'MUIRCHEARTAIGH, 1986, Whitecaps and the passive remote sensing of the ocean surface, *International Journal of Remote Sensing*, **7**, no. 5, 627–642.

[15] C. SIMMER, 1994, *Satellitenfernerkundung hydrologischer Parameter der Atmosph"are mit Mikrowellen.* Verlag Dr. Kovac, Hamburg.

[16] U. HARGENS, C. SIMMER, and E. RUPRECHT, 1994, Remote sensing of cloud liquid water, *accepted for publication in Meteorology and Atmosph. Physics.*

[17] K. NIEKAMP, 1994. personal communication.

Microwave Dielectric Model of Leaves

Christian Mätzler
Institute of Applied Physics, University of Bern
Sidlerstr. 5, 3012 Bern, Switzerland

Key words: Complex dielectric constant, permittivity, leaves, Debye relaxation, temperature dependence

Summary: (This note summarizes a contribution which will be published elsewhere [2]. In addition a further test is presented here.)

The complex permittivity, $\varepsilon = \varepsilon' + i\varepsilon''$, of vegetation plays a central role in the coupling between the electromagnetic properties of a vegetation canopy and its physical properties. A semi-empirical formula for broad leaves of different plants was recently developed from a comparison of reported measurements covering the frequency range from 1 to 100 GHz (Mätzler, 1994). The explicit parameters are the dry-matter fraction, m_d = (dry mass)/(fresh mass), of the leaf, and the complex permittivity, ε_{sw}, of saline water (e.g. Ulaby et al. 1986). Implicit parameters are temperature, salinity and frequency as determined by ε_{sw}. For most plants the salinity S is close to 1%. The physical part of the formula is its basis on ε_{sw}, while the empirical part is its linearity with m_d. The formula is applicable to fresh leaves; their m_d value is in the range $0.1 < m_d < 0.5$ while their density is near $1 g/cm^3$. The explicit form is given [2] by

$$\varepsilon = 0.522(1 - 1.32m_d)\varepsilon_{sw} + 0.51 + 3.84m_d; \qquad S \cong 1\%; \qquad (1)$$

Model curves for $m_d = 0$ and 0.5 together with extrapolated measurements to these m_d values are presented in Figure 1.

In a first test of Equation (1) the computed temperature dependence of the optical depth of a beech tree at 4.9 GHz was compared with actual measurements [2]. Here another test is presented based on experimental data of both the real part ε' and the imaginary part ε''. The test frequency is 4 GHz, the data are from El Rayes and Ulaby, 1987 [1], and the object is a fresh fatshedera leaf. The temperature dependence is shown in Figure 2. Note that the experimental data were obtained by reducing the temperature from 30°C to -5°C; freezing occurred only at still lower temperatures. The data agree with the model within about 10%. The decrease of ε'' with increasing temperature is mainly due to the change of the relaxation frequency, f_0, increasing from 9 GHz at 0°C to about 22 GHz at 30°C. Further tests with the same data, but at 1 and 8 GHz show similar results as Figure 2.

The simplicity of Equation (1) and the validity over a large spectrum covering 1 to 100 GHz makes this new formula attractive for the electromagnetic modeling of vegetation canopies.

References:

1. El Rayes M.A. and F.T. Ulaby (1987): Microwave dielectric behavior of vegetation material, NASA GSFC Report, Contract NAG 5-480, Jan. 1987.
2. Mätzler C. (1994): "Microwave (1-100 GHz) dielectric model of leaves", IEEE Trans. Geoscience and Remote Sensing, in press.
3. Ulaby F.T., R.K. Moore and A.K. Fung (1986): Microwave Remote Sensing, Active and Passive, Vol. 3, Appendix E-2, Artech House, Dedham, MA-U.S.A.

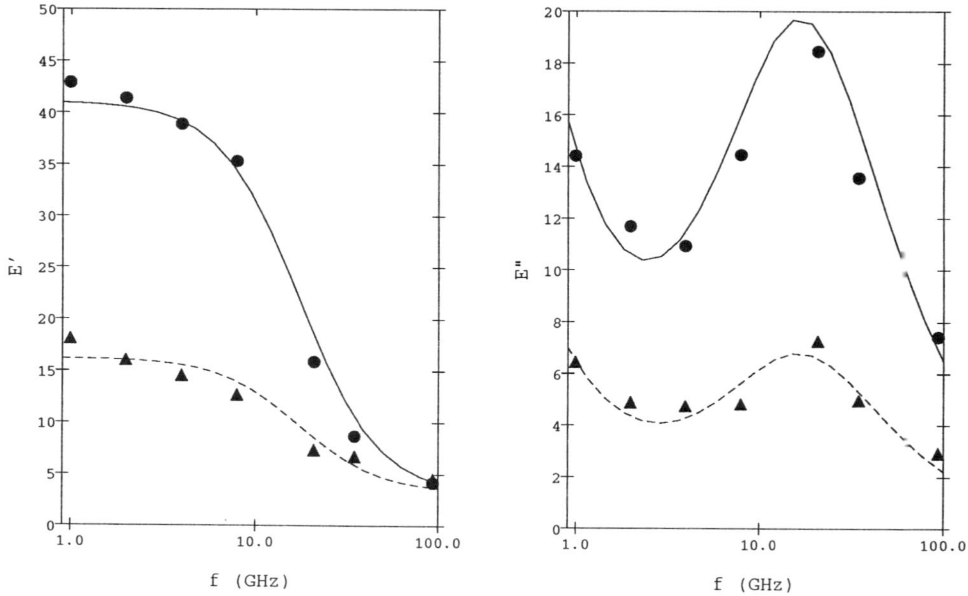

Figure 1: Real parts (left), and imaginary parts (right) of Equation (1) for m_d=0 (solid lines) and m_d=0.5 (dashed) for T=20°C together with experimental data at 1, 2, 4, 8, 21, 35, 94 GHz extrapolated to these m_d values [2].

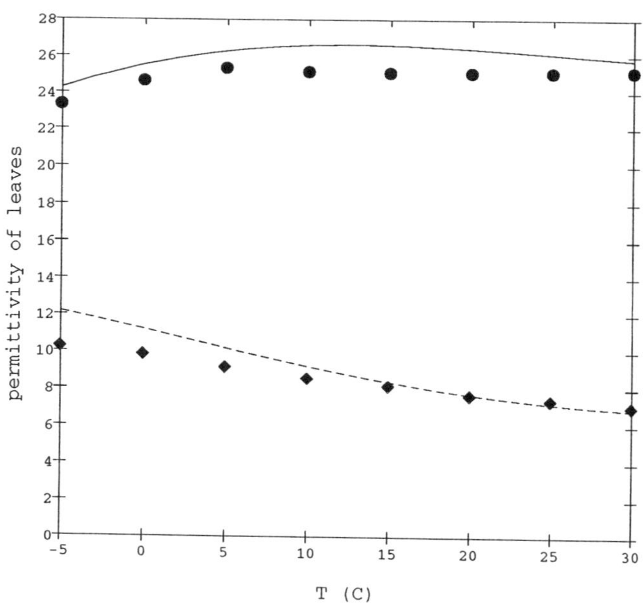

Figure 2: Measured (data points, from [2]) and computed (lines, Equation 1) complex permittivity versus temperature from -5°C to +30°C for unfrozen fatshedera leaves at f=4 GHz, m_d =0.27. The upper curves are the real part, lower curves the imaginary part.

Potential of multifrequency techniques in microwave radiometry of crops

P. FERRAZZOLI[1], L. GUERRIERO[1], S. PALOSCIA[2] and P. PAMPALONI[2]

[1] *Dipartimento di Ingegneria Elettronica, Università Tor Vergata Via della Ricerca Scientifica, 00133 Roma (I)*
[2] *Istituto di Ricerca sulle Onde Elettromagnetiche - CNR Via Panciatichi 64, 50127 Firenze (I)*

Abstract Brightness temperatures measured at L, X and K_a band over various kinds of crops are compared with theoretical values predicted by a discrete model based on the Radiative Transfer theory. A substantial agreement between experimental and theoretical results is noted. By properly combining data acquired at different frequencies and polarizations a satisfactory sensitivity to crop type and crop growth may be achieved.

Keywords: Microwave emission, Crops, Modeling, Multifrequency techniques

1. INTRODUCTION

Detection of vegetation properties is one of the most important goals of the future remote sensing systems aiming towards monitoring global changes and understanding cycles of water, biogeochemicals and energy. Much theoretical and experimental work has been done on studying microwave emission from land surfaces. Agricultural fields located in various countries, characterized by different climatic conditions, have been observed by microwave radiometers and sensor responses have been correlated to crop conditions [1-4]. Moreover, theoretical models have been developed, aimed at simulating microwave emission behaviour of soil and vegetation at various frequencies [5-8].

It is recognized that microwave emission from agricultural fields depends on the properties of both soil (moisture, roughness) and crop (type, stage of growth, wetness). In general, soil properties play a dominant role at low frequencies and low observation angles (typically L band, $\theta <\sim 20^o$), while vegetation properties are important at high frequencies and high observation angles. However, when information about a single parameter are required, the effects of the other parameters must be taken into account, since they are not completely negligible. In order to separate the effects of the required parameter from those of the others, it is necessary to optimize system configuration and processing of the measured data. The separation capability is generally poor if a single frequency is used, but it can substantially be improved by a multifrequency configuration.

In this paper, the problem of optimizing the radiometer parameters for detecting crop type and crop growth is analyzed. Brightness temperatures measured at L (1.4 GHz), X

(9.6 GHz) and K_a (36 GHz) band over Italian agricultural fields in the summer season are shown. Measurements were carried out over bare soils and various kinds of crops (mainly corn, sunflower, wheat and alfalfa) at different stages of growth. The experimental data are combined in different ways, by using polarization and frequency indexes, and the potential of each index in the retrieval of crop type and crop growth is discussed.

In parallel with the analysis based on experimental results, a theoretical study is carried out by using an electromagnetic model [8-10] which is based on the Radiative Transfer theory and describes the agricultural field as a half-space with rough interface (representing the soil) overlaid by an ensemble of discrete scatterers (representing the vegetation elements). Scatterers are selected to be discs and cylinders and their dimensions, orientations and locations are chosen according to the particular geometry of the modeled crop.

Experimental and theoretical results are in substantial agreement to each other and indicate that, by properly combining the brightness temperatures measured at L, X and K_a band, a satisfactory sensitivity to crop type and crop growth may be achieved.

2. MATERIALS AND METHODS

2.1 *Field measurements*

In the summers of 1987 and 1988 a profile multichannel microwave radiometer (Microwave Sensor Package - MSP) was flown over two agricultural areas at different periods of crop growth, so that several types of soils and crops were observed with an incidence angle θ varying between 10 and 40 degrees from nadir. Ground truth measurements were also performed according to tested procedures [2]. The MSP, purposely designed for land studies, included microwave radiometers at L (1.4 GHz, V or H polarizations), X (9.6 GHz, V and H pol.) and K_a (36 GHz, V and H pol.) band, a thermal infrared sensor (8-14 μm) and a colour TV camera used to geolocate data.

Both X and K_a band radiometers were able to measure almost simultaneously H and V polarization components of the brightness temperatures. L band data were available only at horizontal polarization and in a limited number of measurements. Bare soil, corn, sunflower, wheat, barley and alfalfa fields were observed during the experiments. Also some X, K_a and IR band brightness temperatures of sugarbeet fields were available. Details concerning the instruments and the airborne campaigns are given elsewhere [11, 12].

Other measurements at X, K_a and IR band had been carried out using ground based radiometers in the summers of previous years over plots of corn, wheat and alfalfa [2].

2.2 *The model*

The model assumes the vegetation medium to be a homogeneous half-space with rough interface, representing the soil, overlaid by an ensemble of discrete lossy scatterers, representing the plant constituents. Details are given elsewhere [8, 10]; only the main aspects are summarized here.

The electromagnetic properties of the soil are described by its bistatic scattering coefficient. The latter is computed by the Small Perturbation approximation at L band, while the Geometrical Optics approximation, with the shadowing function derived by Smith, is used at X and K_a band [1]. The semiempirical model proposed in [1] is used to compute the soil permittivity.

The electromagnetic properties of the scatterers, which represent the plant constituents, are described by the absorption cross sections $\sigma_a(\theta, \phi)$ and by the bistatic scattering cross sections $\sigma(\theta, \phi; \theta_s, \phi_s)$. θ and θ_s are the incidence and scattering off-normal angles, while ϕ and ϕ_s are the incidence and scattering azimuth angles. Dielectric elements of simple shape, such as discs and cylinders, are assumed. Discs represent small leaves or parts of large leaves. At L band the disc cross-sections are computed by means of the Rayleigh-Gans approximation [13], which assumes the disc to be an oblate ellipsoid with one axis small with respect to the wavelength, so that the inner field may be considered homogeneous along the small dimension. At X and K_a band the disc cross sections are computed by using the Physical Optics approximation [14], which assumes the internal field to be the same as that of an infinitely extended slab. Cylinders may represent stems or ears of wheat and barley. For these kinds of scatterers the theory proposed in [15] is used, which assumes the internal field to be the same as that of an infinite length cylinder.

The dimensions of the scatterers (radius and thickness of discs, radius and length of cylinders) are chosen in such a way as to represent, with reasonable accuracy, the real plant constituents. The dielectric constant ϵ_r is expressed as a function of the plant moisture, by empirically fitting the experimental data published by El-Rayes and Ulaby [16]. Once the absorption and bistatic scattering cross sections of the scatterers have been computed for a discrete set of incidence and scattering directions, the electromagnetic behaviour of the ensemble of scatterers is obtained. To this end, the numerical algorithm described in [8] and [10] is used.

Three canopy structures are considered in this paper.

Corn and sunflower crops are described by upper randomly oriented discs, representing the leaves, lower vertical cylinders, representing the stalks, and a half-space with rough interface, representing the soil. Outputs are used for comparisons with brightness temperatures measured over corn and sunflower crops. At X and K_a band the emission contribution of vertical cylinders is low [9], so that the comparisons at these bands can be extended to sugarbeet crops, although they do not show long vertical stems.

Wheat and barley crops are described by upper vertical cylinders, representing the ears, lower randomly oriented discs, representing the leaves, and a half-space.

Alfalfa crops are modeled as a layer of vertical cylinders surmounted by alternate layers of inclined discs and cylinders and assuming that the latter elements may be oriented according to a $cos^2\beta$ distribution law (β is the cylinder axis orientation angle with respect to the normal).

Table 1 shows the model inputs used for the three structures. Further details are available in [9] and [10].

3. RESULTS

3.1 *Normalized temperatures*

The normalized temperatures are defined as follows.

- K_a band: $KT_n = KT_b/IRT_b$
- X band: $XT_n = XT_b/IRT_b$
- L band: $LT_n = LT_b/IRT_b$

IRT_b, KT_b, XT_b and LT_b are brightness temperatures at IR, K_a, X and L band, respectively. In the theoretical data, the model computed emissivities have been corrected by a

Table 1. Model Inputs. FG: Full Growth maximum value.

	Wheat, Barley	Alfalfa	Corn, Sunflower
Disc radius $[cm]$	0.5	0.5	0.5÷ 2.5
Disc thickness $[mm]$	0.15	0.15	0.25
Disc density (FG) $[m^{-2}]$	$6.3 \cdot 10^4$	$5 \cdot 10^4$	$3.3 \cdot 10^3 \div 8.2 \cdot 10^4$
Vertical cylinder radius $[cm]$	0.5	0.2	1.4
Inclined cylinder radius $[cm]$		0.05÷0.1	
Vertical cylinder height $[cm]$	8	20	128
Inclined cylinder length $[cm]$		5÷10	
Vertical cylinder density $[m^{-2}]$	700 (FG)	500	7
Inclined cylinder density $[m^{-2}]$		$1.5 \cdot 10^4$	
Plant water content (FG)$[kg/m^2]$	5	2.5	8
Leaf Area Index (FG)	5	4	6.5
Plant moisture content [%]	70	80	85
ϵ_r (L-band)	23-j1.3	34.6-j2	43.8-j2.7
ϵ_r (X-band)	18-j6.8	27-j10.5	34-j14
ϵ_r (K_a-band)	6.5-j6	9-j9.7	11-j12.8

factor equal to $0.9^{1/4}$ for bare soils and $0.98^{1/4}$ for well developed crops, representing the estimated ratio between infrared and thermodynamic temperatures.

Results concerning KT_n and XT_n have been shown in [10], where they have been plotted as a function of Plant Water Content (PWC) $[kg/m^2]$. Experimental data had been obtained by airborne radiometers at $\theta = 40^o$. The main aspects are summarized here. For bare soil samples, KT_n is higher than XT_n and tends towards values higher than 0.95. Well developed wheat and barley crops approximate fairly well black-bodies at both bands. Relatively low KT_n and XT_n values are observed in samples of wide leaf crops (corn, sunflower and sugarbeet). Alfalfa samples are located in an intermediate position, with $KT_n \simeq XT_n$. Both KT_n and XT_n values are fairly well correlated with the PWC; the best sensitivities are noted for wheat and barley at X band, with a positive trend, and for wide leaf crops at K_a band, with a negative trend.

Also model predicted values of KT_n and XT_n versus PWC have been plotted in [10]. All the general tendencies shown in the experimental results are confirmed by theoretical results. In particular, the model confirms that developed wheat and barley crops behave similarly to black-bodies, while emission from wide leaf crops tends towards lower values. An electromagnetic interpretation of these different behaviours is given in [10] in terms of bistatic scattering patterns of elements located on top. For vertical cylinders, modeling wheat and barley ears, the upward scattered power is very low, so that the emissivity is close to the unity. For discs, modeling corn and sunflower top located leaves, the upward scattered power is appreciable, which contributes to lower the emissivity.

Although the model well reproduces the experimental trends, some moderate discrepancies are observed in the absolute values. Normalized temperatures of well developed wide leaf crops are slightly underestimated; this problem may be due to many different reasons: difficulties in matching discs dimensions and orientations to the real complex geometrical structure of leaves, presence of non-planar elements (like flowers), "clustering" effects pointed out in [17]. Also the bare soil emission is underestimated by the Geometrical Optics model, unless high slopes (≥ 0.7) are assumed.

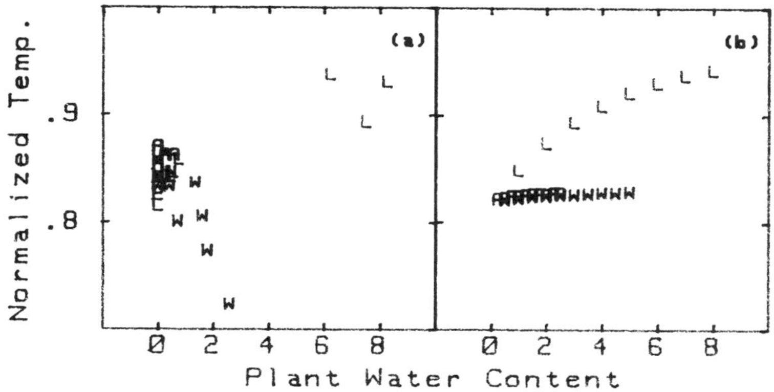

Figure 1. L band normalized temperature vs. PWC $[kg/m^2]$. Aircraft experimental results (a) and model simulations (b). Horizontal polarization, $\theta = 40^o$. Wide leaf crops (L), wheat and barley (W), alfalfa (A).

As far as LT_n is concerned, Figure 1 shows the aircraft measured (a) and model predicted (b) values at horizontal polarization, $\theta = 40^o$.

Both data sets indicate that at L band, differently from the case of higher frequencies, the normalized temperatures of well developed wide leaf crops are higher than those of bare soils. This behaviour finds theoretical explanation in the high values assumed by absorption cross-sections of leaves and stems; on the other side, the corresponding scattering cross-sections are still relatively low. As far as wheat and alfalfa are concerned, the model predicts these crops to be almost transparent at L band, as the flat trend of LT_n vs. PWC indicates. Indeed, the measured LT_n of wheat decreases when increasing PWC; however, this effect could be due to the soil, rather than to the vegetation, since we have noted that the samples with the higher PWC show the higher soil moisture content, too.

3.2 *Microwave indexes*

Results shown in [10] and 3.1 indicate that KT_n, XT_n and (partially) LT_n are sensitive to crop type and crop growth. However, at least in the cases of bare soils and moderately developed crops, the brightness temperature is also influenced by soil moisture and soil roughness. In this section, in order to separate as far as possible the dependencies on the different ground parameters, data measured and computed at $\theta = 40^o$ are combined in different ways, leading to the definitions of the following indexes.

- X band polarization index:

$$XPI = 2\frac{XT_{bV} - XT_{bH}}{XT_{bV} + XT_{bH}}$$

- K_a-X band frequency index:

$$KXFI = 2\frac{KT_{bH} - XT_{bH}}{KT_{bH} + XT_{bH}}$$

- K_a-L band frequency index:

$$KLFI = 2\frac{KT_{bH} - LT_{bH}}{KT_{bH} + LT_{bH}}$$

- X-L band frequency index:

$$XLFI = 2\frac{XT_{bH} - LT_{bH}}{XT_{bH} + LT_{bH}}$$

The V and H suffixes indicate vertical and horizontal polarization, respectively. Only the X band polarization index has been considered, since experimental LT_{bV} data are not available, while the K_a polarization index demonstrated to rapidly saturate even at low PWC's [12]. The horizontal polarization, for which experimental data are available at all frequencies, has been considered in the frequency indexes definition.

Figure 2 shows aircraft measured (a) and model predicted (b) XPI's. Both sets of data indicate a decreasing trend vs. PWC, with moderate differences among the different crop species. The sensitivity of this index to crop growth, already described and discussed in previous papers (see, for example, [2] and [18]) is here confirmed. However, in the experimental results an appreciable dispersion of values is noted for bare soils (PWC=0) since XPI is also lowered by decreasing soil moisture or increasing soil roughness. This effect is not noted in theoretical data, since they have been computed with constant soil parameters.

Other experimental XPI's, obtained by ground-based measurements, are shown in Figures 3a (vegetated soils) and 3b (bare soils). Figure 3a confirms XPI to decrease with PWC in vegetated fields. Figure 3b indicates that, in moisty bare soils (samples with soil moisture content $>\sim$ 20% have been considered) XPI is also much influenced by surface standard deviation σ_z: very high XPI's (~ 0.5) are observed for the case of smooth soils, while rough soils show values similar to those of well developed crops. It can be concluded that XPI contains information on both PWC and soil parameters; the two effects can be separated if some a-priori information is available.

Figure 4 shows aircraft measured (a) and model predicted (b) KXFI's. Both the experimental and the theoretical trends show general characteristics similar to those of XPI (see Figure 2). Indeed, KXFI decreases with vegetation growth; the model slightly overestimates the differences among the various vegetation species, but confirms the general decreasing effect. For bare soils KT_{bH} is mostly contributed by the thin top layer, which is generally dry, while at X band the emission is contributed by a soil volume which extends up to a depth of some centimeters [11] and is appreciably influenced by moisture and roughness. It follows that, for bare soils, KT_{bH} is higher than XT_{bH}, but the difference varies according to soil conditions. For developed crops $KT_{bH} \simeq XT_{bH}$, with moderate differences among the species. Results of Figure 4 are confirmed by ground-measured KXFI's, as it is shown in Figure 5, where the effects of crop growth (a) and soil roughness standard deviation (b) are evident.

It may be concluded that both XPI and KXFI show some peculiar properties.

- High values are associated to bare, smooth, moisty soils.
- Soil moisture reduction, soil roughness increase and crop growth lower both indexes.
- Differences among crop types are moderate.

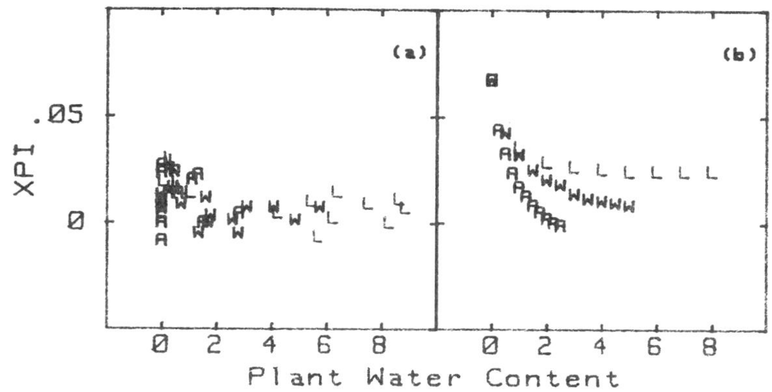

Figure 2. X band polarization index vs. PWC $[kg/m^2]$. Aircraft experimental results (a) and model simulations (b). $\theta = 40^o$. Wide leaf crops (L), wheat and barley (W), alfalfa (A).

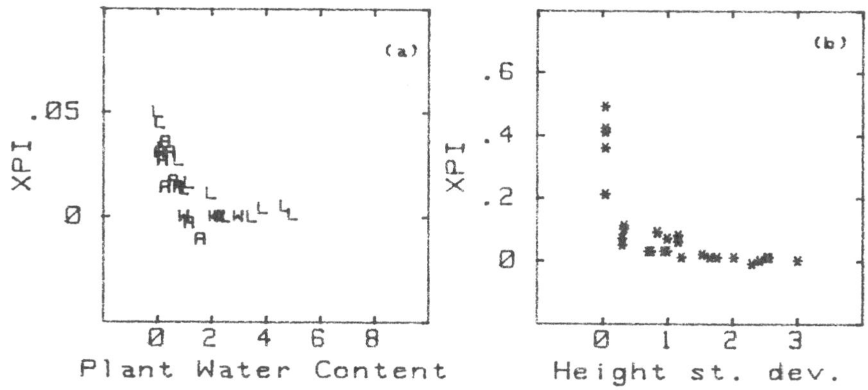

Figure 3. X band polarization index vs. PWC $[kg/m^2]$ (a) and surface height standard deviation $[cm]$ (b). Ground-based experimental results. $\theta = 40^o$. Wide leaf crops (L), wheat and barley (W), alfalfa (A), bare soils (*).

Experimental and theoretical diagrams of KLFI and XLFI vs. PWC are shown in Figures 6 and 7, respectively.

Both indexes, differently from KXFI, are much influenced by crop type: increasing trends are noted in wheat samples, while negative trends are noted for wide leaf crops. Variations associated with crop growth are relatively large, since vegetation effects on L band emission (see Figure 1) are opposite with respect to those observed on X and K_a band emission (see [10]). On the contrary, the versus of emissivity variation due to soil parameters is the same at all frequencies, although with different intensities. The range of variation within soil samples is therefore relatively low with respect to variations associated to crop growth. The theoretical data are in general agreement with experimental data, although some high values of KLFI and XLFI, noted in the experimental data are only partially confirmed by the model; probably, they are due to residual soil moisture effects on L band emission, as pointed out in 3.1.

In synthesis, the following characteristics are shown by KLFI and XLFI.

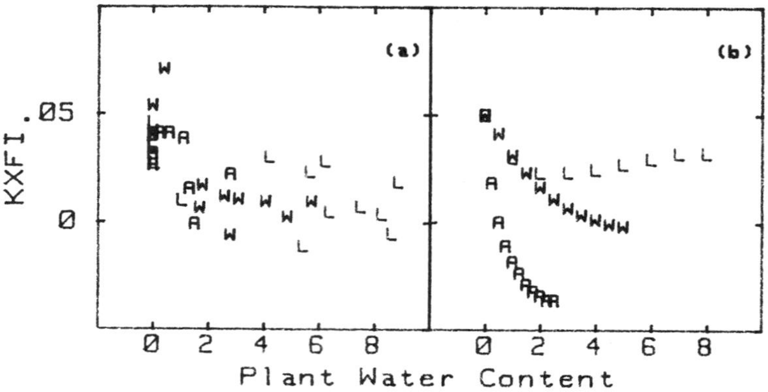

Figure 4. K_a-X band frequency index vs. PWC $[kg/m^2]$. Aircraft experimental results (a) and model simulations (b). $\theta = 40^o$. Wide leaf crops (L), wheat and barley (W), alfalfa (A).

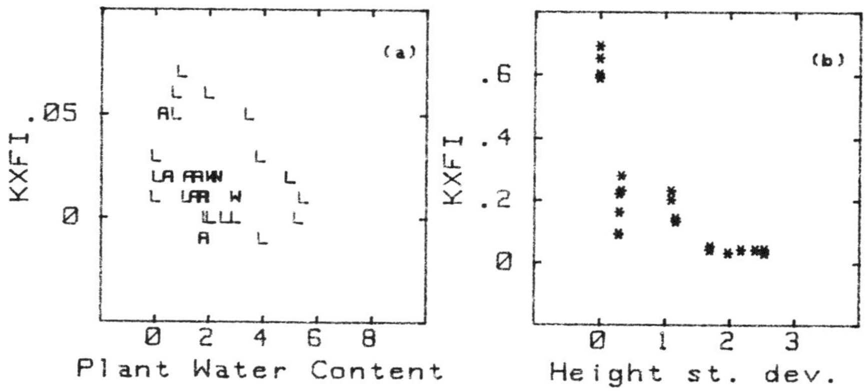

Figure 5. K_a-X band frequency index vs. PWC $[kg/m^2]$ (a) and surface height standard deviation $[cm]$ (b). Ground-based experimental results. $\theta = 40^o$. Wide leaf crops (L), wheat and barley (W), alfalfa (A), bare soils (*).

- Variations associated with crop growth are relatively higher than those observed within bare soils.
- Crop growth effects are considerably different depending on crop type.

4. CONCLUSIONS

Experimental K_a, X and L band brightness temperatures of bare soils and various crops at different stage of growth have been shown and compared with theoretical predictions of a model based on the Radiative Transfer theory. Experimental and theoretical results are in substantial agreement and indicate that a multifrequency radiometric system is sensitive to crop growth and crop type.

If the brightness data are suitably combined in polarization and frequency indexes, the

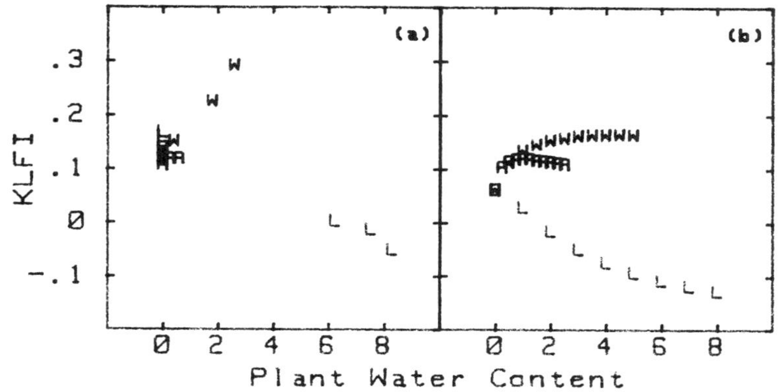

Figure 6. K_a-L band frequency index vs. PWC $[kg/m^2]$. Aircraft experimental results (a) and model simulations (b). $\theta = 40^o$. Wide leaf crops (L), wheat and barley (W), alfalfa (A).

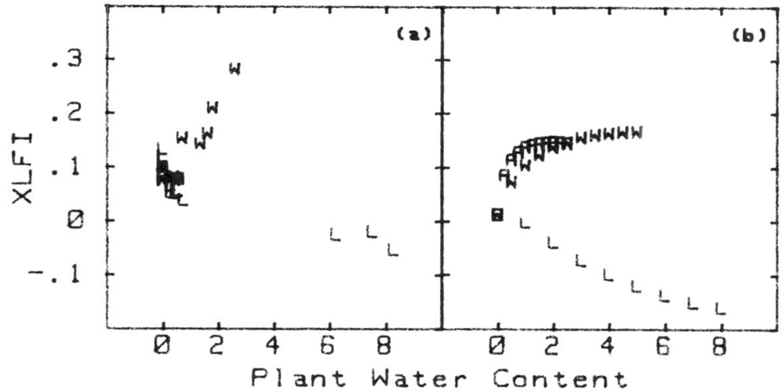

Figure 7. X-L band frequency index vs. PWC $[kg/m^2]$. Aircraft experimental results (a) and model simulations (b). $\theta = 40^o$. Wide leaf crops (L), wheat and barley (W), alfalfa (A).

sensitivities to single parameters of vegetation (PWC, type) and underlying soil (moisture, roughness) may be improved.

Acknowledgments
Work partially supported by ASI, Agenzia Spaziale Italiana.

REFERENCES

1. Ulaby, F. T., R. K. Moore and A. K. Fung, 1982. *Microwave Remote Sensing: Active and Passive*, Vol.II, Reading, Massachussets: Addison-Welsey.
2. S. Paloscia and P. Pampaloni, 1992. Microwave vegetation indexes for detecting biomass and water conditions of agricultural crops, *Remote Sens. Environ.*, **40**, 15-26.
3. T. J. Jackson and T. J. Schmugge, 1991. Vegetation effects on the microwave emission of soils, *Remote Sens. Environ.*, **36**, 203-212.

4. C. Maetzler, 1990. Seasonal evolution of microwave radiation from an oat field, *Remote Sens. Environ.*, **31**, 161-173.

5. L. Tsang, J.A. Kong and R.T. Shin, 1985. *Theory of Microwave Remote Sensing.* New York: Wiley-Interscience.

6. T. Mo, B. Choudury, T. Schmugge, J. Wang and T. Jackson, 1981. A model for microwave emission from vegetation-covered fields, *J. Geophys. Res.*, **87**, 11229-11237.

7. A.K. Fung and H.J. Eom, 1981. Emission from a Rayleigh layer with irregular boundaries, *J. Quant. Spectrosc. Radiat. Transfer*, **26**, 397-409.

8. P. Ferrazzoli, L. Guerriero and D. Solimini, 1991. Numerical model of microwave backscattering and emission from terrain covered with vegetation, *J. Appl. Comp. Electromagn. Soc.*, **6**, 175-191.

9. P. Ferrazzoli and L. Guerriero, 1994. Modeling microwave emission from vegetation-covered surfaces: a parametric analysis, in *Proc. ESA/NASA International Workshop on Passive Microwave Remote Sensing Research Related to Land-Atmosphere Interactions*, Saint-Lary (France).

10. P. Ferrazzoli, L. Guerriero, S. Paloscia and P. Pampaloni, 1994. Modeling X and Ka band emission from leafy vegetation, *Journal of Electromagnetic Waves and Applications*, (in press).

11. S. Paloscia, P. Pampaloni, L. Chiarantini, P. Coppo, S. Gagliani and G. Luzi, 1993. Multifrequency passive microwave remote sensing of soil moisture and roughness, *Int. J. Remote Sensing*, **14**, 467-483.

12. P. Ferrazzoli, S. Paloscia, P. Pampaloni, G. Schiavon, D. Solimini and P. Coppo, 1992. Sensitivity of microwave measurements to vegetation biomass and soil moisture content: a case study, *IEEE Trans. Geosci. Remote Sensing*, **GE-30**, 750-756.

13. R. Schiffer and K.O. Thielheim, 1979. Light scattering by dielectric needles and disks, *J. Appl. Phys.*, **50**, 2476-2483.

14. D.M. Le Vine, R. Meneghini, R.H. Lang and S.S. Seker, 1983. Scattering from arbitrarily oriented dielectric disks in the physical optics regime, *J. Opt. Soc. Am.*, **73**, 1255-1262.

15. M.A. Karam and A.K. Fung, 1988. Electromagnetic scattering from a layer of finite length, randomly oriented, dielectric, circular cylinders over a rough interface with application to vegetation, *Int. J. Remote Sensing*, **9**, 1109-1134.

16. M.A. El-Rayes and F.T. Ulaby, 1987. Microwave dielectric spectrum of vegetation - Part I: Experimental observations, *IEEE Trans. Geosci. Remote Sensing*, **GE-25**, 541-549.

17. L. Tsang, J. A. Kong and G. Zhang. Analytical wave theory and Monte Carlo simulations of scattering in microwave remote sensing, this issue.

18. P. Ferrazzoli, L. Guerriero, S. Paloscia, P. Pampaloni and D. Solimini, 1992. Modeling polarization properties of emission from soil covered with vegetation, *IEEE Trans. Geosci. Remote Sensing*, **GE-30**, 157-165.

A model and experiments for Microwave Radiometry of forests

L. GUERRIERO[1] and C. MÄTZLER[2]

[1] *Dipartimento di Ingegneria Elettronica, Università Tor Vergata*
Via della Ricerca Scientifica, 00133 Roma (I)
[2] *Institut für Angewandte Physik, Universität Bern*
Sidlerstrasse 5, 3012 Bern (CH)

Abstract A model which describes the electromagnetic properties of vegetation has been recently developed at Tor Vergata University. It has been applied to simulate the microwave emissivity of various tree canopies observed by two radiometer systems developed at the University of Bern. These instruments recorded the brightness temperature observed below a single beech and coniferous and deciduous forests at different frequencies (from 1 to 100 GHz) and during several years.

The two approaches complement each other: a theoretical sensitivity analysis, as a function of frequency and of the physical tree characteristics (LAI and wood volume) will be used as a tool to interpret experimental data; on the other hand, the latter will ensure the validity of the model and of the electromagnetic representation.

Keywords: Microwave emission, Forests, Sensitivity analysis, Model

1. INTRODUCTION

Monitoring of forest ecosystem is one of the most challenging goals of Remote Sensing, both from the observing method point of view, and from that of tree parameter retrieval. While much attention has been given to scattering properties of trees, few works have been devoted to the study of microwave emission of forested areas. This paper intends to contribute in the filling of this gap reporting both experimental and theoretical results.

During the last decade many ground-based experiments have been performed by the University of Bern (Switzerland) that designed and developed two radiometer systems: the RASAM (RAdio-ScAtteroMeter) and the PAMIR (Passive and Active Microwave Infrared Radiometer), whose technical characteristics are reported in Tables 1 and 2. Tree observations have been carried out to study the transmission properties of forestal vegetation. To this end the measurement configuration with sky background has been chosen: the instrument is installed below the tree canopy and looks upward to the vegetation. The vegetation brightness temperature, together with the sky brightness temperature and the vegetation physical temperature, is then measured. These quantities have been connected to each other to give the transmissivity coefficient, with the aim of representing the fraction of energy which reaches the ground after penetrating the vegetation. Results

Table 1. RASAM channel and antenna parameters.

Frequency [GHz]	1.5	2.5	3	4.6	7.2	10.2	11
Bandwidth [GHz]	0.3	0.5	0.6	0.6	1	1	1.6
3-dB beamwidth [°]							
E-plane	38.1	23.8	18.1	11.8	7.9	5.4	5.2
H-plane	42.1	20.2	17.5	12.7	8.9	6.8	6.4

Table 2. PAMIR channel and antenna parameters.

Frequency [GHz]	4.9	10.4	21	35	94
Bandwidth [GHz]	0.2	1.2	0.8	0.8	2
3-dB beamwidth [°]	10	9	9	9	9

of these studies have been reported elsewhere, relating transmissivity variability to tree parameters [1], wind, wetness, temperature and forest damage parameters [2].

It is widely recognized that model application is necessary to interpret experimental data and to give them physical basis. In this paper, we will use the model developed at Tor Vergata University. This model is based on the Radiative Transfer Theory and describes vegetation as an ensemble of lossy scatterers overlying a homogeneous half-space. While we refer to literature [3] for a detailed description of the model, we would like to recall here its main advantages: it covers the whole microwave spectrum, it yields backscattering coefficient and emissivity through a unified approach, it includes multiple scattering effects of any order. The model is also applicable to a wide spectrum of vegetation types. The numerical algorithm can be, in fact, subdivided in two main steps: first, the electromagnetic characterization of the elements composing the medium; second, the combination of single contributions. This allows to select the shapes, dimensions and the relative positions of the canopy elements in the most suitable way to represent the physical and geometrical peculiarities of the vegetation under study. In this book, an application to agricultural cases is also described [4], here we will focus on forests.

2. THE MODEL SIMULATION

The Tor Vergata Model applies the energy conservation relation [3] to get the vegetation emissivity, therefore the bistatic scattering coefficient is needed. To this end, the Rayleigh-Gans approximation [5] is used for discs (deciduous leaves) and needles (coniferous leaves) at P, L and C band. According to this theory, the inner field of an object with at least one dimension small with respect to the wavelength can be approximated by a homogeneous field along this dimension, while in the other two dimensions phase differences must be taken into account. At higher frequencies, the Physical Optics approximation [6], which substitutes the field inside a thin disc with the one of a slab with the same thickness, has been applied to compute disc electromagnetic properties. The infinite length approximation [7], which approximates the field inside a finite cylinder with that of an infinite one, is used for cylindrical objects like twigs, branches and trunks.

An important parameter which controls scattering from vegetation is the dielectric constant. It has been computed through the semi-empirical formula, given in [8], which needs the water content and the wet to dry density ratio as inputs. An updated version

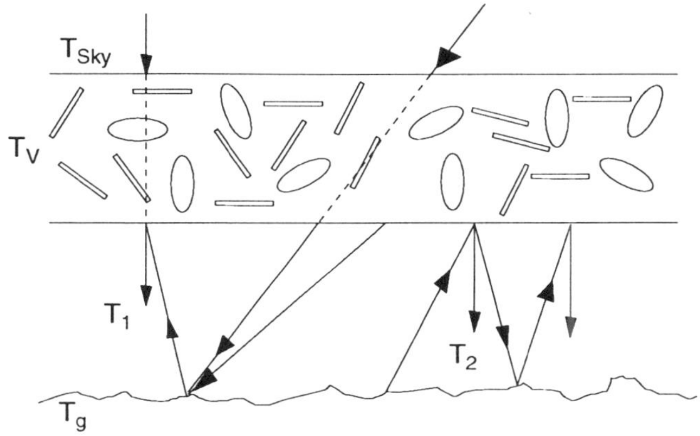

Figure 1. Illustration of the brightness temperature contributions (T_1 and T_2) computed by the model. T_v and T_g are, respectively, the vegetation and ground physical temperatures. T_{sky} is the sky brightness temperature.

of the model which includes an alternative formula for the permittivity of fresh leaves [9], tested over the whole 1-100 GHz range, is now under study.

The soil is described as a homogeneous half-space. Out of the several theories proposed for soil scattering we have applied the Small Perturbation approximation, which seems to be, at present, the most suitable at P and L band, and the Geometrical Optics approximation from 5 GHz on [10]. The soil permittivity is computed according to the semiempirical formula proposed in [11].

Of the large data base available from the RASAM and PAMIR measurements, only a small subset will be used in this paper: data collected in the summer 1986 by RASAM below beech and fir forests, and those collected by PAMIR below a 100-year-old beech over the whole year from August 1987 to August 1988. Besides, the model simulation of PAMIR and RASAM experiments has been carried out for zenith observation only. In this situation, trunk contributions can be neglected and the medium azimuthal symmetry, required by the model, is more easily met in the crown.

Beside geometrical and physical parameters of the tree canopy (such as leaf shape and dimensions, branch size and orientations, vegetation and soil water content, etc.), inputs required by the model are the sky brightness temperature and the vegetation physical temperature. The output brightness temperature T_B is computed as the sum $T_1 + T_2$ (fig. 1). The first term represents the vegetation brightness temperature combined with the sky brightness temperature attenuated by vegetation. Together with direct contributions, multiple scattering phenomena, between vegetation and soil, are considered. The second term is the soil brightness temperature scattered by vegetation into the observation angle.

3. TREE ARCHITECTURE

As already pointed out, trunk contributions will be neglected in the simulation of RASAM and PAMIR zenith data, so that the forest medium is described as sketched in fig. 2 and 3. The crown is filled with different elements, positioned at different height, according to the properties of the tree species to be modeled, as will be pointed out in the following section.

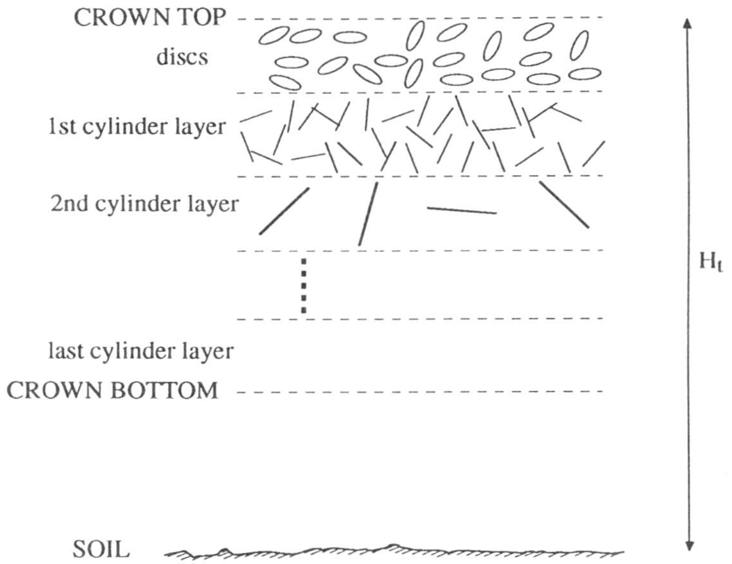

Figure 2. Model representation of a deciduous forest. H_t=tree height. Dimensions have been chosen according to available ground measurements, as reported in Table 3.

The deciduous tree crown is subdivided into several layers (Figure 2). The top layer is filled with randomly oriented discs representing leaves. Lower layers are filled with cylinders, representing branches, whose dimensions increase from crown top to crown bottom. For each layer of equal cylinders, the density is inversely proportional to the single cylinder volume, so that the total filled volume is the same for all the layers. The cylinder radius in the uppermost layer is 0.2 cm, then it increases by 0.2 cm in every subsequent layer until the lowest one where it assumes a value that is a function of the tree height. The branch length-to-radius ratio is fixed to 100. Finally, the smaller cylinders (radius$\leq$1 cm) are randomly oriented, while larger cylinders have random orientation in azimuth and are assumed to follow a uniform distribution, between 30° and 60°, in the inclination angle from the vertical.

In coniferous forests, generally the primary branches are present up to the tree top and the structure is less ramified than in the case of deciduous forests. We have described the crown as a uniform layer, in which branches of different dimensions and needles are uniformly located at different heights (Figure 3). A maximum branch radius is assumed also in this case, and supposed to be a function of the tree height. Again, for each value of branch radius, the number of branches is inversely proportional to the single branch volume.

In situ measurements, when available, were used as model input, otherwise average values found in literature were chosen. Table 3 reports those quantities for each of the three simulations that will be described in the following section.

3. RESULTS

A sensitivity analysis upon tree age and species has been carried out by the model, and results have been compared with RASAM and PAMIR data.

In figure 4, either data collected below a beech forest and theoretical curves obtained

Table 3. Input data for model simulations of Figure 1, 2 and 3. Sky Temperature is frequency dependent as from [1].

	Fig. 1	Fig. 2	Fig. 3
Disc radius and	2.		2
thickness [cm]	0.01		0.01
Needle radius and		0.03	
length [cm]		7	
Leaf moisture content [%]	70	70	70
Leaf orientation	random	random	random
LAI	3–6	3–5	Summer 7,8 Winter 0
Branch radius a[cm]	$0.2 \leq a \leq 2$	$0.2 \leq a \leq 2$	$0.2 \leq a \leq 2$
Length-to-radius ratio	100	100	100
$a < 1$ branch orientation	random	random azimuth 40 ≤inclination≤ 70	random
$a \geq 1$ branch orientation	random azimuth 30 ≤inclination≤ 60	random azimuth 40 ≤inclination≤ 70	random azimuth 30 ≤inclination≤ 60
Branch moisture content [%]	20	20	20
Tree height [m]	10–20	15–30	23,25
Crown Wood volume [m^3/ha]	55–175	55–140	150,190
Vegetation Temperature [°K]	295	295	Summer 295 Winter 280
Soil roughness standard deviation [cm]	1	1	1
Soil moisture content [% by volume]	5	5	5

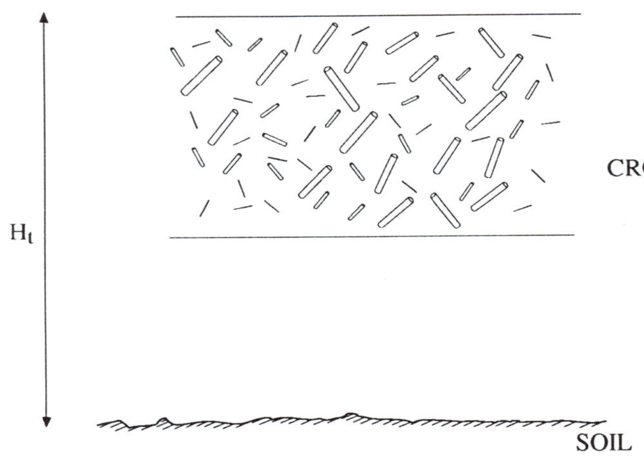

Figure 3. Model representation of a coniferous forest

through the model are reported. Since zenith observation is concerned, polarization differences are absent in the model, while the experimental values have been averaged to give point data in the plots.

RASAM observations have been performed below various spots of a beech forest whose older trees were about 20 m high, and the younger ones were about 10 m. In the model simulation, an increase in tree age has been translated in an increase in crown wood volume which, on its turn, can be related to a growth either in height and/or density. Also the Leaf Area Index is supposed to increase when the tree is getting older. The four solid lines in fig. 4 correspond to a linear increase from 55 m^3/ha to 175 m^3/ha, in crown wood volume, and from 3 to 6 in LAI. The spectral behaviour of beech brightness temperature is fairly well reproduced by the model, as well as the trend with age. We can observe that the brightness temperature dynamic range — between younger and older trees — remains quite constant from low to higher frequencies, so that a discrimination between different tree age seems to be possible in the whole frequency range here considered.

Applying the coniferous tree architecture, a model simulation has also been performed for the case of firs which, from a visual inspection, looked about 30 m high. However, fir branches are less dense, so that, despite the larger height, a linear increase from 75 m^3/ha to 140 m^3/ha in the crown wood volume has been taken into account in this case. LAI=3 in younger trees, and 5 in older ones. Unfortunately, experimental data pertaining to young firs are not available, so that the model prediction can not be confirmed. With respect to the deciduous case, the spectral trend (fig. 5) is steeper and the variability range of brightness temperature seems to be frequency dependent, making tree age discrimination easier at lower frequencies. Indeed, needle emissivity is higher than that of deciduous leaves and, consequently, at about 10 GHz the brightness temperature dynamic range is largely reduced compared to the corresponding one of deciduous trees.

The PAMIR observation of the old beech was performed all-year-round and allows to separately investigate the emissivity of leaves and that of branches. Data collected in spring and summer were averaged, and so were the autumn and winter data, to be compared with the model simulation of foliated and defoliated conditions. Results are

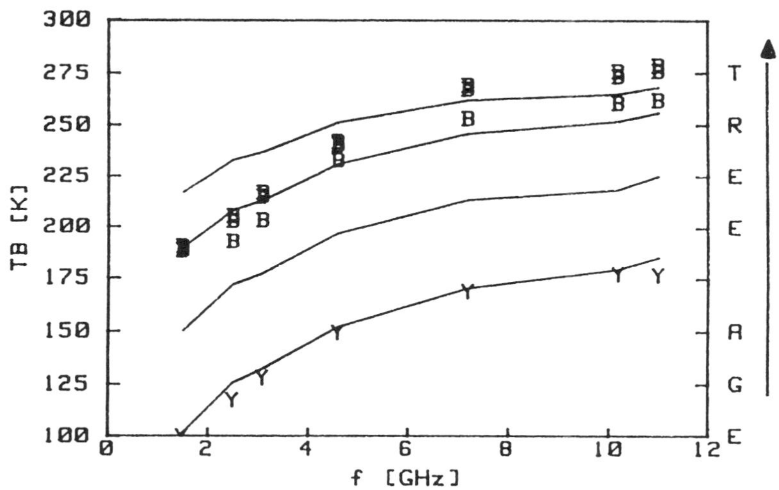

Figure 4. Brightness temperature vs. frequency. RASAM data collected below beeches of different ages: B=old Beech forest, Y=Young beech. The solid lines are the model simulations corresponding to increasing crown volumes and LAI, from bottom to top.

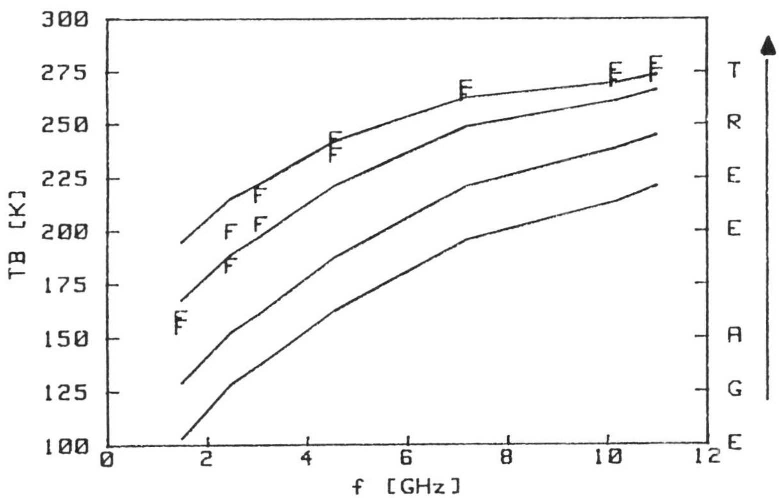

Figure 5. As in fig. 4, but for the coniferous case. Data collected by RASAM: F=Fir forest.

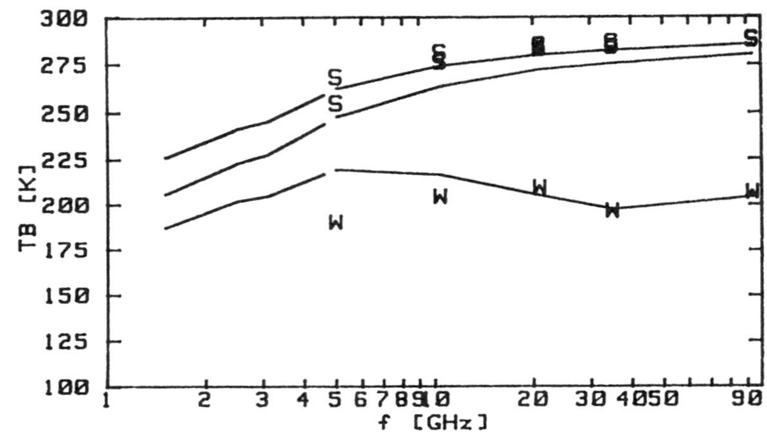

Figure 6. Data collected by PAMIR below an old beech during Summer 1987 (the lower S points), Winter 1987/88 (W) and Summer 1988 (higher S's). Theoretical curves have been extended to lower frequencies

reported in fig. 6, where a theoretical extension to lower frequencies is also given. Tree growth, from one summer to the following, has been represented by an increase in LAI (from 7 to 8, as results from visual inspection [2]) and in crown wood volume (from 145 m^3/ha to 190 m^3/ha). Intermediate winter condition has been simulated by a tree canopy with the lower wood volume and LAI=0. The average measured physical temperatures, used as input, are: $T_v = 295°K$ for the summer cases, and $T_v = 280°K$ for winter. Besides verification of the model applicability even at 95 GHz, the graph shows that leaf contribution becomes important above 5 GHz. Toward lower frequencies, branch emissivity is predominant since the two lower curves (corresponding to the same crown wood volume) tend to converge.

4. CONCLUSIONS

In this paper theoretical data from Tor Vergata electromagnetic Model have been compared with ground-based data collected by two experimental apparatus of the University of Bern. Results show fairly good agreement, both quantitatively and qualitatively, over a large frequency range and for different tree species canopies. Although our observation perspective was from below the canopy, some considerations may be drawn which should still be valid for aircraft or satellite point of view.

It appears that multifrequency systems are suitable for tree classification. They can take advantage of the emissivity dependence on leaf geometry at high frequencies, and on branch orientation at lower frequencies.

Radiometer sensitivity to tree age, or woody biomass, was also pointed out. At frequencies lower than 5 GHz, where contributions coming from branches are the most important, the brightness temperature shows a large dynamic range as the wood volume varies. Therefore, discrimination between young forests, with a small amount of woody biomass, and older ones seems to be possible. On the other hand, at C and X band contributions from leaves become effective, so that information about the foliation state can be drawn.

Acknowledgments
Work partially supported by ASI, Agenzia Spaziale Italiana, and the Italian National Research Council.

REFERENCES

1. A. Sume, C. Mätzler, R. Hüppi, and E. Schanda, 1988. *Microwave Radiometer and Scatterometer measurements of vegetation*, FOA Report C30494-3,2. Swedish Defense Rersearch Establishment, Linköping, Sweden.

2. C. Mätzler, 1994. Microwave transmissivity of forest canopy: experiments made with a beech. In press in *Remote Sens. Environ.*

3. P. Ferrazzoli, L. Guerriero, and D. Solimini, 1991. Numerical model of microwave backscattering and emission from terrain covered with vegetation, *J. Appl. Comp. Electromagn. Soc.*, **6**, 175–191.

4. P. Ferrazzoli, L. Guerriero, S. Paloscia, and P. Pampaloni, 1994. Potential of multifrequency techniques in microwave radiometry of crops, in *this book*.

5. R. Schiffer, and K.O. Thielheim, 1979. Light scattering by dielectric needles and disks, *J. Appl. Phys.*, **50**, 2476–2483.

6. D. M. Le Vine, R. Meneghini, R. H. Lang, and S. S. Seker, 1983. Scattering from arbitrarily oriented dielectric disks in the physical optics regime, *J. Optical Soc. America*, **73**, 1255–1262.

7. M. A. Karam, and A. K. Fung, 1988. Electromagnetic scattering from a layer of finite length, randomly oriented, dielectric, circular cylinders over a rough interface with application to vegetation, *Int. J. Remote Sens.*, **9**, 1109–1134.

8. M. A. El-Rayes, and F. T. Ulaby, 1987. Microwave dielectric spectrum of vegetation - Part I: Experimental observations, *IEEE Trans. Geosci. Remote Sens.*, **GE-25**, 541–549.

9. C. Mätzler, 1994. Microwave (1-100 GHz) dielectric model of leaves, in press in *IEEE Trans. Geosci. Remote Sens.*

10. F. T. Ulaby, R. K. Moore, A. K. Fung, 1982. *Microwave Remote Sensing: Active and Passive*, Vol.II-*Surface Scattering and Emission Theory*, Addison-Wesley, Reading, Massachusetts.

11. F. T. Ulaby, R. K. Moore, A. K. Fung, 1986. *Microwave Remote Sensing: Active and Passive*, Vol.III-*From Theory to Applications*, Artech House, Dedham, Massachusetts.

Microwave radiometry as a tool for forest fires detection: model analysis and preliminary experiments

[1]G. LUZI , [1] P.COPPO, [2] P. FERRAZZOLI, [1] S. GAGLIANI, [3]T. MAZZONI

[1] *Fondazione Scienza per l'Ambiente*
Viale Galileo 32, 50125 Firenze - Italy
Tel. +39 55 2337738/ Fax +39 55 224076

[2] *Dip. Ingegneria Elettronica, Università Tor Vergata*
Via della Ricerca Scientifica, 00133 Roma - Italy
Tel. +39 6 7259 4401/ Fax +39 6 2020519

[3] *Officine Galileo Viale Einstein, 34*
Campi Bisenzio (FI) - Italy
Tel. +39 55 89501/ Fax +39 55 8950600

Abstract
The early warning performance of the present optical remote sensing systems for forest fires detection is often reduced by the screening effect of leaves, branches, smokes and vapours. In this paper the potential of microwave radiometry to improve the efficiency of forest fires monitoring is evaluated. A simple model has been used to simulate the physical and operational limitations of this application; theoretical simulations have shown that, although vegetation attenuation increases with frequency, the system sensitivity could be acceptable at least up to 10 GHz, provided the resolution is adequate. Some experimental results obtained at C band confirmed also that at microwave frequencies the presence of a fire spot can be easily detected as a sharp increase of the brightness temperature

Keywords
Fires, Forest, Microwave, Radiometry, Temperature.

INTRODUCTION

Nowadays, forest fires detection systems mainly use the optical portion of the electromagnetic spectrum; surface reflectance variations at visible band [1], or surface radiance increases [2] at IR band, are measured. The performance of these systems is heavily reduced by the effects of elements like leaves, branches, smokes and vapours which interpose themselves between the soil and the sensor. These elements introduce both attenuation and noise, which can delay the warning and increase the probability of false alarms occurrence. At microwave (MW) frequencies the above mentioned problems are considerably less dramatic for a large part of the spectrum. According to Planck's formula, the thermal emission of a natural isothermal medium at the MW band is much lower than at the IR band; however the typical sensitivity of a MW radiometer is sufficient to detect changes in natural media emission due to temperature or emissivity variations. The potential usefulness of MW radiometry for fire detection consists in sensing the beginning phase of forest fires, i.e. when smokes, vapours and flames are masked to optical sensors by the canopy. Besides MW radiometers can be also used as a prevention system as recently discussed in [3]. Due to the complementary aspects of MW and optical sensors the best solution could probably be represented by an integrated multiband system.

Experimental investigations carried out in the ex-USSR [4] indicated that MW emission in the presence of fires is appreciably higher than background MW emission. Recently a simulation study [5] has been done on the fire detection capabilities of microwave radiometers considering both instrumental and electromagnetic aspects. Results showed that the main factors affecting the sensitivity of the sensor to the presence of fire are: the coarse spatial resolution, due to operational configuration needs at the lower frequencies, and the attenuation of the vegetation at the higher ones. To acquire experimental data about fire emission at lower frequencies, the thermal emission of a small confined artificial fire at C band, looking at a fixed observation angle and polarisation, has been measured.

In this paper an overview of our theoretical and experimental studies is given. Results of the model simulations, aimed at predicting the radiometric system performance at various microwave frequencies including vegetation effects, are discussed. Also an analysis of the experimental results collected at C band is made; the choice of the best instrumental configuration is analysed and some aspects to be further investigated are indicated.

MODEL ANALYSIS

The fire is considered to be in the beginning phase and it is supposed that the thermal properties of the atmosphere and of the vegetation layer over the fire are not changed with respect to the absence of a fire. The thermal behaviour of burning matter on the soil surface is evaluated by considering that in the first phase a typical temperature required to sustain the exothermic reactions in cellulostic fuels is in the range 470-550 K [2]. The hot source is modelled as a half-space with the same dielectric properties as those of the unburned soil, at 293 K, and an equivalent temperature, Ths, which is variable in the range between 293-500 K. Ths is the temperature of an isothermal homogeneous medium which represents the heterogeneous situation of the physical behaviour of a fire with the high spatial and time variability of its thermal and dielectric properties due to loss of water, and heating of the soil. For instance the overestimation of the fire effects related to assuming the temperature to be constant within the whole half-space is compensated by assuming lower temperature values with respect to those of burning matter (typical glowing combustion temperature is about 800 K [2]).

Many studies deal with the MW emission of soil and vegetation under natural conditions which is affected by many physical and geometrical factors. The emission of bare natural soils strongly depends on its temperature and dielectric constant (DC), which is markedly affected by the presence of liquid water, and on surface parameters like the height standard deviation and the horizontal correlation length. An overlaying vegetation canopy does not only attenuate the soil emission, but it itself also generates emission, which is related to both absorption and multiple scattering effects. So the global emission of the soil-vegetation system is affected by some vegetation parameters such as the total plant water content, the shape and the dimensions of the vegetation components (leaves, branches and trunks). The apparent brightness temperature in the observation direction is due to the sum of many different contributions; the most important of them are: the direct emission of the soil attenuated through the vegetation layer, and the direct contribution of the vegetation.

To estimate the vegetation effects, the model described in [6] is used, which describes the canopy as an ensemble of dielectric scatterers of simple shape (disks, cylinders), computes the scatterer cross-sections and combines the various contributions by means of a numerical algorithm named Matrix Doubling. In order to predict the capability of radiometric systems to detect fires, the brightness temperature of a forested area must be estimated, in the two different situations of presence and absence of fire. The study considers four frequencies: 5, 10, 35, and 94 GHz. Atmospheric attenuation is neglected: this hypothesis is well satisfied at the two lower frequencies (5 and 10 GHz) while it will need further investigations at the higher frequencies.

The effect of soil roughness is taken into account by the surface geometrical optics model [7] with a 0.3 average surface slope . The soil permittivity is computed by a semi-empirical model at 5 and 10 GHz [7], while the refractive model [8] is used at 35 and 94 GHz. The soil moisture is supposed to be equal to 10.% by volume. The following permittivity values are obtained: 5.8 - j 0.6 at 5 GHz, 5.5 - j 1. at 10 GHz, 3.8 - j 1.1 at 35 GHz, 3.2 - j 0.7 at 94 GHz. The relevant parameters of the vegetation canopy have been taken by ground measurements carried out in the Netherlands in the framework of the Maestro-1 Campaign [9] over poplar stands, which are rather common in the European environment.

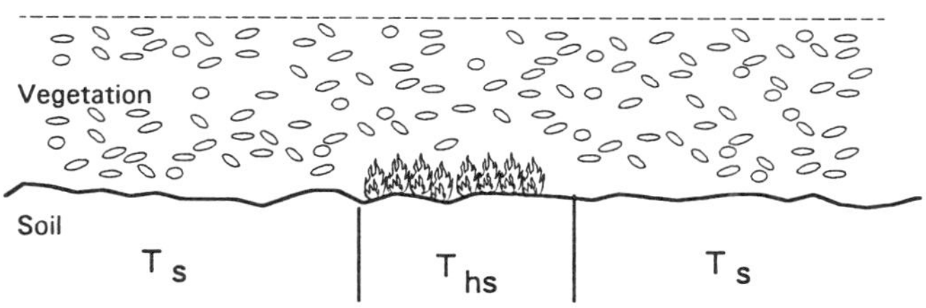

Figure 1 : *Scheme of the model used for fire presence simulation*

A scheme of the model is shown in fig. 1. The leaves are modelled as randomly oriented disks. Branch and trunk contributions are neglected. According to results shown in [10] this approximation is acceptable at 10 GHz and above, while further refinements are needed at 5 GHz

MODEL RESULTS

A representative result of the calculations implemented to estimate the brightness temperature upwelling from the scenario of a small confined hot source below a vegetation layer is shown in fig. 2: the increase in the brightness temperature coming from the whole forest environment due to an increase in Ths is shown. Results are given for 4 frequencies (5,10,35,94 GHz), two polarisations (V, H) and an observation angle of 45°: due to the slant path, this choice represents a worst condition, but it can be better compared to the experimental data shown in the following. Vegetation parameters are taken from [9] for poplar. The linear trends of T_b vs Ths, very close to linearity, suggest the definition of a constant sensitivity parameter $d(T_b)/d(Ths)$. Also a polarisation dependence can be observed: T_b is higher at V polarisation than at H polarisation, even if the difference decreases when the frequency is increased. A strong slope decrease, with increasing frequency, can be appreciated due to the higher vegetation effect.

To include operational aspects, the brightness temperature upwelling from the observed scenario is convoluted with the antenna pattern to estimate the antenna temperature increase ΔTB. The presence of a fixed circular hot spot with a 3 meter radius under the vegetation (without wind) is simulated. The antenna configuration takes into account operational realisation aspects; the brightness temperature measured by a sensor mounted aboard a small helicopter flying at about 120 meters altitude, and observing at 10° from nadir is simulated. The antenna pattern is simulated with a Gaussian function and the 3dB aperture corresponds to that of a 75 cm diameter reflector and changes with the observing wavelength (see further details in [5]). In fig.3 the profile of the brightness temperature increase with respect to background (absence of fire), ΔTB, corresponding to a flight passage over the fire covered by vegetation, is shown. The antenna temperature increments due to the presence of a fire are plotted by simulating a space/time profile measured from a platform flying over the monitored zone. The equivalent temperature Ths, which has a strong influence on brightness contrast, is assumed to be equal to 450K, while the unburned soil temperature, Ts, is 292 K. Computations have been carried out at vertical polarisation.

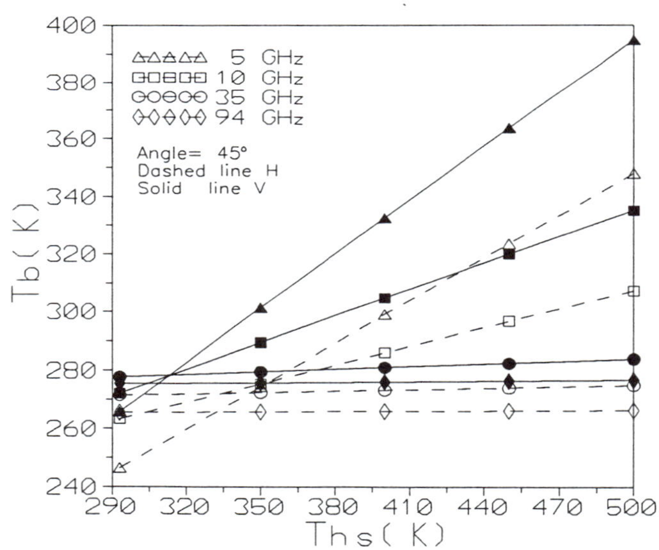

Figure 2: *Simulated brightness temperature as a function of the equivalent temperature of the hot source Ths*

The more sensitive frequency is 10 GHz: these results can be explained by the relatively low vegetation attenuation and the good spatial resolution (3dB angle c.a. 3°) attainable at this frequency. In fact considering a fixed antenna diameter, the attainable spatial resolution of the system deteriorates at lower frequencies. For a given brightness contrast between the hot source and the background, the corresponding antenna temperature will be reduced according to the ratio between footprint and hot source dimensions. The T_b increase is high enough to be distinguished from background fluctuations, which is typically in the order of some K degrees. For this reason, and for the purpose of reducing the attenuating path through the vegetation, low observation angles are convenient.

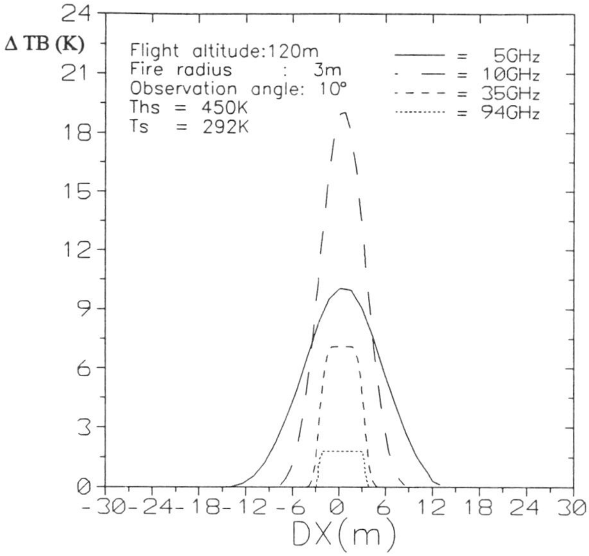

Figure 3 : *Brightness temperature (V polarisation) increase obtained by simulating a flight passage over a fire below the vegetation layer*

EXPERIMENTS

The experimental investigation described in [4] underlined that the entity of brightness increase due to the fire depends on the hot source temperature, on the observed frequency and on the phase and kind of fire; in the case of uncovered and well developed fires the most sensitive wavelength corresponds to 0.8 cm, even if the maximum is not strongly marked. If the fire has strongly blown out, the high temperature of the flame (800-1000 K) gives rise to brightness increases up to 150-200 K at 0.8 cm wavelength.

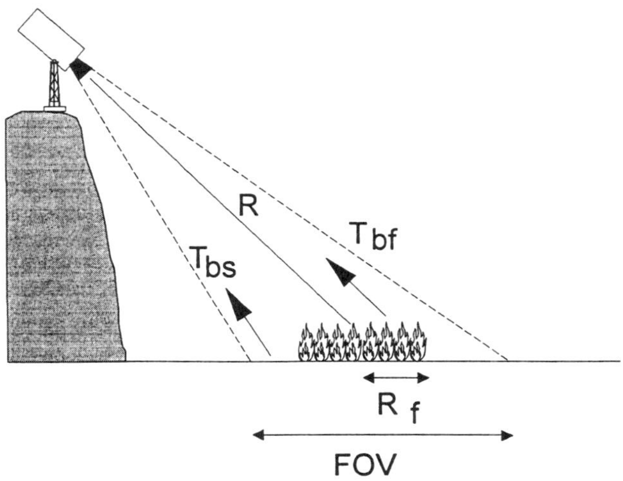

Figure 4 : *Experimental set-up.*

Finally it is noted that the region corresponding to T_b increase is larger than the burning zone. Other experiments related to fire of hydrocarbon, based on propagation measurements, are described in [11]; here data collected at various frequencies are shown and flames appeared to be rather transparent at lower frequencies.

Some measurements of the MW emission of an artificial fire in a natural-like environment were carried out in a background situation typical of the high risk period in an Italian region, near the city of Siena (Italy), on the 27th and 28th of August, 1992. The aim was to detect and quantify the increase of brightness temperature due to the fire over the scenario background. A microwave radiometer, working at C band (5.5 GHz), observing at a fixed angle (53°) with respect to nadir and at horizontal polarisation, was used to measure the thermal emission upwelling from a scenario which included the background soil emission and the contribution of a confined artificial fire. A representative parameter related to the spatial resolution of the system is the "filling factor", Q, i.e. the fraction of the area viewed by the radiometer occupied by the fire. Fig. 4 shows the simple scheme of the geometry of the measurement. By changing the diameter of the fire spot, and observing the fire from different distances, various values of the filling factor can be obtained. While the matter was burning, the radius of the flames was generally smaller than R_f; anyhow, from now on we will speak of fire radius referring to the dimension of the starting "fuel" as specified. The brightness temperature measured by a radiometer, T_{bm}, can be evaluated considering the two contributions: the one due to the background, T_{bs} ,and that one due to the fire, T_{bf}, through a weighted summation expressed by the following simple relationship (eq.1):

$$T_{bm}=(1-Q)T_{bs}+QT_{bf} \qquad \textbf{(eq.1)}$$

The measurements were carried out during the late morning of the 27th and 28th of August 1992. The sensitivity of the radiometer, with an integration time of .1 s, is lower than 1. K. The filling factor is calculated with reference to the antenna 3dB beam width. The used antenna is a pyramidal horn (HPBW=12.8°):

During the 27th of August, measurements with a low filling factor value (less than 5%) were carried out. The brightness temperature, T_{bm}, measured in correspondence with the flames showed an increase of some K with respect to the background, clearly detectable by the sensor. During the measurements the soil and air temperatures were very high and the soil surface was very dry. The fuel consisted of typical wood and leaves of the Mediterranean bush. The soil surface was very rough.

On the 28th of August the radius of the fire was about 1. m and the smaller distance between the instrument and the fire, about 7 meters, made the filling factor very high (near 100%): hence in this case the main contribution came from the burning area. The brightness temperature profile vs time obtained on the 28th of August is shown in fig. 5. It is easy to distinguish the different phases of the fire; a difference of about 60 K between the flame phase (point f) and the background is clearly noticeable, as well as a quick peak, probably due to the high flames temperature. Considering the different phases and in particular when the fire lights up, a characteristic slope seems to be identified which is repeated every time the fire is rekindled (points 1, 2 and 3). The presence of a slow decrease over the long time scale, can be probably due to soil thermal inertia effects. Physical considerations about absorption properties of the burning gas of the flame and some previous experiments [11] suggest that the main responsible factor for the emission should be the warmed soil, even if this consideration needs further investigations. If 27th and 28th of August data are compared, the linear dependence on Q is confirmed.

Also a simple test of vegetation attenuation was carried out. A screen made with the same kinds of vegetation used as fire fuel, thick enough to be opaque to visible and infrared radiation, was inserted between the fire and the radiometer (point * in fig.5). Observing the T_b profile it can be noted that in correspondence with this point the T_b is reduced, but a difference of about 10 K with respect to background is still present.

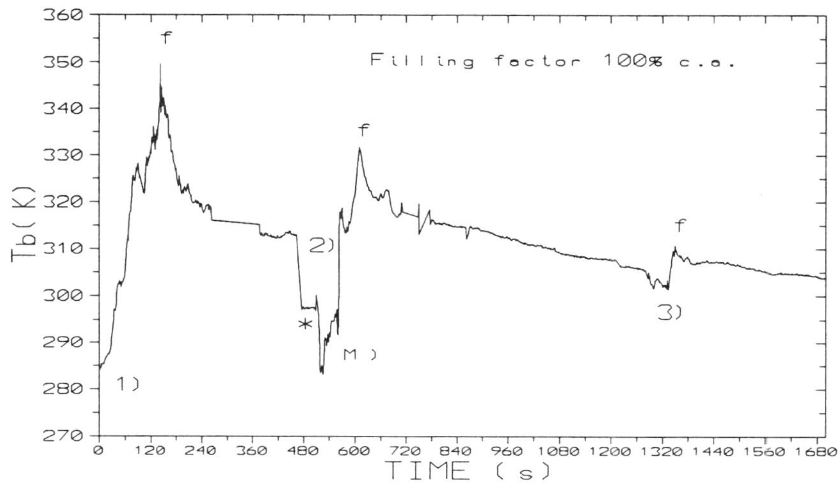

Figure 5 : *Brightness temperature measured the 28th of August 1992 vs time. 1) ignition, 2) and 3) rekindling, f) maximum of the flame phase, *) presence of the vegetation screen, M) noise due to experimental adjustments.*

CONCLUSIONS

The potential of MW radiometry to monitor forest fires has been investigated. Even if some simplifications have been used in modelling the electromagnetic and thermal behaviour of the fire, the obtained results may be considered a first basis to discuss operational aspects. Theoretical simulations showed that, although vegetation attenuation increases with frequency, the system sensitivity could be acceptable at least up to 10 GHz, provided the resolution is adequate.

The experiment carried out, aimed at measuring the brightness temperature of burning matter at C band, confirms some peculiarities of thermal emission at MW, previously found by other researchers at higher frequencies; the brightness temperature, at C band, of forest burning matter over soil has been estimated to be about 50-60 K over background. Further experiments, using multifrequency systems, are needed to evaluate the best configuration for the detection of forest fires. The analysis must be extended also to atmospheric sensitive frequencies where the role of heated water vapour could be important.

Acknowledgements

Thanks are due to the Siena Province Administration and to the staff of the local State Forestry Corps (Corpo Forestale dello Stato) for their support.

REFERENCES

1. Cappellini V., L. Matii, A. Mecocci, 1989. An intelligent system for automatic fire detection in forests, *Proceedings of the 9th EARSeL Symposium*, Espoo-Finland, 216-266.

2. Robinson J. M., Fire from space: global fire evaluation using infrared remote sensing, 1991. *Int. J. Remote Sensing*, **12**, 1, 3-24.

3. Shestopalov Y.K. , 1993. The Statistical Treatment of Passive Microwave Data, *IEEE Trans. Geosci. and Remote Sensing*, **31**, 5 ,1060-1065.

4. Armand A., A. E. Basharinov, L.F. Borodin, E. N. Zotova, A.M. Shutko, 1985. Radiophysical Methods in Remote Sensing of Environment: detection of forest and peat fires, in *Problems of Modern Radio Engineering and Electronics*, V. A. Kotelnikov editor, Moscow (URSS), 136-139.

5. Coppo P., S. Fini, G. Luzi, T. Mazzoni , P. Ferrazzoli, L. Guerriero, G. Lio, 1991. Microwave Radiometry applied to Forest Fires: operational aspects and model analysis, *Proceeedings of the 11th EARSeL Symposium*, Graz (Austria) 1991, 259-265.

6. Ferrazzoli P. , L. Guerriero, D. Solimini, 1991. Numerical model of microwave backscattering and emisson from terrain covered with vegetation, *J. Appl. Comp. Electromagnetic Soc.* , **6**, 175-191.

7. Ulaby F. T. , R. K. Moore, A. K. Fung, 1986. *Microwave Remote Sensing: Active and Passive*, **3**, Artech House , Norwood MA

8. Birchak R.J., G.C. Gardner, J. E. Hipp, J.M. Victor, 1974. High dielectric constant microwave probes for Sensing soil moisture, *Proceedings of the IEEE*, **62**, 1, 93-98.

9. Droesen W.J., D.K. Hoeckman, H.J.C. Van Leeuwen, J.J. Van der Sanden, A.M. Bouman, D. Uenk, A.M. Vissers, G.G. Lemoine, 1990. *Maestro 89 ground data collection Hosterwold / Speuldeerbos /Flevoland*, Wageningen Agricultural University Tech. Rep., 1990.

10. Ferrazzoli P. , L. Guerriero, 1993. Modeling Microwave Emission from Vegetation -covered surface: a parametric analysis, *Proceedings of ESA/NASA International Workshop on Passive Microwave Remote Sensing Research Related to Land- Atmosphere Interactions*, held in St.Lary (France), Utrecht VSP Press,1994 .

11. Gibbins C.J. and R. Chadha, 1987, Millimetre wave propagation through hydrocarbon flame, *IEEE Proceedings*, **134**, 2, 169-173.

Polarimetric Scattering and Emission from a Layer of Random Clusters of Small Spheroids and Dense Spheres

YA-QIU JIN
Department of Electronic Engineering, Fudan University
Shanghai, China 200433

Abstract Employing the volume integral equation formulation with self-interaction term for internal fields, the scattering amplitude functions and the phase matrix for random clusters of small spheroids are derived. Clustering enhances the internal fields and the scattering amplitude functions. Hence, the effective size of clustered scatterers tends to be larger. Substituting newly-derived phase matrix into the Mueller matrix in the approach of vector radiative transfer for a layer of random clusters of small spheroids, fully polarimetric co-polarized and depolarized backscattering are numerically calculated. Using the enhanced effective size in scattering and emission model for a layer of random clusters of dense spheres, the polarized brightness temperatures are numerically calculated. Clustering effects of a layer of random spheroids and dense spheres are quantitatively illustrated, and functional dependencies on parameters are discussed.

Keywords: random clusters, spheroids, scattering, emission

INTRODUCTION

In recent years, there has been a growing interest in study of electromagnetic scattering from clusters of particles, such as branched chains or coagulations of particles.[1-4] These studies have many applications including the effects on the propagation of millimetre and radar waves by aerosol coagulations, detection of smokes and powders emitted from automobile or coal fired exhausts, laser diagnostic and radiative transfer of biological cells, polymers and many others. Purcell and Pennypacker (P-P)[2], by using the method of multi-polar expansion, obtained the scattering field from a cluster of spheres. Each primary sphere is assumed to have a uniform internal field and is treated as a dipole. Iskander et al.(ICP)[3], based on the volume integral equation and the method of moments, calculated scattering from a cluster of chained spheres. The resulting solutions for electrical fields is actually the same as P-P except for the addition of a self-interaction term, i.e. the interaction of the field in a primary sphere with itself. Ku et al.[4] numerically compared several approaches of the Refs.[1-3] for a cluster of chained spheres in Rayleigh limit, and pointed out that the approach by ICP seems to be better than the others. However, scattering from random clusters of non-spherical or dense spherical particles remained to be done. Most recently, Vargas et al.[5] used a T-matrix to calculate the internal field of a cluster of two spheroids in long wavelength regime. T-matrix is a strong tool to deal with scattering from complex scatterers. However, it seems not tractable yet to be applicable to random clusters of spatially oriented, non-spherical scatterers due to very intensive computational works.

During recent years, the development of SAR imaging technology and study of polarimetry have become of importance. Clustering effect from leaves and branches of vegetation canopies, agglomerated ice particles of snowpack, etc. has remained as a intractable problem for both active and passive remote sensing. In this paper, we extend the volume integral equation formulation of ICP to the randomly oriented small spheroids, and present an approach to polarimetric scattering from a layer of random clusters of small spheroids. By using the volume integral equation formulation with a self interaction term for small spheroid in Rayleigh-Debye approximation, we obtain the internal field of the spheroid in a cluster. Then, we derive all elements of the scattering amplitude functions and the phase matrix for vector radiative transfer, which take account of clustering, spatial orientation of random spheroids, and boundary reflection. Since all of these terms are explicitly formulated and are tractable for programming, we then can construct the Mueller matrix[6,7], and numerically calculate co-polarized and depolarized backscattering for any polarized incidence. Clustering enhances the scattering amplitude functions and, in some way, increases the effective size of scatterers in the scattering model. We discuss the effective size for a layer of snowpack which consists of random clusters of dense and size-distributed spherical ice particles. Using the effective size enhanced by clustering, the polarized brightness temperatures from a layer of snow are calculated. The clustering effect and functional dependence on the parameters can be discussed.

SCATTERING AMPLITUDE FUNCTIONS OF A CLUSTER

As a plane wave $\hat{e}E_0\exp(i\vec{k}_i\cdot\vec{r})$ incident on a small spheroid with semi-axes (a,b,c) and local coordinates $(\hat{x}_b,\hat{y}_b,\hat{z}_b)$, the internal field of the spheroid can be written as[8]

$$\vec{E}_{int}(\vec{r}') = \left[\frac{\hat{x}_b\hat{x}_b}{1+(\epsilon_{sr}-1)g_1} + \frac{\hat{y}_b\hat{y}_b}{1+(\epsilon_{sr}-1)g_2} + \frac{\hat{z}_b\hat{z}_b}{1+(\epsilon_{sr}-1)g_3}\right]\cdot \hat{e}E_0 e^{i\vec{k}_i\cdot\vec{r}'}$$

$$\equiv \overline{\overline{A}}_b\cdot \hat{e}E_0\exp(i\vec{k}_i\cdot\vec{r}) \tag{1}$$

where

$$\overline{\overline{A}}_b = \sum_{n=1}^{3}\frac{\hat{x}_n\hat{x}_n}{1+(\epsilon_{sr}-1)g_n}, \quad \hat{x}_1=\hat{x}_b,\ \hat{x}_2=\hat{y}_b,\ \hat{x}_3=\hat{z}_b \tag{2a}$$

$$g_1 = \frac{1}{2}abc\int_0^\infty \frac{1}{(s+a^2)R_s}ds, \qquad R_s = [(s+a^2)(s+b^2)(s+c^2)]^{1/2} \tag{2b}$$

and g_2, g_3 are the same as g_1 with the term of $(s+a^2)$ replaced by $(s+b^2)$, $(s+c^2)$, respectively, and $g_1+g_2+g_3=1$. In the case of spheres,

$$\overline{\overline{A}}_b \rightarrow \frac{3}{\epsilon_{sr}+2}\overline{\overline{I}} \tag{2c}$$

The transformation between $(\hat{x}_b,\hat{y}_b,\hat{z}_b)$ and the principle coordinates $(\hat{x},\hat{y},\hat{z})$ is expressed by the functions of the Euler angles (α,β,γ) of the spheroid

$$[\ \hat{x}_b,\ \hat{y}_b,\ \hat{z}_b\ ]^T = \overline{\overline{a}}(\alpha,\beta,\gamma)\cdot[\ \hat{x},\ \hat{y},\ \hat{z}\ ]^T \tag{3}$$

where the superscript T denotes transpose, and the matrix $\overline{\overline{a}}$ can be found in section 4.2 of Ref.[8]. Substituting the internal field Eq.(1) into the following equation, the scattering field can be written as

$$\begin{aligned}\vec{E}_s &= \frac{e^{ikr}}{r}(\hat{v}_s\hat{v}_s+\hat{h}_s\hat{h}_s)\cdot\frac{1}{3}k^2abc\int_{all\ v_0} d\vec{r}'(\epsilon_{sr}-1)\cdot\vec{E}_{int}(\vec{r}')e^{-i\vec{k}_s\cdot\vec{r}'}\\ &= \frac{e^{ikr}}{r}\overline{\overline{f}}\cdot\hat{e}E_0\end{aligned} \tag{4}$$

where v_0 is the volume of a single scatterer, and in the Rayleigh approximation the scattering amplitude functions f_{pq}, $p,q=v,h$, is defined as

$$f_{pq}(\theta,\phi;\theta',\phi') = \hat{p}\cdot\overline{\overline{T}}\cdot\hat{q}' \tag{5a}$$

$$\overline{\overline{T}} = \sum_{n=1}^{3} T_n\ \hat{x}_n\hat{x}_n\ ,\quad T_n = \frac{k^2}{3}abc(\epsilon_{sr}-1)\frac{1}{1+(\epsilon_{sr}-1)g_n} \equiv \frac{3}{2k}t_n \tag{5b}$$

In the case of spheres, Eq.(5) becomes

$$f_{pq}(\theta,\phi;\theta',\phi) = k^2a^3\frac{\epsilon_{sr}-1}{\epsilon_{sr}+2}\ \hat{p}\cdot\hat{q}' \tag{6}$$

To ensure the optical theorem be satisfied, an additional term will be kept[8], and t_n in Eq.(5b) is written as $t_n(1+it_n)$ and

$$T_n = \frac{3}{2k}t_n(1+it_n) \tag{7}$$

In the Rayleigh-Debye approximation, T_n of Eq.(7) should be multiplied by a factor

$$\mu_{is} = \int_{v_0} d\vec{r}'e^{i(\vec{k}_i-\vec{k}_s)\cdot\vec{r}'} \approx \begin{cases}\dfrac{2J_1(qa)}{qa}, & a>>c\\[2ex] \dfrac{\sin[kc(\cos\theta_i+\cos\theta_s)]}{kc(\cos\theta_i+\cos\theta_s)}, & c>>a\end{cases} \tag{8}$$

where $q=k[\sin^2\theta_s+\sin^2\theta_i-2\sin\theta_s\sin\theta_i\cos(\phi_s-\phi_i)]$, J_1 is the first-order Bessel function.

We extend the volume integral equation formulation of ICP[3,4] to the case of n_0 spheroids in a cluster. The internal field in the i-th spheroid of a cluster is

$$\vec{E}_i = \overline{\overline{A}}_b \cdot \hat{e}E_0 e^{i\vec{k}_i\cdot\vec{r}'} + \sum_{j\neq i}^{n_0} \frac{\exp(kR_{ij})}{R_{ij}^3}\overline{\overline{G}} \cdot \vec{E}_j + \overline{\overline{s}}_i \cdot \vec{E}_i \tag{9}$$

where the elements G_{mn} $m,n=1,2,3$ of the matrix $\overline{\overline{G}}$ are written as

$$G_{nn}=\frac{1}{3}abc\frac{(\epsilon_{sr}-1)}{1+(\epsilon_{sr}-1)g_n}\left(k^2R_{ij}^2-(1-ikR_{ij})-\frac{X_n^2}{R_{ij}^2}[k^2R_{ij}^2-3(1-ikR_{ij})]\right) \tag{10a}$$

$$G_{mn}= -\frac{1}{3}abc\frac{(\epsilon_{sr}-1)}{1+(\epsilon_{sr}-1)g_n}\frac{X_mX_n}{R_{ij}^2}[\ k^2R_{ij}^2 - 3(1-ikR_{ij})], \quad m\neq n \tag{10b}$$

where

$$X_1\equiv X_{ij}=x_i-x_j,\ X_2\equiv Y_{ij}=y_i-y_j,\ X_3\equiv Z_{ij}=z_i-z_j,\ R_{ij}=(\sum_{n=1}^{3} X_n^2)^{1/2}$$

The clustering effect is from the second and third terms of the RHS of Eq.(9). The self interaction term with $\overline{\overline{s}}_i$ is a diagonal integral over the ith-scatterer self

$$\overline{\overline{s}} \cdot \vec{E}_i = P.V. \int_{i-th\ self} d\vec{r}'\ \overline{\overline{G}}(\vec{r}-\vec{r}')\cdot[\epsilon_{sr}(\vec{r}')-1]\vec{E}(\vec{r}') \equiv [\ P_{ij}\ ]\cdot \vec{E}_i \tag{11}$$

where P.V. denotes the principle integral. Let $\vec{r}=\vec{r}_0$, $\vec{r}'=\vec{r}_0+\vec{s}$. Substituting Eqs.(10a,b) into the integral of the self term Eq.(11), the ij-th elements of the integral becomes

$$P_{ij}=\int_{self} d\vec{s}\frac{e^{iks}}{4\pi s}[k^2(\delta_{ij}-\tilde{s}_i\tilde{s}_j)-(\delta_{ij}-3\tilde{s}_i\tilde{s}_j)(s^{-2}-iks^{-1})] \tag{12}$$

where $\tilde{s}_i$ $i=1,2,3$ is the i-th Cartesian components of the unit vector $\hat{s}$. Since we use the centre point of the scatter cell as the lattice point $\vec{r}_0$, the above integral has symmetric limits and volumetric symmetry.[3] Therefore, $\tilde{s}_i\tilde{s}_j$,$-\tilde{s}_i\tilde{s}_j$ will cancel each other. It yields

$$\int d\vec{s}\ \tilde{s}_i\tilde{s}_j = \frac{1}{3}\int d\vec{s}\ \delta_{ij} \tag{13a}$$

and leads to

$$P_{ij}=\frac{k^2}{6\pi}\delta_{ij}\int_{self} d\vec{s}\ \frac{e^{iks}}{s} \approx \frac{k^2}{6\pi}\delta_{ij}[\int_{self} d\vec{s}\ (\frac{1}{s}+ik)] \tag{13b}$$

If the scatter cell is a sphere, then Eq.(13b) yields

$$P_{ij}=\frac{1}{3}\delta_{ij}(k^2a^2+i\frac{2}{3}k^3a^3) \tag{14a}$$

$$\overline{\overline{s}}=\frac{(\epsilon_{sr}-1)}{(\epsilon_{sr}+2)}\overline{\overline{I}}\ (k^2a^2+i\frac{2}{3}k^3a^3) \tag{14b}$$

If the scatterer is an oblate spheroid, we obtain the integral

$$\int_{spheroid} d\vec{s}\frac{1}{s} = \frac{2\pi a^2c}{\sqrt{a^2-c^2}}\arctan\frac{\sqrt{a^2-c^2}}{c} \tag{15}$$

$$\overline{\overline{s}} = \frac{(\epsilon_{sr}-1)}{3}\overline{\overline{A}}_b\left[\frac{k^2a^2c}{\sqrt{a^2-c^2}}\arctan\frac{\sqrt{a^2-c^2}}{c} + i\frac{2}{3}k^3a^2c\right] \tag{16}$$

If let $c\to a$, Eq.(16) reduces to Eq.(14b).

Now the internal fields of Eq.(9) can be numerically solved. To make the amplitude functions in an explicit form, we assume that all internal fields in n_0 spheroids of a cluster are equal, i.e. $\vec{E}_j=\vec{E}_i$, $i,j\in n_0$ as a first-order approximation, then the internal field of Eq.(9) can be simply written as

$$\vec{E}_i = \left[\overline{\overline{I}} - \sum_{j\neq i}^{n_0}\frac{\exp(ikR_{ij})}{R_{ij}^3}\overline{\overline{G}} - \overline{\overline{s}}_i\right]^{-1}\cdot\overline{\overline{A}}_b\cdot\hat{e}E_0e^{i\vec{k}_i\cdot\vec{r}'} \tag{17}$$

Substituting Eq.(17) into Eq.(4), the scattering amplitude functions can be obtained as

$$\begin{aligned} f_{pq} &= \hat{p}\cdot(\overline{\overline{B}}\cdot\overline{\overline{T}})\cdot\hat{q}' \\ &= \sum_{m=1}^{3}\sum_{n=1}^{3} b_{mn}\ T_n(\hat{p}\cdot\hat{x}_m)(\hat{x}_n\cdot\hat{q}') \end{aligned} \tag{18}$$

where

$$\overline{\overline{B}} = \left[\overline{\overline{I}} - \sum_{j\neq i}^{n_0}\frac{\exp(ikR_{ij})}{R_{ij}^3}\overline{\overline{G}} - \overline{\overline{s}}_i\right]^{-1} \equiv [b_{mn}] \tag{19}$$

Comparing Eq.(18) with Eq.(5), or Eq.(17) with Eq.(1), it may be seen that the clustering factor b_{mn} enhances the scattering amplitude functions and the internal fields.

We can now calculate the forward scattering amplitude function $\langle f_{pq}(\theta,\phi;\theta,\phi)\rangle$ for calculation of extinction coefficient[9], where the angular brackets $\langle . \rangle$ denotes the orientation average of the Euler angles (α,β,γ).

ELEMENTS OF PHASE MATRIX AND EFFECTIVE SIZE

We consider a layer of random clusters of small spheroids as shown in Fig.1. By using iterative method to solve the vector radiative transfer equation and boundary conditions, the first-order Mueller matrix can be obtained[6],

$$[\ I_{vs}\ ,\ I_{hs}\ ,\ U_s\ ,\ V_s\ ]^T = \overline{\overline{M}} \cdot [\ I_{vi}\ ,\ I_{hi}\ ,\ U_i\ ,\ V_i\ ]^T \tag{20}$$

The Mueller matrix $\overline{\overline{M}}$ is a 4*4 real matrix, which takes account of scattering phase matrix and boundary reflections. By using the functions f_{pq} of Eq.(18) to construct the phase matrix and Mueller matrix, we derive all elements of the phase matrix, where $\phi=\pi$, $\phi'=0$ always assumed. We have obtained following ten terms[9]:

$$<|f_{hh}|^2>,\quad <|f_{vv}|^2>,\quad <|f_{hv}|^2>,\quad <|f_{vh}|^2>,\quad <f_{vv}f_{hh}^*>,$$

$$<f_{vh}f_{hv}^*>,\quad <f_{hh}f_{hv}^*>,\quad <f_{vv}f_{vh}^*>,\quad <f_{hh}f_{vh}^*>,\quad <f_{vv}f_{hv}^*>$$

Then we readily compose the phase matrix and the Mueller matrix. From Eq.(20), we obtain co-polarized back-scattering coefficient σ_c and depolarized back-scattering coefficient σ_d, and the polarization degree m_s [6,7].

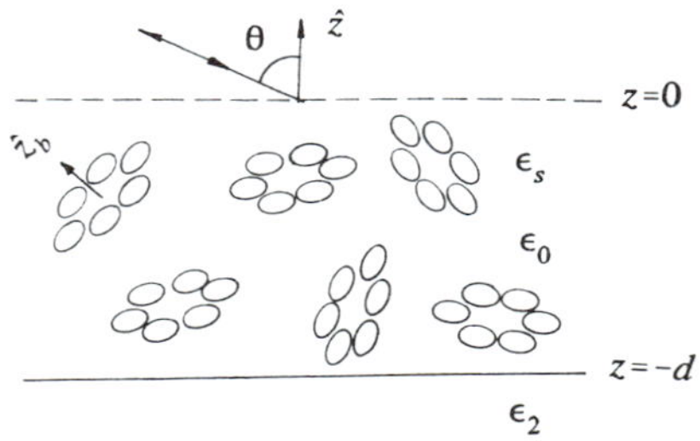

Figure 1 Geometry of the problem

Snowpack is a dense scatter medium, where the ice particle fraction can be larger than 0.2. Non-spherical ice particles were usually modeled by uniformly and size-distributed spherical particles. It has been discussed[10] that the effective radius of the size-distributed particles is always larger than the average or mode radii in the scattering model since the dominant scattering effect is caused by larger scatterers, especially. Actually, dense ice particles of snowpack are also clustered. The scattering amplitude functions enhanced by clustering may increase the effective size further. Consider a randomly oriented cluster of n_0+1 closely aligned spheres. Averaging over uniform orientation from Eqs.(10a,b), we may have $G_{mn}=0$ $(m\neq n)$ and

$$G_{nn} = \frac{2}{3}k^2a^3\frac{\epsilon_{sr}-1}{\epsilon_{sr}+2}\ R_{ij}^2 \tag{21}$$

Since the amplitude functions are proportional to the cubic size, substituting Eq.(21) into Eq.(19) to obtain the enhancement factor b_{mn} , the effective size of clustered dense spheres should become as

$$a_e = (b_{nn}^{1/3}) a_{e0} \tag{22}$$

where a_{e0} is the effective size of uniformly random dense spheres.

NUMERICAL RESULTS OF POLARIMETRIC SCATTERING

We suppose to have a layer of random clusters of hexad oblate spheroids as shown in Fig.1. The centres of each spheroids compose a hexagon with length R. Then,

$$R_{0j} = 2R\sin(j\frac{\pi}{n_0}),\ X_{0j} = R_{0j}\sin(j\frac{\pi}{n_0}),\ Y_{0j} = R_{0j}\cos(j\frac{\pi}{n_0}),\quad n=1,\cdots,5$$

are given. The number of clusters per unit volume is N, then the fractional volume $f_s=(4/3)\pi abcn_0N$. Figs.2a,b show the co-polarized back-scattering σ_c and depolarized back-scattering σ_d versus the polarization angles (χ,ψ) of the incidence for a layer of random clusters of spheroids. The orientations of β, γ are distributed over the regions (0,45°) and (0,360°), respectively. As a comparison, Fig.3a,b give σ_c, σ_d vs (χ,ψ) for uniformly random spheroids without clustering. It can be seen that clustering significantly enhances the back-scattering, especially for linear polarizations.

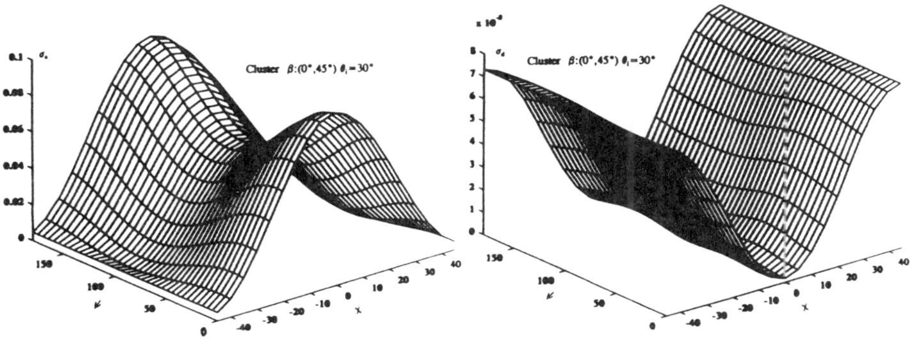

Figures 2a,2b σ_c, σ_d vs (χ,ψ) for a layer of random clusters frequency=1.225 GHz, d=2.5m, $\epsilon_s=(6+i0.5)\epsilon_0$, $\epsilon_2=(15+i2)\epsilon_0$, $f_s=0.0055$, a=b=2.5cm, c=0.5cm, R=5cm, β:(0,45°), $\theta_i=30°$;

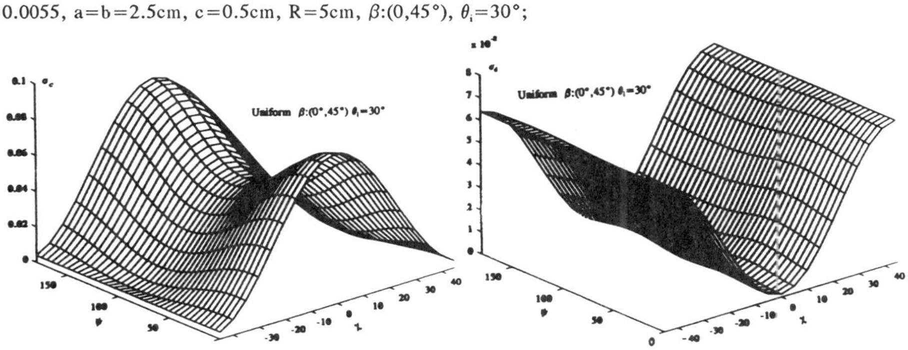

Figures 3a,3b σ_c, σ_d vs (χ,ψ) for a layer of uniformly random spheroids

Fig.4a,b give the comparisons of polarization degree m_s for a layer of clusters and uniformly random scatterers for β over (0,45°). Non-uniformly oriented clusters cause very strong polarization degree.

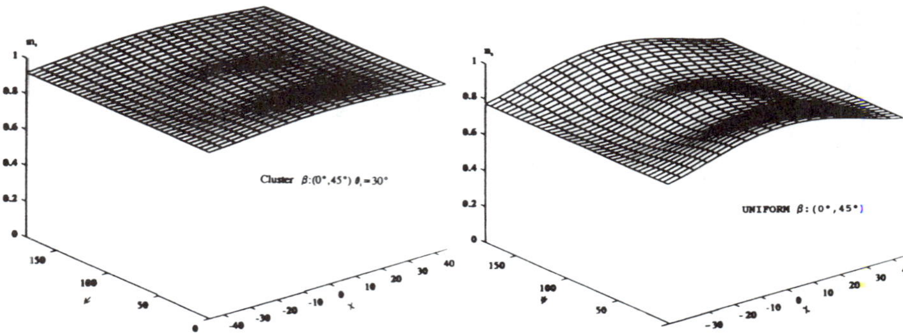

Figures 4a,4b Polarization degree m_s vs (χ,ψ)

Fig.5 gives the brightness temperature vs observation angle at 19.35 GHz for a layer of snow. Let $n_0+1=5$ for a cluster of closely aligned spheres with a_{e0}=0.1cm , we may obtain $b_{nn}\approx 1.16$ and a_e=0.105cm from Eqs.(19) and (22). The dense medium radiative transfer (DMRT) equation for respective a_e, a_{e0} is numerically solved by the discrete-ordinate and eigenanalysis method. The solid lines show the results of the clustered dense spheres with the enhanced effective size a_e , and the dashed lines are the results of uniformly random dense spheres with a_{e0} . At low frequency, scattering is proportional to particle size to the sixth power. At high frequencies, Mie scattering effects are important. The size in the scattering model is always a very sensitive parameter. Clustering yields a larger effective size of scatterers, and may enhance scattering and darken emission.

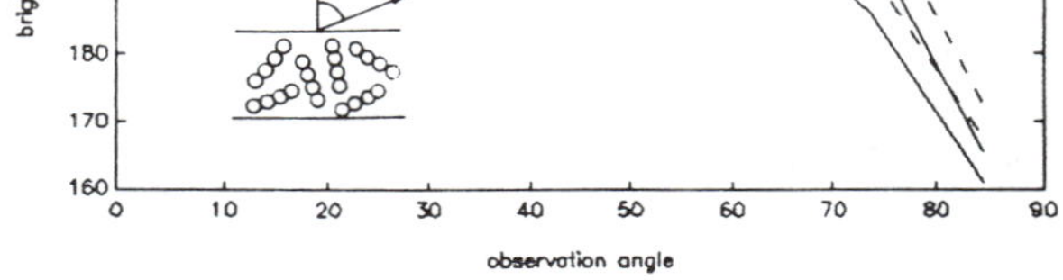

Figure 5 Brightness temperature vs angle
d=60cm, f_s=0.3, $T_1=T_2$=273K, $\epsilon_s=(3.2+0.0032)\epsilon_0$, $\epsilon_2=(5+i0.1)\epsilon_0$

CONCLUSIONS

We have extended the volume integral equation formulation to the case of random small spheroids and calculated the internal fields in the Rayleigh-Debye approximation. The scattering amplitude functions and all elements of the phase matrix for vector radiative transfer are explicitly derived. The Mueller matrix for a layer of random clusters can be readily constructed. By using the Mueller matrix approach, fully polarimetric co-polarized and depolarized back-scattering, and polarization degree are numerically calculated. Clustering enhances the effective size in the scattering model. Using the enhanced effective size, the brightness temperatures from a layer of clustered dense spheres are numerically calculated. Numerical results show the enhanced polarized scattering and darken emission due to clustering. Comparisons with uniformly random scatterers and some functional dependencies on physical parameters are also discussed.

ACKNOWLEDGEMENT

This work was supported by UK Royal Society K.C. Wong Fellowship and the National Natural Science Foundation of China.

REFERENCES

1. A.R. Jones, "Scattering efficiency factors for agglomerates of small spheres", *J. Phys. D: Appl. Phys.*, Vol.12, pp1661-1672, 1979.
2. E.M. Purcell and C.R. Pennypacker, "Scattering and absorption of light by nonspherical dielectric grains", *The Astrophys. Journal*, vol.186, pp705-714. 1973.
3. M.F. Iskander, H.Y. Chen, and J.E. Penner, "Optical scattering and absorption by branched chains of aerosols", *Applied Optics*, vol.28, pp3083-3091, 1989.
4. J.C. Ku and K.H. Shim, "A comparison of solutions for light scattering and absorption by agglomerated or arbitrarily-shaped particles", *J. Quant. Spectrosc. Radiat. Transfer*, vol.47, pp201-220, Mar. 1992.
5. W. Vargas, L. Cruz, L.F. Fonseca, and M. Gomez, "T -matrix approach for calculating local fields around clusters of rotated spheroids", *Applied Optics*, vol.32, pp2164-2170, April 1993.
6. L. Tsang, K.H. Ding, and B. Wen, "*Polarimetric Remote Sensing*", ed. by J.A. Kong, pp75-141, 1990, Elsevier.
7. Y.Q. Jin, "A Mueller matrix approach to complete polarimetric scattering from a layer of non-uniformly oriented, non-spherical scatterers", *J. Quant. Spectrosc. Radiat. Transfer*, Vol.48, pp295-306, Sep. 1992.
8. L. Tsang, J.A. Kong, and R. Shin, "*Theory of Microwave Remote Sensing*", John Wiley and Sons., N.Y. 1985.
9. Y.Q. Jin, "Polarimetric scattering from a layer of random clusters of small spheroids", to be published in *IEEE Trans. Ant. Propag. 1994.*
10. Y.Q. Jin, "Size parameters of snow grains in scattering and emission models", in *Proceedings of the ESA/NASA International Workshop*, ed. by P. Pampaloni, forthcoming.

Microwave Radiometry of Vegetation: Recent Advances

G. MACELLONI, S. PALOSCIA, P. PAMPALONI, R. RUISI and C. SUSINI
Istituto di Ricerca sulle Onde Elettromagnetiche - CNR
via Panciatichi 64, 50127, Firenze (Italy) - Fax: 39 55 480193

Abstract

Microwave radiometers have a high sensitvity to water content of soil and vegetation, and their use is of great interest in land studies. This paper reports the results of recent investigations on emission features of vegetation, on different spatial scales, and at those frequencies of SSM/I and MIMR bands 10, 19 and 37 GHz where the atmospheric attenuation is still modest. The study has shown that discrimination between bare and vegetation covered soils can be accomplished from satellite using the polarization index at 19 GHz. Furthermore field measurements allowed us to quantify the contribution of single plant constituents.

Keywords: land emission, vegetation, plant water content.

INTRODUCTION

Microwave measurement from space or aircraft of the thermal emission from the Earth's surface has proven to be an important method for investigating atmosphere and land surface properties. The need for global data sets derived from remote sensing observations is clear in many disciplines. For example, monitoring of vegetation biomass on a global and regional scale is very important for a better understanding of chemical and hydrological processes and for improving climate change assessments.

Experimental investigations and theoretical analysis, carried out over the past years by our group at CNR in Florence and by other scientists worldwide, showed that microwave emission is sensitive to vegetation biomass (e.g.: 1-4). Most of the experiments were carried out with ground based or airborne systems and only a few significant studies for discriminating land surfaces or estimating quantitative parameters have been conducted using satellite data from the Scanning Multichannel Microwave Radiometer (SMMR) (5) and the Special Sensor Microwave Imager (SSM/I) (6,7). The electromagnetic features of many crop types have been studied on a frequency range between 1.4 GHz (L-band) and 37 GHz (Ka-band) and several relationships between brightness temperatures and plant growth have been found (8). The next generation satellite sensors such as the ESA Multi-frequency Imaging Microwave Radiometer, with considerably improved performances (number of frequency channels, sensitivity and ground resolution), will allow significant progress towards operational use in several fields, provided adequate efforts are carried out to develop algorithms and models for interpreting emissivity data.

A simple approach to studying microwave emission from a canopy covered soil is to model vegetation as a uniform absorbing and scattering layer overlaying a homogeneous half space (9, 10). According to this model the variation of brightness temperature Tb as vegetation grows depends on optical depth τ, single scattering albedo ω, and scattering phase function, which in turn are functions of frequency and plant type. When the scattering is mainly forward the formal solution of radiative transfer equation can be highly simplified, provided τ and ω

are replaced by the equivalent τ' and ω'according to the transformation introduced in (9). If the single scattering albedo is low Tb increases as vegetation grows, whereas the opposite occurs when scattering plays a significant role.

Our experiments have confirmed that the increase of vegetation biomass is associated with a monotonic variation of the emissivity when the observation wavelength is much smaller (Ka-band) or much larger (L-band) than the dimensions of plant constituents. At X-band (10 GHz), microwave emission is significantly influenced by crop type; it increases with leaf area index for small leaf crops and decreases for wide leaves (11).

According to both theoretical and experimental results, we can remark that microwave radiation from vegetation canopies is influenced by both the total biomass and the plant geometry. At a certain frequency the emission may be dominated by a specific geophysical parameter such as leaf dimensions, plant water content or soil moisture, but in general it is affected simultaneously by several factors and the extraction of geophysical parameters from microwave emission requires the use of appropriate combinations of frequencies and polarizations (8, 12). For example, it has been demonstrated that significant information on vegetation biomass is obtained from the polarization index (PI) at X band (3, 13) which has a maximum value on smooth bare soil and decreases as vegetation grows. However, surface roughness, similarly to vegetation growth, reduces PI, so that, without the aid of additional information it may be difficult to separate the effects.

In general the highest contrast between emission from bare and vegetated soils is obtained at relatively low frequencies (5 - 10 GHz). On the other hand, the need for reasonable ground resolution in vegetation monitoring from satellite, is very important and requires the use of higher frequencies to avoid impractical antenna dimensions. Hence, there is interest in further exploration of the emission features of vegetation, on different spatial scale, and at the frequencies of SSM/I and MIMR bands like 19 and 37 GHz where the atmospheric attenuation is still modest. Two research lines have been initiated to achieve this goal:

- to investigate the capability of microwave data from spaceborne sensors in monitoring land surface parameters in the European and Mediterranean landscape, where spatial and temporal variations may be very high.

- to study microwave polarimetric features of plant constituents and evaluate in detail the contribution of leaves, stalks and fruits to the total emission from the canopy.

This paper reports some significant results of these investigations which are still in progress.

EXPERIMENTAL RESULTS

Emission from vegetation covers

The potential of microwave radiometry from space in detecting land surface features in the European landscape, where spatial and temporal variations are high, is being investigated using multitemporal data from the SMM/I. This sensor measures brightness temperature at 19.3, 22.2, 37 and 85.5 GHz (V and H pol.), with ground resolution of 43, 40, 29 and 13 km respectively. The preliminaries of the study have been carried out on 10 selected 'test' areas with different surface and climatic characteristics.

The area dimensions have been determined as a compromise between the minimum number of measurements necessary for a statistical analysis which is driven by the big resolution cell of the low frequency channel, and the homogeneity of the area itself, better characterized by the higher resolution of the 85 GHz channel. The areas usuallyinclude more than 100 measurements at the lower frequencies, with a standard deviation of the brightness temperature lower than 5 K at the highest frequency. Data have not been corrected for atmospheric attenuation and this may represent a significant uncertainty especially for the 85 GHz channel. However a check for the eventual presence of precipitations has been performed by using the same data and the standard algorithms suggested in scientific literature and in SSM/I User's Guide (7, 13). To reduce the effects of the surface temperature variation on the measured microwave radiation, data collected in the morning have been separated from those collected in the afternoon. All groups of data have been averaged over a four day period. The mean value, standard deviation σ and histograms of the single channel brightness temperatures computed for each test area have allowed a first evaluation of the homogeneity characteristics of the surface and an adjustment of the area limits. For each test area the analysis has been based on the frequency spectra of vertical (T_{bv}) and horizontal (T_{bh}) polarization components and on certain combinations of frequency and polarization channels, which, according to previous investigations (7), appear suitable for identifying the surface type or its physical conditions. In particular we have used the Polarization Index [PI = 2 $(T_{bv}-T_{bh})/(T_{bv}+T_{bh})$] as a basic parameter for identifying vegetated areas (3 - 14). Indeed it has been proven that this index is very high on desert and smooth bare soils while it decreases on agricultural regions to almost zero on deep vegetation (5).

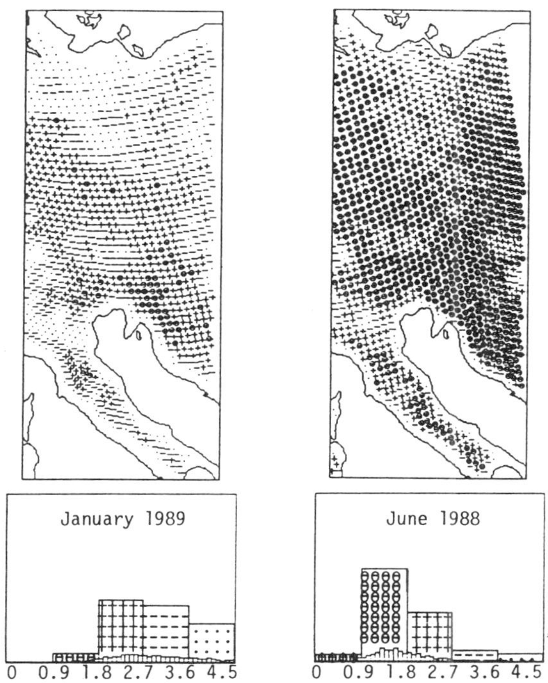

Figure 1 - Maps and histograms of Polarization index (PI) at 19 GHz measured in January 1989 and June 1988 over a portion of Central Europe. Darker areas represent lower values of PI.

Fig. 1 shows the polarization index at 19 GHz measured in January 1989 and June 1988 over a section of Central Europe. Darker areas represent lower values of PI and correspond to more densely vegetated areas. The histograms confirm the shift of PI toward lower values as vegetation grows from January to June.

The average polarization index at 19 and 37 GHz of the Po valley in Northern Italy, is represented as a function of time in Fig. 2. In this region (about 80 Km x 80 Km wide) the agricultural landscape is dominant, but important urban and industrial installations are also present which cannot be separated due to the coarse SSM/I resolution. The agricultural areas are partially covered by mixed crops whose amount varies noticeably according to the season and their microwave emission is essentially the combination of the two main components from soil and vegetation. As expected the polarization index changes according to the vegetation cycle, being maximum in winter and minimum in June. We see that, in spite of the lower resolution, the 19 GHz channel has a wider PI variation than the 37 GHz. The T_b frequency spectrum exhibits a slightly positive slope at both vertical and horizontal polarization.

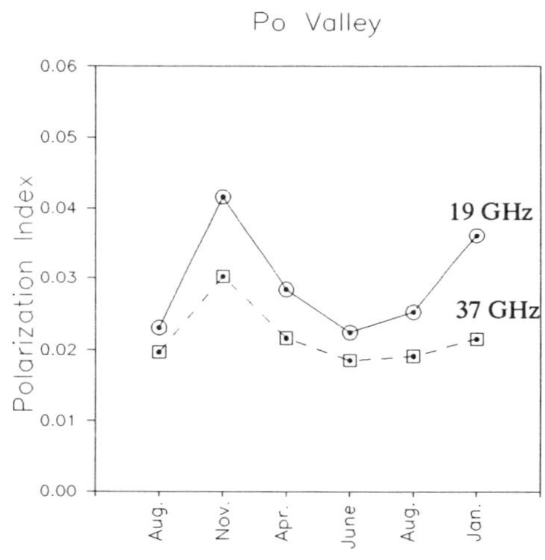

Figure 2 - Polarization Index (at 19 GHz and 37 GHz) of an agricultural area in Northern Italy as a function of time .

Data collected over an area located in the northern part of Scandinavia (69.5° N, 26° E - 68° N, 22° E) covered by short grass (tundra) show that the brightness temperature measured in summer (July) slightly increases with frequency as in the case of moderately vegetated areas (Fig. 3), and the polarization index is quite low. On the other hand, according to data collected in April, this area has the typical spectral trend of a snow cover; indeed as the frequency increases the volume scattering of snowpack increases and the brightness temperature decreases. The polarization index, much higher in winter (by about 5 times) than in summer, is consistent with the spectrum trend and confirms that the mechanisms which influence emission in winter are different from those prevailing in summer (Fig. 4).

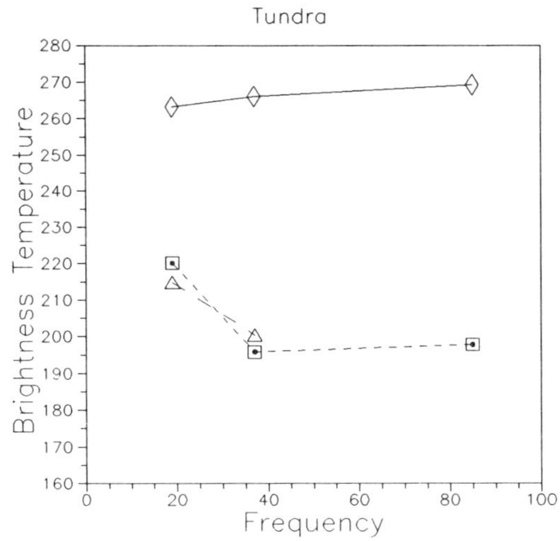

Figure 3 - Brightness temperature spectra(horizontal polarization) of tundra region in summer (◇), spring ((□) and winter (△).

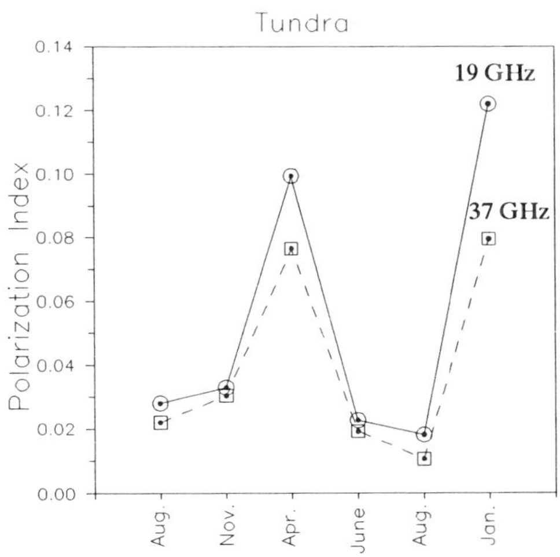

Figure 4 - Polarization Index of Tundra as a function of time

A similar behavior has been found in a site located in Russia (60° N, 52° E - 59° N, 50° E) and mostly covered by a coniferous forest. The frequency spectrum, measured in summer, is almost flat at both horizontal and vertical polarizations, which is typical for densely vegetated areas, and the polarization index is small (0.013 at 19 GHz). On the contrary, in January the spectrum has the negative slope characteristic of snow cover.

Emissivity of plant constituents

Field measurements have been carried out using a dual polarized (V and H polarizations) microwave radiometer at 36.5 GHz and a thermal IR sensor (8 - 14 μm) (15). The microwave instrument is a portable, battery operated system, which can be easily used in field measurements as well as from an aircraft. The instrument is based on a frequently calibrated total power radiometer with two reference loads at different temperatures. It uses two corrugated conical horns, one for each polarization channel, and the signal is switched between the two horns at a frequency close to 10 Hz. The detected signal is digitized with a voltage to frequency converter and all the functions of the radiometer (integration time, calibration and observation sequences, etc.) are programmable with an internal microprocessor. Finally, row and calibrated data are recorded by means of a notebook computer. The achieved measurement accuracy (repeatability) is in the order of ± 0.5 K, with an integration time of 0.1 sec. The infrared sensor is a commercial type hand-held radiometer with an accuracy (repeatability) of ± 0.5 K.

The experiments included measurements of emission from:

- well developed standing 'hard' wheat, in natural condition and after sequential cut of fruits, leaves, and stems. The ears of this wheat variety were included long hairlike fibers called awns.
- single plant constituents of wheat and alfalfa (fruit, leaves, stems) placed on a reflecting plate.

The microwave and infrared brightness temperatures have been measured at an observation angle $\theta = 45°$ and, for plant elements, at various azimuth angles (ϕ). Total biomass and geometrical characteristics of plant constituents have been measured with standard methods.

During the measurements wheat was in the flowering stage and had a global plant water content (PWC) equal to 1.0 Kg/m^2, the fraction due to ears (with awns), leaves and stems was respectively of 0.2, 0.15, and 0.65 Kg/m^2 . The leaf area index was 1.2 m^2/m^2 and the average height 70 cm. The PWC of alfalfa, which was well developed (76 cm height), was 1.4 Kg/m^2.

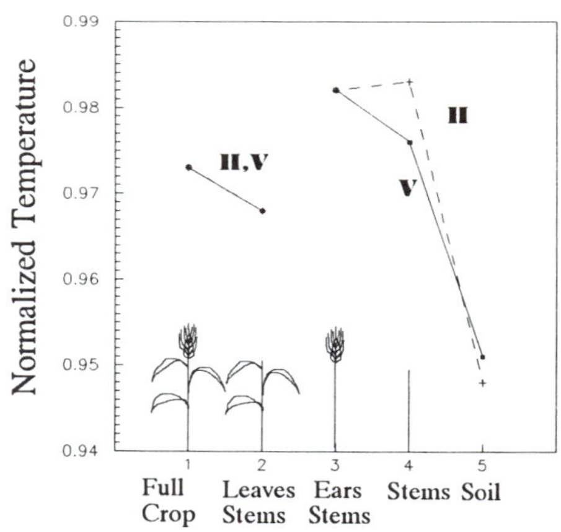

Figure 5 - The normalized temperature Tn (f = 37 GHz, $\theta = 45°$) of a wheat canopy: 1. Full crop, 2. without ears, 3. without leaves, 4. only stems, 5. bare soil. (V = and H = vertical and horizontal polarizations)

Fig. 5 shows a sequence of measurements on well-developed hard wheat, observed as full crop with a PWC = 1 Kg/m^2 and after cutting different plant constituents, i.e.: stems with leaves and without ears (PWC = 0.8 Kg/m^2), stems with ears and without leaves (PWC = 0.85 Kg/m^2), stems only (PWC = 0.65 Kg/m^2), and soil after the cutting of plants (gravimetric soil moisture = 12%) with a small residue of vegetation biomass.

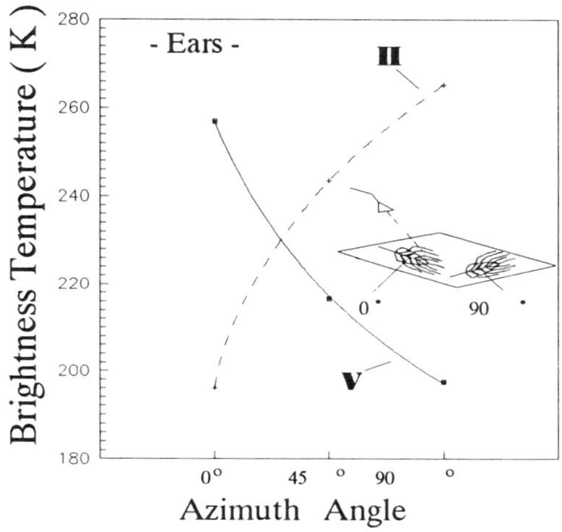

Figure 6 - The brightness temperature at V and H pol. (f = 37 GHz, θ = 45°) of a layer of wheat ears (PWC = 0.28 Kg/m^2) as a function of azimuth angle ϕ

The decrease of normalized temperature T_n (i.e. the ratio between the brightness temperature measured by the microwave radiometer and the radiance surface temperature monitored by the thermal infrared radiometer) when the ears are cut can be interpreted in terms of the bistatic scattering cross section of single elements. Indeed, assuming energy conservation, the emissivity of a half space of absorbing and scattering elements can be written as 1 minus the fraction of power incident on the surface from a given direction (at a specific polarization and frequency) scattered in the upper half space (16, 17). The pattern of the bistatic scattering cross-sections of an ensemble of vertical cylinders simulating ears indicates that the incoming radiation (at θ = 45°)is either absorbed or scattered downwards for both vertical and horizontal polarizations (18). It follows that the total upwards scattered power is very low and the emissivity is close to 1 at both polarizations. On the other hand, the scattering pattern of thin discs simulating leaves is almost symmetrical with respect to the horizontal plane, since each scatterer gives a significant contribution to the upwards scattered power, and the emissivity is lowered. Thus, emissivity seems to be significantly influenced by ears with awns for full crop and by leaves when ears are cut. For an ensemble of ears and thin stems, the downwards scattered power is further attenuated by the lower plant constituents; the contribution to the upward scattered radiation is even lower than for full crop and the emissivity is higher. When soil is covered with a layer of long and thin (with respect to the wavelength) vertical stems, the screening effect due to scattering from thin cylinders is stronger for V polarization and emissivity can result higher at H polarization. Finally, when vegetation is cut, emission is scarcely affected by the Brewster angle effect due to surface roughness and the residue of cut stems, and the vertical polarized component is only slightly higher than the horizontal.
The effects of plant geometry on the canopy emission have been further investigated by

measuring the emission of single elements placed on a reflecting plate. Fig. 6 shows the brightness temperature (measured at Θ = 45°) of a layer of wheat ears as a function of the azimuth angle ϕ. The ears are placed close together in order to form an almost uniform layer.

When the main axis of ears is in the incidence plane (ϕ=0°), the V component of the emission is much higher than the H one, while the reverse is obtained for ϕ = 90°. A similar behavior is shown by the stems and the leaves (15). The measurements confirm the anisotropic emission of these elements, the emissivity being much higher when the main axis of plant constituents is aligned with the electric field of the antenna. When the medium becomes optically thick, as in three layers of leaves, the anisotropy tends to disappear and the normalized temperature is almost independent of azimuth angle, with a vertical polarized component higher than the horizontal one similar to the Brewster angle effect typical of surface scattering (15). Consequently, and due to the almost random distribution of leaves in the standing crop, the anisotropic effect is significantly smoothed when plants are observed in field conditions.

The alfalfa plant is characterized by a ramified structure where separation of the small quasi circular leaves results tedious. Fig. 7 shows the brightness temperature of layered composite elements consisting of stems and leaves of an alfalfa crop, with a total plant water content equal to 0.70 Kg/m^2. We see that the difference between V and H polarizations is small, even for one single layer; moreover the increase of emission with increasing biomass is rapid and attains saturation from the second layer.

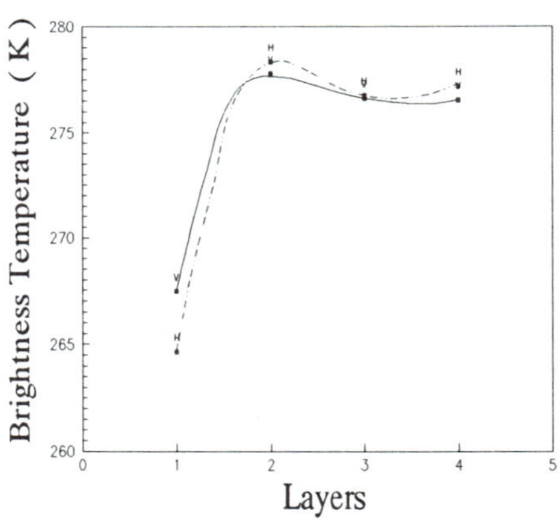

Figure 7 - The brightness temperature (f = 37 GHz, θ = 45°) of layered stems with leaves of alfalfa plants. (V = vertical pol., H = horizontal pol.)

SUMMARY

In summary, the preliminary investigation based on SSM/I data confirms that at least some dominant features of land surfaces can be identified using multi-frequency radiometric data from SSM/I, even on relatively small areas. Indeed different surfaces may have the same brightness at the same frequency, but in general the spectrum and polarization state is different for various land categories. In particular, discrimination between bare and vegetation covered soils can be accomplished using the polarization index at 19 GHz which, in spite of the coarser resolution, results more sensitive than that at 37 GHz.

The results of field measurements, aiming at detecting and quantifying the contribution of single plant constituents and eventual anisotropic effects, have shown that microwave emission from wheat in the flowering stage is significantly affected by ears. All the single plant elements have an intrinsic anisotropic emission behavior which is smoothed out by the geometrical distribution of the plants in the field and by multiple scattering. On the contrary, for alfalfa emission from both plants, which are characterized by small quasi circular leaves and ramified structure, and single plant elements was found almost isotropic.

Acknowledgments
This work has been partly supported by the Italian Space Agency (ASI). G. Macelloni gratefully acknowledges Alenia Spazio Spa for funding his research.

REFERENCES

1. Kirdiashev K., A. Chuklantsev and A. Shutko, 1979. Microwave radiation of the Earth's surface in the presence of vegetation cover *Radio Engin. Electron. Phys.* Engl. Transl., *IEEE Trans. Geosci. Remote Sensing,* **20** (1982), 27 - 29.
2. Ulaby F. T., M. Razani, and C. Dobson, 1983. Effects of vegetation cover on the microwave radiometric sensitivity to soil moisture" *IEEE Trans. Geosci. Remote Sensing,* **21**, 51-61.
3. Pampaloni P. and S. Paloscia, 1985. Experimental relationships between microwave emission and vegetation features, *Int. J. Remote Sensing*, **6**,. 315-323
4. Huppi R. , E. Stotzer, and E. Schanda, 1985. Calibrated microwave signature measurements of soil and wheat, in *Proc. 3rd Intern. Coll. on Spectral Signatures of Objects in Remote Sensing*, Les Arcs, France. ESA SP-247, 351-355.
5. Choudhury, B., 1989, Monitoring global land surface using the Nimbus-7 37 GHz data: theory and observations, *Int. J. Remote Sensing*, **10**, 1579-160
6. Ferraro, R.R, N.C. Grody, and J.A. Kogut, 1986. Classification of geophysical parameters using passive microwve satellite measurements, *IEEE Trans. Geosci. Remote Sensing,* **24**, 1008-1013
7. Neal C. M., M.J. McFarland, K. Chang, 1990, Land-Surface-Type Classification Using Microwave Brightness Temperatures from the Special Sensor Microwave/Imager," IEEE Trans. Geosci. Remote Sensing, **28**,. 829-838..
8. Paloscia S., and P. Pampaloni, 1992, Microwave vegetation indexes for detecting biomass and water conditions of agricultural crops, *Remote Sens. Environ.*,. **40**, 15-26.
9. Mo T., B. Choudhury, T. Schmugge, J. Wang and T. Jackson, 1982, "A model for microwave emission from vegetation covered fields, *J. Geophys. Res.* **87**, C13, 11229-11237
10. Pampaloni P., S. Paloscia, 1986. Microwave emission and plant water content: a comparison between field measurement and theory, *IEEE Trans. on Geoscience and Remote Sensing*, **24**, 900-904.
11. Pampaloni P., 1992. Microwave radiometry of soil and vegetation, *Proc. of Specialist Meeting on Mi rowave Radio etry and Re ote Sen ing Appli ation* , Boulder, CO, E. Westwater ED., 42-49.
12. Ferrazzoli P., L. Guerriero, S. Paloscia, P. Pampaloni, 1994. Potential of multifrequency techniques in microwave radiometry of crops, *Proc. of the IV Specialist Meeting on Microwave Radiometry and Remote Sensing*, Roma(Italy), February 15-17, 1994.

13. Hollinger J., R. Lo, G. Poe, R. Savage, and J. Pierce, 1987. *Special Sensor Microwave Imager User's Guide*, Washington, DC: Naval Res. Lab..
14. Paloscia P. and P. Pampaloni, 1988, Microwave polarization index for monitoring vegetation growth, *IEEE Trans. Geosci. Remote Sensing*, **26**, 617-621.
15. Macelloni G., S. Paloscia, P. Pampaloni, R. Ruisi and C. Susini, 1994, Microwave emissivity of plant constituents at 37 GHz, *Proc. VI Intern. Symp. Physical Measurements and Signatures in Remote Sensing*, Val d'Isere (France), 535-540.
16. Peake W.H., 1959, Interaction of Electromagnetic Waves with Some Natural Surfaces, *IRE Trans. Antennas and Propagation,* **7**, S324-S329.
17. Ferrazzoli P., G. Luzi, S. Paloscia, P. Pampaloni, G. Schiavon, and D. Solimini, 1989, Comparison Between Microwave Emissivity and Backscatter Coefficient of Crops, *IEEE Trans. Geosci. Remote Sensing*, **27**, 772-778.
18. Ferrazzoli P., L. Guerriero, S. Paloscia, P. Pampaloni, 1994, Modeling X and Ka emission from leafy vegetation, *J. Electromagnetic Waves and Appl.* (in press)

4. NEW RADIOMETRIC SYSTEMS

OVERVIEW

- **Synthetic aperture radiometers**

  The effort to remove the substantial limitation of the coarse ground resolution of spaceborne microwave radiometers has led to the development of synthetic aperture radiometers. These systems are now in different stages of progress, ranging from aircraft prototype (*ESTAR*) to laboratory demonstration model (*TUD SARad*), to bread-board design (*MIRAS*). Experiments conducted with *ESTAR* have confirmed the calibration of the instrument and successfully produced promising soil moisture maps.

- **MIMR**

  Real aperture radiometers with enhanced performances are under development to be ready for flight on the next generation platforms.

  MIMR is an imaging radiometer with six dual-polarization frequency channels in the range 7 to 92 GHz. The total power configuration, the stability constraints, and the wide bandwidth require a sophisticated hot target for the radiometric calibration of the instrument. Also the configuration of the antenna system is rather challenging, due to the number of channels, the external calibration requirements, and the overall antenna specifications.

- **AMSU**

  The Advanced Microwave Sounding Unit (*AMSU*) is being developed to fly on the new generation of NOAA orbiters for identification of precipitation, snow and ice, and, principally, for global sounding of atmospheric temperature and humidity. *AMSU-B* is a five channel radiometric system operating towards the high end of the microwave frequency range (88 to 191 GHz), i.e., in two atmospheric windows and on a water vapor absorption line. The high frequency of operation and the stringent specifications make the antenna design, the test of the system and the calibration rather demanding.

- **Ground-based**

  Ground-based radiometers, in their turn, are undergoing improvements both from the hardware and from the software point of view. Examples of this activity are the Multifrequency mm-wave Radiometer (*MFR*) and the new portable microwave profiling radiometer (*MTP*). *MFR* is a seven frequency instrument with two channels on the low frequency water absorption line, one channel in the atmospheric window, and four channels in the 60 GHz oxygen absorption complex. Its profiling capabilities benefit from the use of a retrieval technique making use of appropriate sets of orthogonal functions. *MTP* has the distinct feature of a highly stable synthesizer which allows the tuning of almost any user-selected ensemble of frequencies within the low frequency shoulder of the oxygen absorption complex. Both instruments are intended for the production of improved data on the atmospheric environment.

Progress in Remote Sensing with Thinned Array Radiometers

D.M. Le Vine[1], C.T. Swift[2], T.J. Jackson[3], A. Griffis[4], M. Kao[5] and P. Gaiser[6]

(1) Goddard Space Flight Center, Greenbelt, MD 20771
(2) University of Massachusetts, Amherst, MA 01003
(3) USDA/ARS Hydrology Lab, Beltsville, MD 20705
(4) Kaman Aerospace Corp., Tucson, AZ 85703
(5) Hughes STX Corp, NASA/GSFC, Greenbelt, MD 20771
(6) Naval Research Laboratory, Washington, D.C.

Abstract: The Electronically Steered Thinned Array Radiometer (ESTAR) represents a new instrument technology for microwave remote sensing from space being developed and validated jointly by the NASA/Goddard Space Flight Center, the University of Massachusetts and the USDA Agricultural Research Service. This interferometric technique permits a substantial reduction in the antenna aperture needed for a given spatial resolution. ESTAR is an L-band prototype built to demonstrate the technique from an aircraft platform. Experiments to validate the instrument's ability to make maps of soil moisture have been successfully conducted at the USDA watersheds at Walnut Gulch in Arizona (1991) and the Little Washita River in Oklahoma (1992). Both experiments indicate that a valid calibration has been obtained for this radiometer.

Keywords: Microwave Radiometer, Synthetic Aperture, Soil Moisture

INTRODUCTION

The Goddard Space Flight Center in cooperation with the University of Massachusetts and the USDA Agricultural Research Service in Beltsville, Maryland have been working together to develop the technique of aperture synthesis (thinned array radiometers) for remote sensing from space at the long wavelength end of the microwave spectrum. Passive microwave remote sensing at long wavelengths offers the potential for measuring parameters such as soil moisture and ocean salinity which are needed for understanding the dynamics of the earth's environment [1]. However, passive microwave remote sensing at long wavelengths requires large antennas to achieve high spatial resolution, and the problems associated with deploying large antennas on satellites are sufficiently severe to have limited the deployment of passive microwave sensors in space.

A possible means for overcoming this limitation is to use an interferometric technique called aperture synthesis to reduce the physical antenna aperture needed in space. In aperture synthesis, the coherent product (correlation) of the output voltage from pairs of antennas is measured at many different baselines. Each baseline produces a sample point in the Fourier transform of the scene and a map of the scene is obtained by inverting the transform. The technique is similar to "earth rotation synthesis" developed in radio astronomy [2]. The advantage of aperture synthesis for microwave remote sensing applications is that one can achieve high spatial resolution with sparse arrays which require no mechanical scanning. Sparse arrays are possible because only one measurement is needed at each baseline [2,3], and mechanical scanning is not needed because the image reconstruction algorithm yields a map of the entire field-of-view (footprint) of the individual antennas. One obtains an instrument with a wide field-of-view by using pairs of small antennas in the interferometer [3,4].

As part of this research program, an aircraft prototype called ESTAR has been built. ESTAR is an L-band radiometer designed for the remote sensing of soil moisture [5]. Initial flights near Goddard's Wallops Flight Facility in Virginia demonstrated that images could be successfully obtained; and recently the remote sensing capability of the instrument has been demonstrated in a series of soil moisture measuring experiments at the USDA watersheds at Walnut Gulch, Arizona (August, 1991) and at the Little Washita River basin near Chickasha, Oklahoma (June, 1992). The instrument is currently undergoing changes to improve internal calibration and thermal control, and is scheduled to participate in another series of flights at the Little Washita River watershed in April, 1994.

INSTRUMENT CONFIGURATION

ESTAR is a hybrid real-and-synthetic aperture microwave radiometer. It uses a real antenna aperture to obtain resolution along track and employs aperture synthesis to obtain resolution in the cross track dimension. The real antennas employed in this hybrid are stick antennas oriented with their long axis in the direction of motion as shown in Figure 1 (top). Each stick antenna produces a fan beam which is narrow in the direction of motion and broad in the cross track dimension. The fan beam is swept along the surface as the sensor moves and it provides resolution along track. Resolution across track is achieved using aperture synthesis. Figure 1 (bottom) illustrates the signal processing employed to do this. The signal from each antenna is filtered, amplified and mixed to a convenient IF in the block labelled "RF front end". Then, the signal from pairs of antennas with different spacing (baselines) are multiplied and averaged in the correlator. This multiplication is done coherently (both phase and amplitude are measured). The products are digitized and stored in the computer for later processing.

The actual hardware is shown in Figure 2. ESTAR is shown at the top of Figure 2 as it was installed on the NASA P-3 aircraft during a series of experiments at GSFC's Wallops Flight Facility during May, 1993. The white structure at the bottom of the ESTAR box is a radome covering the stick antennas and inside the box is the RF circuitry, correlator and digitizer. The computer and power supplies are mounted on racks inside the airplane's passenger compartment. At the bottom of Figure 2 ESTAR is shown with the radome removed (and upside down) so that the individual stick antennas can be seen. Each stick is a linear array of eight horizontally polarized dipoles. This is in contrast to an earlier configuration [5,6] in which crossed dipoles with elliptical polarization were employed. The new antennas are described in detail by Griffis [7]. The 5 stick antennas are spaced in multiples of $\lambda/2$ and with the configuration shown in Figure 2 it is possible to obtain all baselines which are multiples of $\lambda/2$ from $N = 0$ to $N = 7$ as illustrated in Figure 1.

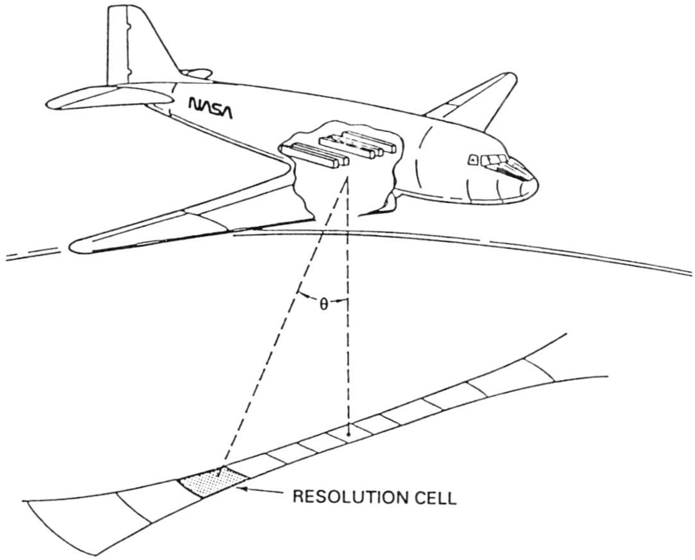

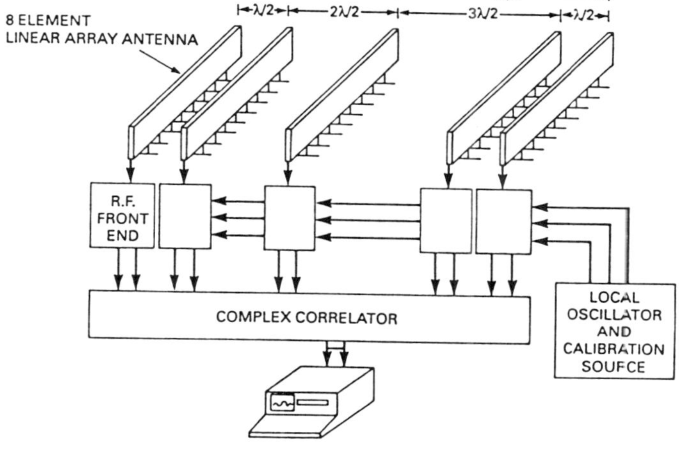

Figure 1: (Top) Aircraft configuration of the ESTAR prototype; (Bottom) Block diagram of the ESTAR instrument.

ESTAR employs a sideband correlator to form the required products at all non-zero spacings. In this correlator, the signal from each antenna is mixed to two IF's separated by 30 MHz (113.5 MHz and 143.5 MHz) and the desired products (one from each antenna pair) are formed by using signals with different IF. These products exit the multiplier at 30 MHz and are then mixed with a reference oscillator (at 30 MHz) to form inphase and quadrature signals at baseband. This sideband technique was adopted because a baseband multiplier was not available. However, this is no longer the case and the instrument is currently being modified to do away with the dual IF scheme [8].

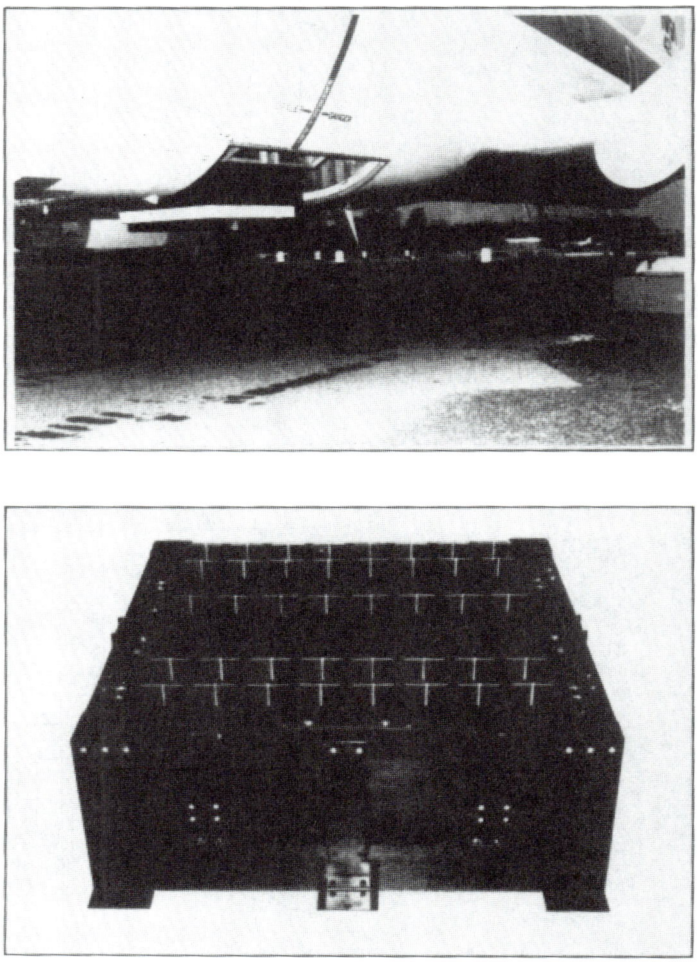

Figure 2: (Top) ESTAR as installed on NASA WFF P-3; (Bottom) ESTAR stick antennas with radome removed.

ESTAR also contains an internal reference source (an oscillator at 1.413 GHz) which is periodically injected at the antenna output so that it passes through the entire RF circuit. The reference oscillator provides a calibration of each of the RF paths relative to each other and is used to correct for long term drifts in gain. This reference signal is also used during diagnostics where it acts as a built-in system check. The reference signal is presently being changed from a CW tone to a wideband noise source [8].

The multiplication at zero baseline must be done differently in order to eliminate system noise which is correlated in this case. The radiometer used at zero spacing is a variation of what is sometimes referred to as a pulsed-noise injection radiometer [9] except that the synchronous (Dicke) detection is not needed. In the radiometer employed in ESTAR, a hybrid adds and subtracts the signal from a reference source (e.g. a thermally controlled matched load) from the sum of the antenna voltage plus injected noise. The outputs from the hybrid are processed as if they were the output from two different antennas in a sideband correlator (i.e. they are mixed to different IF and multiplied and averaged). A feedback loop adjusts the injected noise until the inphase product is zero. In this scheme, the noise in each channel of the correlator is uncorrellated and therefore does not appear in the product. This radiometer was developed for ESTAR so that the same RF circuitry could be used for both the zero spacing and the non-zero baselines. Variations of this radiometer have been employed successfully in a number of sensor systems developed at the Microwave Remote Sensing Laboratory of the University of Massachusetts.

RESULTS

The first successful ESTAR image was obtained during initial flight tests over the Delmarva peninsula south of NASA/GSFC's Wallops Flight Facility [5]. Research immediately shifted to calibrating the instrument and demonstrating that measurements of soil moisture could be made. This was successfully done in 1991 during a series of flights over a highly instrument USDA watershed near Tombstone, Arizona [10]. The following season, June, 1992, ESTAR participated in a major experiment (WASHITA-92) as the primary sensor for mapping soil moisture. Results from these experiments are presented below to illustrate instrument performance.

A. Walnut Gulch

During the summer, 1991, measurements were made at the USDA's Walnut Gulch watershed near Tombstone, Arizona to validate the image reconstruction and calibration procedure in a soil moisture measuring experiment. The objective was to see how well the ESTAR images corresponded to patterns of surface soil moisture and to make direct comparison of ESTAR derived brightness temperature with measurements of soil moisture at selected sites on the surface.

The watershed is classified as semiarid rangeland and the vegetation cover was sparse grass on the eastern portion and desert shrub on the western side [10]. Surface soil was mostly sandy loam with varying rock fraction. Soil moisture samples were collected at 10 sites within the watershed. Soil moisture was sampled in the upper 5 cm of soil and measurements were made within about an hour of the ESTAR overflights. The location of the sampling sites plus maps of the ESTAR brightness temperature made on August 1 and 3 are shown in Jackson et al [10].

The program of aircraft flights and ground sampling employed with ESTAR during the summer of 1991 are essentially identical to those employed in an experiment performed the previous year (1990) using the Push Broom Microwave Radiometer (PBMR). Furthermore, the PBMR and ESTAR are similar in many respects: both radiometers operate at L-band, both are horizontally polarized and both have similar effective beam widths. The radiometers are physically about the same size and are deployed in the aircraft (the NASA/ARC C-130 in this case) in the same manner. PBMR differs from ESTAR in that it has 4 real beams oriented across track at ±8 and ±24 degrees with respect to nadir whereas ESTAR achieves its resolution across track synthetically and in effect scans its beam across-track in 2 degree intervals. During the summer of 1990, the PBMR flew the same flight lines over the Walnut Gulch watershed as ESTAR and ground samples of soil moisture were collected in the same manner. Eight of the sites sampled in 1991 were the same as those used in 1990 for the PBMR. Also, the range of soil moisture conditions encountered in 1990 [10] were nearly the same as occurred during the experiments with ESTAR in 1991. Examples of the ESTAR images and a comparison with rainfall during the experiment have been presented by Jackson et al [10].

Figure 3 shows a comparison of the ESTAR data obtained in 1991 with the data obtained by the PBMR the previous year. The open circles are the PBMR data, the solid

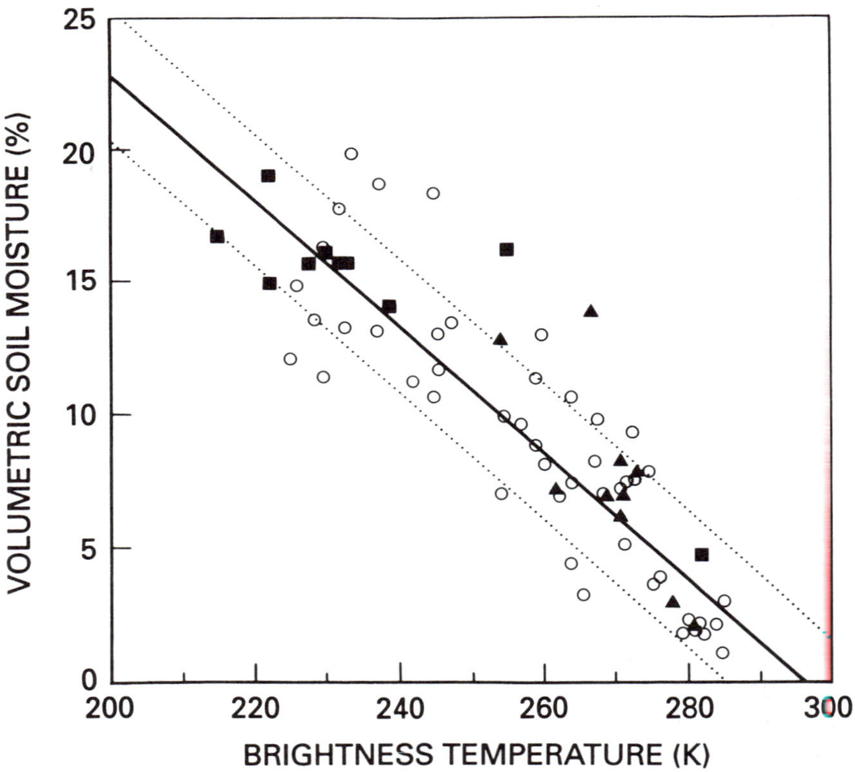

Figure 3: Comparison of ESTAR and PBMR data. The open circles are PBMR data obtained in 1990 and the solid squares and triangles are the ESTAR data obtained on August 1 and 3, respectively. The solid line is a best fit to the PBMR data and the dashed line is standard error for this fit.

line is a best fit to the PBMR data and the dashed lines are the standard deviation of this fit. The solid symbols are the ESTAR data obtained in 1991. Notice that for the most part the ESTAR data fall well within the standard deviation of the best fit to the PBMR data. The data clearly indicate that ESTAR is yielding brightness temperatures which are equivalent to those obtained with the real aperture microwave radiometer.

B. Washita River Basin

A subsequent series of measurements were made during June, 1992, at the USDA watershed at the Little Washita River near Chickasha, OK, which is about 50 miles southwest of Oklahoma City. This is an area of moderately rolling hills with a climate classified as "sub-humid" with average annual rainfall of about 75 cm. Land use is largely rangeland and pasture but there are significant areas of winter wheat and other crops in the floodplain and western portions of the watershed. Soils include a wide range of textures with higher percentages of loam on the eastern and western ends of the watershed and a higher percentage of sand in the center [11].

ESTAR flew on the NASA/ARC C-130 in the same configuration as at Walnut Gulch the year before. In this experiment, measurements were made at 7,000 ft and at 500 ft. The high altitude flights were designed to map a large portion of the watershed to give an overview of moisture patterns. A rectangular area about 20 km wide and 42 km

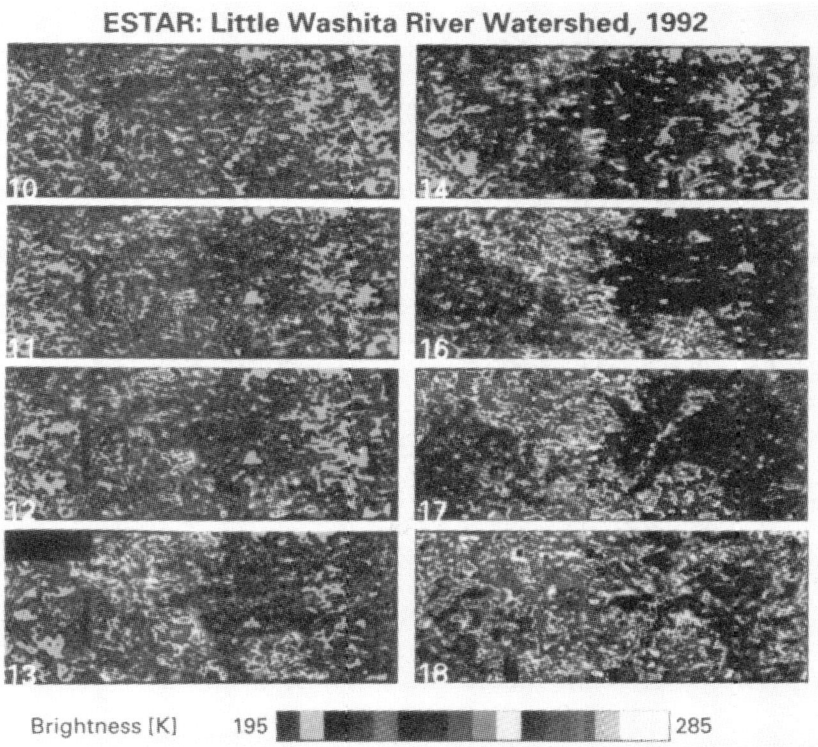

Figure 4: ESTAR brightness temperature over the Washita River watershed. Images are from 7,000 ft and include eight days from June 10 (upper left) through June 18 (lower right).

long was mapped. The low level flights were made to obtain high resolution data over selected sites with varying ground cover where detailed sampling of surface soil moisture and temperature was done.

The measurements began on June 10 at the end of a very wet period which left the watershed in a nearly saturated condition with substantial areas of standing water. Flights were made each day for the next eight days except for a day off on June 15. The last flight was on June 18. No rain occurred over the watershed during this period and the watershed slowly dried.

Figure 4 is a composite showing the ESTAR brightness temperature maps of the watershed obtained on each of the 8 flights at 7,000 ft. Several features of this composite indicate that the ESTAR calibration and image reconstruction algorithms are working well. First, the drying which took place during this experiment is clearly evident in the 8 images. The image obtained on the first flight (June 10; upper left) yielded brightness temperatures indicative of very wet soil, and as is evident in Figure 4 the brightness temperatures become increasingly warm as the experiment progresses. Also, notice that the center of each image is somewhat warmer than the eastern (right) and western (left) portions. This pattern of brightness temperatures is consistent with the distribution of soil type for the watershed [11]. In fact, the ESTAR images are in excellent agreement with the soils map, reflecting the sandier soils in the center and the loamier content on the eastern and western extremes. Finally, there are several landmarks in the watershed which are clearly evident in the ESTAR images and which have been used to check the consistency of the image reconstruction. An example is Lake Birche, located in the north-central portion of the image, which has a distinctive kidney bean shape. It is consistently imaged with the correct shape and location on each of the eight days.

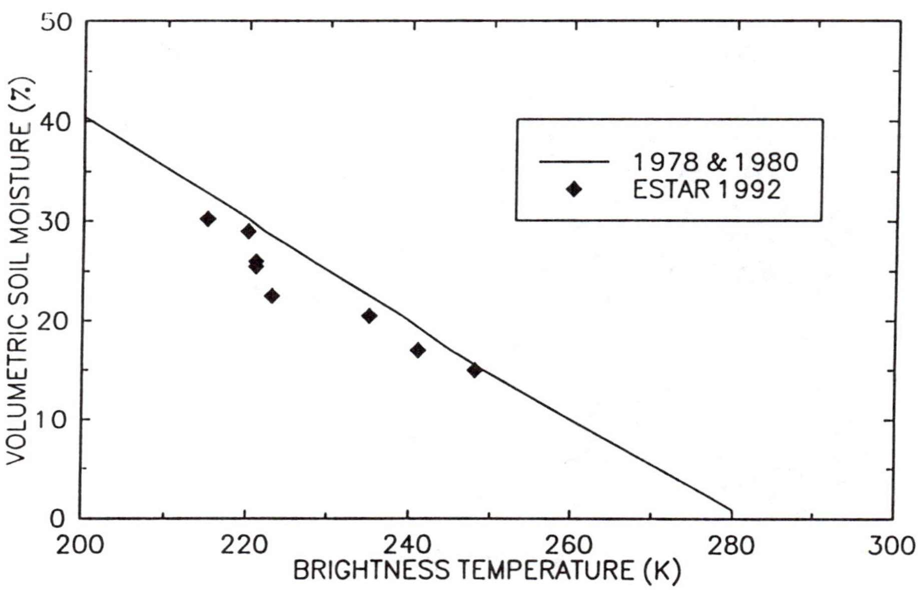

Figure 5: Comparison of ESTAR data averaged over the watershed (diamonds) with measurements made at a nearby site in 1978 and 1980 by Jackson et al [12].

A more quantitative validation is presented in Figure 5 in which a comparison is made with measurements made at a nearby site (just off the upper right-hand corner of the ESTAR images) in 1978 and 1980 [12]. The solid diamonds are ESTAR data averaged over the watershed. Soil moisture was sampled during the flights at about 35 sites distributed across the watershed. These sites were located in the ESTAR images and a brightness representative of the site computed. Then on each day, the brightness temperatures and the measured soil moisture for all the sites were averaged to obtain a single soil moisture and brightness temperature for that day. The results are the solid diamonds in Figure 5. The straight line is a regression line representative of the data obtained in 1978 and 1980 (Ref 12, Table 3). Although the ESTAR data represent an average over a much greater area than observed in 1978 and 1980, the ESTAR data clearly are close to this regression line and exhibit a similar behavior as a function of soil moisture.

REFERENCES

1. Swift, C.T. et al, 1993, ESTAR-The Electronically Scanned Thinned Array Radiometer for Remote Sensing Measurement of Soil Moisture and Ocean Salinity, NASA Technical Mem. 4523, July.

2. Thompson, A.R., J.M. Moran and G.W. Swenson, 1986, Interferometry and Synthesis in Radio Astronomy, J. Wiley & Sons, Chap 4, pg 82.

3. Le Vine, D.M., 1990, The Sensitivity of a Synthetic Aperture Radiometer for Remote Sensing Applications from Space, *Radio Science*, **25 (#4)**, pp 441-453, July.

4. Le Vine, D.M., T.T. Wilheit, R.E. Murphy and C.T. Swift, 1989, A Multifrequency Microwave Radiometer of the Future, *IEEE Trans. GRS*, **27(#2)**, pp 193-199, March.

5. Le Vine, D.M., M. Kao, A.B. Tanner, C.T. Swift, A. Griffis, 1990, Initial Results in the Development of a Synthetic Aperture Microwave Radiometer, *IEEE Trans. GRS*, **28(#4)**, pp 614-619, July.

6. Tanner, A.B., 1990, Aperture Synthesis For Passive Microwave Remote Sensing: The Electronically Scanned Thinned Array Radiometer, Ph-D dissertation, Univ of Mass, February, *Avail UMI* #AAD90-22749.

7. Griffis, A., 1993, Earth Remote Sensing with an Electronically Scanned Thinned Array Radiometer, Ph-D dissertation, Univ. of Mass, February, *Avail UMI* #AAD93-16656.

8. Gaiser, P.W., 1993, The Development of a Second Generation Electrically Scanned Thinned Array Radiometer, Ph-D dissertation, Univ. of Mass, May, *Avail UMI* #AAD93-29609.

9. Skou, N., 1989, Microwave Radiometer Systems: Design and Analysis, Artech House, Chapter 4.

10. Jackson, T.J., D.M. Le Vine, A. Griffis, D.C. Goodrich, T.J. Schmugge, C.T. Swift, P.E. O'Neill, 1993, Soil Moisture and Rainfall Estimation over a Semiarid Environment with the ESTAR Microwave Radiometer, *IEEE Trans. GRS*, **31**, pp 836-841, July.

11. Allen, P.B. and J.W. Naney, 1991, Hydrology of the Little Washita River Watershed, Oklahoma, USDA/Agricultural Research Service, *Tech memo ARS-90*, December.

12. Jackson, T.J., T.J. Schmugge and P.E. O'Neill, 1984, Remote Sensing of Soil Moisture from an Aircraft Platform, *Remote Sens. Environ.*, **14**, pp. 135-151.

The TUD Synthetic Aperture Radiometer Demonstration Model

BRIAN LAURSEN, HENRIK M. PEDERSEN, and NIELS SKOU

Electromagnetics Institute, B 348
Technical University of Denmark
DK 2800 Lyngby, Denmark

Abstract - The TUD Synthetic Aperture Radiometer demonstration model consists of a 2-channel X-band correlation radiometer with two horn antennas and an antenna mounting structure enabling the horns to be mounted in relevant positions within a certain aperture. The cross correlation of the signals from the 2 antennas produces a sample of the visibility function, which is the Fourier transform of the brightness temperature distribution of the target area. A total aperture synthesis is obtained by sequentially placing the two antenna elements in all required pairs of positions and measuring the corresponding samples of the visibility function followed by an image reconstruction procedure which is based on the inverse Fourier transform.
The system has been used to demonstrate two-dimensional synthetic aperture imaging from outdoor ground experiments. To enable a practical distance between the aperture and the target the system is focused by having the antenna elements mounted on a curved rail. By rotating this rail around a vertical axis the full aperture can be covered.

INTRODUCTION

The Synthetic Aperture Radiometer (SARad) systems are fundamentally different from traditional radiometer systems in that they measure in the spatial frequency domain and an inverse Fourier transformation is needed to obtain the normal brightness temperature distribution in the spatial domain. Aperture synthesis involves interferometric measurements made by cross-correlating the signals from two spatially separated antennas having overlapping field-of-view.

The cross-correlation of the signals from the two antennas (having passed through two properly designed radiometers) produces a sample of the visibility function corresponding to the relative positions of the antenna elements. The visibility function is the Fourier transform of the brightness temperature distribution of the target area. By making a new measurement with another relative position of the antennas, another sample of the visibility function is found. Only a limited number of measurements i. e. positions are required to adequately sample the visibility function. In a practical imaging system this is done by having a proper number of (small) antenna elements placed properly within the aperture and then simultaneously cross-correlate all possible pairs to generate the visibility function.

The TUD (Technical University of Denmark) demonstration model consists of a 2-channel X-band correlation radiometer with two horn antennas and an antenna mounting structure enabling the horns to be mounted in relevant positions within a certain aperture. A total aperture synthesis is obtained by sequentially placing the two antenna elements in all possible/required pairs of positions and measuring the corresponding samples of the visibility function. The demonstration model thus requires the test scene to remain stable for a considerable time - typically a few hours.

The purpose of such a 2-channel demonstration model is twofold: with relatively cheap instrumentation the synthetic aperture radiometer concept can be demonstrated and investigated, and specific configurations for air- or spaceborne systems can be simulated and evaluated while they are still in the design phase.

SARad THEORY

The general theory concerning synthetic aperture radiometry, is published in (1) and (2), and discussed with Roger A. Davidheiser (TRW, USA). The discussion in this section is specific for the case where the scene (target / object) is placed in the near field of the synthetic aperture.

Samples of the visibility function $V(\bar{a}_1, \bar{a}_2)$ are found as the complex cross correlation between signals from two antennas. Let the position of the antennas be given by the vectors $\bar{a}_1 = (x_{A1}, y_{A1}, z_{A1})$ and $\bar{a}_2 = (x_{A2}, y_{A2}, z_{A2})$ in accordance with Figure 1, and the visibility function is determined by

$$V(\bar{a}_1, \bar{a}_2) = \int_0^{2\pi}\int_0^{\pi} \frac{T_B(\theta,\phi)}{4\pi} D(\theta,\phi)\, \rho(\Delta R)\, e^{j\frac{2\pi}{\lambda}\Delta R} \sin(\theta) d\theta d\phi. \tag{1}$$

$T_B(\theta,\phi)$	brightness temperature from a point in direction (θ,ϕ),
$D(\theta,\phi)$	antenna element directivity,
$\rho(\Delta R)$	$= \mathcal{F}^{-1}\{ H_1(f - f_C)\, H_2^*(f - f_C) \}$,
f_C	centre frequency
ΔR	difference of distance from the two antennas to a point in direction (θ,ϕ), i.e. $\Delta R = R_2 - R_1$,
λ	wavelength.

$\rho(\Delta R)$ is a function that takes into consideration the difference in time between the signals which are correlated. When we are looking at the correlation between signals from a direction far away from the centre, and when we at the same time are measuring a visibility sample with a large distance between the two antennas, the correlation between the signals will decrease as a function of the system bandwidth. $H_1(f)$ and $H_2(f)$ are the system filter transfer functions of the two particular channels. In case of a rectangular filter transfer function, with a system bandwidth B_{SYS}, it is well known that $\rho(\Delta R) = \mathrm{sinc}(\pi B_{SYS} \Delta R / c)$. For other filter transfer functions this decorrelation effect will increase.

In the following, it will be assumed that signals from behind the aperture can be neglected, and a subscript $()_m$ as main lobe indicates signals from the half sphere in front of the aperture; signals from behind the aperture cannot be distinguished from signals from a direction in front of the aperture. It is also assumed that the system is nadir looking on a flat ground z_G=h; the (x,y,z) and the spherical co-ordinate systems are given in Figure 1.

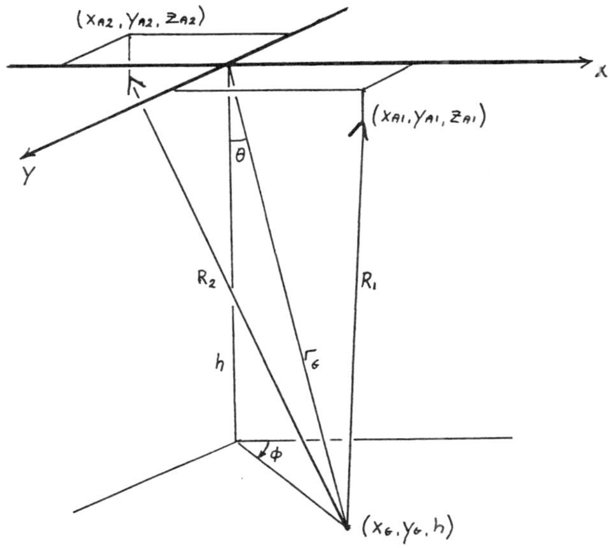

Figure 1. Two dimensional nadir looking synthetic aperture interferometer.

The distance from the two antennas to a point (x_G,y_G,h), denoted respectively R_1 and R_2, are given by

$$R_1 = \sqrt{(x_G - x_{A1})^2 + (y_G - y_{A1})^2 + (h - z_{A1})^2}, \tag{2}$$

$$R_2 = \sqrt{(x_G - x_{A2})^2 + (y_G - y_{A2})^2 + (h - z_{A2})^2}. \tag{3}$$

Placing the antennas on a spherical surface with the centre of the sphere at the centre of the scene, and by introducing $r_G = \sqrt{x_G^{\,2} + y_G^{\,2} + h^2}$, (2) and (3) can be written as

$$R_1 = r_G\left[1 + \frac{(-1)}{2r_G^{\,2}}(2x_G x_{A1} + 2y_G y_{A1}) - \frac{(-1)^2}{8r_G^{\,4}}(2x_G x_{A1} + 2y_G y_{A1})^2 + \sigma_1\right], \tag{4}$$

$$R_2 = r_G\left[1 + \frac{(-1)}{2r_G^{\,2}}(2x_G x_{A2} + 2y_G y_{A2}) - \frac{(-1)^2}{8r_G^{\,4}}(2x_G x_{A2} + 2y_G y_{A2})^2 + \sigma_2\right], \tag{5}$$

where σ is the remainder after a second order Taylor approximation. It is now clear that the antennas should be placed symmetrically about the same point for all measurements of the spatial frequencies, x_{A2}=-x_{A1} and y_{A2}=-y_{A1}. This has an advantage when the antenna element co-ordinates are separated from the ground co-ordinates since the second order term and all other terms of even order cancel, when the difference of distances from a ground pixel to the two antennas is determined

$$\Delta R = R_2 - R_1 \approx 2x_{A1} \frac{x_G}{\sqrt{x_G^2 + y_G^2 + h^2}} + 2y_{A1} \frac{y_G}{\sqrt{x_G^2 + y_G^2 + h^2}}. \tag{6}$$

Introducing the spherical co-ordinate system (r,θ,ϕ) and defining the spatial frequencies (u,v) as the distance between the antenna elements in x-axis and y-axis directions measured in wavelength $u=2x_{A1}/\lambda$ and $v=2y_{A1}/\lambda$, (6) can be written as

$$\frac{\Delta R}{\lambda} \approx u \sin(\theta)\cos(\phi) + v \sin(\theta)\sin(\phi). \tag{7}$$

The same equation is achieved in case of an object placed in the far field of the synthetic aperture, but without the restriction that the synthetic aperture should be a spherical surface and that the antenna pairs should be placed symmetrically. The differentiation criterion between near field (aperture is a spherical surface) and far field (flat aperture) is

$$r_{\text{far field}} > 20 \frac{D_{SYN}^2}{\lambda}, \tag{8}$$

D_{SYN} diameter (size) of the synthetic aperture.

Thus in general, ground based systems must use a spherical aperture while airborne and spaceborne systems can use a flat aperture.

Recalling that only signals from directions in front of the aperture are considered, the (ξ,η) co-ordinates can be defined as

$$\begin{aligned} \xi &= \sin(\theta)\cos(\phi), \\ \eta &= \sin(\theta)\sin(\phi), \end{aligned} \tag{9}$$

and the visibility function can be written as

$$V(u,v) = \iint_{\xi^2+\eta^2<1} \frac{T_{B,m}(\xi,\eta)}{4\pi} D_m(\xi,\eta)\rho_m(u,v,\xi,\eta) \frac{1}{\sqrt{1-\xi^2-\eta^2}} e^{j2\pi(u\xi+v\eta)}\, d\xi\, d\eta. \tag{10}$$

The image reconstruction can now be based on an inverse Fourier transform if the system filter transfer functions are chosen so the influence of $\rho_m(u,v,\xi,\eta)$ can be neglected

$$T_B(\xi,\eta) = \frac{1}{D_{N,m}(\xi,\eta)} \sqrt{1-\xi^2-\eta^2}\ \mathcal{F}^{-1}\{V(u,v)\}, \tag{11}$$

$D_{N,m}(\xi,\eta)$ normalised antenna element directivity.

TUD X-BAND SARad MEASUREMENTS

The antenna mounting structure

The TUD demonstration model consists of a 2-channel X-band correlation radiometer with two horn antennas and an antenna mounting structure enabling the horns to be mounted in relevant positions within a certain aperture. A total aperture synthesis is obtained by sequentially placing the two antenna elements in all possible/required pairs of positions and measuring the corresponding samples of the visibility function. The demonstration model thus requires the test scene to remain stable for a considerable time - typically a few hours.

The demonstration model is a nadir looking synthetic aperture radiometer. The experiments are carried out outdoors, and the scene consists of metal targets reflecting the sky (low brightness temperature) against a warm uniform background.

To enable a practical distance between the aperture and the target, the antenna element patterns are focused by having the antenna elements mounted on a curved rail, see Figure 2.

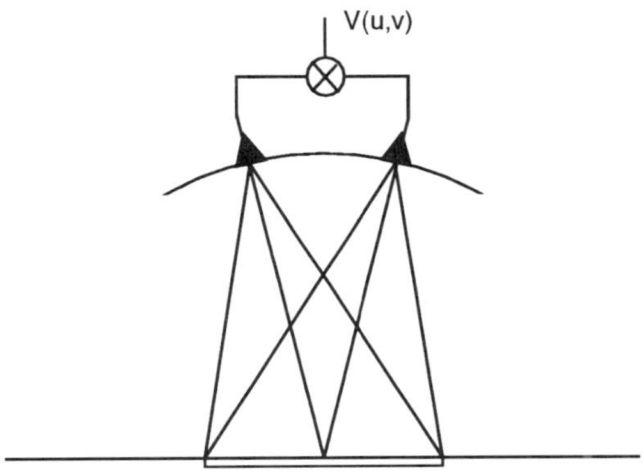

Figure 2. Focused antennas in near field synthetic aperture.

By rotating this rail around a vertical axis, the full aperture can be covered. When the scene is placed in the near field of the synthetic aperture, the pixel size in the reconstructed image, is small compared to the size of the synthetic aperture. The requirements to the accuracy of the circular curve are determined, not by the pixel size, but by a fraction of the period of the fastest fringe pattern on ground; the fringe pattern is determined by the complex exponential function in equation (10). The circular curve makes sure that all spatial frequencies are exactly centred, and this is necessary when an inverse Fourier transform is used in the image reconstruction on data from a near field measurement. A photo, taken looking upwards, of the antenna mounting structure is shown in Figure 3.

Finally, in case the scene is a flat ground it is previously explained (SARad THEORY) that placing the antenna on a circular curve is the optimum in relation to the shift of co-ordinate system from (θ,ϕ) to (ξ,η); the referred shift of co-ordinate system is necessary when an inverse Fourier transform is used.

The visibility function is sampled with a 2 λ spacing on the rail. When the rail is rotated, the angular sampling distance corresponds to a 2 λ spacing between visibility samples on the largest circle in the (u,v) plane, see Figure 4.

Figure 3. X-band antenna element mounting structure.

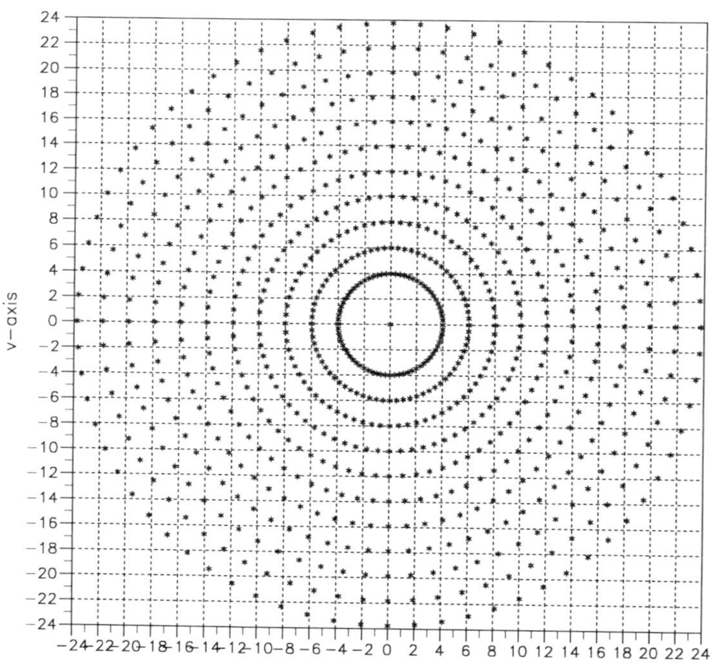

Figure 4. Measured (u,v) frequencies and a rectangular grid (the grid line intersections corresponds to the sampling points needed for an inverse Fourier transform).

The lowest visibility sample cannot be measured because of the size of the antenna element. The samples on the ring nearest to the centre are therefore set to zero.

The antenna mounting structure also opens up the possibility of a $\lambda/2$ and a λ antenna element spacing. Using these options the total time of a measurement will increase and a reduction of the synthesised aperture size will be necessary. The combination of a poorer ground resolution, a wider image size (swath) and a constant number of pixels are normally not attractive for the near field test situation.

The X-band SARad block diagram

The TUD X-band synthetic aperture radiometer consists of two single sideband super heterodyne receivers, see Figure 5. The RF centre frequency is 10.3 GHz and the IF centre frequency is 300 MHz with a 150 MHz system bandwidth. Horn antennas are used as antenna elements and phase stable flexible microwave cables are used between the antennas and receivers. In SARad techniques all local oscillator signals must be phase locked to measure the I & Q correlation between input signals. In a two channel demonstration model this is achieved simply by using a common LO. The phase shifter between the LO and the mixer in channel 2 is used to compensate for any difference in electrical delay in the two receivers.

The power divider network used to feed the analogue correlators is also shown in Figure 5. In each channel it consists of a quadrature hybrid followed by two in-phase 2-way power dividers. The analogue correlator is made by a four quadrant analogue multiplier and an integrator. Finally the correlation radiometer also includes a detection channel used for measuring the spatial DC frequency and used for calibration.

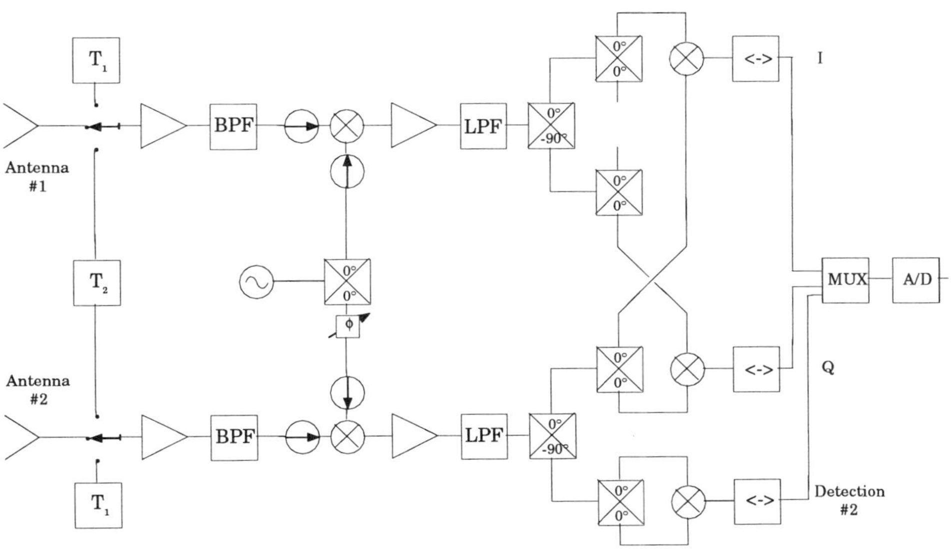

Figure 5. Block diagram of TUD X-band SARad.

The electronics require calibration of the gain and noise figure every 10 - 15 minutes, and two reference temperatures are used. The sky is used as a cold reference temperature by pointing the antenna element at zenith. As a warm reference temperature the antenna element is pointed at absorbers.

To achieve identical electrical delays in the two channels used for measuring the real part of a visibility sample, the adjustment of the phase of the local oscillator in one of the channels is necessary. The electrical delays in the two channels used for measuring the imaginary part of a visibility sample will then be 90° out of phase as a result of the quadrature hybrid, see Figure 5.

To adjust the phase, two correlated input signals are necessary. These are generated from a common noise source and a Wilkinson power divider; T_2 on Figure 5. The reference noise temperature must be different from the physical temperature of the power divider to obtain two correlated output signals of the power divider. The reason for this is the noise from the internal resistor in the power divider; the resistor that ensures isolation between output ports. The thermal noise from the internal resistor is 180° out of phase on the output ports, leading to a correlation between output signals corresponding to the physical temperature of the power divider, but with a negative (opposite) sign. When the reference noise temperature from the load impedance equals the temperature of the power divider, the effect of the two noise signals cancels in correlation.

The reference temperatures T_1 are used as uncorrelated input signals in the two channels.

TUD X-BAND SARad IMAGE RECONSTRUCTION

Each visibility sample needs a correction of the error corresponding to the output of the correlator when the inputs in the two particular channels (antenna elements) are uncorrelated. This error signal is caused by the fact that the two radiometer channels are using the same local oscillator and the same power supply and thus some common - correlated - noise is always present.

The visibility samples must be interpolated from the measured (u,v) grid to a grid suitable for an inverse Fourier transform. In this case data are interpolated to a rectangular Fourier transform, see Figure 4. Visibility samples at the first square of the rectangular (u,v) grid are set to zero because of the missing frequencies. Visibility samples on the inner rings are oversampled because of the constant angular samplings distance. On the basis of the missing frequencies at the first ring and the oversampling on the following rings, a nearest neighbour interpolation routine is used to interpolate visibility samples on the second square of the rectangular (u,v) grid. For all other rectangular (u,v) samples within a radius corresponding to synthesised aperture, a third order Lagrange interpolation routine is used. This interpolation routine is found as the best in this case among the following routines: nearest neighbour, radix, and Lagrange interpolation routines.

The visibility function is inverse Fourier transformed to obtain the brightness distribution of the scene. To have a continuous function describing the measured brightness temperature distribution, the size of the inverse Fourier transform is increased. Zeros are placed at spatial frequencies larger than the highest measured frequency. After multiplication with an appropriate window the visibility function is inverse Fourier transformed. The choice of window depends on the requirements to the spatial resolution and the beam efficiency of the synthesised beam.

The image reconstruction is based on equation (11)

$$T_B(\xi,\eta) = \frac{1}{D_{N,m}(\xi,\eta)} \sqrt{1-\xi^2-\eta^2}\ \mathcal{F}^{-1}\{V(u,v)\}.$$

The output of the inverse Fourier transform is the brightness temperature in (ξ,η) co-ordinates multiplied by two functions. The first one is the antenna element directivity and the second one is the Jacobi determinant corresponding to the shift of co-ordinate system from (θ,ϕ) to (ξ,η). The influence of these functions must be removed.

Finally a ground projection is performed to obtain a true image of the measured brightness temperature distribution as a function of ground co-ordinates (x_G,y_G).

$$x_G = h\frac{\xi}{\sqrt{1-\xi^2-\eta^2}},\qquad y_G = h\frac{\eta}{\sqrt{1-\xi^2-\eta^2}}. \tag{12}$$

TUD X-BAND SARad IMAGES

Metal targets reflecting the sky temperature (low brightness temperature) against a warm uniform background (absorbers) have been successfully imaged. The main specifications are shown in Table 1.

Frequency	10.3 GHz
Bandwidth	150 MHz
Noise figure	5 dB
Altitude	1.429 m
Aperture	0.66 m
Synthetic beam width	3°
Antenna element type	Horn
Spatial frequency sampling	2 λ
Measured samples of the visibility function	545
Image size	10 *10 pixels

Table 1 TUD X-band 2 channel demonstration model

Metal letters spelling "93" and "EMI" have been used as targets; see the following page. Low brightness temperatures, 100 K and below, are coded as blue and warm brightness temperatures are coded as red.

In the "93" image a large peak , 160 K above the mean image temperature, is present 1.8 cm from the centre. It is caused by a constant error signal in the measured visibility samples of +0.5 K and -0.8 K in the respective channels measuring the real and the imaginary part of a visibility sample; recalling that the negative spatial frequencies are found as the complex conjugated of the measured positive spatial frequencies, the error is complex conjugated as well when the negative spatial frequencies are used in the inverse Fourier transform. The peak

X-Band SARad Image

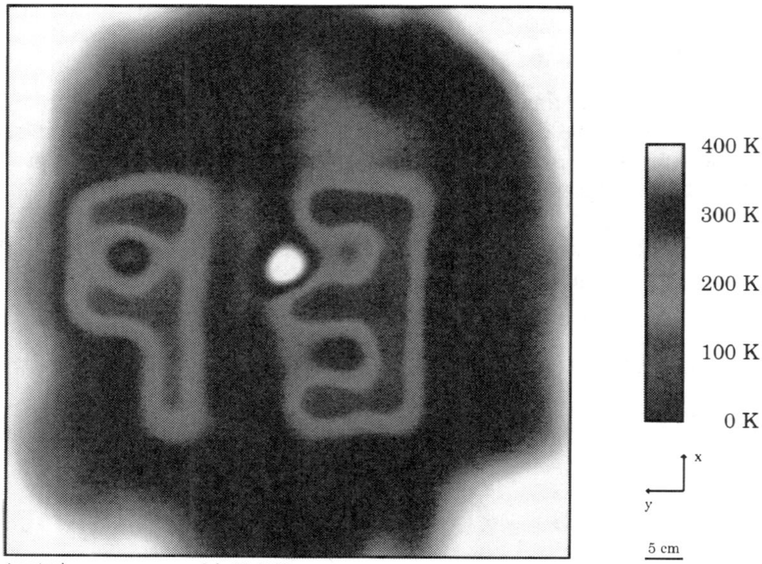

Acquired: July 27, 1993
Processed: Jan. 21, 1994

Electromagnetics Institute, Technical University of Denmark

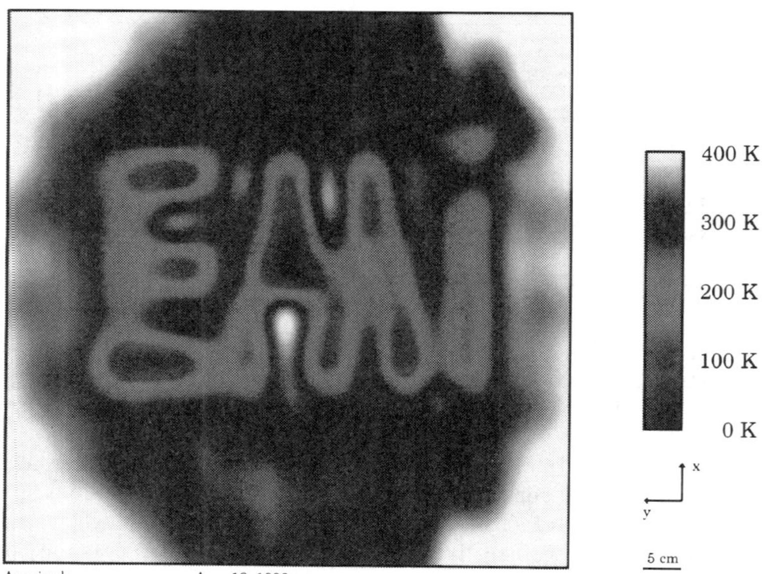

Acquired: Aug. 18, 1993
Processed: Jan. 21, 1994

Electromagnetics Institute, Technical University of Denmark

is followed by a warm (dark red) tail in the positive x-axis direction and a cold (green) tail in the negative x-axis direction caused by the error in the imaginary part of a visibility sample. The reason for this error signal is an only partly corrected error caused by the fact that the two radiometer channels are using the same local oscillator and the same power supply and thus some common - correlated - noise is present.

For SARad systems using more than two receivers the error will not be identical in all the visibility samples. It is then important that the effect is reduced by proper changes in the hardware design. Software error correction may be necessary, but only possible in case the error can be specified.

In the "EMI" image the error is seen as a cold black tail from the centre in the positive x-axis direction, and a warm white tail in the negative x-axis direction. It correspond to an error of less than 0.1 K in the channels measuring the real part of a visibility sample, and +0.4 K in the channel measuring the imaginary part.

Also, it is seen that the temperature rise to 400 K, and above, at the edge of the image and especially in the corners. This is a simple aliasing effect in the spatial domain (the brightness temperature domain) caused by an undersampling in the spatial frequency domain (the visibility function domain) ; undersampling compared to the Nyquist sampling theorem. The required sampling distance in the spatial frequency domain is determined by the antenna element radiation pattern. It is well known that a small aliasing only influence the edge of the image, and not the centre, as it is seen on the images.

CONCLUSIONS

Using rather simple hardware - the microwave components are cheap DBS receiver components and the correlators are 4 quadrant analog multipliers - it has been possible to demonstrate the 2-dimensional synthetic aperture radiometer concept and to experience some of the problem areas. A next generation system is presently under development using professional equipment. Great emphasis will be put to long term stability recalling the duration of a typical measurement. The new system will be used to assess the problem areas like calibration, and it will be used to simulate specific system configurations being investigated for future spaceborne use.

REFERENCES

(1) Ruf C.S, Swift C.T., Tanner A.B., Le Vine D.M.: "Interferometric Synthetic Aperture Microwave Radiometry for the Remote Sensing of the Earth", in IEEE Trans. Geoscience and Remcte Sens., vol.26 no.5; p.597-611;sept. 1988.

(2) Thompson A.R., Moran J.M., Swenson G.W.: "Interferometry and Synthesis in Radio Astronomy", JOHN WILEY & SONS, 1986.

MIRAS:preliminary concept of a two-dimensional L-band aperture synthesis radiometer

J.M. GOUTOULE 1 , U. KRAFT 2 , M. MARTIN -NEIRA 3

1 MATRA MARCONI SPACE 31 RUE DES COSMONAUTES 31077 TOULOUSE CEDEX FRANCE
2 DASA OTTOBRÜN P.O. BOX 801169 D-8000 MUNCHEN 80 GERMANY
3 EUROPEAN SPACE AGENCY P.O. BOX 299 2200 AG NORDWICK THE NETHERLANDS

Abstract : The passive imaging of soil moisture and ocean salinity from low orbit platforms with 1 to 3 days revisit time, requires microwave radiometers at 1400 MHz frequency (wavelength =21cm) with large antenna apertures (up to 20 m) and wide angle continuous scan capabilities . Rotating Reflectors instrument, and even push broom instruments (fixed reflector with 30/40 beams) are un-realistic today for that mission.The only valuable concept is based on interferometers principles,also called synthetic aperture . In MIRAS the large antenna reflector is replaced by 133 , 18cm diameter antennas,arranged on three coplanar beams .After low noise amplification the output signals are cross-multiplied and integrated,providing the 8776 'visibility functions'.The data are sent to ground for image reconstruction .MIRAS images the earth with 890 km swath , 20 km ground resolution at NADIR and 35km on swath edge.In the current contract for ESA a space-borne instrument has been studied , and a representative bread-board is planned for mid 1995.It will be flown on an aeroplane by the end of 1995.

Key-words :microwave interferometer , microwave radiometer , synthetic aperture , passive remote sensing,soil moisture measurement

INTRODUCTION

The passive imaging of soil moisture at L band (λ=21cm) from space requires large antenna apertures (up to 20 m) for which thinned array using synthetic aperture principles are the only valuable candidates by comparison with mechanically and electronically steered antennas and pushbrooms.ESTAR [1] , a one dimensional L band synthetic aperture radiometer , has been investigated for NASA , a bread-board instrument being built , tested on ground and then on an aircraft.As an alternative a two dimensional instrument compatible with a European low orbit platform has been proposed by ESA [2] , and is currently studied by European industries and research labs. A two dimensional interferometer approach for earth observation is described by Kraft [3] , the elementary antenna being regularly placed on 2 orthogonal directions , in a T or U shape . MIRAS takes into account the approach of radio astronomers , as in the Very Large Array [4],which leads to a three coplanar beams ,Y shape structure , with equally spaced elementary antennas.

MIRAS MISSION AND SPECIFICATION

The scientific needs for MIRAS have been established Yan Kerr (LERTS ,CNRS,France) taking into account various scientific works (and its owns) and the conclusions of the "Workshop on Advanced Microwave radiometer Techniques" held in Copenhagen in 1991. Those scientific wishes have are the starting point to specify a 200 kg /200w instrument at 799 km altitude with 890 km swath , the spatial resolution being limited by the mass and power allocations . The specification set of MIRAS is given on table1.

Table 1 : MIRAS specifications

Parameter	Allocation (BASELINE)	Allocation (OPTION)
Spatial Coverage	global, ±85° NS	global, ±85° NS
Spatial Resolution	20 - 60 km	20 - 60 km
Temporal Coverage	full year	full year
Revisit Time	3.2 days	3.2 days
Basic Measurement Geometry	Conical Scan	Nadir Scan
Nominal Incidence Angles	45-55°	0°-35°
Polarisation	H and V	V
Swath	≥ 880 km	≥ 880 km
Nominal Off-Nadir Angle	31.3° in-flight	0°
IFOV Cross-Track	≥ 24.5°	≥ 28.6°
IFOV On-Track	≤ 24.5°	≤ 28.6°
Orbit Type	sun-synchronous	sun-synchronous
Mean Orbit Altitude	799.8 km	799.8 km
Orbit Inclination	98.55	98.55
Orbit Eccentricity	0.001165	0.001165
Nodal Crossing Time	10:00 hour AM	10:00 hour AM
Sensor Frequency	1.4 GHz	1.4 GHz
Sensor Bandwidth	≤ 27 MHz	≤ 27 MHz
Radiometric Accuracy	≤ 3.0 K	≤ 3.0 K
Radiometric Sensitivity	≤ 1.0 K	≤ 1.0 K
Mission Lifetime	≥ 4 years	≥ 4 years
Total Antenna Dimensions	≤ 20 m	≤ 20 m
Total Sensor Mass	≤ 200 kg	≤ 200
Total Sensor Power Consumption	≤ 300 W (average)	≤ 300 W (average)
Total Data Rate	≤ 10 Mbps	≤ 10 Mbps

MIRAS CONCEPT AND DESIGN

Array geometry

Extensive work have been published on the optimum configuration of linear interferometer arrays achieving a complete Fourier coverage with the lowest redundancy level.In the case of two dimensions interfrometers,the antennas arrangement is not straightforward.For ground observation a dense set of visibility functions in the so called u-v plane is required : equally spaced Elementary Radiating Elements (ERE) may be installed on antenna arms arranged in T or U shape achieving a rectangular and orthogonal gridding.Radio-astronomers,who do not need a so dense ERE array,generally use a Y shape antenna in their interferometers as in the Very Large Array. Such a shape , but with equally spaced ERE , is investigated for MIRAS (fig1).In first simulations 15% improvement in aliasing is obtained which means that the Y shape saves 15% ERE by comparison with rectangular grid approaches:Lannes and Anterrieu [5].Moreover up to 70% improvement in spatial resolution may be observed when the large base-lines (corresponding to largely spaced ERE pairs) are used in the image reconstruction process.

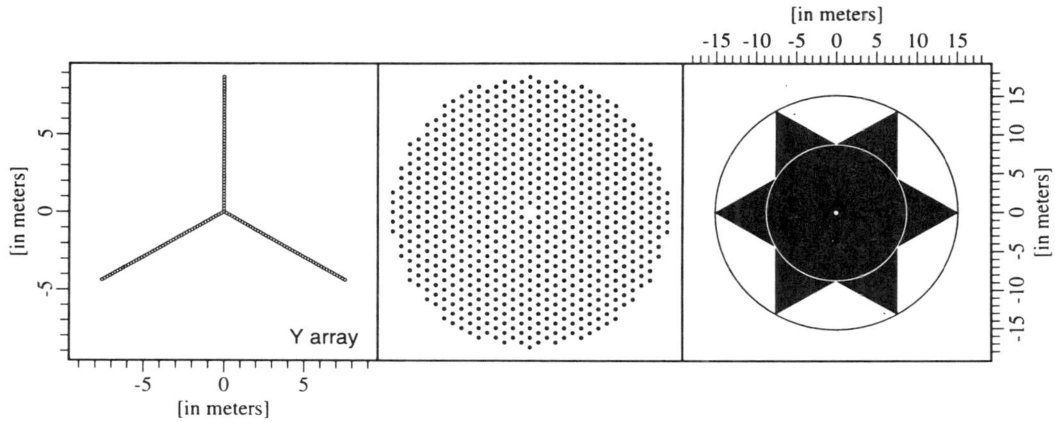

fig 1 : MIRAS array geometry , centre region of the uv plane, complete uv plane

The instrument is constituted of 133 ERE operating in dual polarisation . The array plane is tilted by 31.2° , in order to observe the earth between 40 and 55 ° incidence , fig 2 , which allows a direct retrieval of the ground temperature. The ERE have a 70° half power beam width .They are spaced by 0.89λ , which rejects aliasig lobes in cold space.

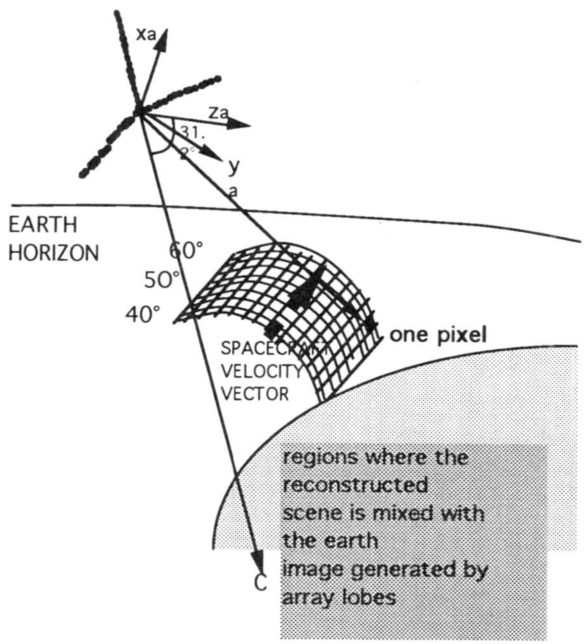

fig 2 :MIRAS field of view : +- 10 ° on track,+- 25 ° cross track.

Miras design and accommodation on a low orbit platform

Each ERE is associated to a single MMIC L band receiver to which both polarisation are alternatively sent. The base-band digital signals are routed to the correlators via optical fibers ,in order to minimise the gain and phase drifts. The 8778 correlations , which are simply the complex multiplication and integration between the receivers outputs, are done by 1bit digital correlators implemented in a single equipment.

The ERE are installed on three carbon fiber deployable beams .The deployed configuration on an ENVISAT platform ,is shown on fig 3 .

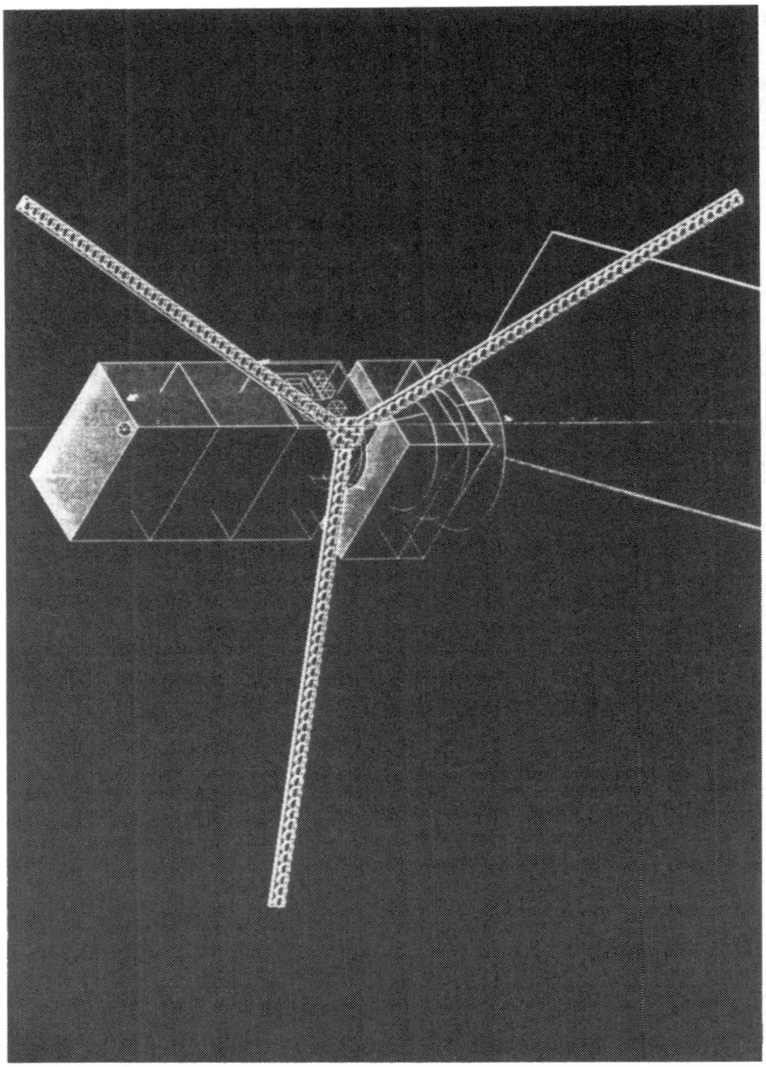

fig 3: :artist view of MIRAS on a European platform (ENVISAT)

MIRAS AIRBORNE DEMONSTRATOR

A demonstrator instrument based on the same principles ,but with a limited number of Elementary Radiating Elements (11) , is being developed , fig 4.It will be tested on a C130 for soil moisture imaging.

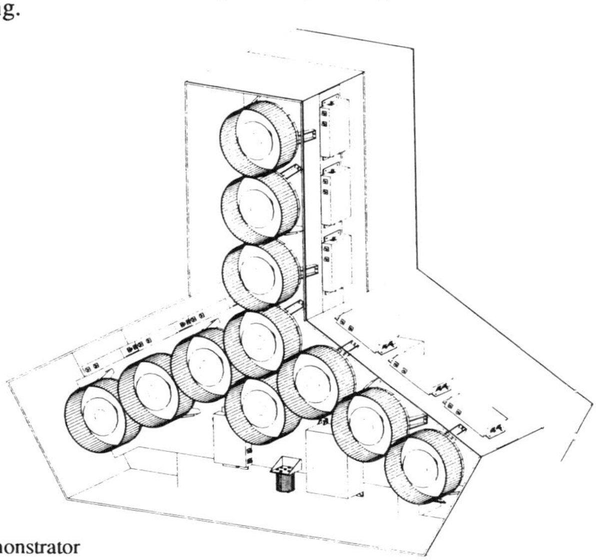

fig 4 : MIRAS Airborne demonstrator

CONCLUSION

A two dimensional passive interferometer has been presented .It operates at L band , is devoted to soil moisture and ocean salinity measurements , and is compatible with low orbit platforms accommodation . An airborne demonstrator bread-board is to be tested in 1995.

REFERENCES

[1] Swift C.T. et al."The Electronically Steered Thinned Array Radiometer (ESTAR)",*Proceedings of IGARSS'86 Symposium* , Zurich ,pp 591-593.

[2] Martin-Neira M. "MIRAS:A Two-Dimensional Passive Aperture Synthesis Radiometer" *Proceedings of Piers'93*

[3] Kraft U.R."Spaceborne Synthetic Aperture Interferometer for Global Monitoring of Earth", *Proceedings of IGARSS'91 ,Helsinski Symposium* pp 971.

[4] Napier P.J. "et al."The Very Large array:design and performances of a modern Synthesis Radio Telescope",*Proceedings of IEEE*,Vol 71,N11,Nov.1983,pp 1295-1320.

[5] Lannes A. Anterieu E. "Image Reconstruction Methods for Remote Sensing by Aperture Synthesis" *Proceedings of IGARSS"94* , Pasadena Symposium .

Multifrequency Imaging Microwave Radiometer: Design, Calibration and Expected Performance.

R. BORDI, M. L'ABBATE and P. SPERA
Alenia Spazio S.p.A. Remote Sensing Directorate, Via Saccomuro 24, 00131 Rome, ITALY

Abstract - The Multifrequency Imaging Microwave Radiometer (MIMR) is an instrument designed and developed for the European Space Agency (ESA) by the European Industry, with Alenia Spazio as Prime Contractor.
MIMR is a conical scanning radiometer which will make an image of the Earth with 6 channels in both horizontal and vertical polarizations.
MIMR output products are of prime importance for weather monitoring of the environment, and will provide useful data for climatology and meteorology with day and night operation capabilities and near global daily coverage.
The experience gained through various programs since 1972 with NIMBUS satellite's series utilizing ESMR, NEMS, SCAMS, SMMR [1] instruments and DMSP-Block 5D-2 satellite with on-board SSM/I instrument [2] led to the MIMR concept with better performance than the instruments already flown. This was required in order to fulfil in a more efficient way and with better accuracy the scientific objectives in the various application fields of atmosphere, ocean, cryosphere, land, ocean/atmosphere interaction.
In this paper, after a brief survey of the instrument design concept, the MIMR radiometric performance prediction are tackled, and emphasis is given to the error analysis/budget and to the design/technology choices adopted in the radiometric receivers.

Key words: Imaging radiometer, error analysis, calibration, receivers technology, expected performance.

1. INTRODUCTION

The MIMR is a 20 channel, 6 frequency radiometric system. It receives both vertically and horizontally polarized radiation at 6.8, 10.65, 18.7, 23.8, 36.5, 89.0 GHz.
MIMR will fly on a spacecraft in a near circular orbit, sun-synchronous, near polar orbit mission with the capability to achieve a daily coverage of the Earth surface of over 90%, and 99.8% coverage after three days. MIMR rotates continuously around an axis parallel to the local spacecraft vertical with an active portion of the scan about 120 deg. centered in the fore direction of the spacecraft and with the antenna system viewing the Earth scene with a nearly constant incidence angle of 55 deg.. A 26 RPM spin rate provides a rotation period of 2.3 sec. after which the sub-satellite point advancement corresponds to a footprint overlap along track for all channels greater than 50 %. Channels 36.5 and 89.0 GHz use a pushbroom technique with the two beams for the former channel and four beams for the latter one.
MIMR instrument (Figure 1) consists of an offset parabolic reflector of 1.6x1.4 m dimensions, illuminated by a cluster of 9 feeds (single horns for the 6.8 and 10.65 GHz,

multiple horns for channels 36.5 and 89.0 GHz, double frequency corrugated horn for 18.7 and 23.8 GHz).
The reflector and antenna feed horns are mounted on a "drum" which contains the receivers, the digital data unit, the mechanical balancing subsystem and power supply. The entire drum assembly is rotated about the axis of the instrument by a coaxially mounted bearings and brushless DC motor.
All data, commands, timing and telemetry signals, and power, pass through the rotating part to the stator via roll rings assembly. The fixed part of the instrument includes an inertial wheel as an angular momentum compensation device and a set of piezoelectric transducers as mechanical unbalance sensors, and allocates the interfacing units with the platform (power supply connected to the platform power bus and a control processing unit which links respectively with the platform command and telemetry bus and the platform science data bus).
A fixed rod through the scan assembly emerges above the rotating drum upper plane providing off axis mechanical support for the two calibration targets feeding the horns with an accurately known radiation power; they consist in a small reflector for the cold sky pointing (2.7 °K) and a hot blackbody load (300 °K), which, once per rotation, illuminates the feed horns obscuring them from the view of the main reflector; this "external calibration" technique allows an overall end-to-end absolute calibration (RF path from feeds to associated receivers is the same either during measurement or during cold and hot calibration).

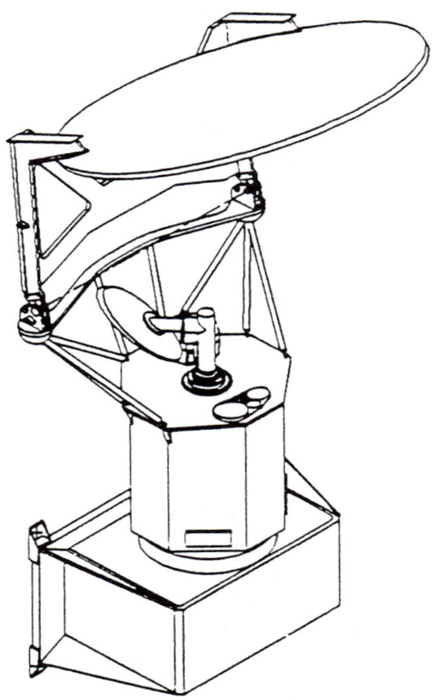

Figure 1. Instrument Configuration (deployed).

2. CALIBRATION

MIMR calibration involves all tasks to be performed on the ground before flight and in flight during instrument operation, in order to meet the required radiometric measurement performance, by keeping under control the uncertainty/drift of parameters involved in the reconstruction of input antenna noise temperature (associated to a brightness temperature scenario) starting from the radiometer measured output voltage.
MIMR calibration tasks aim at deriving the instrument correction factors which are needed in the ground processing in order to compensate for bias or drift in measurement data. These correction factors may be transformed in look-up tables defining the parameters for the ground processing correction algorithms [3]. In addition MIMR calibration is needed to establish a data base of support parameters to be used in ground pre-processing to allow exploitation/interpretation of the on-board instrument output data. In principle the calibration data are gathered both as ground characterization and as in orbit operation (pre-launch and in-flight calibration respectively).
Figure 2 shows how ground characterization and in-flight calibration intervene in the logics of antenna temperature reconstruction.
Flight calibration implies, apart from a 1/3 scan period involving Earth scene data acquisition, a calibration time for hot and cold reference sources observation to be properly allocated inside the remaining ~2/3 scan period. The allocation of the two (cold and hot) "calibration windows" within the scan period engages 2.5 deg. of angular range. This limitation is imposed by the focusing constraints of the cold reflector and by the hot load size regarding the widest (6.8 GHz) feed.
Within the framework of the above flight calibration, constraints on MIMR functional and hardware design are pointed out, with particular emphasis to the calibration targets accuracy in providing known (and stable) reference noise temperatures at feed input to continuously update the current transfer function of the radiometer during flight operation.

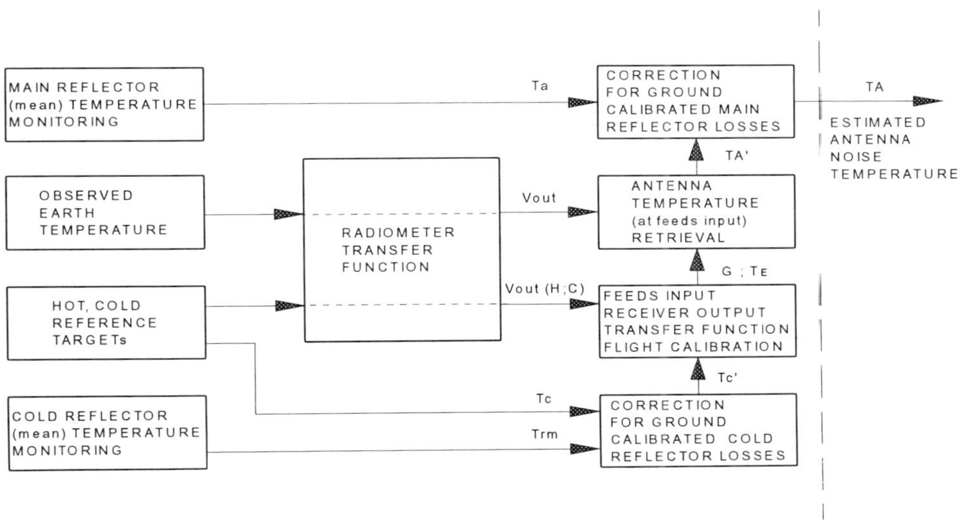

Figure 2. Antenna temperature reconstruction block-diagram

3. ERROR ANALYSIS FOR PERFORMANCE PREDICTION

An error analysis has been developed to evaluate the system performance of the MIMR radiometer in brightness temperature measurements, aiming to identify the critical aspects of the design both at system and subsystem level. Therefore, a parametric analysis of performance versus source of errors, including calibration errors, has been carried out.
Here below the main items are summarized.

Definitions
The temperature transfer equation is the relationship that relates the apparent brightness temperature scenario at antenna input with the output voltage of the receiver. Under the following assumptions, it could be possible to retrieve without error the apparent brightness temperature seen by the antenna:

-Reference temperatures: stable and well known;
-Perfect linear relationship between input power and output voltage;
-Continuos calibration;
-External calibration;
-Very low receiver noise;
-No gain variations and drift;
-Negligible quantization error;
-No reflector ohmic losses.

The radiometric measurement performance is based upon definition and specification of:

- Sensitivity: minimum detectable change in the radiometric antenna temperature.
- Accuracy: representative of deterministic error on parameters, which affect with a bias the retrieved antenna temperature.
- Stability: representative of the short term and long term variations in the antenna temperature retrieval due to changes of environmental parameters (e.g. physical temperatures).

It is worth considering the radiometric chain as divided into:

-Non calibrated path.
-Flight calibrated path.

As shown in Figure 3, one can distinguish three different non calibrated paths, one for the measurement mode and two for the calibration modes.

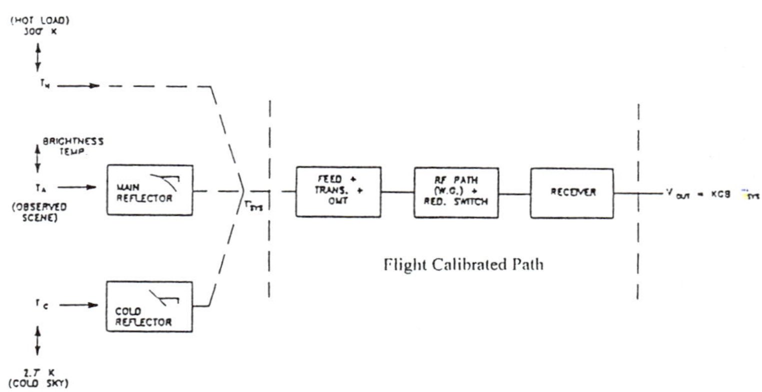

Figure 3. Radiometric chain

The non calibrated path can be characterized by pre-flight calibration measurements, and by flight monitoring (telemetry) of the physical parameters of its RF devices. The flight calibrated path is the part of the radiometric chain common to the three modes (measurements, hot calibration, cold calibration) and starts from the feeds input: this is a key advantage of the "external calibration" technique with respect to the "internal" one, where three different RF paths introduce uncontrolled sources of error between calibrations and measurement. The periodic occurrence of cold reflector and hot load in front of the feed input allows one to calibrate at every scanning period the feed input-receiver output transfer function.

The schematization used for the radiometric chain is important for the correct development and structuring of the error analysis (see Figure 4), in fact one can state that the errors present in the non calibrated paths are:

-measurement error during ground calibration (characterization);
-effects of ageing on characterized parameter figures;
-errors in flight monitoring of physical temperature;
-errors in the knowledge of the actual hot/cold calibration input to feed horns.

On the other hand, the errors present in the flight calibrated path are residual errors after flight calibration due to:

-receiver transfer function instability during intercalibration period;
-receiver transfer function non-linearity;
-A/D conversion error;
-different reflection properties between main reflector, cold reflector and hot load.

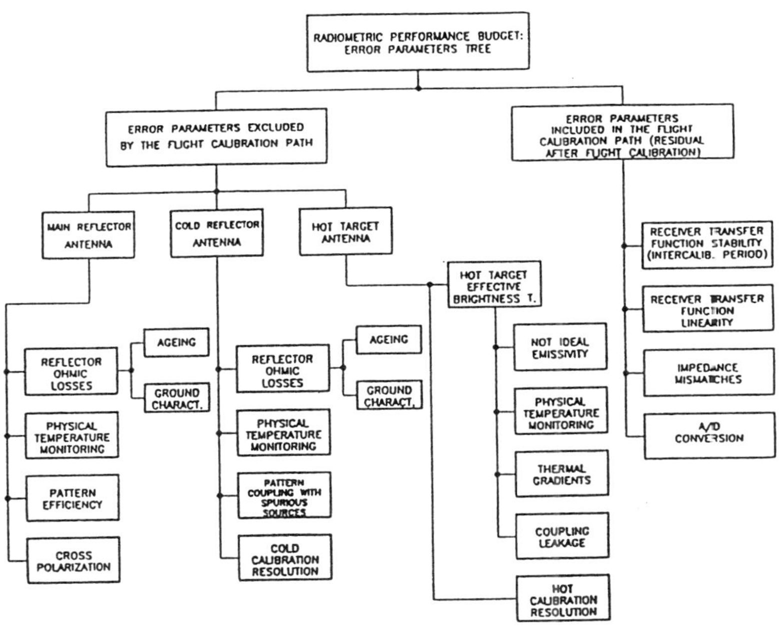

Figure 4. Error parameters tree

Error Parameters included in the non calibrated path

Error in ground ohmic loss characterization, flight ageing, and physical temperature monitoring of reflectors (main and cold). For a reflector with an ohmic loss L_A and physical temperature T_{pa} , the antenna noise temperature T_A is related to the noise temperature measured at feeds input T_A' through the following equation:

$$T_A = L_A \cdot (T_A' - T_{pa}) + T_{pa} \qquad \textbf{(1)}$$

It is not possible to relate T_A and T_A' without errors, in fact the L_A value is affected by an uncertainty due to pre-flight characterization error (plus ageing effects at end of life) , and one must consider also monitoring accuracy on T_{pa}.
A value less than 0.1 dB over the MIMR frequency range is required for reflector ohmic losses characterization, which means a residue of $\delta\varepsilon/\varepsilon$ less than 4% on the emissivity measurement accuracy.
The number of thermal sensors, positions and their measurement accuracy are assumed to give a maximum error in reflector mean physical temperature flight monitoring is less than 1 °K.

Crosspolarization effects. The antenna noise temperature retrieved after ground processing differs from the true brightness temperature of the observed scene, notably due to antenna spillover plus sidelobes contribution and crosspolarization coupling. Therefore the antenna noise temperature must be corrected because of these effects.
Additional corrections must be made for second-order cause of polarization coupling:
- Faraday rotation;
- Polarization rotation angle.

After ground processing corrections there will be a residual uncertainty in the crosspolarization. This uncertainty will degrade the MIMR radiometric performance for some application. In fact the effects of crosspolarization coupling is a magnification of instrument unaccuracy. The required characterization accuracy on the crosspolarization factor is less than 10%.

Main reflector antenna pattern knowledge. The antenna patterns at all MIMR frequencies shall be accurately measured in order to perform the deconvolution process of measured data and thus retrieve the brightness temperature distribution. High quality test range allow directivity to be assessed with accuracy better than 0.03 dB (1σ), while a -32 dB sidelobe can be measured to within better than 0.3 dB .

Cold reflector antenna: coupling with spurious sources. Differently from the ideal case, where the brightness temperature scenario collected by the cold reflector and weighted by its radiation pattern gives the noise power generated by the 2.7 °K cold sky blackbody, in the solid angle $\Omega=4\pi$ are also captured the unwanted radiations from Sun, Moon, Earth and platform appendices, which fall within the main lobe and near sidelobes affecting the measured cold sky reference brightness temperature. In the above respect the cold sky antenna design has been optimized from the configuration point of view (in terms of boresight elevation above the normal to rotation axis and azimuth from flight direction) and by integrating the directivity function with the simulated brightness temperature environment accounting for the mission/orbit parameters, a less than 0.2°K spurious contribution has been detected.

Hot target effective brightness temperature. The knowledge of the hot load brightness temperature is given by its physical temperature value with an associated uncertainty given by:
-error due to non ideal emissivity of the target;
-error due to coupling leakage between the feeds-target system and the external environment;
-error due to thermal gradients through the target;
-thermal temperature measurement error.
These errors can be considered as independent. The hot load breadboard manufactured during development phase has been tested/analyzed from e.m. and thermal points of view, with the result that the accuracy achievable on the hot reference temperature (300 °K) is equal to 0.2 °K.

Calibration (hot; cold) resolution. The standard deviation of the receiver output voltage corresponding to hot/cold reference temperatures, depends on the calibration window width(calibration integration time during one scan). In spite of the fact that the calibration window should be the widest possible to get a good resolution of the calibration measurement, other design aspects drive to minimize the calibration duration (e.g. focusing limitation of cold reflector; hot load size; etc). Limited duration acts as uncertainty on the reference temperature. The corresponding contribution errors can be reduced by performing ground processing running averages over a certain number (e.g. nine) of calibration samples taken during consecutive scans (it is required the instrument stability over the number of scans to be averaged).

Error parameters included in the flight calibrated path

Receiver transfer function instability during intercalibration period. It is here intended "short term" stability of the receiver transfer function referred to the intercalibration period, i.e. to the time base of less than 1 sec order between the completion of calibration data collection and the Earth data acquisition range, during which the radiometer transfer function shouldn't vary w.r.t. the calibrated one to have ensured that the earth radiation measured point lies on it, thus permitting input brightness temperature to be retrieved.
Although calibration and measurement occurr consecutively within each scan, intrinsic fluctuations at higher frequency than the calibration rate (e.g. flicker noise, conducted EMC aspects, supply voltage ripples) and temperature dependent fluctuations (which involve drifts in receiver gain, noise equivalent bandwidth, noise figure), induce a contributing term to the error budget.
For analytical simplicity, one can gather all the receiver transfer function fluctuations into receiver gain fluctuations during the intercalibration period, without loss of generality. The intrinsic gain variation requirement issued at receiver subsystem level is $\Delta G/G \leq 10^{-4}$ over one second.

Receiver transfer function non linearity. The antenna brightness temperature retrieval algorithm is based on a two points calibration, under the assumption that the instrument transfer function (one for each receiver) is linear.
Non linearities typically arise with the signal amplification and detection; in order to verify the linearity, it is necessary during pre-flight verification to stimulate the instrument/receiver with a variable temperature target able to cover the input range: 77 ÷ 330 °K, making use of liquid nitrogen target: this is not however a limitation as non linearities are not expected in the lower portion (2.7°÷ 77°K) of the dynamic. The linearity will be calculated as the peak departure from a best fit line on the data. As for short term stability we simply refer, without loss of generality, to gain linearity aspects. It is applied to receiver designer a figure of $\Delta G/G < 5x10^{-4}$ as gain linearity performance.

Effect of impedance mismatches. Impedance mismatches [4] in the radiometric chain can be seen in the two following aspects:

- The equivalent losses of the passive network (RF path) increase, and therefore the system temperature rises (worse performance in radiometric sensitivity).
- The reflection coefficient viewed into OMT output port might change among the three different antenna configurations (hot, cold, measurement), hence a calibration error may be present. Assuming a VSWR always better than 30 dB, at OMT output, the calibration error gives a negligible contribution.

Effect of quantization error. The effect of quantization error is a contribution to sensitivity performance. The current guideline is a 16 bit A/D converter and the quantization error can be neglected.

4. RECEIVER DESIGN/TECHNOLOGY

Previous paragraph has evidentiated some key performances which shall be considered as the receiver design driving elements, in particular the Noise Equivalent Temperature, the input impedance matching (VSWR), the frequency response, the short term gain variations (over 2.3 sec. mean time, antenna rotation period), gain linearity, gain stability over temperature and ageing, ADC quantization noise.
Further design driving requirements [5] are imposed to receivers as flight space hardware, in particular for minimum power consumption, mass, dimensions, wide temperature ranges under vacuum conditions, high reliability, exposure to radiations, electro-magnetic and radio frequency compatibility with the instrument and the spacecraft environment, mechanical vibrations and shocks, use of proven and already qualified (or at least qualifiable) technologies for space applications.
In the following major emphasis will be put on the trade-offs and design choices driven by radiometric performance, with some basic considerations regarding the construction, environment and technological aspects, as far as performance are concerned.

Receiver general concepts
As the required radiometric stability and accuracy performances are achieved through the external calibration technique described above, MIMR receivers are based on the "total

power" concept, thus providing the best possible radiometric sensitivity performance when compared to Dicke or noise injection concepts.
The detected output voltage is linear with respect to the input mean noise power collected from the antenna, i.e. the detection is based on the quadratic mode, with the detector diode operating within the quadratic region of its characteristics.
Two main receiver architectures may be considered : the superheterodyne and the direct detection.
The superheterodyne offers the inherent advantage of flexibility, as by proper selection of the Local Oscillator (LO) frequency, the band defining filtering, as well as most of the needed amplification, can be allocated in a convenient Intermediate Frequency (IF) band, where adequate components and technologies can be selected to satisfy the requirements. On the other hand, the direct detection approach offers the advantage of simplicity, as no LO is necessary and the receiver components are reduced, thus achieving potential mass and power consumption reduction, as well as increased reliability.

Filtering

The key elements to be considered in the trade-offs are the filtering, amplification, and the stability aspects.The filtering of a radiometric chain is a key element, since the output voltage is :

$$\mathrm{Vod} = \mathrm{k} \times (\mathrm{Trec} + \mathrm{Ta}) \times \mathrm{Bn} \times \mathrm{G} \qquad \textbf{(2)}$$

where k is the Boltzmann constant, Trec is the receiver noise temperature, Ta is the antenna noise temperature, G is the receiver gain and Bn is the noise equivalent bandwidth. This means that all the energy collected in Bn will be detected as brightness temperature coming from the Earth scene being observed, and therefore the Bn shall be as much as possible closed to the allocated bandwidth for the radiometric measurement of that channel (the filter shape shall be as much as possible rectangular).
MIMR requirements impose a rejection of at least 40 dB in a band which is 1.3 times the allocated band for that channel, which means that the energy collected outside the 1.3 band is 1/10000 with respect to the in-band energy. Also the Bn variations over temperature and life will be seen as gain variations, and therefore the selected filter shall be very stable.
The best suited filters are of the Elliptical and Chebyshev type, where excellent out of band attenuation characteristics are achievable at frequencies of 1.3 times the pass band frequency, with acceptable pass band ripple and order number (acceptable filter complexity).

Architecture

Depending from the selected architecture, superheterodyne or direct detection [7], the frequency where the filtering and most of the amplification will be performed can vary significantly, as well as the fractional bandwidth, which might affect the receiver filtering characteristics and other performance such as the Noise Figure, the VSWR, the frequency stability, the short term gain variations, the linearity.
In case a direct detection topology was chosen for each of the MIMR channels, fractional bandwidths in the range 0.9% (at 10.65 GHz) to 3.3% (at 89.0 GHz) would result.
In these conditions to achieve the required rejection at 10.65 GHz, a difficult cavity filter with at least 8 dual or triple mode cavities is needed. An additional constraint is represented by the required center frequency stability, 2.5 MHz over temperature and life, which leads to the development of an extremely stable Invar filter or similar.

The drawbacks in this case are represented by the difficulty of achieving the required cut-off band attenuation and frequency stability, but also by the mass and dimensions of the filter itself.
If a superheterodyne concept was chosen, the IF frequency could be selected in order to achieve a fractional bandwidth between 5 and 15%, for example 700 MHz, where a lumped elements filter or a SAW filter may be used with excellent performances and higher level of miniaturisation.
Frequency stability would be higher due to the lower frequency (as it is a percentage of the centre frequency) and by proper selection of the LO.
A larger fractional bandwidth, exceeding the 20%, might rise some difficulties to achieve the required rejection , as the "Q" factor is lower and the required shape factor of the filter is more stringent. Moreover a worse impedance matching would result, with consequent degradation of the in-band ripple. Also due to the larger bandwidth, the Noise Figure, the gain variation over temperature and gain linearity performance will be degraded, as the variation of gain with frequency will be an additional effect .
The superheterodyne concept seems to offer more benefits for MIMR application. The highest frequency channels, 36.5 and 89 GHz, as well as the 6.8 GHz one, might be suitable for direct detection, from the fractional band point of view. In addition the higher frequency would significantly reduce the dimensions of the needed filters. Nevertheless the problem of achieving the needed amplifications directly at RF is presently the blocking point for the application of this topology at 89 GHz, where only experimental amplifiers have been developed but with not adequate performances, especially from the noise figure and achievable gain per stage point of view.
Figure 5 and Figure 6 show the application of the superheterodyne concept respectively for the 6.8/36.5 GHz and 89 GHz channels. The direct detection topology would be achieved by suppression of the mixer and LO, implementing all the circuitry up to detector directly at RF.
In the 89 GHz channel a DSB (Double Side Band) concept is applied, in order to improve the receiver Noise Figure (lower Noise Temperature).
The required pre-detection band-width is 3 GHz; therefore by extending the RF band to 6 GHz and by choosing an LO frequency equal to the centre frequency (89 GHz), a base band of 0÷3 GHz would result at the mixer output: the energy in the 3 GHz base band will include the energy of the full 6 GHz RF band, which means a noise power 3 dB higher (as the "negative" frequency side-band folds-up to the positive one). Consequently, from the radiometric point of view, the mixer insertion losses and hence the receiver Noise Figure, is improved by a 3dB factor. This characteristic of the DSB architecture makes it very attractive when compared to a Single Side Band (SSB) one or to a direct detection one.
The disadvantage is, of course, the double RF bandwidth to be considered, which will degrade the mixer input VSWR, the Noise Figure and the in band ripple (in this case by an acceptable amount compared to the benefits).

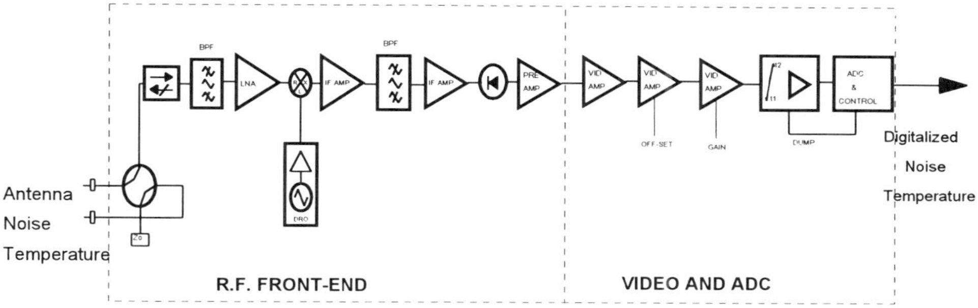

Figure 5. 6.8/36.5 GHz Receiver architecture

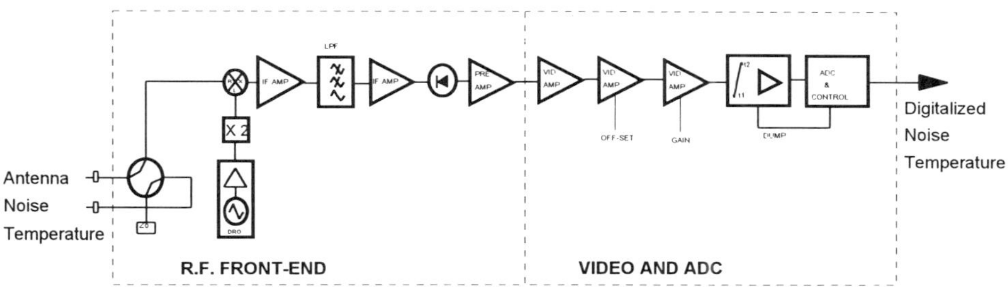

Figure 6. 89.0 GHz Receiver architecture

Receiver's components

The main sections of the receiver are the front-end, the local oscillator, the IF channel, the detector, video and ADC.

Receiver Front-end. The front-end includes both active and passive components, e.g. isolators, filters, Low Noise Amplifiers (LNA) and mixers.
Isolators and filters, ahead of the LNA, produce an increase of the Noise Figure, and their use should be therefore controlled, but they avoid back-reflection towards antenna of microwave energy, or to reject undesired signals which may corrupt the useful signal.
A key element is the trade-off for the optimization of the LNA noise figure with respect to its input matching, which are, for physical reasons, not achievable at the same time.
Concerning active devices of the LNA, several alternatives are available today: the most suited devices are low noise pseudo-morphic HEMT (High Electron Mobility Transistors). By use of these devices, amplifiers in hybrid MIC (Microwave Integrated Circuits) technology (thin film on Alumina substrates) can be developed with good Noise performances up to the Ka band, by use of 0.5 or 0.25µ gate length devices.
MMIC amplifier (Monolithic MIC) devices are currently under evaluation for use on MIMR, but their main limitation is the achievable noise figure when compared to the

discrete HEMT solution; these devices are very interesting for space application due to their lower dimensions, power and mass, but the current technologies aren't fully proven for space application.
For the 89 GHz receiver, the key component of the front-end is the mixer-LNA assembly. Good noise figure characteristics are achieved by use of a balanced waveguide mixer, with a suspended microstrip circuit which allocates a pair of beam lead schottky diodes; the LNA is integrated in the same assembly, very closed to the mixer output port, in order to improve the critical matching with the mixer. Alternatively a mixer/LNA/LO assembly in MMIC technology is being evaluated for MIMR, but the achievable Noise Figure is degraded due to mixer intrinsic characteristics and also due to the ohmic loss of the transition from the waveguide to the microstrip circuit where the MMIC chip are assembled.
The achieved noise figures are in the range of 1.4 dB at 6.8/10.65 GHz, 4 dB at 36.5 GHz, and 5 dB at 89 GHz for the front-ends including all the needed passive components .

Amplifier's chain. The post LNA amplification, either at IF or RF, is less critical from the Noise Figure stand point, and, depending from the frequency, many solutions may be considered, such as hybrid MIC based on discrete HEMT or MESFET, or conventional GAsFET devices, or MMIC which represents a good alternative in particular at lower frequency (up to 3GHz) due to their good performances.
The amplifier chain shall be properly sized to achieve good linearity characteristics, since the last stages of the strip will be closed to saturation. Therefore a good back-off margin between the operating point of the dynamic and the 1dB gain compression point of the amplifier shall be implemented.
Stability over temperature is usually achieved by use of temperature compensation circuits acting on these amplifiers.

Local Oscillator (LO). Several alternatives are today available to develop a microwave solid state oscillator. The characteristics of radiometric receivers are such that a direct microwave oscillator approach is preferred.
Both Gunn diode oscillators and Dielectricallly stabilised Oscillators (DRO) may be suited for the application.
High power levels are easily achieved by Gunn types, but with lower power efficiency compared to DRO type. Given the reasonable needed power level which is between 0 and +5 dBm, the DRO can be used.
The key parameter which drives the selection of the DRO type is its frequency stability: 12 MHz accuracy/stability have been achieved by MIMR 89 GHz LO, in this configuration.
The DRO characteristics offer better performances in terms of output power stability over temperature and AM and phase noise: these characteristics are very important, as they affect the mixer conversion loss (critical especially at 89 GHz) and the short term gain stability, as AM noise modulate the mixer conversion losses itself; in general noise generated by the LO can fall inside the IF band, thus degrading receiver sensitivity.
The detector may be based on a a biased or unbiased diode, of the Schottky or of the Tunnel type, depending from the required sensitivity and stability (the former one is more sensitive, the latter one is more stable). Planar doped diodes, which is a relatively advanced product, provides intermediate characteristics between Schottky and Tunnel detector diodes.

Video section. The video section is the analog processing part of the radiometer, and is based on the video filter which determines the instrument characteristics in terms of sensitivity (noise bandwidth) and aliasing (integration and sampling time).
The ideal video filter should have the smallest possible Noise Bandwidth (rectangular filter) from the sensitivity point of view, while its pass band characteristics should be as much as possible matched to the antenna spatial frequency response, which, depending from the rotation speed and from the pattern characteristics of the antenna, defines the signal bandwidth at detector output, weighting the observed Earth scene.
The adopted solution in MIMR is the "Integrate and Dump" filter.

Analog to Digital Converter (ADC). The ADC selection is a design driver, as several errors may be introduced by this component. Quantization noise is degraded by the ADC : the higher is the number of bits (resolution) the lower results this error and vice versa.
Quantization noise can also be reduced by adapting the signal dynamics in order to occupy at the maximum extent the ADC dynamics, i.e. by use of gain and off-set control functions as shown in fig. 5 and 6.
Linearity is especially important for radiometric application as the accuracy error is compensated by the radiometric end to end calibration . The most important characteristics is the integral linearity error, as it defines the straightness of the input to output characteristics compared to the ideal one achieved by the best fit function.
Linearity error is usually in the order of a few LSB (Least Significant Bit), and therefore receiver linearity might be improved by increasing the ADC resolution.

5. EXPECTED RADIOMETRIC PERFORMANCE

Here after are summarized the MIMR predicted radiometric performance based on the outcome of current design activity and, where available, test results on breadboard hardware items.

Sensitivity Performance
In order to give a certain flexibility to the processing performed on the detected voltage, in MIMR design it is foreseen to implement a video filtering on-board and to complete the averaging process on-ground applying dedicated algorithms.
The parameters affecting radiometric resolution [8] ΔT_N are:

- System temperature T_{sys};
- Receiver equivalent noise bandwidth B_n (before detection);
- Global integration time τ_{INT} (on-board video filter + on-ground data pre-processing);
- Intrinsic gain variations $\dfrac{\Delta G}{G}$.

$$\Delta T_N = \sqrt{T_{SYS}^2 \cdot \left[\frac{1}{B_n \cdot \tau_{INT}} + \left(\frac{\Delta G}{G} \right)^2 \right]} \qquad (3)$$

For the intrinsic gain variation , it is assumed a figure of 2 x10^{-4}.

Table 1. Predicted Sensitivity performance

Channel [GHz]	IF Channel Bandwidth [MHz]	Integration Time after processing [msec]	System Temperature at feeds input [°K]	Predicted Sensitivity performance [°K]	Sensitivity Goal Specification [°K]
6.8	200	20	345.6 (1)	**0.18**	0.2
10.65	100	10	327.0 (1)	**0.33**	0.4
18.7	200	10	793.1 (2)	**0.58**	0.5
23.8	400	10	898.7 (2)	**0.48**	0.5
36.5	1000	5	993.0 (2)	**0.48**	0.5
89.0	2700	2.5	1573.4 (2)	**0.68**	0.7

(1) T_A=150 °K
(2) T_A=300 °K

Accuracy & Stability Performance
Tables 2 and 3 report the current budget of accuracy and stability (as defined in chapter 3), including breakdown of the various error contributions.
Given stability figures are computed over one year and include ageing effects.

Table 2. Predicted Accuracy performance

Channel [GHz]	error in main reflect. ohmic losses char. [°K]	cross-pol. error [°K]	error in cold refl. ohmic losses charact. [°K]	coupling of cold refl. with spurious sources [°K]	receiver non linearities [°K]	error due to imp. mism. [°K]	main refl. ohmic losses drift [°K]	cold refl. ohmic losses drift [°K]	Total RSS (1);(2) [°K]	Accuracy Goal Spec. (no T_A ref.) [°K]
6.8 (1)	0.15	0.086	0.12	0.15	0.182	0.28	0.46	0.38	**0.73**	1
10.65 (1)	0.15	0.086	0.12	0.15	0.163	0.28	0.46	0.38	**0.73**	1
18.7 (2)	0.05	0.086	0.04	0.05	0.376	0.051	0.16	0.13	**0.45**	1.5
23.8 (2)	0.05	0.086	0.04	0.05	0.419	0.051	0.16	0.13	**0.49**	1.5
36.5 (2)	0.05	0.086	0.04	0.05	0.47	0.051	0.16	0.13	**0.53**	1.5
89.0 (2)	0.05	0.086	0.04	0.05	0.742	0.051	0.16	0.13	**0.78**	1.5

(1) T_A=150 °K
(2) T_A=250 °K

Table 3. Predicted Stability performance

Channel	error in main refl. ohmic losses char.	flight monitor -ing of main refl. phys. temp.	hot load effecti -ve bright. temp.	hot load temp. resolu -tion	error in cold refl. ohmic losses char.	monito- ring of cold refl. phys. temp.	cold calibr. resol.	error due to imp. mism.	main refl. ohmic losses drift	cold refl. ohmic. losses drift	Tot. RSS	Goal Spec.
[GHz]	[°K]	[°K]	[°K]	[°K]	[°K]	[°K]	[°K]	[°K]	[°K]	[°K]	[°K]	[°K]
6.8 (1)	0.05	0.047	0.1	0.1	0.04	0.01	0.06	0.028	0.13	0.1	**0.24**	0.2
10.65 (1)	0.05	0.047	0.1	0.18	0.04	0.01	0.06	0.028	0.13	0.1	**0.28**	0.4
18.7 (2)	0.05	0.047	0.16	0.34	0.01	0.003	0.04	0.028	0.13	0.03	**0.42**	0.5
23.8 (2)	0.05	0.047	0.16	0.28	0.01	0.003	0.034	0.028	0.13	0.03	**0.36**	0.5
36.5 (2)	0.05	0.047	0.16	0.2	0.01	0.003	0.03	0.028	0.13	0.03	**0.30**	0.5
89.0 (2)	0.05	0.047	0.16	0.22	0.01	0.003	0.036	0.028	0.13	0.03	**0.32**	0.7

(1) T_A=150 °K
(2) T_A=250 °K

6. REFERENCES

1. E.G. Njoku, J.M. Stacey and F.T. Barath, The Seasat scanning multichannel microwave radiometer (SMMR): Instrument Description and Performance, IEEE J. Oceanic Eng., Vol.OE-5, No.2 1980
2. J.P. Hollinger, J.L. Peirce and G.A. Poe, SSM/I Instrument evaluation, IEEE Tr. on Geos. and R.S. , Vol.28, No.5, Sept. 1990
3. P.N. Swanson and A.L. Riley, The Seasat scanning multichannel microwave radiometer (SMMR): Radiometric calibration algorithm development and performance, IEEE J. Oceanic Eng., Vol.OE-5, No.2 1980
4. F.T.Ulaby,R.K.Moore,A.K.Fung:" Microwave Remote Sensing", Vol.1 , Artech House 1981.
5. Alenia Spazio MIMR Phase B/ Phase B Rider design and breadboard development reports.
6. G. Evans and C.W. Mc Leish, RF Radiometer Handbook, Artech House 1977
7. N. Skou, Microwave Radiometer System: Design and Analysis, Artech House 1989
8. M.S. Hershman and G.A. Poe, Sensitivity of the total power radiometer with periodic absolute calibration, IEEE Tr. on MTT, Vol. MTT-29, No.1 Jan. 1981

RADIOMETER ABSOLUTE CALIBRATION: THE SOLUTION IMPLEMENTED IN MIMR

E. BATTISTELLI[1], R. BORDI[2],

[1]*GF-GALILEO SMA S.r.l, via Einstein 35, 50013 Campi Bisenzio (FI), Italy*
[2]*Alenia Spazio S.p.A, via Saccomuro 24, 00131 Rome, Italy*

Abstract. While operating with micro-mm-wave radiometers there is the need of associating values of brightness temperature to the instrument voltage outputs. At a subsequent time these brightness temperature values will form the input of theoretical algorithms used to extract the physical parameters of interest.
In order to optimize the system accuracy the radiometric chain has to be continuously calibrated on-board: to this end the external Calibration technique appears as the most suitable procedure, permitting the exploitation of the high Sensitivity performance of Total Power Radiometer configuration. External Calibration, as implemented on-board in MIMR, foresees through the instrument rotation the periodical interception of the feeds beam path by a small reflector pointing the cold sky (2.7 K) and by a blackbody device, thermally controlled at 300 K (Hot Load).
A very interesting task in MIMR project is the development of the Hot Load, taking into account the 0.2 K required accuracy on brightness temperature over the wide bandwidth (15:1 range).
In this paper a description of the Hot Load designed, manufactured and tested by SMA, at breadboard level, is reported.
A tight thermal control, both passive and active, is applied to Hot Load in order to keep it at the nominal 300 K temperature. A set of thermal sensors permits the continuous monitor of the Hot Load physical temperature, allowing the functionality of the Calibrator also in the case of thermal control failure.
Hot Load exhibits high electromagnetic performance: a minimum figure of 0.9997 for Emissivity has been obtained over the entire MIMR band.

Key words: Radiometer, Calibration, Blackbody.

INTRODUCTION

The work we describe herebelow is relevant to the Target developed by SMA for In-Flight Calibration of MIMR.
MIMR is a multichannel radiometer devoted to provide scientific data for remote sensing of atmosphere, ocean, cryosphere, land.
MIMR instrument consists of an offset parabolic reflector (main dimension 1.6 m), illuminated by a cluster of 9 horns arranged as close as possible the reflector focal point. The receiving section is composed by a set of 20 Total Power radiometric channels covering both H and V polarizations in the 6.7÷92 GHz.
MIMR will fly on a Polar Platform in a circular sun synchronous near polar orbit.

Accuracy of MIMR radiometric data strongly benefits of the External Calibration strategy, based on the observation, at each antenna scan, of the cold sky at 2.7 K and a Hot Reference Target at 300 K.
The Hot Target is substantially a blackbody panel in a thermomechanical structure kept at a fixed temperature by an active thermal control.
Active thermal control works by means of heating resistors; heat radiative exchange with external environment is controlled by a passive external insulation. A dedicated circuitry permits the continuous thermal monitor by means of embedded platinum resistance sensors.

We like to remark the very challenging task in designing the Hot Target, considering the very tight mass constraint (3 Kg overall), the need to match with a cluster composed by 9 feed, covering a 15:1 bandwidth. Present requirements are very constraining if compared with other similar projects, for instance SSM/I [1] [2], where the bandwidth was 19÷86 GHz (5:1) and the active thermal control was not required.

SMA system analysis on MIMR Hot Target for external calibration started at half 1990; design activity started at half 1991 and finally the breadboard of Hot Target was tested at half 1992.

SYSTEM REQUIREMENTS

The Hot Target main requirement is to provide a stable and known blackbody source at 300 K for all the MIMR frequency bands during the time allocated for In-Flight Hot Calibration, i.e. at each rotation of the drum containing the feed cluster.
The mechanical allocation on MIMR is shown in Figure 1.

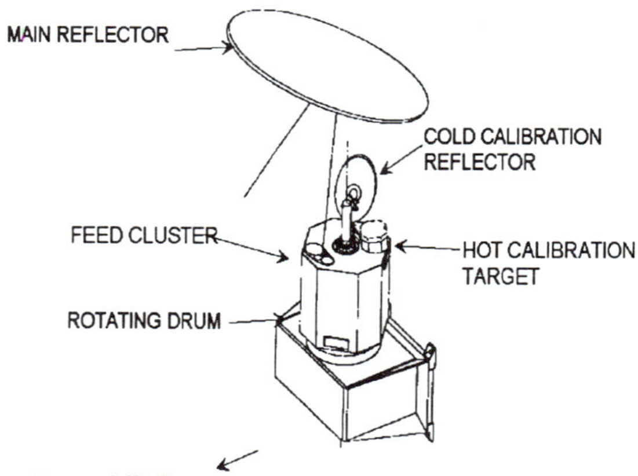

Figure 1. *Hot Target location on MIMR*

MIMR frequency bands are reported in Table 1.
The time interval reserved to Hot Calibration of each RF channel (H+V polarization) is 16 ms, corresponding to 2.5° angular width.

The main thermal and electrical requirements are summarized in a single parameter, the Accuracy that is the overall uncertainty in the knowledge of brightness temperature of the Hot Target; required figure is 0.2 K.

Table 1. MIMR Frequency Bands

Center Frequency (GHz)	Bandwidth (MHz)
6.8	200
10.65	100
18.7	200
23.8	400
36.5	1000
89.0	6000

Main electromagnetic requirements are summarized in Table 2.

Table 2. Main e/m requirements

Emissivity	≥0.9997, equivalent to
Return LossL_R	≥35 dB
Coupling Loss L_C	40 dB as goal
Frequency range	6÷92 GHz
Polarization	H and V

Other main constraints are the mass and the power consumption, required to be less than 3 Kg and 3 W, respectively.

ACCURACY OF TARGET BRIGHTNESS TEMPERATURE

Sources of uncertainty on brightness temperature are: non ideal Emissivity, Coupling Loss, accuracy on temperature monitoring, thermal gradients in the blackbody volume.
In the case of Emissivity ε of the Target less than 1, a reduction of the radiated power is expected, associated to the reflection of a fraction of power radiated toward the Target.
The difference from the received brightness temperature and the ideal one is composed by two terms: the first is retrieved by the on-ground characterization and temperature monitoring, while the second is originated by the error on characterization and monitoring In other words, for Return Loss:

$$T_R = T_T(\varepsilon) + T_X(1-\varepsilon) \quad (1)$$

where:
T_T is the Target physical Temperature;
T_X is brightness temperature radiated toward the Target from the receiving chain;
ε is the Emissivity at μ-mm-wave and it is related to measurable Return Loss R_L by:

$$\varepsilon=1-1/R_L \quad (2)$$

The spurious contribution due to non-infinite Return Loss is given by:

$$\Delta T = T_R - T_T = (T_X - T_T) / R_L = (T_X - T_T)(1-\varepsilon) \quad \textbf{(3)}$$

The term ΔT_R is given by the sum of a known part (measured or monitored) and an error part; this latter term is relevant to the uncertainties in measuring T_X, T_T and R_L. The known part of ΔT_R can be used for correction, so the final uncertainty on Target brightness temperature results substantially lower then the figure obtained from Eq.(3).

<u>Coupling Loss</u> is the fraction of electromagnetic power that feeds receive from external environment instead of the Target.
A similar assessment as above can be made for Coupling Loss L_C;

$$T_R = T_T (1 - 1/L_C) + T_E / L_C \quad \textbf{(4)}$$

where T_E is the worst case environment brightness temperature.
Previous equation leads to a bias term:

$$\Delta T = T_R - T_T = (T_E - T_T) / L_C \quad \textbf{(5)}$$

Again the term ΔT is composed by a known term that can be removed thus reducing the final uncertainty on Target brightness temperature.

Another source of uncertainty on Target brightness temperature is the error in evaluating its physical temperature; this error is related to:
- the accuracy of thermal monitor system;
- thermal gradients inside the blackbody volume.

Effects of gradients inside the Target can be minimized by embedding several thermal sensors and averaging their output.

DESIGN

The high level requirement for Hot Target is to provide an accurate RF blackbody source at 300 K for the In-Flight Calibration of MIMR receivers.
The design of the Hot Target is restricted by both thermal design and strict mass limitations.
Main areas of concern in designing the Hot Target were the following:

- Mass constraints;
- Emissivity performance;
- Thermal uniformity;
- Active thermal control.

In the following we report the design activity for each important item, altough the final design is derived from a extensive trade-off activity covering in a overall way all the design aspects.
The basic scheme of Hot Target is shown in Figure 2.

Mechanical and mass constraints.
Low reflection performance (i.e. high Emissivity) is expected for Target dimension of the order of several wavelenghts, in the direction of field propagation.
Actually the mass requirement imposes small dimensions for the Target, making critical the design task regarding Emissivity.
Footprint of Target have to cope with the feed cluster geometry, see Figure 3. In order to fully intercept the radiation beam of each feed a large Target would be required, possibly with a negligible gap between the Target and the rotating drum underneath it. Practically the gap and footprint dimensions of the Target are limited by mechanical constraints, thus the feeds / Target coupling is expected non ideal.

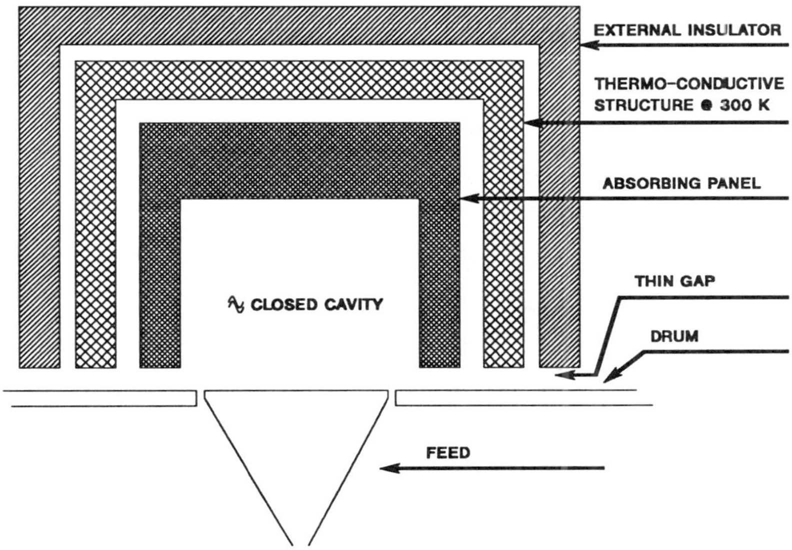

Figure 2. Basic scheme of Hot Target

To save mass the Target is dimensioned for the sequential calibration of the nine feeds, when each one is looking at the central portion of Target.

Blackbody concept
The selected solution for Hot Target design is based on the "closed oven" concept: an absorbing panel is contained in a thermomechanical structure in order to form a quasi-closed cavity. High IR emissivity inside the cavity assures temperature uniformity.
The absorbing panel is composed by a planar array of pyramids; this proven geometry has been considered for its favourable electromagnetic behaviour and for thermal reasons: thermal control can be applied on the back wall of the panel very uniformly.

Absorbing material

Selection of absorbing material has been constrained by several needs: electromagnetic performance, space suitability, thermal conductivity and mass.

Thermal conductivity is a very important item for the minimization of gradients along the Target. A preliminary tradeoff has involved foam materials against solid ones; foam materials (pyramidal, convoluted or flat with gradients in the dielectric constant) have been discarded for they very low thermal conductivity and also because only a few foam materials meet the outgassing requirement for use in space.

After an accurate analysis of solid absorbing materials the selected one was an epoxy resin loaded with magnetic particles.

Its thermal conductivity is not as high as the one of aluminum, by which the structure is made, anyway it leads to a satisfactory thermal performance (low gradients).

Specific weight of the absorbing material results as a compromise between the lightweight requirement and the need for a high content of magnetic metal, that shall assure the electromagnetic performance.

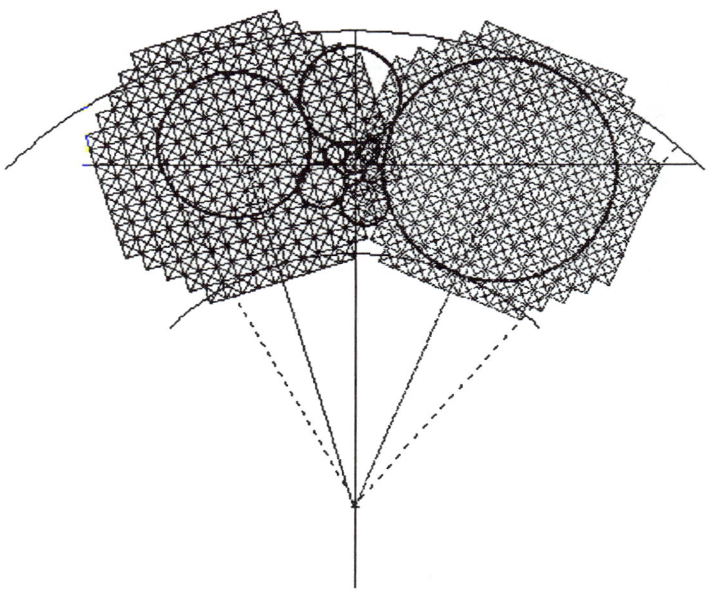

Figure 3. Feed cluster with Target footprint

E/M design

The challenging electromagnetic requested performances are related to the geometry and RF behaviour of the absorbing material.

The electromagnetic design has been constrained by several factors, in particular the wide bandwidth (15:1), to be obtained keeping the overall mass below 3 Kg. Mass requirement impacts on the length of pyramids, thus limiting the Return Loss / Emissivity performance at the low frequency. The pyramidal panel geometry applied to the selected material appears very effective for high values of the ratio l/λ (l is the pyramid lenght). In our case the mass

constraint limits l/λ to a figure less than 2, so a critical performance were expected at the lowest frequency.
In order to define the height to base ratio of pyramids for optimum e/m performance we considered two different guidelines depending on frequency:
- at higher frequencies the pyramid dimensions are larger than wavelenght, so a quasi optical approach is valid: the radiations rays are subjected to multiple reflections / scattering on absorbing walls of pyramids, thus originating the expected very good absorbitive performance of the panel;
- at the lowest frequencies the pyramids dimensions, including its lenght, are less than or comparable with the wavelenght, so a different approach has been followed. For the sake of simplicity we consider a plane wave inciding on the pyramidal panel; this one can be considered as a continuum medium with its (complex) dielectric constant smootly varying from the one of vacuum (at pyramid tip) to the one of absorbing resin (at pyramid bottom). The volume averaged dielectric constant and permeability follow a parabolic behaviour given by the increase of pyramid section, see Figure 4. The effectiveness of presented geometry is based on the very smooth transition from vacuum to solid absorber, that strongly reduces the reflected wave.

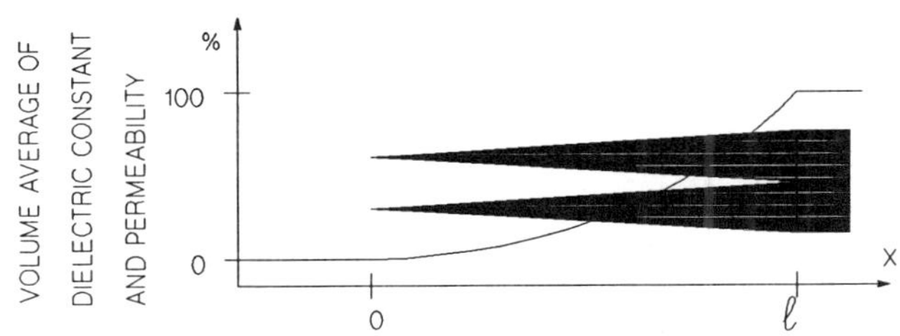

Figure 4. e/m constants average behaviour

In the limits for which the pyramidal panel can be considered a continuum, i.e. the pyramids are sufficiently thin and consequently their number high, the behaviour of averaged dielectric constant and permeability is independent from the vertex angle of the tip. Following the above modelization, the ratio l/lambda impacts directly on the smoothness of transition from vacuum to the lossy medium.

Thermal design.
Thermal design of Hot Target has covered several areas, the main are the analysis of radiative heat fluxes through external walls, the minimizations of gradients along the pyramids, the active thermal control and monitoring subsystems.
Depending on the MIMR in-orbit position and season, the Hot Target will be exposed to several radiative conditions, ranging from a partial sun exposure to eclipse.
A suitable insulation (thermal blanket) has been designed in order to reduce the overall radiative thermal exchange, that results to be 0.54 W towards the cold space and 0.18 W received from the sun.

During operations the absorbing panel will be faced to a surface (the MIMR drum) at a different temperature with respect to the nominal one of Hot Target (300 K), so a radiative exchange take place; the worst case amount of this flux has been evaluated to be 0.19 W. A heat flux along the pyramids produce a temperature gradient that gives us some uncertainty about the real brightness temperature of each pyramid.
In order to minimize gradient along pyramids an aluminum core has been inserted in each pyramid, thermally connected to the aluminum plate at pyramid base. The physical temperature gradient has been evaluated through thermal analysis, then a weighed average has been performed in order to evaluate the final uncertainty on brightness temperature providing a worst case result of 0.16 K.

Thermal control is performed by means of five independent thermal controls, one for the pyramids base and the others for the lateral walls. Each control reads the physical temperature through a platinum resistance sensor, processes the signal and drives a set of heating resistors. Thermofoil heaters are fixed on external walls of the Target.

Thermal monitoring is performed by means of a set of six thermometers, each one composed by a platinum sensor and a conditioning circuit. Four sensors are located in four selected pyramids, the remaining two are placed on two opposite lateral walls.

MANUFACTURED HOT TARGET

The manufactured Hot Target is shown in Figure 5.

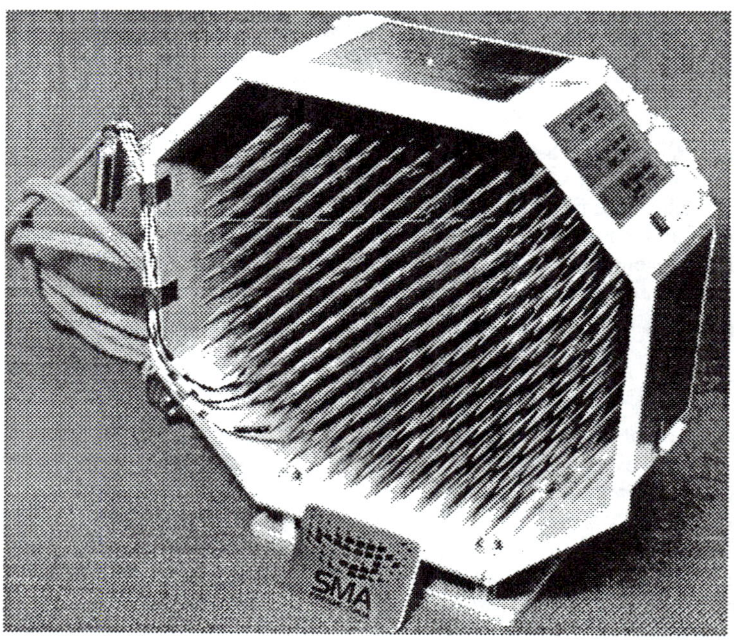

Figure 5. SMA Hot Load Breadboard

ELECTROMAGNETIC TEST

Here we report the electromagnetic tests that have been carried out on the manufactured Hot Target.

Emissivity / Return Loss
Emissivity of Hot Target at MIMR frequency bands has been verified by means of Return Loss measurement, as per Eq.(2).
Return Loss R_L is usually measured illuminating the device under test with a feed and measuring the reflected power.
The test bench is composed by a RF sweeper source, a micro/millimeter-wave test set with the horn pattern as close as possible to MIMR feeder one and a scalar Network Analyzer to process and to display the RF reflection coefficient. A mechanical support has been manufactured in order to keep the Target in a correct position in front of the feeder and to assure a configuration similar to the real one.
The expected Target Return Loss (>35 dB) is substantially higher of test feed one (typical 25 dB), making the former difficult to precisely detect.
For this reason a non standard method has been used: the reflected power has been measured at different distances between Target and test feed.
The resultant Reflection Coefficient detected by the Network Analyzer is composed by two contributes: one due to the reflection at the mouth of the feeder and the other to the true reflection from the panel. This two contributes sum themselves: in phase, when the difference of their paths is an integer number of wavelengths, giving a maximum of VSWR; and out of phase when this difference is an integer number of half-wavelengths, giving a minimum of VSWR displayed on the Network Analyzer.

In terms of VSWR, if X_{min} and X_{max} are the VSWR values in the extremes of the modulation, we have:

$$\text{VSWR(horn)} * \text{VSWR(panel)} = X_{max}[\text{in phase}] \tag{6}$$

and

$$\text{VSWR(horn)} / \text{VSWR(panel)} = X_{min}[\text{out of phase}] \tag{7}$$

Solving the equations system we obtain the value of the intrinsic VSWR of the Hot Target.

A dedicated mechanical test equipment has been used to perform the relative Target / feed movement, following the scheme in Figure 6.

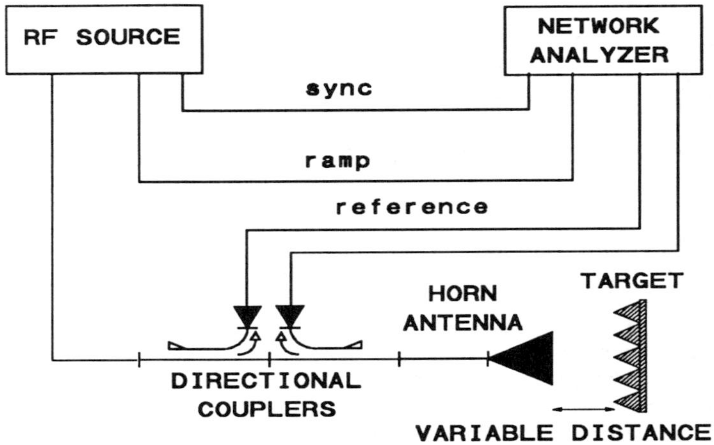

Figure 6. Return Loss Test set-up

Result of Tests are displaied in Figure 7 and Figure 8 where the measured Return Loss and Emissivity, retrieved through Eq.(2) are shown with respect to the test frequency. Maximum uncertainty for each measure is reported.

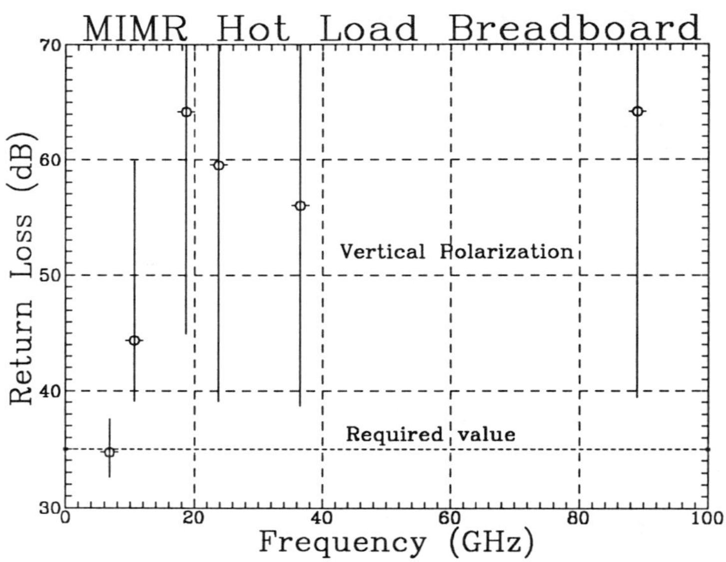

Figure 7. Return Loss Test results

Results for a single Polarization are shown, having the other Polarization an identical behaviour.

Test instrumentation depends on the frequency and it is reported herebelow:

• 6.8 GHz, 10.65 GHz, 18.7 GHz.
Measurements has been performed by means a HP8510B network analyzer with HP8341B synthetized sweeper, HP8514B S-parameter test set and HP calibration set for the considered frequency bands.

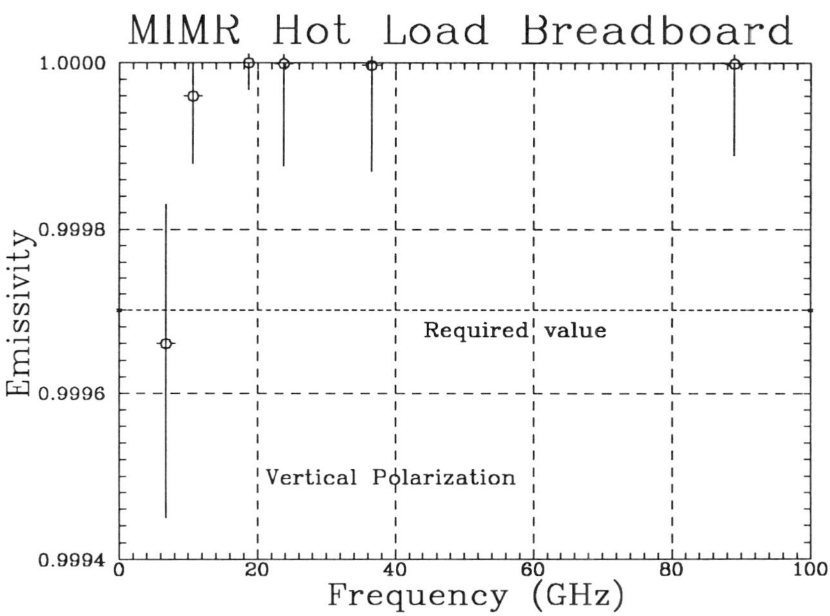

Figure 8. Emissivity performance

• 23.8 GHz instruments.
The 23.8 GHz measurements are based on the HP8757C scalar network analyzer with the extension up to 40 GHz.

• 36.5 GHz.
Tests at 36.5 GHz have been performed through a set-up formed by:
-Network analyzer: WAVETEK N.A. 1038-N10
-Sweep oscillator: HP8350B/HP83572 Plug-In
-Directional couplers: HP R752D
-Detectors:PACIFIC MEASUREMENTS INC 15850

• 89 GHz
The measurements at 89 GHz have been performed by means of the HP8510B network analyzer with the relevant W-band extension.

Coupling Loss.
The task is to measure the power loss, due to the radiation coming from the thin gap between the Hot Target box and the drum plane, with respect to the total energy collected by the antenna feeders.
The linear thin gaps (about 20mm height), in correspondance of the Target walls, can be considered as antennas, so the Coupling Loss can be evaluated by measuring their radiation pattern on the whole solid angle.

The measurement of the amount of Coupling Leakage between feeders and external environment during Hot Calibration appeared a complex task, in particular for the low power level we dealed with.
With reference to Figure 9, Coupling Leakage has been measured placing a transmitting horn faced to Hot Target in a position similar to the MIMR operative one; then a receiving horn has been used to collect the leakage power over the semisphere around the Target.

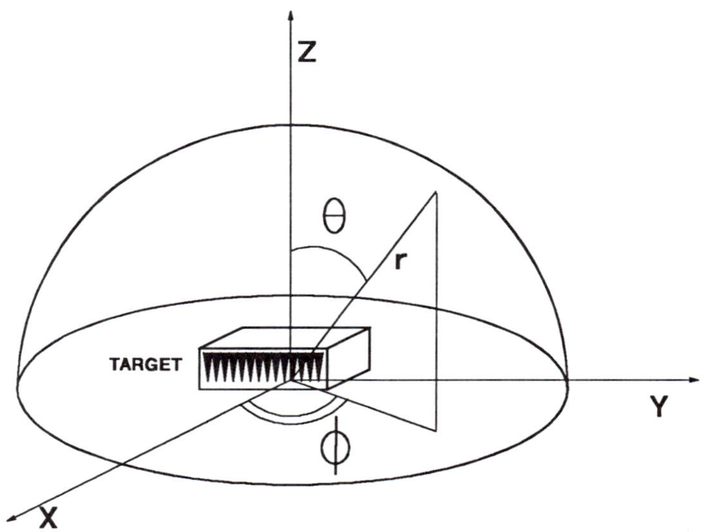

Figure 9. *Coupling Loss geometry*

Tests have been carried out at four frequencies: 6.8, 18.7, 36.5 and 89.0 Ghz.
Two different approaches have been followed depending on frequency. At 6.8 and 18.7 the total radiated power has been evaluated through power level measurements taken by sampling the semisphere in 24 points, at a distance of about 30 cm from the Target center.
At 36.5 and 89 GHz the measurement has been performed by means of field probing in proximity of the coupling slot.
The near field measurement has been performed by means of small horns, repeating the coupled power reading at adjacent locations in front of the slot, in order to fully cover the slot area.

The distance of receiving horn was selected in order to meet the required sensitivity of the receiver.
For each of the two polarizations of the transmitting feed under the Target, the measurement has been repeated for both the two polarizations of the receiving horn.
The total coupled power has been evaluated through the integration of all the measured contribution.

Test results are reported in Table 3, for a single Polarization (results for the other Polarization are substantially the same).

Table 1. Coupling Loss Test Results

Frequency (GHz)	Coupling Loss (dB)
6.8	25.1
18.7	32.2
36.5	34.6
89.0	40.4

As expected, a certain amount of leakage is present at the minimum frequency, where the feed aperture is very near the Target edge, allowing the feed to couple with external environment.
It is important to remark that Coupling Loss has not measured with the real MIMR feed cluster, so in the real case the test results could be quite different.

CONCLUSIONS

As result of performed activity a very effective Blackbody Target for In-Flight Calibration of MIMR has been designed, manufactured and tested by an European Industry.

REFERENCES

[1] J. Hollinger, 1989, *DMSP SPECIAL SENSOR MICROWAVE/IMAGE CALIBRATION/VALIDATION*, Naval Research Laboratory, Washington DC.

[2] J. P. Hollinger, J. L. Pierce, G. A. Poe, 1990, SSM/I Instrument Evaluation, *IEEE TRANSACTIONS ON GEOSCIENCE AND REMOTE SENSING*, VOL. 28, NO. 5, SEPTEMPER 1990, 781-790

ANTENNA ELECTRICAL DESIGN FOR THE MULTIFREQUENCY IMAGING MICROWAVE RADIOMETER : M.I.M.R.

S. Contu, F. M. Marinelli,
Alenia Spazio S.p.A.
Via Saccomuro, 24 - 00131 Rome , Italy

P. Rinous,
E.S.T.E.C. : European Space Research and Technology Centre
Keplerlaan, 1 - Noordwijk - The Netherlands

ABSTRACT

This paper describes the characteristics of the antenna system of the radiometer M.I.M.R. which has the objective to provide Earth and atmosphere remote sensing observations.
The guide lines followed in the design of the two antennas of the existing system are described with also the future possible developpements of it.

1. INTRODUCTION

The Multifrequency Imaging Microwave Radiometer, M.I.M.R., is an instrument being developed with the sponsorship of the European Space Agency, E.S.A., for which Alenia Spazio is the prime contractor.
In the frame of its responsability as prime contractor , Alenia Spazio has established the radiometer antenna electrical design, the related breadboarding activity to ascertain the design technical validity and to asses the design improvements to meet the requirements given by E.S.A.
The radiometer M.I.M.R. measures the brightness temperature which is related to the emissivity of the Earth surface and its atmosphere. The emissivity of the Earth depends upon dielectric properties of the surface materials and moreover atmospheric propagation effects, all the measurements and all these quantities are frequency and incidence angle dependent. By reconstituting the brightness temperature field over the Earth at selected frequencies one can retrieve geophysical parameters.
The M.I.M.R. scientific objectives are the monitoring and prediction of the global environment, the monitoring of climate and its changes, the resource managment (snow, ice, forestry), the meteorology with the prediction of convection, precipitations and surface/atmosphere interactions.
The radiometer will make image of the Earth at 6 different frequencies or channels in both horizontal and vertical polarisations as indicated in table 1. The Earth image is provided by a conical scanning antenna system positioned on a rotating drum above which are placed calibration devices and which contains in itself the instrument receivers.
The M.I.M.R. daily coverage of the Earth surface, due to its polar orbit, must be better than 92% over one day, 98.8% over two days, 99.8 after three days giving rise to a total earth coverage, with the exception of two small caps over the poles representing 0.2% of the Earth surface area.

The M.I.M.R. antenna system is composed by a main reflector and a cold calibration reflector which are illuminated in successive time periods by the same feed cluster. The main reflector and the feed cluster constitute the instrument antenna and are placed on the rotating drum in a adequate position to look toward the Earth surface. The instrument antenna rotation around a scan axis in the drum centre allows the conical scanning and brightness temperature imaging of the Earth surface in an appropriate solid angle called field of view (F.O.V).
The cold calibration reflector (C.C.R.) is fixed to the scan axis and oriented toward the cold space to permit the instrument calibration with the cold space brightness temperature of 2.7 degree Kelvin. During the cold calibration the C.C.R. obscures the main reflector with respect to the feed cluster in an adequate time period or angular range.
The feed cluster rotation around the scan axis allows also the illumination of an hot load placed above the drum (reference to figure 1). This load is constituted by a metallic box filled with absorbing material and heaters aiming to produce a temperature of 300 degree Kelvin with a behaviour similar to a black body.
With these 2 devices, C.C.R. and hot load, an accurate instrument calibration is reached through the same RF chain.
The M.I.M.R. instrument and antenna system layout is depicted in figures 1 and 2.
The feed cluster is composed by 9 feeds which illuminate both antennas, main and cold calibration reflectors, at 6 frequencies.
Channel 1 and 2 are provided by two distinct feeds. Channels 3 and 4 are derived from the same feed giving rise to two secondary patterns with the same antenna pointing.
To provide the required Earth coverage, because of the narrow secondary patterns beamwidths at 36.5 and 89 [Ghz], two feeds are present for channel 5 and four feeds for channel 6. The feed cluster layout is depicted in figure 3 which shows the feed apertures disposition around the antenna focus, whilst figure 4 and figure 5 show the feed cluster breadboard assembly.
Each feed is associated to an orthomode transducer (O.M.T.) to separate the two polarisations, with the exception of channels 3 and 4 which are equipped with a quadruplexer to separe two frequencies and two polarisations.

2. ANTENNA ELECTRICAL REQUIREMENTS

Several M.I.M.R. antenna requirements are expressed in an integral form which is the most adequate to take into account the possible error sources coming from the Earth surface monitoring.
For the main reflector the beam efficiency is defined as the ratio of power from the desired polarisation received within the region of interest to the total power received by the antenna. This beam efficiency requirement has been tightened from a value of 91% to a new value of 95% over the main beam region (2.5 x 3 dB beamwidth) and 96% over the main beam and its near sidelobe region (7.5 x 3 dB beamwidth) for all channels and all beam positions. The efficiency value over the main beam region can be obtained only with an appropriate feed and antenna geometry design.
Another integral requirement is the polarisation purity that must be better than 21 dB for the instrument antenna. This purity is expressed as the integral ratio of the crosspolar power in the angular region of interest to the total power in the same angular region. An antenna pattern with a high crosspolar component can give rise to a brightness temperature measurement error especially for some Earth surfaces and for angles out to 20 or 30 degrees from the main beam.

The two requirements above described give to the antenna designer a rather clear objective. The radiometer antenna must be designed so that sidelobes, cross-polarisation lobes and backlobes do not pick up unwanted contributions
The instrument antenna beamwidth is an other important requirement for the desired spatial resolution as depicted in table 2 and it is defined as the -3 dB pattern contour.
The spillover due to the feedhorn pattern not subtended by the main reflector must not exceed 2% of the total normalized feedhorn gain. This parameter is part of the beam efficiency loss budget and can be obtained with a carefully feed design taking into account the antenna geometry in which it is inserted.
The antenna halfpower beamwidth before described and the spillover requirements are, to some extent, in contrast one with the other and the antenna designer must find a compromise to satisfy at the same time these two specifications.
Another parameter of interest for the instrument antenna is the footprint overlap along the track coincident with the flight direction. The overlap between two consecutive footprints (due to the field of view conical scanning and the satellite motion along the orbit) at mid swath must be better than 50% at channels 5 and 6. This value has been tightened from values of 32% for channel 5 and 19% for channel 6 that were applicable for the initial M.I.M.R. design phases . This requirement influences the feed apertures layout as explained in a following paragraph.
Concerning the C.C.R., its efficiency is defined as the ratio of the antenna pattern power integral over cold space and the same integral over all space.
Even if this efficiency value depends from the antenna pattern shape, it is not a parameter that can be given without system considerations because it depends first of all from the definition of cold space and on the satellite position in its orbit around the Earth and on the cold antenna position during its rotation around the scan axis.
The efficiency of the cold reflector defined in this way must be 99.9% or better.
The calibration reflector polarisation purity must be better than 10 dB.

3. ANTENNA GEOMETRY: INSTRUMENT ANTENNA AND COLD CALIBRATION REFLECTOR

The instrument antenna is an offset parabolic reflector with a projected aperture of 1485 [mm], an offset angle of 41.23 degrees, a focal length of 1350 [mm] and a clearance of 340 [mm].
The projected aperture value has been chosen to meet the 3 dB halfpower beamwidth requirements for the different frequencies, taking into account that these beamwidths depend also on the primary patterns illuminations which in turn determine the antenna beam efficiency.
The offset angle value 41.23 degrees and the associated clearance of 340 [mm] have been chosen to allow the main reflector rotation around the scan axis without mechanical interferences with the C.C.R. placed under it in a fixed position.
Secondary pattern simulations have shown that the offset angle of 41.23 degrees does not constitute a problem to satisfy the polarisation purity requirement for the instrument antenna and does not give rise to a disymmetric secondary pattern shape. In these simulations all the crosspolarisation contributions due to reflector , feed and O.M.T. are included.
The cold calibration reflector is an offset parabolic reflector with an offset angle of 85 degrees, a projected aperture of 294.65 [mm] and a focal length of 130 [mm]. The compact cold reflector dimensions allow its positioning below the main reflector thus permitting its rotation without mechanical interferences. The calibration reflector is

positioned in such a way that its boresight is at 137 degrees with respect to the flight direction almind not to reduce the instrument field of view.
The 85 degrees offset angle permits the C.C.R. secondary pattern orientation toward the cold sky. In spite of its large value, the offset angle gives crosspolarisation levels that are not in contrast with the C.C.R. polarisation purity requirement of 10 dB.
The high offset angle gives rise to a parabola with its major axis extended in a direction perpendicular to the drum plane.

4. FEED CLUSTER DESCRIPTION

For each of the 9 feeds which constitute the feed cluster a design has been made to satisfy the beam efficiency requirements and a spillover requirement of 2%.
A feed cluster breadboard was implemented based on the previous requirement of 91% beam efficiency over the main beam region. Secondary pattern predictions, performed using the measured feed patterns, give predicted beam efficiencies in the range 93% to 95%, thus making reasonable a tightening up of the relevant requirement.
A new design of the feed cluster is being developed to cope with this new requirement with the objective of obtaining a primary pattern with most part of its energy intercepted by the main reflector and with an adequate shape to give a reduced spillover and an increased antenna beam efficiency. A primary pattern with a raised edge taper with respect to the main reflector half illumination angle gives the required spillover value with low sidelobes in the secondary pattern but also widens the antenna -3 dB halfpower beamwidth degrading the spatial resolution.
A correct compromise must be taken into account in the feed design.
Potter horns are employed at all frequencies, with the exception of 6.8 [Ghz] (channel 1) where a skyrt horn is employed to satisfy the length limitation of 300 [mm] and of 18.7 [Ghz] and 23.8 [Ghz] (channels 3 and 4) where a corrugated horn has been adopted.
In the skyrt horn a choke allows the right phase combination between the two modes TE11 and TM11 without impacts on feed length.
A corrugated horn for channels 3 and 4 has been chosen to obtain good performances through the wide frequency band.
For channels 1,2 and 5 each horn is connected with a conventional O.M.T. where the polarisations are separated by two rectangular ports in mutual orthogonal positions and by two metallic plates which improve the port isolation.
The channels 3 and 4 are separated in frequency and polarisation by a wide band O.M.T. and two diplexers which are connected to the O.M.T. by waveguide sections. The advantage of a similar configuration is the positioning flexibility for the diplexers.
At channel 6 an O.M.T. without lamina has been developed to avoid, at this high frequency (89 [Ghz]), mechanical tolerance problems. Advantage is taken from multiple reflections between the two orthogonal ports to achieve acceptable matching and isolation.
An adequate feed apertures disposition in the feed cluster allows one to obtain the required footprint overlap along track of 50% for the two adjacent beams at 36.5[Ghz] and of 50% for the four adjacent beams at 89 [Ghz]. The only way to obtain the 50 % at 89 [Ghz] is to place the 4 feed apertures in a straight line at a given angle.

5. CONCLUSIONS

The M.I.M.R. antenna design has been presented.
At this stage of the design it has been shown, through secondary pattern simulations, that a new beam efficiency requirement of 95% over the main beam region can be approached even if the feeds were designed to provide an efficiency of 91% in this angular region. For this reason a new feed cluster design activity will start to meet the new efficiency requirement.
This design activity will concern first the channel 6 (89 [Ghz]) that has the broader bandwidth, 6.7 %, with respect to the other channels. At this frequency a corrugated feed has not been taken into account because of the hazards in constructing grooves with a depth in the order of 0.84 [mm]. A pottern horn with multiflare sections will be adopted in which the length and angle of each section will be varied to maximize the antenna beam efficiency.
This 89 [Ghz] design will be scaled or adapted to the other channels with the exception of channels 3 and 4 for which a corrugated feed will be considered.
To improve the return loss at 89 [Ghz], a feed layout with 8 feeds at this frequency could be considered. Four feeds will provide one polarisation, the remaining four the other. It is possible in this way to remove the O.M.T.'s and to obtain better return loss performances with a high degree of reproducibility at this high frequency.

CHANNEL	1	2	3	4	5	6
C. FREQ. [Ghz]	6.8	10.65	18.7	23.8	36.5	89
BAND [Mhz]	200	100	200	400	1000	6000

TABLE 1: M.I.M.R CENTRE FREQUENCIES AND BANDWIDTHS.

CHANNEL	1	2	3	4	5	6
-3 dB HPBW	2.158	1.382	0.797	0.732	0.416	0.174

TABLE 2: INSTRUMENT ANTENNA BEAMWIDTHS

BIBLIOGRAPHY: F. T. Ulaby, R. K. Moore and A. K. Fung, Microwave Remote Sensing, ***(1981) Addison-Wesley Publishing Company.***

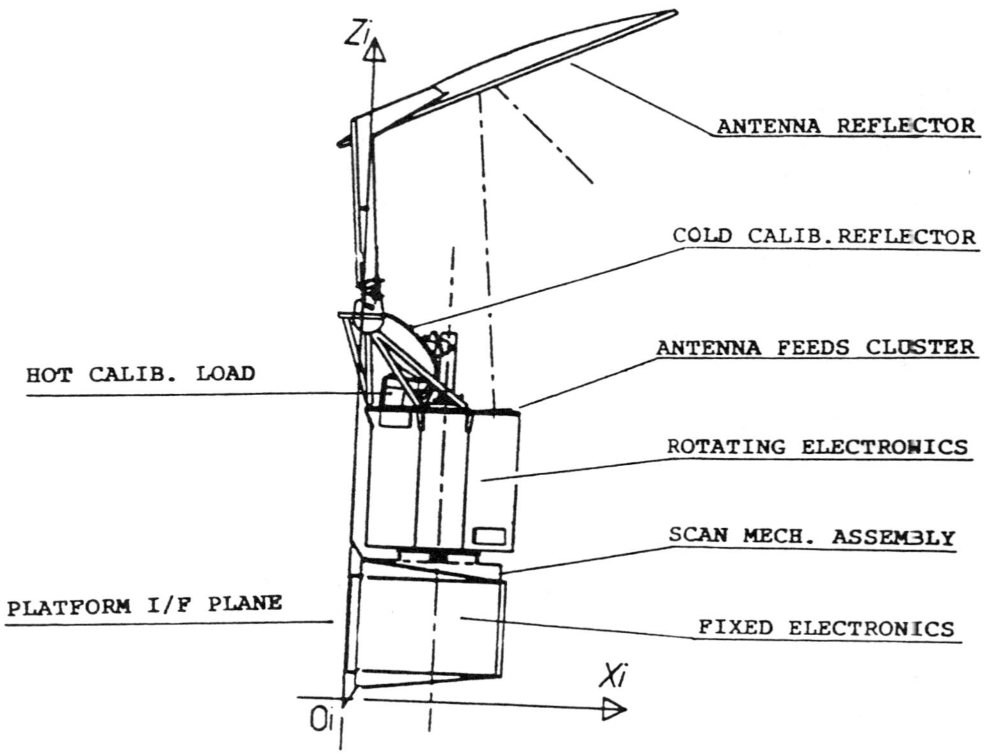

FIGURE 1

M.I.M.R. ANTENNA SYSTEM ON THE ROTATING DRUM

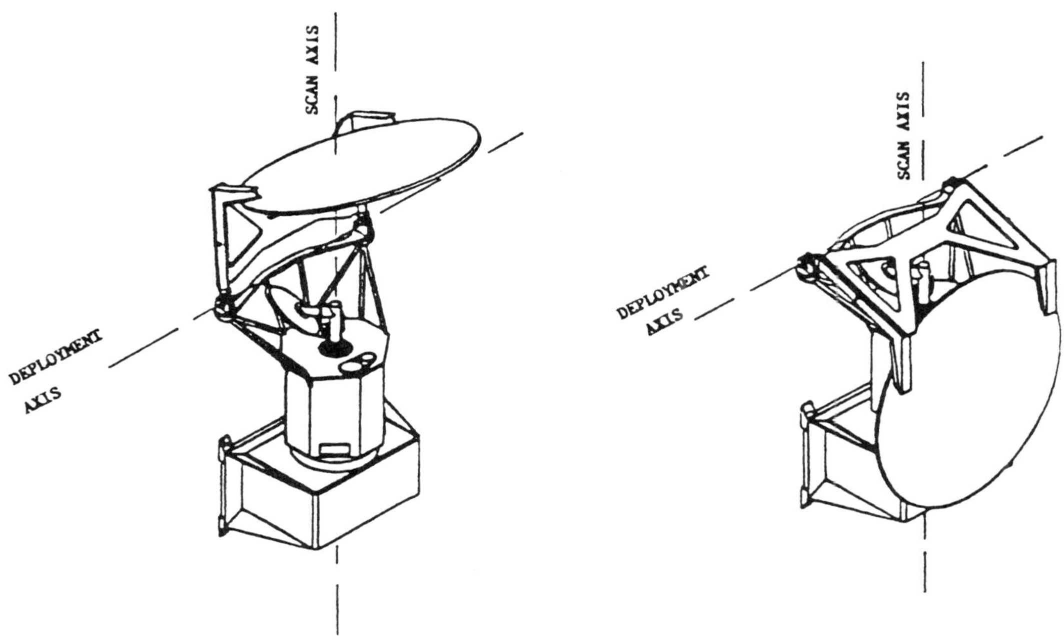

DEPLOYED STOWED

FIGURE 2

M.I.M.R. ANTENNA IN STOWED AND DEPLOYED POSITION

feed cluster layout

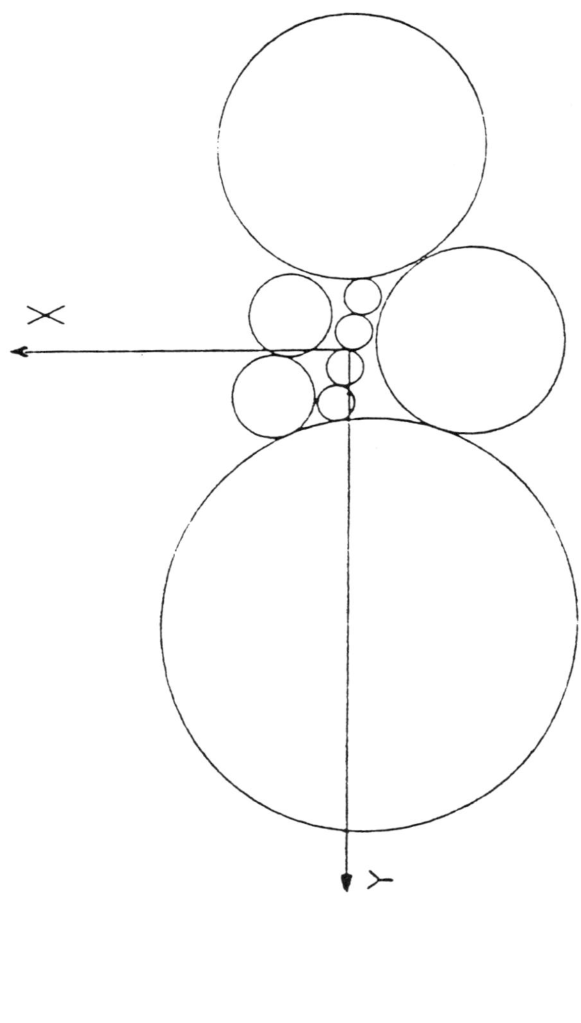

FIGURE 3

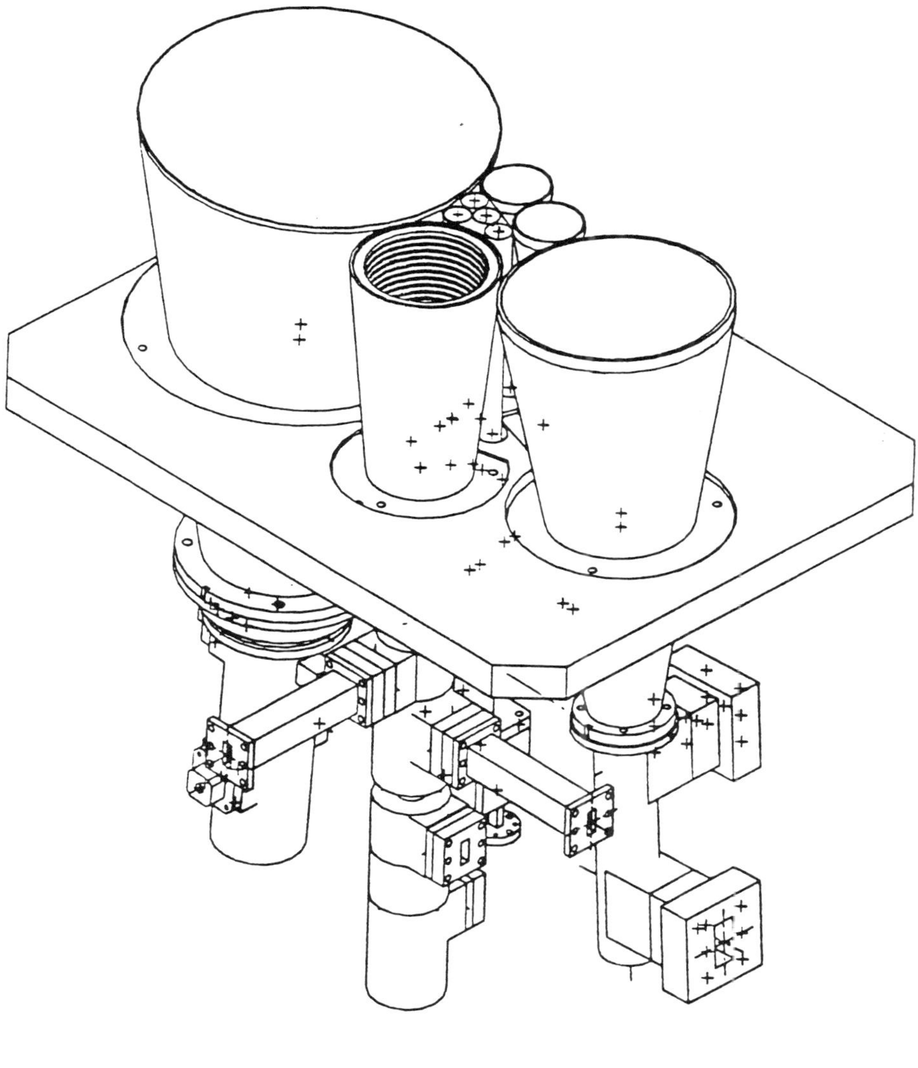

FIGURE 4 - MIMR FEED CLUSTER ASSEMBLY DRAWING

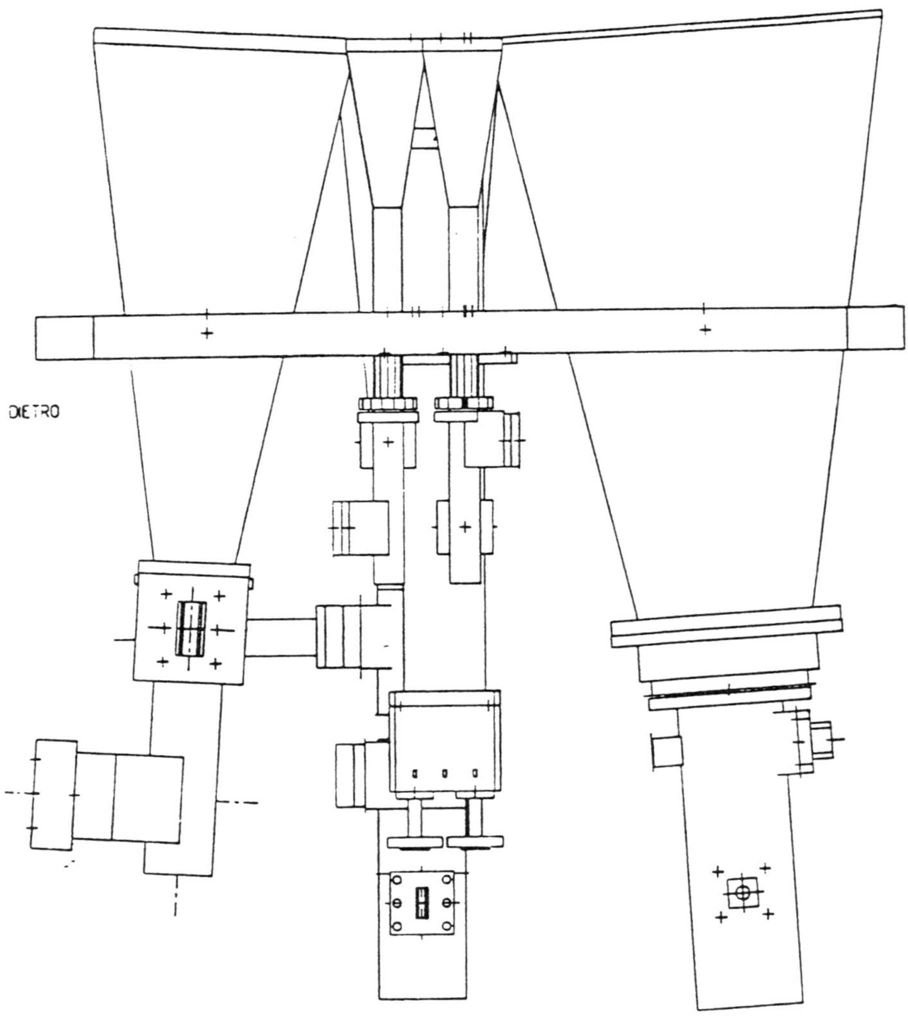

FIGURE 5 - MIMR FEED CLUSTER ASSEMBLY DRAWING

AMSU-B ANTENNA TEST RESULTS

T.J. HEWISON

Remote Sensing Instrumentation
U.K. Meteorological Office
Y70, D.R.A., Farnborough
Hants GU14 6TD, U.K.
Tel:+44-252-395781 Fax:+44-252-515523
e-mail: tjhewison@email.meto.govt.uk

Abstract. Antenna patterns of the Advanced Microwave Sounding Unit (AMSU-B) were measured on the Compact Antenna Test Range at Queen Mary & Westfield College, London. This paper presents the results of the UK Met. Office analysis of the measurements for the three AMSU-B flight models and discusses their implications to the operational calibration of AMSU-B.

Keywords: AMSU, microwave, radiometer, antenna

INTRODUCTION

The UK Meteorological Office (UKMO) has procured the high frequency component of the Advanced Microwave Sounding Unit (AMSU-B). One engineering model and three flight models of this instrument were produced under contract by British Aerospace, for use on the US meteorological polar orbiting satellites: NOAA-K,-L and -M.

AMSU-B consists of a five channel microwave radiometer. Its channels are centred on 89GHz, 150GHz, 183.31±1GHz, 183.31±3GHz, and 183.31±7GHz, known as Channels 16, 17, 18, 19 and 20 respectively. The first two of these channels fall in an atmospheric window (affected mainly by the water vapour continuum), the three high frequency channels are centred on a water vapour absorption line. These channels will allow the identification of precipitation, snow and ice, and, when used in conjunction with AMSU-A, will allow global sounding of atmospheric temperature and humidity.

British Aerospace carried out a comprehensive measurement programme of the AMSU-B antenna beam shapes on the compact antenna test range at Queen Mary and Westfield College, London. The results of the analysis carried out by the UKMO on these measurements for the three AMSU-B flight models are reported here.

AMSU-B Antenna Description

The instrument's antenna scans across its track, sampling 90 earth views within ±48.95° of nadir, each separated by 1.10°, refered to as Pixels 1 to 90. It also views an internal calibration target and space in one of four directions, which are used as radiometric calibration points. The period of one full scan is $\frac{8}{3}$s.

AMSU-B's antenna system consists of a offset paraboloidal main reflector, housed in a scanning beryllium shroud. This directs radiation on to a secondary hyperbolic mirror and into the instrument's quasi-optical system, where it is divided into three separate corrugated feed-horns by two dichroic filter plates in a series of beam folding mirrors. In this way, all channels are multiplexed onto a common beam axis and all have the same far-field beamwidth [1].

The polarisation vector rotates with antenna scan angle from "vertical" at nadir (ie. when the electric field vector is perpendicular to the direction of satellite motion) to ±50° at the scan's extremes.

AMSU-B Antenna Specification

In order to perform adequately, the antenna system of all channels of each AMSU-B model must meet a series of specifications, defined by the UKMO, based on simple criteria.

The full-width half-power beamwidth of all AMSU-B channels in any plane should be within ±10% of 1.10°. This will provide a horizontal resolution of 15km at the sub-satellite point.

The main beam efficiency must be at least 95%. This is defined as the fraction of the total that is received from the main beam (ie within a cone of half-angle equal to 1.25 times the mean beamwidth). Similarly, more than 99% of the received power must originate from a cone of half-angle of 4°. This is defined as the wide beam efficiency.

The cross-polarisation is defined as the fraction of total power within the main beam that is detected in the orthogonal polarisation. This must be less than 2% for the window channels (16 and 17) and 10% for the broadest of the channels centred on the water vapour line (Channel 20). It is unspecified for the other two channels, as these are not sensitive to polarisation dependent surface effects.

COMPACT ANTENNA TEST RANGE

The use of a far-field range is not accurate for this instrument because of the variable absorption due to atmospheric water vapour, which would affect any outdoor test range. Given the size of the antenna (21.9cm) then at 89GHz the far field is beyond 30m. Each model of AMSU-B was tested on a specially modified Compact Antenna Test Range (CATR) at Queen Mary & Westfield College, London, to verify that they comply with the above specifications. In this facility, a test source, mounted in a feed tower, irradiates a 3m aperture offset paraboloidal reflector. This forms a plane wave over the 1m³ "quiet zone", where the antenna under test is positioned. The CATR is shown in Figure 1 and described in more detail by Parini and Olver in [2].

The test source can be rotated to locate the plane of polarisation of the antenna under test and to allow its response to be measured in two orthogonal polarisations. Channel 16 of AMSU-B was tested in its lower sideband, using a 87.9GHz Gunn oscillator. A 74.15GHz Gunn with a frequency doubler was used to test Channel 17 in its lower sideband. Channel 19 was selected to test the high frequency channels using a doubled 93.15GHz Gunn oscillator in its upper sideband.

The parabolic reflector consists of 18 panels, each with a measured RMS surface accuracy of 15μm. However, there are 2-3mm gaps between these panels, which produce a diffraction pattern in the measured antenna response (see next section for treatment of these "range effects").

The instrument is mounted on an azimuth over elevation turntable in the quiet zone. It is enclosed in a tent of transparent balloon material to protect it from the dust produced by the RAM in the anechoic chamber. The instrument's antenna is parked in the view to be measured and the output from the channel under test is recorded while the turntable maps out the antenna pattern by sampling it at fixed elevation intervals in consecutive lines of azimuth scans.

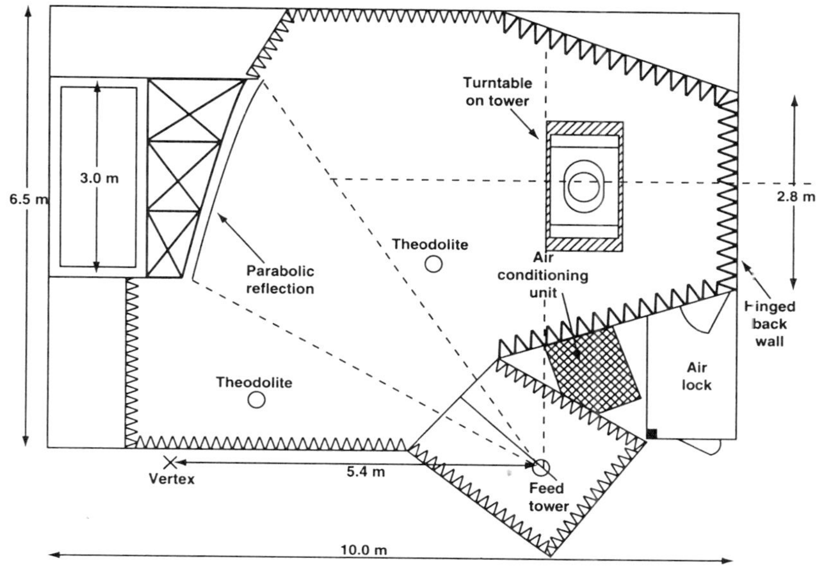

Figure 1: Plan of QMW Compact Antenna Test Range.

ANTENNA MEASUREMENTS

Although different sets of antenna patterns were measured for each model, coarse patterns were always measured in both polarisations with the antenna in nadir position (Pixel 45). These covered $\pm 35°$ in azimuth and elevation with a resolution of $0.25°$, and are used to estimate the total power received by the antenna for all antenna positions and to check no serious side-lobes are present outside the main beam.

Azimuth cuts along AMSU's scan direction were extended to $\pm 90°$ to ensure no sidelobes were present beyond the sampled antenna pattern.

Figure 2 shows the azimuth cut through the beam axis of the co-polar coarse pattern measured for the first flight model (PFM) Channel 17 at nadir. Note the apparent sidelobes at $\pm 6°$, with magnitudes of –35dB. By repeating this measurement with the instrument in a slightly different position within the quiet zone, it was confirmed that these are due to the diffraction pattern produced by the gaps in the CATR reflector.

Prior to further processing, the first order range effects are removed from the co-polar coarse antenna patterns. The positions of maxima beyond the main beam are found and linear interpolation across $2°$ wide strips, centred on these maxima is used to remove them. The result of this process on the same data-set is also shown in Figure 2.

Finer resolution patterns $(0.05°)$ were also measured, covering only a $\pm 2°$ square centred on the beam axis at Pixels 1, 45 and 90. From these, the beamwidths are calculated in two planes through the beam axis. These data are also used to calculate the power within the main beam area and the cross polarisation, where both polarisation patterns were measured.

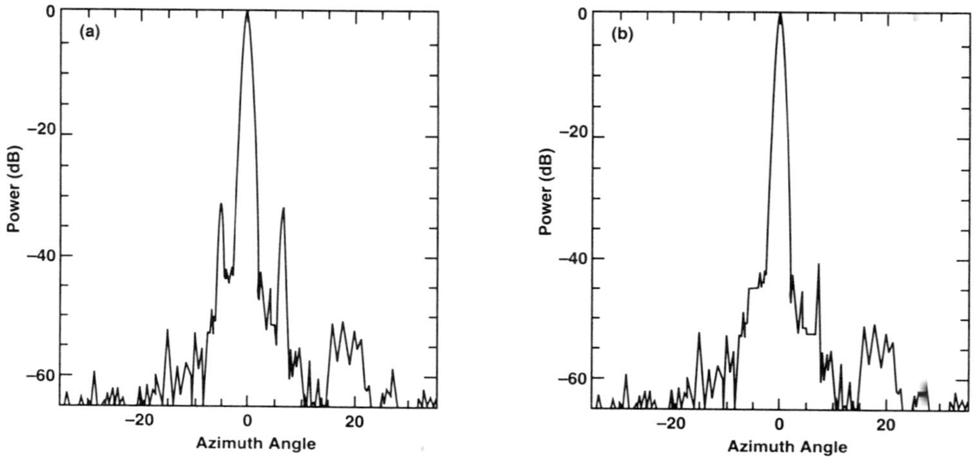

Figure 2: Antenna pattern of PFM Channel 17 Pixel 45 (a) Before and (b) After removal of Range Effects

THEORY OF CALCULATIONS

The measured radiation pattern, $F_n(\theta',\phi')$ comprises two orthogonally polarised components.

$$F_n(\theta',\phi') = F_n^{co}(\theta',\phi') + F_n^{x}(\theta',\phi') \quad (1)$$

where F_n^{co} and F_n^{x} are the co- and cross-polarised components of the pattern respectively, normalised with respect to the co-polar maximum, and θ' and ϕ' are the angles in elevation and azimuth planes of the CATR turntable with respect to the beam axis.

Note these are not the usual spherical coordinates, but are relative to the beam axis, where both θ' and ϕ' are zero. Elevation, θ' ranges from $-90°$ to $+90°$ and azimuth, ϕ' from $-180°$ to $+180°$. Hence, the elemental solid angle, $d\Omega = cos(\theta').d\theta'.d\phi'$

These radiation patterns are corrected to remove the range effects as described in the previous section.

Beamwidths

The half-power beamwidths, θ'_{bw} and ϕ'_{bw} in the elevation and azimuth planes through the beam axis are defined as the angular width between two angles at which the magnitude of $F_n(\theta',\phi')$ is equal to half its peak value.

$$F_n(\theta'_{bw}/2, 0) = F_n(0, \phi'_{bw}/2) = 0.5 \quad (2)$$

Linear interpolation of the fine antenna patterns is used to calculate the beamwidths shown in Table 1.

Beam Efficiencies

The antenna pattern solid angle, Ω_p, is defined by Ulaby [3] thus:

$$\Omega_p = \iint_{4\pi} F_n(\theta',\phi')d\Omega \quad (3)$$

In CATR coordinates, the elemental solid angle, $d\Omega = cos(\theta').d\theta'.d\phi'$

In this analysis this is approximated by:

$$\Omega_p \sim \int_{-35^\circ}^{+35^\circ} \int_{-35^\circ}^{+35^\circ} F_n(\theta', \phi').cos(\theta')d\theta' d\phi' \tag{4}$$

which can be calculated from the co- and x-polar coarse pattern data, substituting the central areas with fine patterns where available.

The main beam solid angle, $d\Omega_{MB}$ is defined as:

$$\Omega_{MB} = \iint_{MB} F_n(\theta', \phi')d\Omega \tag{5}$$

where MB is the main beam area defined as a cone of half-angle equal to 1.25 times the mean beamwidth.

This can be calculated from the fine pattern measurements and thus allows the calculation of the main beam efficiency, η_{MB}, defined as the ratio of the main beam to pattern solid angles:

$$\eta_{MB} = \frac{\Omega_{MB}}{\Omega_p} \tag{6}$$

The wide beam efficiency, η_{WB}, is similarly defined as the ratio of the wide beam solid angle, Ω_{WB}, to the pattern solid angle. This must be calculated by combining the fine and coarse patterns. It is defined as:

$$\Omega_{WB} = \iint_{-4^\circ}^{+4^\circ} F_n(\theta', \phi')d\Omega \tag{7}$$

where WB is the wide beam area, defined as a cone of half-angle equal to 4°.

The wide beam efficiency is then defined as,

$$\eta_{WB} = \frac{\Omega_{WB}}{\Omega_p} \tag{8}$$

Cross Polarisation
The cross polarisation efficiency, η_x, is defined as the ratio of the measured power in the main beam of the cross-polarisation component to the total power in the main beam :

$$\eta_x = \frac{\iint_{MB} F_n^x(\theta', \phi')d\Omega}{\Omega_{MB}} \tag{9}$$

This is calculated from fine resolution patterns measured in both polarisations when available.

RESULTS

Table 1. AMSU-B Half-power Full Beamwidths [degrees]

Channel	Plane	Pixel	PFM	FM2	FM3	Specification
		1	1.15	1.15	1.18	1.10
16	Az	45	1.10	1.12	1.13	±0.11
		90	1.13	1.13	1.17	
		1	1.11	1.11	1.15	1.10
16	El	45	1.14	1.13	1.17	±0.11
		90	1.12	1.12	1.16	
		1	1.10	0.99	0.99	1.10
17	Az	45	1.17	1.16	1.12	±0.11
		90	1.05	1.04	1.09	
		1	1.02	1.01	1.02	1.10
17	El	45	0.99	0.99	0.96	±0.11
		90	1.03	1.05	1.00	
		1	1.20	1.06	1.05	1.10
19	Az	45	1.18	1.14	1.08	±0.11
		90	1.02	0.99	1.06	
		1	1.08	1.03	1.03	1.10
19	El	45	1.07	0.96	0.97	±0.11
		90	1.05	1.08	1.00	

The measurements described in the previous section were analysed both by the UKMO and British Aerospace and the results show good agreement. The results presented here are from the UKMO analysis.

Table 1 summarises the measured beamwidths and shows that Channel 16 is consistently broader than the nominal 1.10°, especially at the scan extremes, though always within the specified limits and symmetric. The beamwidths of Channels 17 and 19 at Pixel 45 in the elevation plane are just narrower than the original specification on the later flight models but the impact on the measured brightness temperatures was calculated to be negligible.

Table 2. AMSU-B Main Beam Efficiencies [%]

Channel	Pixel	PFM	FM2	FM3	Specification
16	45	94.4	94.4	94.9	>95.0
17	45	95.1	94.9	94.9	>95.0
19	45	96.9	96.2	96.5	>95.0

Table 3. AMSU-B Wide Beam Efficiencies [%]

Channel	Pixel	PFM	FM2	FM3	Specification
16	45	98.36	98.43	98.33	>99.00
17	45	99.45	99.53	99.56	>99.00
19	45	99.64	99.15	99.15	>99.00

Tables 2 and 3 show the results of the calculation of main and wide beam efficiencies for all AMSU-B flight models. Channel 16 just fails to meet original specifications, though these were the most technically difficult to achieve, due to the broader quasi-optic beam of this channel. From the extensive knowledge of performance acquired through these tests, operational corrections may be applied to achieve acceptable results (see next section).

The Engineering Model was measured in both Pixels 1 and 45, and showed no significant difference in beam efficiency along the scan.

Table 4. AMSU-B Cross Polarisation Efficiencies [%]

Channel	Pixel	PFM	FM2	FM3	Specification
16	1	-	-	-	<2.0
	45	0.2	0.3	0.2	
	90	-	-	-	
17	1	2.4	2.3	2.3	<2.0
	45	0.5	0.8	0.6	
	90	1.0	1.1	1.3	
19	1	1.0	0.7	0.9	<10.0 for Ch.20
	45	0.5	0.3	0.2	
	90	-	1.7	1.6	

The results, in Table 4, show that all views tested meet the original cross polarisation specifications, except Channel 17 in Pixel 1. An impact analysis showed this would produce a worst case error of 0.4K at the edge of the swath which is a small uncertainty that can be tolerated.

Channel 16 of the Engineering Model was measured in both Pixels 1 and 45 and showed low cross polarisation figures in both cases.

ANTENNA TEMPERATURE PREDICTIONS

The implications of the antennas' actual performance are assessed in this section.

Theory

The antenna radiometric temperature, T_A, can be calculated by convolving the antenna pattern, $F_n(\theta', \phi')$, with an assumed far-field apparent temperature distribution, $T_{AP}(\theta', \phi')$ [3].

$$T_A = \frac{\iint_{4\pi} T_{AP}(\theta, \phi) F_n(\theta, \phi) d\Omega}{\iint_{4\pi} F_n(\theta, \phi) d\Omega} \tag{10}$$

This calculation can be used to predict the extent of the bias of the measured antenna temperature due to antenna sidelobes being exposed to contrasting scenes. Such systematic errors will affect all Earth and space views to variable degrees. T_A is estimated using the coarse resolution antenna patterns, which are available only over a $\pm 35°$ square centred at nadir. A map of the apparent temperature distribution is shown in Figure 3. This is based on a crude projection of other instrumentation onboard the NOAA-K, -L and -M satellites and the silhouette of the Earth, against the cosmic microwave background of 2.73K. The satellite instrumentation is assumed to have surfaces reflecting the Earth radiance but any secondary effects of reflected solar radiation are not included. It is important to note that the above assumes that all radiation sources are in the far field. The satellite instrumentation seen will be in the near field so the results of the calculations presented below are subject to some uncertainty due to this approximation.

It is more appropriate to perform this calculation in terms of radiances, by applying the

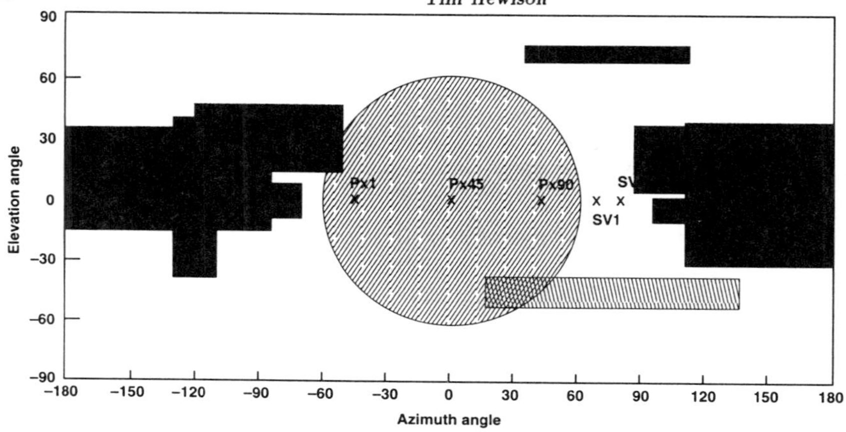

Figure 3: Map of Apparent Temperature Distribution

Planck function, $B(\nu, T)$, rather than brightness temperatures, as suggested earlier. This is because the Rayleigh-Jeans approximation breaks down for brightness temperatures of 2.73K at AMSU-B frequencies. The resulting antenna temperature is expressed in terms of temperature, by application of the inverse Planck function, $B^{-1}(\nu, T)$:

$$T_A = B_f^{-1}\left\{\frac{\int_{-35^\circ}^{+35^\circ}\int_{-35^\circ}^{+35^\circ} B_f(\nu, T_{AP}(\theta', \phi'))F_n(\theta', \phi').cos(\theta')d\theta' d\phi'}{\Omega_p}\right\} \tag{11}$$

where $B_f(\nu, T)$ and $B_f^{-1}(\nu, T)$ are the Planck function and its inverse, and ν is the channel's centre frequency.

It is convenient to write this integration as a combination of the efficiency with which the antenna detects the Earth, η_E, the satellite platform, η_P, and space, η_S. These efficiencies vary according to which Earth or space view, v, is being considered,

$$T_A = B^{-1}\left\{\eta_E(v).B(\overline{T_E}) + \eta_P(v).B(\overline{T_P}) + \eta_S(v).B(\overline{T_S})\right\} \tag{12}$$

where $\overline{T_E}$, $\overline{T_P}$ and $\overline{T_S}$ are the average apparent temperatures of the Earth, the satellite platform and space respectively.

Results of Antenna Temperature Calculation for Space and Earth Views

All views are subject to contamination due to the exposure of antenna sidelobes to the satellite platform and Earth or space. Figure 4 shows the efficiencies with which the Earth, platform and their union are detected in all four possible Space views and 90 Earth pixels, as calculated by performing the integration (11) on PFM Channel 16 Pixel 45 antenna patterns. Also shown are the exponential functions used as approximations.

Note that the third space view is the most efficient in this example. This is true for all cases. Assuming both Earth and platform radiate as black bodies at 250K. PFM channel 16 would measure the equivalent of 3.66K in space view 3 and 249.3K for pixel 90. There is a large systematic error in the estimate used of the scene apparent temperature distribution, but this may be reduced by comparing the apparent radiances in each of the four space views to produce an improved estimate of $\overline{T_S}$.

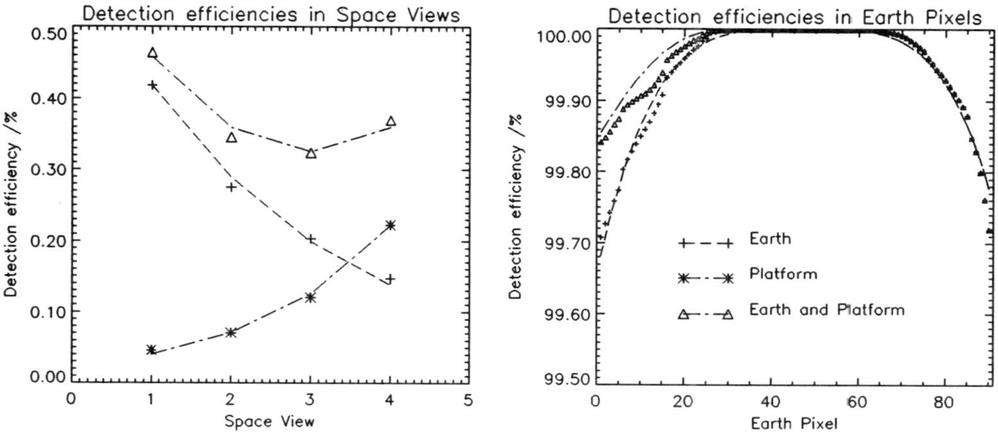

Figure 4: Detection efficiencies for Space and Earth Views

CONCLUSIONS

The antenna patterns of all AMSU-B flight models were measured on the Compact Antenna Test Range at Queen Mary & Westfield College. A selection of these measurements will be made available to the user community. All flight models performed well, meeting most of the original specifications. The main exceptions were Channel 16 not meeting the most stringent target, the main beam efficiency, and Channel 17 exceeding 2% cross polarisation at one extreme of the scan. Our extensive knowledge of AMSU-B's antenna performance based on the measurements allows corrections to be applied.

Operationally, corrections may be applied to apparent radiances to allow for the antenna sidelobes being exposed to scenes of high contrast. The radiances in the four possible space views can be compared and, combined with efficiencies calculated in this report, an estimate of the apparent temperature of the satellite platform can be inferred. This is required to produce the corrections to Earth and space view radiances.

Based on the assumptions in this report, Space View 3 is recommended for operational calibration, because it suffers least from sidelobe contamination. Ideally, a better thermal model of the spacecraft is required together with a near field analysis (to see how good the far field approximation is) to refine these calculations.

Acknowledgements
Many thanks to Bill Simpson and the other BAe and QMW staff involved in the AMSU-B antenna tests and their help in processing the data.

REFERENCES

1. R.J.Martin & W.J.Hall, 1993. Three Dimensional Design of Quasi-Optical Systems at. BAe Space Systems, *SPIE International Symposium - Aerospace and Remote Sensing*, **SPIE Volume 1935**, 178-188
2. C.G.Parini & A.D.Olver, 1992. Millimetrewave Compact Antenna Test Range, *Pro-. ceedings of ISAP '92, Sapporo, Japan*, 989-992
3. Ulaby, Moore & Fung, 1981. Microwave Remote Sensing Active and Passive Volume. I, **Addison Wesley**, 94-106 and 203-205

THE RADIOMETRIC CHARACTERISATION OF AMSU-B

R.W. SAUNDERS, T.J. HEWISON, S.J. STRINGER & N.C. ATKINSON

Remote Sensing Instrumentation
U.K. Meteorological Office
Building Y70, D.R.A.
Farnborough, Hants GU14 6TD
U.K.
tel:+44-252-395717 fax:+44-252-515523
e-mail: rwsaunders@email.meto.govt.uk

Abstract. The Advanced Microwave Sounding Unit, AMSU, is being developed to fly on the new generation of NOAA polar orbiters due to be launched in the latter half of the 1990s. The U.K. Meteorological Office (UKMO) are procuring the high frequency component of AMSU (AMSU-B) with five channels in the range 88–191GHz. To determine the radiometric performance of AMSU-B an extensive series of tests have been performed by the UKMO on the engineering and three flight models. The instruments were placed in a 3m thermal-vacuum chamber where their temperature could be controlled over the full range expected in orbit and an Earth target and a space target could be viewed. For the first flight model the measured NeΔT values were all <1.1K at the nominal instrument temperature using a 300K target. Absolute calibration accuracy and linearity in response were measured to be well within the required limits.

Keywords: AMSU, microwave, radiometer, calibration

INTRODUCTION

The next major step forward in temperature/humidity sounding of the atmosphere from space is the introduction of a new generation of microwave radiometers. They will provide improved accuracy in temperature retrievals, especially in the stratosphere, and much improved humidity retrievals especially in the lower troposphere. They will also improve the coverage obtained, by being less affected by cloud cover than the infrared sounders particularly over the extensive sheets of maritime stratocumulus cloud.

The current operational sounder TOVS (TIROS Operational Vertical Sounder), on the NOAA polar orbiting satellites, consists of a 20 channel infrared sounder (HIRS-2) for temperature (surface to 20hPa) and humidity (surface to 400 hPa), a 3 channel infrared stratospheric sounding unit (SSU) for temperature (1 – 20 hPa) and a 4 channel microwave sounding unit (MSU) for temperature retrievals through cloud up to 70 hPa. TOVS will be upgraded to the ATOVS (Advanced TOVS) system due for launch on NOAA's K, L & M in the mid 90s. The new generation of instruments which will make up the ATOVS are an infrared sounder, HIRS-3, (similar to the previous version) and a new 20 channel Advanced Microwave Sounding Unit (AMSU). AMSU is comprised of three separate instruments: AMSU-A1 & AMSU-A2, which have the 15 lower frequency channels (23.8–89.0 GHz) primarily for temperature sounding (surface to 2 hPa), and AMSU-B which has 5 higher frequency channels (89–190 GHz) primarily for humidity sounding (surface to 200 hPa). AMSU-A has a nominal field of view of 3.3° (45 km on surface at nadir) and AMSU-B a field of view of 1.1° (15 km on surface at nadir). AMSU-A and AMSU-B sample 30 and 90 Earth views respectively, covering ±48.95° from the sub-satellite point.

The U.K. Meteorological Office (UKMO) is providing 3 AMSU-B flight instruments through a co-operative agreement with NOAA. Before launch the instruments are being comprehensively tested to determine their actual antenna, spectral, thermal and radiometric characteristics. This paper describes the radiometric tests which are being carried out by the UKMO and presents the results for the first two flight models. Based on these test data, recommendations for the in-orbit calibration procedure for AMSU-B are made.

FUNCTIONAL DESCRIPTION OF AMSU-B

AMSU-B is a total power microwave radiometer with channels nominally centred at 89, 150 GHz and 3 centred on the 183.31 GHz water vapour line at 183.31±1, ±3 and ±7 GHz. More details of the spectral characteristics of AMSU-B are given in Table 1. The exact values of the central frequencies for each instrument and their variation with temperature in vacuum have been measured and are described in more detail elsewhere [1]. The passband sensitivity and limits have also been measured for each flight model [2]. Once every 2.667s AMSU-B scans through 90 Earth views, 4 space views and 4 internal black body target views. There are 4 possible viewing direction options for the space view, selectable by ground command, which view either closer to the limb of the Earth or closer to the spacecraft. During the commissioning phase of the spacecraft, investigations will be carried out to determine which of the 4 space view options give the lowest stable radiances and hence the most reliable calibration. The Earth view samples are numbered sequentially such that sample 1 is at the edge of the swath where the scan line starts and samples 45 & 46 are centred 0.55° either side of nadir. Measurements of the antenna pattern of the AMSU-B flight models have been made on a compact antenna test range and are described by an accompanying paper in this report [3].

Table 1: Nominal centre frequencies, passbands and polarization angles for AMSU-B.

Channel number designation	Centre Frequency of channel (GHz)	No. of Pass Bands	Channel Passband Limits Lower GHz	Channel Passband Limits Upper GHz	Polarization Angle†
16	89.0	2	87.6–88.6	89.4–90.4	$90-\theta$
17	150.0	2	148.6–149.6	150.4–151.4	$90-\theta$
18	183.31±1.00	2	182.06–182.56	184.06–184.56	$90-\theta$
19	183.31±3.00	2	179.81–180.81	185.81–186.81	$90-\theta$
20	183.31±7.00	2	175.31–177.31	189.31–191.31	$90-\theta$

† The polarization angle is defined as the angle from horizontal polarization (i.e. electric field vector parallel to satellite track) where θ is the angle from nadir.

The Earth view radiances are derived from the measured counts received in the spacecraft telemetry (counts are proportional to the receiver output voltage averaged over the scene integration time of 18ms) and the calibration coefficients inferred from the internal target and space view data.

The internal target is comprised of a magnesium alloy substrate with pyramidal tines (aspect ratio 4:1) covered by a 1.3mm layer of Eccosorb CR114. There are 7 two wire Platinum Resistance Thermometers (PRTs) in the internal target to monitor its temperature. The emissivity of a similar target has been radiometrically measured to be > 0.99999 [4]. The temperature of the internal target is passively controlled. It is located in a particularly benign thermal environment and is well isolated against thermal instabilities and gradients, facilitated by a large thermal inertia. It is predicted to be around 20°C in orbit.

THE U.K. MET. OFFICE CALIBRATION FACILITY

The microwave instrument calibration facility allows the radiometric performance of AMSU-B to be determined under the full range of operating conditions using a 3m thermal vacuum chamber. The facility is also used to temperature cycle the instrument in vacuum to qualify it before launch.

The 90 possible Earth views for AMSU-B are simulated by a movable black body target, hereafter referred to as the 'Earth target', which for all the tests described here was placed in a nadir view (taken as sample 46), but can be moved to be in any desired Earth viewing direction. It is temperature controlled over the range 80–330K so that temperature gradients across the surface are less than 50mK. It must also remain at a stable temperature (< 50mK variation) for longer than 5 minutes (110 scans). The Earth target temperature is measured to an absolute accuracy of better than 0.1K by regularly calibrating the PRTs with a 'standard' PRT traceable to a national standard. A cylindrical 'shroud' surrounds the target which is maintained at a temperature as close to the target as possible (always within 1K) and a 'snout' (with reflective interior) closely couples the target to the antenna. A similar black body target is used for the space view, hereafter referred to as the 'space target'. This target is maintained at around 84K for all the tests. This obviously gives a much higher space view radiance than experienced in orbit (i.e. 2.73K) and so the instrument performance in space has to be extrapolated from these measurements.

AMSU-B calibration

AMSU-B has been designed to allow a two point calibration, for each scan line, from the internal target and space views. The instrument response is assumed to be linear between these two calibration points. This allows the radiance from the antenna to be given by,

$$R_{Earth}(i) = a(i) \times (C_{Earth}(i) - C_o(i)) \qquad \mu W/(cm^{-1}.ster.m^2) \tag{1}$$

where $C_{Earth}(i)$ is the measured count for channel i whilst viewing the Earth scene. Calibration parameters (slope $a(i)$ & offset $C_o(i)$) are determined for every scan line from the internal target and space view counts. Equation 1 is written in this form because $C_o(i)$ is then directly related to receiver noise. The radiance units of $\mu W/(cm^{-1}.ster.m^2)$ are chosen as this scales to convenient values at these frequencies. The slope $a(i)$ (the reciprocal of the instrument gain) for channel i is given by:

$$a(i) = \frac{B(\nu, T'_{BB}) - B(\nu, T_{SP})}{(\overline{C}_{BB}(i) - \overline{C}_{SP}(i))} \qquad \mu W/(cm^{-1}.ster.m^2)/count \tag{2}$$

where $B(\nu, T)$ is the Planck function for a frequency ν corresponding to the channel central frequency and temperature T. The mean internal target temperature, T_{BB}, is computed from the 7 individual PRT temperatures.

For channels 19 and 20 of AMSU-B the monochromatic assumption breaks down (e.g. channel 20 spans 16GHz leading to errors of 0.4K if a frequency of 183.31GHz is assumed) and so band correction coefficients have to be applied. These coefficients modify T_{BB} to give an effective temperature T'_{BB},

$$T'_{BB} = b + c \times T_{BB} \qquad degK \tag{3}$$

which is then used in the Planck function in (2) to give an accurate radiance.

In the chamber the space view has to be simulated by a cold target placed in the space view. T_{SP} is then the mean temperature of the space target (which is also converted to effective temperature for channels 19 and 20 using (3)). $\overline{C}_{BB}(i)$ and $\overline{C}_{SP}(i)$ are mean internal target and space view counts respectively. The means are computed from the four consecutive samples which view the internal target and space view for each scan line. In addition values from scan lines before and after the current scan line can also be included in the average to reduce the random noise in the calibration views. This is achieved using a triangular function which is convolved with the mean space view counts, $\overline{C}_{SP}(t)$, and internal target counts, $\overline{C}_{BB}(t)$, for each scan line. This averages up to n scan lines ahead of and behind the current scan line, giving less weight to the lines further from the current scan line. For instance this convolution for the internal target counts can be written as,

$$\overline{C}_{BB} = \frac{1}{n+1} \sum_{j=-n}^{+n} \left(1 - \frac{|j|}{n+1}\right) \overline{C}_{BB}(t_j) \qquad \textit{counts} \tag{4}$$

where t_j is the time of the scan line before or after the current scan line, that is $t_j = t + j \times 8/3$ secs.

The zero radiance offset, $C_o(i)$, for channel i is given by,

$$C_o(i) = \overline{C}_{BB}(i) - \frac{B(\nu, T'_{BB})}{a(i)} \qquad \textit{counts} \tag{5}$$

$C_o(i)$ is related to the receiver noise (see below) and the DC offset applied to the video amplifier.

The radiance computed from (1), $R_{Earth}(i)$, can be converted into effective brightness temperature using the inverse Planck function $B^{-1}(\nu, R_{Earth})$,

$$T'_{Earth}(i) = B^{-1}(\nu, R_{Earth}(i)) \qquad K \tag{6}$$

The scene brightness temperature, $T_{Earth}(i)$, is then computed using the inverse of (3). Note that $T_{Earth}(i)$ is strictly the antenna temperature which hereafter is referred to as brightness temperature.

RADIOMETRIC TESTS

A series of tests were carried out to fully characterise the radiometric response of the instrument. Most of them were carried out at several instrument operating temperatures to characterise the instrument behaviour over the range of temperatures expected in orbit. For the purposes of these tests the 'instrument temperature' was defined as the channel 18/19/20 mixer temperature which is available in the spacecraft telemetry and so can be used to relate to the temperature in-orbit. It was found that this temperature closely followed the mean receiver temperature determined from several thermocouples placed near all three receivers on the engineering model (EM). A full set of measurements were made at instrument temperatures of 16°C, 26°C and 36°C which represent the minimum, nominal and maximum instrument temperatures expected in orbit. In addition a few radiometric checks were made at extreme instrument temperatures of 6°C and 46°C to extend the measurement range of any temperature dependent parameters.

(A) Temperature Sensitivity

The temperature sensitivity or noise equivalent temperature (NeΔT) for each channel is a measure of the minimum change in antenna temperature detectable by the receiver and is primarily a function of the input referred system noise T_{sys}. Other factors such as noise generated by the electronics, short term variations in gain and noise in the measured calibration counts also influence the sensitivity.

The sensitivity measurement is performed with the instrument at a stable temperature and the antenna scanning. The Earth target is at a temperature close to 300K. 100 scans (~4.4 minutes) of nadir Earth target counts are extracted, converted into brightness temperature (as described above) and the standard deviation of the 100 brightness temperatures is

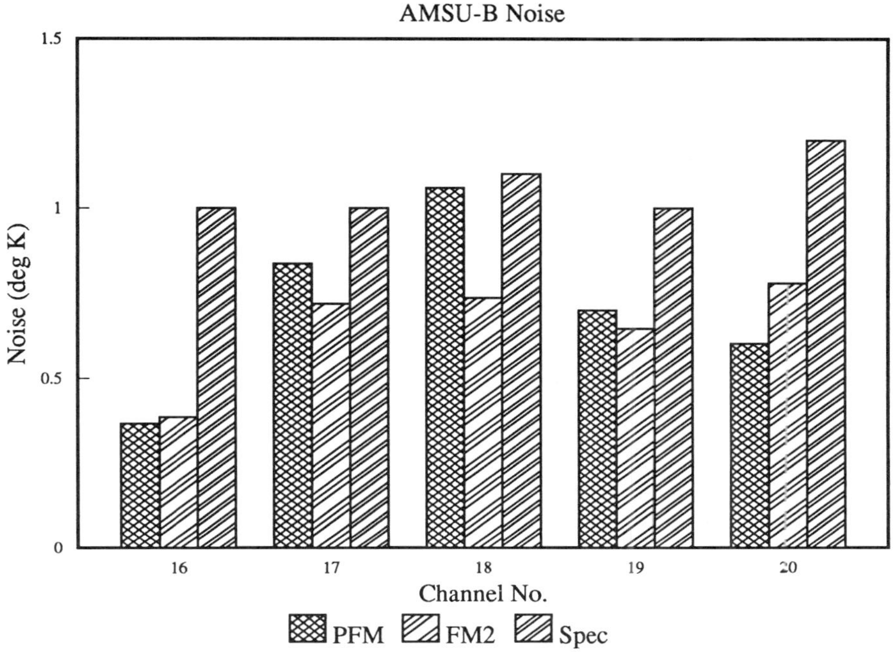

Figure 1: The measured NeΔT values for the AMSU-B PFM and FM2 flight models at 26°C. The specification for each channel is also plotted.

computed. More than ten such runs are performed and the standard deviation values from each run are averaged to give the NeΔT value for each channel. For the results presented here the calibration parameters were all averaged over 7 scan lines (i.e. $n = 3$ in equation 4).

The NeΔT results for both PFM and the second flight model, FM2, are shown in Figure 1 which also plots the specification for AMSU-B. All channels are within specification at the nominal instrument temperature of 26°C. Figure 2 demonstrates how the NeΔT values are reduced by about 0.1K when the number of scan lines included in the weighted average of the calibration parameters is increased from 1 up to 7. Beyond this, the low frequency (> 20secs) 1/f noise starts to influence the means and the NeΔT starts to rise again. An optimum value for n in (4) appears to be 2 or 3 (i.e. 5 or 7 scan lines) for AMSU-B. Note that the values plotted in Figure 1 assume a value for n of 3.

To better quantify AMSU-B's low frequency (periods > 20 secs) receiver noise characteristics a power spectrum of the radiometer's output, whilst viewing a thermally stable target, is required. High frequencies are dominated by white noise, with a uniform power spectral density. In this regime, the noise on the calibration parameters (a and C_0, defined in section 4) will decrease as the number of calibration data samples averaged increases. For a uniformly weighted average, the noise would be inversely proportional to the square root of the number of scans averaged. Lower frequencies are dominated by 1/f noise, with a power spectral density increasing rapidly as the frequency decreases, and hence the calibration noise will start to increase as the number of scans averaged exceeds a certain limit.

During AMSU-B radiometric testing, the internal and Earth targets are held in a thermally stable environment for extended periods (>1 hr). Power spectra were calculated using Fast Fourier Transform (FFT) routines on the radiometric data. Periods of 1024 scans were taken when the instrument was scanning across an Earth target held at a con-

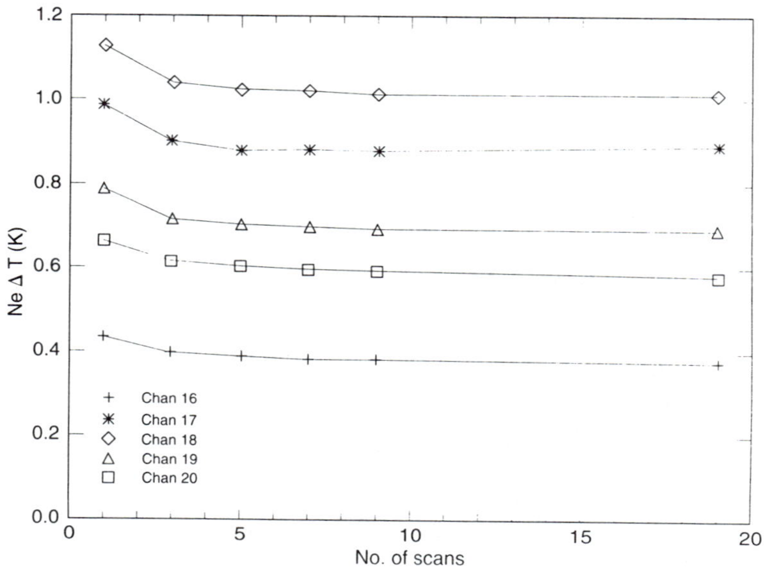

Figure 2: The variation of $Ne\Delta T$ with number of scan lines included in the average of the space view and internal target counts for the PFM.

stant temperature of around 300K. These tests showed no consistent changes in spectra with instrument operating temperature. Values related to the maximum time calibration parameters can be averaged over, before 1/f noise starts to increase the error in the average can be inferred from these power spectra. The lowest value for the maximum averaging period was 34s (13 scan lines) for channel 17. This is consistent with the small increase in channel 17 NeΔT values after averaging 19 scan lines, shown in Figure 2.

(B) Receiver Linearity

As implied by equation 2, a two point calibration is employed for AMSU-B (i.e. space and internal target). For all scenes with radiances in between these two calibration points it is assumed that the instrument response is linear. To determine to what level this assumption holds the Earth target was placed in the nadir view and radiance measurements were made for Earth target temperatures ranging from 85K to 330K (with the space target at ~84K). This encompasses the full range of Earth scene temperatures but leaves the range from 2.7K to 85K unmeasured. Ideally knowledge of linearity over this range is necessary when the instrument is operating in orbit in order to correctly relate the space view measured counts to radiance. The linearity of the individual electronic components has been measured at unit level for input signals corresponding to near zero radiance and their response was found to be linear. The increased difficulty of maintaining a target at 2.7K and making target temperature measurements which are representative of the radiating surface ruled out this option in the vacuum chamber.

The measurements of linearity were made with the antenna in normal scan mode at three instrument temperatures 16°C, 26°C and 36°C. The instrument temperature was required to vary by less than 0.1K during these measurements. The space target temperature was kept roughly constant at around 83K throughout. For each Earth target temperature at least ten 100 scan line 'runs' were made during which the temperatures of both external targets were required to vary by less than 0.05K and the gradients across the Earth and Space targets to be less than 0.1K.

Having converted the measured counts into brightness temperatures using the calibration procedure described above, the differences between the AMSU-B calculated brightness temperatures and the actual target temperatures computed from the target PRTs can then be calculated. These differences are plotted as a function of target temperature in Figure 3. This plot demonstrates the absolute calibration accuracy (see below) as well as the linearity of the receiver. The specification for AMSU-B is that the peak departure from linearity should not exceed 0.3× NeΔT (i.e. 0.3K for channels 16, 17 & 19, 0.33K for channel 18 and 0.36K for channel 20). In fact the results for the PFM shown in Figure 3 show peak deviations no greater than ±0.1K. No evidence is seen of any significant changes with instrument temperature. The channel 16 response does suggest a quadratic fit might be appropriate for this channel due to the close resemblance to the inherently non-linear properties of square law detectors but the departure from linearity is small.

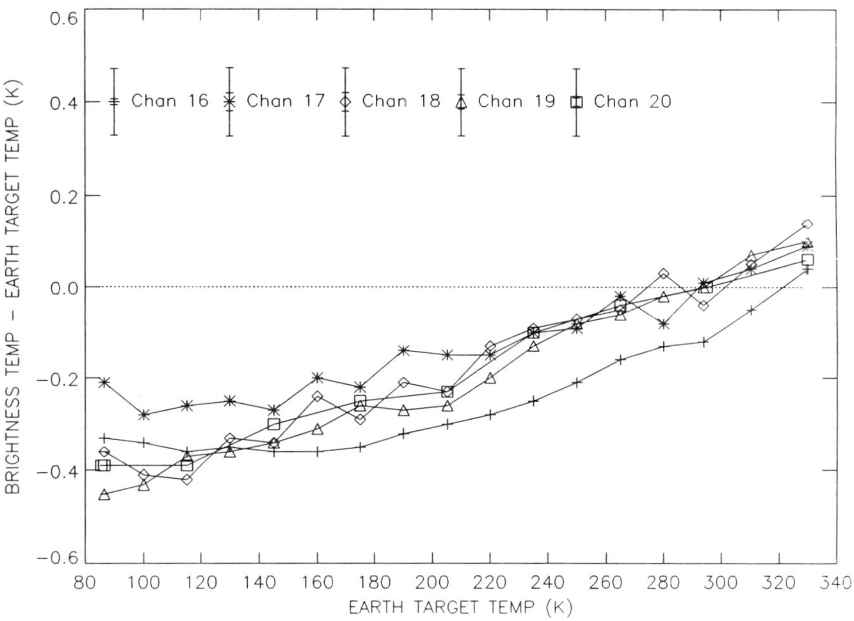

Figure 3: Difference between measured brightness temperature of the Earth target and the measured target temperatures for an instrument temperature of 26°C and a nadir view. The relative and absolute measurement uncertainties are shown by the smaller and larger error bars along the top for each channel.

(C) Absolute calibration accuracy

The absolute calibration accuracy is defined as the difference between the 'measured' brightness temperature and the actual calculated brightness temperature of a target determined from PRTs on the target and a knowledge of the target emissivity. The bias can only be estimated with confidence for the internal target calibration point (i.e. warm bias). The mean of the 100 scan line averages of nadir view brightness temperatures, when the Earth target is at the same temperature as the internal target, allows the warm bias to be determined. The standard deviation of the 100 scan line averages, which is the randomly varying component of the bias, is also computed.

An error analysis for the warm bias measurements has been carried out which suggests the absolute uncertainty in the Earth target temperature is ±0.07 K and the standard error in the mean AMSU-B brightness temperatures ranges from ±0.006K for channel 16 to ±0.028K for channels 18 & 20. Given these uncertainties there is no significant warm bias measured except for channel 16 which is 0.12K colder. This may be caused by cooler stray radiation entering around the Earth target. As the specification for AMSU-B is for the uncertainty in the mean bias to be less than ±0.98K the instrument is well within this limit. The randomly varying part of the bias for the various channels ranged from 0.04K to 0.10K. This is also well within the limit of 0.2K for AMSU-B.

Estimating the bias at the cold calibration point is more difficult as stray radiation has a bigger effect on both Earth and Space Targets. Figure 3 shows that there is an apparent cold bias in the measured brightness temperatures of up to –0.5K. The reason for this cold bias is still under investigation to determine if it is a function of the instrument or if it is caused by stray warm radiation leaking into the space target.

(D) Variation of Gain with instrument temperature

It is important to be aware of any significant changes in the instrument gain either with instrument temperature or time. During the duration of the characterisation tests the gains were continuously monitored during six thermal profiles of the instrument between 6°C and 46°C. Table 2 lists the gains (i.e. $1/a(i)$) of each channel for the five instrument temperatures during the tests on the PFM.

Table 2: AMSU-B PFM gains in units of $Counts/\mu W/(m^2.ster.cm^{-1})$.

	Instrument Temperature				
Channel	6°C	16°C	26°	36°C	46°C
16	320.9	324.8	347.1	333.4	341.3
17	95.0	94.4	92.4	89.5	82.1
18	62.83	60.94	61.31	55.26	50.73
19	60.42	56.07	54.20	49.64	45.87
20	59.54	59.24	54.40	50.34	42.15

To simulate the variations in temperature that AMSU-B will experience around an orbit a test was carried out in the chamber where the instrument temperature was forced to vary in a similar manner to that predicted by a thermal model for orbital conditions. The thermal model itself has been verified by measuring the thermal behaviour of the EM in a solar simulation chamber which was also able to simulate orbital conditions. The receiver temperature variations achieved were 0.8K around an orbit which was 0.1K greater than that predicted by the thermal model. However one orbit took 2 hours whereas the real orbital period is 100 mins. The PFM was subjected to ten orbital cycles. The effect on the gains of each channel around a simulated orbit was small (<0.4%). There was no evidence of the imposed orbital temperature variations influencing the measured brightness temperatures in any of the channels.

RECOMMENDATIONS FOR CALIBRATION OF IN-ORBIT DATA

Based on the results described above, this section outlines how the calibration coefficients for AMSU-B once in-orbit should be calculated and the way they are used for computing the required Earth view radiances. The recommendations given below are based on the PFM test results and so there may be differences for other flight models. The calibration procedure outlined above is still valid for the in-orbit data but the following points should be borne in mind.

(A) Black body temperatures and calibration counts

The test results have shown that for the calculation of the internal target temperature no significant warm bias correction has to be applied. However a cold space temperature correction factor ΔT_{SP} will be required for the cold calibration point,

$$T_{SP} = 2.73 + \Delta T_{SP} \qquad degK \tag{7}$$

where ΔT_{SP} is an estimate of the amount of radiation detected by AMSU-B in the space view in addition to the cosmic background at 2.73K. This radiation originates in the side lobes of the antenna pattern from the spacecraft and the Earth's limb. ΔT_{SP} will initially be pre-calculated for each channel and possible space view using a thermal model of the spacecraft and the measured antenna patterns [3]. The optimal values for ΔT_{SP} will need to be determined post-launch during the spacecraft commissioning phase.

As described in section 4, the noise in the calibration parameters can be reduced by applying a triangular function (4) which is convolved with the mean space ($\overline{C}_{SP}(t)$) and internal target ($\overline{C}_{BB}(t)$) counts for each scan line. The recommended value for n based on the PFM test data is 3 (i.e. a weighted average over 7 scan lines).

(B) Calibration coefficients and radiance adjustments

Having calculated the mean calibration counts and radiances the calibration coefficients $a(i)$ and $C_o(i)$ can now be computed using equations (2) and (5). The coefficients $a(i)$ and $C_o(i)$ for the scan line of interest, are then used in (1) to give the required Earth scene radiances for samples 1 to 90 of that scan line. If desired these radiances can be converted into equivalent black body brightness temperatures using (6).

A final adjustment which could be applied is to correct for cold space beyond the limb of the Earth seen in the antenna sidelobes which reduces the measured Earth radiances. This effect is largest for Earth viewing angles at the edges of the swath. Calculations using the measured antenna patterns [3] suggest in the worst case this effect reduces the brightness temperatures by 0.7K.

CONCLUSIONS

An extensive series of radiometric tests on the AMSU-B flight models has been successfully carried out in the UKMO test facility at Farnborough. The results show that the PFM is within specification in all aspects of its radiometric performance. The receiver has a linear response for scene temperatures between 85 and 330K obviating the need for a non-linear correction factor except possibly for channel 16 where a small correction could be applied. The absolute calibration of the instrument shows no significant biases for the internal target calibration point. The changes in gain with instrument temperature are relatively small (<16% over the full range of temperatures expected in orbit). Subjecting the instrument to the temperature variations expected during one orbit did not affect the measured brightness temperatures. Before the launch of NOAA-K the gains, offsets and NeΔT figures for PFM will be routinely monitored to check there are no sudden changes in the instrument performance.

The second AMSU-B flight model (FM2) exhibits characteristics similar to the PFM but with lower NeΔT values (except for channel 20). A more comprehensive paper describing these results for the PFM is in preparation. It is planned to test the AMSU-A1 engineering model in the UKMO chamber which will provide a comprehensive radiometric characterisation of that instrument and allow a better cross-calibration of both AMSU-A1 and AMSU-B which have a common channel at 89GHz. A technical document describing the recommended calibration procedure in detail for both AMSU-A and AMSU-B and giving the coefficients required, for each flight model, is in preparation by NOAA and UKMO.

Acknowledgements

British Aerospace Space Systems Ltd are the prime contractor for building the AMSU-B and were closely involved in formulating the radiometric tests to ensure the instrument met the specification. Aerojet ESD provided the receivers for AMSU-B. The efforts of both B. Greener's team (UKMO) and the D.R.A. personnel in Y60 are acknowledged in carrying out the tests in a timely fashion.

REFERENCES

1. Hewison T.J., 1992 Monitoring AMSU-B Local Oscillators. *RSI Branch Working Paper no.* **35** *1992.* (Available from Remote Sensing Instrumentation, Y70, DRA, Farnborough, Hants, UK.)
2. Jarrett M.L. and J.E. Charlton, 1993 Characterisation of the channel passband response of AMSU-B. *Microwave Instrumentation for Remote Sensing of the Earth, SPIE* **1935** 169–177; Orlando, Apr 13–14, 1993
3. Hewison T.J. 1994 AMSU-B Antenna Test Results. *Accompanying paper in this report*
4. Hewison T.J. 1991 Reflectivity tests of microwave black body targets. *RSI Branch Working Paper no.* **27** *1991.* (Available from Remote Sensing Instrumentation, Y70, DRA, Farnborough, Hants, U.K.)

A Multi-Frequency mm-wave Radiometer (MFR): instrument description and retrieval algorithm

E. BATTISTELLI[1], C. CAPITANI[1], A. CULEBRAS[2], F. DEL FRATE[3], G. SCHIAVON[3], B. ARBESSER-RASTBURG[4], J. P. V. POIARES BAPTISTA[4]
[1]*S.M.A azienda della GF-GALILEO SMA S.r.l, Florence, Italy*
[2]*R.Y.M.S.A., Madrid, Spain*
[3]*Dipartimento di Ingegneria Elettronica, Tor Vergata University, Rome, Italy*
[4]*ESA/ESTEC, Noordwijk, The Netherlands*

Abstract This paper describes a ground based Multi-Frequency mm-wave Radiometer (MFR) system designed to provide high performance measurements of atmospheric brightness temperatures in the 20-60 GHz range of the microwave spectrum.
These measurements allow the accurate retrieval of parameters of interest in earth-satellite propagation studies, space geodesy and meteorology .
In the following the key instrument features, such as radiometer stability and sensitivity, are presented, and the algorithm developed for retrieving profiles of atmospheric quantities is described.

Keywords: Microwave radiometry, Propagation, MFR.

INTRODUCTION

The Multi-Frequency mm-wave Radiometer (MFR) system has been conceived to cope with the requirements for improved performance that have been steadily increasing in the fields of application of ground-based radiometry, such as telecommunications and meteorology. This need concerns both instrumental aspects and algorithms for the retrieval of the physical parameters.
In particular satellite communication studies applied to payloads available in OLYMPUS and ITALSAT need the analysis of electromagnetic propagation at frequencies ranging from 12.5 to 50 GHz. Radiometric measurements in this range permit obtaining estimates of the atmospheric attenuation at the foreseen frequencies, of the columnar water vapor and liquid water content, and of the temperature profile.
To carry out this task, the MFR system will be operating at seven frequency channels in the 20-60 GHz range consisting of two channels in the water vapor absorption line at 22.235 GHz, one channel in the troposphere transmission window at 31.6 GHz and four channels in the lower frequency wing of the 60 GHz oxygen complex.
To fully exploit retrieval capabilities offered by the availability of seven frequency channels, algorithms for the retrieval of the atmospheric parameters have been developed at Tor Vergata University (Rome).

INSTRUMENT DESCRIPTION

The MFR system is composed by two complete modules, one for both K and Ka band and the other for the 50-60 GHz channels.
Both modules are contained in a thermally controlled enclosure mounted on a positioner for complete elevation and azimuth movement.
The whole radiometric system is strictly temperature controlled. A three stage thermostabilization, has been adopted corresponding to main housing, receiver section, and reference loads.
All critical points are also temperature monitored; temperature measurements can then be used either for diagnostic or for small correction of the radiometric output data.
The instrument is equipped with meteorological sensors for ground measurements of pressure, temperature and humidity.
Calibration philosophy foresees both the tipping curve method of measuring radiometer output versus elevation angle, and the switching of receiver input on hot/cold reference loads placed in the receiving chain following the antenna.
The system is required to operate continuously and autonomously in open air and all weather conditions.
All functions concerning system control, data acquisition and processing are fulfilled by a remote control unit.
A schematic diagram of the MFR radiometer is shown in Fig. 1.
Main design specifications are listed in table 1.

Table 1. MFR main requirements

Frequencies	22.235 +/- 0.150 GHz
	23.870 +/- 0.130 GHz
	31.650 +/- 0.150 GHz
	51.25 +/- 0.200 GHz
	52.85 +/- 0.200 GHz
	53.85 +/- 0.200 GHz
	54.85 +/- 0.200 GHz
Antenna	
HPBW	2+/-0.25 deg.
Sidelobe level	< -30 dB
Beam efficiency	90% at first null
	99% (goal) at 10° from boresight
Beam alignment	≤ 0.25deg
Radiometer	
Sensitivity for τ=1s	better than 0.5 K
Accuracy	0.5 K
Repeatability	0.5 K
Thermal control	
Main housing	300 ± 1°K
Receiver section	300 ± 0.1°K
References	280 ± 0.1°K, 360 ± 0.1°K

Antenna

The antenna subsystem consists of two offset fed reflectors, one for each module.

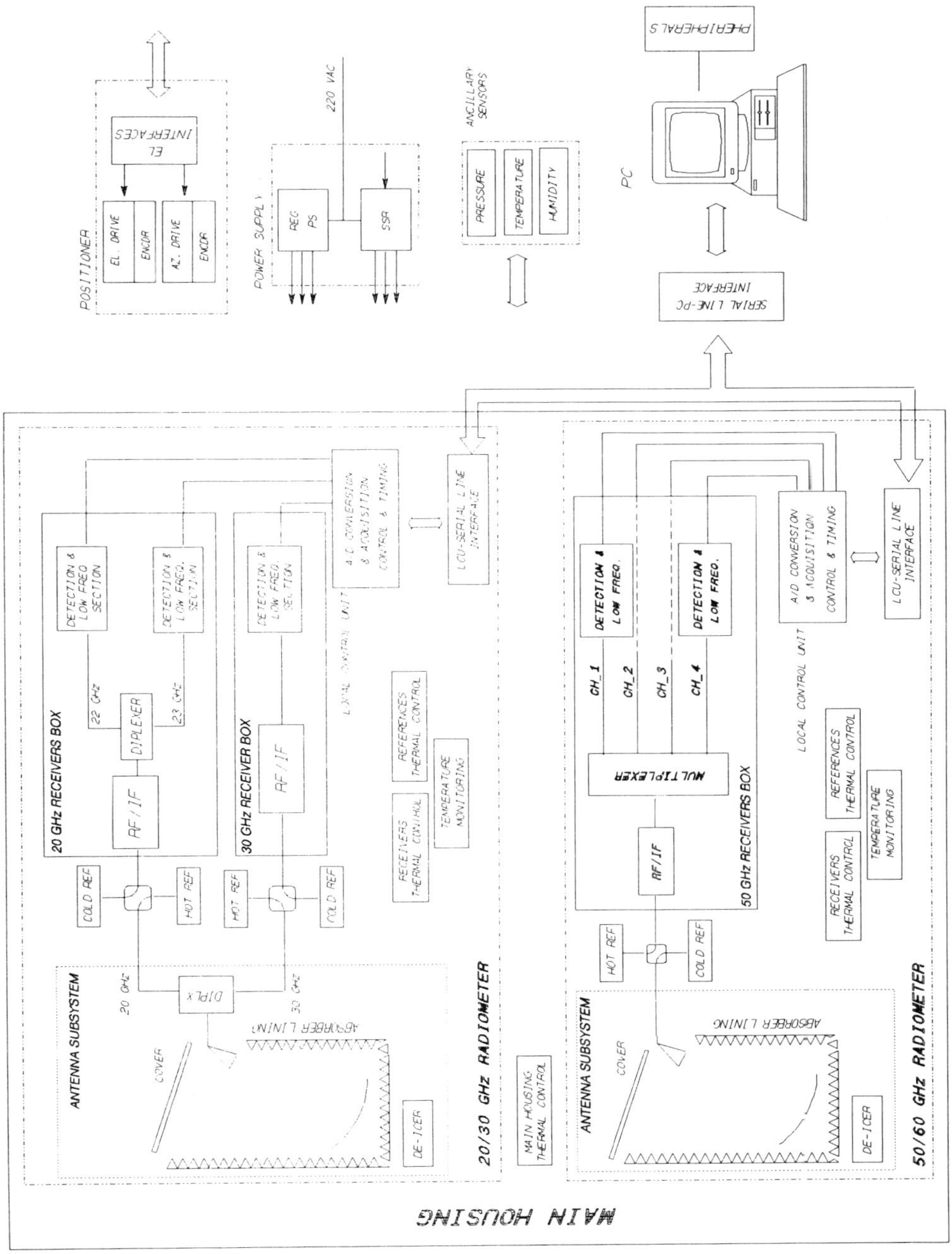

Figure 1. MFR schematic diagram

The 20/30 GHz antenna is a light weight Carbon Fibre skin reflector, while the reflector for the Q band antenna is made in aluminium. Due to the large bandwidth required, both reflectors are fed by corrugated horns.
Band separation for the antenna operating at 20/30 GHz is accomplished by means of a waveguide frequency diplexer.
The most important parameter for characterizing antenna performance is the main beam efficiency, defined as the percentage of energy received within the main beam of the radiation pattern to the energy in the whole pattern. The far field radiation pattern is the combination of the radiation patterns of both the feed-horn antenna and the reflector.
It can be seen (Ref.1) that a low beam efficiency results in a degradation of radiometric accuracy. The effect of an efficiency lower than 1 is not only the reduction of the power entering the antenna, but also the increased contribution of extraneous sources, mainly radiation from the scene averaged on the secondary lobes of the antenna pattern and on the spillover lobes of the horn (i.e. at angles higher than the illumination one). A reduction of this effect can be achieved by means of a careful design of the antenna system, with a particular care to spillover. Feed spillover is reduced by placing a thermally controlled wall of absorbing material around the reflector periphery.
Table 2 presents antenna main features as a function of channel frequency.

Table 2. Antenna e/m features

Frequency (GHz)	Beamwidth (deg)	First sidelobe level (dB)	Beam efficiency at 6° from boresight (%)
22.235	1.90	-30.80	95
23.87	1.95	-30.40	
31.65	2.12	-30.40	
51.25	2.05	-34.8	96
52.85	2.02	-36.0	
53.85	2.00	-36.3	
54.85	1.97	-36.8	

Receivers

The receiving section is based on the balanced Dicke principle. Balance is accomplished using Automatic Gain Control (AGC) techniques (Refs. 2, 3). In the scheme proposed by Hach, the receiver input, during the half Dicke cycle connected to the reference channel, is further switched between two reference sources at different temperature. If these sources are highly stable, the radiometer response is made insensitive to variation in system gain and receiver noise temperature.
Single side band (SSB) heterodyne configuration is adopted for the K and Q bands receivers; channels separation is done at IF level. The drawback of wide band RF components in the front end assembly, is recovered at 20 GHz using a RF low noise amplifier, common to both channels, as front end. For the four Q band channels the degradation in sensitivity is overcome by a more compact and efficient thermal/mechanical design. The Ka channel was designed for double side band (DSB) operation.
Figs. 2 through 4 show plots of the theoretical radiometric sensitivity versus antenna temperature. These curves are sensitive to the values of the noise temperatures of the two reference loads. A good balance (best sensitivity) is obtained for temperatures chosen according to the scene brightness temperatures.

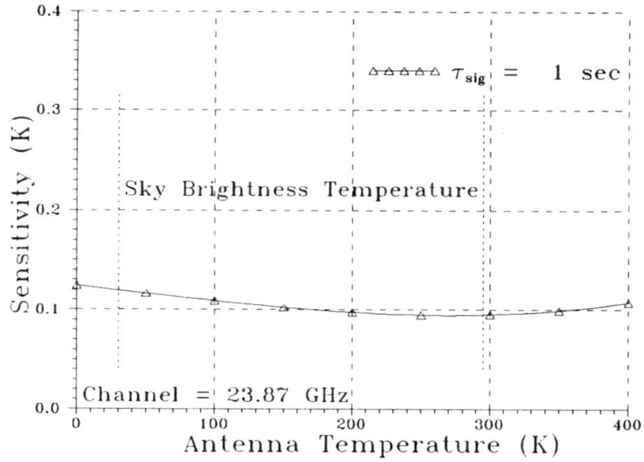

Figure 2. Radiometric sensitivity for one channel in the K band

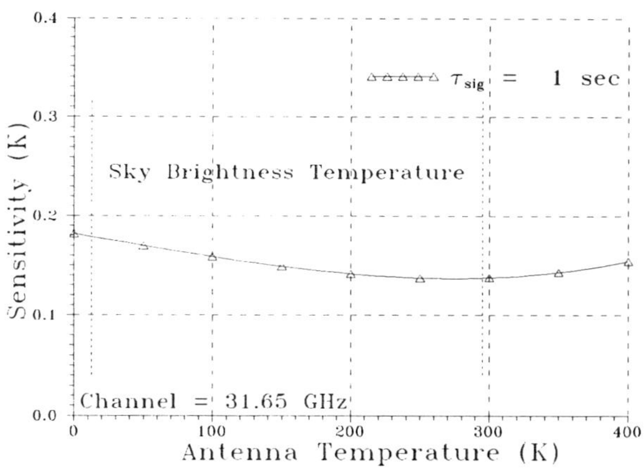

Figure 3. Radiometric sensitivity for the Ka band

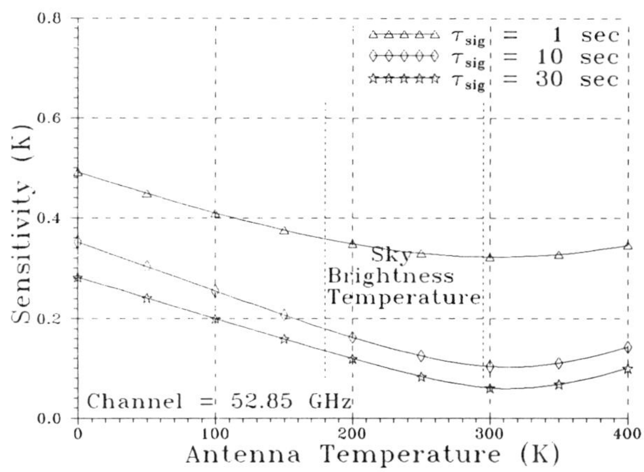

Figure 4. Radiometric sensitivity for one channel in the Q band

Expected sky noise temperatures range from a few tens of Kelvin up to 300 Kelvin across the seven MFR frequency channels, low frequency to high frequency respectively.
This wide dynamic range made necessary a tradeoff between the balance requirements and the feasibility of a tight thermal control of the reference sources.
The temperatures selected for the two AGC reference loads are 280 K and 360 K respectively.
The stability of AGC receivers is mainly related to the thermal stability of the reference loads and to the stability of DC processing circuitry (demodulators, AGC amplifiers and final filtering).
Reference loads temperature is strictly controlled and continuously monitored. This implies that the residual overall stability is equal to the total drift of the monitoring circuit. This drift is evaluated to be less than 0.1°C.
Also DC processing section is temperature controlled; its stability is related to the aging of electronic components

RETRIEVAL ALGORITHM

The algorithm developed to be used with the radiometer described in the preceding sections aims at retrieving profiles of the atmospheric quantities, namely temperature, water vapor and cloud liquid. Once these profiles have been retrieved, the use of a suitable atmospheric propagation model allows the evaluation of propagation parameters too.
Fig.5 shows a block diagram of the developed algorithm. The measured brightness temperature and the ground measurements are fed, on one hand, to a linear estimator to evaluate integrated vapor and liquid and mean temperature. On the other hand, they are used, in combination with these last integrated quantities, to retrieve the profiles of temperature, vapor and cloud liquid. The attenuation and path delays at the frequencies of interest are subsequently evaluated by means of Liebe's MPM propagation model (Ref. 4).

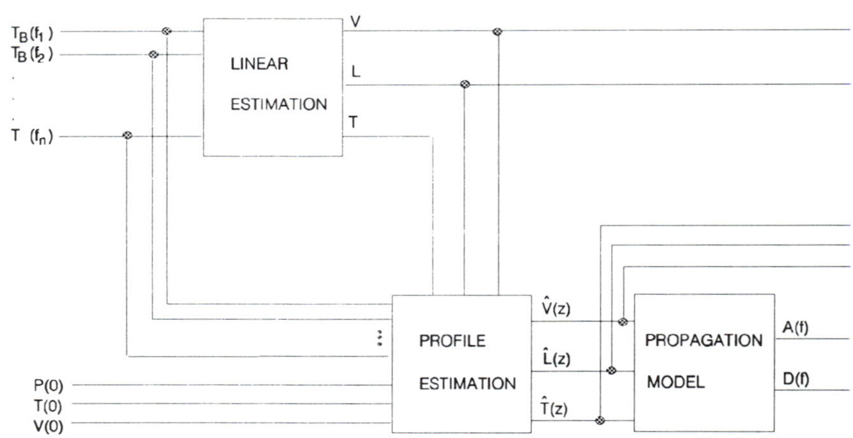

Figure 5 . Block diagram of the developed algorithm

The retrieval method is based on the expansion of the atmospheric profiles on a basis of natural orthogonal functions (Ref. 5). These functions, which are the eigenfunctions of the covariance function of the profiles, can be evaluated from a statistical set of atmospheric realizations.

A linear statistical regression was used to relate the coefficients of the expansion on the natural orthogonal function basis to the brightness temperatures.
As an example of results Fig.6 reports a sample of retrieved profiles of temperature, water vapor and cloud liquid. They compare well with those recovered by other techniques.

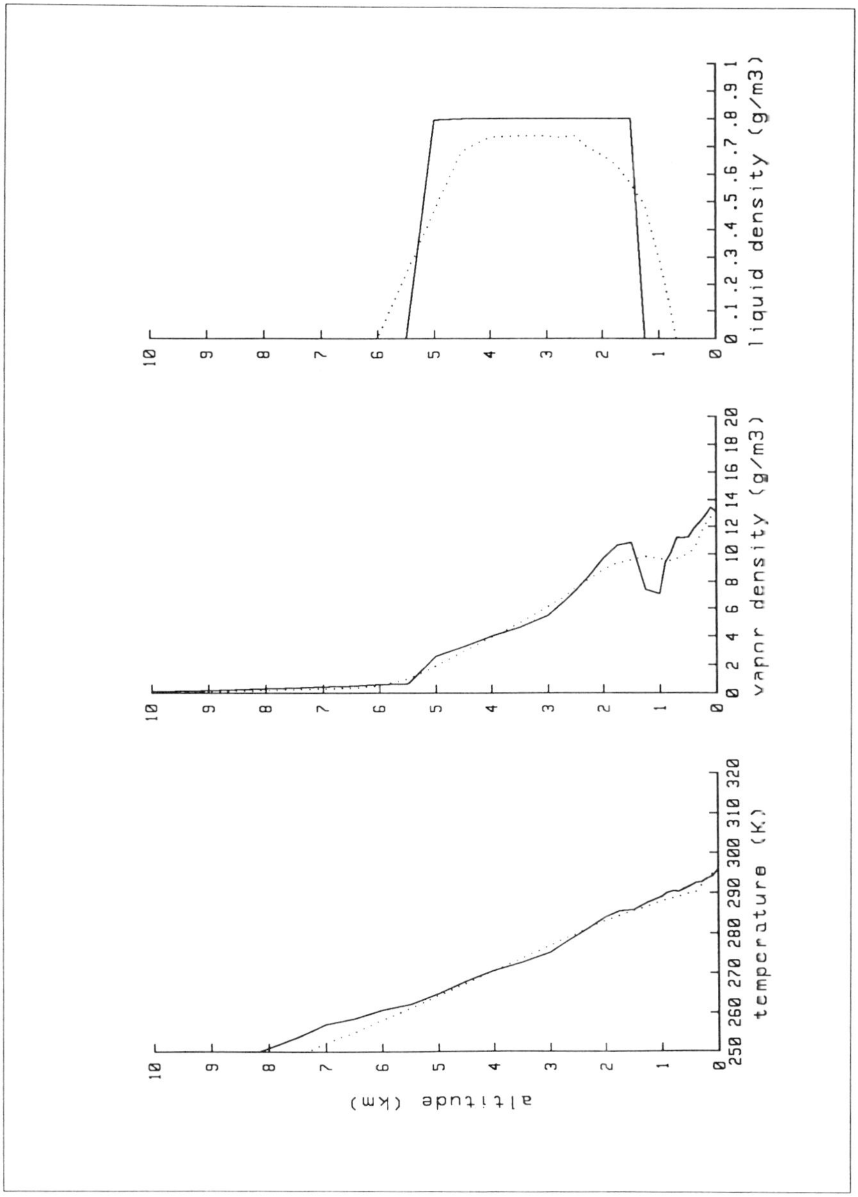

Figure 6. Sample of actual (solid line) and retrieved (dotted line) profiles of temperature, vapor and liquid

CONCLUSION

The radiometric system described in this paper is presently under manufacturing and testing at SMA and RYMSA. It is expected to be fully operative by the end of next summer.

Acknowledgment
The work presented in this paper is being carried out under ESA contract 9960/92/NL/GS(SC).

REFERENCES

1. Battistelli E. et al. 1990, *Study of multifrequency mm-wave radiometer, Final Report* ESA contract # 8809/90/NL/PM(SC), pp. 110-116

2. Hach J.P., 1968, A very sensitive airborne microwave radiometer using two reference temperatures, IEEE Trans. on Microwave Theory and Techniques, Vol MTT-16, 9

3. Ulaby.F. T., Moore R. K., Fung A. K., Microwave remote sensing, Addison-Wesley, 1981

4. Liebe H. J., Hufford G. A., Cotton M. G., Propagation modeling of moist air and suspended water/ice particles at frequencies below 1000 GHz, AGARD 52nd Specialist Meeting of the Electromagnetic Wave Propagation Panel, Spain, pp.3-1--3-10, 1993

5. Obukhov A. M., The statistically orthogonal expansion of empirical functions, Izvestiya Acad. Sci. USSR, Geophys. Sci., pp.288-291, 1960.

New Designs for Portable Microwave Temperature Profilers and for Water Vapor Profilers

FREDRICK SOLHEIM[1] and STEPHEN J. KEIHM[2]

[1] Radiometrics Corporation,
2760 29th Street, Boulder, CO USA 80301

[2] Ball Corporation, under contract to JPL
California Institute of Technology
Pasadena, CA

Abstract A temperature profiling radiometer based on a highly stable synthesizer, rather than traditionally used Gunn oscillators, has been designed and fabricated under contract to White Sands Missile Range. This profiler is capable of tuning a user-selected ensemble of frequencies in the range of 52.85 to 58.8 GHz as well as fixed water vapor and water channels at 23.8 and 31.4 GHz. This instrument weighs about 30 kg and consumes about 200 watts. Because of the frequency stability of the receiver, and because of the large number of observing frequencies possible, improved profile accuracy and resolution is expected over Gunn-based temperature profilers.

An inexpensive, simple, and accurate calibration target system that includes the antenna system has been developed for this profiler. The calibration target can be loaded with liquid nitrogen or other cryogenic liquids. Tests with a Radiometrics water vapor radiometer have demonstrated stability of 0.1K over 6 hours. The cryogenic liquid temperature can be known by its boiling point to within several hundredths of a Kelvin by a simple barometric pressure measurement.

Preliminary design work has also been accomplished on a portable (35 kg) water vapor profiling radiometer. This radiometer is to utilize a stable synthesizer to map pressure broadening by tuning across the 22 GHz water vapor line. Based on prior theoretical instrument analysis of R. Peter and N. Kämpfer[1] of the University of Bern, we expect accuracies of better than 0.8 mm total water vapor and 0.04 mm total liquid water, with correspondingly good accuracies along the profiles.

Key words: *water vapor, profiling, calibration, temperature profiles*

Introduction

Microwave radiometers provide a means of passively and continuously measuring atmospheric parameters remotely. Satellite, airborne, and ground based microwave radiometers are increasingly being used to remotely and continuously measure atmospheric temperature profiles.

Temperature profiling can be accomplished utilizing the oxygen resonance lines at 60 GHz (Westwater[2]). The oxygen line at 59.4 GHz is actually a superposition of a fine structure of about 40 lines, nearly symmetrically (but not evenly) spaced from the line center. These lines are pressure broadened into one feature at sea level.

The density of atmospheric oxygen with pressure height is quite constant in time. The zenith attenuation of oxygen on the wing of the 60 oxygen resonance feature ranges from about 1 dB (around 50 GHz), to about 100 dB at the line center (59.4 GHz). Because of this dependence of absorption upon frequency, and because of the predictable concentration of atmospheric oxygen, this atmospheric resonance feature is useable for temperature profiling. Superposed across this band is a monotonically increasing brightness continuum due to any water and water vapor present in the atmosphere.

Ground-based sky brightness observed near the central peak of this group (near 60 GHz) is due to emissions in the lower atmosphere, and brightness at the 50 GHz wing of this band tends to originate at higher altitudes. This results in surface observations of emissions that are weighted with altitude. By appropriately inverting observations at frequencies within the 50 to 60 GHz waveband, and correcting for water and water vapor emissions, the temperature profile of the atmosphere can be inferred.

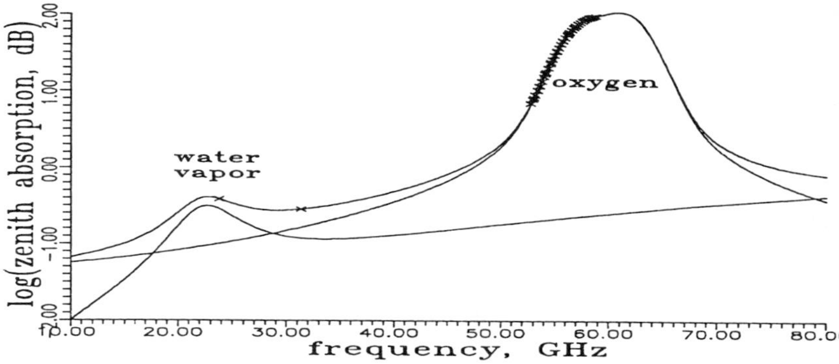

Figure 1. Zenith atmospheric absorption. The atmospheric spectrum from 10 to 80 GHz is dominated by the water vapor line at 22.235 GHz and the oxygen feature at 60 GHz. The MTP described herein receives at the fixed frequencies 23.8 and 31.4 GHz, and can tune in .25 MHz steps across 52.8 to 58.8 GHz.

Microwave temperature profiler error sources

1. Instrument errors include uncertainties in calibration, pointing, and the effects of beam smearing. Beam pointing errors can be effectively diminished with care in construction of the radiometer and setup of the instrument. Finite beam effects can be modelled out if the antenna pattern is known (Lundquist and Potash[3]; Jordan[4]). For zenith observations, for which the variations of brightness temperature with elevation are minimum, neither of these effects are large relative to other error sources. Calibration errors, including calibration target temperatures, receiver non-linearities, and receiver frequency drift, can be significant in current radiometer designs.

Current technology in ground-based microwave temperature profiling radiometers (MTPs) include the use of Gunn diode local oscillators in the microwave receivers. Receiver gain is commonly determined with Dicke-switched hot and ambient loads in the antenna system. Because sky brightness observations are made on the sloping shoulder and wings of the 60 GHz oxygen feature, fluctuations in the receiver frequency shift the observation up or down the line and can produce significant errors in the retrieved profile. The frequency excursions of a free-running Gunn are shown in figure 2.

These excursions are as great as 10 MHz peak-to-peak. The effect of such 10 MHz shifts upon the retrieved temperature profile are shown for the NOAA 4 channel MTP in figure 3. Note that the 53.85 channel has the greatest effect upon profile errors; the slope of the oxygen line is greatest at this frequency. Near the peak of the oxygen feature at 58.8 GHz the receiver is insensitive to frequency drifts. Because we are using a synthesizer in the MTP, rather than Gunn oscillators, frequency accuracy and stability is maintained to several KHz, eliminating any significant profile error due to frequency drift.

The most common calibration scheme for current MTPs is the use of hot (400-500 K) and ambient (~300 K) loads for gain monitoring. Errors resulting from uncertainties in target temperatures and receiver non-linearity are amplified for the lowest antenna temperatures (~100 K) which fall far below the calibration target range. In our present design a cryogenic cold target calibration is implemented which effectively spans the range of expected antenna temperatures. This calibration scheme, and its importance for the reduction of profiling errors, is discussed in detail in the next section.

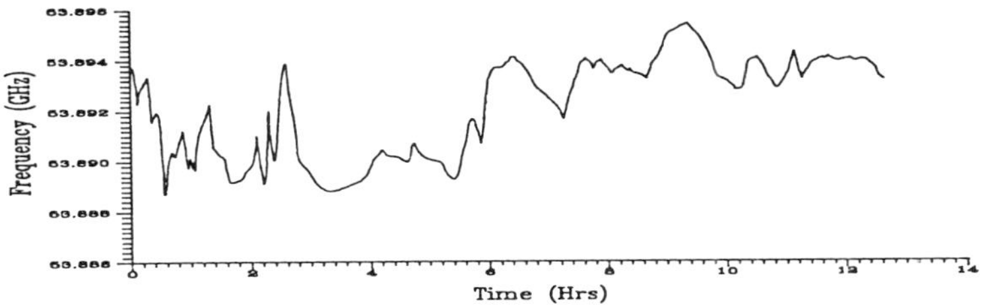

Figure 2. Frequency excursion over time of a temperature stabilized 53.85 GHz Gunn diode oscillator over 12 hours. Note the 7 MHz p-p drift in this example. This drift would result in as much as 0.6K shift in the inferred temperature profile at midlevels. From James Jordan, NOAA/WPL.

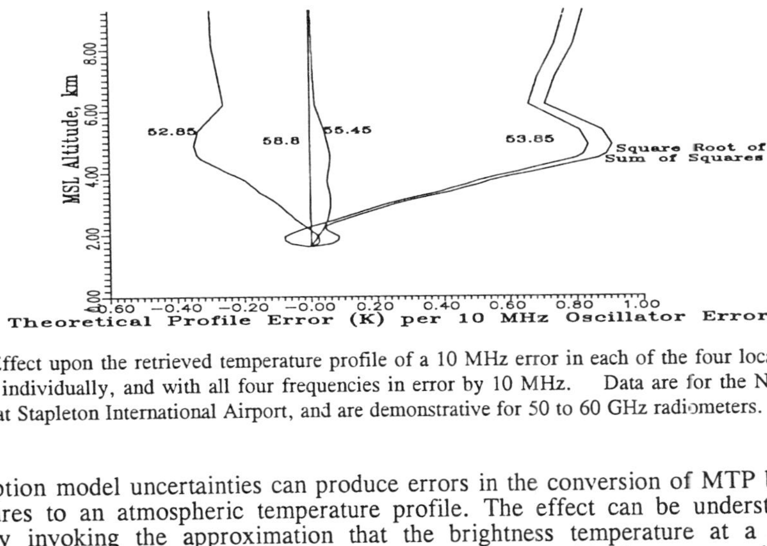

Figure 3. Effect upon the retrieved temperature profile of a 10 MHz error in each of the four local oscillator frequencies individually, and with all four frequencies in error by 10 MHz. Data are for the NOAA/WPL instrument at Stapleton International Airport, and are demonstrative for 50 to 60 GHz radiometers.

2. Absorption model uncertainties can produce errors in the conversion of MTP brightness temperatures to an atmospheric temperature profile. The effect can be understood most clearly by invoking the approximation that the brightness temperature at a particular frequency is representative of the physical temperature at an altitude corresponding to the 1/e folding distance produce by the oxygen absorption. Thus, for example, at 54 GHz, which effectively measures the temperature at ~ 2 km altitude, a 10% absorption model error would produce a 200 m error in the assigned altitude. For a 7 K/km nominal lapse rate, this corresponds to a 1.4 K error in the retrieved temperature.

The retrieval algorithm model we have adopted for oxygen absorption is that of Liebe et al.[5] which is based on extensive laboratory measurements. The model reproduces the laboratory measurements with accuracies better than 2% for lower troposphere conditions over the 50-70 GHz spectral range. At higher altitudes, where the lines are resolved, the absolute model accuracies are at the 0.01 neper/km level. These accuracies are sufficient to restrict the resultant profiling errors to 0.3 K or less which is insignificant relative to the inherent retrieval algorithm errors discussed below.

3. Retrieval algorithm errors result due to limits on the vertical resolution of any MTP observable set and the non-unique mapping of brightness temperatures to the atmospheric temperature distribution. The number of receiver channels desired for optimum temperature profiling using frequencies in the 50-60 GHz band is a tradeoff between accuracy of the profile (more frequencies) and simplicity/ economy (fewer frequencies). Diminishing returns are reached because adjacent observations tend to be correlated, especially if observational errors are large. Measure et al.[6] find from eigenvalue analysis (after Twomey[7]) that, at most, 5-7 frequency measurements are independent. NOAA/WPL researchers (Schroeder et al.[8]) find from their models that more than four channels is essentially redundant in the context of their instrument design. J. Askne[9] of Chalmers University of Technology at Goteberg, however, finds that 11 channels yield a better result than 4 channels.

Our current MTP instrument allows observation at any of 6000 frequencies between 52.8 and 58.8 GHz (1 MHz tuning stepsize). We utilize the 11 "windows" in this bandwidth that fall between the individual oxygen lines in the 60 GHz feature. Observations at each channel can be scanned in elevation which yields some improvement in retrieval accuracy in the lower 300 meters. We are currently conducting field tests with co-located radiosondes to optimize channel selection for best performance. It is clear that both instrument calibration and retrieval algorithm uncertainties dominate the temperature profiling error budget. In terms of frequency selection, the quality of measurement is as important as the number of channels. A higher number of channels will not improve the accuracy of the inversion unless calibration errors are reduced below the current design limits. Minimizing the instrument error sources effectively decorrelates observations adjacent in frequency and elevation angle, allowing improved profiles from a greater number of observations.

The cryogenic cold target calibration system

Because brightness temperature is the fundamental radiometer measurement, the most meaningful and direct calibration is to input radiation from sources of precisely known temperature into the radiometer system, and record the radiometer output parameter. Several variations of this method are implemented in radiometry; most implementations have deficiencies.

Unlike radiometers operating in less opaque wavebands, MTP radiometers cannot utilize tipping curves for calibration. Hot load/ambient load calibration systems are generally implemented; this method requires extrapolation to the range of sky observables (100 to 300 Kelvins) As shown in figure 4, receiver nonlinearity due to signal compression and non square law behavior of the detector can thus induce calibration errors. Additionally, this calibration method is incomplete in that these references are switched into the radiometer system behind, and therefore exclude, the antenna system. We have therefore developed a cold target/ambient target system that spans and closely brackets the temperature range of the sky observables, and includes the antenna system.

Errors in the calibration of the 4 channel radiometer manifest themselves as temperature profile errors as shown in figure 5 below. These errors are based on Denver retrievals.

Two requirements of radiometric target calibration are that the blackbody temperature of the target be accurately known, and that antenna sidelobe signal be effectively accommodated or modeled. This target accomplishes both.

The target consists of a Styrofoam container into which open cell convoluted blackbody foam is placed and immersed in liquid nitrogen. The target is viewed from underneath, upward through the Styrofoam container. The open cell structure allows bubbles from the liquid nitrogen to pass through and avoid target scintillation. Convolutions capture reflections from the surface of the foam.

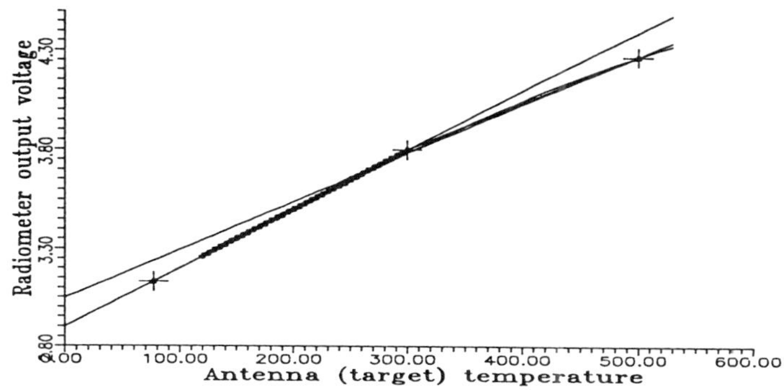

Figure 4. Comparison of hot load/cold calibration to ambient load/cryogenic load calibration. Hot/cold calibration points are shown at 300 and 500K. Receiver nonlinearity induces an error in the range of sky observables (shown as 120 to 300K) when this calibration is extrapolated. The cryogenic target at 77K allows a calibration that spans and closely brackets the sky observables.

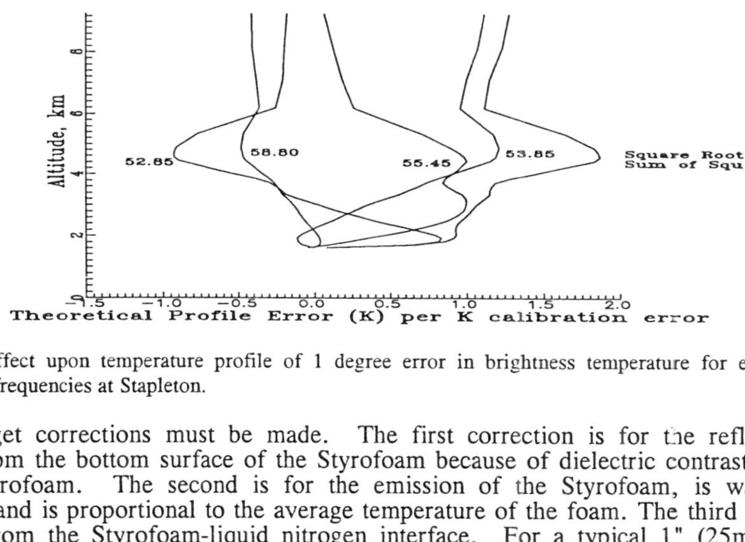

Figure 5. Effect upon temperature profile of 1 degree error in brightness temperature for each of the NOAA/WPL frequencies at Stapleton.

Several target corrections must be made. The first correction is for the reflection of radiation from the bottom surface of the Styrofoam because of dielectric contrast between air and Styrofoam. The second is for the emission of the Styrofoam, is wavelength dependent, and is proportional to the average temperature of the foam. The third is for the reflection from the Styrofoam-liquid nitrogen interface. For a typical 1" (25mm) thick Styrofoam container of density 1.5 pounds/cubic foot, the insertion loss is about 2.2 Kelvins at 60 GHz. Typical target corrections are given below. These corrections are relatively small, and errors in the corrections therefore result in diminishingly small errors in target temperature estimation:

air-Styrofoam interface	-0.2K
Styrofoam insertion loss	-2.2K
Styrofoam-LN2 interface	-1.6K

Side lobes are captured by metal blinders that ensure that the more significant antenna side lobes are reflected into the blackbody.

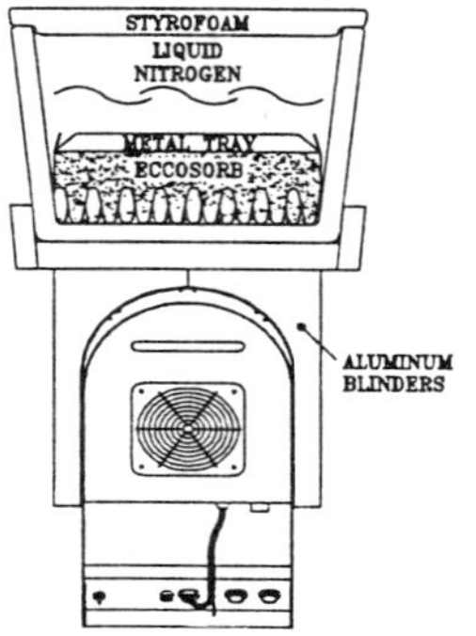

Figure 6. The cryogenic blackbody target. A saddle holds the Styrofoam cooler on top of the radiometer. The blinders ensure that antenna sidelobes terminate in the blackbody. The target is valid for a broad range of radiometer frequencies.

Many cryogenic fluids can be implemented in this target. Liquid helium is a consideration for deep space radiometers operating below 10 GHz. Liquid nitrogen at 77K is suitable for the MTP operating at 60 GHz. Other cryogens could utilized to measure receiver linearity. The boiling point of liquid nitrogen is a weak function of atmospheric pressure, and can be calculated from the equation below:

$$\text{LN2 temperature} = 68.23 + 0.009037 * \text{pressure(mb)}$$

A simple barometric measurement therefore determines the target temperature. We estimate this target to be accurate to about 0.2K. This accuracy could be increased to 0.1K with some additional care in determining the target corrections and in capturing sidelobes into the blackbody.

The synthesizer-based MTP

We have completed construction of the MTP based on a stable and tuneable synthesizer under contract to the Atmospheric Sciences Laboratory at White Sands Missile Range. This portable total integrated water, water vapor, and temperature profiling radiometer is shown in figure 7.

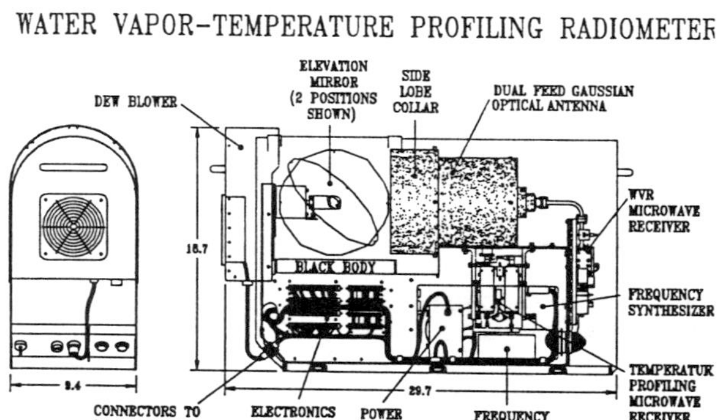

Figure 7. The synthesized MTP. The instrument is about 30"x19"x10", weighs about 30 kg, and consumes 200 watts. The instrument is controlled by a notebook computer.

We are completing the software, and therefore have limited field data. Our first profile comparison, performed at the National Weather Service site in Denver Colorado, is shown in figure 8.

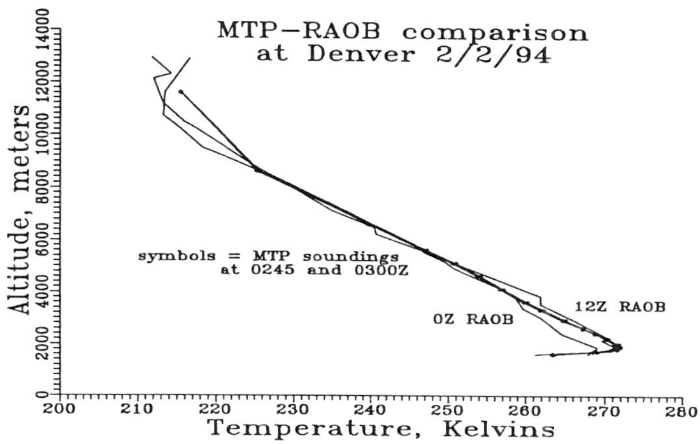

Figure 8. A comparison of the MTP profiles (symbols) with co-located RAOBs at Denver. The RAOBS span the MTP profiles in time.

The MTP profiles were taken intermediate in time to the RAOBs. The agreement with the 1200 UT RAOB is excellent with rms differences of 0.7K in the inversion layer of the first kilometer. The 1200UT radiosonde comparison is more meaningful because the MTP and RAOB physical surface temperature measurements more closely match for this post-sunset condition.

Simulated Performance of the Synthesized MTP

Because we are able to significantly diminish the instrumental and measurement errors below those of Gunn-based MTPs, we expect more accurate and resolute profiles. The calibration and frequency stability errors of conventional profilers were discussed above, and can be 1K or more.

Our analysis, based Denver 1989 RAOBs, show that significant improvement in profile quality can be achieved by bringing instrumental and measurement errors down to the 0.5K level we expect with the synthesized MTP and cryogenic calibration. The expected rms errors are shown in figure 9.
For comparison to the 11 channel result, the 4 channel error profile for 0.5K error is shown.

The estimated errors in figure 9 for Gunn-based 4 channel profilers are between 1 and 2K, assuming Gunn drift and hot load calibration errors. The bulk of residual errors with 11 channels and cryogenic calibration are due to inherent retrieval algorithm uncertainties.

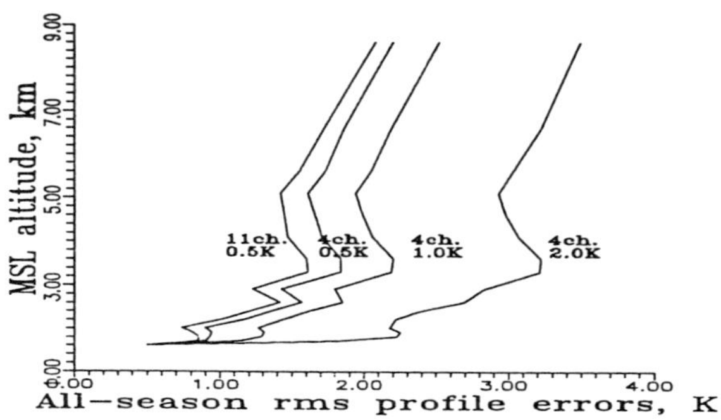

Figure 9. Profile errors based on one year of RAOBs, stratified into four seasons, at Denver Colorado. The errors expected from the synthesized MTP are 0.5K or less. Three cases are given for the 4 channel MTP: 0.5, 1, and 2K instrument+calibration errors.

Profiling atmospheric water vapor

We are proposing to develop a synthesizer-based water vapor profiler as a modification to the existing total integrated WVR-MTP.

Although several technologies have demonstrated the ability to profile atmospheric water vapor (Lidars, near infrared Fourier transformation spectrometers), all methods degrade severely in the presence of cloud. RAOBs are still the profiling standard, but are expensive (about $200 per release) and are only point-in-time measurements.

R. Peter and N. Kämpfer of the Institute of Applied Physics of the University at Bern

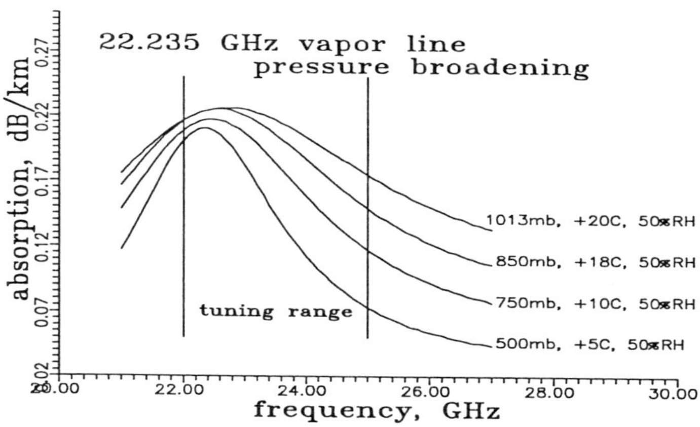

Figure 10. Pressure broadening of the 22.235 water vapor line. We are proposing to map this line from 22 to 25 GHz, and to invert the line shape for the water vapor profile.

have demonstrated profiling of tropospheric water vapor by making observations across the pressure broadened water vapor line at 22.235 GHz. This method demonstrates the ability to define cloud boundaries. Their instrument was designed and developed to measure stratospheric water vapor and is not optimum for tropospheric measurements, but demonstrates the ability to measure tropospheric water vapor profiles.

We will scan this vapor resonance line over a greater width of the line (3 GHz) to improve the profile resolution, especially at low altitudes. Elevation scans are also possible. The instrument design is a modification to an existing commercial dual channel total integrated water and water vapor radiometer.

Acknowledgement: This study and construction of the synthesized MTP was funded under U.S. Army Atmospheric Sciences Laboratory contracts DAAD-07-90-C-158 and DAAL-06-92-C-0013.

References

1 Peter, R., and N. Kämpfer, *Radiometric Determination of Water Vapor and Liquid Water and Its Validation With Other Techniques,* JGR 97, No. D16, pp 19173-18183, November 20, 1992

2 Westwater, E.R., *Ground-Based Passive Probing Using the Microwave Spectrum of Oxygen*, Radio Science **69D**, No. 9, September 1965.

3 Lundquist, G.L. and R.I. Potash (1985), *WVR brightness temperature corrections due to finite beam width,* Interferometrics/GSFC Internal Working Paper, August 30, 1985

4 Jordan, J. (1988), *Finite Beamwidth Error,* NOAA/WPL Internal Memo, October 17, 1988

5 Liebe, H.J., P.W. Rosenkranz, and G.A. Hufford (1992), *Atmospheric 60-GHz oxygen spectrum: New laboratory measurements and line parameters*, J. Quant. Spectrosc. Radiat. Transfer, vol. 48, pp. 629-643.

6 Measure, E., T.L. Barber, and Y.P. Lee (1986), *Number of Nonredundant frequencies for Ground-Based Microwave Radiometric Temperature Profiling,* JOAT **6**, October 1989.

7 Twomey, Sean (1977), *Introduction to the Mathematics of Inversion in Remote Sensing and Indirect Measurements*. G70.4 T86, Elsevier Scientific Publishing.

8 Schroeder, J.A., J.R. Jordan, M.T. Decker, *Design Considerations for a Network of Thermodynamic Profilers*, Bulletin of the American Meteorological Society, 1989

9 Askne, J.I.H., and Ed R. Westwater, *A Review of Ground-Based Remote Sensing of Temperature and Moisture by Passive Microwave Radiometers,* IEEE Trans. on Geosci. and Remote Sensing, **GE-34**, No. 3, May 1986.

AUTHOR INDEX

ALBER, C. 65
ARBESSER-RASTBURG, B. 529
ATKINSON, N.C. 519
BAPTISTA, J.P.V. 321, 529
BARBALISCIA, F. 139
BASILI, P. 229
BASIST, A.N. 309
BATTISTELLI, E. 485, 529
BORDI, R. 469, 485
BORGEAUD, M. 321
BOSISIO, A.V. 129
BURNS, B.A. 265
CAPITANI, C. 529
CAPSONI, C. 129
CHANG, L.A. 33
CIOTTI, P. 229
CONTU, S. 499
COPPO, P. 411
CULEBRAS, A. 529
D'AURIA, G. 229
DAVIS, J.L. 69
DEBLONDE, G. 43
DEL FRATE, F. 529
ELGERED, G. 69
ENGLISH, S.J. 357, 369
FERRARE, R.A. 3
FERRARO, R.R 309
FERRAZZOLI, P. 391, 411
FIONDA, E. 139
FLENDER, F. 265
FUHRHOP, R. 165
GAGLIANI, S. 411
GAISER, P.W. 441
GARAND, L. 43
GAUTHIER, P. 43
GOUTOULE, J.M. 463
GRANDELL, J. 333
GRASSOTTI, C. 43
GRIFFIS, A. 441
GRODY, N.C. 281, 309
GUERRIERO, L. 391, 401
GUILLOU, C. 369
HALLIKAINEN, M. 333, 347
HAN, Y. 3
HASHIMOTO, S. 53
HEWISON, T.J. 509, 519
HORNBOSTEL, A. 197, 239
JACKSON, T.J. 441
JARLEMARK, P.O.J. 119
JIN, Y.Q. 419
JOHANSSON, J.M. 69
JONES, D.C. 357
KAO, M. 441
KEIHM, S.J. 537
KRAFT, U. 463
KUMMEROW, C. 209
KURVONEN, L. 333, 347
KUTUZA, B.G. 175, 197, 239
L'ABBATE, M. 469
LAURSEN, B. 451
LE VINE, D.M. 441
LIBERTI, G.L. 297
LILJEGREN, J.C. 107
LIU, Q. 219, 379
LUZI, G. 411
MACELLONI, G. 429
MARINELLI, F.M. 499
MARTIN-NEIRA, M. 463
MARZANO, F. 229, 253
MASULLO, P. 139

MÄTZLER, C. 389, 401
MAZZONI, T. 411
MELFI, S.H. 3
MUGNAI, A. 253
NOLL, J. 321
OLMI, L. 79
PALOSCIA, S. 391, 429
PAMPALONI, P. 391, 429
PEDERSEN, H.M. 451
PETER, R. 95
PIERDICCA, N. 229, 253
PRIGENT, C. 155, 369
RACETTE, P. 33
RAYER, P.J. 23
RINOUS, P. 499
ROBERTI, L. 209
ROCKEN, C. 65
ROGNES, A. 321
RUISI, R. 429
RUPRECHT, E. 165
SAUNDERS, R.W. 357, 519
SCHIAVON, G. 529
SCHRADER, M. 379
SCHROTH, A. 197, 239
SEZE, G. 155
SIMMER, C. 219
SKOU, N. 451
SMITH, E.A. 253
SNIDER, J.B. 3, 13
SOLHEIM, F. 65, 537
SPERA, P. 469
STRINGER, S.J. 519
SUSINI, C. 429
SWIFT, C.T. 441
TJEMKES, S. 147
TURK, J. 187, 253
VAN HOVE, T. 65
VISSER, M. 147
VIVEKANANDAN, J. 187, 253
WANG, J.R. 33
WARE, R. 65
WENG, F. 281
WESTWATER, E.R. 3

SUBJECT INDEX

A
Absorption:
  in clouds, 175
  models, 3, 13, 23, 175
AMPR, 253
AMSU
  antenna of AMSU-B, 509
  characteristics, 23, 357
  performance of AMSU-B, 519
Antenna
  of AMSU-B, 509
  of MFR, 529
  of MIMR, 499
ARM, 107
ASTEX, 357, 369
Atmospheric turbulence, 79
Attenuation, 129, 139

B
Black body 485, 519

C
CaPE, 253
Cloud
  3-D model, 209, 219
  classification, 229
  effects, 165, 197
  microphysics, 187, 229
Cloud liquid
  absorption, 175
  retrieval, 95, 107, 119, 155, 281, 357
  spatial variations, 139, 197

D
Debye formula, 175
Dielectric constant
  of leaves, 389
Drop size distribution, 239, 253

E
Emission, 13, 23
Emissivity
  of clusters, 419
  of crops, 391
  of forest, 401, 411
  of rough surfaces, 369, 379
  of vegetation, 429
ESTAR, 441

F
Fires
  detection, 411
Forest, 401
  fires, 411

G
GENLN2, 23
GPCP-AIP/2, 297
GOES, 43
GPS, 65, 69

H
Hydrological cycle, 147
HIRS, 147

I
Imaging radiometer, 33, 469
Interferometers, 79, 441, 451, 463
IRAM, 79
Ice
  in precipitation, 187
  retrieval from SSM/I, 281
  sea ice
    mapping, 347
    retrieval from SSM/I, 309, 333

L
Land surface
  emission, 321

M
MARSS, 357
METEOSAT, 155
MFR, 529
Millimeter-wave radioastronomy, 79
MIMR
  antenna, 499
  calibration, 485, 537
  characteristics, 321
  performance, 469
MIR, 33
MIRAS, 463
MTP, 537
Multi-temporal data, 321

N
Nonspherical particles, 187

O
OPEX, 239

P
PAMIR, 401
PBMR, 441
Phase fluctuations, 79
PIPOR, 347
Plant
  biomass, 391
  constituents, 429

Polarization, 187, 239
 in dense media, 419
 index, 391, 429
Propagation, 129, 139, 529

R

Radar meteorology, 253
Radiative transfer, 155, 165
 in dense media, 419
 in precipitation, 187, 197, 209, 219, 229, 239, 281
 in vegetation, 391, 401
 three-dimensional, 209, 219
Radiometers:
 airborne, 33, 253, 347, 357, 369, 391, 441
 dual-channel, 69, 107
 ground-based, 3, 13, 95, 107, 119, 239, 529, 537
 millimeter-wave, 33, 529
 single-frequency, 119
Radiometric systems
 antenna, 499, 509, 529
 calibration, 485, 499, 519
 receiver, 469, 519, 529
Rainfall rate, 197, 219, 239, 253, 265, 281, 309
Raman lidar, 3
RASAM, 401
Refractive index, 79
Retrieval methods:
 cloud ice , 281
 cloud liquid, 357
 cloud parameters, 229, 281
 errors, 53, 165
 intercomparison, 297
 Kalman filtering, 119
 nonlinear algorithms, 95, 155, 357
 precipitable water, 165
 precipitation profile, 253
 profiles, 529, 539
 rain, 265, 281, 297, 309
 sea ice, 333, 347
 snow cover, 309, 321, 33
 statistical, 107, 229
 two-beamwidth antenna, 53
 using GPS, 65
 wind speed, 379
Rough surfaces
 of sea, 369, 379
RTTOV, 23

S

SARad, 451
Sea
 foam, 379
 waves, 369
Snow
 cover, 309, 321, 351
 water equivalent, 321, 333
Soil moisture, 441
Solar extinction, 53
Space geodesy, 119
SSM/I data for
 cloud ice, 281
 cloud liquid, 147, 155, 165, 281
 rain retrieval, 265, 281, 297, 309
 snow cover, 309, 321, 333
 vegetation, 429
 water vapor, 43, 147, 155, 165
 wind speed, 369, 379
Synthetic aperture
 radiometers, 441, 451, 463

T

Telecommunications, 129, 139, 529
Temperature:
 effect on absorption, 175
 profiles, 95, 529

V

Vegetation
 crops, 391
 measurements, 429
 models of emission, 391, 401
 permittivity, 389
VLBI, 69

W

Water vapor:
 integrated, 43, 53, 65, 69, 95, 107, 119, 155, 165
 profiles, 3, 33, 95, 529, 537
 space variations, 139, 147, 197
 time variations, 69, 79, 197
Weather prediction, 43
Wet delay, 65, 69, 119
Wind speed
 retrieval, 155, 357, 369, 379